$\mathcal{P}_n$	polynomials of degree n or less, 203
$\ell_j(x)$	Lagrange polynomials, 207
δ_{ij}	Kronecker delta, 207
$\Delta f(x)$	forward difference, 216
$\Delta^k f(x)$	kth forward difference, 217
$\binom{k}{i}$	binomial coefficients, 217
$\mathrm{Spr}\{y_i\}_{i=1}^n$	spread of $\{y_1, y_2, \ldots, y_n\}$, 224
$T_k(x)$	Chebyshev polynomials, 227
$\langle f, g \rangle$	inner product of functions, 257
$I(f)$	weighted integral, 289
$Q_n(f)$	interpolatory quadrature, 290
$T_N(f)$	composite trapezoidal rule, 302
$S_N(f)$	composite Simpson's rule, 303
B_{2j}	Bernoulli numbers, 310
$C_{2\pi}$	continuous periodic functions, 341
$\phi(x_i, y_i; h)$	increment function, 358
τ_i	local truncation error, 372
$\nabla f(x, y)$	gradient, 508

Numerical Analysis

Second Edition

Lee W. Johnson

R. Dean Riess

Virginia Polytechnic Institute
and State University

Addison-Wesley Publishing Company

Reading, Massachusetts
Menlo Park, California
London Amsterdam
Don Mills, Ontario Sydney

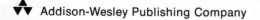

To our wives,
ROCHELLE and JAN

Library of Congress Cataloging in Publication Data

Johnson, Lee W.
Numerical analysis.

Includes index.
1. Numerical analysis. I. Riess, R. Dean
(Ronald Dean), 1940- . II. Title.
QA297.J63 1982 519.4 81-15019
ISBN 0-201-10392-3 AACR2

ISBN 0-201-10392-3
FGHIJ-DO-898765

Preface

In this second edition we have substantially rewritten Chapters 2 and 3 (numerical linear algebra) and Chapter 7 (numerical solution of ordinary differential equations). We have also included a new chapter on the numerical solution of partial differential equations and added some new material to Chapters 4, 5, and 6. In writing the second edition we have tried to emphasize many of the recent developments in the fast-changing area of mathematical software; specifically, we have expanded our coverage of error-sensing/error-control devices and of adaptive algorithms. We have also greatly increased the number of exercises and added quite a few that are relatively easy and have a computational flavor.

Briefly, these are the major changes in the second edition.

Chapter 2 An expanded discussion of LU-decomposition is based in part on the ideas employed in LINPACK, and includes practical ways of estimating from the LU-decomposition of A the condition number and the accuracy of the computed solution to $A\mathbf{x} = \mathbf{b}$.

Chapter 3 The material on the eigenvalue problem has been reordered and rewritten. The inverse power method and reduction to Hessenberg form using Householder transformations are emphasized. A brief introduction to the QR algorithm is also included.

Chapter 5 Included is a section on the divided difference form for the interpolating polynomial, and the material on interpolation at equally spaced points has been streamlined. A new section on multivariate interpolation and bicubic splines has been added as well as a section on B-splines and their application in the Rayleigh-Ritz procedure.

Chapter 6 The material on adaptive quadrature has been expanded to include a detailed description of an adaptive Simpson's rule, and a new section on multiple integrals has been added.

Chapter 7 In this completely rewritten chapter, the sections covering one-step methods include an extended and complete discussion of variable-step Runge-Kutta methods along with their implementation and error-control mechanisms. A comprehensive treatment of relative stability and absolute stability for multistep methods and one-step methods has been added, along with a discussion of stiff systems.

Chapter 8 This new chapter contains material on finite-difference methods and a brief introduction to finite-element methods. Topics include Crank-Nicolson, line (block) methods, ADI, cyclic ADI, and introductory to stability considerations.

Blacksburg, Virginia L.W.J.
January 1982 R.D.R.

Suggestions on the Use of This Text

Our primary goal in writing this book has been to provide a *textbook* that is well suited for use in an introductory numerical analysis course at the undergraduate level. We view the essential ingredients of such a text to be a proper range of topics, a cohesive and understandable presentation, numerous examples that stress insight into the topics, and large selections of exercises that both reinforce the material of the text and encourage further investigation.

The list of topics is quite extensive, ranging from important classical material to a number of relatively modern concepts. However, the text is by no means an encyclopedia of all possible numerical methods. Rather the text presents a selection of commonly used basic procedures together with a comparative analysis of their strengths and weaknesses. Each topic is introduced in its simplest and most understandable form and then developed to a point at which the reader has a sound and fundamental background in the subject. With this background, the student may independently delve deeper into the advanced aspects of problems of interest. The examples, for the most part, are kept as simple and direct as possible in order to illustrate a point without obscuring it. There are a few fairly intricate examples, however, which should help the student appreciate the complex nature of typical real-world problems. The exercises include both theoretical and computational problems ranging from the routine to the challenging.

Although this book is intended for use in the classroom, we believe it will serve as a valuable reference book as well since it includes an introduction to several modern topics not usually found in many such texts. These topics include solutions for overdetermined linear systems, spline approximation, the fast Fourier transform, adaptive quadrature, collocation methods for differential equations, and an introduction to some optimization techniques such as quasi-Newton methods, Lagrange multipliers, and linear programming. The text should appeal to engineers, scientists, and mathematicians alike in that it presents computationally efficient and practical methods and also includes sufficient mathematical theory for a thorough understanding of each method presented. The theoretical material is kept at a minimum and is presented in an expository and intuitive fashion, but is still sufficiently comprehensive so that the reader can understand why each technique works, how efficient it is, and, possibly most important, what can cause it to fail.

The text can be used either for a one-term course in introductory numerical methods or for a full-year numerical analysis course at the junior-senior level. These two uses are possible since the more elementary material is placed at the beginning of each chapter while the more sophisticated material is near the end of the chapter where it can be omitted without loss of continuity. We have given some illustrations at the end of this section as to how the text may be adapted to fulfill the purposes of either type of course. There are other alternatives to our suggestions as the chart of chapter dependencies shows. For example, the instructor of a one-term course may wish to omit entirely the material on eigenvalues and concentrate more on interpolation, quadrature, or differential equations. In either type of course, the instructor can present the material in a different order from that in the text. For example, interpolation can be covered first if so desired. The starred sections are more sophisticated than the others and can be either skimmed or omitted. Varying the coverage of these special sections is one way of adjusting the level of the course. The broad range of exercises provides additional flexibility in adjusting the level of the course.

The prerequisites for either type of course are basic calculus and a familiarity with the ideas of a matrix and a determinant. The fundamental results from calculus that we use frequently are listed in Chapter 1 and most of the fundamentals of matrix theory are reviewed in Chapter 2. We occasionally introduce material that is not always covered in a freshman/sophomore calculus course; for example, norms, inner products, and eigenvalues. In these instances, we give a careful exposition, and the reader should have little difficulty understanding the material and its relevance to the particular topic or method being discussed. Whenever possible, material that requires some mathematical maturity is left to the later sections of a chapter or included in the starred (optional) sections.

Although some computational experience can be gained on a hand calculator, it is useful for a student to have a modest programming ability in a language such as FORTRAN. To aid the student in gaining computational experience, we have included a number of fairly simple programs for some of the numerical methods. These programs are in the form of subroutines, which are documented and easy to understand since they are written to parallel the statement of the algorithms. In the interest of clarity, we have not tried to include all-purpose, fool-proof codes that handle every possible contingency. Comprehensive, well-tested programs of this sort are available in the literature and in most computer libraries. As we develop techniques to avoid pitfalls that can occur in a particular method, the reader should be able to expand the simple programs given in the text to accommodate these techniques.

The first chapter presents basic material on rounding errors and floating-point arithmetic. These topics are discussed further and illustrated as they pertain to particular methods in succeeding chapters. However, since a thorough understanding of the effects of rounding and ill-conditioning requires a considerable theoretical background in mathematics and statistics, we primarily present these sources of error intuitively and illustrate them in the methods in which they can cause the greatest difficulties.

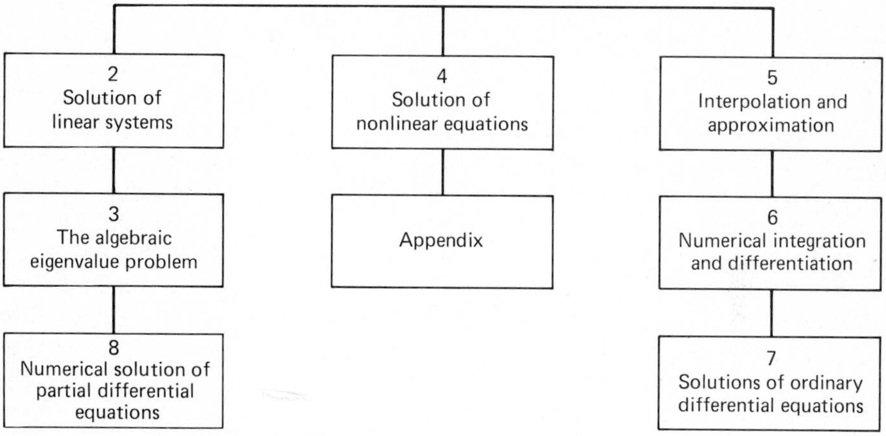

Fig. S.1 Relationship of the chapters.

Chapter 2: Sections 1., 2., 2.1, 2.2, 2.3, 2.4
Chapter 3: Sections 1., 2.
Chapter 4: Sections 1., 2., 3., 3.1, 3.2, 3.3, 3.4, 4., 4.1
Chapter 5: Sections 1., 2., 2.1, 2.2, 2.3, 2.4
Chapter 6: Sections 1., 2., 2.1, 2.2, 2.3, 2.4
Chapter 7: Sections 1., 2., 2.1, 2.2, 2.3, 2.4, 3., 4., 5.

Fig. S.2 Suggested list of sections for a self-contained one-term course in numerical methods.

The chart in Fig. S.1 shows the relationships among the chapters of this book. For example, Chapters 5 and 6 are prerequisite to Chapter 7; but Chapter 7 does not depend on the material in Chapters 2, 3, and 4. (In a few instances, some starred sections and starred problems depend on other chapters.) Figure S.2 is a list of sections that are appropriate for a comprehensive and self-contained one-term course on numerical methods. The instructor may wish to modify this suggested set of topics, delete some, and include others that we have not listed. Furthermore, there are sufficient topics that the instructor of a full-year sequence can selectively omit or skim some while treating others more carefully.

L.W.J
R.D.R

Contents

1

Computational and Mathematical Preliminaries

1.1 INTRODUCTION

In this text we shall concentrate on that portion of numerical analysis that is concerned with the solution of scientific problems utilizing modern high-speed computers. Even from this possibly narrow point of view, numerical analysis is quite widely interdisciplinary. It involves engineering and physics in converting a physical phenomenon into a mathematical model; numerical analysis involves mathematics in developing techniques for the solution (or approximate solution) of the mathematical equations describing the model; finally numerical analysis involves computer science for the implementation of these techniques in an optimal fashion for the particular computer available. These three processes are not independent and we should not lose sight of this dependence as we consider a particular aspect of a problem. For example, often certain idealistic restrictions must be placed on a physical system in order to obtain a tractable mathematical model. Furthermore, it is possible that mathematical techniques that are derived to "solve" the equations of the model and that serve very well in theory cannot be practically implemented on a particular computer. The task of the numerical analyst is therefore to synthesize these processes and obtain "acceptable" numerical answers. (It is also part of the task to be able to determine when and why "acceptable" answers for a particular problem on a particular computer *cannot* be attained.)

In this text we shall concentrate on the latter two phases of this threefold process in order to cover more topics, but we will try not to lose sight of the "real world" origins of the problems. In this chapter we shall discuss some basic concepts such as computer representation of numbers, floating-point arithmetic, and rounding errors. We shall also list from the calculus some basic mathematical results that will be useful in the analysis of the numerical methods.

1.2 ERRORS IN COMPUTATIONS

In analyzing the accuracy of numerical results, the numerical analyst should be aware of the possible sources of error in each stage of the computational process and of the extent to which these errors can affect the final answer. There are three types of errors that occur in a computation. First, there are errors that we call "initial data" errors. When the equations of the mathematical model are formed, these errors arise because of idealistic assumptions made to simplify the model, inaccurate measurements of data, miscopying of figures, or the inaccurate representation of mathematical constants (for example, if the constant π occurs in an equation, we must replace π by 3.1416 or 3.141593, etc.). Another class of errors, "truncation" errors, occurs when we are forced to use mathematical techniques that give approximate, rather than exact, answers. For example, suppose we use the Maclaurin's series expansion to represent e^x so that $e^x = 1 + x + x^2/2! + \cdots + x^n/n! + \cdots$. If we want a number that approximates e^β for some β, we must terminate the expansion in order to obtain $e^\beta \approx 1 + \beta + \beta^2/2! + \cdots + \beta^k/k!$. Thus $e^\beta = 1 + \beta + \beta^2/2! + \cdots + \beta^k/k! + E$ where E is the truncation error introduced in the calculation. Truncation errors in numerical analysis usually occur because many numerical methods are iterative in nature, with the approximations theoretically becoming more accurate as we take more iterations. As a practical matter, we must stop the iteration after a finite number of steps, and thus introduce a truncation error. The last type of error we shall consider, "round-off" or "rounding" errors, is due to the fact that a computer has a finite word length. Thus most numbers and the results of arithmetic operations on most numbers cannot be represented exactly on a computer. Even though the computer is capable of representing numerical values and performing operations on them, we should be aware of how this is accomplished so that we can understand the error that is produced by inexact representation.

Initial data errors and truncation errors are dependent mostly on the particular problem we are examining, and we shall deal with them as they arise in the context of the different numerical methods we derive throughout the text. The total effect of round-off errors is sometimes dependent on the particular problem in the sense that the more operations we perform, the more we can probably expect the round-off error to affect the solution. The individual round-off error caused by any individual number representation or arithmetic operation is dependent, however, on the particular computer being used; and thus we shall examine the possible sources of this error in the next section before we introduce any specific numerical methods. We emphasize that our ultimate concern is the effect of total error from any and all sources. For example, we shall later see problems in which a "small" error (no matter from what source) can cause a "large" error in the final solution. Problems of this type are called *ill-conditioned* and must be treated very carefully to obtain an acceptable computed answer.

There are two ways to measure the size of errors. In analyzing the error of a computation, if we let $\tilde{x}$ represent the "computed approximation" to the "true solution" x, then we define the *absolute error* to be $(x - \tilde{x})$ and the *relative error* to be $(x - \tilde{x})/x$. (If the true solution x is zero, then we say the relative error is undefined.) We shall consider these concepts again in later sections, but we briefly pause to mention here that the relative error is usually more significant than the absolute error, and hence we shall try to establish bounds for the relative error whenever possible. To illustrate this point, suppose that in "Computation A" we have $x = 0.5 \times 10^{-4}$ and $\tilde{x} = 0.4 \times 10^{-4}$; in "Computation B" we have $x = 5000$ and $\tilde{x} = 4950$. The absolute errors are 0.1×10^{-4} and 50, respectively; but the relative errors are 0.2 and 0.01, respectively. Stated differently, Computation A has a 20% error; Computation B has only a 1% error.

In investigating the effect of the total error in various methods, we shall often mathematically derive an "error bound," which is a limit on how large the error can be. (This limit applies to both absolute and relative errors.) It is important that the reader realize that the error bound can be much larger than the actual error and in practice often is. Any mathematically derived error bound must account for the worst possible case that can occur and is often based upon certain simplifying assumptions about the problem, assumptions which in many particular cases cannot be actually tested. For the error bound to be used in any practical way, the user must have a good understanding of how the error bound was derived in order to know how crude it is; i.e., how likely it is to overestimate the actual error. Of course, whenever possible, our goal is to eliminate or lessen the effects of errors, rather than to estimate them after they occur.

1.3 ROUNDING ERRORS AND FLOATING-POINT ARITHMETIC

The first type of rounding error that we encounter in performing a computation evolves from the fact that most real numbers cannot be represented exactly on a computer. Readers are probably not surprised by this statement since they are aware that irrational numbers such as π or e have an infinite nonrepeating decimal expansion. Thus they know that even for a hand computation they must use an approximation such as 3.14159 for π or 2.718 for e, and they carry as many digits in their approximations as they feel are necessary for a particular computation. Nevertheless, they realize that once these approximations have been used, an error that can never exactly be corrected has been introduced into the calculation.

Since only a finite number of digits can be represented in computer memory, each number x must be represented in some fashion that uses only a fixed number of digits. One of the most common forms is the "floating-point" form, in which one position is used to identify the sign of x, a prescribed number of digits are used to represent the "mantissa" or fractional form of x, and an integer is used to represent the "exponent" or "characteristic" of x with respect to the base b of the representation. (Modern, preferred terminology uses "significand" for mantissa and "exrad" for exponent.) Thus each x can be thought of as being represented by a number $\bar{x}$ of the form $\pm(0.a_1a_2 \ldots a_m) \times (b^c)$ where m is the number of digits allowed in the mantissa, b is the base of the representation, and c is the exponent. Additionally there are two machine-dependent constants, μ and M, such that $\mu \leq c \leq M$. We shall consider three bases: (i) $b = 10$ (decimal), with which the reader is familiar and which is used on some machines; (ii) $b = 16$ (hexadecimal), which is common to the IBM 360 and 370 series; and (iii) $b = 2$ (binary), which is, in a sense, the most fundamental of the three. Proper form requires the mantissa, $a \equiv 0.a_1a_2 \ldots a_m$, to satisfy $|a| < 1 = b^0$ and each a_i, $1 \leq i \leq m$, to be an integer such that $0 \leq a_i \leq b - 1$ with $a_1 \neq 0$ (unless $\bar{x} = 0$). The floating-point representation, $\bar{x}$, is then said to be "normalized."

The floating-point decimal form ($b = 10$) should be familiar to the reader. For example, the decimal number 150.623 is the same as 0.150623×10^3 and can also be regarded as

$$(1 \times 10^2) + (5 \times 10^1) + (0 \times 10^0) + (6 \times 10^{-1}) + (2 \times 10^{-2}) + (3 \times 10^{-3}).$$

Likewise, the binary number 110.011 equals

$$(1 \times 2^2) + (1 \times 2^1) + (0 \times 2^0) + (0 \times 2^{-1}) + (1 \times 2^{-2}) + (1 \times 2^{-3}),$$

and the hexadecimal number 15F.A03 equals

$$(1 \times 16^2) + (5 \times 16^1) + (F \times 16^0) + (A \times 16^{-1}) + (0 \times 16^{-2}) + (3 \times 16^{-3}).$$

Note that there are only two digits, 0 and 1, in the binary system; and there are sixteen digits in the hexadecimal system; 0, 1, 2, 3, 4, 5, 6, 7, 8, 9, A, B, C, D, E, F where the decimal equivalents of A, B, C, D, E, F are 10, 11, 12, 13, 14, 15, respectively. Also note that the hexadecimal system is a natural extension of the binary system since $2^4 = 16$, and hence there is precisely one hexadecimal digit for each group of four binary digits ("bits") and vice versa [$0 = (0000)_2$, $7 = (0111)_2$, $A = (1010)_2$, $F = (1111)_2$, etc.].

The conversion of an integer from one system to another is fairly simple and can probably best be presented in terms of an example. Let $k = 275$ in decimal form; that is, $k = (2 \times 10^2) + (7 \times 10^1) + (5 \times 10^0)$. Now $(k/16^2) > 1$, but $(k/16^3) < 1$; so in hexadecimal form k can be written as $k = (\alpha_2 \times 16^2) + (\alpha_1 \times 16^1) + (\alpha_0 \times 16^0)$. Now $275 = 1(16^2) + 19 = 1(16^2) + 1(16) + 3$; and so the decimal integer 275 can be written in hexadecimal form as 113; that is, $(275)_{10} = (113)_{16}$. The reverse process is even simpler. For example, $(5C3)_{16} = 5(16^2) + 12(16) + 3 = 1280 + 192 + 3 = (1475)_{10}$. Conversion of a hexadecimal fraction to a decimal is similar. For example, $(0.2A8)_{16} = (2/16) + (A/16^2) + (8/16^3) = (2(16^2) + 10(16) + 8)/16^3 = (680)/4096 = (0.166)_{10}$ (carrying only three digits in the decimal form). Conversion of a decimal fraction to hexadecimal (or binary) proceeds as in the following example. Consider the number $r_1 = 1/10 = 0.1$ (decimal form). Then there exist constants $\{\alpha_k\}_{k=1}^{\infty}$ such that

$$r_1 = 0.1 = \alpha_1/16 + \alpha_2/16^2 + \alpha_3/16^3 + \alpha_4/16^4 + \cdots .$$

Now $16r_1 = 1.6 = \alpha_1 + \alpha_2/16 + \alpha_3/16^2 + \alpha_4/16^3 + \cdots$. Thus $\alpha_1 = 1$ and $r_2 \equiv 0.6 = \alpha_2/16 + \alpha_3/16^2 + \alpha_4/16^3 + \cdots$. Again $16r_2 = 9.6 = \alpha_2 + \alpha_3/16 + \alpha_4/16^2 + \cdots$, so $\alpha_2 = 9$ and $r_3 \equiv 0.6 = \alpha_3/16 + \alpha_4/16^2 + \cdots$. From this stage on we see that the process will repeat itself, and so we have $(0.1)_{10}$ equals the infinitely repeating hexadecimal fraction, $(0.1999\ldots)_{16}$. Since $1 = (0001)_2$ and $9 = (1001)_2$, we also have the infinite binary expansion

$$r_1 = (0.1)_{10} = (0.1999\ldots)_{16} = (0.0001\ 1001\ 1001\ 1001\ldots)_2.$$

From the example above we begin to discern one problem of number representation. Not only do we have problems with irrational numbers and infinite repeating decimal expansions such as $1/3 = 0.333\ldots$, but also we see that an m-digit terminating fraction with respect to one base may not have an n-digit terminating representation in another base. (If we were performing an iteration on a hexadecimal machine, the example above suggests that we should probably choose a step size of $h = 1/16$ over $h = 1/10$, $h = 1/1024 = 1/2^{10}$ over $h = 1/1000$, etc., if at all possible.) In Table 1.1 we have taken several integers k, formed the reciprocals, $1/k$, and added the reciprocal to itself k times. The theoretical result of each computation should, of course, equal 1. The calculations were performed on a six-digit hexadecimal machine.†

TABLE 1.1

k	Sum(1/k)	k	Sum(1/k)	k	Sum(1/k)
2	0.1000000E 01	9	0.9999999E 00	16	0.1000000E 01
3	0.9999999E 00	10	0.9999996E 00	⋮	
4	0.1000000E 01	11	0.9999997E 00	1000	0.9999878E 00
5	0.9999999E 00	12	0.9999998E 00	1006	0.9999912E 00
6	0.9999998E 00	13	0.9999999E 00	1012	0.9999843E 00
7	0.9999999E 00	14	0.9999995E 00	1018	0.9999678E 00
8	0.1000000E 01	15	0.9999999E 00	1024	0.1000000E 01

We must, of course, realize that part of the error in Table 1.1 is due to the round-off from addition. We shall discuss this error momentarily, but for now the reader should notice the relative accuracies for the values of k that are powers of 2. For example, compare $k = 1000$ to $k = 1024 = 2^{10}$.

We next consider how numbers are represented in an m-digit machine, particularly those numbers for which m digits are not sufficient to represent the

†Most of the computer programs in this text were run on an IBM 370/158.

number exactly. For example, how might a number like $\frac{1}{3}$ be represented on a 5-digit decimal computer since $\frac{1}{3} = 0.3333333\ldots$. As a general example, suppose a real number x is given exactly by

$$x = \pm(0.\tilde{a}_1\tilde{a}_2 \ldots \tilde{a}_m\tilde{a}_{m+1} \ldots) \times 10^c, \qquad \tilde{a}_1 \neq 0$$

and suppose we want to represent x in an m-digit decimal computer. There are two common ways of representing x. First, we can simply let $a_k = \tilde{a}_k$, $1 \leq k \leq m$, discard the remaining digits, and let the computer representation be

$$\bar{x} = \pm(0.a_1a_2 \ldots a_m) \times 10^c.$$

This method is known as "chopping." The other representation is the familiar "symmetric rounding" process, which is equivalent to adding $5 \times 10^{c-m-1}$ to x and then chopping (we think of adding 5 to $\tilde{a}_{m+1}$). Of course, if $c < \mu$ or $c > M$, the number lies outside the range of admissible computer representations. If $c < \mu$, (underflow), it is common to regard x as zero, notify the user, and continue further computation. If $c > M$, then overflow results. It is usually deemed not worth the added expense to use symmetric rounding and most machines simply chop. Whether a representation is obtained via chopping or symmetric rounding, we shall hereafter refer to the machine representation, $\bar{x}$, as being "rounded." Note that the illustration above was in decimal form for simplicity; analagous results are valid for other bases. For example, if x has the hexadecimal form

$$x = \pm(0.\tilde{a}_1\tilde{a}_2 \ldots \tilde{a}_m\tilde{a}_{m+1} \ldots) \times 16^c, \qquad \tilde{a}_1 \neq 0,$$

then the "decimal" point is really a "hexadecimal" point; and

$$x = \left(\sum_{k=1}^{\infty} (\tilde{a}_k \times 16^{-k})\right) \times 16^c.$$

The chopped form is still obtained by deleting $\tilde{a}_k$, $k \geq m + 1$. Symmetric rounding is done by adding $8 \times 16^{c-m-1}$ to x and then chopping.

Returning to decimal form for simplicity, we first consider the error in symmetric rounding. If $\tilde{a}_{m+1} \geq 5$, then

$$x - \bar{x} = (\pm 0.\tilde{a}_1 \ldots \tilde{a}_m\tilde{a}_{m+1} \ldots) \times 10^c - [(\pm 0.\tilde{a}_1 \ldots \tilde{a}_m) + (\pm 10^{-m})] \times 10^c.$$

If $\tilde{a}_{m+1} < 5$, then

$$x - \bar{x} = (\pm 0.\tilde{a}_1 \ldots \tilde{a}_m\tilde{a}_{m+1} \ldots) \times 10^c - (\pm 0.\tilde{a}_1 \ldots \tilde{a}_m) \times 10^c.$$

In either case $|x - \bar{x}| \leq 0.5 \times 10^{c-m}$, and this is a bound on the absolute error. To get a bound on the relative error, we note that since $\tilde{a}_1 \neq 0$, then $|x| \geq 0.1 \times 10^c$. Thus the relative error satisfies

$$\frac{|x - \bar{x}|}{|x|} \leq \frac{0.5 \times 10^{c-m}}{0.1 \times 10^c} = 5 \times 10^{-m} = 0.5 \times 10^{-m+1}.$$

In a similar manner we can show that the relative error bound from chopping is 10^{-m+1}. It is interesting to note that neither of these relative error bounds depends on the magnitude of x, that is, the size of c. Rather they depend only on the value of m, which is thus said to be the number of "significant digits" of the representation. For example, the IBM System 360 and 370 are hexadecimal with a mantissa of $m = 6$. The relative error from chopping in number representation thus does not exceed 1×16^{-5}. To compare this with the size of relative errors we expect in the decimal mode, we set $1 \times 16^{-5} = 1 \times 10^{-m+1}$. Solving for m yields $m \approx 7$. Thus we have approximately seven-significant-digit decimal accuracy with respect to the relative error of the representation. This statement does *not* mean that any seven-digit decimal number can be represented exactly on this machine as the previous example of $x = 1/10$ shows. The statement says, however, that $|x - \bar{x}|/|x| \leq 10^{-6}$ (approximately, since $m \approx 7$).

In the discussion above, we analyzed the error made by replacing the true value of x by its machine representation $\bar{x}$. Now we wish to assess the effect of each arithmetic operation $(+, -, \cdot, \div)$ to see how errors propagate. Let us use $\bar{x}$ to denote the machine approximation to the true value x where the error, $e(x) = x - \bar{x}$, includes *all* errors that have been made in going from x to $\bar{x}$; that is, $e(x)$ not only includes the error of representation but also includes errors from previous calculations, initial data errors, etc. In other words, $e(x)$ includes all errors from the beginning of the calculation that have led to any discrepancies between x and $\bar{x}$. With $e(y)$ defined similarly for any quantity y, we investigate the error resulting from the "machine addition" of x and y. Now,

$$x + y = (\bar{x} + e(x)) + (\bar{y} + e(y)) = (\bar{x} + \bar{y}) + (e(x) + e(y)).$$

At first glance it seems that the error of the addition is merely the sum of the individual errors. However, this is not always true since even though $\bar{x}$ and $\bar{y}$ can be represented exactly on the machine, it is *not* necessarily true that their sum can also be; i.e., it does not follow that $\bar{x} + \bar{y} = \overline{\bar{x} + \bar{y}}$. For example, consider a four-digit floating-point machine and let $\bar{x} = 0.9621 \times 10^0$ and $\bar{y} = 0.6732 \times 10^0$. Then $\bar{x} + \bar{y} = 1.6353 \times 10^0$. Now it is not uncommon for an m-digit machine to perform arithmetic operations in a $2m$-digit accumulator and then to round the answer. Assuming this situation to be true, we have $\overline{\bar{x} + \bar{y}} = 0.1635 \times 10^1$. Returning to our general analysis, we see that where $z = x + y$, the true error, $z - \bar{z}$, is actually $(e(x) + e(y))$ *plus* the error between $\bar{x} + \bar{y}$ and $\overline{\bar{x} + \bar{y}}$.

Before examining the other three arithmetic operations, let us consider addition somewhat further. We can see immediately that addition can lead to overflow; for instance, $\bar{x} = 0.9621 \times 10^M$ and $\bar{y} = 0.6732 \times 10^M$; thus $\bar{x} + \bar{y}$ does not fall within the range of the computer. Another factor that we must consider is that before a machine can perform an addition, it must align the decimal points. For example, again let $m = 4$ with $\bar{x}_1 = 0.5055 \times 10^4$ and $\bar{x}_2 = \bar{x}_3 = \cdots \bar{x}_{11} = 0.4000 \times 10^0$. To perform the addition $\bar{x}_1 + \bar{x}_2$, the computer must shift the decimal four places to the left in $\bar{x}_2$ and form $\bar{x}_1 + \bar{x}_2 = (0.5055 \times 10^4) + (0.00004$

$\times\ 10^4) = (0.50554 \times 10^4)$, which rounds to $\overline{\bar{x}_1 + \bar{x}_2} = 0.5055 \times 10^4 = \bar{x}_1$. Continuing, we see that

$$\overline{(\cdots(((\bar{x}_1 + \bar{x}_2) + \bar{x}_3) + \bar{x}_4) + \cdots + \bar{x}_{11})} = \bar{x}_1,$$

but

$$\overline{(\cdots(((\bar{x}_{11} + \bar{x}_{10}) + \bar{x}_9) + \bar{x}_8) + \cdots + \bar{x}_1)} = 0.5059 \times 10^4,$$

which is the correct answer. Thus the machine calculates $\Sigma_{i=0}^{10} \bar{x}_{11-i}$ correctly, but not the sum $\Sigma_{i=1}^{11} \bar{x}_i$. This example illustrates the rule of thumb that if we have several numbers of the same sign to add, we should add them in ascending order of magnitude to minimize the propagation of round-off error. The mathematical foundation underlying this statement is that machine addition is not associative; i.e., it can happen that

$$\overline{\overline{(\bar{x} + \bar{y})} + \bar{z}} \neq \overline{\bar{x} + \overline{(\bar{y} + \bar{z})}}.$$

The following numerical examples were run to illustrate this phenomenon. (We used a hexadecimal machine with $m = 6$.) With $x = 1048576 = 16^5 = \bar{x}$ and $y = z = 1/2 = 8/16 = \bar{y} = \bar{z}$, our machine results were

$$(\bar{x} + \bar{y}) + \bar{z} = 0.1048576\text{E}07$$

and

$$\bar{x} + (\bar{y} + \bar{z}) = 0.1048577\text{E}07$$

as our analysis above leads us to expect. Letting $w_0 = 1048576 = 16^5 = \overline{w_0}$ and $w_k = 1/16 = \overline{w_k}$, $1 \leq k \leq 256$, we also obtained the following results (adding in the order indicated by the sum):

$$\sum_{k=0}^{256} w_k = 0.1048576\text{E}07 = w_0$$

but

$$\sum_{k=0}^{256} w_{256-k} = 0.1048592\text{E}07,$$

the latter being the correct result (which we can easily check by hand). As a final example we computed the sum $S \equiv \Sigma_{k=1}^{999} (1/k(k+1)) \equiv 0.999$ by adding forwards and backwards and obtained

$$S \approx 0.9989709\text{E}00 \text{ (forwards)}, \qquad S \approx 0.9989992\text{E}00 \text{ (backwards)}.$$

A concept that is useful in numerical methods is that of *machine epsilon*. Machine epsilon is the smallest positive machine number, ϵ, such that

$$\overline{1 + \epsilon} > 1.$$

(Machine epsilon varies from computer to computer and from compiler to compiler, but a simple program can be written to determine machine epsilon for a given machine and compiler.) As an example to clarify machine epsilon, let us consider a machine (such as the IBM 370) using 6-digit hexadecimal arithmetic. Consider the machine numbers 1 and δ where

$$1 = .100000 \times 16^1 \qquad \delta = .000001 \times 16^1.$$

Clearly the machine addition of $1 + \delta$ will produce $\overline{1 + \delta} = .100001 \times 16^1$. Therefore $\overline{1 + \delta} > 1$, and thus machine epsilon is at least as large as $\delta = 16^{-5} = .95367432 \times 10^{-6}$. If the machine arithmetic rounds by chopping, then machine epsilon is exactly equal to δ, for the next smallest machine number is

$$\delta' = .000000F \times 16^1$$

and (assuming the machine chops) $\overline{1 + \delta'} = 1$.

To see the significance of machine epsilon, consider the following frequently used approximation to $f'(x)$:

$$f'(x) \simeq \frac{f(x + h) - f(x)}{h}.$$

In problems in which such an approximation would be used, we would be able to program the machine to evaluate f at any point; and by choosing h smaller and smaller, we would expect to get better and better estimates of $f'(x)$. However if h is small relative to the size of x, then it is quite possible that the machine number x and the machine number $x + h$ are the same. If $x + h = x$ in the machine, then $f(x + h) = f(x)$ in the machine; and we have passed the point at which we can improve our estimate to $f'(x)$ by choosing h smaller. Thus if the approximation is to be reasonable, we need to have (if x and h are machine numbers and $h > 0$)

$$\overline{x + h} > x;$$

or (assuming $x > 0$) we need to have

$$\overline{1 + \frac{h}{x}} > 1.$$

For the approximation to be valid, we need the machine representation for h/x to be at least as large as machine epsilon. In general, machine epsilon gives us the limit to which we can resolve or distinguish two numbers (relative to the size of the numbers).

The error analysis for subtraction is much the same as for addition in that

$$x - y = (\bar{x} - \bar{y}) + (e(x) - e(y)),$$

but now we have the problem that subtraction can cause loss of significant digits. This problem occurs when x and y are nearly equal. For example, if $x = 0.6532849 \times 10^2$, $y = 0.6531212 \times 10^2$, and $m = 4$, then $\bar{x} = 0.6532 \times 10^2$ and $\bar{y} = 0.6531 \times 10^2$ (rounding by chopping). Now $\bar{x} - \bar{y} = 0.1000 \times 10^{-1}$, which is the machine representation for $\overline{\bar{x} - \bar{y}}$, that is, $\bar{x} - \bar{y} = 0.1000 \times 10^{-1}$. However, $x - y = 0.1637 \times 10^{-1}$. The problem we encounter here is not with this computation itself, but in its effect on later computations since the zeros in $\overline{\bar{x} - \bar{y}}$ have no significance whatsoever and are, in effect, just filling up spaces. The answer is smaller than x or y; and hence errors already present, $e(x)$ and $e(y)$, can have a drastic effect on the relative error of any further calculations since the zeros will be treated thereafter as though they were significant. This source of error is known as *subtractive cancellation* and leads to serious errors in many computations.

When this type of error occurs, we may try to locate the "sensitive" subtraction and eliminate it, if possible, by analytical manipulation. For example, consider the function $f(x) = (1 + x - e^x)/x^2$. Even if we have a very precise way of evaluating e^x, if x is "small," the evaluation of $f(x)$ will suffer from subtractive cancellation in the numerator, which will be magnified by the division by a number very close to zero. One way to help alleviate this problem is to use the Maclaurin series expansion, $e^x = \sum_{k=0}^{\infty} x^k/k! = 1 + x + x^2/2! + x^3/3! + \cdots$. Then $1 + x - e^x = -x^2/2! - x^3/3! - x^4/4! - \cdots$, and we see that the factor $[(1 + x) - (1 + x)]$, which was essentially causing the trouble, is eliminated. Furthermore, $f(x) = -1/2! - x/3! - x^2/4! - \cdots$; thus this analytical technique also eliminates the problem of "division by zero." Since we are considering the evaluation of $f(x)$ for small values of x, this series should be fairly rapidly convergent as evidenced in the following computer results. However, for large values of x, the series is no longer rapidly convergent and we experience considerable difficulties when the series is truncated after a finite number of terms.

We note from the convergent series expansion for $f(x)$ above that $f(0) = -1/2$. In Table 1.2 the series evaluation was truncated after the seventh term; such truncation should theoretically give us at least seven-decimal-place accuracy for the values of x that were used. Table 1.2 demonstrates that the series evaluations yield good answers whereas direct substitution becomes worse as x decreases.

TABLE 1.2

x	$f(x)$ (Direct Substitution)	$f(x)$ (Series Evaluation)
0.1	−0.5171781E 00	−0.5170917E 00
0.01	−0.5054476E 00	−0.5016708E 00
0.001	−0.9536749E 00	−0.5001667E 00
0.0001	−0.9536745E 02	−0.5000166E 00
0.00001	0.0	−0.5000016E 00

Analysis of multiplication yields

$$x \cdot y = (\bar{x} + e(x)) \cdot (\bar{y} + e(y)) = \bar{x}\bar{y} + \bar{x}e(y) + \bar{y}e(x) + e(x)e(y).$$

We can usually assume that the term $e(x)e(y)$ is negligible with respect to the other terms and we omit it. Furthermore, the product of the m-digit numbers, $\bar{x}$ and $\bar{y}$, is at most a $2m$-digit number and can be represented exactly in the $2m$-digit mode that computers commonly use for individual operations. The product will be rounded to m digits, of course, to form its machine representation, $\overline{\bar{x}\bar{y}}$. Thus, even though $\bar{x}\bar{y} \neq \overline{xy}$ (necessarily), the error is merely that of rounding $\bar{x}\bar{y}$ to its machine representation. Thus the relative error of the difference between $\bar{x}\bar{y}$ and $\overline{\bar{x}\bar{y}}$ is bounded by $0.5 \times 10^{-m+1}$ for symmetric rounding and by $1 \times 10^{-m+1}$ for chopping.

It is common, especially in matrix methods, to have to compute a quantity of the form, $\Sigma_{j=1}^{n} \bar{x}_j\bar{y}_j$, called an "inner product." The multiplications, $\bar{x}_j\bar{y}_j$, are individually done in the $2m$-digit mode, but often the addtion is not. In that case, it is often a simple task in a high-level language to program the machine to do the additions of this type in double precision as well. This method is called "double precision accumulation of inner products," and serves to lessen the errors resulting from the multiplications, $\bar{x}_j\bar{y}_j$, $1 \leq j \leq n$.

Finally, we analyze the division operation

$$\frac{x}{y} = \frac{\bar{x} + e(x)}{\bar{y} + e(y)} = \left(\frac{x + e(x)}{\bar{y}}\right)\left(\frac{1}{1 + e(y)/\bar{y}}\right).$$

Assuming that $e(y)$ is "small" with respect to $\bar{y}$, we recall the geometric series where for $|r| < 1$

$$\frac{1}{1 + r} = 1 - r + r^2 - r^3 + \cdots.$$

Letting $r \equiv e(y)/\bar{y}$ in the expression above for x/y, we have

$$\frac{x}{y} = \left(\frac{\bar{x} + e(x)}{\bar{y}}\right)(1 - e(y)/\bar{y} + e(y)^2/\bar{y}^2 - e(y)^3/\bar{y}^3 + \cdots) \simeq \frac{\bar{x}}{\bar{y}} + \frac{e(x)}{\bar{y}} - \frac{\bar{x}e(y)}{\bar{y}^2}$$

where we assume the omitted terms in the expansion are negligible with respect to the first three.

Division is much the same as multiplication in that the difference between $\bar{x}/\bar{y}$ and its machine representation $(\overline{\bar{x}/\bar{y}})$ is merely that of rounding the representation for $\bar{x}/\bar{y}$. The preceding expression clearly illustrates the disastrous effect on the total error that division by a value of $\bar{y}$ very close to zero can have. As mentioned earlier, this problem can sometimes be avoided by alternate approaches that analytically circumvent "zero divisions."

PROBLEMS, SECTION 1.3

1. a) Convert the following decimal numbers to hexadecimal form:

   i. 1023 ii. 1025

   iii. 278.5 iv. 14.09375

   v. 0.1240234375

   b) Convert the answers above to binary form.

2. Convert the following hexadecimal numbers to decimal form.

a) 1023, b) 100, c) 1A4.C, d) 6B.1C, e) FFF.118

3. Convert the numbers π, e, and 1/3 into hexadecimal form with a mantissa of six hexdigits using

a) symmetric rounding,

b) chopping.

***4.** Prove that any proper fraction that has a terminating hexadecimal expansion also has a terminating decimal expansion. Explain in general terms why the converse of this statement is not valid.

5. For the computer example that computed the sums of reciprocals, compute the relative errors for the operations corresponding to $k = 1000, 1006, 1012, 1018$.

6. a) If x is a real number and $\bar{x}$ is its chopped machine representation on a computer with base 10 and mantissa length m, show that the relative error, $|x - \bar{x}|/|x|$, is bounded by 10^{-m+1}.

b) If $\delta = 10^{-m+1}$ (in the case of chopping) or $\delta = 0.5 \times 10^{-m+1}$ (in the case of symmetric rounding), show that $\bar{x} = x(1 + \varepsilon)$ where $|\varepsilon| \leq \delta$.

c) Given a function $F(x)$, F' continuous, we have from the mean-value theorem that in evaluating F at x the relative error from this source alone is

$$\frac{F(\bar{x}) - F(x)}{F(x)} = \frac{F(x + \varepsilon x) - F(x)}{F(x)} = \frac{F'(\xi)[(x + \varepsilon x) - x]}{F(x)} \simeq \varepsilon x \frac{F'(x)}{F(x)}.$$

Analogously, $F(\bar{x}) - F(x) \simeq \varepsilon x F'(x)$. Use these formulas to give estimates of absolute and relative error if

i. $F(x) = x^k$, for various values of k and x;

ii. $F(x) = e^x$, for large values of x;

iii. $F(x) = \sin(x)$, for small values of x;

iv. $F(x) = x^2 - 1$, for x near 1;

v. $F(x) = \cos(x)$, for x near $\pi/2$.

7. Write a program to compute the value of $S_N = \sum_{k=1}^{N} 1/k$ for various values of N. Adapt the program so that it sums forwards and backwards, and compare the results as N increases. Find the value of N such that $S_n = S_N$ for all $n \geq N$ for the forward summation. How is this value reconciled with the fact that theoretically $\lim_{N\to\infty} S_N = \infty$? Would there be such a number N for any modern digital computer and would the number be the same on all such machines?

8. Consider a computer with mantissa length $m = 3$, base $b = 10$, and exponent constraints $\mu = -2 \leq c \leq 2 = M$. How many real numbers will this computer represent exactly?

9. For any positive integer N and a fixed constant, $r \neq 1$, we recall the formula for the geometric sum:

$$G_N \equiv 1 + r + r^2 + \cdots + r^N = (1 - r^{N+1})/(1 - r) \equiv Q_N.$$

Write a computer program to compute G_N and Q_N for arbitrary values of r and N. Let r be chosen close to 1 and note the discrepancy. In which of the two calculations do you have the most faith? Use values of r for input for which you can easily check the exact value of Q_N by hand calculation (for example, $r = 1 - 10^{-k}$ for some integer $k \geq 0$). Calculate the relative and the absolute errors of these computations.

1.4 REVIEW OF FUNDAMENTAL MATHEMATICAL RESULTS

In this section we shall briefly list some basic mathematical results, mainly from the calculus, that are useful in the development and investigation of numerical procedures in subsequent chapters. We are, of course, assuming familiarity with the rudiments of calculus such as the real number system, functions, and at least an intuitive understanding of the concept of limits (the latter being especially true with respect to sequences since many numerical methods are iterative in nature and thus generate a sequence of approximations to the true solution). A simple example is the classical Newton's method applied to finding the square root of an arbitrary positive real number, α. Given an initial guess, x_0, for $\sqrt{\alpha}$, Newton's method generates the sequence: $x_{n+1} = \frac{1}{2}(x_n + \alpha/x_n)$, for $n = 0, 1, 2, \ldots$. Three questions immediately come to mind if we are to be able to use this algorithm successfully in practice. (1) Is the sequence generated by the algorithm convergent to $\sqrt{\alpha}$ and $\sqrt{\alpha}$ alone? (2) How is the convergence related to the choice of the initial guess x_0? (3) How fast does the method converge, or how large is the error, $x_n - \sqrt{\alpha}$, after the nth iteration? These are questions involving the concept of convergence of sequences. There are, of course, other important questions to answer. (4) How does rounding error affect the method? (5) Is it competitive with other methods for finding $\sqrt{\alpha}$? (6) How well can the method be implemented on the particular computer available and what cost is necessary to obtain an acceptable answer?

Other basic ideas necessary from calculus are the concepts of continuity, differentiation, and integration. We have assumed in Chapters 2 and 3 that the reader is able to evaluate determinants. We have, however, tried to keep the material involving determinants minimal, and the other necessary material from matrix theory and linear algebra is self-contained. We have also assumed that the reader is aware of the importance of differential equations in modeling and solving modern technological problems although we assume very little background in the mathematical theory of differential equations in Chapter 7.

We list here some fundamental theorems of calculus and algebra that are used in later chapters and with which the reader should already be familiar. We will present other results only as they are needed to treat particular topics, and the results will be presented only in the context of the particular topic being treated. The following results and their proofs can be found in almost any sophomore-level calculus text.

Theorem 1.1. Rolle's Theorem

If $f(x)$ is a continuous function in the closed interval $[a, b]$ and is differentiable in (a, b), and if $f(a) = f(b)$, then there exists at least one point ξ in (a, b) such that $f'(\xi) = 0$.

Theorem 1.2. Mean Value Theorem

If $f(x)$ is a continuous function in the closed interval $[a, b]$ and is differentiable in (a, b), then there exists at least one number ξ in (a, b) such that

$$f(b) - f(a) = (b - a)f'(\xi).$$

Theorem 1.3. First Mean Value Theorem for Definite Integrals

If $f(x)$ is continuous for $a \leq x \leq b$, then there exists a number ξ, $a < \xi < b$, such that

$$\int_a^b f(x)\,dx = f(\xi)(b - a).$$

Theorem 1.4. Second Mean Value Theorem for Definite Integrals

Let $g(x)$ be an integrable function that does not change sign on the interval $[a, b]$. If $f(x)$ is continuous on $[a, b]$, then there exists a number ξ in $[a, b]$ such that

$$\int_a^b f(x)g(x)\,dx = f(\xi)\int_a^b g(x)\,dx.$$

Theorem 1.5

Let f be a continuous function on the closed finite interval $[a, b]$. Then there exist points, x_m and x_M, in $[a, b]$ such that $f(x_m) \leq f(x) \leq f(x_M)$ for any x in $[a, b]$; that is, "a continuous function attains its maximum and minimum on a closed interval."

Corollary 1.5.1

Under the hypotheses of Theorems 1.3 and 1.5,

$$\left|\int_a^b f(x)\,dx\right| \leq (b - a)\left\{\max_{a \leq x \leq b} |f(x)|\right\}.$$

Theorem 1.6. Intermediate Value Theorem

If $f(x)$, x_m, and x_M are given as in Theorem 1.5, and if $m \equiv f(x_m)$ and $M \equiv f(x_M)$, then for any value y^* such that $m \leq y^* \leq M$, there exists at least one number x^* in $[a, b]$ such that $f(x^*) = y^*$.

Theorem 1.7. Taylor's Theorem

Let $f(x)$ be a function such that its $(n + 1)$st derivative, $f^{(n+1)}(x)$, is continuous on the interval (a, b). If x and x^* are any two points in (a, b), then

$$f(x) = f(x^*) + \frac{f'(x^*)}{1!}(x - x^*) + \frac{f''(x^*)}{2!}(x - x^*)^2 + \cdots$$
$$+ \frac{f^{(n)}(x^*)}{n!}(x - x^*)^n + R_{n+1}(x^*; x)$$

where there exists a number ξ between x and x^* such that

$$R_{n+1}(x^*; x) = \frac{f^{(n+1)}(\xi)}{(n+1)!}(x - x^*)^{n+1}.$$

(Note that Theorem 1.2 is a special case of this theorem with $n \equiv 0$.)

Theorem 1.8. Fundamental Theorem of Algebra

Let $p(x) \equiv a_n x^n + a_{n-1}x^{n-1} + \cdots + a_1 x + a_0$ be an nth degree polynomial with $n \geq 1$, $a_n \neq 0$. Then there exist uniquely n constants, $\{r_j\}_{j=1}^{n}$, ("zeros of $p(x)$") such that

$$p(x) = a_n(x - r_1)(x - r_2) \cdots (x - r_n)$$

where the r_j's need not all be distinct or all real. (This is not quite the usual statement of the Fundamental Theorem of Algebra, but this form is more suited to our purposes.)

2

Solution of Linear Systems of Equations

2.1 INTRODUCTION

One of the most basic and important problems in science and engineering is the accurate and efficient simultaneous solution of a system of m linear equations in n unknowns. This problem is written in the form

$$\begin{aligned} a_{11}x_1 + a_{12}x_2 + \cdots + a_{1n}x_n &= b_1 \\ a_{21}x_1 + a_{22}x_2 + \cdots + a_{2n}x_n &= b_2 \\ \vdots \qquad\qquad\qquad\qquad & \ \ \vdots \\ a_{m1}x_1 + a_{m2}x_2 + \cdots + a_{mn}x_n &= b_m \end{aligned} \tag{2.1}$$

where each a_{ij}, $1 \le i \le m$, $1 \le j \le n$, and each b_i, $1 \le i \le m$, are known values, and x_j, $1 \le j \le n$, are the unknowns for which we must solve. Problems of this type arise in almost all disciplines and most elementary algebra courses introduce the elimination of variables techniques for the solution of (2.1).

For the study of various numerical methods for solving (2.1) and their implementation on the computer, we will first review some of the basic ideas associated with matrices. In the following we use the common notation that any $(k \times \ell)$ matrix C will be written as "$C = (c_{ij})$" where c_{ij} is the element in the ith row and jth column of C and where it is understood that $1 \le i \le k$ and $1 \le j \le \ell$. Let $R = (r_{ij})$ be an $(m \times n)$ matrix and let $S = (s_{ij})$ be an $(n \times p)$ matrix. Then the *product* of R and S is defined and is an $(m \times p)$ matrix $T = (t_{ij})$ where $t_{ij} = \Sigma_{k=1}^{n} r_{ik}s_{kj}$. We write $T = RS$ to denote the product of R and S. If α is any real number, then the product αR is formed by multiplying each entry of R by the number α. If $Q = (q_{ij})$ is any $(m \times n)$ matrix, then the *sum* of R and Q is defined to be the $(m \times n)$ matrix $V = (v_{ij})$ where each element v_{ij} of V is given by $v_{ij} = r_{ij} + q_{ij}$. We write $V = R + Q$ to denote the sum of the two matrices R and Q. Note that the product of R and Q cannot be defined if $m \ne n$, and the sum of R and S cannot be defined unless R and S both have the same number of rows and columns. Also the product SR is not defined unless $m = p$. Moreover, even for $m = p = n$, SR need not equal RS. It is true, however, that if $R + Q$ is defined, then so is $Q + R$ and also $R + Q = Q + R$. [We say that two $(m \times n)$ matrices are *equal* if and only if each of their corresponding entries are equal.]

EXAMPLE 2.1. Let

$$R = \begin{bmatrix} 1 & 2 & 1 & 1 \\ -1 & 0 & 1 & 1 \\ 1 & 0 & 2 & 3 \end{bmatrix}, \quad S = \begin{bmatrix} 1 & 1 & 1 \\ 1 & 0 & 1 \\ -1 & 1 & -1 \\ 2 & 0 & 3 \end{bmatrix}, \quad \text{and} \quad Q = \begin{bmatrix} 1 & 0 & 1 \\ -1 & 1 & -1 \\ 2 & 1 & 0 \end{bmatrix}.$$

Then

$$RS = \begin{bmatrix} 4 & 2 & 5 \\ 0 & 0 & 1 \\ 5 & 3 & 8 \end{bmatrix}, \quad SR = \begin{bmatrix} 1 & 2 & 4 & 5 \\ 2 & 2 & 3 & 4 \\ -3 & -2 & -2 & -3 \\ 5 & 4 & 8 & 11 \end{bmatrix},$$

but RQ is not defined and $RS \neq SR$. Also,

$$RS + 2Q = \begin{bmatrix} 6 & 2 & 7 \\ -2 & 2 & -1 \\ 9 & 5 & 8 \end{bmatrix},$$

but $R + Q$ and $S + Q$ are not defined.

Using matrix notation, we may write the linear system (2.1) in the compact form

$$A\mathbf{x} = \mathbf{b} \tag{2.2}$$

where $A = (a_{ij})$ represents the $(m \times n)$ *coefficient matrix* for the system and where

$$\mathbf{x} = \begin{bmatrix} x_1 \\ x_2 \\ \vdots \\ x_n \end{bmatrix} \quad \text{and} \quad \mathbf{b} = \begin{bmatrix} b_1 \\ b_2 \\ \vdots \\ b_m \end{bmatrix} \tag{2.3}$$

are $(n \times 1)$ and $(m \times 1)$ column vectors, respectively.

EXAMPLE 2.2. Consider the linear system

$$\begin{aligned} x_1 \qquad + x_3 &= 1 \\ -x_1 + x_2 - x_3 &= 0 \\ 2x_1 + x_2 \qquad &= 0 \end{aligned} \quad \text{(or in matrix form)} \quad \begin{bmatrix} 1 & 0 & 1 \\ -1 & 1 & -1 \\ 2 & 1 & 0 \end{bmatrix} \begin{bmatrix} x_1 \\ x_2 \\ x_3 \end{bmatrix} = \begin{bmatrix} 1 \\ 0 \\ 0 \end{bmatrix}.$$

This system is easily solved by elimination of variables. Adding the first and second equations yields $x_2 = 1$. Then the third equation gives $x_1 = -1/2$, and the first equation gives $x_3 = 3/2$.

In the following sections we will be primarily concerned with solving (2.1) in the special case in which A is $(n \times n)$ and $\mathbf{b}$ is $(n \times 1)$. In this case, the number of equations equals the number of unknowns and A is said to be a *square* matrix. (If $m \neq n$, A is said to be *rectangular*.) If one considers (2.1) with $m =$

n, the coefficient matrix A is called *nonsingular* if there always exists a unique solution, $\mathbf{x}$, to (2.1) for any choice of $b_1, b_2, \ldots, b_n$. It can be shown in this case that there exists a unique $(n \times n)$ matrix, A^{-1}, called the *inverse of* A such that $A^{-1}A = AA^{-1} = I$ where I is the $(n \times n)$ *identity matrix*:

$$I = \begin{bmatrix} 1 & 0 & 0 & \cdots & 0 \\ 0 & 1 & 0 & \cdots & 0 \\ \vdots & & & & \\ 0 & 0 & 0 & \cdots & 1 \end{bmatrix}. \tag{2.4}$$

It is easily seen that if B is any $(n \times n)$ matrix, then $IB = BI = B$. Thus if we knew A^{-1}, we could multiply both sides of (2.2) by A^{-1} and obtain the solution $\mathbf{x}$ by

$$A^{-1}(A\mathbf{x}) = (A^{-1}A)\mathbf{x} = I\mathbf{x} = \mathbf{x} = A^{-1}\mathbf{b}.$$

It is often very useful to have an alternative way of expressing the product $A\mathbf{x}$ where $A = (a_{ij})$ is an $(m \times n)$ matrix, and $\mathbf{x}$ is an $(n \times 1)$ column vector as given in (2.3). Let $\mathbf{A}_k$ denote the kth column of A:

$$\mathbf{A}_k = \begin{bmatrix} a_{1k} \\ a_{2k} \\ \vdots \\ a_{mk} \end{bmatrix}, \qquad \text{for } 1 \le k \le n.$$

Then each $\mathbf{A}_k$ is an $(m \times 1)$ column vector. By the rules of matrix multiplication, the jth component of the vector $A\mathbf{x}$ is given by $c_j = \sum_{k=1}^{n} a_{jk}x_k$ for each j, $1 \le j \le m$. We claim that each c_j is also the jth component of the $(m \times 1)$ vector

$$x_1\mathbf{A}_1 + x_2\mathbf{A}_2 + \cdots + x_n\mathbf{A}_n.$$

This claim is easily seen since the jth component of each vector $x_k\mathbf{A}_k$ is equal to x_ka_{jk}, and adding these for $1 \le k \le n$ yields $\sum_{k=1}^{n} x_ka_{jk} = c_j$. Thus we have shown that $A\mathbf{x}$ may be written in the form $A\mathbf{x} = x_1\mathbf{A}_1 + x_2\mathbf{A}_2 + \cdots + x_n\mathbf{A}_n$. (We will often find it useful to express the matrix A symbolically in column form as $A = [\mathbf{A}_1, \mathbf{A}_2, \ldots, \mathbf{A}_n]$.)

Similarly from the definition of matrix multiplication, it is easily verified that if A and B are any two $(n \times n)$ matrices, then for any j, $1 \le j \le n$, the jth column of the product AB is equal to $A\mathbf{B}_j$ where $B = [\mathbf{B}_1, \mathbf{B}_2, \ldots, \mathbf{B}_n]$. This observation shows how we can compute A^{-1}. To be specific, let $\mathbf{e}_j$, for $1 \le j \le n$, be the $(n \times 1)$ column vector that has a 1 in the jth component and zeros in all other components. Thus $\mathbf{e}_j$ is the jth column vector of I, or $I = [\mathbf{e}_1, \mathbf{e}_2, \ldots, \mathbf{e}_n]$. If A is nonsingular, then for $1 \le j \le n$, the jth column of AA^{-1} equals $\mathbf{e}_j$ since $AA^{-1} = I$. We can therefore find the jth column of A^{-1}, $\mathbf{x}^{(j)}$, by solving the linear system

$$A\mathbf{x}^{(j)} = \mathbf{e}_j.$$

We can use any of the numerical methods described in the following sections to find A^{-1} by solving each of the n systems: $A\mathbf{x}^{(j)} = \mathbf{e}_j$, $1 \le j \le n$, as in Example 2.3.

EXAMPLE 2.3. Consider the two linear systems

$$\begin{aligned} x_1 \quad\quad + x_3 &= 0 \\ -x_1 + x_2 - x_3 &= 1 \\ 2x_1 + x_2 \quad\quad &= 0 \end{aligned} \qquad \text{and} \qquad \begin{aligned} x_1 \quad\quad + x_3 &= 0 \\ -x_1 + x_2 - x_3 &= 0 \\ 2x_1 + x_2 \quad\quad &= 1. \end{aligned}$$

Proceeding exactly as in Example 2.2, we obtain $x_1 = -1/2$, $x_2 = 1$, $x_3 = 1/2$ as the solution to the first system and $x_1 = 1/2$, $x_2 = 0$, $x_3 = -1/2$ as the solution to the second. Now, the coefficient matrix of all three systems in Example 2.2 and Example 2.3 is the matrix Q of Example 2.1. Taking these three solutions as column vectors, we get

$$Q^{-1} = \begin{bmatrix} -\frac{1}{2} & -\frac{1}{2} & \frac{1}{2} \\ 1 & 1 & 0 \\ \frac{3}{2} & \frac{1}{2} & -\frac{1}{2} \end{bmatrix},$$

and we may easily verify that $QQ^{-1} = Q^{-1}Q = I$. Thus, if $\mathbf{b} = \begin{bmatrix} 1 \\ 1 \\ 1 \end{bmatrix}$, for example, then the solution of $Q\mathbf{x} = \mathbf{b}$ is given by

$$\mathbf{x} = Q^{-1}\mathbf{b} = \begin{bmatrix} -\frac{1}{2} & -\frac{1}{2} & \frac{1}{2} \\ 1 & 1 & 0 \\ \frac{3}{2} & \frac{1}{2} & -\frac{1}{2} \end{bmatrix} \begin{bmatrix} 1 \\ 1 \\ 1 \end{bmatrix} = \begin{bmatrix} -\frac{1}{2} \\ 2 \\ \frac{3}{2} \end{bmatrix}.$$

The general problem of solving several linear systems with the same coefficient matrix will be discussed in more detail in Section 2.2.4, with an emphasis on efficient programming for a digital computer.

Several *theoretical* tests from the theory of matrices can be used to determine whether a square matrix is nonsingular or not. Two of the most familiar characterizations of nonsingular matrices are these.

1) Let θ denote the $(n \times 1)$ column vector that has all components equal to zero. (θ is called the *zero vector*.) A is nonsingular if and only if θ is the only solution of $A\mathbf{x} = \theta$. (Note: $\mathbf{x} = \theta$ is always a solution of $A\mathbf{x} = \theta$.)
2) A is nonsingular if and only if the determinant of A is nonzero; that is, $\det(A) \ne 0$. (We assume here that the reader is familiar with the rules for calculating determinants of square matrices.)

EXAMPLE 2.4. Let

$$A = \begin{bmatrix} 1 & 2 & 3 & 1 \\ -1 & 0 & 2 & 1 \\ 2 & 1 & -3 & 0 \\ 1 & 1 & 2 & 1 \end{bmatrix}.$$

Recall that the evaluation of det(A) by cofactor expansion of the first row yields

$$\det(A) = \begin{vmatrix} 0 & 2 & 1 \\ 1 & -3 & 0 \\ 1 & 2 & 1 \end{vmatrix} - 2\begin{vmatrix} -1 & 2 & 1 \\ 2 & -3 & 0 \\ 1 & 2 & 1 \end{vmatrix} + 3\begin{vmatrix} -1 & 0 & 1 \\ 2 & 1 & 0 \\ 1 & 1 & 1 \end{vmatrix} - \begin{vmatrix} -1 & 0 & 2 \\ 2 & 1 & -3 \\ 1 & 1 & 2 \end{vmatrix}$$

$$= \left\{0\begin{vmatrix} -3 & 0 \\ 2 & 1 \end{vmatrix} - 2\begin{vmatrix} 1 & 0 \\ 1 & 1 \end{vmatrix} + \begin{vmatrix} 1 & -3 \\ 1 & 2 \end{vmatrix}\right\} - 2\left\{-\begin{vmatrix} -3 & 0 \\ 2 & 1 \end{vmatrix} - 2\begin{vmatrix} 2 & 0 \\ 1 & 1 \end{vmatrix} + \begin{vmatrix} 2 & -3 \\ 1 & 2 \end{vmatrix}\right\}$$

$$+ 3\left\{-\begin{vmatrix} 1 & 0 \\ 1 & 1 \end{vmatrix} - 0\begin{vmatrix} 2 & 0 \\ 1 & 1 \end{vmatrix} + \begin{vmatrix} 2 & 1 \\ 1 & 1 \end{vmatrix}\right\} - \left\{-\begin{vmatrix} 1 & -3 \\ 1 & 2 \end{vmatrix} - 0\begin{vmatrix} 2 & -3 \\ 1 & 2 \end{vmatrix} + 2\begin{vmatrix} 2 & 1 \\ 1 & 1 \end{vmatrix}\right\}$$

$$= (0 - 2 + 5) - 2(3 - 4 + 7) + 3(-1 - 0 + 1) - (-5 - 0 + 2) = -6.$$

Thus A is nonsingular since $\det(A) \neq 0$. (Note that the evaluation above could be facilitated by elementary row and column operations, but it does not serve our purpose to go into this here.) Similarly it is easy to verify that if

$$\begin{bmatrix} 1 & 2 & 3 & 1 \\ -1 & 0 & 2 & 1 \\ 2 & 1 & -3 & 0 \\ 1 & 1 & 2 & 1 \end{bmatrix} \begin{bmatrix} x_1 \\ x_2 \\ x_3 \\ x_4 \end{bmatrix} = \begin{bmatrix} 0 \\ 0 \\ 0 \\ 0 \end{bmatrix},$$

then $x_1 = x_2 = x_3 = x_4 = 0$.

We hasten to add that while determinants and inverses are of great importance to a theoretical analysis of the solution of (2.1), they are rarely recommended for actual use in computational procedures. We will return to this point later and confine ourselves for the moment to saying that procedures of this type, such as the well-known Cramer's Rule for solving (2.1) or computing A^{-1} and forming $\mathbf{x} = A^{-1}\mathbf{b}$, are not competitive methods.

For theoretical purposes it is also important to recall that if A and B are $(n \times n)$ nonsingular matrices, then the product (AB) is nonsingular and $(AB)^{-1} = B^{-1}A^{-1}$ (Problem 12). In the case that A is singular, then A has no inverse and the equation $A\mathbf{x} = \mathbf{b}$ will have either no solution or infinitely many solutions. It will be demonstrated later that a direct computational method such as Gauss elimination (Section 2.2.1) can be implemented to display conveniently all of the solutions (if any exist) of $A\mathbf{x} = \mathbf{b}$ even if A is rectangular or if A is square but singular. (See Example 2.6.)

In the following sections on numerical procedures for solving the equation $A\mathbf{x} = \mathbf{b}$, we will be working extensively with *triangular* matrices. A square matrix $U = (u_{ij})$ is said to be *upper triangular* if $u_{ij} = 0$ whenever $i > j$ (that is, all entries below the main diagonal are zero). Similarly, a square matrix $L = (\ell_{ij})$ is called *lower triangular* if all entries above the main diagonal are zero (that is, $\ell_{ij} = 0$ whenever $i < j$). A square matrix $D = (d_{ij})$ is called *diagonal* if $d_{ij} = 0$ whenever $i \neq j$. Thus a diagonal matrix is simultaneously upper triangular and lower triangular. Finally, we wish to recall the definition of the *transpose* of a matrix. If $A = (a_{ij})$ is an $(m \times n)$ matrix, then the transpose of A, denoted A^T, is an $(n \times m)$ matrix where the ijth entry of A^T is a_{ji}; $1 \leq j \leq n$, $1 \leq i \leq m$; that is,

A^T is formed from A by interchanging the rows and columns of A. It is easily verified (Problem 13) that $(AB)^T = B^T A^T$ whenever the product of AB is defined and that $(A + B)^T = A^T + B^T$ whenever the sum is defined.

EXAMPLE 2.5. Let

$$A = \begin{bmatrix} 2 & 1 & 3 \\ 6 & 0 & 4 \end{bmatrix}; \qquad \text{then} \qquad A^T = \begin{bmatrix} 2 & 6 \\ 1 & 0 \\ 3 & 4 \end{bmatrix}.$$

For an example of a triangular matrix, let L be given by

$$L = \begin{bmatrix} 1 & 0 & 0 \\ 2 & 4 & 0 \\ 1 & 3 & 5 \end{bmatrix}.$$

Then L is lower triangular and, in addition, L^T is upper triangular. Transposes are used frequently in connection with column and row vectors. For example, let

$$\mathbf{x} = \begin{bmatrix} 1 \\ 3 \\ 0 \end{bmatrix} \qquad \text{and} \qquad \mathbf{y} = \begin{bmatrix} 2 \\ 1 \\ 2 \end{bmatrix}.$$

Then $\mathbf{x}$ and $\mathbf{y}$ are (3×1) column vectors [or (3×1) matrices if we choose to think of them in that fashion]. So, $\mathbf{x}^T = (1, 3, 0)$ and $\mathbf{y}^T = (2, 1, 2)$. According to the rules of matrix multiplication, we can form both of the products $\mathbf{x}^T\mathbf{y}$ and $\mathbf{x}\mathbf{y}^T$:

$$\mathbf{x}^T\mathbf{y} = 5, \qquad \mathbf{x}\mathbf{y}^T = \begin{bmatrix} 2 & 1 & 2 \\ 6 & 3 & 6 \\ 0 & 0 & 0 \end{bmatrix}.$$

Many other fundamental concepts from matrix theory could be mentioned here: *rank*, *augmented matrix*, *echelon matrix*, etc. However since we will be primarily concerned with the computational solution of $A\mathbf{x} = \mathbf{b}$ where A is $(n \times n)$ and nonsingular, a review of this additional material is not necessary for our purposes. Some additional concepts (such as *symmetric matrix*, *positive definite*, *diagonally dominant*, etc.) are introduced as they become necessary, but they are explained and illustrated in detail so that the reader with a calculus background plus some very rudimentary experience with matrices should have no difficulty with the material.

PROBLEMS, SECTION 2.1

1. Using the definition of matrix multiplication, write a computer program to form the product of two arbitrary $(n \times n)$ matrices where $n \leq 10$.
2. In Problem 1, let A and B be the two $(n \times n)$ matrices to be multiplied and suppose that $C = AB$ denotes the product of A and B. It is most natural to store A and B as doubly subscripted arrays [so, for example, the ijth entry of A is $A(I, J)$]. In

applications involving large matrices (or small computers), storage requirements can sometimes present problems. In Problem 1 if A, B, and C are all saved, then $3n^2$ locations are needed. Suppose that A is no longer needed once the product C has been formed. How might the program in Problem 1 be modified so that only $2n^2 + n$ locations are needed to form C? [Hint: Once the first row of C is formed, the first row of A is no longer needed.]

3. How many multiplications must be performed to form the product of two $(n \times n)$ matrices?

4. How many multiplications must be performed to form the product $L\mathbf{x}$ when L is an $(n \times n)$ lower triangular matrix and $\mathbf{x}$ is an $(n \times 1)$ column vector? (Do not count multiplications by zero.)

***5.** Write a computer program to evaluate the determinant of an arbitrary (5×5) matrix, using a cofactor expansion along the first row. (This is a fairly challenging programming problem.)

6. How many multiplications must be performed to evaluate the determinant of an $(n \times n)$ matrix according to the procedure of Problem 5 when $n = 3$, when $n = 4$, when $n = 5$, and for arbitrary n? (For $n = 10$, more than 3,000,000 multiplications are required. If a person were able to multiply two numbers and record the result at a rate of one per second, it would require 126 eight-hour days to find the determinant of a (10×10) matrix using a cofactor expansion.)

7. Let $L = (\ell_{ij})$ be a lower triangular (5×5) matrix such that $\ell_{11}\ell_{22}\ell_{33}\ell_{44}\ell_{55} \neq 0$. By considering the equation $L\mathbf{x} = \theta$, show that L is nonsingular. [Hint: Testing $L\mathbf{x} = \theta$ is criterion (1) for nonsingularity.] Extend your proof to an arbitrary $(n \times n)$ lower triangular matrix L such that $\ell_{11}\ell_{22} \ldots \ell_{nn} \neq 0$.

8. Let L be as in Problem 7, but this time suppose that $\ell_{44} = 0$ and $\ell_{55} \neq 0$. Show there is a nonzero vector $\mathbf{x}$ such that $L\mathbf{x} = \theta$.

***9.** Let $L = (\ell_{ij})$ be an $(n \times n)$ lower triangular matrix. Show that L is singular if and only if $\ell_{11}\ell_{22} \ldots \ell_{nn} = 0$.

10. Let T and S be $(n \times n)$ upper triangular matrices. Use the definition of matrix multiplication to show that the product ST is also upper triangular.

***11.** Let A be a lower (upper) triangular nonsingular matrix. Show that A^{-1} is also lower (upper) triangular. [Hint: Consider the equation $AA^{-1} = I$, entry by entry.]

12. If A and B are $(n \times n)$ nonsingular matrices, show that AB and BA are also nonsingular. Furthermore, show that $(AB)^{-1} = B^{-1}A^{-1}$.

***13.** a) If A is $(r \times s)$ and B is $(r \times s)$, show that $(A + B)^T = A^T + B^T$.

b) If A is $(r \times s)$ and B is $(s \times p)$, show that $(AB)^T = B^T A^T$.

c) Show $(A^T)^T = A$.

2.2 DIRECT METHODS

The term *direct method* refers to a numerical procedure that can be executed in a finite number of steps. Direct methods are in contrast to *iterative methods*, which generate an infinite sequence of approximations that (it is hoped) converge to the solution. We will discuss iterative methods in Section 2.4. The two most common types of direct methods for solving the system (2.1) are elimination methods and factorization methods.

2.2.1. Gauss Elimination

Gauss elimination is the familiar variable elimination technique whereby the variables are eliminated one at a time to reduce the original system to an equivalent triangular system. The first step in this procedure is to replace the ith equation by the equation that is the result of multiplying the first equation by $(-a_{i1}/a_{11})$ and adding it to the original ith equation. Proceeding thus for $i = 2, 3, \ldots, n$, we obtain a system of equations equivalent to (2.1):

$$\begin{aligned} a_{11}x_1 + a_{12}x_2 + \cdots + a_{1n}x_n &= b_1 \\ a_{22}^{(1)}x_2 + \cdots + a_{2n}^{(1)}x_n &= b_2^{(1)} \\ &\vdots \\ a_{n2}^{(1)}x_2 + \cdots + a_{nn}^{(1)}x_n &= b_n^{(1)}. \end{aligned} \tag{2.5}$$

(We assume here that $a_{11} \neq 0$. If $a_{11} = 0$, we find an $a_{i1} \neq 0$, interchange the first and ith rows, and proceed in the same fashion.) We continue in this manner until the system is in an equivalent triangular form:

$$\begin{aligned} a'_{11}x_1 + a'_{12}x_2 + a'_{13}x_3 + \cdots + a'_{1,\,n-1}x_{n-1} + a'_{1n}x_n &= b'_1 \\ a'_{22}x_2 + a'_{23}x_3 + \cdots + a'_{2,\,n-1}x_{n-1} + a'_{2n}x_n &= b'_2 \\ a'_{33}x_3 + \cdots + a'_{3,\,n-1}x_{n-1} + a'_{3n}x_n &= b'_3 \\ &\vdots \\ a'_{n-1,\,n-1}x_{n-1} + a'_{n-1,\,n}x_n &= b'_{n-1} \\ a'_{n,\,n}x_n &= b'_n. \end{aligned} \tag{2.6}$$

The solution to this triangular system is now easily found by *backsolving*; that is, by solving the last equation for x_n, then the $(n-1)$st equation for x_{n-1}, and continuing until each x_k where $k = n, n-1, n-2, \ldots 1$, is determined.

EXAMPLE 2.6. As a specific case to demonstrate Gauss elimination, consider the linear system

$$\begin{aligned} 5x_1 + 7x_2 + 6x_3 + 5x_4 &= 23 \\ 7x_1 + 10x_2 + 8x_3 + 7x_4 &= 32 \\ 6x_1 + 8x_2 + 10x_3 + 9x_4 &= 33 \\ 5x_1 + 7x_2 + 9x_3 + 10x_4 &= 31. \end{aligned}$$

Proceeding as indicated by (2.5), multiply the first equation by $-7/5$ and add the result to the second equation to eliminate x_1 in the new second equation. Proceeding similarly to eliminate x_1 from the third and fourth equations, we obtain an equivalent reduced linear system

$$\begin{aligned} 5x_1 + 7x_2 + 6x_3 + 5x_4 &= 23 \\ 0.2x_2 - 0.4x_3 &= -0.2 \\ -0.4x_2 + 2.8x_3 + 3x_4 &= 5.4 \\ 3x_3 + 5x_4 &= 8 \end{aligned}$$

that corresponds to (2.5). Eliminating x_2 from the third equation gives the equivalent linear system

$$\begin{aligned} 5x_1 + 7x_2 + 6x_3 + 5x_4 &= 23 \\ 0.2x_2 - 0.4x_3 \qquad &= -0.2 \\ 2x_3 + 3x_4 &= 5 \\ 3x_3 + 5x_4 &= 8. \end{aligned}$$

Finally by eliminating x_3 from the fourth equation, we obtain a linear system in the triangular form of (2.6):

$$\begin{aligned} 5x_1 + 7x_2 + 6x_3 + 5x_4 &= 23 \\ 0.2x_2 - 0.4x_3 \qquad &= -0.2 \\ 2x_3 + 3x_4 &= 5 \\ 0.5x_4 &= 0.5. \end{aligned}$$

Backsolving the triangular linear system, we find from the fourth equation that $x_4 = 1$. Because we have x_4, the third equation yields $x_3 = 1$; with x_3 and x_4 known, the second equation says $x_2 = 1$; and finally from the first equation, we obtain $x_1 = 1$. The coefficient matrix for the original system is

$$A = \begin{bmatrix} 5 & 7 & 6 & 5 \\ 7 & 10 & 8 & 7 \\ 6 & 8 & 10 & 9 \\ 5 & 7 & 9 & 10 \end{bmatrix}$$

This matrix is due to T.S. Wilson (see Todd, 1962). The matrix has several interesting properties and will serve as one example when we discuss topics such as error analysis and condition numbers.

Continuing this example of solution by variable elimination, consider the (3×4) linear system

$$\begin{aligned} 5x_1 + 7x_2 + 6x_3 + 5x_4 &= 23 \\ 7x_1 + 10x_2 + 8x_3 + 7x_4 &= 32 \\ 6x_1 + 8x_2 + 10x_3 + 9x_4 &= 33. \end{aligned}$$

Proceeding exactly as above, we arrive at the equivalent system

$$\begin{aligned} 5x_1 + 7x_2 + 6x_3 + 5x_4 &= 23 \\ 0.2x_2 - 0.4x_3 \qquad &= -0.2 \\ 2x_3 + 3x_4 &= 5. \end{aligned}$$

Here we see that either x_3 or x_4 can be chosen completely arbitrarily. For example, let x_4 be arbitrary; then $x_3 = (1/2)(5 - 3x_4)$. Continuing, we obtain $x_2 = 4 - 3x_4$ and $x_1 = 5x_4 - 4$, and we may substitute any value for x_4 and have a corresponding solution of the system. In particular if we let $x_4 = 1$, we obtain $x_3 = x_2 = x_1 = 1$ as above. This example of a (3×4) system shows how Gauss elimination can be used for rectangular systems as well as for square systems.

In the simple (4×4) example above, no row interchanges were necessary. When Gauss elimination is programmed for use on a digital computer, we will clearly have to include a statement to check whether we might be dividing by zero. Thus, given (2.1) to solve, the first step would be to test a_{11} to determine whether or not $a_{11} = 0$, and then to perform a row interchange if $a_{11} = 0$. Having obtained the first reduced system (2.5), we would again test $a_{22}^{(1)}$ to see if $a_{22}^{(1)} = 0$ and perform a row interchange if necessary. Example 2.7 below illustrates the sort of numerical results that a simple Gauss elimination program will produce.

EXAMPLE 2.7. The basic steps of Gauss elimination that are outlined above were programmed to solve the system in Example 2.6. The program was executed in single precision; and as can be seen, the errors in the computed solution range from 0.000583 to 0.00009. For purposes of comparison, the final triangular form of the matrix was printed out as shown and can be compared with the exact triangular form displayed in Example 2.6.

SOLUTION VECTOR
0.999417E 00
0.100035E 01
0.100015E 01
0.999910E 00

THE TRIANGULARIZED MATRIX

0.500000E 01	0.700000E 01	0.600000E 01	0.500000E 01
0.000000E 00	0.200003E 00	−0.399998E 00	0.190735E-05
0.000000E 00	0.000000E 00	0.200002E 01	0.300000E 01
0.000000E 00	0.000000E 00	0.000000E 00	0.500038E 00

This computer example illustrates the effects of round-off error, and we note that this computer solution would probably not be satisfactory in most practical problems. As we shall see later, the coefficient matrix of Example 2.6 is moderately "ill-conditioned" (see Section 2.3.3). Thus even a few small computer-round-off errors will induce relatively large errors in the computed solution.

This program was executed again, in double precision, and gave the solution vector correct to as many places as are listed:

SOLUTION VECTOR
0.10000000D 01
0.10000000D 01
0.10000000D 01
0.10000000D 01

THE TRIANGULARIZED MATRIX

0.500000D 01	0.700000D 01	0.600000D 01	0.500000D 01
0.000000D 00	0.200000D 00	−0.400000D 00	0.444089D-15
0.000000D 00	0.000000D 00	0.200000D 01	0.300000D 01
0.000000D 00	0.000000D 00	0.000000D 00	0.500000D 00

Since linear systems of any appreciable size are normally solved on the computer, round-off errors can cause problems and the technique of Gauss elimination must be examined more carefully. In addition to the question of rounding errors, other aspects must be considered for practical and efficient implementation of numerical procedures on the machine. Three aspects that can influence the choice of an algorithm are storage requirements, round-off errors, and execution time.

2.2.2. Operations Counts

For the system (2.1), storage requirements are clearly linked to the size of the system, n. In order to give a feeling for the execution time and rounding errors of direct methods, an *operations count* is customary. An operations count is a calculation of the number of arithmetic operations necessary to carry out the direct method. One rationale for an operations count is that execution time and the error caused by round-off in the computed solution are both related to the total number of arithmetic operations. For Gauss elimination, multiplication and addition are (essentially) the only arithmetic operations performed. (We consider "division" as a multiplication and "subtraction" as an addition.) To illustrate how an operations count is made, we will count the multiplications necessary to solve the system (2.1) by Gauss elimination. The reader can quickly verify that the necessary number of additions is about the same as the number of multiplications.

To transform (2.1) to the equivalent system (2.5) requires $(n - 1)(n + 1)$ multiplications. This result is seen by noting that to eliminate the variable x_1 in the ith equation of (2.1), we multiply the first equation $a_{11}x_1 + a_{12}x_2 + \cdots + a_{1n}x_n = b_1$ by the scalar $-a_{i1}/a_{11}$ and add the resulting multiple of the first equation to the ith equation. It takes one multiplication to form the scalar $-a_{i1}/a_{11}$ (we don't count multiplications by -1) and an additional n multiplications to form each of the products.

$$\frac{-a_{i1}}{a_{11}}a_{12}, \frac{-a_{i1}}{a_{11}}a_{13}, \ldots, \frac{-a_{i1}}{a_{11}}a_{1n}, \qquad \frac{-a_{i1}}{a_{11}}b_1.$$

Note that there is no need to form the product $(-a_{i1}/a_{11})a_{11}$ since we *already* know we are going to have 0 as the scalar multiplying x_1 in the new ith equation. Thus multiplying the first equation by $-a_{i1}/a_{11}$ and adding the result to the ith equation to produce the new ith equation

$$a_{i2}^{(1)}x_2 + a_{i3}^{(1)}x_3 + \cdots + a_{in}^{(1)}x_n = b_i^{(1)} \tag{2.7}$$

takes $(n + 1)$ multiplications (and n additions). We have to perform this variable elimination in rows $2, 3, \ldots, n$ and hence require a total of $(n - 1)(n + 1)$ multiplications to produce the equivalent reduced system (2.5).

Given the system (2.5), the same analysis shows that n multiplications are needed to eliminate x_2 in equations $3, 4, \ldots, n$ of (2.5). Thus the next step of Gauss elimination that results in

$$\begin{aligned}
a_{11}^{(2)}x_1 + a_{12}^{(2)}x_2 + a_{13}^{(2)}x_3 + \cdots + a_{1n}^{(2)}x_n &= b_1^{(2)} \\
a_{22}^{(2)}x_2 + a_{23}^{(2)}x_3 + \cdots + a_{2n}^{(2)}x_n &= b_2^{(2)} \\
a_{33}^{(2)}x_3 + \cdots + a_{3n}^{(2)}x_n &= b_3^{(2)} \\
\vdots \qquad\qquad \vdots \quad &\quad \vdots \\
a_{n3}^{(2)}x_3 + \cdots + a_{nn}^{(2)}x_n &= b_n^{(2)}
\end{aligned} \tag{2.8}$$

takes $(n - 2)n$ multiplications. Continuing until the triangular system (2.6) is obtained, we have performed a total of $(n - 1)(n + 1) + (n - 2)n + \cdots + 1 \cdot 3$ multiplications. To evaluate this sum, we write it symbolically as

$$\sum_{j=1}^{n-1} j(j + 2) = \sum_{j=1}^{n-1} j^2 + 2 \sum_{j=1}^{n-1} j. \tag{2.9}$$

In this form, using the well-known formulas for the sum of the first $(n - 1)$ integers and the sum of the first $(n - 1)$ integers squared, we find that a total of

$$\frac{n(n - 1)(2n - 1)}{6} + n(n - 1) = \frac{n(n - 1)(2n + 5)}{6} \tag{2.10}$$

multiplications are needed to produce an equivalent triangular system from a square $(n \times n)$ system.

That essentially $n^3/3$ multiplications are needed to triangularize (2.1) turns out to be quite an important observation. For the moment, however, we have not completed the operations count of Gauss elimination. We have yet to count the multiplications necessary to backsolve the triangular system (2.6). We leave it to the reader to show that $n(n + 1)/2$ multiplications are required to solve (2.6). Thus the bulk of the computation involved in Gauss elimination is in triangularizing the coefficient matrix.

2.2.3. Pivoting and Scaling in Gauss Elimination

One modification of Gauss elimination is relatively easy to program and should in most cases reduce the effects of round-off error. To illustrate, we first consider the following example from Forsythe (1967); for simplicity, we consider a three-digit floating-decimal machine. [Recall, we assumed in Chapter 1 that on such a machine the arithmetic operations are done in a six-digit mode and then rounded to three digits. For example, this machine will record $(1 - 10000)$ as -10000.]

EXAMPLE 2.8.*

$$\begin{aligned} 0.0001x + 1.00y &= 1.00 \\ 1.00x + 1.00y &= 2.00 \end{aligned}$$

The true solution rounded to five decimals is $x = 1.00010$ and $y = 0.99990$. If we proceed as above, without rearranging the equations, we obtain $-10000y = -10000$ and so we get $y = 1.00$ and $x = 0.00$, quite different from the true solution. However if we rewrite

---

*Reprinted with permission from G. E. Forsythe, "Today's computational methods of linear algebra," SIAM Review, Vol. 9, 1967. Copyright © 1967 by Society for Industrial and Applied Mathematics.

the system as

$$\begin{aligned} 1.00x + 1.00y &= 2.00 \\ 0.0001x + 1.00y &= 1.00, \end{aligned}$$

then we obtain the answer $x = 1.00$ and $y = 1.00$, a remarkable improvement

This example demonstrates one source of round-off error; namely, rounding errors can become an important factor in adding two numbers if one is much larger than the other, and the significant digits of the smaller one are essentially lost. A technique to compensate for this loss is called *pivoting* and is explained below. In ordinary Gauss elimination if $|a_{11}|$ is relatively small in comparison with some $|a_{i1}|$, then the multiplicative factor $(-a_{i1}/a_{11})$ is a relatively larger number in magnitude. Let $\lambda = -a_{i1}/a_{11}$; then the result of multiplying the first equation by λ and adding to the ith equation is the new ith equation

$$(a_{i2} + \lambda a_{12})x_2 + \cdots + (a_{in} + \lambda a_{1n})x_n = b_i + \lambda b_1.$$

If λ is large, then significant rounding errors can occur and can be quite harmful. In particular if λ is large, then adding a multiple of λ times the first equation to the ith equation may destroy most of the information in the ith equation (if λa_{1j} is large relative to a_{ij} for $2 \leq j \leq n$, the effect is almost the same as replacing the ith equation by a multiple of the first equation). If λa_{1j} is small relative to a_{ij}, rounding can still occur. However in this case the effect will be minimal in that the information in the ith equation will be left nearly intact.

With these ideas in mind, we modify the steps of Gauss elimination as follows. Find a row index I such that

$$|a_{I1}| = \max_{1 \leq i \leq n} |a_{i1}|,$$

and rewrite the system so that the Ith equation becomes the first equation. (Note that this altered system still has the same solution as the original system.) Now the multiplicative constant to eliminate the coefficients of x_1 is $(-a_{i1}/a_{I1})$, which is the relatively smallest factor we could obtain in this fashion. Thus the effect of the round-off error is reduced in the remaining equations. This process is repeated for each successive variable elimination until the triangularization is completed. This technique is known as *partial pivoting*.

Further reduction of round-off error can be accomplished in the following manner. As the first step, find $|a_{IJ}| = \max_{1 \leq i,\, j \leq n} |a_{ij}|$. Retain the Ith equation and eliminate the coefficient of x_J in the ith equation, $1 \leq i \leq n$, $i \neq I$, by replacing the ith equation by $(-a_{iJ}/a_{IJ})$ times the Ith equation plus the original ith equation. Repeat this process until the final system can be backsolved in a manner similar to that for a triangular system. This process, called *total pivoting*, reduces the effect of round-off error over partial pivoting but is more complicated to program since the order in which the variables are eliminated must be recorded.

In Fig. 2.1 we list a FORTRAN subroutine that uses Gauss elimination with partial pivoting to solve a square system $A\mathbf{x} = \mathbf{b}$. For convenience of computa-

```
      SUBROUTINE GAUSS(A,B,X,N,MAINDM,IERROR,RNORM)
      DIMENSION A(MAINDM,MAINDM),B(MAINDM),X(MAINDM)
      DIMENSION AUG(50,51)
      NM1=N-1
      NP1=N+1
C
C  SET UP THE AUGMENTED MATRIX FOR AX=B.
C
        DO 2 I=1,N
          DO 1 J=1,N
          AUG(I,J)=A(I,J)
    1     CONTINUE
        AUG(I,NP1)=B(I)
    2   CONTINUE
C
C  THE OUTER LOOP USES ELEMENTARY ROW OPERATIONS TO TRANSFORM
C  THE AUGMENTED MATRIX TO ECHELON FORM.
C
        DO 8 I=1,NM1
C
C  SEARCH FOR THE LARGEST ENTRY IN COLUMN I, ROWS I THROUGH N.
C  IPIVOT IS THE ROW INDEX OF THE LARGEST ENTRY.
C
        PIVOT=0.
          DO 3 J=I,N
          TEMP=ABS(AUG(J,I))
          IF(PIVOT.GE.TEMP)  GO TO 3
          PIVOT=TEMP
          IPIVOT=J
    3     CONTINUE
        IF(PIVOT.EQ.0.)  GO TO 13
        IF(IPIVOT.EQ.I)  GO TO  5
C
C  INTERCHANGE ROW I AND ROW IPIVOT.
C
          DO 4 K=I,NP1
          TEMP=AUG(I,K)
          AUG(I,K)=AUG(IPIVOT,K)
          AUG(IPIVOT,K)=TEMP
    4     CONTINUE
C
C  ZERO ENTRIES (I+1,I), (I+2,I),....,(N,I) IN THE AUGMENTED MATRIX.
C
    5   IP1=I+1
          DO 7 K=IP1,N
          Q=-AUG(K,I)/AUG(I,I)
          AUG(K,I)=0.
            DO 6 J=IP1,NP1
            AUG(K,J)=Q*AUG(I,J)+AUG(K,J)
    6       CONTINUE
    7     CONTINUE
    8   CONTINUE
      IF(AUG(N,N).EQ.0.)  GO TO 13
C
C  BACKSOLVE TO OBTAIN A SOLUTION TO AX=B.
C
      X(N)=AUG(N,NP1)/AUG(N,N)
        DO 10 K=1,NM1
        Q=0.
          DO 9 J=1,K
          Q=Q+AUG(N-K,NP1-J)*X(NP1-J)
    9     CONTINUE
        X(N-K)=(AUG(N-K,NP1)-Q)/AUG(N-K,N-K)
   10   CONTINUE
```

Figure 2.1 Subroutine GAUSS.

```
C
C   CALCULATE THE NORM OF THE RESIDUAL VECTOR, B-AX.
C   SET IERROR=1 AND RETURN.
C
      RSQ=0.
        DO 12 I=1,N
        Q=0.
          DO 11 J=1,N
          Q=Q+A(I,J)*X(J)
   11     CONTINUE
        RESI=B(I)-Q
        RMAG=ABS(RESI)
        RSQ=RSQ+RMAG**2
   12   CONTINUE
      RNORM=SQRT(RSQ)
      IERROR=1
      RETURN
C
C   ABNORMAL RETURN --- REDUCTION TO ECHELON FORM PRODUCES A ZERO
C   ENTRY ON THE DIAGONAL.  THE MATRIX A MAY BE SINGULAR.
C
   13 IERROR=2
      RETURN
      END
```

Figure 2.1 (continued)

tion we perform Gauss elimination on the *augmented* matrix $[A\,\vdots\,\mathbf{b}]$. [Recall that if A is $(n \times n)$, then the augmented matrix for the system $A\mathbf{x} = \mathbf{b}$ is the $(n \times (n + 1))$ matrix obtained by attaching the column $\mathbf{b}$ to the $(n \times n)$ array A so that $\mathbf{b}$ is the $(n + 1)$st column of $[A\,\vdots\,\mathbf{b}]$. Using Gauss elimination to transform A to triangular form means, in the language of linear algebra, that we are trying to reduce the augmented matrix $[A\,\vdots\,\mathbf{b}]$ to "echelon" form.] Like others listed later, the program in Fig. 2.1 was written with an emphasis on simplicity and readability, with a minimum of special features and options. Also, in the interests of clarity, we include only enough comments to highlight the main computational segments of the program.

The inputs required to use Subroutine GAUSS to solve an $(n \times n)$ system $A\mathbf{x} = \mathbf{b}$ are these: A, an $(N \times N)$ matrix; B, an $(N \times 1)$ vector; N, the size of the system $A\mathbf{x} = \mathbf{b}$; MAINDM, the declared dimension of the array A in the calling program. Subroutine GAUSS will return these: X, an $(N \times 1)$ array containing the machine solution to $A\mathbf{x} = \mathbf{b}$ if Gauss elimination is successful; IERROR, an error flag set to 1 if the elimination is successful and set to 2 if elimination cannot proceed because of a zero pivot; RNORM, a value that measures the size of the residual vector (we will discuss this in Section 2.3). Fig. 2.2 lists a simple program that reads in a system $A\mathbf{x} = \mathbf{b}$ and calls GAUSS to solve the system.

This section concludes with a brief discussion of a concept known as scaling. From the discussion above on pivoting one might expect that if the magnitudes of the elements of the coefficient matrix vary greatly in size, then the solution of $A\mathbf{x} = \mathbf{b}$ is more susceptible to rounding error. In practice this

```
      DIMENSION A(20,20),B(20),X(20)
      MAINDM=20
    1 READ 100,N
      IF(N.LE.1)  STOP
        DO 2 I=1,N
        READ 101,(A(I,J),J=1,N),B(I)
    2   CONTINUE
      CALL GAUSS(A,B,X,N,MAINDM,IERROR,RNORM)
      PRINT 102,IERROR
      IF(IERROR.EQ.2)  GO TO 1
      PRINT 103,RNORM
      PRINT 104,(X(I),I=1,N)
  100 FORMAT(I2)
  101 FORMAT(21F4.0)
  102 FORMAT(8H IERROR=,I3)
  103 FORMAT(7H RNORM=,E20.6)
  104 FORMAT(1H ,6E16.6)
      GO TO 1
      END
```

Figure 2.2

expectation is usually correct. This variation in magnitudes can be countered by multiplying the rows and columns of A by scaling constants to produce a matrix of the form $B = D_r A D_c$ where D_r and D_c are diagonal matrices for the row scaling and the column scaling, respectively, with the scaling constants as their diagonal elements. The system $A\mathbf{x} = \mathbf{b}$ is then equivalent to the system $(D_r A D_c)(D_c^{-1}\mathbf{x}) = D_r \mathbf{b}$, which is solved by successively solving $B\mathbf{z} = D_r \mathbf{b}$ and $\mathbf{x} = D_c \mathbf{z}$. The problem of constructing a practical computational scheme for scaling a matrix A is not well understood at this time, and an extensive discussion can be found in Young and Gregory (1973) and in the LINPACK User's Guide (1979). (LINPACK is an excellent library of mathematical software for numerical linear algebra.)

One frequently used approach to scaling is to divide each entry in each row by the largest (in magnitude) entry in that row. This approach uses row scaling only; so in the context of the discussion above, $D_c = I$. In particular, for each i, $1 \leq i \leq n$, define

$$d_i = \max_{1 \leq j \leq n} |a_{ij}|,$$

and let $1/d_i$ be the diagonal elements of D_r. This method produces $B = (b_{ij})$ where $B = D_r A$ and where $\max_{1 \leq j \leq n} |b_{ij}| = 1$ for $1 \leq i \leq n$. Unfortunately this technique can introduce rounding error in every element of B. An alternative, called *machine-base scaling*, is to replace $1/d_i$ by $1/b^{k_i}$ where b is the base of the machine arithmetic and k_i is an integer such that $d_i \simeq b^{k_i}$. The resulting divisions are performed exactly using an exponent shift so that no rounding results (for example, on a hexadecimal machine such as an IBM 370, row i is scaled by a power of 16).

Another sort of scaling that is frequently employed is called *implicit pivoting*; here, the term "implicit" is used because the scaling multiplications are not actually performed, but rather the Gauss elimination algorithm selects the pivot element at each stage of the elimination as if scaling had taken place. In particular, let $A^{(k)} = (a_{ij}^{(k)})$ be the coefficient matrix at the kth step of Gauss elimination. For $A^{(k)}$, compute values $d_i^{(k)}$ as above, find the row index I that satisfies

$$\frac{|a_{Ik}^{(k)}|}{d_I^{(k)}} = \max_{k \le i \le n} \frac{|a_{ik}^{(k)}|}{d_i^{(k)}}.$$

We use row I as the pivot row; so we interchange row k with row I, b_k with b_I, and then proceed with the elimination as usual.

2.2.4. Implementation of Gauss Elimination and *LU*-decomposition

Besides pivoting, another important consideration is the efficient programming of Gauss elimination and we wish to mention one aspect of this general organizational problem. Frequently in practice, one is confronted with solving a succession of linear systems that have the same coefficient matrix. That is, solve

$$A\mathbf{x} = \mathbf{b}_i, \qquad i = 1, 2, \ldots, m. \tag{2.11}$$

(For example, such was the case when we were computing A^{-1} in the first section.) If we employ a Gauss elimination routine to solve, say, $A\mathbf{x} = \mathbf{b}_1$, then we first obtain an equivalent triangular system [such as (2.6)]

$$A'\mathbf{x} = \mathbf{b}'_1; \tag{2.12}$$

and this system is backsolved. If we next employ the same routine to solve $A\mathbf{x} = \mathbf{b}_2$, we obtain, as above,

$$A'\mathbf{x} = \mathbf{b}'_2; \tag{2.13}$$

and (2.13) is then backsolved.

A moment's reflection shows that instead of calling the same routine repeatedly to solve the m systems in (2.11), it would be better if we could store A' and just have a procedure to generate $\mathbf{b}'_i$ from $\mathbf{b}_i$. If such could be done, the approximately $n^3/3$ operations necessary to triangularize A would have to be performed only once. This programming problem can be solved by considering how the triangular system (2.6) is obtained from the original system (2.1). In particular as the system $A\mathbf{x} = \mathbf{b}$ is transformed to $A'\mathbf{x} = \mathbf{b}'$, the vector $\mathbf{b}$ is transformed to the vector $\mathbf{b}'$ by the same sequence of scalar multiples that transforms A to the triangular matrix A'. Thus what is needed is some way to store the sequence of the scalar multiples of the triangularization. Rather than trying to elaborate abstractly on this method, we will give an example that makes the procedure clear.

EXAMPLE 2.9. Consider the linear system

$$\begin{aligned} x_1 + 2x_2 + x_3 &= b_1 \\ 3x_1 + 4x_2 \quad\quad &= b_2. \\ 2x_1 + 10x_2 + 4x_3 &= b_3 \end{aligned} \tag{2.14}$$

We are seeking the scalar multiples to transform $\mathbf{b}$ into $\mathbf{b}'$; so we first determine how A is transformed to A'. That is, we focus our attention upon how to triangularize the matrix

$$A = \begin{bmatrix} 1 & 2 & 1 \\ 3 & 4 & 0 \\ 2 & 10 & 4 \end{bmatrix} \tag{2.15}$$

by adding multiples of one row to the next. Multiplying the first row by -3 and adding the result to the second row, then multiplying the first row by -2 and adding the result to the third row, we obtain

$$A^{(1)} = \begin{bmatrix} 1 & 2 & 1 \\ 0 & -2 & -3 \\ 0 & 6 & 2 \end{bmatrix}, \qquad M_{21} = -3, \qquad M_{31} = -2 \tag{2.16}$$

where M_{21} and M_{31} are respective multiples needed to introduce zeros in the first column below the main diagonal. Continuing this simple example, if we multiply the second row of $A^{(1)}$ by 3 and add to the third row, we obtain A' where

$$A' = \begin{bmatrix} 1 & 2 & 1 \\ 0 & -2 & -3 \\ 0 & 0 & -7 \end{bmatrix}, \qquad M_{32} = 3. \tag{2.17}$$

If we had performed Gauss elimination on the system (2.14), the coefficient matrix of the equivalent triangular system would clearly be A' in (2.17).

Moreover, the multiples M_{21}, M_{31}, and M_{32} tell us how to transform $\mathbf{b}$. To be specific, we first obtain

$$\mathbf{b}^{(1)} = \begin{bmatrix} b_1 \\ b_2 + M_{21}b_1 \\ b_3 + M_{31}b_1 \end{bmatrix} = \begin{bmatrix} b_1^{(1)} \\ b_2^{(1)} \\ b_3^{(1)} \end{bmatrix}, \tag{2.18}$$

and finally

$$\mathbf{b}' = \begin{bmatrix} b_1^{(1)} \\ b_2^{(1)} \\ b_3^{(1)} + M_{32}b_2^{(1)} \end{bmatrix} = \begin{bmatrix} b_1' \\ b_2' \\ b_3' \end{bmatrix}. \tag{2.19}$$

Clearly, in terms of efficient storage, we can overlay the scalar multiples M_{ij} in the zero entries of A' that they introduce during the triangularization.

Thus we proceed as indicated by the arrows:

$$A = \begin{bmatrix} 1 & 2 & 1 \\ 3 & 4 & 0 \\ 2 & 10 & 4 \end{bmatrix} \to \left[\begin{array}{c|cc} 1 & 2 & 1 \\ \hline -3 & -2 & -3 \\ -2 & 6 & 2 \end{array}\right] \to \left[\begin{array}{cc|c} 1 & 2 & 1 \\ \hline -3 & -2 & -3 \\ \hline -2 & 3 & -7 \end{array}\right] = A''. \tag{2.20}$$

(Note: no additional locations are needed to store M_{21}, M_{31}, and M_{32}.) Moreover, the positions of the multiples in A'' describe how they are to operate on vector $\mathbf{b}$ to transform $\mathbf{b}$ to $\mathbf{b}'$. As an illustration, consider

$$\begin{aligned} x_1 + 2x_2 + x_3 &= 3 \\ 3x_1 + 4x_2 &= 3 \\ 2x_1 + 10x_2 + 4x_3 &= 10, \end{aligned} \tag{2.21}$$

which we know from above is equivalent to

$$\begin{aligned} x_1 + 2x_2 + x_3 &= b_1' \\ -2x_2 - 3x_3 &= b_2' \\ -7x_3 &= b_3' \end{aligned} \tag{2.22}$$

where we get the coefficients of (2.22) from the upper triangle portion of A''. To find b_1', b_2', b_3', we change $\mathbf{b}$ to $\mathbf{b}'$ by the sequence

$$\mathbf{b} = \begin{bmatrix} 3 \\ 3 \\ 10 \end{bmatrix} \to \begin{bmatrix} 3 \\ 3 + (-3)\cdot 3 \\ 10 + (-2)\cdot 3 \end{bmatrix} = \begin{bmatrix} 3 \\ -6 \\ 4 \end{bmatrix} \to \begin{bmatrix} 3 \\ -6 \\ 4 + 3\cdot(-6) \end{bmatrix} = \begin{bmatrix} 3 \\ -6 \\ -14 \end{bmatrix} \tag{2.23}$$

where the numbers below the main diagonal in A'' define the sequence. Thus from A'', we can read that (2.21) is equivalent to

$$\begin{aligned} x_1 + 2x_2 + x_3 &= 3 \\ -2x_2 - 3x_3 &= -6 \\ -7x_3 &= -14; \end{aligned} \tag{2.24}$$

and backsolving, we obtain $x_3 = 2$, $x_2 = 0$, and $x_1 = 1$.

The basic ideas in Example 2.9 are fairly easy to program. One sees in (2.23) how solving several systems of the form $A\mathbf{x} = \mathbf{b}_k$, $1 \le k \le m$, can be accommodated by simple transformations of the vector on the right-hand side. In this simple example no row interchanges were necessary, but accounting for interchanges is also an easy bookkeeping task, one which we will be discussing shortly.

One might suppose if m in (2.11) is very large, that the most efficient way of solving the m linear systems, $A\mathbf{x} = \mathbf{b}_k$, $1 \le k \le m$, is first to calculate A^{-1} and then generate the kth solution vector, $\mathbf{y}_k = A^{-1}\mathbf{b}_k$, for each k. However since A^{-1} is an $(n \times n)$ matrix, it follows that n^2 multiplications are required to compute $A^{-1}\mathbf{b}_k$. By contrast, Problem 9 shows that a total of $n(n-1)/2$ multiplications are necessary to generate $\mathbf{b}_k'$ from $\mathbf{b}_k$ and a total of $n(n+1)/2$ multiplications are needed to solve $A'\mathbf{x} = \mathbf{b}_k'$. Thus if we know the triangular matrix A' and the $n(n-1)/2$ scalar multiples needed to generate $\mathbf{b}_k'$, we can solve $A\mathbf{x} = \mathbf{b}_k$ with precisely as many multiplications as it takes to form $A^{-1}\mathbf{b}_k$. Clearly then, the computation of A^{-1} for solving $A\mathbf{x} = \mathbf{b}_i$, $i = 1, 2, \ldots, m$ is not justified since in order to find A^{-1}, we must first solve n systems of equations of the form $A\mathbf{x} = \mathbf{e}_i$, $i = 1, 2, \ldots, n$ (see Section 2.1).

At this point, we would like to offer as a guideline the following: in a practical problem it is seldom necessary to compute either the inverse of a matrix, the determinant of a matrix, or the power of a matrix. In the formulation of a problem it is often convenient to use inverses, powers, and determinants in an abstract mathematical sense (e.g., the solution of $A\mathbf{x} = \mathbf{b}$ is given by $\mathbf{x} = A^{-1}\mathbf{b}$). However, it is rare to encounter a problem in which the *solution* requires the computation of inverses, powers, or determinants. Thus for example, the solution to $A\mathbf{x} = \mathbf{b}$ is found efficiently by using Gauss elimination and is only *represented mathematically* by $A^{-1}\mathbf{b}$. Exceptions to the guidelines above do occur. For example, in certain problems in statistical analysis, the inverse matrix is of significance and may have to be calculated.

Although the basic ideas in Example 2.9 are fairly simple to program, they require examination in more detail and expression in matrix terms for some later analyses. Briefly, we will see that Gauss elimination with partial or implicit pivoting applied to $A\mathbf{x} = \mathbf{b}$ amounts (in effect) to constructing a nonsingular matrix S and then forming an equivalent system $SA\mathbf{x} = S\mathbf{b}$ where $SA = U$ is an upper-triangular matrix. There are primarily two reasons for wanting to view Gauss elimination in matrix terms. First of all, such a view provides a logical and conceptually convenient framework in which to organize an efficient Gauss elimination package. Perhaps more important, this framework will serve to show how to obtain estimates to the "condition number" of A (see Section 2.3.3). Such estimates are quite important because they give us a way to assess the accuracy of the machine solution to a problem $A\mathbf{x} = \mathbf{b}$.

As we stated above, using Gauss elimination to solve $A\mathbf{x} = \mathbf{b}$ amounts *formally* to two steps (the "factor" step and the "solve" step).

1. Find a nonsingular matrix S such that $SA = U$ where U is upper triangular.
2. Solve the equivalent triangular system $U\mathbf{x} = S\mathbf{b}$.

To see this two-step procedure in matrix terms, let A and L_1 be the matrices

$$A = \begin{bmatrix} a_{11} & a_{12} & a_{13} & \cdots & a_{1n} \\ a_{21} & a_{22} & a_{23} & \cdots & a_{2n} \\ a_{31} & a_{32} & a_{33} & \cdots & a_{3n} \\ \vdots & & & & \vdots \\ a_{n1} & a_{n2} & a_{n3} & \cdots & a_{nn} \end{bmatrix} \qquad L_1 = \begin{bmatrix} 1 & 0 & 0 & \cdots & 0 \\ m_{21} & 1 & 0 & \cdots & 0 \\ m_{31} & 0 & 1 & \cdots & 0 \\ \vdots & & & & \vdots \\ m_{n1} & 0 & 0 & \cdots & 1 \end{bmatrix}$$

where $m_{i1} = -a_{i1}/a_{11}$, $2 \le i \le n$. Forming the product $A_1 = L_1 A$ will produce the matrix

$$A_1 = \begin{bmatrix} a_{11} & a_{12} & a_{13} & \cdots & a_{1n} \\ 0 & a'_{22} & a'_{23} & \cdots & a'_{2n} \\ 0 & a'_{32} & a'_{33} & \cdots & a'_{3n} \\ \vdots & & & & \vdots \\ 0 & a'_{n2} & a'_{n3} & \cdots & a'_{nn} \end{bmatrix};$$

and in fact, forming $A_1 = L_1A$ is the same as adding a multiple of m_{i1} times the first row of A to the ith row of A, $2 \le i \le n$. In other words, forming $A_1 = L_1A$ is equivalent to carrying out the first stage of Gauss elimination on A if no row interchange is made. Similarly if L_2 is the matrix

$$L_2 = \begin{bmatrix} 1 & 0 & 0 & \cdots & 0 \\ 0 & 1 & 0 & \cdots & 0 \\ 0 & m_{32} & 1 & \cdots & 0 \\ \vdots & & & & \vdots \\ 0 & m_{n2} & 0 & \cdots & 1 \end{bmatrix}$$

where $m_{i2} = -a'_{i2}/a'_{22}$, $3 \le i \le n$, then the matrix $A_2 = L_2A_1$ is what would be produced by the second stage of Gauss elimination, again if no row interchange is made. If no row interchanges are made at any stage, then Gauss elimination amounts (in matrix terms) to constructing an upper-triangular matrix U where U is given by

$$U = L_{n-1}L_{n-2} \cdots L_2L_1A.$$

Setting $S = L_{n-1}L_{n-2} \cdots L_2L_1$, we have $U = SA$, which completes the factor step. (Note that S is lower triangular; hence $S^{-1} = L$ is also lower triangular. Thus since $U = SA$, we have in effect factored A into a product of the form $A = LU$ where L is lower triangular and U is upper triangular. Such a factorization is called an LU-decomposition; we return to this topic later.) Having $U = SA$, we can now solve a problem $A\mathbf{x} = \mathbf{b}$ by solving the equivalent triangular system $U\mathbf{x} = S\mathbf{b}$.

In the discussion above we did not account for the possibility of row interchanges; and as we know, Gauss elimination with partial or implicit pivoting might require a row interchange at each stage of the elimination. We can express row interchanges in matrix terms by using permutation matrices. In particular if an $(n \times n)$ matrix P is derived from the identity by interchanging rows i and j of I, then P is called a permutation matrix. Next, it is easy to verify that forming the product PA gives the same result as interchanging the ith and jth rows of A. Thus permutation matrices can be used to incorporate partial or implicit pivoting into a matrix-theoretic description of Gauss elimination. Suppose P_k is the permutation matrix that describes the row interchange at the kth stage of Gauss elimination where we set $P_k = I$ if no interchange is made. Then in matrix terms the first step of Gauss elimination amounts to forming an upper-triangular matrix U where

$$U = (L_{n-1}P_{n-1})(L_{n-2}P_{n-2}) \cdots (L_2P_2)(L_1P_1)A. \tag{2.25a}$$

As before, we set $S = L_{n-1}P_{n-1}L_{n-2}P_{n-2} \cdots L_2P_2L_1P_1$ so that $U = SA$; and this completes the factor step.

The solve step now amounts to calculating $S\mathbf{b}$ from

$$S\mathbf{b} = L_{n-1}P_{n-1}L_{n-2}P_{n-2} \cdots L_2P_2L_1P_1\mathbf{b} \tag{2.25b}$$

and then solving the triangular system $U\mathbf{x} = S\mathbf{b}$. Clearly if we have several systems $A\mathbf{x} = \mathbf{b}_i$, $i = 1, 2, \ldots, m$, then we perform the factor step once to obtain $U = SA$. Next, for each i, we form $S\mathbf{b}_i$ and solve $U\mathbf{x} = S\mathbf{b}_i$, $1 \le i \le m$. The factor step takes on the order of $n^3/3$ operations, and each application of the solve step requires n^2 operations.

Giving Gauss elimination in matrix terms is conceptually useful and also useful for purposes of error analyses. However, a Gauss elimination program would not actually calculate the matrix S in the factor step nor form the product $S\mathbf{b}$ in the solve step; instead the program would be structured along the lines of Example 2.9. In particular, an elementary Gauss elimination program should consist of at least two subroutines, FACTOR and SOLVE; and we ask the reader to modify the Gauss elimination program in Fig. 2.1 to accomplish this task (see Problem 11). As illustrated in Example 2.9, the FACTOR routine accepts a matrix A and generates a triangular array of multipliers $m_{21}, m_{31}, \ldots m_{n1}, m_{32}, m_{42}, \ldots$ and also generates the upper-triangular matrix U. The array of multipliers and the matrix U can be written over A, or stored in a separate $(n \times n)$ array if we do not wish to destroy A in FACTOR. In addition, FACTOR must keep a record of row interchanges; and this record can be stored in a vector IPIVOT of length $n - 1$ where IPIVOT $(K) = J$ means that row k and row j were interchanged at the kth stage of the elimination.

Given a system $A\mathbf{x} = \mathbf{b}$, subroutine SOLVE is passed the vector $\mathbf{b}$, along with the array of multipliers, the triangular matrix U, and the vector IPIVOT. The calculation of $S\mathbf{b}$ in SOLVE is done in $n - 1$ stages as was illustrated in Example 2.9. In the first stage if IPIVOT(1) $= J$, then the entries b_1 and b_j are interchanged in $\mathbf{b}$, and the interchange produces a new vector, $\tilde{\mathbf{b}}_1 = [\tilde{b}_1, \tilde{b}_2, \ldots, \tilde{b}_n]^T$. Then the entries of $\tilde{\mathbf{b}}_1$ are modified to produce $\mathbf{b}_1$ where the ith entry of $\tilde{\mathbf{b}}_1$ is changed to $\tilde{b}_i + m_{i1}\tilde{b}_1$, $2 \le i \le n$. In terms of (2.25b),

$$\tilde{\mathbf{b}}_1 = P_1\mathbf{b} \quad \text{and} \quad \mathbf{b}_1 = L_1P_1\mathbf{b}.$$

At the second stage, if IPIVOT(2) $= K$, then the second and kth entries of $\mathbf{b}_1$ are switched to produce $\tilde{\mathbf{b}}_2$. The entries of $\tilde{\mathbf{b}}_2$ are then modified to produce $\mathbf{b}_2$, by adding a multiple of m_{i2} times the second entry of $\tilde{\mathbf{b}}_2$ to the ith entry of $\tilde{\mathbf{b}}_2$, for $3 \le i \le n$. In terms of (2.25b),

$$\tilde{\mathbf{b}}_2 = P_2\mathbf{b}_1 \quad \text{and} \quad \mathbf{b}_2 = L_2P_2\mathbf{b}_1.$$

The remaining stages follow analogously, each stage consists of an interchange followed by the multiplier operations. The end result is the vector $S\mathbf{b}$ in (2.25b), and SOLVE can complete its assigned task by backsolving $U\mathbf{x} = S\mathbf{b}$.

EXAMPLE 2.10. To illustrate the factor and solve steps, we consider a simple (3×3) system $A\mathbf{x} = \mathbf{b}$. For brevity we let v denote the two-dimensional pivot vector; so $v_1 =$ IPIVOT(1) and $v_2 =$ IPIVOT(2). We also write the multipliers and the matrix U over the entries of A although it is not necessarily desirable to destroy A in the factor step.

$$A = \begin{bmatrix} 1 & -1 & 0 \\ 2 & -1 & 1 \\ 2 & -2 & -1 \end{bmatrix} \qquad \mathbf{b} = \begin{bmatrix} 2 \\ 4 \\ 3 \end{bmatrix}.$$

Factor step:

$$A \xrightarrow[R_1 \longleftrightarrow R_2]{(v_1 = 2)} \begin{bmatrix} 2 & -1 & 1 \\ 1 & -1 & 0 \\ 2 & -2 & -1 \end{bmatrix} \xrightarrow[m_{31} = -1]{m_{21} = -\frac{1}{2}} \begin{bmatrix} 2 & -1 & 1 \\ -\frac{1}{2} & -\frac{1}{2} & -\frac{1}{2} \\ -1 & -1 & -2 \end{bmatrix}$$

$$\xrightarrow[R_2 \longleftrightarrow R_3]{(v_2 = 3)} \begin{bmatrix} 2 & -1 & 1 \\ -\frac{1}{2} & -1 & -2 \\ -1 & -\frac{1}{2} & -\frac{1}{2} \end{bmatrix} \xrightarrow{m_{32} = -\frac{1}{2}} \begin{bmatrix} 2 & -1 & 1 \\ -\frac{1}{2} & -1 & -2 \\ -1 & -\frac{1}{2} & \frac{1}{2} \end{bmatrix}$$

$$\mathbf{v} = \begin{bmatrix} 2 \\ 3 \end{bmatrix} \quad \text{(pivot vector).}$$

Solve step:

$$\mathbf{b} \xrightarrow[R_1 \longleftrightarrow R_2]{(v_1 = 2)} \begin{bmatrix} 4 \\ 2 \\ 3 \end{bmatrix} \xrightarrow[m_{31} = -1]{m_{21} = -\frac{1}{2}} \begin{bmatrix} 4 \\ 0 \\ -1 \end{bmatrix} \xrightarrow[R_2 \longleftrightarrow R_3]{(v_2 = 3)} \begin{bmatrix} 4 \\ -1 \\ 0 \end{bmatrix} \xrightarrow{m_{32} = -\frac{1}{2}} \begin{bmatrix} 4 \\ -1 \\ \frac{1}{2} \end{bmatrix} = S\mathbf{b}.$$

Backsolving $U\mathbf{x} = S\mathbf{b}$ yields $x_3 = 1$, $x_2 = -1$, $x_1 = 1$.

As we mentioned earlier, finding a nonsingular matrix S such that $SA = U$ is equivalent to factoring A as $A = S^{-1}U$. A related idea is that of an *LU-decomposition* for A. Specifically, suppose we can factor the $(n \times n)$ matrix A into a product of two matrices L and U so that

$$A = LU$$

where L is an $(n \times n)$ lower-triangular matrix and U is an $(n \times n)$ upper-triangular matrix.

If A has a factorization $A = LU$, then solutions of the system $A\mathbf{x} = \mathbf{b}$ are found by solving $LU\mathbf{x} = \mathbf{b}$. To solve $LU\mathbf{x} = \mathbf{b}$, we proceed in two stages.

1. Solve $L\mathbf{y} = \mathbf{b}$.
2. Solve $U\mathbf{x} = \mathbf{y}$.

In this fashion if $U\mathbf{x} = \mathbf{y}$, then $LU\mathbf{x} = L\mathbf{y} = \mathbf{b}$; moreover, both systems $L\mathbf{y} = \mathbf{b}$ and $U\mathbf{x} = \mathbf{y}$ are triangular and hence easy to solve. If Gauss elimination proceeds with no row interchanges, then as we noted before, the matrix S in (2.25a) is lower triangular and therefore $S^{-1} = L$ is also lower triangular. Thus Gauss elimination without pivoting can sometimes be used to produce an LU-decomposition for A. Another procedure is to consider the matrix equation $A = LU$:

$$\begin{bmatrix} a_{11} & a_{12} & \cdots & a_{1n} \\ a_{21} & a_{22} & \cdots & a_{2n} \\ \vdots & & & \vdots \\ a_{n1} & a_{n2} & \cdots & a_{nn} \end{bmatrix} = \begin{bmatrix} \ell_{11} & 0 & 0 & \cdots & 0 \\ \ell_{21} & \ell_{22} & 0 & \cdots & 0 \\ \vdots & & & & \vdots \\ \ell_{n1} & \ell_{n2} & & \cdots & \ell_{nn} \end{bmatrix} \begin{bmatrix} u_{11} & u_{12} & \cdots & u_{1n} \\ 0 & u_{22} & \cdots & u_{2n} \\ \vdots & & & \vdots \\ 0 & 0 & \cdots & u_{nn} \end{bmatrix}. \qquad (2.26)$$

Writing (2.26) out at length and equating both sides coordinatewise gives n^2 equations, which can then be (possibly) solved to determine L and U. We do not intend to do an extensive analysis of the general case represented by (2.26), but we do want to observe that as (2.26) is written, L and U both contain $n(n+1)/2$ undetermined entries, a total of $n^2 + n$ undetermined entries. Since we have just n^2 equations, it seems feasible that n quantities may be chosen as we wish. In practice, these free parameters are normally specified by either

$$\ell_{11} = \ell_{22} = \cdots = \ell_{nn} = 1$$

or

$$\ell_{11} = u_{11}, \qquad \ell_{22} = u_{22}, \ \ldots, \ \ell_{nn} = u_{nn}.$$

The following example should serve to illustrate LU-decomposition.

Given

$$A = \begin{bmatrix} 3 & 1 & 2 \\ 6 & 3 & 3 \\ -3 & 2 & -3 \end{bmatrix},$$

find an LU-decomposition of the form

$$\begin{bmatrix} 3 & 1 & 2 \\ 6 & 3 & 3 \\ -3 & 2 & -3 \end{bmatrix} = \begin{bmatrix} 1 & 0 & 0 \\ \ell_{21} & 1 & 0 \\ \ell_{31} & \ell_{32} & 1 \end{bmatrix} \begin{bmatrix} u_{11} & u_{12} & u_{13} \\ 0 & u_{22} & u_{23} \\ 0 & 0 & u_{33} \end{bmatrix}. \tag{2.27}$$

The most natural way to set up the equations that must be solved is in a row-by-row fashion:

$$\begin{array}{rlrlrl} u_{11} = & 3 & u_{12} = & 1 & u_{13} = & 2 \\ \ell_{21}u_{11} = & 6 & \ell_{21}u_{12} + u_{22} = & 3 & \ell_{21}u_{13} + u_{23} = & 3 \\ \ell_{31}u_{11} = & -3 & \ell_{31}u_{12} + \ell_{32}u_{22} = & 2 & \ell_{31}u_{13} + \ell_{32}u_{23} + u_{33} = & -3. \end{array} \tag{2.28}$$

Solving these equations successively, we obtain

$$\begin{bmatrix} 3 & 1 & 2 \\ 6 & 3 & 3 \\ -3 & 2 & -3 \end{bmatrix} = \begin{bmatrix} 1 & 0 & 0 \\ 2 & 1 & 0 \\ -1 & 3 & 1 \end{bmatrix} \begin{bmatrix} 3 & 1 & 2 \\ 0 & 1 & -1 \\ 0 & 0 & 2 \end{bmatrix}. \tag{2.29}$$

Not all matrices have an LU-decomposition. As a simple example, the nonsingular matrix

$$A = \begin{bmatrix} 0 & 1 \\ 1 & 0 \end{bmatrix} \tag{2.30}$$

can be shown (Problem 19) to have no LU-decomposition; the singular matrix $I + A$ does have an LU-decomposition. It *is* known that if a matrix A is nonsingular, then there is some rearrangement of the rows of A such that the rearranged matrix has an LU-decomposition. See Forsythe and Moler (1967), for a proof.

To investigate the relations between various LU-decompositions of A, consider the two factorizations $A = L_1U_1$ and $A = L_2U_2$. If A is nonsingular, then the factors are nonsingular; and $L_1U_1 = L_2U_2$ implies that $L_2^{-1}L_1 = U_2U_1^{-1}$. Since the products and inverses of lower- (upper-) triangular matrices remain lower (upper) triangular, then $L_2^{-1}L_1 = U_2U_1^{-1}$ means that $L_2^{-1}L_1$ is a diagonal matrix. Suppose $D = L_2^{-1}L_1$ where $D = (d_{ij})$, $L_1 = (b_{ij})$, and $L_2 = (c_{ij})$; then $L_1 = L_2D$ implies $b_{ii} = c_{ii}d_{ii}$, $1 \le i \le n$. Thus if the diagonal elements of L_1 and L_2 are the same, then $D = I$ and $L_1 = L_2$. Therefore, setting $\ell_{ii} = 1$, $1 \le i \le n$, in a factorization yields the same factorization as Gauss elimination without pivoting (Problem 16). The solution of $A\mathbf{x} = \mathbf{b}$ by this factorization is called the Doolittle method. Analogously, setting $u_{ii} = 1$, $1 \le i \le n$, is known as the Crout method. An advantage of such factorizations (called "compact" methods) of A into LU is that the elements ℓ_{ij} and u_{ij} are computed in terms of inner products. One can form these inner products in double precision and increase the accuracy of L and U. Any similar use of limited double precision in the usual form of Gauss elimination is not effective [see Atkinson (1978) for a more detailed discussion].

A similar compact method is often used when A is symmetric and positive definite; that is, $A^{\mathrm{T}} = A$ and $\mathbf{x}^{\mathrm{T}}A\mathbf{x} > 0$ for all nonzero vectors $\mathbf{x}$. If we set $\ell_{ii} = u_{ii}$, $1 \le i \le n$, then the resulting factorization yields $U = L^{\mathrm{T}}$ and $A = LL^{\mathrm{T}}$. Recursively the elements of L are given by

$$\ell_{ij} = \Big(a_{ij} - \sum_{k=1}^{j-1} \ell_{ik}\ell_{jk}\Big)\Big/\ell_{jj}, \qquad j = 1, 2, \ldots, i-1$$

$$\ell_{ii} = \Big(a_{ii} - \sum_{k=1}^{i-1} \ell_{ik}^2\Big)^{\frac{1}{2}}.$$

Since A is positive definite, the arguments of the square roots for ℓ_{ii} are positive. Since A is symmetric and $U = L^{\mathrm{T}}$, fewer operations are necessary to compute the factorization. Again inner products may be performed in double precision. This technique, known as Cholesky's method, can be shown to be effective without pivoting or scaling.

There are special sorts of problems for which factorization methods are very natural to use. One such problem that occurs in many different applied settings is that of solving a *tridiagonal* linear system:

$$\begin{aligned}
a_{11}x_1 + a_{12}x_2 \qquad\qquad\qquad\qquad\qquad &= b_1 \\
a_{21}x_1 + a_{22}x_2 + a_{23}x_3 \qquad\qquad\qquad &= b_2 \\
a_{32}x_2 + a_{33}x_3 + a_{34}x_4 \qquad\quad &= b_3 \\
a_{43}x_3 + a_{44}x_4 + a_{45}x_5 &= b_4 \\
\ddots \qquad\qquad \ddots \qquad &\ \ \vdots \\
a_{n,\,n-1}x_{n-1} + a_{nn}x_n &= b_n
\end{aligned} \tag{2.31}$$

Such systems are called tridiagonal because the coefficient matrix A for the system (2.31) has three *diagonals;* or more precisely, $a_{ij} = 0$ if $j < i - 1$ or if $j > i + 1$. These systems commonly arise, for example, when numerical methods are used to solve two-point boundary value problems or when cubic spline approximations are used to fit data (see Chapters 7 and 5). In most practical situations, a factorization of the form (2.32) below is possible:

$$LU = \begin{bmatrix} 1 & 0 & 0 & 0 & \cdots & 0 \\ \ell_{21} & 1 & 0 & 0 & \cdots & 0 \\ 0 & \ell_{32} & 1 & 0 & \cdots & 0 \\ 0 & 0 & \ell_{43} & 1 & \cdots & 0 \\ \vdots & & & & & \vdots \\ 0 & 0 & 0 & 0 & \cdots & 1 \end{bmatrix} \begin{bmatrix} u_{11} & a_{12} & 0 & 0 & \cdots & 0 \\ 0 & u_{22} & a_{23} & 0 & \cdots & 0 \\ 0 & 0 & u_{33} & a_{34} & \cdots & 0 \\ 0 & 0 & 0 & u_{44} & \cdots & 0 \\ \vdots & & & & & \vdots \\ 0 & 0 & 0 & 0 & \cdots & u_{nn} \end{bmatrix}. \tag{2.32}$$

If A is the coefficient matrix of (2.31), we can express A as $A = LU$ by solving recursively:

$$\begin{aligned} u_{11} &= a_{11} \\ \ell_{i,\, i-1} &= a_{i,\, i-1}/u_{i-1,\, i-1} \qquad i = 2, 3, \ldots, n \\ u_{ii} &= a_{ii} - \ell_{i,\, i-1}a_{i-1,\, i} \end{aligned} \tag{2.33a}$$

Having the factorization above, we can solve (2.31) by first solving $L\mathbf{y} = \mathbf{b}$ and then $U\mathbf{x} = \mathbf{y}$. The explicit equations are

$$\begin{aligned} y_1 &= b_1 \\ y_i &= b_i - \ell_{i,\, i-1}y_{i-1}; \qquad i = 2, 3, \ldots, n \\ x_n &= y_n/u_{nn} \\ x_{n-i} &= (y_{n-i} - a_{n-i,\, n+1-i}x_{n+1-i})/u_{n-i,\, n-i}; \qquad i = 1, 2, \ldots, n-1. \end{aligned} \tag{2.33b}$$

PROBLEMS, SECTION 2.2.4

1. Use Subroutine GAUSS to solve the (3×3) system of Example 2.2, Section 2.1.
2. Repeat Problem 1 for the (3×3) system of Example 2.14, Section 2.3.3.
3. Write a program using Subroutine GAUSS to find the inverse of a nonsingular $(n \times n)$ matrix and to find the solution of $A\mathbf{x} = \mathbf{b}$ by forming $\mathbf{x} = A^{-1}\mathbf{b}$. Use this program on the system in Problem 2 and compare answers of both problems to the exact solution: $x_1 = 1$, $x_2 = 3$, $x_3 = 2$.
4. In Example 2.8 replace .0001 by 10^{-n}. Use three-digit floating-decimal calculations for solving the system without pivoting, and determine the positive integer values of n for which the computed solution is "significantly" different from the true solution.
5. Repeat Problem 4 for the system

$$\begin{aligned} 10^{-n}x_1 + x_2 &= 3 \\ x_1 - x_2 &= -2. \end{aligned}$$

6. Use the program in Problem 3 to find A^{-1} for the (4×4) matrix A in Example 2.6 (the exact inverse is given in Problem 7 at the end of Section 2.3.4). In theory, AA^{-1}

$= A^{-1}A$; but because of round-off errors, we cannot expect this equality in practice. To illustrate the problem, calculate and print the two products AA^{-1} and $A^{-1}A$, using the machine version of A^{-1} found by the program in Problem 3.

7. Small changes in the right-hand side of $A\mathbf{x} = \mathbf{b}$ may lead to relatively large changes in the solution. As an illustration, use GAUSS to solve $A\mathbf{x} = \mathbf{b}$ where A is the matrix in Problem 6 and $\mathbf{b}$ is either of the vectors

$$\text{a)} \quad \mathbf{b} = \begin{bmatrix} 23.01 \\ 31.99 \\ 32.99 \\ 31.01 \end{bmatrix} \qquad \text{b)} \quad \mathbf{b} = \begin{bmatrix} 23.1 \\ 31.9 \\ 32.9 \\ 31.1 \end{bmatrix}.$$

8. Let A_1 and A_2 be matrices obtained from the (4×4) matrix in Problem 6 by replacing the (1, 1) entry of A by 5.01 and 4.99, respectively. As in Problem 6, calculate A_1^{-1} and A_2^{-1}. Compare your results with A^{-1}.

9. Given the system of equations (2.6), show that $n(n + 1)/2$ multiplications are required to solve for $x_n, x_{n-1}, \ldots, x_2, x_1$.

10. Construct a (3×3) coefficient matrix where the implicit-scaling row interchanges are different from the row interchanges of partial pivoting.

11. Write FACTOR and SOLVE subroutines along the lines described in Section 2.2.4. If you write these subroutines in FORTRAN and use execution-time dimensioning, they should be of a form similar to

```
SUBROUTINE FACTOR(A, SA, IPIVOT, N, MAINDM, IERROR)
SUBROUTINE SOLVE(SA, B, X, IPIVOT, N, MAINDM)
```

In this version of FACTOR, the array SA contains U in the upper-triangular portion and the multipliers in the lower-triangular portion; the array A is not destroyed in FACTOR. To test your program (if you use partial pivoting), put some temporary print statements in FACTOR and SOLVE and then verify that you duplicate each stage of the problem worked in Example 2.10. Subroutine GAUSS can be used as a rough model for designing these two routines.

12. Let the $(n \times n)$ matrix P be derived from the identity matrix by interchanging the ith and jth rows. If $\mathbf{x}$ is any $(n \times 1)$ vector, argue that $P\mathbf{x}$ is the same as $\mathbf{x}$ with the ith and jth components interchanged. Use this result to describe PA where A is any $(n \times n)$ matrix. Show that $P = P^{\mathrm{T}} = P^{-1}$.

13. If L_1 and P_1 are given as in (2.25), show that L_1P_1A is the resulting matrix of the first step of Gauss elimination with partial pivoting on A.

14. Argue that the matrix $S = L_{n-1}P_{n-1} \ldots L_2P_2L_1P_1$ in (2.25) is nonsingular. (Hint: First show that AB is nonsingular if and only if both A and B are nonsingular.)

15. The entries of the $(n \times n)$ Hilbert matrix $H = (h_{ij})$ are given by $h_{ij} = 1/(i + j - 1)$. Hilbert matrices are considered to be badly behaved or *ill-conditioned* (see Section 2.3.3), and are often used to test solution routines. Write a program using the subroutines of Problem 11 to find the inverse of the (4×4) Hilbert matrix. Test your answer by forming the products HH^{-1} and $H^{-1}H$. Do the same with the program in Problem 3 and compare results. (Compute the entries h_{ij} rather than reading them as data.)

16. Noting that Gauss elimination in Example 2.6 proceeds without row interchanges, use only the computations given in this example to give an LU-decomposition for the (4×4) matrix A. Use this decomposition to solve the system $A\mathbf{x} = \mathbf{b}$ where $\mathbf{b} = [-1, -1, -8, -11]^T$.
17. If L is a lower-triangular matrix equal to I except that its jth column is $[0, \ldots, 0, 1, M_{j+1,j}, \ldots, M_{nj}]^T$, show that L^{-1} is of the same form with jth column $[0, \ldots, 0, 1, -M_{j+1,j}, \ldots, -M_{nj}]^T$.
18. The Hilbert matrices of Problem 15 are known to be symmetric and positive definite. Program Cholesky's method to yield an LU-decomposition of the (4×4) Hilbert matrix. Use this decomposition to compute H^{-1} and again compare results with Problem 15.
19. Show that the nonsingular matrix

$$A = \begin{bmatrix} 0 & 1 \\ 1 & 0 \end{bmatrix}$$

has no LU-decomposition, but the singular matrix $A + I$ does have. Give the permutation matrix P such that PA has an LU-decomposition.
20. Apply the Doolittle factorization to the (3×3) coefficient matrix in Problem 1 and use the factorization to solve the system. Use the Cholesky method on the system in Problem 16.

2.3 ERROR ANALYSIS AND NORMS

There are many different types of linear systems of equations each having their own special characteristics. Thus we can hardly expect any particular direct method, like Gauss elimination, to be the best possible method to use in all circumstances. Moreover if we do use a direct method, our computed solution will almost certainly be incorrect because of round-off error. Therefore we need some way to determine the size of the error of any computed solution and also some way to improve this computed solution. In this section, we will briefly develop a little of the theoretical background that is needed both for the analysis of errors and for the analysis of the "iterative" algorithms of Section 2.4 that generate a sequence of approximate solutions to $A\mathbf{x} = \mathbf{b}$.

The subject of error analysis must be approached somewhat carefully since a particular computed solution (say $\mathbf{x}_c$) to $A\mathbf{x} = \mathbf{b}$ may be considered badly in error or quite acceptable depending on how we intend to use the vector $\mathbf{x}_c$. For example, let $\mathbf{x}_t$ denote the "true" solution of $A\mathbf{x} = \mathbf{b}$ and let $\mathbf{r} = A\mathbf{x}_c - \mathbf{b}$ be the *residual vector*. Then $\mathbf{r} = A\mathbf{x}_c - \mathbf{b} = A\mathbf{x}_c - A\mathbf{x}_t$ or

$$\mathbf{x}_c - \mathbf{x}_t = A^{-1}\mathbf{r}. \tag{2.34}$$

If A^{-1} has some very large entries, then Eq. (2.34) demonstrates that the residual vector, $\mathbf{r}$, might be small, and yet $\mathbf{x}_c$ might be quite far from $\mathbf{x}_t$. Depending on the context of the problem that gave rise to the equation $A\mathbf{x} = \mathbf{b}$, we might be happy with having $A\mathbf{x}_c - \mathbf{b}$ small or we might need $\mathbf{x}_c$ to be near $\mathbf{x}_t$.

2.3.1. Vector Norms

In the discussion above, we were forced in a natural way to use words like "large" and "small" to describe the size of a vector and words like "near" and "far" to describe the proximity of two vectors. Also, in subsequent material we will be developing methods that generate *sequences* of vectors, $\{\mathbf{x}^{(k)}\}_{k=1}^{\infty}$, which one hopes *converge* to some vector $\mathbf{x}$. In these methods we will need to have some idea of how "close" each $\mathbf{x}^{(k)}$ is to $\mathbf{x}$ in order that we may know how large k must be so that $\mathbf{x}^{(k)}$ is an acceptable approximation to $\mathbf{x}$. Thus, we must have some meaningful way to measure the size of a vector or the distance between two vectors. To do this, we extend the concept of *absolute value* or *magnitude* from the real numbers to vectors. For a vector

$$\mathbf{x} = \begin{bmatrix} x_1 \\ x_2 \\ \vdots \\ x_n \end{bmatrix} \tag{2.35}$$

we already know the Euclidean length $|\mathbf{x}| = \sqrt{(x_1)^2 + (x_2)^2 + \cdots + (x_n)^2}$ as a measure of size. It turns out, as we shall see, that other measures of size for a vector are also *practical* to use in many computational problems. This realization leads us to the definition of a *norm*.

To make the setting precise, let R^n denote the set of all n-dimensional vectors with real components

$$R^n = \left\{ \mathbf{x} \middle| \mathbf{x} = \begin{bmatrix} x_1 \\ x_2 \\ \vdots \\ x_n \end{bmatrix} \quad x_1, x_2, \ldots, x_n \text{ real} \right\}. \tag{2.36}$$

A norm on R^n is a real-valued function $\|\cdot\|$ defined on R^n and satisfying the three conditions of (2.37) below (where, as before, θ denotes the zero vector in R^n):

$$\|\mathbf{x}\| \geq 0, \text{ and } \|\mathbf{x}\| = 0 \text{ if and only if } \mathbf{x} = \theta; \tag{2.37a}$$

$$\|\alpha\mathbf{x}\| = |\alpha|\,\|\mathbf{x}\|, \text{ for all scalars } \alpha \text{ and vectors } \mathbf{x}; \tag{2.37b}$$

$$\|\mathbf{x} + \mathbf{y}\| \leq \|\mathbf{x}\| + \|\mathbf{y}\|, \text{ for all vectors } \mathbf{x} \text{ and } \mathbf{y}. \tag{2.37c}$$

As noted above, the quantity $\|\mathbf{x}\|$ is thought of as being a measure of the size of the vector $\mathbf{x}$, and double bars are used to emphasize the distinction between the norm of a vector and the absolute value of a scalar. Three useful examples of norms are the so-called ℓ_p norms, $\|\cdot\|_p$, for R^n, $p = 1, 2, \infty$:

$$\begin{aligned} \|\mathbf{x}\|_1 &= |x_1| + |x_2| + \cdots + |x_n| \\ \|\mathbf{x}\|_2 &= \sqrt{(x_1)^2 + (x_2)^2 + \cdots + (x_n)^2} \\ \|\mathbf{x}\|_\infty &= \max\{|x_1|, |x_2|, \ldots, |x_n|\}. \end{aligned} \tag{2.38}$$

EXAMPLE 2.11. By way of illustration for R^3,

$$\mathbf{x} = \begin{bmatrix} 1 \\ -2 \\ -2 \end{bmatrix}, \qquad \|\mathbf{x}\|_1 = 5, \qquad \|\mathbf{x}\|_2 = 3, \qquad \|\mathbf{x}\|_\infty = 2. \tag{2.39}$$

To emphasize further the properties of norms, we now show that the definition of the function $\|\mathbf{x}\|_1$ in (2.38) satisfies the three conditions of (2.37). Clearly for each $\mathbf{x} \in R^n$, $\|\mathbf{x}\|_1 \geq 0$. The rest of (2.37a) is also trivial, for if $\mathbf{x} = \theta$, then

$$\mathbf{x} = \begin{bmatrix} 0 \\ 0 \\ \vdots \\ 0 \end{bmatrix} \quad \text{and} \quad \|\mathbf{x}\|_1 = 0.$$

Conversely if

$$\mathbf{x} = \begin{bmatrix} x_1 \\ x_2 \\ \vdots \\ x_n \end{bmatrix} \quad \text{and} \quad \|\mathbf{x}\|_1 = 0,$$

then $|x_1| + |x_2| + \cdots + |x_n| = 0$ and hence $x_1 = x_2 = \cdots = x_n = 0$, or $\mathbf{x} = \theta$. For part (2.37b),

$$\alpha\mathbf{x} = \begin{bmatrix} \alpha x_1 \\ \alpha x_2 \\ \vdots \\ \alpha x_n \end{bmatrix};$$

so $\|\alpha\mathbf{x}\|_1 = |\alpha x_1| + |\alpha x_2| + \cdots + |\alpha x_n| = |\alpha| \, \|\mathbf{x}\|_1$. If

$$\mathbf{y} = \begin{bmatrix} y_1 \\ y_2 \\ \vdots \\ y_n \end{bmatrix}, \quad \text{then} \quad \mathbf{x} + \mathbf{y} = \begin{bmatrix} x_1 + y_1 \\ x_2 + y_2 \\ \vdots \\ x_n + y_n \end{bmatrix};$$

and therefore

$$\begin{aligned} \|\mathbf{x} + \mathbf{y}\|_1 &= |x_1 + y_1| + |x_2 + y_2| + \cdots + |x_n + y_n| \\ &\leq (|x_1| + |x_2| + \cdots + |x_n|) + (|y_1| + |y_2| + \cdots + |y_n|) \\ &= \|\mathbf{x}\|_1 + \|\mathbf{y}\|_1. \end{aligned}$$

Similarly (Problem 5) it is easy to show that $\|\cdot\|_\infty$ is a norm for R^n. The remaining ℓ_p norm, $\|\cdot\|_2$, is handled not quite so easily. In this case, the triangle inequality in condition (2.37c) does not follow immediately from the triangle inequality for absolute values, and (2.37c) is usually demonstrated with an application of the Cauchy-Schwarz inequality (see Section 2.6).

Having the concept of a norm, we can now make precise quantitative statements about size and distance, and can say that $A\mathbf{x}_c - \mathbf{b}$ is small if $\|A\mathbf{x}_c - \mathbf{b}\|$ is small and that $\mathbf{x}_c$ is near $\mathbf{x}_t$ if $\|\mathbf{x}_c - \mathbf{x}_t\|$ is small. The definition of a norm also has some flexibility. For example, we might have reason to insist that the first coordinate of the residual vector $\mathbf{r} = A\mathbf{x}_c - \mathbf{b}$ is quite critical and must be small, even at the expense of growth in the other coordinates. In this case we might select a norm like the one below where $\mathbf{x}$ is as in (2.35)

$$\|\mathbf{x}\| = \max\{10|x_1|, |x_2|, |x_3|, \ldots, |x_n|\}.$$

A norm such as this emphasizes the first coordinate. For example, saying that $\|\mathbf{x}\| \leq 10^{-5}$ implies that $|x_i| \leq 10^{-5}$ for $i = 2, 3, \ldots, n$ and $|x_1| \leq 10^{-6}$. Weightings of this sort are quite common in problems that involve fitting curves to data and we shall see some examples when we discuss topics such as least-squares fits.

An idea related to norms that weight different coordinates differently is the concept of *relative error*. This is the simple and practical notion that when the size of the error $\mathbf{x}_c - \mathbf{x}_t$ is measured, we should take into account the size of the components of $\mathbf{x}_t$. That is, if $\mathbf{x}_t$ has components of the order of 10^4, then an error of 0.01 is probably acceptable; if $\mathbf{x}_t$ has components that are generally of the order of 10^{-4}, then an error like 0.01 is disastrous. Thus if $\|\cdot\|$ is a norm for R^n, we define $\|\mathbf{x}_c - \mathbf{x}_t\|$ to be the *absolute error* and define the quantity $\|\mathbf{x}_c - \mathbf{x}_t\|/\|\mathbf{x}_t\|$ to be the *relative error*. We will usually be more interested in the relative error than in the absolute error.

EXAMPLE 2.12. Consider the two vectors

$$\mathbf{x}_t = \begin{bmatrix} 0.000397 \\ 0.000214 \\ 0.000309 \end{bmatrix} \quad \text{and} \quad \mathbf{x}_c = \begin{bmatrix} 0.000504 \\ 0.000186 \\ 0.000342 \end{bmatrix}.$$

Then the vector $\mathbf{x}_c - \mathbf{x}_t$ is given by

$$\mathbf{x}_c - \mathbf{x}_t = \begin{bmatrix} 0.000107 \\ -0.000028 \\ 0.000033 \end{bmatrix}.$$

One measure of the absolute error is $\|\mathbf{x}_c - \mathbf{x}_t\|_1 = 0.000168$; so $\mathbf{x}_c$ seems a reasonably good approximation to $\mathbf{x}_t$. However, checking the relative error, we see that

$$\frac{\|\mathbf{x}_c - \mathbf{x}_t\|_1}{\|\mathbf{x}_t\|_1} = \frac{0.000168}{0.000920} \approx 0.183,$$

which more nearly reflects the true state of affairs, i.e., that our approximation $\mathbf{x}_c$ is in error by nearly 20 percent. Relative errors become particularly important in computer programs in which a criterion is needed for terminating an iteration.

2.3.2. Matrix Norms

In Eq. (2.34) we see that in order to measure the size of $\mathbf{x}_c - \mathbf{x}_t$, we must also be able to measure the size of the *vector* $(A^{-1}\mathbf{r})$ since $\mathbf{x}_t$ is not known. If we knew A^{-1}, then we could simply multiply A^{-1} times $\mathbf{r}$ and measure the size of $(A^{-1}\mathbf{r})$ by one of the vector norms of the previous section. However since A^{-1} is not known, we must use a different approach since it would be quite inefficient to compute A^{-1} accurately just to check the accuracy of $\mathbf{x}_c$. We are thus led to consider a way of measuring the "size" or "norms" of matrices as well as of vectors. We shall do this measurement in such a way that we are able to estimate the "size" of A^{-1} by knowing the "size" of A and then to estimate the norm of $(A^{-1}\mathbf{r})$. The concept of matrix norms is not limited to this particular problem, i.e., estimating $\|\mathbf{x}_c - \mathbf{x}_t\|$, but is practically indispensable in deriving computational procedures and error estimates for many other problems as we shall see presently.

By way of notation, let M_n denote the set of all $(n \times n)$ matrices and let $\mathcal{O}$ denote the $(n \times n)$ zero matrix. Then a *matrix norm* for M_n is a real-valued function $\|\cdot\|$ which is defined on M_n and will satisfy for all $(n \times n)$ matrices and A and B

$$\|A\| \geq 0 \text{ and } \|A\| = 0 \text{ if and only if } A = \mathcal{O} \tag{2.40a}$$

$$\|\alpha A\| = |\alpha|\,\|A\| \text{ for any scalar } \alpha \tag{2.40b}$$

$$\|A + B\| \leq \|A\| + \|B\| \tag{2.40c}$$

$$\|AB\| \leq \|A\|\,\|B\|. \tag{2.40d}$$

The addition of condition (2.40d) should be noted. Thus matrix norms have a triangle inequality for both addition and multiplication.

Just as there are numerous ways of defining specific vector norms, there are also numerous ways of defining specific matrix norms. We will concentrate on three norms that are easily computable and are intrinsically related to the three basic vector norms discussed in the previous section. Specifically if $A = (a_{ij}) \in M_n$, we define

$$\begin{aligned}
\|A\|_1 &\equiv \max_{1\le j\le n}\left[\sum_{i=1}^{n} |a_{ij}|\right] \text{— (Maximum absolute column sum)}\\
\|A\|_\infty &\equiv \max_{1\le i\le n}\left[\sum_{j=1}^{n} |a_{ij}|\right] \text{— (Maximum absolute row sum)}\\
\|A\|_E &\equiv \sqrt{\sum_{i=1}^{n}\sum_{j=1}^{n} (a_{ij})^2}.
\end{aligned} \tag{2.41}$$

EXAMPLE 2.13. Let

$$A = \begin{bmatrix} 0 & 0 & 10 & 0 \\ 1 & 1 & 5 & 1 \\ 0 & 1 & 5 & 1 \\ 0 & 0 & 5 & 1 \end{bmatrix};$$

then $\|A\|_1 = \text{Max}\,\{1, 2, 25, 3\} = 25$, $\|A\|_\infty = \text{Max}\,\{10, 8, 7, 6\} = 10$, and $\|A\|_E = \sqrt{181} \approx 13.454$.

These three norms are stressed here because they are *compatible* with the ℓ_1, ℓ_∞, and ℓ_2 vector norms, respectively. Thus given any matrix $A \in M_n$, then for *all* vectors $\mathbf{x} \in R^n$ it is true that

$$\|A\mathbf{x}\|_1 \le \|A\|_1\|\mathbf{x}\|_1, \quad \|A\mathbf{x}\|_\infty \le \|A\|_\infty\|\mathbf{x}\|_\infty, \quad \text{and} \quad \|A\mathbf{x}\|_2 \le \|A\|_E\|\mathbf{x}\|_2. \tag{2.42}$$

(The reader should be careful to distinguish between vector norms and matrix norms since they bear the same subscript notation in two of the three cases. This distinction is clear from the usage. For example, $A\mathbf{x}$ is a vector; so $\|A\mathbf{x}\|_1$ denotes the use of the $\|\cdot\|_1$ vector norm, whereas $\|A\|_1$ denotes use of the matrix norm. The reader should also note that the pairs of matrix and vector norms are compatible only in the orders given by (2.42) and cannot be mixed. For example, let A be given as in the example above and let $\mathbf{x} = (0, 0, 1, 0)^T$. Then $\|\mathbf{x}\|_1 = 1$, but $\|A\mathbf{x}\|_1 = 25$ and $\|A\|_\infty\|\mathbf{x}\|_1 = 10$. That is, we cannot expect to have the inequality $\|A\mathbf{x}\|_1 \le \|A\|_\infty\|\mathbf{x}\|_\infty$ or $\|A\mathbf{x}\|_1 \le \|A\|_\infty\|\mathbf{x}\|_1$.)

Compatibility is a property that connects vector norms and matrix norms. For example, in (2.34) we had $(\mathbf{x}_c - \mathbf{x}_t) = A^{-1}\mathbf{r}$. Using the idea of compatible vector and matrix norms, we can estimate $\|\mathbf{x}_c - \mathbf{x}_t\|_1$ in terms of $\|A^{-1}\|_1$ and $\|\mathbf{r}\|_1$:

$$\|\mathbf{x}_c - \mathbf{x}_t\|_1 = \|A^{-1}\mathbf{r}\|_1 \le \|A^{-1}\|_1\|\mathbf{r}\|_1.$$

As we shall see in Section 2.4, compatibility is crucial also to a clear understanding of iterative methods.

Thus far we have not shown that the three matrix norms satisfy the compatibility properties, (2.42), or even that their definitions given by (2.41) satisfy the necessary norm properties given by (2.40). For the sake of brevity we shall supply only the necessary proofs for the $\|\cdot\|_1$ norm and leave the $\|\cdot\|_\infty$ and $\|\cdot\|_E$ norms to the reader.

Let $A = (a_{ij}) \in M_n$; and write A in terms of its column vectors, $A = [\mathbf{A}_1, \mathbf{A}_2, \ldots, \mathbf{A}_n]$. Let $\mathbf{x} = (x_1, x_2, \ldots, x_n)^{\mathrm{T}}$ be any vector in R^n, and recall from Section 2.1 that $A\mathbf{x}$ may be written as $A\mathbf{x} = x_1\mathbf{A}_1 + x_2\mathbf{A}_2 + \cdots + x_n\mathbf{A}_n$. Since $A\mathbf{x}$ is an $(n \times 1)$ vector, we use (2.37) to get

$$\begin{aligned} \|A\mathbf{x}\|_1 &= \|x_1\mathbf{A}_1 + x_2\mathbf{A}_2 + \cdots + x_n\mathbf{A}_n\|_1 \\ &\le \|x_1\mathbf{A}_1\|_1 + \|x_2\mathbf{A}_2\|_1 + \cdots + \|x_n\mathbf{A}_n\|_1 \\ &= |x_1|\,\|\mathbf{A}_1\|_1 + |x_2|\,\|\mathbf{A}_2\|_1 + \cdots + |x_n|\,\|\mathbf{A}_n\|_1 \\ &\le (|x_1| + |x_2| + \cdots + |x_n|)\Big(\max_{1\le j\le n}\|\mathbf{A}_j\|_1\Big) \equiv \|A\|_1\|\mathbf{x}\|_1. \end{aligned} \tag{2.43}$$

Thus we have shown compatibility for the $\|\cdot\|_1$ norm.

It is trivial to see that parts (2.40a) and (b) hold for $\|\cdot\|_1$. Since the ith column of $A + B$ is precisely the ith column of A plus the ith column of B, part (2.40c) is also easily seen to be true. For part (2.40d), we recall from Section 2.1 that the ith column of AB equals $A\mathbf{B}_i$ where $\mathbf{B}_i$ is the ith column of B. By compatibility, $\|A\mathbf{B}_i\|_1 \le \|A\|_1\|\mathbf{B}_i\|_1$. Now choose i such that $\|\mathbf{B}_i\|_1 \ge \|\mathbf{B}_j\|_1$ for $1 \le j \le n$. Then $\|B\|_1 = \|\mathbf{B}_i\|_1$, and

$$\|AB\|_1 = \max_{1\le j\le n} \|A\mathbf{B}_j\|_1 \le \max_{1\le j\le n} \|A\|_1\|\mathbf{B}_j\|_1 = \|A\|_1\|\mathbf{B}_i\|_1 = \|A\|_1\|B\|_1;$$

and thus part (2.40d) is satisfied. We have therefore shown that $\|A\|_1 = \max_{1\le j\le n} \sum_{i=1}^{n} |a_{ij}|$ is a matrix norm and that it is compatible with the $\|\cdot\|_1$ vector norm.

We conclude this material with an observation on matrix norms, and consider the three numbers K_p, $p = 1, 2, \infty$, where if $A \in M_n$ is given, then

$$K_p = \inf\{K \in R^1 \colon \|A\mathbf{x}\|_p \le K\|\mathbf{x}\|_p, \text{ for all } \mathbf{x} \in R^n\} \tag{2.44}$$

(where "inf" denotes *infimum* or *greatest lower bound*). It can be shown that $K_1 = \|A\|_1$ and $K_\infty = \|A\|_\infty$ although the demonstration goes beyond our purposes. We introduce these numbers, however, since $K_2 \ne \|A\|_E$; and this explains why we use the subscript "E" instead of "2." Thus although we have $\|A\mathbf{x}\|_2 \le \|A\|_E\|\mathbf{x}\|_2$ (compatibility), there is a matrix norm $K_2 \equiv \|A\|_2$, smaller for most matrices than $\|A\|_E$, and such that $\|A\mathbf{x}\|_2 \le \|A\|_2\|\mathbf{x}\|_2$ for all $\mathbf{x} \in R^n$. $\|A\|_2$ is rather unwieldy in computations and involves some deeper theory to derive, and so we are satisfied to use the easily computable and compatible $\|A\|_E$ in place of $\|A\|_2$ (see Theorem 3.7).

2.3.3. Condition Numbers and Error Estimates

In this section, we use the ideas of matrix and vector norms to provide some more information that is useful in helping to determine how good a computed (approximate) solution to the system $A\mathbf{x} = \mathbf{b}$ is. The norms used below can be any pair of compatible matrix and vector norms; for convenience we have omitted the subscripts. The reader can tell from the context whether a particular norm is a matrix norm or a vector norm. The following theorem provides valuable information with respect to the relative error.

Theorem 2.1.

Suppose $A \in M_n$ is nonsingular and $\mathbf{x}_c$ is an approximation to $\mathbf{x}_t$, the exact solution of $A\mathbf{x} = \mathbf{b}$ where $\mathbf{b} \ne \theta$. Then for any compatible matrix and vector norms

$$\frac{1}{\|A\|\,\|A^{-1}\|}\,\frac{\|A\mathbf{x}_c - \mathbf{b}\|}{\|\mathbf{b}\|} \le \frac{\|\mathbf{x}_c - \mathbf{x}_t\|}{\|\mathbf{x}_t\|} \le \|A\|\,\|A^{-1}\|\,\frac{\|A\mathbf{x}_c - \mathbf{b}\|}{\|\mathbf{b}\|}. \tag{2.45}$$

Proof. Again by (2.34), $\mathbf{x}_c - \mathbf{x}_t = A^{-1}\mathbf{r}$ where $\mathbf{r} = A\mathbf{x}_c - \mathbf{b}$. Thus by the compatability conditions, (2.42), $\|\mathbf{x}_c - \mathbf{x}_t\| \le \|A^{-1}\| \|\mathbf{r}\| = \|A^{-1}\| \|A\mathbf{x}_c - \mathbf{b}\|$. Now $A\mathbf{x}_t = \mathbf{b}$; so $\|A\| \|\mathbf{x}_t\| \ge \|\mathbf{b}\|$, and $\|A\|/\|\mathbf{b}\| \ge 1/\|\mathbf{x}_t\|$. Thus,

$$\frac{\|\mathbf{x}_c - \mathbf{x}_t\|}{\|\mathbf{x}_t\|} \le \|A^{-1}\| \|A\mathbf{x}_c - \mathbf{b}\| \frac{\|A\|}{\|\mathbf{b}\|},$$

establishing the right-hand side of (2.45). Now

$$\|A\mathbf{x}_c - \mathbf{b}\| = \|\mathbf{r}\| = \|A\mathbf{x}_c - A\mathbf{x}_t\| \le \|A\| \|\mathbf{x}_c - \mathbf{x}_t\|;$$

and since $\|A\| > 0$,

$$\|\mathbf{x}_c - \mathbf{x}_t\| \ge \|A\mathbf{x}_c - \mathbf{b}\|/\|A\|.$$

Also $\mathbf{x}_t = A^{-1}\mathbf{b}$; so $\|\mathbf{x}_t\| \le \|A^{-1}\| \|\mathbf{b}\|$, or $1/\|\mathbf{x}_t\| \ge 1/\|A^{-1}\| \|\mathbf{b}\|$. Combining these last two inequalities establishes the left-hand side of 2.45). ■

Note the appearance of the term $\|A\| \|A^{-1}\|$ in both the upper and the lower bounds for the relative error. This term is called the *condition number* and is denoted by $\kappa(A)$. If one lets $\epsilon = \|A\mathbf{x}_c - \mathbf{b}\|/\|\mathbf{b}\|$, Eq. (2.45) becomes

$$\frac{\epsilon}{\kappa(A)} \le \frac{\|\mathbf{x}_c - \mathbf{x}_t\|}{\|\mathbf{x}_t\|} \le \epsilon\kappa(A).$$

It is easy to show that $\kappa(A) \ge 1$ (Problem 6), and that the closer $\kappa(A)$ is to 1, the more accurate ϵ becomes as a measurement of the relative error. If $\kappa(A) >> 1$, we are alerted to the possibility that the relative error may not be small even if ϵ is small. This situation occurs when the system $A\mathbf{x} = \mathbf{b}$ is ill-conditioned; that is, when small changes in input data can cause large variations in the solution $\mathbf{x}$. The ultimate in ill-conditioning occurs when A is singular. In this case there are infinitely many vectors $\mathbf{u} \ne \theta$ such that $A\mathbf{u} = \theta$, and hence $\mathbf{x}_t$ and $\mathbf{x}_t + \mathbf{u}$ are both solutions of $A\mathbf{x} = \mathbf{b}$ even when A and $\mathbf{b}$ undergo no change at all. To show the connection between the size of $\kappa(A)$ and how close A is to being singular, it can be proved that if A is nonsingular, then [see Conte and deBoor (1980)]

$$\frac{1}{\kappa(A)} = \min\left\{\frac{\|A - B\|}{\|A\|} : B \text{ is singular}\right\}. \tag{2.46}$$

Hence A can be well approximated (in a relative sense) by a singular matrix if and only if $\kappa(A)$ is large. To elaborate, we observe that if $\mathbf{x}_c$ is the computed solution to $A\mathbf{x} = \mathbf{b}$ and if $\mathbf{x}_c \ne \mathbf{x}_t$, then we have in effect solved a perturbed problem of the form $(A + E)\mathbf{x} = \mathbf{b}$ where E is a matrix that accounts for the errors in the computation. From (2.46) we see that if $\kappa(A)$ is large, then $A + E$ could be singular or almost singular even if E is small. With $A + E$ almost singular we need not expect $\mathbf{x}_c$ to be close to $\mathbf{x}_t$. Although we cannot justify it here, a basic rule of thumb is that if $\kappa(A) = 10^k$ and we are working in d-decimal arithmetic, then we should not expect more than $(d - k)$ accurate figures in $\mathbf{x}_c$ if A is properly scaled. [See the *LINPACK User's Guide* (1979) for a more complete discussion. If $d \leq k$ we say that A is "singular to working precision."]

From the discussion above one sees that a Gauss elimination package should not only provide a machine estimate $\mathbf{x}_c$ to the solution of $A\mathbf{x} = \mathbf{b}$, but also give an estimate for $\kappa(A)$ to test the reliability of $\mathbf{x}_c$. Since $\|A\|$ can be calculated immediately, the crux of the problem is to estimate $\|A^{-1}\|$. Obviously it would be extremely inefficient to try to compute A^{-1} accurately just for this purpose. An alternative is provided by the following. For any nonzero vector $\mathbf{x}$, $\mathbf{x} = A^{-1}A\mathbf{x}$, $\|\mathbf{x}\| \leq \|A^{-1}\| \, \|A\mathbf{x}\|$; and so

$$\|A^{-1}\| \geq \frac{\|\mathbf{x}\|}{\|A\mathbf{x}\|}.$$

Hence we can estimate $\|A^{-1}\|$ by trying to make the quotient $\|\mathbf{x}\|/\|A\mathbf{x}\|$ as large as possible. It can be shown that there exist infinitely many vectors $\tilde{\mathbf{x}}$ such that $\|A^{-1}\| = \|\tilde{\mathbf{x}}\|/\|A\tilde{\mathbf{x}}\|$. In practice we do not actually try to compute such an $\tilde{\mathbf{x}}$ precisely; we merely look for an efficient way of making the quotient $\|\mathbf{x}\|/\|A\mathbf{x}\|$ reasonably large. One reason is that the upper bound for the relative error in (2.45) is usually very conservative; that is, the bound is usually too large, and so a smaller estimate for $\|A^{-1}\|$ is probably acceptable in analyzing the relative error.

Although a detailed theoretical development goes beyond the scope of this text, the *LINPACK* authors recommend the following way of obtaining an estimate for $\|A^{-1}\|$. Let $\mathbf{c}$ be a vector whose components are each ± 1, solve $A^{\mathrm{T}}\mathbf{y} = \mathbf{c}$, and then solve $A\mathbf{x} = \mathbf{y}$. Use $\|\mathbf{x}\|/\|\mathbf{y}\|$ as the estimate for $\|A^{-1}\|$; since $\mathbf{y} = A\mathbf{x}$, it follows that $\|A^{-1}\| \geq \|\mathbf{x}\|/\|\mathbf{y}\|$. We discuss the choice of ± 1 in $\mathbf{c}$ momentarily, but for now we consider the solution process for $A^{\mathrm{T}}\mathbf{y} = \mathbf{c}$ with respect to the computations already performed in Gauss elimination with partial pivoting in solving $A\mathbf{x} = \mathbf{b}$. In particular, the factor step produces matrices S and U such that $SA = U$ where

$$S = L_{n-1}P_{n-1} \ldots L_2P_2L_1P_1.$$

Thus since $A = S^{-1}U$, $A^{\mathrm{T}}\mathbf{y} = \mathbf{c}$ becomes $U^{\mathrm{T}}(S^{-1})^{\mathrm{T}}\mathbf{y} = \mathbf{c}$. Setting $\mathbf{z} = (S^{-1})^{\mathrm{T}}\mathbf{y}$, we can solve $A^{\mathrm{T}}\mathbf{y} = \mathbf{c}$ by solving $U^{\mathrm{T}}\mathbf{z} = \mathbf{c}$ and then setting $\mathbf{y} = S^{\mathrm{T}}\mathbf{z}$. Since U^{T} is lower triangular, $U^{\mathrm{T}}\mathbf{z} = \mathbf{c}$ is simply solved by successive variable substitution. The matrix S is characterized by the elimination multiples M_{ij} and by the vector IPIVOT that records the necessary row interchanges of the pivoting. Without actually forming S or S^{T}, we can compute $S^{\mathrm{T}}\mathbf{z}$ as follows. The permutation matrices P_k are symmetric, $P_k^{\mathrm{T}} = P_k$, and so $S^{\mathrm{T}} = P_1L_1^{\mathrm{T}}P_2L_2^{\mathrm{T}} \ldots P_{n-1}L_{n-1}^{\mathrm{T}}$. Each L_k^{T} is the same as the identity matrix I except the kth row is

$$[0, \ldots, 0, 1, M_{k+1,\,k}, M_{k+2,\,k}, \ldots, M_{nk}].$$

Hence for $\mathbf{w} = [w_1, w_2, \ldots, w_n]^{\mathrm{T}}$, $L_k^{\mathrm{T}}\mathbf{w}$ equals $\mathbf{w}$ except in the kth component, which is given by

$$(L_k^{\mathrm{T}}\mathbf{w})_k = w_k + M_{k+1,\,k}w_{k+1} + M_{k+2,\,k}w_{k+2} + \cdots + M_{nk}w_n. \qquad (2.47)$$

Thus to form $S^{\mathrm{T}}\mathbf{z}$, we successively multiply by L_k as indicated above, and then perform the interchange specified by P_k.

As a simple numerical example, consider the solution of $A^{\mathrm{T}}\mathbf{y} = \mathbf{c}$ where A is as in Example 2.10; and arbitrarily select $\mathbf{c} = [1, -1, -1]^{\mathrm{T}}$. The matrix U is displayed in Example 2.10, and it is routine to solve $U^{\mathrm{T}}\mathbf{z} = \mathbf{c}$ to obtain

$$\mathbf{z} = \begin{bmatrix} \frac{1}{2} \\ \frac{1}{2} \\ -1 \end{bmatrix}.$$

The multipliers and the pivot vector $\mathbf{v} = [2, 3]^{\mathrm{T}}$ are also given in Example 2.10; so $\mathbf{y} = S^{\mathrm{T}}\mathbf{z}$ is calculated [see (2.47)] by

$$\mathbf{z} = \begin{bmatrix} \frac{1}{2} \\ \frac{1}{2} \\ -1 \end{bmatrix} \xrightarrow{m_{32} = -\frac{1}{2}} \begin{bmatrix} \frac{1}{2} \\ 1 \\ -1 \end{bmatrix} \xrightarrow[R_2 \leftrightarrow R_3]{v_2 = 3}$$

$$\begin{bmatrix} \frac{1}{2} \\ -1 \\ 1 \end{bmatrix} \xrightarrow[m_{31} = -1]{m_{21} = -\frac{1}{2}} \begin{bmatrix} 0 \\ -1 \\ 1 \end{bmatrix} \xrightarrow[R_1 \leftrightarrow R_2]{v_1 = 2} \begin{bmatrix} -1 \\ 0 \\ 1 \end{bmatrix} = \mathbf{y}.$$

In the example above, the vector $\mathbf{c}$ was selected at random to illustrate how the solution of $A^{\mathrm{T}}\mathbf{y} = \mathbf{c}$ is organized. What is really wanted is an algorithm that forms a vector $\mathbf{c}$ (whose entries are ± 1) such that the quantity $\|\mathbf{x}\|/\|\mathbf{y}\|$ is a good lower estimate to $\|A^{-1}\|$ and where, as before, $A^{\mathrm{T}}\mathbf{y} = \mathbf{c}$ and $A\mathbf{x} = \mathbf{y}$. The reason for choosing $\mathbf{c}$ to have components ± 1 is beyond the scope of this text and is developed in Cline, Moler, Stewart, and Wilkinson (1979). However, the system $A^{\mathrm{T}}\mathbf{y} = \mathbf{c}$ is solved as above where the specific choices for components of $\mathbf{c}$ are predicated on making $\mathbf{z}$ as large as possible, where $U^{\mathrm{T}}\mathbf{z} = \mathbf{c}$ and $\mathbf{y} = S^{\mathrm{T}}\mathbf{z}$. Part of the explanation for this choice of $\mathbf{c}$ lies in the fact that pivoting usually implies that any ill-conditioning in A is reflected in a corresponding ill-conditioning in U and that S is usually not ill-conditioned. (The bulk of the explanation lies in the "singular-value decomposition of A," the eigenvalues of $A^{\mathrm{T}}A$, and their relation to $\|A\|_2$. See the references above.)

Denoting $U = (u_{ij})$, $\mathbf{z} = (z_i)$, and $\mathbf{c} = (c_i)$ where $c_i = \pm 1$, the kth step of the solution of $U^{\mathrm{T}}\mathbf{z} = \mathbf{c}$ is solving the equation

$$u_{kk}z_k = c_k - (u_{1k}z_1 + \cdots + u_{k-1,k}z_{k-1}).$$

A first strategy at this point is to choose the sign of c_k to be opposite to the sign of $q_k \equiv (u_{1k}z_1 + \cdots + u_{k-1,k}z_{k-1})$ in order to maximize the size of z_k. Setting $c_1 = 1$ and doing this for $2 \le k \le n$ is one way to determine the vector $\mathbf{c}$. The authors mentioned above advocate a second, more sophisticated strategy, which uses some information from the last $n - k$ equations as well as the kth equation. Again we consider the kth stage (choosing c_k) where we assume that

$z_1, \ldots, z_{k-1}$ and $c_1, \ldots, c_{k-1}$ have been already determined. For $k \le i \le n$ we have that the ith equation of $U^T\mathbf{z} = \mathbf{c}$ is

$$u_{ii}z_i = -p_i - (u_{ki}z_k + \cdots + u_{i-1,i}z_{i-1}) + c_i \tag{2.48a}$$

where $p_i = (u_{1i}z_1 + u_{2i}z_2 + \cdots + u_{k-1,i}z_{k-1})$. Note that p_i is independent of the choice of c_k (p_i is determined by the choices $c_1, c_2, \ldots, c_{k-1}$ that have already been made) and note also that the kth equation is $u_{kk}z_k = -p_k + c_k$ in this notation. The choice of $c_k = 1$ or $c_k = -1$ will determine z_k, but it will also influence the last components $z_{k+1}, z_{k+2}, \ldots, z_n$ of $\mathbf{z}$. To make a choice for c_k, let z_k^+ and z_k^- denote the solution of (2.48a) for $i = k$ and for $c_k = 1$ and $c_k = -1$, respectively. Given the two possible choices for c_k, we can rewrite (2.48a) for $k + 1 \le i \le n$ as either

$$u_{ii}z_i = -p_i^+ - (u_{k+1,i}z_{k+1} + \cdots + u_{i-1,i}z_{i-1}) + c_i$$

or (2.48b)

$$u_{ii}z_i = -p_i^- - (u_{k+1,i}z_{k+1} + \cdots + u_{i-1,i}z_{i-1}) + c_i$$

where $p_i^+ = p_i + u_{ki}z_k^+$ and $p_i^- = p_i + u_{ki}z_k^-$. Since we do not know c_j or z_j for $j \ge k + 1$, we temporarily set these equal to zero on the right-hand side of (2.48b); select $c_k = +1$ if

$$|-p_k + 1| + \sum_{i=k+1}^{n} |p_i^+| \ge |-p_k - 1| + \sum_{i=k+1}^{n} |p_i^-|,$$

and choose $c_k = -1$ otherwise. Essentially this procedure is trying to maximize $\|\mathbf{z}\|$ by forcing the absolute sum of the terms on the right-hand side of (2.48a) to be as large as possible when we neglect z_j and c_j for $k + 1 \le j \le n$. For $k = 1$, we can choose $c_k = +1$ to start the selection algorithm.

Using this strategy on the matrix A in Example 2.10,

$$A = \begin{bmatrix} 1 & -1 & 0 \\ 2 & -1 & 1 \\ 2 & -2 & -1 \end{bmatrix} \quad \text{and} \quad U = \begin{bmatrix} 2 & -1 & 1 \\ 0 & -1 & -2 \\ 0 & 0 & \frac{1}{2} \end{bmatrix},$$

we obtain $\mathbf{c} = [1, 1, -1]^T$, $\mathbf{z} = [1/2, -3/2, -9]^T$, and $\mathbf{y} = [-9, 2, 3]^T$ where $U^T\mathbf{z} = \mathbf{c}$, $\mathbf{y} = S^T\mathbf{z}$, and $A^T\mathbf{y} = \mathbf{c}$. Then $A\mathbf{x} = \mathbf{y}$ where $\mathbf{x} = [32, 41, -21]^T$, and so $\|A^{-1}\|_1 \ge \|\mathbf{x}\|_1/\|\mathbf{y}\|_1 = 94/14 \simeq 6.7$. Actual calculation of A^{-1} yields $\|A^{-1}\|_1 = 9$. The matrix above is not ill-conditioned. The strategy was also applied to the (3×3) Hilbert matrix H_3 (Problem 15, Section 2.2.4) with the result that $\|H_3^{-1}\|_1 \ge 371.35$ where $\|H_3^{-1}\|_1 = 408$. (We leave it to the reader to verify this result.)

EXAMPLE 2.14. Consider the linear system

$$\begin{aligned} 6x_1 + 6x_2 + 3.00001x_3 &= 30.00002 \\ 10x_1 + 8x_2 + 4.00003x_3 &= 42.00006 \\ 6x_1 + 4x_2 + 2.00002x_3 &= 22.00004, \end{aligned}$$

which has the unique solution $x_1 = 1$, $x_2 = 3$, $x_3 = 2$. This example illustrates the notion of an ill-conditioned system as well as how matrix norms can be used to analyze the results of a computational solution. (Recall that a system of linear equations is called ill-conditioned if "small" changes in the coefficients produce "large" changes in the solution.) Before solving this system, note that the perturbed system below

$$\begin{aligned} 6x_1 + 6x_2 + 3.00001x_3 &= 30 \\ 10x_1 + 8x_2 + 4.00003x_3 &= 42 \\ 6x_1 + 4x_2 + 2.00002x_3 &= 22 \end{aligned}$$

has a unique solution $x_1 = 1$, $x_2 = 4$, $x_3 = 0$, which is substantially different from the solution of the first system, even though the coefficient matrices are the same and the constants on the right-hand side are the same through six significant figures. Therefore we call these two systems ill-conditioned.

The first system demonstrates also that we should be cautious about how we determine whether an estimate to a solution is a good estimate or not. If we try $x_1 = 1$, $x_2 = 4$, and $x_3 = 0$ in the first system, we obtain the residual vector

$$\mathbf{r} = \begin{bmatrix} -2. \times 10^{-5} \\ -6. \times 10^{-5} \\ -4. \times 10^{-5} \end{bmatrix}$$

whereas the actual error in using this estimate is on the order of 10^5 times as large as the residual vector would indicate. By (2.34) it follows that the inverse of this coefficient matrix has large entries.

To show what happens when Gauss elimination is used to solve the first system of equations, a single-precision Gauss elimination routine was employed and found

$$\begin{bmatrix} x_1 \\ x_2 \\ x_3 \end{bmatrix} = \begin{bmatrix} 0.9999898 \\ 1.907699 \\ 4.184615 \end{bmatrix}, \qquad \mathbf{r} = \begin{bmatrix} 0.000019 \\ 0.000076 \\ 0.000049 \end{bmatrix}, \qquad \mathbf{x}_t - \mathbf{x}_c = \begin{bmatrix} 0.000010 \\ 1.092301 \\ -2.184615 \end{bmatrix}.$$

These results are typical if this system is solved on any digital device that has six- to eight-place accuracy. A double-precision computation on a computer would give almost exactly the correct answer, but going to double precision is obviously not a cure-all, for there are simple examples like the one above for which double-precision arithmetic is not sufficient.

Although this example is somewhat contrived (so that the essence of the problem is not hidden in a mass of cumbersome calculations), many real-life problems, particularly in areas such as statistical analysis and least-squares fits, are ill-conditioned. Thus it is appropriate that a person who must deal with numerical solutions of linear equations have at hand as many tools as possible in order to test computed results for accuracy. Often a very reliable test is simply a feeling for what the computed results are supposed to represent physically; i.e., whether the answers fit the physical problem. In the absence of a physical intuition for what the answer should be, or in terms of a tool for a mathematical analysis of the significance of the computed results, the estimates of Theorem 2.1 provide a beginning.

We continue now with Example 2.14 to illustrate how the $\|\cdot\|_1$ and $\|\cdot\|_\infty$ norms can be readily used to analyze errors in computed solutions. Thus $A\mathbf{x} = \mathbf{b}$ has $\mathbf{x}_t$ as its solution vector where

$$A = \begin{bmatrix} 6 & 6 & 3.00001 \\ 10 & 8 & 4.00003 \\ 6 & 4 & 2.00002 \end{bmatrix}, \qquad \mathbf{b} = \begin{bmatrix} 30.00002 \\ 42.00006 \\ 22.00004 \end{bmatrix}, \qquad \mathbf{x}_t = \begin{bmatrix} 1 \\ 3 \\ 2 \end{bmatrix}. \tag{2.49}$$

In order not to obscure the point of this example, let us suppose that a computed estimate to the solution, $\mathbf{x}_c$ and hence the residual $\mathbf{r} = A\mathbf{x}_c - \mathbf{b}$ are given by

$$\mathbf{x}_c = \begin{bmatrix} 1 \\ 4 \\ 0 \end{bmatrix} \quad \text{and} \quad \mathbf{r} = \begin{bmatrix} -0.00002 \\ -0.00006 \\ -0.00004 \end{bmatrix}. \tag{2.50}$$

For an analysis of $\mathbf{x}_c - \mathbf{x}_t$, let us use the ℓ_∞ norm and carry only three significant figures. Thus

$$\frac{\|A\mathbf{x}_c - \mathbf{b}\|_\infty}{\|\mathbf{b}\|_\infty} \approx \frac{6 \times 10^{-5}}{42} \approx 1.43 \times 10^{-6}. \tag{2.51}$$

Clearly $\|A\|_\infty = 22.00003 \approx 22$, and therefore by Theorem 2.1

$$\frac{\|\mathbf{x}_c - \mathbf{x}_t\|_\infty}{\|\mathbf{x}_t\|_\infty} \le (22.)(1.43 \times 10^{-6})\|A^{-1}\|_\infty \le (3.15 \times 10^{-5})\|A^{-1}\|_\infty. \tag{2.52}$$

Hence an estimate for $\|A^{-1}\|_\infty$ is in order. With the following vector $\mathbf{x}'$ we obtain

$$\mathbf{x}' = \begin{bmatrix} 0 \\ -1 \\ 2 \end{bmatrix}, \qquad A\mathbf{x}' = \begin{bmatrix} 0.00002 \\ 0.00006 \\ 0.00004 \end{bmatrix}. \tag{2.53}$$

Thus a lower bound for $\|A^{-1}\|_\infty$ is provided by

$$\frac{\|\mathbf{x}'\|_\infty}{\|A\mathbf{x}'\|_\infty} = \frac{2}{6. \times 10^{-5}} \approx 3.33 \times 10^4 \le \|A^{-1}\|_\infty. \tag{2.54}$$

This estimate for $\|A^{-1}\|_\infty$ makes the estimate of (2.52) more meaningful, for now the upper bound for the relative error in (2.52) is at least as large as $(3.15 \times 10^{-5})(3.33 \times 10^4) \approx 1.05$; and in fact,

$$\frac{\|\mathbf{x}_c - \mathbf{x}_t\|_\infty}{\|\mathbf{x}_t\|_\infty} = \frac{2}{3}.$$

Subsequent sections will make further use of matrix norms in such topics as iterative procedures to solve $A\mathbf{x} = \mathbf{b}$, methods for finding eigenvalues of matrices, methods for solving nonlinear systems, and methods of optimization.

2.3.4. Iterative Improvement

If an error analysis indicates that the computed solution is unacceptable, either we can start again with a different method (such as an indirect method or even Gauss elimination with higher precision), or we can try to improve the answer we have. Iterative improvement (sometimes known as the "method of residual correction") is a procedure that uses a limited amount of double precision to try to refine a computed solution $\mathbf{x}_c$ of the system $A\mathbf{x} = \mathbf{b}$. The procedure is relatively simple to implement and frequently will yield improved estimates of the solution. In order to describe this method, we again let $\mathbf{x}_t$ denote the true solution of $A\mathbf{x} = \mathbf{b}$ and use $\mathbf{r} = A\mathbf{x}_c - \mathbf{b}$ to denote the residual vector [where $\mathbf{x}_c$ is a computed (approximate) solution to $A\mathbf{x} = \mathbf{b}$]. If we let $\mathbf{e} = \mathbf{x}_c - \mathbf{x}_t$, then $A\mathbf{e} = A\mathbf{x}_c - A\mathbf{x}_t = A\mathbf{x}_c - \mathbf{b} = \mathbf{r}$. Thus if we could solve $A\mathbf{e} = \mathbf{r}$, we would be able to find $\mathbf{x}_t$ from the equation $\mathbf{x}_t = \mathbf{x}_c - \mathbf{e}$. The method of iterative improvement is precisely the implementation of this idea. We immediately see the difficulty inherent in such a procedure since if we cannot solve $A\mathbf{x} = \mathbf{b}$ exactly, we cannot expect to solve $A\mathbf{e} = \mathbf{r}$ exactly.

Iterative improvement, when implemented properly on the computer, attempts to overcome the objection noted above by calculating the residual, $\mathbf{r}$, in double precision. To be precise, the steps in iterative improvement are these.

1. Calculate $\mathbf{r} = A\mathbf{x}_c - \mathbf{b}$ in double precision.
2. Solve $A\mathbf{e} = \mathbf{r}$ and let $\mathbf{e}_c$ be the computed solution to this system.
3. Let $\mathbf{x}_c' = \mathbf{x}_c - \mathbf{e}_c$.

Solving $A\mathbf{e} = \mathbf{r}$ is greatly facilitated by retaining the factored form of A from the previous computation since then one need only form $S\mathbf{r}$ and backsolve the equivalent system $U\mathbf{e} = S\mathbf{r}$. We hope that $\mathbf{x}_c'$ is a better approximation to $\mathbf{x}_t$ than was $\mathbf{x}_c$. In order to analyze the method a bit more carefully, let $\mathbf{e}_t$ denote the true solution to $A\mathbf{e} = \mathbf{r}$. First we can obtain some qualitative information from the steps outlined above by noting that

$$\frac{\|\mathbf{e}_t\|}{\|\mathbf{x}_t\|} = \frac{\|\mathbf{x}_c - \mathbf{x}_t\|}{\|\mathbf{x}_t\|}.$$

Thus we can hope that given the numbers we have, namely $\|\mathbf{e}_c\|$ and $\|\mathbf{x}_c\|$, the ratio

$$\frac{\|\mathbf{e}_c\|}{\|\mathbf{x}_c\|}$$

might give an indication of the relative error.

We next note that the relative error of solving $A\mathbf{e} = \mathbf{r}$ is given by

$$\frac{\|\mathbf{e}_c - \mathbf{e}_t\|}{\|\mathbf{e}_t\|}.$$

From Step 3 we have that $\mathbf{e}_c = \mathbf{x}_c - \mathbf{x}_c'$; so since $\mathbf{e}_t = \mathbf{x}_c - \mathbf{x}_t$, it follows that

$$\frac{\|\mathbf{e}_c - \mathbf{e}_t\|}{\|\mathbf{e}_t\|} = \frac{\|\mathbf{x}_t - \mathbf{x}_c'\|}{\|\mathbf{x}_t - \mathbf{x}_c\|}.$$

Therefore if we can solve $A\mathbf{e} = \mathbf{r}$ with a relative error less than 1, then $\mathbf{x}_c'$ is a better approximation to $\mathbf{x}_t$ than is $\mathbf{x}_c$.

Finally we recall from (2.45) that an error bound for the relative error is of the form

$$\frac{\|\mathbf{x}_c - \mathbf{x}_t\|}{\|\mathbf{x}_t\|} \le M\|A\mathbf{x}_c - \mathbf{b}\| \tag{2.55a}$$

where $M = \|A\|\,\|A^{-1}\|/\|\mathbf{b}\|$. We can find a similar bound for $\|\mathbf{x}_c' - \mathbf{x}_t\|/\|\mathbf{x}_t\|$ as follows. We let $\mathbf{r}_c = A\mathbf{e}_c - \mathbf{r}$ so that $A^{-1}\mathbf{r}_c = \mathbf{e}_c - \mathbf{e}_t$, and hence $\|A^{-1}\|\,\|\mathbf{r}_c\| \ge \|\mathbf{e}_c - \mathbf{e}_t\|$. But, as above, $\|\mathbf{e}_c - \mathbf{e}_t\| = \|\mathbf{x}_t - \mathbf{x}_c'\| = \|\mathbf{x}_c' - \mathbf{x}_t\|$, so $\|\mathbf{x}_c' - \mathbf{x}_t\| \le \|A^{-1}\|\,\|A\mathbf{e}_c - \mathbf{r}\|$. Just as in Theorem 2.1, $\|A\|\,\|\mathbf{x}_t\| \ge \|\mathbf{b}\|$; so we get $1/\|\mathbf{x}_t\| < \|A\|/\|\mathbf{b}\|$. Using these two inequalities, we find

$$\frac{\|\mathbf{x}_c' - \mathbf{x}_t\|}{\|\mathbf{x}_t\|} \le \frac{\|A\|\,\|A^{-1}\|}{\|\mathbf{b}\|}\|A\mathbf{e}_c - \mathbf{r}\|,$$

which is an inequality of the form

$$\frac{\|\mathbf{x}_c' - \mathbf{x}_t\|}{\|\mathbf{x}_t\|} \le M\|A\mathbf{e}_c - \mathbf{r}\| \tag{2.55b}$$

where M is the same constant that appears in (2.55a). We can calculate both the numbers $\|A\mathbf{x}_c - \mathbf{b}\|$ and $\|A\mathbf{e}_c - \mathbf{r}\|$; and if $\|A\mathbf{e}_c - \mathbf{r}\| < \|A\mathbf{x}_c - \mathbf{b}\|$, we expect that $\mathbf{x}_c'$ might be a relatively better approximation to $\mathbf{x}_t$ than is $\mathbf{x}_c$. Notice that since $\mathbf{x}_c'$ is an approximation to $\mathbf{x}_t$ just as was $\mathbf{x}_c$, we can apply iterative improvement again and hope to improve $\mathbf{x}_c'$ by solving $A\mathbf{e} = A\mathbf{x}_c' - \mathbf{b}$. If the corrections, $\mathbf{e}_c^{(i)}$, that we obtain in this fashion do not decrease in size, then we have a strong indication that the matrix A is ill-conditioned.

To illustrate iterative improvement, the method was used on the system in Example 2.7. The approximate solution, $\mathbf{x}_c$, was calculated using a single precision Gauss elimination routine. Displayed below are the results of two applications of iterative improvement. That is, $\mathbf{x}_c'$ is the result of solving $A\mathbf{e} = \mathbf{r}'$ where $\mathbf{r}' = A\mathbf{x}_c - \mathbf{b}$; and $\mathbf{x}_c''$ is the result of solving $A\mathbf{e} = \mathbf{r}''$ where $\mathbf{r}'' = A\mathbf{x}_c' - \mathbf{b}$.

$$\mathbf{x}_c = \begin{bmatrix} 0.999417\text{E } 00 \\ 0.100035\text{E } 01 \\ 0.100015\text{E } 01 \\ 0.999910\text{E } 00 \end{bmatrix} \qquad \mathbf{x}_c' = \begin{bmatrix} 0.100004\text{E } 01 \\ 0.999977\text{E } 00 \\ 0.999991\text{E } 00 \\ 0.100000\text{E } 01 \end{bmatrix}$$

$$\mathbf{x}_c'' = \begin{bmatrix} 0.100000\text{E } 01 \\ 0.100000\text{E } 01 \\ 0.100000\text{E } 01 \\ 0.100000\text{E } 01 \end{bmatrix}$$

Thus two steps of the iterative improvement algorithm give the answer correct to as many places as are printed.

PROBLEMS, SECTION 2.3

1. For the matrix A of Problem 7, find a (4×1) vector $\mathbf{x}$, $\mathbf{x} \neq \theta$, such that $\|A\mathbf{x}\|_\infty = \|A\|_\infty\|\mathbf{x}\|_\infty$. [Hint: Consider only vectors $\mathbf{x}$ such that $\|\mathbf{x}\|_\infty = 1$.] Find a (4×1) vector $\mathbf{y}$, $\mathbf{y} \neq \theta$, such that $\|A\mathbf{y}\|_1 = \|A\|_1\|\mathbf{y}\|_1$. Use A^{-1} and repeat this problem.
2. Let I be the $(n \times n)$ identity matrix. What is $\|I\|_E$? Is there any $(n \times 1)$ vector $\mathbf{x}$, $\mathbf{x} \neq \theta$, such that $\|I\mathbf{x}\|_2 = \|I\|_E\|\mathbf{x}\|_2$?
3. Let $\mathbf{x}$ be any vector in R^n as given by (2.36), and let $\{\mathbf{x}^{(j)}\}_{j=1}^{\infty}$ be any sequence of vectors in R^n.
   a) Show that $\|\mathbf{x}\|_1 \geq \|\mathbf{x}\|_2 \geq \|\mathbf{x}\|_\infty \geq (1/n)\|\mathbf{x}\|_1$ for all $\mathbf{x} \in R^n$.
   b) From the definition of convergence of a sequence of real numbers, use Part (a) to show that if $\lim_{j\to\infty} \|\mathbf{x}^{(j)}\|_p = 0$ for $p = 1, 2$, or ∞, then $\lim_{j\to\infty} \|\mathbf{x}^{(j)}\|_q = 0$ for $q = 1$, 2, or ∞, $q \neq p$.
   c) For p given as 1, 2, or ∞, show that $\lim_{j\to\infty} \|\mathbf{x}^{(j)} - \mathbf{x}\|_p = 0$ implies that
   $$\lim_{j\to\infty} \|A(\mathbf{x}^{(j)} - \mathbf{x})\|_p \equiv \lim_{j\to\infty} \|A\mathbf{x}^{(j)} - A\mathbf{x}\|_p = 0.$$
4. For each norm, $\|\cdot\|_p$, $p = 1, 2, \infty$, graph the set of points $\mathbf{x} = [x_1, x_2]^T$ in R^2 such that $\|\mathbf{x}\|_p = 1$. Compare this result with the results of Problem 3a.
5. If $\|\cdot\|_\infty$ is given by (2.38), show that it satisfies all three properties of (2.37) and thus is a norm for R^n.
6. Let $\|\cdot\|_m$ be any matrix norm on M_n. Show that $\|A\|_m\|A^{-1}\|_m \geq 1$. [Hint: Consider $\|I^2\|_m = \|I\|_m$ and use the norm properties (2.40).]
7. Let A be the (4×4) coefficient matrix of the system in Example 2.6. Then A^{-1} is given by
$$A^{-1} = \begin{bmatrix} 68 & -41 & -17 & 10 \\ -41 & 25 & 10 & -6 \\ -17 & 10 & 5 & -3 \\ 10 & -6 & -3 & 2 \end{bmatrix}.$$
   Find the condition number $\|A\|_\infty\|A^{-1}\|_\infty$ of this matrix. Use the inequality (2.45) of Theorem 2.1 and the computer results of Example 2.7 to estimate the relative error $\|\mathbf{x}_c - \mathbf{x}_t\|_\infty/\|\mathbf{x}_t\|_\infty$. What is the true relative error?
8. Let A_n be the (2×2) matrix given by
$$A_n = \begin{bmatrix} 1 & 2 \\ 2 & 4 + 1/n^2 \end{bmatrix}.$$
   Find A_n^{-1} and the condition number $\|A_n\|_\infty\|A_n^{-1}\|_\infty$. Let $n = 100$ so that
$$A_n = \begin{bmatrix} 1 & 2 \\ 2 & 4.0001 \end{bmatrix},$$
   and let
$$\mathbf{b} = \begin{bmatrix} 1 \\ 2 - 1/n^2 \end{bmatrix} = \begin{bmatrix} 1 \\ 1.9999 \end{bmatrix}.$$
   Solve $A_{100}\mathbf{x} = \mathbf{b}$ mathematically and call the answer $\mathbf{x}_t$. Let $\mathbf{x}_c$ be given by
$$\mathbf{x}_c = \begin{bmatrix} 1 \\ 0 \end{bmatrix}.$$

Find $\mathbf{r} = A_{100}\mathbf{x}_c - \mathbf{b}$ and check the error bound $\|\mathbf{x}_c - \mathbf{x}_t\|_\infty \le \|A^{-1}\|_\infty \|\mathbf{r}\|_\infty$. One might expect from this problem and from Example 2.14 that ill-conditioned matrices must have small determinants. Find the determinant of the matrix A in Problem 7 to see that this suspicion is not always valid.

9. For the matrix A_n in Problem 8, find a (2×2) singular matrix B as in (2.46) such that $(\|A_n - B\|/\|A_n\| - 1/\kappa(A_n)) < 1/n^2$.

10. Let D_n be the $(n \times n)$ diagonal matrix with diagonal entries all equal to 0.1. Compute $\det(D_n)$ and $\|D_n\|_\infty \|D_n^{-1}\|_\infty$. Among all singular $(n \times n)$ matrices B what is about the smallest that the number $\|B - D_n\|_\infty$ can be? For large n, $\det(D_n) \simeq 0$; does this mean that D_n is "almost singular"? [See (2.46).] Is D_n ill-conditioned?

11. In the lines of the computations following (2.47), use the previously computed information on the factorization of A from Example 2.10 to solve the (3×3) system $A^{\mathrm{T}}\mathbf{y} = \mathbf{b}$ where $\mathbf{b}$ is given by

(a) $[-5, 3, -1]^{\mathrm{T}}$ (b) $[8, -8, -3]^{\mathrm{T}}$ (c) $[1, -2, -2]^{\mathrm{T}}$

12. Let

$$A'' = \begin{bmatrix} 1 & -1 & 4 \\ -\frac{1}{2} & -1 & 4 \\ \frac{1}{2} & \frac{1}{4} & -3 \end{bmatrix}, \qquad \mathbf{v} = \begin{bmatrix} 3 \\ 3 \end{bmatrix},$$

$$\text{and } \mathbf{b} = \begin{bmatrix} -2 \\ 4 \\ 2 \end{bmatrix}.$$

Suppose that A'' contains the upper-triangular part of a (3×3) matrix A after Gauss elimination, and the lower triangular part of A'' contains the elimination multiples; and suppose that $\mathbf{v}$ has recorded the row interchanges. Follow the lines of Section 2.3.3.

a) Solve $A\mathbf{x} = \mathbf{b}$. b) Solve $A^{\mathrm{T}}\mathbf{y} = \mathbf{b}$. c) Compute an estimate for $\|A^{-1}\|_1$, as in Problem 13.

d) Reconstruct the original matrix A and use it to check parts (a) and (b).

13. Using the first strategy [described prior to (2.48)] for the choice of the vector $\mathbf{c}$, calculate an estimate for $\|H_3^{-1}\|_1$ where H_3 is the (3×3) Hilbert matrix.

14. Repeat Problem 13 and use the second strategy [described following (2.48)] for the choice of the vector $\mathbf{c}$.

15. In describing the technique for solving $A^{\mathrm{T}}\mathbf{y} = \mathbf{b}$, we used the fact that $(S^{-1})^{\mathrm{T}} = (S^{\mathrm{T}})^{-1}$; prove this.

16. Following the technique described in Section 2.3 and making use of the subroutines FACTOR and SOLVE, write a subroutine to estimate the condition number of an $(n \times n)$ matrix A. (Use either the first or the second strategy for choosing the vector $\mathbf{c}$ and either the 1 or ∞ norms.)

17. Use the subroutine in Problem 16 to estimate $\|H_3^{-1}\|_1$ and $\|A^{-1}\|_1$ where H_3 is the (3×3) Hilbert matrix and A is the (4×4) matrix in Example 2.6.

18. If an $(n \times n)$ matrix A has an LU-decomposition where L and U are known, what are the two necessary steps for solving the system $A^{\mathrm{T}}\mathbf{y} = \mathbf{b}$? Find an LU-decomposition of the (3×3) coefficient matrix in Example 2.2, Section 2.1, and use the decomposition in this manner to solve $A^{\mathrm{T}}\mathbf{y} = \mathbf{e}_1$.

2.4 ITERATIVE METHODS

There are instances in which direct methods such as Gauss elimination and LU-decomposition may not be the best methods to use for solving the system of linear equations $A\mathbf{x} = \mathbf{b}$. There is an alternative class of methods that can be used for this problem; namely, the *iterative* methods. We will first develop the concept of an iterative method and give some particular examples. Later we will discuss the relative merits of iterative and direct methods.

The methods presented here are called iterative because each method is designed to generate a sequence of vectors (*iterates*), $\{\mathbf{x}^{(k)}\}_{k=0}^{\infty}$, which converge to the true solution, $\mathbf{x}_t$, of $A\mathbf{x} = \mathbf{b}$. The basic idea of iterative methods can be described as follows.

1. The matrix A is written as the difference of two matrices N and P so that $A = N - P$. This decomposition of A is called a *splitting*.
2. An initial guess $\mathbf{x}^{(0)}$ is made for the solution vector $\mathbf{x}_t$.
3. A sequence $\mathbf{x}^{(1)}, \mathbf{x}^{(2)}, \mathbf{x}^{(3)}, \ldots,$ of estimates to $\mathbf{x}_t$ is generated by the formula

$$N\mathbf{x}^{(k+1)} = P\mathbf{x}^{(k)} + \mathbf{b}, \qquad k = 0, 1, 2, \ldots. \tag{2.56}$$

Since the idea of iteration is probably not too familiar, we present an example below before discussing iterative methods further.

EXAMPLE 2.15. Let

$$A = \begin{bmatrix} 4 & 1 & 0 \\ 2 & 5 & 1 \\ -1 & 2 & 4 \end{bmatrix}, \qquad \mathbf{b} = \begin{bmatrix} 1 \\ 0 \\ 3 \end{bmatrix}, \text{ and}$$

let $A\mathbf{x} = \mathbf{b}$ be the linear system to be solved. Let the splitting be given by

$$N = \begin{bmatrix} 4 & 0 & 0 \\ 0 & 5 & 0 \\ 0 & 0 & 4 \end{bmatrix} \quad \text{and} \quad P = \begin{bmatrix} 0 & -1 & 0 \\ -2 & 0 & -1 \\ 1 & -2 & 0 \end{bmatrix}$$

so that $A = N - P$. If we denote the vector $\mathbf{x}^{(k)}$ by $\mathbf{x}^{(k)} = \begin{bmatrix} x_k \\ y_k \\ z_k \end{bmatrix}$,

then writing out formula (2.56) yields the equations

$$\begin{aligned} 4x_{k+1} &= -y_k + 1 \\ 5y_{k+1} &= -2x_k - z_k \\ 4z_{k+1} &= x_k - 2y_k + 3, \end{aligned}$$

which define $\mathbf{x}^{(k+1)}$ for $k = 0, 1, 2, \ldots$. As an initial guess, let

$$\mathbf{x}^{(0)} = \begin{bmatrix} 1 \\ 1 \\ 1 \end{bmatrix}.$$

Then the first few iterates are

$$\mathbf{x}^{(0)} = \begin{bmatrix} 1 \\ 1 \\ 1 \end{bmatrix}, \qquad \mathbf{x}^{(1)} = \begin{bmatrix} 0 \\ -\frac{3}{5} \\ \frac{1}{2} \end{bmatrix}, \qquad \mathbf{x}^{(2)} = \begin{bmatrix} \frac{2}{5} \\ -\frac{1}{10} \\ \frac{21}{20} \end{bmatrix}.$$

[The true solution in this example is $\mathbf{x}_t = (1/3, -1/3, 1)^T$.]

Let us return now to the general procedure outlined in steps (1) to (3). The first thing to observe is that if $A\mathbf{x}_t = \mathbf{b}$ then $N\mathbf{x}_t = P\mathbf{x}_t + \mathbf{b}$, and vice versa (that is, solving $A\mathbf{x} = \mathbf{b}$ is equivalent to solving $N\mathbf{x} = P\mathbf{x} + \mathbf{b}$). In order to get some idea of what might constitute a good choice for N and P, consider formula (2.56). This formula says that if we have $\mathbf{x}^{(k)}$, then we can get the next iterate $\mathbf{x}^{(k+1)}$ provided we can solve the linear system $N\mathbf{x}^{(k+1)} = \mathbf{h}^{(k)}$ where the vector $\mathbf{h}^{(k)}$ is given by $\mathbf{h}^{(k)} = P\mathbf{x}^{(k)} + \mathbf{b}$. Thus it is clear that we must require N to be nonsingular in order to be assured that we can implement the iteration. Furthermore, for an iterative procedure to be efficient, N should be chosen so that $N\mathbf{x}^{(k+1)} = \mathbf{h}^{(k)}$ is quite easy to solve. This is the case if, for instance, N is chosen to be a triangular matrix (or a diagonal matrix as in Example 2.15 above).

Last, the question of convergence to $\mathbf{x}_t$ must be considered. It is fairly easy to make a start at answering this question. Let $\mathbf{e}^{(k)} = \mathbf{x}^{(k)} - \mathbf{x}_t$ denote the error vector at the kth step. As noted above, $N\mathbf{x}_t = P\mathbf{x}_t + \mathbf{b}$; so from this and (2.56) it follows that $N(\mathbf{x}^{(k+1)} - \mathbf{x}_t) = P(\mathbf{x}^{(k)} - \mathbf{x}_t)$ or $N\mathbf{e}^{(k+1)} = P\mathbf{e}^{(k)}$. Since we have required that N be nonsingular, we can multiply by N^{-1} to obtain $\mathbf{e}^{(k+1)} = N^{-1}P\mathbf{e}^{(k)}$ for $k = 0, 1, 2, \ldots$. If we set $M \equiv N^{-1}P$, then a fundamental relationship among the error vectors has been established:

$$\mathbf{e}^{(k+1)} = M\mathbf{e}^{(k)}, \qquad k = 0, 1, 2, \ldots. \tag{2.57}$$

If M is in some sense a "small" matrix, then (2.57) would indicate that the errors are diminishing, or that $\{\mathbf{x}^{(k)}\} \to \mathbf{x}_t$. This statement is made more precise in Theorem 2.2 below. In this theorem, we use the ℓ_∞ vector and matrix norms. It is evident from the proof, however, that any compatible vector and matrix norms could have been used.

Theorem 2.2

Suppose $A = N - P$ and suppose $\|N^{-1}P\|_\infty \le \lambda < 1$. Then

1. A is nonsingular;
2. if $\mathbf{x}_t$ is the solution of $A\mathbf{x} = \mathbf{b}$ and if $\{\mathbf{x}^{(j)}\}$ is given by (2.56), then $\lim_{j\to\infty} \mathbf{x}^{(j)} = \mathbf{x}_t$; and
3. $\|\mathbf{x}^{(j)} - \mathbf{x}_t\|_\infty \le \lambda^j \|\mathbf{x}^{(0)} - \mathbf{x}_t\|_\infty$ (or $\|\mathbf{e}^{(j)}\|_\infty \le \lambda^j \|\mathbf{e}^{(0)}\|_\infty$).

Proof. Before establishing Theorem 2.2, it is worth observing that a theorem of this type has a lot of practical computational value when it is possible to verify the hypotheses. The theorem

1. guarantees a solution to the problem,
2. shows that the numerical method will converge to the solution for any initial guess $\mathbf{x}^{(0)}$, and
3. indicates to the user of the numerical method how many steps should be taken to attain a desired accuracy.

Part (1) is easiest to prove by contradiction. If A is singular, then there must be a nonzero vector $\mathbf{y}$ such that $A\mathbf{y} = \theta$. Therefore $(N - P)\mathbf{y} = \theta$ or $\mathbf{y} = N^{-1}P\mathbf{y}$. From the compatibility properties of the ℓ_∞ vector and matrix norms, it follows that

$$\|\mathbf{y}\|_\infty = \|N^{-1}P\mathbf{y}\|_\infty \le \|N^{-1}P\|_\infty\|\mathbf{y}\|_\infty \le \lambda\,\|\mathbf{y}\|_\infty. \tag{2.58}$$

Since $\mathbf{y} \ne \theta$, then $\|\mathbf{y}\|_\infty > 0$; and hence the inequality (2.58) means that $1 \le \lambda$. This statement is a contradiction of the hypothesis; so it must be that there is no nonzero vector $\mathbf{y}$ such that $A\mathbf{y} = \theta$. Hence A is nonsingular.

We next establish part (3). If we set $M = N^{-1}P$ and $\mathbf{e}^{(j)} = \mathbf{x}^{(j)} - \mathbf{x}_t$, then a repeated application of (2.57) gives

$$\mathbf{e}^{(j)} = M\mathbf{e}^{(j-1)} = M(M\mathbf{e}^{(j-2)}) = M^2\mathbf{e}^{(j-2)} = \cdots = M^j\mathbf{e}^{(0)}.$$

Since $\|\mathbf{e}^{(j)}\|_\infty \le \|M^j\|_\infty\,\|\mathbf{e}^{(0)}\|_\infty \le \|M\|_\infty^j\,\|\mathbf{e}^{(0)}\|_\infty \le \lambda^j\|\mathbf{e}^{(0)}\|_\infty$, part (3) of the theorem is proved. Since $\lim_{j\to\infty}\|\mathbf{x}^{(j)} - \mathbf{x}_t\|_\infty = 0$ and since $\|\mathbf{x}^{(j)} - \mathbf{x}_t\|_\infty$ is the absolute value of the largest component of the vector $\mathbf{x}^{(j)} - \mathbf{x}_t$, it follows that $\lim_{j\to\infty}\mathbf{x}^{(j)} = \mathbf{x}_t$. This proves part (2). (We note from Problem 3, Section 2.3, that $\|\mathbf{x}^{(j)} - \mathbf{x}_t\|_2 \to 0$ and $\|\mathbf{x}^{(j)} - \mathbf{x}_t\|_1 \to 0$ as well). ■

Theorem 2.2 together with the observations made previously allows us to summarize the properties that a splitting should have in order to define a useful iterative method. For $A = N - P$, the properties are these.

1. N should be nonsingular.
2. The equation $N\mathbf{x} = \mathbf{h}$ should be easy to solve.
3. $\|N^{-1}P\|$ should be less than 1 for some matrix norm.

In this context, condition (1) assures us that the sequence $\{\mathbf{x}^{(j)}\}$ given by formula (2.56) can be generated. Condition (2) assures us that the sequence $\{\mathbf{x}^{(j)}\}$ can be generated efficiently (after all, we do not want to work as hard to perform one step of an iterative method as we would to solve $A\mathbf{x} = \mathbf{b}$ by a direct method). Last, condition (3) assures us that the sequence we generate will in fact converge to the solution $\mathbf{x}_t$.

2.4.1. Basic Iterative Methods

The first and simplest iterative method described in this section is the Jacobi method. For this method N is taken to be a diagonal matrix with its main diagonal entries equal to a_{ii}. The matrix P is then determined by $P = N - A$. We note that P can be visualized as the sum of a lower-triangular matrix and an upper-triangular matrix. For the purposes of analyzing the Jacobi method and the ensuing Gauss-Seidel method, it is quite convenient to think of P in this way. To be specific, let $A = (a_{ij})$ be an $(n \times n)$ matrix. Define L, D, and U to be the lower-triangular, diagonal, and upper-triangular parts of A:

$$L = \begin{bmatrix} 0 & 0 & 0 & \cdots & 0 \\ a_{21} & 0 & 0 & \cdots & 0 \\ a_{31} & a_{32} & 0 & \cdots & 0 \\ \vdots & & & & \vdots \\ a_{n1} & a_{n2} & a_{n3} & \cdots & 0 \end{bmatrix}, \quad D = \begin{bmatrix} a_{11} & 0 & 0 & \cdots & 0 \\ 0 & a_{22} & 0 & \cdots & 0 \\ 0 & 0 & a_{33} & \cdots & 0 \\ \vdots & & & & \vdots \\ 0 & 0 & 0 & \cdots & a_{nn} \end{bmatrix}$$

$$U = \begin{bmatrix} 0 & a_{12} & a_{13} & \cdots & a_{1n} \\ 0 & 0 & a_{23} & \cdots & a_{2n} \\ 0 & 0 & 0 & \cdots & a_{3n} \\ \vdots & & & & \vdots \\ 0 & 0 & 0 & \cdots & 0 \end{bmatrix}.$$

Thus $A = L + D + U$, and so the Jacobi splitting is given by $N = D$ and $P = -(L + U)$.

The Jacobi method for solving $A\mathbf{x} = \mathbf{b}$ (that is, the splitting defined above) is this:

$$D\mathbf{x}^{(k+1)} = -(L + U)\mathbf{x}^{(k)} + \mathbf{b}. \tag{2.59}$$

The matrix $M_J = -D^{-1}(L + U)$ is called the *Jacobi matrix*. In actual computation, Eq. (2.59) would have to be written out elementwise. Suppose the vector $\mathbf{x}^{(k)}$ is given by

$$\mathbf{x}^{(k)} = \begin{bmatrix} x_1^{(k)} \\ x_2^{(k)} \\ \vdots \\ x_n^{(k)} \end{bmatrix}, \qquad k = 0, 1, 2, \ldots . \tag{2.60}$$

Then Eq. (2.59) leads to the following iteration for the ith component of $\mathbf{x}^{(k+1)}$:

$$x_i^{(k+1)} = \frac{-1}{a_{ii}}\left[\sum_{\substack{j=1 \\ j \neq i}}^{n} a_{ij}x_j^{(k)} - b_i\right], \qquad i = 1, 2, \ldots, n. \tag{2.61}$$

This formula shows that the Jacobi iteration is quite easy to program. The only real problem is to determine an efficient test for terminating the iteration. Also, it is obvious from (2.61) or from (2.59) that in order for the Jacobi method to be used, the diagonal elements of A must all be nonzero. In practice, this requirement causes no real difficulty. If A is the coefficient matrix of the system $A\mathbf{x} = \mathbf{b}$ and if $a_{ii} = 0$, then the ith equation can be interchanged with another equation that will give a coefficient matrix with a nonzero diagonal entry in the

ith row. Thus the situation with the Jacobi method is similar to that of Gauss elimination in which the possibility of zero pivot elements must be guarded against.

Finally, we note that when D^{-1} exists, it is relatively easy (in comparison to a direct method) to carry out each step of the iteration. Thus in those cases in which $\|-D^{-1}(L + U)\| < 1$, the Jacobi method provides an alternative to direct methods.

Examination of (2.61) reveals that each component of the vector $\mathbf{x}^{(k+1)}$ is computed entirely from the vector $\mathbf{x}^{(k)}$. If $x_j^{(k+1)}$ is assumed to be closer to the true answer than $x_j^{(k)}$, the estimate for $x_i^{(k+1)}$ should be improved by replacing $x_j^{(k)}$ by $x_j^{(k+1)}$ whenever $j < i$. That is, we should use our most recent information as soon as it becomes available. The implementation of this idea leads to the procedure known as the Gauss-Seidel method.

If we use the new information as soon as it is available in (2.61), we obtain (after multiplication by a_{ii}) this equation:

$$a_{ii}x_i^{(k+1)} = -\sum_{j=1}^{i-1} a_{ij}x_j^{(k+1)} - \sum_{j=i+1}^{n} a_{ij}x_j^{(k)} + b_i, \qquad i = 1, \ldots, n, \tag{2.62}$$

(in which we interpret the first sum as zero when $i = 1$). We can write this equation in matrix form, using $A = L + D + U$ as in the Jacobi method, and obtain

$$D\mathbf{x}^{(k+1)} = -L\mathbf{x}^{(k+1)} - U\mathbf{x}^{(k)} + \mathbf{b}. \tag{2.63}$$

Putting this in the standard form Eq. (2.56) for an iterative method, we have

$$(D + L)\mathbf{x}^{(k+1)} = -U\mathbf{x}^{(k)} + \mathbf{b}. \tag{2.64}$$

The matrix $M_G = -(D + L)^{-1}U$ is called the *Gauss-Seidel matrix*. Since the Gauss-Seidel method is refinement of the Jacobi method, the former usually (but not always) converges faster. For deeper results on convergence and comparison of rates of convergence, see the Ostrowski-Reich and Stein-Rosenberg Theorems in Varga (1962). Note that the choice of the starting vector $\mathbf{x}^{(0)}$ is not particularly critical, and one natural choice is $\mathbf{x}^{(0)} = \theta$. We will have more to say of this choice in Section 3.4.

EXAMPLE 2.16. As an example of the sorts of computational results that the Jacobi and Gauss-Seidel methods give, consider the linear system

$$\begin{aligned} 3x_1 + x_2 + x_3 &= 5 \\ 2x_1 + 6x_2 + x_3 &= 9 \\ x_1 + x_2 + 4x_3 &= 6 \end{aligned} \qquad \text{with solution vector } \begin{bmatrix} 1 \\ 1 \\ 1 \end{bmatrix}.$$

With $\mathbf{x}^{(0)} = \theta$, we obtain Tables 2.1 and 2.2. The coefficient matrix of the system is *diagonally dominant*, a condition that is sufficient to guarantee convergence of the Jacobi and Gauss-Seidel iterations (see Theorem 2.3).

TABLE 2.1 Jacobi iteration.

k	$x_1^{(k)}$	$x_2^{(k)}$	$x_3^{(k)}$
1	0.166667E 01	0.150000E 01	0.150000E 01
2	0.666667E 00	0.694445E 00	0.708333E 00
3	0.119907E 01	0.115972E 01	0.115972E 01
4	0.893518E 00	0.907022E 00	0.910301E 00
5	0.106089E 01	0.105044E 01	0.104986E 01
6	0.966564E 00	0.971392E 00	0.972166E 00
7	0.101881E 01	0.101578E 01	0.101551E 01
8	0.989568E 00	0.991144E 00	0.991350E 00
9	0.100584E 01	0.100492E 01	0.100482E 01
10	0.996753E 00	0.997251E 00	0.997312E 00
11	0.100181E 01	0.100153E 01	0.100150E 01
12	0.998991E 00	0.999146E 00	0.999165E 00
13	0.100056E 01	0.100047E 01	0.100047E 01
14	0.999687E 00	0.999735E 00	0.999741E 00
15	0.100017E 01	0.100015E 01	0.100014E 01
16	0.999903E 00	0.999918E 00	0.999919E 00
17	0.100005E 01	0.100005E 01	0.100004E 01
18	0.999970E 00	0.999974E 00	0.999975E 00
19	0.100002E 01	0.100001E 01	0.100001E 01
20	0.999991E 00	0.999992E 00	0.999992E 00

TABLE 2.2 Gauss-Seidel iteration.

k	$x_1^{(k)}$	$x_2^{(k)}$	$x_3^{(k)}$
1	0.166667E 01	0.944445E 00	0.847222E 00
2	0.106944E 01	0.100231E 01	0.982060E 00
3	0.100521E 01	0.100125E 01	0.998385E 00
4	0.100012E 01	0.100023E 01	0.999913E 00
5	0.999953E 00	0.100003E 01	0.100000E 01
6	0.999989E 00	0.100000E 01	0.100000E 01
7	0.999998E 00	0.100000E 01	0.100000E 01
8	0.100000E 01	0.100000E 01	0.100000E 01

As an example in which iteration is not so successful, consider the (4×4) linear system of Example 2.6 (solved by Gauss elimination in Example 2.7). This coefficient matrix is *positive-definite* and hence the Gauss-Seidel iteration will converge (see Theorem 2.4); but as can be seen, convergence is exceedingly slow. (See Table 2.3.) The question of how fast an iterative procedure will converge is considered in Section 3.4. Through the theory of the above-mentioned section it can be shown that the Jacobi method will not converge for the (4×4) system above.

TABLE 2.3

k	$x_1^{(k)}$	$x_2^{(k)}$	$x_3^{(k)}$	$x_4^{(k)}$
1	0.460000E 01	−0.199982E−01	0.556000E 00	0.313602E 00
2	0.364720E 01	−0.173599E−01	0.843327E 00	0.529559E 00
3	0.308276E 01	−0.327911E−02	0.976370E 00	0.682185E 00
4	0.275076E 01	0.158432E−01	0.102290E 01	0.792917E 00
5	0.255742E 01	0.364410E−01	0.102277E 01	0.875290E 00
6	0.244637E 01	0.566254E−01	0.999118E 00	0.937971E 00
7	0.238381E 01	0.754578E−01	0.965172E 00	0.986620E 00
8	0.234954E 01	0.925567E−01	0.928278E 00	0.102499E 01
9	0.233149E 01	0.107840E 00	0.892337E 00	0.105566E 01
10	0.232256E 01	0.121379E 00	0.859268E 00	0.108042E 01
⋮				
50	0.220322E 01	0.276739E 00	0.694661E 00	0.117948E 01
51	0.219950E 01	0.278992E 00	0.695580E 00	0.117894E 01
52	0.219578E 01	0.281235E 00	0.696501E 00	0.117840E 01
⋮				
100	0.203333E 01	0.378932E 00	0.737633E 00	0.115422E 01
101	0.203012E 01	0.380858E 00	0.738446E 00	0.115374E 01
102	0.202693E 01	0.382777E 00	0.739256E 00	0.115326E 01
⋮				
196	0.176445E 01	0.540536E 00	0.805902E 00	0.111409E 01
197	0.176208E 01	0.541962E 00	0.806505E 00	0.111373E 01
198	0.175972E 01	0.543384E 00	0.807103E 00	0.111338E 01
199	0.175736E 01	0.544801E 00	0.807701E 00	0.111303E 01
200	0.175501E 01	0.546213E 00	0.808298E 00	0.111268E 01

The question of convergence for iterative methods is sometimes hard to analyze, for convergence depends on $M = N^{-1}P$ being small in some sense as is shown in (2.57). For the Jacobi method, it is easy to display the form of the Jacobi matrix M_J. From (2.59) it follows quickly that

$$M_J = -\begin{bmatrix} 0 & \frac{a_{12}}{a_{11}} & \frac{a_{13}}{a_{11}} & \cdots & \frac{a_{1n}}{a_{11}} \\ \frac{a_{21}}{a_{22}} & 0 & \frac{a_{23}}{a_{22}} & \cdots & \frac{a_{2n}}{a_{22}} \\ \vdots & & & & \vdots \\ \frac{a_{n1}}{a_{nn}} & \frac{a_{n2}}{a_{nn}} & \frac{a_{n3}}{a_{nn}} & \cdots & 0 \end{bmatrix}. \tag{2.65}$$

An explicit representation for M_G for a general matrix A is obviously more complicated and it is of little use to find M_G.

While Theorem 2.2 provides a general criterion for convergence, it would be helpful to have some convergence theorems that related specifically to the Jacobi and Gauss-Seidel methods. In particular, the ideal theorem would be one that guaranteed convergence for all matrices in some particular class or for all matrices that satisfy some particular property. Below, we illustrate two theorems of this type, theorems that guarantee convergence for *diagonally dominant* matrices and for *symmetric positive-definite* matrices. Diagonally dominant matrices and symmetric positive-definite matrices, two important types of matrices that occur frequently in practical problems, are defined below.

We say a matrix $A = (a_{ij})$ is diagonally dominant if

$$|a_{ii}| > \sum_{\substack{j=1 \\ j \neq i}}^{n} |a_{ij}|, \text{ for } i = 1, 2, \ldots, n. \tag{2.66}$$

That is, each main diagonal entry is larger in absolute value than the sum of the absolute values of all the off-diagonal entries in that row. An $(n \times n)$ matrix $A = (a_{ij})$ is called *symmetric* if

$$A^{\mathrm{T}} = A. \tag{2.67}$$

Clearly, an equivalent way of stating that A is symmetric is to say that $a_{ij} = a_{ji}$ for $1 \le i \le n$, $1 \le j \le n$. Finally, an $(n \times n)$ matrix $A = (a_{ij})$ is said to be *symmetric* and *positive-definite* if

1. A is symmetric and
2. $\mathbf{x}^{\mathrm{T}} A\mathbf{x} > 0$ for all $\mathbf{x} \in R^n$, $\mathbf{x} \neq \theta$. (2.68)

[Note that since A is $(n \times n)$ and $\mathbf{x}$ is $(n \times 1)$, then $A\mathbf{x}$ is $(n \times 1)$. Thus since $\mathbf{x}$ is $(n \times 1)$, $\mathbf{x}^{\mathrm{T}}$ is $(1 \times n)$ and so we have that $\mathbf{x}^{\mathrm{T}} A\mathbf{x}$ is a scalar quantity. Most readers will recognize $\mathbf{x}^{\mathrm{T}} A\mathbf{x}$ as the familiar "dot product" of the vector $\mathbf{x}$ and the vector $A\mathbf{x}$.]

EXAMPLE 2.17. Consider the simple (2×2) matrix A, $A = \left[\begin{smallmatrix} 2 & 1 \\ 1 & 3 \end{smallmatrix}\right]$. Clearly A is diagonally dominant and also symmetric. To determine whether A is positive-definite, we consider

$$\mathbf{x}^{\mathrm{T}}A\mathbf{x} = [x_1, x_2]\begin{bmatrix} 2 & 1 \\ 1 & 3 \end{bmatrix}\begin{bmatrix} x_1 \\ x_2 \end{bmatrix} = [x_1, x_2]\begin{bmatrix} 2x_1 + x_2 \\ x_1 + 3x_2 \end{bmatrix}.$$

Therefore $\mathbf{x}^{\mathrm{T}}A\mathbf{x} = 2x_1^2 + x_1x_2 + x_2x_1 + 3x_2^2 = (x_1 + x_2)^2 + x_1^2 + 2x_2^2$. Hence, we see that A is positive-definite since $\mathbf{x}^{\mathrm{T}}A\mathbf{x} > 0$ whenever $\mathbf{x} \neq \theta$. If we modify A and consider $B = \left[\begin{smallmatrix} 2 & 1 \\ 1 & -3 \end{smallmatrix}\right]$, then we still have that B is symmetric and diagonally dominant. However, B is not positive-definite since for $\mathbf{x} = \left(\begin{smallmatrix} 0 \\ 1 \end{smallmatrix}\right)$, $\mathbf{x}^{\mathrm{T}}B\mathbf{x} = (0, 1)\left(\begin{smallmatrix} 1 \\ -3 \end{smallmatrix}\right) = -3 < 0$.

Having the concepts of diagonally dominant and positive-definite, we can now state two convergence theorems.

Theorem 2.3.

If A is diagonally dominant, then A is nonsingular and the sequence $\{\mathbf{x}^{(k)}\}$ defined by the Jacobi method (2.59) converges for any initial guess $\mathbf{x}^{(0)}$.

Proof. Dividing both sides of inequality (2.66) by $|a_{ii}|$ for $i = 1, 2, \ldots, n$, we obtain

$$1 > \sum_{\substack{j=1\\ j\neq i}}^{n} \frac{|a_{ij}|}{|a_{ii}|} \quad \text{for } i = 1, 2, \ldots, n.$$

Referring to (2.65), we see that this inequality means that $\|M_j\|_\infty < 1$. Consequently, by Theorem 2.2, A is nonsingular and the Jacobi method will converge. ■

The fact that a diagonally dominant matrix is nonsingular will be important also in our study of cubic splines in Chapter 5. The proof of Theorem 2.4 is more difficult and we defer it to Section 3.4.

Theorem 2.4.

If A is symmetric and positive-definite, then the sequence $\{\mathbf{x}^{(k)}\}$ defined by the Gauss-Seidel method (2.64) converges for any initial guess $\mathbf{x}^{(0)}$.

Both of these theorems are quite useful in various computational settings. For example, some numerical procedures result in large linear systems $A\mathbf{x} = \mathbf{b}$, which must be solved, where A is diagonally dominant. Theorem 2.3 guarantees that the Jacobi method will work on this system. In Chapter 8 we consider some finite-difference approximations to partial differential equations which give rise to systems with symmetric, positive-definite coefficient matrices.

2.4.2. Implementation of Iterative Methods

There are many other important iterative methods besides Jacobi and Gauss-Seidel. Two other important, but rather more complicated methods are successive overrelaxation (SOR) and alternating-direction implicit methods (ADI) (see Chapter 8). A good advanced reference is Varga (1962). Iterative methods are normally used when it can easily be shown in advance that they will converge and that they will require less computation than a direct method. The numerical solution of partial differential equations provides the most common area of application for iterative methods. When finite difference or finite element schemes are used, the end result is to replace the partial differential equation by a large matrix equation $A\mathbf{x} = \mathbf{b}$. The matrix equation is an approximation to the partial differential equation, and the solution of $A\mathbf{x} = \mathbf{b}$ gives approximate values for the function that satisfies the partial differential equation. These matrices have a structure that is known in advance; so it is normally easy to see whether or not an iterative procedure will converge. Moreover, these matrices are often quite large [say (1000×1000) perhaps] and *sparse* (that is, there are relatively few nonzero entries in the matrix). In the following, we indicate how this "sparseness" can be exploited by an iterative method.

Note that an iterative method written in the form

$$\mathbf{x}^{(k+1)} = M\mathbf{x}^{(k)} + \mathbf{g} \tag{2.69}$$

where M is $(n \times n)$ requires just n^2 multiplications per iteration. This observation forms a part of the reason why iterative methods are valuable since a direct method like Gauss elimination takes about $(n^3/3)$ multiplications. Thus if we are satisfied with the approximate solution $\mathbf{x}^{(k)}$ where $k < n/3$, we have saved machine time. In the context of partial differential equations, since the equation $A\mathbf{x} = \mathbf{b}$ is itself an approximation, it is not necessary to ask for an extremely good estimate of the true solution $\mathbf{x}_t$. Another point to be made is that if the matrix A is sparse [a typical example is when A is (1000×1000) with five nonzero entries per row], then iterative methods can take advantage of the fact that multiplication by zero is not necessary. For example in (2.61) or (2.62), if only four of the coefficients a_{ij}/a_{ii} are nonzero for each i, then only 5 multiplications are required to form $x_i^{(k+1)}$. Thus forming $\mathbf{x}^{(k+1)}$ in the Jacobi method (or the Gauss-Seidel method) would take $5n$ multiplications. The total number of multiplications needed to form $\mathbf{x}^{(j)}$ from $\mathbf{x}^{(0)}$ is then $5nj$. So for $j \leq n^2/15$, less effort is needed to compute $\mathbf{x}^{(j)}$ than is needed to solve the system by Gauss elimination. A final consideration is that of storage since a matrix with only five nonzero entries per row needs only $5n$ storage locations. If, however, Gauss elimination is used on the same sparse system, (as a whole) the system may become less sparse as the elimination proceeds. Thus Gauss elimination might require many more than $5n$ storage locations. We should also remark that sparse matrices arising from practical problems may have such a regular pattern that programming this information is easy.

EXAMPLE 2.18. Consider a system of linear equations with the coefficient matrix

$$\begin{bmatrix} 1 & 2 & 3 & 4 & 5 & 6 & 7 & 8 & 9 & 10 \\ 2 & 1 & 0 & 0 & 0 & 0 & 0 & 0 & 0 & 0 \\ 3 & 0 & 1 & 0 & 0 & 0 & 0 & 0 & 0 & 0 \\ 4 & 0 & 0 & 1 & 0 & 0 & 0 & 0 & 0 & 0 \\ 5 & 0 & 0 & 0 & 1 & 0 & 0 & 0 & 0 & 0 \\ 6 & 0 & 0 & 0 & 0 & 1 & 0 & 0 & 0 & 0 \\ 7 & 0 & 0 & 0 & 0 & 0 & 1 & 0 & 0 & 0 \\ 8 & 0 & 0 & 0 & 0 & 0 & 0 & 1 & 0 & 0 \\ 9 & 0 & 0 & 0 & 0 & 0 & 0 & 0 & 1 & 0 \\ 10 & 0 & 0 & 0 & 0 & 0 & 0 & 0 & 0 & 1 \end{bmatrix}. \tag{2.70}$$

This provides a graphic example of a sparse matrix that becomes "dense" as Gauss elimination is used. To illustrate our point, this example is admittedly contrived in that we could interchange the first and last rows, and solution by Gauss elimination becomes almost immediate. The point of the example, however, is well taken and remains valid in less contrived situations.

PROBLEMS, SECTION 2.4

1. Write a computer program to implement the Jacobi method.
2. Write a computer program to implement the Gauss-Seidel method.
3. Test the Gauss-Seidel and Jacobi routines of Problem 1 and Problem 2 on the (2×2) system $A^TA\mathbf{x}^* = A^T\mathbf{b}$ in Example 2.19 of Section 2.5.
4. Let $A = (a_{ij})$ be the (4×4) diagonally dominant matrix given by $a_{ii} = 10$, $i = 1, 2, 3, 4$, and $a_{ij} = 2$, for $i \neq j$. Use the Gauss-Seidel and Jacobi methods to solve $A\mathbf{x} = \mathbf{b}$ where $\mathbf{b}^T = (16, 24, 16, 8)$. [The exact solution is $\mathbf{x}^T = (1, 2, 1, 0)$.]
5. Let $A = N - P$ where

$$N = \begin{bmatrix} 5 & 0 & 0 \\ 0 & 4 & 0 \\ 0 & 0 & 5 \end{bmatrix}, \qquad P = \begin{bmatrix} -1 & -1 & 1 \\ -2 & -1 & 2 \\ -1 & -1 & -1 \end{bmatrix}.$$

   a) Show that the iterative method for this splitting converges to the solution of $A\mathbf{x} = \mathbf{b}$.

   b) Use (2.57) to argue why you expect the Jacobi method to converge faster than this method.
6. For $k = 0, 1, \ldots$, let $\mathbf{x}^{(k)} = [x_k, y_k, z_k]^T$ be generated by the equations

$$\begin{aligned} 4x_{k+1} &= -x_k - y_k + z_k + 2 \\ 6y_{k+1} &= 2x_k + y_k - z_k - 1 \\ -4z_{k+1} &= -x_k + y_k - z_k + 4. \end{aligned}$$

   Show that the vector sequence $\{\mathbf{x}^{(k)}\}_{k=1}^{\infty}$ is convergent for any initial vector $\mathbf{x}^{(0)}$, and determine the vector $\mathbf{x}$ to which it converges. [Hint: Find A, N, P, and $\mathbf{b}$.]
7. How many multiplications and divisions are required for one iteration of the Gauss-Seidel method? How many iterations may be performed before this number exceeds the operations count of Gauss elimination?
8. If the $(n \times n)$ matrix A is nonsingular, then there is at least one nonzero entry in the first column. (Why?) Suppose now that no matter what row interchange is made to create a nonzero $(1, 1)$ entry, the resulting $(2, 2), \ldots, (n, 2)$ entries are zero. Examine the first two columns and explain why this situation contradicts the nonsingularity of A.
9. Let A be an $(m \times n)$ matrix. Using Problem 13, Section 2.1, show that A^TA is an $(n \times n)$ symmetric matrix.
10. If $\mathbf{y}$ is any $(n \times 1)$ vector, $\mathbf{y} \neq \theta$, show that $\mathbf{y}^T\mathbf{y} > 0$.
11. Suppose A is an $(m \times n)$ matrix such that $A\mathbf{y} \neq \theta$ for any $(n \times 1)$ vector $\mathbf{y}$ where $\mathbf{y} \neq \theta$. Using Problems 9 and 10, show that A^TA is a symmetric, positive-definite matrix.
12. Using Problem 11, verify for the (4×2) matrix A in Example 2.19 that A^TA is symmetric and positive-definite. (From Theorem 2.4, the results of this problem have a bearing on Problem 3.)
13. For the matrix A^TA of Example 2.19, verify that the Jacobi matrix $M_J = -D^{-1}(L + U)$ is given by

$$M_J = -\begin{bmatrix} 0 & \frac{3}{7} \\ \frac{3}{2} & 0 \end{bmatrix} \quad \text{and} \quad M_J^2 = \begin{bmatrix} \frac{9}{14} & 0 \\ 0 & \frac{9}{14} \end{bmatrix}.$$

Using $M_J^4 = M_J^2 M_J^2$, $M_J^6 = M_J^2 M_J^2 M_J^2$, etc., show that the entries of M_J^{2n} all tend to zero as $n \to \infty$. From this information, give a bound for $\|M_J^i\|_\infty$; and using (2.57), show that the Jacobi method must converge for any initial guess $\mathbf{x}^{(0)}$. Note that A^TA is not diagonally dominant so that Theorem 2.3 does not apply. Additionally, $\|M_J\| > 1$ for the three matrix norms of (2.41) so that Theorem 2.2 does not apply either. However, the analysis of this problem shows why the Jacobi method is convergent in Problem 3.

14. As another example of a slowly converging iterative method, use the Gauss-Seidel iteration to solve the (2×2) system $A_{100}\mathbf{x} = \mathbf{b}$ of Problem 8, Section 2.3.

15. Assume that an original system of equations has been altered to an equivalent system $A\mathbf{x} = \mathbf{b}$ where each of the diagonal entries of A equals one ($a_{ii} = 1$). From (2.64) the splitting for the Gauss-Seidel method is $(I + L)\mathbf{x} = -U\mathbf{x} + \mathbf{b}$. Multiplying this equation by a scalar ω and adding $\mathbf{x}$ to both sides lead to the iterative method

$$(I + \omega L)\mathbf{x}^{(k+1)} = [(1 - \omega)I - \omega U]\mathbf{x}^{(k)} + \omega\mathbf{b}.$$

(Verify these steps.) This method is called successive overrelaxation (SOR) and ω is called the relaxation factor. How would one use (2.57) to choose a value of ω to make this iteration faster than Gauss-Seidel? Program this method and try to find an ω that improves the convergence on the system of Table 2.3. (SOR methods are frequently used on systems arising from partial differential equations. The choice of ω is a difficult and delicate matter, and the value of ω is sometimes changed with each step of the iteration.)

2.5 LEAST-SQUARES SOLUTION OF OVERDETERMINED LINEAR SYSTEMS

We now return our attention to the linear system (2.1) [or Eq. (2.2) in matrix form] in which we have m linear equations in n unknowns; and we consider the case in which $m > n$. Since we have more equations than unknowns, we do not normally expect a solution, $\mathbf{x}$, of $A\mathbf{x} = \mathbf{b}$ to exist. (As pointed out previously, if such an $\mathbf{x}$ does exist, Gauss elimination can be implemented to find it.) However, in this section we shall assume that $A\mathbf{x} = \mathbf{b}$ does *not* have a solution. The reader may thus feel that this is necessarily an unimportant problem. Quite the opposite is true, however, for we can reformulate the problem to ask for the vector $\mathbf{x}^*$ that somehow "minimizes" the vector expression $(A\mathbf{x}^* - \mathbf{b})$. That is, find $\mathbf{x}^*$ such that $\|A\mathbf{x}^* - \mathbf{b}\|$ is minimized for some vector norm $\|\cdot\|$. In the case in which we use the Euclidean norm, $\|\cdot\|_2$, this problem becomes precisely the important statistical problem of finding the best *least-squares solution* of $A\mathbf{x} = \mathbf{b}$. We shall see later that this problem is a special case of the famous Fourier series approximation problem, but for now we shall be content to solve it in this one particular context.

Let A be any given $(m \times n)$ matrix, and define Y to be the set of vectors $\mathbf{y}$ in R^m such that

$$Y = \{\mathbf{y} \in R^m\colon \mathbf{y} = A\mathbf{x}, \text{ for some } \mathbf{x} \in R^n\}.$$

Thus Y is in reality the "range of A" and is a subset (actually a "subspace") of R^m. Since we have assumed that there is no $\mathbf{x} \in R^n$ such that $A\mathbf{x} = \mathbf{b}$, then $\mathbf{b}$ does not belong to Y in this case. (However, the following procedures would be valid even if $\mathbf{b} \in Y$.) Let $\mathbf{x} = (x_1, x_2, \ldots, x_n)^T$ be arbitrary in R^n, and write $A = [\mathbf{A}_1, \mathbf{A}_2, \ldots, \mathbf{A}_n]$ where each $\mathbf{A}_k$ is the kth $(m \times 1)$ column vector of A, $1 \le k \le n$. Recall from Section 2.1 that $A\mathbf{x}$ can be written as

$$A\mathbf{x} = x_1\mathbf{A}_1 + x_2\mathbf{A}_2 + \cdots + x_n\mathbf{A}_n.$$

Thus we see that Y can be equivalently written as

$$Y = \{\mathbf{y} \in R^m \colon \mathbf{y} = x_1\mathbf{A}_1 + x_2\mathbf{A}_2 + \cdots + x_n\mathbf{A}_n\}.$$

Therefore we can see that our problem of minimizing $\|A\mathbf{x} - \mathbf{b}\|_2$ can be restated thus: find $\mathbf{y}^* \in Y$ such that $\|\mathbf{y}^* - \mathbf{b}\|_2 \le \|\mathbf{y} - \mathbf{b}\|_2$ for all $\mathbf{y} \in Y$, and then find $\mathbf{x}^* \in R^n$ such that $A\mathbf{x}^* = \mathbf{y}^*$. This problem can be formally solved easily by use of the following theorem.

Theorem 2.5

Assume that there exists a vector $\mathbf{y}^* \in Y$ such that $(\mathbf{b} - \mathbf{y}^*)^T\mathbf{y} = 0$ for all $\mathbf{y} \in Y$. Then $\|\mathbf{b} - \mathbf{y}^*\|_2 \le \|\mathbf{b} - \mathbf{y}\|_2$ for all $\mathbf{y} \in Y$.

Proof. We observe that if $\mathbf{z}$ is any vector in R^m, then $\mathbf{z}^T\mathbf{z} = \|\mathbf{z}\|_2^2$ and $\mathbf{x}^T\mathbf{z} = \mathbf{z}^T\mathbf{x}$. Let $\mathbf{y}$ be any vector in Y and remember in the following that $(\mathbf{b} - \mathbf{y}^*)^T\mathbf{y} = 0$.

$$\begin{aligned} 0 \le \|\mathbf{b} - \mathbf{y}\|_2^2 &= (\mathbf{b} - \mathbf{y})^T(\mathbf{b} - \mathbf{y}) = \mathbf{b}^T\mathbf{b} - 2\mathbf{b}^T\mathbf{y} + \mathbf{y}^T\mathbf{y} \\ &= \mathbf{b}^T\mathbf{b} + 2(\mathbf{b} - \mathbf{y}^*)^T\mathbf{y} - 2\mathbf{b}^T\mathbf{y} + \mathbf{y}^T\mathbf{y} \\ &= \mathbf{b}^T\mathbf{b} + 2\mathbf{b}^T\mathbf{y} - 2\mathbf{y}^{*T}\mathbf{y} - 2\mathbf{b}^T\mathbf{y} + \mathbf{y}^T\mathbf{y} \\ &= \mathbf{b}^T\mathbf{b} - \mathbf{y}^{*T}\mathbf{y}^* + (\mathbf{y}^{*T}\mathbf{y}^* - 2\mathbf{y}^{*T}\mathbf{y} + \mathbf{y}^T\mathbf{y}) \\ &= \|\mathbf{b}\|_2^2 - \|\mathbf{y}^*\|_2^2 + (\mathbf{y}^* - \mathbf{y})^T(\mathbf{y}^* - \mathbf{y}). \end{aligned}$$

Thus $\|\mathbf{b} - \mathbf{y}\|_2^2 = \|\mathbf{b}\|_2^2 - \|\mathbf{y}^*\|_2^2 + \|\mathbf{y}^* - \mathbf{y}\|_2^2$, and obviously this last expression is minimized if and only if $\mathbf{y} = \mathbf{y}^*$. ■

The geometric rationale underlying this theorem is as follows: suppose that Y is a plane in R^3 (in our particular case it is either a plane or a line through the origin when $m = 3$). If $\mathbf{b}$ is not in this plane, then the closest point on the plane to $\mathbf{b}$ is $\mathbf{y}^*$ where $\mathbf{y}^*$ is such that the projection vector, $(\mathbf{b} - \mathbf{y}^*)$, is perpendicular to every $\mathbf{y}$ in the plane. Since $(\mathbf{b} - \mathbf{y}^*)^T\mathbf{y}$ is the usual "dot product" of $(\mathbf{b} - \mathbf{y}^*)$ and $\mathbf{y}$, $(\mathbf{b} - \mathbf{y}^*)^T\mathbf{y} = 0$ means that $(\mathbf{b} - \mathbf{y}^*)$ and $\mathbf{y}$ are perpendicular. So this theorem is a natural analytical extension of this geometric concept.

Later, when we investigate this problem in a more general setting, we will have the necessary mathematical tools to prove that $\mathbf{y}^*$, as given in the above theorem, always exists and is unique. For now we will merely assume that this statement is always true. Under this assumption, there is always at least one set of scalars, $\{x_1^*, x_2^*, \ldots, x_n^*\}$, such that $\mathbf{y}^* = x_1^*\mathbf{A}_1 + x_2^*\mathbf{A}_2 + \cdots + x_n^*\mathbf{A}_n \equiv$

$A\mathbf{x}^*$ where $\mathbf{x}^* \equiv (x_1^*, x_2^*, \ldots, x_n^*)^{\mathrm{T}}$. Since every $\mathbf{y} \in Y$ can be written as $\mathbf{y} = x_1\mathbf{A}_1 + x_2\mathbf{A}_2 + \cdots + x_n\mathbf{A}_n$, then $(\mathbf{b} - \mathbf{y}^*)^{\mathrm{T}}\mathbf{y} = 0$ for all $\mathbf{y} \in Y$ is equivalent to saying that $(\mathbf{b} - \mathbf{y}^*)^{\mathrm{T}}\mathbf{A}_i = \mathbf{A}_i^{\mathrm{T}}(\mathbf{b} - \mathbf{y}^*) = 0$ for $1 \le i \le n$, or $\mathbf{A}_i^{\mathrm{T}}\mathbf{y}^* = \mathbf{A}_i^{\mathrm{T}}\mathbf{b}$, $1 \le i \le n$. This statement gives a set of n simultaneous equations, which written in vector form becomes

$$\begin{bmatrix} \mathbf{A}_1^{\mathrm{T}}(A\mathbf{x}^*) \\ \mathbf{A}_2^{\mathrm{T}}(A\mathbf{x}^*) \\ \vdots \\ \mathbf{A}_n^{\mathrm{T}}(A\mathbf{x}^*) \end{bmatrix} = \begin{bmatrix} \mathbf{A}_1^{\mathrm{T}}\mathbf{b} \\ \mathbf{A}_2^{\mathrm{T}}\mathbf{b} \\ \vdots \\ \mathbf{A}_n^{\mathrm{T}}\mathbf{b} \end{bmatrix}. \tag{2.71}$$

since $\mathbf{y}^* = A\mathbf{x}^*$. It is easily verified by the basic rules of matrix multiplication that if $\mathbf{z}$ is any vector in R^m, then since the rows of A^{T} are the columns of A,

$$A^{\mathrm{T}}\mathbf{z} = \begin{bmatrix} \mathbf{A}_1^{\mathrm{T}}\mathbf{z} \\ \mathbf{A}_2^{\mathrm{T}}\mathbf{z} \\ \vdots \\ \mathbf{A}_n^{\mathrm{T}}\mathbf{z} \end{bmatrix}. \tag{2.72}$$

Therefore by (2.72), (2.71) reduces to the simple expression

$$A^{\mathrm{T}}A\mathbf{x}^* = A^{\mathrm{T}}\mathbf{b}. \tag{2.73}$$

This is an $(n \times n)$ linear system commonly known as the *normal equations*, and a *least-squares solution*, $\mathbf{x}^*$, may be obtained by Gauss elimination or Gauss-Seidel. (We note here that if $A\mathbf{x} = \mathbf{b}$ has an exact solution, $\mathbf{x}^*$, that is, $\mathbf{b} \in Y$, then $\mathbf{x}^*$ is still a solution of $A^{\mathrm{T}}A\mathbf{x}^* = A^{\mathrm{T}}\mathbf{b}$.)

A familiar problem that arises in this setting is the following: assume that a function $f(x)$ is known at the m distinct points $\{x_0, x_1, \ldots, x_{m-1}\}$, and we wish to find an $(n-1)$st degree polynomial $p(x) = a_0x^{n-1} + a_1x^{n-2} + \cdots + a_{n-2}x + a_{n-1}$ that minimizes the expression $\Sigma_{i=0}^{m-1}(f(x_i) - p(x_i))^2$ where $m > n$ [$p(x)$ is the $(n-1)$st degree best discrete least-squares polynomial fit for $f(x)$]. The equations, $f(x_i) = p(x_i) = a_0x_i^{n-1} + a_1x_i^{n-2} + \cdots + a_{n-1}$ and $0 \le i \le m-1$, yield the $(m \times n)$ linear system

$$\begin{bmatrix} 1 & x_0 & \cdots & x_0^{n-1} \\ 1 & x_1 & \cdots & x_1^{n-1} \\ \vdots & & & \\ 1 & x_{m-1} & \cdots & x_{m-1}^{n-1} \end{bmatrix} \begin{bmatrix} a_{n-1} \\ a_{n-2} \\ \vdots \\ a_0 \end{bmatrix} = \begin{bmatrix} f(x_0) \\ f(x_1) \\ \vdots \\ f(x_{m-1}) \end{bmatrix}, \tag{2.74}$$

and the least-squares solution for $\{a_i\}_{i=0}^{n-1}$ is obtained via (2.73).

EXAMPLE 2.19. Suppose we wish to draw the straight line in the plane that comes closest to "fitting" the points (0, 1), (1, 2.1), (2, 2.9), (3, 3.2). As we indicated above, if we set $f(0) = 1$, $f(1) = 2.1$, $f(2) = 2.9$ and $f(3) = 3.2$, then we want a_0 and a_1 to minimize $\Sigma_{i=0}^{3}[f(x_i) - p(x_i)]^2$ where $p(x) = a_0x + a_1$ and $x_i = i$, $i = 0, 1, 2, 3$. Hence we are led to the overdetermined system

$$\begin{aligned} a_1 &= 1. \\ a_0 + a_1 &= 2.1 \\ 2a_0 + a_1 &= 2.9 \\ 3a_0 + a_1 &= 3.2 \end{aligned} \qquad \text{or} \qquad \begin{bmatrix} 0 & 1 \\ 1 & 1 \\ 2 & 1 \\ 3 & 1 \end{bmatrix} \begin{bmatrix} a_0 \\ a_1 \end{bmatrix} = \begin{bmatrix} 1 \\ 2.1 \\ 2.9 \\ 3.2 \end{bmatrix}.$$

Then

$$A^{\mathrm{T}}A\mathbf{x}^* = A^{\mathrm{T}}\mathbf{b} \quad \text{is} \quad \begin{bmatrix} 14 & 6 \\ 6 & 4 \end{bmatrix} \begin{bmatrix} a_0 \\ a_1 \end{bmatrix} = \begin{bmatrix} 17.5 \\ 9.2 \end{bmatrix}.$$

Thus we find $a_1 = 1.19$ and $a_0 = 0.74$. (See Fig. 2.3.)

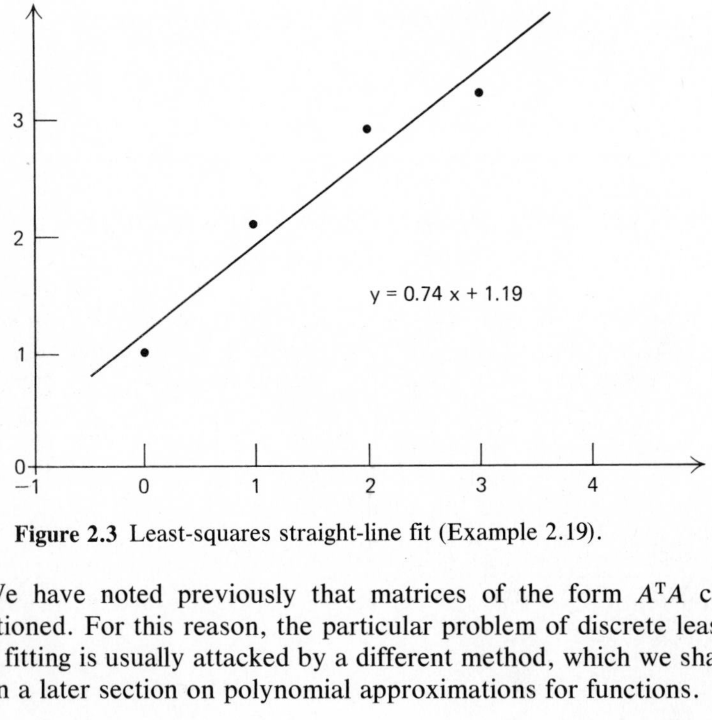

Figure 2.3 Least-squares straight-line fit (Example 2.19).

We have noted previously that matrices of the form $A^{\mathrm{T}}A$ can be ill-conditioned. For this reason, the particular problem of discrete least-squares curve fitting is usually attacked by a different method, which we shall investigate in a later section on polynomial approximations for functions.

PROBLEMS, SECTION 2.5

1. Use the methods of this section to find the best least-squares solution of the system

$$\begin{aligned} x_1 + x_2 &= 1 \\ 2x_1 + x_2 &= 0 \\ x_1 - x_2 &= 0. \end{aligned}$$

2. If A denotes the coefficient matrix of the (3×2) system of Problem 1, show that $A^{\mathrm{T}}A$ is symmetric and positive-definite as in Problem 12, Section 2.4.

3. Consider the (3 × 3) system

$$\begin{aligned} x_1 + 3x_2 + 7x_3 &= 1 \\ 2x_1 + x_2 - x_3 &= 1 \\ x_1 + 2x_2 + 4x_3 &= 1. \end{aligned}$$

Show the system has no solution (so that coefficient matrix A must be singular). Next, find the best least-squares solution by the techniques of this section. Note that even though A^TA is singular, the equation corresponding to (2.73) is solvable.

4. For the matrix A in Problem 3, find a nonzero vector $\mathbf{x}$ such that $\mathbf{x}^T(A^TA)\mathbf{x} = 0$, and thus conclude that A^TA cannot be positive-definite.

5. If $y = a_0x + a_1$, choose a_0 and a_1 such that this straight line is the best least-squares fit to these (x, y) data points:

   a) (1, 1), (4, 2), (8, 4), (11, 5)

   b) (−1, 0), (0, 1), (1, 2), (2, 4)

   c) (−2, 2), (−1, 1), (1, 0), (2, −1).

6. If $y = a_0x^2 + a_1x + a_2$, choose a_0, a_1, and a_2 such that this quadratic is the best least-squares fit to these (x, y) data points:

   a) (−2, 2), (−1, 1), (1, 1), (2, 2)

   b) (0, 0), (1, 0), (2, 1), (3, 2).

*7. Find c_0, c_1, and c_2 so that the expression $y = c_0 + c_1 \cos x + c_2 \sin x$ is a best least-squares fit to the (x, y) data points $(-\pi/2, 1)$, $(0, 0)$, $(\pi/2, 1/2)$, $(\pi, 1)$.

8. An $(n \times n)$ matrix B is called *positive semidefinite* if $\mathbf{x}^TB\mathbf{x} \geq 0$ for all $(n \times 1)$ vectors $\mathbf{x}$. Show that any matrix B of the form $B = A^TA$ is positive semidefinite.

9. Write a computer program to find a fourth-degree polynomial approximation to $f(x) = \cos(x)$, for $0 \leq x \leq \pi$, as follows.

   a) Let $x_j = j\pi/20$; $j = 0, 1, \ldots, 20$.

   b) Set up Eq. (2.74) with $m = 21$ and $n = 5$, to find a best fourth-degree least-squares approximation to $\cos(x)$ on $x_0, x_1, \ldots, x_{20}$.

   c) Having (2.74), form the system (2.73) and use a Gauss elimination routine to solve for $\mathbf{x}^*$.

   d) Let $p(x)$ denote the resulting best fourth-degree polynomial approximation. Print a table, listing $p(y_i), f(y_i), |p(y_i) - f(y_i)|/|f(y_i)|$ for $y_i = (0.01)j\pi; j = 0, 1, 2, \ldots, 100; j \neq 50$.

*2.6 THE CAUCHY-SCHWARZ INEQUALITY.†

Let $\mathbf{x}$ and $\mathbf{y}$ be vectors in R^n, with $\mathbf{x} = (x_1, x_2, \ldots, x_n)^T$ and $\mathbf{y} = (y_1, y_2, \ldots, y_n)^T$. We recall the familiar *scalar* or *dot* product, given by

$$\mathbf{x}^T\mathbf{y} = x_1y_1 + x_2y_2 + \cdots + x_ny_n.$$

---

†This brief section is included for the sake of completeness and may be omitted without loss of continuity. The ambitious reader will find that the ideas contained in this section have wide-ranging application.

We next note that the scalar product satisfies the following conditions for all vectors $\mathbf{x}$, $\mathbf{y}$ and $\mathbf{z}$ in R^n and for all scalars α:

$$\mathbf{x}^T\mathbf{y} = \mathbf{y}^T\mathbf{x} \tag{2.75a}$$
$$(\alpha\mathbf{x})^T\mathbf{y} = \alpha(\mathbf{x}^T\mathbf{y}) \tag{2.75b}$$
$$(\mathbf{x} + \mathbf{y})^T\mathbf{z} = \mathbf{x}^T\mathbf{z} + \mathbf{y}^T\mathbf{z} \tag{2.75c}$$
$$\mathbf{x}^T\mathbf{x} \geq 0 \text{ where } \mathbf{x}^T\mathbf{x} = 0 \text{ if and only if } \mathbf{x} = \theta. \tag{2.75d}$$

The idea of a scalar product can be extended to vector spaces other than R^n, including even infinite-dimensional vector spaces (see Problem 9). The four properties (2.75a, b, c, and d) serve to define this extension, which is usually called an *inner product* rather than a scalar product. Some of the most fruitful areas in modern analysis, such as Hilbert spaces and Fourier analysis, use inner products as a fundamental tool. Returning to the scalar product $\mathbf{x}^T\mathbf{y}$, we next note (Problem 1) that from formula (2.38),

$$(\mathbf{x}^T\mathbf{x})^{1/2} = \|\mathbf{x}\|_2.$$

Having made this observation, we now are ready to give the main result of this section.

Theorem 2.6. Cauchy-Schwarz.

Given any $\mathbf{x}$ and $\mathbf{y}$ in R^n, $|\mathbf{x}^T\mathbf{y}| \leq \|\mathbf{x}\|_2 \|\mathbf{y}\|_2$.

Proof. If either $\mathbf{x} = \theta$ or $\mathbf{y} = \theta$, the proof is trivial; so we assume $\mathbf{x} \neq \theta$, $\mathbf{y} \neq \theta$. For any scalar α, we have by property (2.75d),

$$0 \leq (\mathbf{x} + \alpha\mathbf{y})^T(\mathbf{x} + \alpha\mathbf{y}). \tag{2.76}$$

Expanding the right-hand side of this inequality and using properties (2.75a, b, and c), we obtain

$$0 \leq \mathbf{x}^T\mathbf{x} + 2\alpha\mathbf{x}^T\mathbf{y} + \alpha^2\mathbf{y}^T\mathbf{y}.$$

Choosing $\alpha' = -\mathbf{x}^T\mathbf{y}/\mathbf{y}^T\mathbf{y}$, the inequality above reduces to $0 \leq \mathbf{x}^T\mathbf{x} + \alpha'\mathbf{x}^T\mathbf{y}$, which is equivalent to $(\mathbf{x}^T\mathbf{y})(\mathbf{x}^T\mathbf{y}) \leq (\mathbf{x}^T\mathbf{x})(\mathbf{y}^T\mathbf{y})$. Since both sides of the inequality are nonnegative, taking square roots of both sides establishes the theorem. ∎

The problems that follow give some indication of the applications of the Cauchy-Schwarz inequality.

PROBLEMS, SECTION 2.6

1. Show that for $\mathbf{x} \in R^n$, $(\mathbf{x}^T\mathbf{x})^{1/2} = \|\mathbf{x}\|_2$ as in formula (2.38).
2. For $\mathbf{x}$ and $\mathbf{y}$ in R^n, show that

$$\|\mathbf{x} + \mathbf{y}\|_2^2 + \|\mathbf{x} - \mathbf{y}\|_2^2 = 2\|\mathbf{x}\|_2^2 + 2\|\mathbf{y}\|_2^2.$$

(This equation is called the parallelogram law, and a picture of two vectors $\mathbf{x}$ and $\mathbf{y}$ in a plane will show why.)

3. Suppose $\mathbf{x}$ and $\mathbf{y}$ are two vectors in R^n such that $\mathbf{x}^T\mathbf{y} = 0$. Show that

$$\|\mathbf{x} - \mathbf{y}\|_2^2 = \|\mathbf{x}\|_2^2 + \|\mathbf{y}\|_2^2.$$

(Again, a picture of two vectors in a plane will show that this result is essentially an extension of the Pythagorean theorem of plane geometry.)

4. Use the Cauchy-Schwarz inequality to show that

$$\|\mathbf{x} + \mathbf{y}\|_2^2 \le \|\mathbf{x}\|_2^2 + 2\|\mathbf{x}\|_2\,\|\mathbf{y}\|_2 + \|\mathbf{y}\|_2^2.$$

Use this inequality to show that the ℓ_2 vector norm satisfies Eq. (2.37c). Complete the verification that $\|\cdot\|_2$ is a vector norm by showing that it satisfies (2.37a) and (2.37b).

5. Show that $\|\mathbf{x}\|_2$ and $\|A\|_E$ are compatible vector and matrix norms for $\mathbf{x} \in R^n$ and $A \in M_n$. [Hint: Use the Cauchy-Schwarz inequality on the vector $A\mathbf{x}$, where we note that the ith component of $A\mathbf{x}$ is given by the scalar product of the ith row of A and $\mathbf{x}$.]

6. Show that equality can hold in the Cauchy-Schwarz inequality if and only if $\mathbf{x} = \beta\mathbf{y}$ for some scalar β. [Hint: If $\mathbf{x} = \beta\mathbf{y}$, equality follows. By analyzing the proof of Theorem 2.6, show that strict inequality in (2.76) leads to strict inequality in the Cauchy-Schwarz inequality.]

***7.** Prove the following stronger form of the Cauchy-Schwarz inequality. For $\mathbf{x} = (x_1, x_2, \ldots, x_n)^T$ and $\mathbf{y} = (y_1, y_2, \ldots, y_n)^T$, show that

$$\sum_{i=1}^{n} |x_i y_i| \le \|\mathbf{x}\|_2\,\|\mathbf{y}\|_2.$$

To carry out the proof, define $\mathbf{z} = (z_1, z_2, \ldots, z_n)^T$ where $z_i = -y_i$ if $x_i y_i < 0$ and $z_i = y_i$ if $x_i y_i \ge 0$. Apply Theorem 2.6 to $\mathbf{x}^T\mathbf{z}$.

***8.** Let $\Sigma_{i=1}^{\infty} \alpha_i$ and $\Sigma_{i=1}^{\infty} \beta_i$ be two infinite series such that $\Sigma_{i=1}^{\infty} \alpha_i^2$ and $\Sigma_{i=1}^{\infty} \beta_i^2$ converge. Using partial sums and Problem 7, show that

$$\sum_{i=1}^{\infty} |\alpha_i \beta_i| \le \sqrt{\sum_{i=1}^{\infty} \alpha_i^2}\,\sqrt{\sum_{i=1}^{\infty} \beta_i^2}.$$

***9.** Suppose $f(x)$ and $g(x)$ are continuous functions defined on the finite interval $[a, b]$. Following the proof of Theorem 2.6, show that

$$\left|\int_a^b f(x)g(x)\,dx\right| \le \sqrt{\int_a^b [f(x)]^2\,dx}\,\sqrt{\int_a^b [g(x)]^2\,dx}.$$

You will need to use the fact that if $g(x)$ is not identically zero, then $\int_a^b [g(x)]^2\,dx > 0$. As a loose guide to see how the proof can be constructed, think of $f^Tg = \int_a^b f(x)g(x)\,dx$ where, of course, the symbol f^Tg is used as a hint and does not really make sense mathematically in the context of this chapter. Thinking of Problem 7, prove a stronger version of this inequality, with $\int_a^b |f(x)g(x)|\,dx$ replacing $\left|\int_a^b f(x)g(x)\,dx\right|$.

3

The Algebraic Eigenvalue Problem

3.1 INTRODUCTION

In this chapter we consider computational procedures for solving the following problem:

> Let $A = (a_{ij})$ be any $(n \times n)$ matrix where each a_{ij} is real. Find all scalars λ(real or complex) and all $(n \times 1)$ nonzero vectors $\mathbf{x}$ (whose components may also be complex) such that the vector equation
>
> $$A\mathbf{x} = \lambda\mathbf{x} \tag{3.1}$$
>
> is satisfied. The scalar λ is called an *eigenvalue* of A and the vector $\mathbf{x}$ is called an *eigenvector* corresponding to λ.

This problem arises in many different physical settings ranging from the social sciences such as economics to the physical sciences such as engineering and physics. Often the problem occurs in the solution of a system of differential equations modeling some set of physical phenomena, but the problem can arise in other areas such as structures analysis. As an example that shows how eigenvalues and eigenvectors relate to differential equations, let us consider this simple system of differential equations:

$$\begin{aligned} u'(x) &= u(x) - 2v(x) \\ v'(x) &= u(x) + 4v(x). \end{aligned} \tag{3.2}$$

Thus, in this example, we are seeking functions $u(x)$ and $v(x)$ that satisfy the relationships given in (3.2). It is natural to write this system in matrix form as

$$\mathbf{y}'(x) = A\mathbf{y}(x) \tag{3.3}$$

where

$$\mathbf{y}(x) = \begin{bmatrix} u(x) \\ v(x) \end{bmatrix} \quad \text{and} \quad A = \begin{bmatrix} 1 & -2 \\ 1 & 4 \end{bmatrix}.$$

The equation $\mathbf{y}'(x) = A\mathbf{y}(x)$ is reminiscent of the single variable problem $y' = \alpha y$, which has solutions of the form $y(x) = ce^{\alpha x}$ where c is any nonzero constant. Thus it seems reasonable that a solution of (3.3) might exist in the form

$$\mathbf{y}(x) = e^{\lambda x}\mathbf{y}_0 \tag{3.4}$$

for some scalar λ and some nonzero constant vector $\mathbf{y}_0$. Substituting (3.4) into (3.3), we obtain

$$\lambda e^{\lambda x}\mathbf{y}_0 = e^{\lambda x}A\mathbf{y}_0$$

or

$$e^{\lambda x}(A - \lambda I)\mathbf{y}_0 = \theta.$$

Since the function $e^{\lambda x}$ never vanishes, we must have $(A - \lambda I)\mathbf{y}_0 = \theta$ or $A\mathbf{y}_0 = \lambda\mathbf{y}_0$, which is the eigenvalue problem as formulated in (3.1). Thus we can find solutions to (3.2) by finding eigenvalues and eigenvectors of A and then using (3.4).

In the example above, it is easy to verify by direct substitution that $\lambda_1 = 2$ and $\lambda_2 = 3$ are eigenvalues of A, with corresponding eigenvectors given by

$$\mathbf{y}_1 = \begin{bmatrix} 2\alpha \\ -\alpha \end{bmatrix}, \qquad \mathbf{y}_2 = \begin{bmatrix} \beta \\ -\beta \end{bmatrix}$$

where α and β are any nonzero scalars. That is, the identities $A\mathbf{y}_1 = \lambda_1\mathbf{y}_1$ and $A\mathbf{y}_2 = \lambda_2\mathbf{y}_2$ hold no matter what α and β are. Hence, the functions $\mathbf{y}_1(x) = \alpha e^{2x}\left[\begin{smallmatrix} 2 \\ -1 \end{smallmatrix}\right]$ and $\mathbf{y}_2(x) = \beta e^{3x}\left[\begin{smallmatrix} 1 \\ -1 \end{smallmatrix}\right]$ are each solutions of (3.3). Furthermore if $\mathbf{y}_1(x)$ and $\mathbf{y}_2(x)$ are solutions to $\mathbf{y}'(x) = A\mathbf{y}(x)$, then so is $\mathbf{y}(x) = \mathbf{y}_1(x) + \mathbf{y}_2(x)$. Thus we find, for $\mathbf{y}(x)$ given by

$$\mathbf{y}(x) = \begin{bmatrix} 2\alpha e^{2x} + \beta e^{3x} \\ -\alpha e^{2x} - \beta e^{3x} \end{bmatrix}$$

that $\mathbf{y}(x)$ is a solution of (3.3). Equivalently, the choice $u(x) = 2\alpha e^{2x} + \beta e^{3x}$ and $v(x) = -\alpha e^{2x} - \beta e^{3x}$ is a solution to our original problem (3.2). The presence of the arbitrary constants, α and β, allow us to specify "initial values" for the solution. For example, if we wished to prescribe that $u(0) = 4$ and $v(0) = -3$, we would be led to the system

$$\begin{aligned} 2\alpha + \beta &= 4 \\ -\alpha - \beta &= -3, \end{aligned}$$

which would determine $\alpha = 1$ and $\beta = 2$.

Systems of the sort given in (3.2) occur frequently in problems in engineering and science. For example, suppose we are given a radioactive substance R, which decays into a radioactive substance S; and suppose that S in turn decays into a stable substance T. Next, let us assume that radioactive decay takes place at a rate proportional to the amount of the parent substance present. Thus if $q_1(t)$ denotes the amount of substance R present at time t, then $dq_1/dt = -c_1q_1(t)$. At any time t, the rate of change of $q_2(t)$ [where $q_2(t)$ is the amount of S present at time t], is equal to the rate of formation of S (from R) minus the rate of decay of S (to T). Thus, $dq_2/dt = c_1q_1(t) - c_2q_2(t)$. If we suppose at time $t = 0$ that we have r_0 grams of R, then the following system of equations serves as a mathematical model for the physical system:

$$\frac{dq_1}{dt} = -c_1q_1(t), \qquad q_1(0) = r_0$$

$$\frac{dq_2}{dt} = c_1q_1(t) - c_2q_2(t), \qquad q_2(0) = 0.$$

Returning now to the topic of this chapter, we should first observe that finding eigenvalues and eigenvectors of matrices is a formidable computational problem. The problem is of such practical importance, however, that effective numerical procedures have been developed for its solution. Before presenting some of these, let us recall a few basic mathematical preliminaries that are essential to understanding the problem and to deriving techniques for its solution. As noted above, solving $A\mathbf{x} = \lambda\mathbf{x}$ is equivalent to solving $(A - \lambda I)\mathbf{x} = \theta$. Obviously $\mathbf{x} = \theta$ is always a (trivial) solution, and so we ignore it and make the stipulation that *eigenvectors* must be nonzero. We do allow zero eigenvalues, however, and it should be clear to the reader that A is singular if and only if A has a zero eigenvalue.

Considering the eigenvalue problem in the form $(A - \lambda I)\mathbf{x} = \theta$, $\mathbf{x} \neq \theta$, we see that λ is an eigenvalue of A if and only if the matrix $(A - \lambda I)$ is singular. Therefore eigenvalues of A are the values of λ that satisfy

$$p(\lambda) \equiv \det(A - \lambda I) = 0. \tag{3.5}$$

(Recall the two equivalent criteria for singularity given in Section 1, Chapter 2.) By a cofactor expansion of $\det(A - \lambda I)$ it can be seen that $p(\lambda)$ is an nth degree polynomial in λ (Problem 4) and, in fact, can be written in the form

$$p(\lambda) = (-1)^n\lambda^n + a_1\lambda^{n-1} + \cdots + a_{n-1}\lambda + a_n.$$

The equation $p(\lambda) = 0$ is called the *characteristic equation* of A. Thus the eigenvalues of A are precisely the zeros of $p(\lambda)$; and since the degree of $p(\lambda)$ equals n, there are exactly n eigenvalues, $\{\lambda_i\}_{i=1}^n$, of A. (This result is a consequence of the well-known Fundamental Theorem of Algebra, which states that any nth degree polynomial has exactly n zeros if we count multiple zeros and allow for the possibility that some of the zeros may be complex, even though the polynomial coefficients are all real. See Section 4.4, for more detail.) Example 3.1 illustrates the most elementary approach for solving the eigenvalue problem. Namely, first find $p(\lambda)$ by expanding the determinant of $(A - \lambda I)$; then find the zeros, $\{\lambda_i\}_{i=1}^n$, of $p(\lambda)$ by some polynomial root finding technique (such as those discussed in the next chapter). For each distinct λ_i, the matrix $(A - \lambda_i I)$ is singular; so there exists at least one nonzero vector $\mathbf{x}_i$ that satisfies $(A - \lambda_i I)\mathbf{x}_i = \theta$. Eigenvectors $\mathbf{x}_i$ can be found, for example, by Gauss elimination. Though quite illustrative, techniques that follow this procedure are generally computationally ineffective and are seldom used in practice, in part because of the problem of *ill-conditioning,* which we shall discuss in detail as we progress, and thus we shall present only one method of this type (see Section 3.3).

EXAMPLE 3.1. Let A be the (3×3) matrix

$$A = \begin{bmatrix} 1 & 2 & 1 \\ 0 & 1 & 3 \\ 2 & 1 & 1 \end{bmatrix}.$$

Then

$$\det(A - \lambda I) = \begin{vmatrix} 1-\lambda & 2 & 1 \\ 0 & 1-\lambda & 3 \\ 2 & 1 & 1-\lambda \end{vmatrix}$$

or

$$\begin{aligned} p(\lambda) &= (1-\lambda)\begin{vmatrix} 1-\lambda & 3 \\ 1 & 1-\lambda \end{vmatrix} + 2\begin{vmatrix} 2 & 1 \\ 1-\lambda & 3 \end{vmatrix} \\ &= [(1-\lambda)^3 - 3(1-\lambda)] + [12 - 2(1-\lambda)]. \end{aligned}$$

Thus $p(\lambda) = -\lambda^3 + 3\lambda^2 + 2\lambda + 8 = -(\lambda - 4)(\lambda^2 + \lambda + 2)$. Therefore the eigenvalues of A are $\lambda = 4$, $\lambda = (-1 + i\sqrt{7})/2$, and $\lambda = (-1 - i\sqrt{7})/2$. Solving the equation $(A - 4I)\mathbf{x} = \theta$ by Gauss elimination yields

$$\begin{aligned} -3x_1 + 2x_2 + x_3 &= 0 \\ -3x_2 + 3x_3 &= 0; \\ 2x_1 + x_2 - 3x_3 &= 0 \end{aligned} \qquad \begin{aligned} -3x_1 + 2x_2 + x_3 &= 0 \\ -3x_2 + 3x_3 &= 0; \\ 7x_2 - 7x_3 &= 0 \end{aligned} \qquad \begin{aligned} -3x_1 + 2x_2 &= -x_3 \\ -3x_2 &= -3x_3 \\ 0 &= 0; \end{aligned}$$

and so we see that the solution satisfies $x_1 = x_2 = x_3$. Thus any vector of the form

$$\begin{bmatrix} \alpha \\ \alpha \\ \alpha \end{bmatrix}, \qquad \alpha \neq 0,$$

is an eigenvector belonging to $\lambda = 4$. Since the other two eigenvalues are complex and A is real, it follows from (3.1) that their corresponding eigenvectors have complex entries.

In Example 3.1, we saw that there were infinitely many eigenvectors that correspond to the eigenvalue 4. That is, every vector $\mathbf{x}$ of the form $\mathbf{x} = [\alpha, \alpha, \alpha]^T$ where α is a nonzero scalar satisfies the equation $A\mathbf{x} = 4\mathbf{x}$ as was to be expected since the matrix $A - 4I$ is singular and hence the equation $(A - 4I)\mathbf{x} = \theta$ has infinitely many solutions. However, all the eigenvectors of A that correspond to $\lambda = 4$ are in some sense the same, because each eigenvector $\mathbf{x}$ is of the form $\mathbf{x} = \alpha\mathbf{u}_1$ where $\mathbf{u}_1 = [1, 1, 1]^T$. In order to consider the important question of how many different eigenvectors a matrix possesses, we must make the following definition: let $\{\mathbf{x}_k\}_{k=1}^{m}$ be a collection of $(n \times 1)$ vectors with either real or complex components. Then the set of vectors $\{\mathbf{x}_k\}_{k=1}^{m}$ is said to be *linearly independent* if for any set of scalars, $\{\alpha_k\}_{k=1}^{m}$ (real or complex), the vector equation

$$\alpha_1\mathbf{x}_1 + \alpha_2\mathbf{x}_2 + \cdots + \alpha_m\mathbf{x}_m = \theta$$

implies that every $\alpha_k = 0$, $1 \le k \le m$. The set of vectors $\{\mathbf{x}_k\}_{k=1}^{m}$ is said to be *linearly dependent* if it is not linearly independent. Thus $\{\mathbf{x}_k\}_{k=1}^{m}$ is linearly dependent if we can find $\beta_1, \beta_2, \ldots, \beta_m$ such that $\beta_1\mathbf{x}_1 + \beta_2\mathbf{x}_2 + \cdots + \beta_m\mathbf{x}_m = \theta$ where not all the scalars β_i are zero. Although the concept of linear independence is extremely important in mathematical analysis, it is sufficient for understanding the material of this section to illustrate the concept with the following example.

EXAMPLE 3.2. As before, let $\mathbf{e}_j$ denote the $(n \times 1)$ vector with 1 in the jth component and zeros elsewhere. If the vector equation $\alpha_1\mathbf{e}_1 + \alpha_2\mathbf{e}_2 + \cdots + \alpha_n\mathbf{e}_n = \theta$ is written componentwise, it is obvious that each $\alpha_j = 0$, $1 \le j \le n$. It is equally obvious that any smaller collection of the $\mathbf{e}_j$'s (say, $\{\mathbf{e}_1, \mathbf{e}_3, \mathbf{e}_5, \ldots, \mathbf{e}_n\}$, for n odd) is also linearly independent.

Now we consider the reverse situation, in which $\{\mathbf{x}_k\}_{k=1}^{m}$ is not linearly independent. Then, by definition, there must exist a set of scalars, $\{\alpha_k\}_{k=1}^{m}$, where at least one α_k (say α_1) is not zero and yet $\alpha_1\mathbf{x}_1 + \alpha_2\mathbf{x}_2 + \cdots + \alpha_m\mathbf{x}_m = \theta$. Thus, we can solve for $\mathbf{x}_1$ and find

$$\mathbf{x}_1 = -\frac{1}{\alpha_1}(\alpha_2\mathbf{x}_2 + \cdots + \alpha_m\mathbf{x}_m).$$

Hence, given any equation involving $\{\mathbf{x}_k\}_{k=1}^{m}$, the equation can be rewritten using only $\{\mathbf{x}_k\}_{k=2}^{m}$, and so $\mathbf{x}_1$ is not really needed. In this sense, $\mathbf{x}_1$ is not "different" from the rest and is thus said to be "linearly dependent" on them. As a specific example of linear dependence, consider

$$\mathbf{x}_1 = \begin{bmatrix} 1 \\ 6 \\ 4 \end{bmatrix}, \quad \mathbf{x}_2 = \begin{bmatrix} 1 \\ 2 \\ 0 \end{bmatrix}, \quad \mathbf{x}_3 = \begin{bmatrix} 1 \\ 14 \\ 12 \end{bmatrix}.$$

Writing the vector equation $\alpha_1\mathbf{x}_1 + \alpha_2\mathbf{x}_2 + \alpha_3\mathbf{x}_3 = \theta$ componentwise, we find that α_1, α_2, and α_3 must satisfy

$$\begin{aligned} \alpha_1 + \alpha_2 + \alpha_3 &= 0 \\ 6\alpha_1 + 2\alpha_2 + 14\alpha_3 &= 0 \\ 4\alpha_1 \phantom{{}+ 2\alpha_2} + 12\alpha_3 &= 0. \end{aligned}$$

Solving by Gauss elimination, we have that $\alpha_2 = 2\alpha_3$ and $\alpha_1 + \alpha_2 = -\alpha_3$. Thus, for example, a nontrivial solution is provided by choosing $\alpha_3 = 1$, $\alpha_2 = 2$, and $\alpha_1 = -3$. From this result we see that $-3\mathbf{x}_1 + 2\mathbf{x}_2 + \mathbf{x}_3 = \theta$ or $\mathbf{x}_1 = \frac{1}{3}(2\mathbf{x}_2 + \mathbf{x}_3)$.

We now return to the concept of eigenvectors of an $(n \times n)$ matrix. Unfortunately, it is not necessarily the case that a given $(n \times n)$ matrix has n different or linearly independent eigenvectors. As a simple example of a (2×2) matrix that does not have 2 linearly independent eigenvectors, let

$$A = \begin{bmatrix} 1 & 1 \\ 0 & 1 \end{bmatrix}.$$

Then the characteristic equation is $(1 - \lambda)^2 = 0$, and we find every eigenvector must have the form

$$\begin{bmatrix} \alpha \\ 0 \end{bmatrix} = \alpha \begin{bmatrix} 1 \\ 0 \end{bmatrix}.$$

Since α can be any nonzero scalar, it would seem that A has an infinite number of eigenvectors with respect to $\lambda = 1$. Since, however, the only eigenvectors of A are scalar multiples of $\left[\begin{smallmatrix}1\\0\end{smallmatrix}\right]$, A has only one linearly independent eigenvector. On the other hand, for

$$B = \begin{bmatrix} 1 & 0 \\ 0 & 1 \end{bmatrix},$$

the characteristic equation is again $(1 - \lambda)^2 = 0$; but two eigenvectors are given by $\alpha\left[\begin{smallmatrix}1\\0\end{smallmatrix}\right]$ and $\beta\left[\begin{smallmatrix}0\\1\end{smallmatrix}\right]$ where α and β are arbitrary. Therefore, B has two linearly independent eigenvectors.

In the examples above both A and B have eigenvalues $\lambda = 1, 1$; i.e., 1 is an eigenvalue of multiplicity two. On the other hand, if A is an $(n \times n)$ matrix with *distinct* eigenvalues, $\{\lambda_j\}_{j=1}^n$, then it can be shown (Problem 9) that any set of eigenvectors, $\{\mathbf{x}_j\}_{j=1}^n$, is linearly independent (we assume, in this context, that $A\mathbf{x}_j = \lambda_j\mathbf{x}_j, j = 1, 2, \ldots, n$). Therefore the only instance in which the eigenvectors of A can fail to be linearly independent is the case in which at least one eigenvalue has multiplicity greater than one. The matrix B above illustrates that even with eigenvalues of multiplicity greater than one, the eigenvectors may still be linearly independent. Mathematically, the situation may be summarized by the following theorem (without proof).

Theorem

Let λ be an eigenvalue of A [so $\det(A - \lambda I) = 0$]. Then the number of linearly independent solutions of the equation $(A - \lambda I)\mathbf{x} = \theta$ (thus the number of linearly independent eigenvectors corresponding to λ) equals k where $k = n - r$ and where r is the number of linearly independent column vectors of the matrix $(A - \lambda I)$. [The number r is known as the *rank* of $(A - \lambda I)$].

PROBLEMS, SECTION 3.1

1. Find the eigenvalues and eigenvectors for these matrices:

$$A = \begin{bmatrix} -1 & 1 \\ 4 & 2 \end{bmatrix}, \quad B = \begin{bmatrix} 0 & 2 & 1 \\ -1 & 3 & 1 \\ -1 & 1 & 2 \end{bmatrix}.$$

2. For the matrix A of Example 3.1, find an eigenvector corresponding to the eigenvalue $\lambda = (-1 + i\sqrt{7})/2$; use Gauss elimination to solve the equation $(A - \lambda I)\mathbf{x} = \theta$.
3. Show that the set of vectors S_1 is linearly independent and the set of vectors S_2 is linearly dependent where

$$S_1 = \left\{ \begin{bmatrix} 1 \\ 1 \\ 1 \end{bmatrix}, \begin{bmatrix} 1 \\ 0 \\ 1 \end{bmatrix}, \begin{bmatrix} 2 \\ 3 \\ 1 \end{bmatrix} \right\}, \qquad S_2 = \left\{ \begin{bmatrix} 1 \\ 3 \\ 2 \end{bmatrix}, \begin{bmatrix} 2 \\ 4 \\ 1 \end{bmatrix}, \begin{bmatrix} 0 \\ 2 \\ 3 \end{bmatrix} \right\}.$$

4. Let A be an $(n \times n)$ matrix. Show that the function $p(\lambda)$ defined by $p(\lambda) = \det(A - \lambda I)$ is a polynomial of degree n. [This demonstration is easily done by mathematical induction, verifying directly that $p(\lambda)$ is a quadratic polynomial when $n = 2$.]
5. Let $A = (a_{ij})$ be a lower-triangular $(n \times n)$ matrix. Show that the eigenvalues of A are $a_{11}, a_{22}, \ldots, a_{nn}$.
6. Using the fact that $\det(B) = \det(B^T)$ for any square matrix B, show that the eigenvalues of A and the eigenvalues of A^T are the same where A is a square matrix. Find a (2×2) matrix A that shows that an eigenvector of A is not necessarily an eigenvector of A^T.

*7. A set $\{\mathbf{x}_1, \mathbf{x}_2, \ldots, \mathbf{x}_k\}$ of vectors in R^n is called *orthogonal* if $\mathbf{x}_i^T\mathbf{x}_j = 0$ when $i \neq j$. Show that any orthogonal set is a linearly independent set.

8. Find the largest possible set of linearly independent eigenvectors for each of these:

$$A_1 = \begin{bmatrix} 1 & 1 & 0 \\ 0 & 1 & 1 \\ 0 & 0 & 1 \end{bmatrix}, \qquad A_2 = \begin{bmatrix} 1 & 1 & 0 \\ 0 & 1 & 0 \\ 0 & 0 & 1 \end{bmatrix}, \qquad A_3 = \begin{bmatrix} 1 & 0 & 0 \\ 0 & 1 & 0 \\ 0 & 0 & 1 \end{bmatrix}.$$

*9. Let $\lambda_1, \lambda_2, \ldots, \lambda_k$ be k distinct eigenvalues for A and let $\mathbf{u}_1, \mathbf{u}_2, \ldots, \mathbf{u}_k$ be corresponding eigenvectors. Show that $\{\mathbf{u}_1, \mathbf{u}_2, \ldots, \mathbf{u}_k\}$ are linearly independent. [Hint: Suppose $\{\mathbf{u}_1, \mathbf{u}_2, \ldots, \mathbf{u}_r\}$ is the largest linearly independent subset of $\{\mathbf{u}_1, \mathbf{u}_2, \ldots, \mathbf{u}_k\}$; so if $r < k$, then $\{\mathbf{u}_1, \mathbf{u}_2, \ldots, \mathbf{u}_{r+1}\}$ is linearly dependent.]

10. Show that if λ is a complex eigenvalue of a real $(n \times n)$ matrix A, then $\bar{\lambda}$ is also an eigenvalue.

3.2 THE POWER METHOD

One of the most widely used procedures to estimate eigenvalues of a matrix A is the power method. Our basic assumptions are that A is a real $(n \times n)$ matrix, its eigenvalues $\lambda_1, \lambda_2, \ldots, \lambda_n$ are numbered so that

$$|\lambda_1| > |\lambda_2| \geq |\lambda_3| \geq \cdots \geq |\lambda_n|, \tag{3.6}$$

and A has n linearly independent eigenvectors, $\{\mathbf{u}_i\}_{i=1}^n$, where $A\mathbf{u}_i = \lambda_i\mathbf{u}_i$, $1 \leq i \leq n$. When the power method works, it gives us an estimate of the dominant eigenvalue, λ_1.

We first need the fact that any $(n \times 1)$ vector can be expressed as a linear combination of the eigenvectors $\mathbf{u}_1, \mathbf{u}_2, \ldots, \mathbf{u}_n$. To see this, let B be the $(n \times n)$

matrix defined such that the kth column vector of B is $\mathbf{u}_k$, $1 \le k \le n$; that is, $B = [\mathbf{u}_1, \mathbf{u}_2, \ldots, \mathbf{u}_n]$. Let $\mathbf{x} = (x_1, x_2, \ldots, x_n)^T$ be any $(n \times 1)$ vector. Recall from Section 2.1 that $B\mathbf{x} = x_1\mathbf{u}_1 + x_2\mathbf{u}_2 + \cdots + x_n\mathbf{u}_n$. Now if $B\mathbf{x} = \theta$, then $x_1\mathbf{u}_1 + x_2\mathbf{u}_2 + \cdots + x_n\mathbf{u}_n = \theta$. But by the linear independence of $\{\mathbf{u}_i\}_{i=1}^n$, we must have that $x_1 = x_2 = \cdots = x_n = 0$, and thus B is nonsingular. [Note that we have essentially just proved the basic result that a square matrix is nonsingular if and only if its column vectors are linearly independent. Since $\det(B) = \det(B^T)$, the same result is true for row vectors.] Now let $\mathbf{x}_0$ be an arbitrary $(n \times 1)$ vector, and consider the linear system of equations $B\mathbf{x} = \mathbf{x}_0$. Since B is nonsingular this system has a unique solution, $\mathbf{x} = (\alpha_1, \alpha_2, \ldots, \alpha_n)^T$; and so $\mathbf{x}_0$ can be uniquely expressed as

$$\mathbf{x}_0 = \alpha_1\mathbf{u}_1 + \alpha_2\mathbf{u}_2 + \cdots + \alpha_n\mathbf{u}_n. \tag{3.7}$$

[To be able to write *any* $\mathbf{x}_0$ uniquely in the form (3.7) is the reason we require $\{\mathbf{u}_i\}_{i=1}^n$ to be linearly independent. Note that some of the α_i's may be complex if some of the $\mathbf{u}_i$'s have complex entries.]

Next, observe that if $A\mathbf{u}_i = \lambda_i\mathbf{u}_i$, then $A(A\mathbf{u}_i) = \lambda_i A\mathbf{u}_i$ or $A^2\mathbf{u}_i = (\lambda_i)^2\mathbf{u}_i$. Continuing, it follows that $A^3\mathbf{u}_i = (\lambda_i)^3\mathbf{u}_i$ or in general

$$A^k\mathbf{u}_i = (\lambda_i)^k\mathbf{u}_i. \tag{3.8}$$

Thus if we form the sequence

$$\begin{aligned}
\mathbf{x}_1 &= A\mathbf{x}_0 \\
\mathbf{x}_2 &= A\mathbf{x}_1, \quad (\mathbf{x}_2 = A^2\mathbf{x}_0) \\
\mathbf{x}_3 &= A\mathbf{x}_2, \quad (\mathbf{x}_3 = A^3\mathbf{x}_0), \\
&\ \ \vdots \qquad\qquad\quad \vdots \\
\mathbf{x}_k &= A\mathbf{x}_{k-1}, \quad (\mathbf{x}_k = A^k\mathbf{x}_0),
\end{aligned}$$

then by (3.7) and (3.8) we have $\mathbf{x}_k = \alpha_1 A^k\mathbf{u}_1 + \alpha_2 A^k\mathbf{u}_2 + \cdots + \alpha_n A^k\mathbf{u}_n$ or

$$\mathbf{x}_k = \alpha_1(\lambda_1)^k\mathbf{u}_1 + \alpha_2(\lambda_2)^k\mathbf{u}_2 + \cdots + \alpha_n(\lambda_n)^k\mathbf{u}_n. \tag{3.9}$$

If one recalls (3.6), it is natural to write $\mathbf{x}_k$ as

$$\mathbf{x}_k = (\lambda_1)^k\left\{\alpha_1\mathbf{u}_1 + \alpha_2\left[\frac{\lambda_2}{\lambda_1}\right]^k\mathbf{u}_2 + \cdots + \alpha_n\left[\frac{\lambda_n}{\lambda_1}\right]^k\mathbf{u}_n\right\}. \tag{3.10}$$

In (3.10), we see for large k that the approximation $\mathbf{x}_k \approx (\lambda_1)^k\alpha_1\mathbf{u}_1$ should be good since $|\lambda_1| > |\lambda_2| \ge \cdots \ge |\lambda_n|$. In order to estimate λ_1, we replace k by $k+1$ in (3.10) to obtain $\mathbf{x}_{k+1} \approx (\lambda_1)^{k+1}\alpha_1\mathbf{u}_1$, or $\mathbf{x}_{k+1} \approx \lambda_1\mathbf{x}_k$. This result suggests a number of ways to approximate λ_1. For example, $\mathbf{x}_k^T\mathbf{x}_{k+1} \approx \lambda_1\mathbf{x}_k^T\mathbf{x}_k$ and thus $\lambda_1 \approx \mathbf{x}_k^T\mathbf{x}_{k+1}/\mathbf{x}_k^T\mathbf{x}_k$. Another common procedure is to divide the largest component of $\mathbf{x}_k$ into the corresponding component of $\mathbf{x}_{k+1}$. If the largest component is the jth component, then this approximation amounts to $\lambda_1 \approx \mathbf{e}_j^T\mathbf{x}_{k+1}/\mathbf{e}_j^T\mathbf{x}_k$.

We wish to be fairly careful about the description of the power method, for this description has two parts, namely the "practical" part, which shows how the method is actually carried out, and the "theoretical" part, which shows why the method works, what can go wrong, and how to interpret the output. First of all, powers of A are never actually computed in the implementation of the power method, nor is it necessary to know the α_i's and/or the $\mathbf{u}_i$'s in (3.7). We merely guess a vector $\mathbf{x}_0$ and form the sequence $\{\mathbf{x}_k\}$ from $\mathbf{x}_k = A\mathbf{x}_{k-1}$; $k = 1, 2, 3, \ldots$. The equation $\mathbf{x}_k = A^k\mathbf{x}_0$ is the theoretical basis for the power method, but all we really need is the vector $\mathbf{x}_k$ (note that it is uneconomical to compute A^k and then form $\mathbf{x}_k$ by $\mathbf{x}_k = A^k\mathbf{x}_0$). The estimates for λ_1 can be found by generating the sequence of scalars $\{\beta_k\}$ from the equation

$$\beta_k = \frac{\mathbf{v}^T\mathbf{x}_{k+1}}{\mathbf{v}^T\mathbf{x}_k} \tag{3.11}$$

where $\mathbf{v}$ is usually $\mathbf{x}_k$ or the rth unit-vector $\mathbf{e}_r$ (where the rth component of $\mathbf{x}_k$ is the dominant one). When programming the power method, it is a little easier to generate the estimate β_k from $\beta_k = \mathbf{x}_k^T\mathbf{x}_{k+1}/\mathbf{x}_k^T\mathbf{x}_k$ than to include a test at each step to determine the maximum component of $\mathbf{x}_k$(moreover, the program will execute faster). The advantage in finding the maximum component of $\mathbf{x}_k$ is that round-off errors will tend to be reduced. Finally, as k increases, the vectors $\mathbf{x}_k$ are usually growing quite large if $|\lambda_1| > 1$ (or quite small if $|\lambda_1| < 1$) as can be seen from (3.9). As we note below, it is desirable to "scale" or "normalize" $\mathbf{x}_k$ to keep it within computational limits. This scaling can be done by dividing $\mathbf{x}_k$ by $\|\mathbf{x}_k\|$ at each step where we would use the ℓ_2 vector norm when $\beta_k = \mathbf{x}_k^T\mathbf{x}_{k+1}/\mathbf{x}_k^T\mathbf{x}_k$ and the ℓ_∞ vector norm when $\beta_k = \mathbf{e}_r^T\mathbf{x}_{k+1}/\mathbf{e}_r^T\mathbf{x}_k$. It is easy to show (Problem 8) that scaling has no effect on the convergence.

In order to study the theoretical convergence of the power method, it is most convenient to consider a variation of the method in which the β_k are calculated in a slightly different fashion from the two ways described above. In this variation, we assume $\mathbf{v}$ is a *fixed* vector such that $\mathbf{v}^T\mathbf{u}_1 \neq 0$. We then define β_k to be given by $\beta_k = \mathbf{v}^T\mathbf{x}_{k+1}/\mathbf{v}^T\mathbf{x}_k$. If we set $\gamma_j = \mathbf{v}^T\mathbf{u}_j$, Eq. (3.10) shows that

$$\beta_k = \lambda_1 \frac{\alpha_1\gamma_1 + \alpha_2\gamma_2\left[\frac{\lambda_2}{\lambda_1}\right]^{k+1} + \cdots + \alpha_n\gamma_n\left[\frac{\lambda_n}{\lambda_1}\right]^{k+1}}{\alpha_1\gamma_1 + \alpha_2\gamma_2\left[\frac{\lambda_2}{\lambda_1}\right]^{k} + \cdots + \alpha_n\gamma_n\left[\frac{\lambda_n}{\lambda_1}\right]^{k}}. \tag{3.12}$$

The theoretical result, (3.12), reveals that the sequence $\{\beta_k\}$ converges to λ_1 at essentially the same rate as $(\lambda_2/\lambda_1)^k$ converges to zero. Equation (3.12) shows also how acceleration procedures like Aitkins Δ^2-process can be used to speed up the convergence of $\{\beta_k\}$ (see Section 4.3.1, Problem 7). We leave as a problem for the reader to show that the sequence $\{\beta_k\}$ generated by $\beta_k = \mathbf{x}_k^T\mathbf{x}_{k+1}/\mathbf{x}_k^T\mathbf{x}_k$ also converges to λ_1; the analysis is similar to that in (3.12). When $\beta_k = \mathbf{e}_r^T\mathbf{x}_{k+1}/\mathbf{e}_r^T\mathbf{x}_k$, a rigorous analysis is more difficult since the vector $\mathbf{e}_r$ may change from step to step.

The theoretical analysis in (3.12) reveals also some of the problems that can occur and how to interpret machine output. While we do not wish to examine

(3.12) extensively, general observations are easy to make. First, even if our assumption that A has n linearly independent eigenvectors holds and even if (3.6) is valid (neither of which we can check in general), it might be that our initial vector $\mathbf{x}_0$ is chosen so that $\alpha_1 = 0$ in (3.7), or so that α_1 is quite small. If $\alpha_1 = 0$ and if $|\lambda_2| > |\lambda_3|$, then mathematically it follows that $\{\beta_k\} \to \lambda_2$. However, the presence of round-off errors tends to contaminate the computation of $\{\mathbf{x}_k\}$ in that the round-off can begin to introduce a nonzero multiple of $\mathbf{u}_1$ in the $\mathbf{x}_k$'s. This error makes it quite likely that some $\mathbf{x}_k$ will not be *only* a linear combination of $\mathbf{u}_2, \mathbf{u}_3, \ldots, \mathbf{u}_n$ but will contain the presence of $\mathbf{u}_1$ as well. Then it is just a matter of time until the dominance of λ_1 asserts itself. The machine output would manifest this dominance by showing that the sequence $\{\beta_k\}$ is apparently converging to some value for a time (λ_2, presumably), and then changing directions and converging to another value (λ_1, this time). As a final (theoretical) point, we note that even if A does not have n linearly independent eigenvectors, it is still possible to establish that the power method is convergent provided that A has a single dominant eigenvalue λ_1. We do not want to go into the details of this point other than to say that every $(n \times n)$ matrix has a set of n linearly independent eigenvectors and generalized eigenvectors $\{\mathbf{v}_1, \mathbf{v}_2, \ldots, \mathbf{v}_n\}$. Any starting vector $\mathbf{x}_0$ can be expressed as a linear combination of $\mathbf{v}_1, \mathbf{v}_2, \ldots, \mathbf{v}_n$ just as in (3.7), and an analysis of the power method proceeds as in (3.10).

As we mentioned above, it is desirable to scale the vectors $\mathbf{x}_k$, and most programs employing the power method do this scaling. The first step is to select a suitable vector norm to use in the scaling. Usually either the ℓ_∞ or ℓ_2 norm is employed, and we illustrate the procedure with the ℓ_2 norm. First an initial vector, $\mathbf{x}_0$, is chosen where $\mathbf{x}_0^T\mathbf{x}_0 = 1$ (that is, $\|\mathbf{x}_0\|_2 = 1$). The power method then proceeds for $i = 1, 2, \ldots$.

1. Let $\mathbf{y}_i = A\mathbf{x}_{i-1}$.
2. Set $\beta_i = \mathbf{y}_i^T\mathbf{x}_{i-1}$.
3. Let $\eta_i = \sqrt{\mathbf{y}_i^T\mathbf{y}_i}$.
4. Set $\mathbf{x}_i = \mathbf{y}_i/\eta_i$ and return to step (1).

Subroutine POWERM, given in Fig. 3.1, employs this scaling device. Note that β_i is the approximation to λ_1 and that each $\mathbf{x}_i$ is an approximate eigenvector [recall (3.10)].

EXAMPLE 3.3. In this example, subroutine POWERM was used to estimate the dominant eigenvalue of

$$A = \begin{bmatrix} 1 & -1 & 2 \\ -2 & 0 & 5 \\ 6 & -3 & 6 \end{bmatrix}.$$

It is easy to verify that the eigenvalues of A are $\lambda_1 = 5$, $\lambda_2 = 3$, and $\lambda_3 = -1$. Three corresponding eigenvectors are

$$\mathbf{u}_1 = \begin{bmatrix} 5 \\ 16 \\ 18 \end{bmatrix}, \quad \mathbf{u}_2 = \begin{bmatrix} 1 \\ 6 \\ 4 \end{bmatrix}, \quad \text{and} \quad \mathbf{u}_3 = \begin{bmatrix} -1 \\ -2 \\ 0 \end{bmatrix}.$$

```
      SUBROUTINE POWERM(A,X,LAMBDA,TOL,N,M,ITERM)
      REAL LAMBDA
      DIMENSION A(20,20),X(20),Y(20)
C
C   SUBROUTINE POWERM USES THE POWER METHOD WITH SCALING TO ESTIMATE THE
C   DOMINANT EIGENVALUE OF A MATRIX A.  THE CALLING PROGRAM MUST SUPPLY
C   THE MATRIX A, AN INITIAL VECTOR X OF EUCLIDEAN LENGTH 1, A TOLERANCE
C   TOL, AN INTEGER N (WHERE A IS (NXN)) AND AN INTEGER M=MAXIMUM NUMBER
C   OF POWER ITERATIONS DESIRED.  THE SUBROUTINE RETURNS WHEN THE
C   DIFFERENCE OF TWO SUCCESSIVE ESTIMATES IS LESS THAN TOL IN ABSOLUTE
C   VALUE OR WHEN M ITERATIONS HAVE BEEN EXECUTED.  IN THE FIRST CASE, A
C   FLAG, ITERM, IS SET TO 1 AND IN THE SECOND CASE ITERM IS SET TO 2.
C   THE APPROXIMATE EIGENVALUE IS RETURNED AS LAMBDA AND AN APPROXIMATE
C   EIGENVECTOR AS X.  LAMBDA MUST BE DECLARED REAL IN THE CALLING
C   PROGRAM
C
      ITR=1
C
C   CALCULATE THE INITIAL EIGENVALUE APPROXIMATION
C
      DO 1 I=1,N
      Y(I)=0.
      DO 1 J=1,N
    1 Y(I)=Y(I)+A(I,J)*X(J)
      TEMP=0.
      YSCALE=0.
      DO 2 I=1,N
      TEMP=TEMP+Y(I)*X(I)
    2 YSCALE=YSCALE+Y(I)*Y(I)
      YSCALE=SQRT(YSCALE)
      ESTOLD=TEMP
C
C   POWER METHOD ITERATION WITH SCALING
C
    3 ITR=ITR+1
      DO 4 I=1,N
    4 X(I)=Y(I)/YSCALE
      DO 5 I=1,N
      Y(I)=0.
      DO 5 J=1,N
    5 Y(I)=Y(I)+A(I,J)*X(J)
      TEMP=0.
      YSCALE=0.
      DO 6 I=1,N
      TEMP=TEMP+Y(I)*X(I)
    6 YSCALE=YSCALE+Y(I)*Y(I)
      YSCALE=SQRT(YSCALE)
      ESTNEW=TEMP
C
C   TEST FOR TERMINATION OF THE POWER METHOD ITERATION
C
      IF(ABS(ESTNEW-ESTOLD).LE.TOL)    GO TO 7
      IF(ITR.GE.M)   GO TO 8
      ESTOLD=ESTNEW
      GO TO 3
    7 ITERM=1
      LAMBDA=ESTNEW
      RETURN
    8 ITERM=2
      LAMBDA=ESTNEW
      RETURN
      END
```

Figure 3.1

For this example, the estimates β_k in Table 3.1 were computed from $\beta_k = \mathbf{x}_k^T\mathbf{x}_{k+1}/\mathbf{x}_k^T\mathbf{x}_k$ and initial vector $\mathbf{x}_0 = \gamma(1, 1, 1)^T$ where $\gamma = 1/\sqrt{3}$.

TABLE 3.1

k	β_k	k	β_k
1	0.4666666E 01	14	0.5002243E 01
2	0.7127658E 01	15	0.5001345E 01
3	0.5770836E 01	16	0.5000809E 01
4	0.5460129E 01	17	0.5000484E 01
5	0.5245014E 01	18	0.5000289E 01
6	0.5142666E 01	19	0.5000171E 01
7	0.5083027E 01	20	0.5000103E 01
8	0.5049140E 01	21	0.5000067E 01
9	0.5029204E 01	22	0.5000038E 01
10	0.5017425E 01	23	0.5000021E 01
11	0.5010424E 01	24	0.5000011E 01
12	0.5006239E 01	25	0.5000006E 01
13	0.5003741E 01	26	0.5000005E 01

We have listed below some representative values of the approximate eigenvectors, $\mathbf{x}_k$. These vectors $\mathbf{x}_k$ all have Euclidean length 1 and so are converging to $\gamma\mathbf{u}_1$ where $\gamma = 1/\|\mathbf{u}_1\|_2$. A quick calculation shows that $\gamma\mathbf{u}_1 = [0.2032789, 0.6504925, 0.7318041]^T$.

k		$\mathbf{x}_k^T$	
6	0.2058211E 00	0.6420183E 00	0.7385463E 00
12	0.2033920E 00	0.6501137E 00	0.7321092E 00
18	0.2032841E 00	0.6504748E 00	0.7318184E 00
24	0.2032790E 00	0.6504918E 00	0.7318048E 00

To illustrate the point made earlier about how round-off error may begin to introduce a nonzero multiple of $\mathbf{u}_1$, the program was run again with an initial vector $\mathbf{x}_0 = \mathbf{y}_0/\|\mathbf{y}_0\|_2$ where $\mathbf{y}_0 = \alpha_2\mathbf{u}_2 + \alpha_3\mathbf{u}_3$ and where $\alpha_2 = 1/3$ and $\alpha_3 = -2/3$. In this case, the iterates begin converging to $\lambda_2 = 3$ and then change direction and begin converging to $\lambda_1 = 5$. (See Table 3.2, next page.)

Finally, (3.12) reveals that if $\lambda_1 = \lambda_2$, and $|\lambda_1| > |\lambda_3| \geq \cdots \geq |\lambda_n|$, then the β_i still converge to λ_1. (That is, multiple eigenvalues do not affect the power method. However, if A does not have a full set of eigenvectors corresponding to λ_1, then convergence may be slow. See Problem 13.) If $\lambda_1 = -\lambda_2$ and $|\lambda_1| > |\lambda_3| \geq \cdots \geq |\lambda_n|$, then the β_i would exhibit a periodic behavior while $\mathbf{v}^T\mathbf{x}_{i+2}/\mathbf{v}^T\mathbf{x}_i$ would provide an estimate for $(\lambda_1)^2$. Dominant complex eigenvalues cause the most problems, for if λ_1 is a complex root of $p(\lambda) = 0$, then so is the complex conjugate, $\bar{\lambda}_1$. One further circumstance that must be accounted for is that $|\lambda_1|$

TABLE 3.2

k	β_k	k	β_k	k	β_k
1	0.1527997E 01	17	0.3002388E 01	33	0.4635723E 01
2	0.3680351E 01	18	0.3003982E 01	34	0.4765195E 01
3	0.2787491E 01	19	0.3006630E 01	35	0.4852588E 01
4	0.3073232E 01	20	0.3011026E 01	36	0.4909041E 01
5	0.2975838E 01	21	0.3018322E 01	37	0.4944480E 01
6	0.3008090E 01	22	0.3030365E 01	38	0.4966341E 01
7	0.2997321E 01	23	0.3050138E 01	39	0.4979676E 01
8	0.3000922E 01	24	0.3082280E 01	40	0.4987756E 01
9	0.2999737E 01	25	0.3133698E 01	41	0.4992640E 01
10	0.3000160E 01	26	0.3213868E 01	42	0.4995575E 01
11	0.3000079E 01	27	0.3333886E 01	43	0.4997344E 01
12	0.3000196E 01	28	0.3502944E 01	44	0.4998403E 01
13	0.3000305E 01	29	0.3721320E 01	45	0.4999041E 01
14	0.3000515E 01	30	0.3973443E 01	46	0.4999423E 01
15	0.3000856E 01	31	0.4229045E 01	47	0.4999656E 01
16	0.3001437E 01	32	0.4456168E 01	48	0.4999791E 01

$= |\lambda_2| = \cdots = |\lambda_k|$, $k \le n$. Adjustments for all of these situations can be made, but we shall not go further into this problem here. [See Faddeev and Faddeeva (1963).]

We conclude this section with one particular variation of the power method called the *Rayleigh quotient method*. For this method we assume that the matrix A is symmetric, that is, $A^T = A$; and that $|\lambda_1| > |\lambda_2| \ge \cdots \ge |\lambda_n|$. We will show in Section 3.4 that when A is symmetric, it has n linearly independent real eigenvectors, $\{\mathbf{u}_i\}_{i=1}^n$, and furthermore the eigenvectors can be chosen such that

$$\mathbf{u}_i^T\mathbf{u}_j = \begin{Bmatrix} 1, & i = j \\ 0, & i \ne j \end{Bmatrix}, \qquad 1 \le i, j \le n.$$

(Later we shall see that a set of vectors with this property is called *orthonormal.*)

Now, making use of this property in the power method and expressing $\mathbf{x}_0$ as in (3.7) with $\mathbf{x}_k = A\mathbf{x}_{k-1}$, $k \ge 1$, we can easily verify from (3.9) that $\mathbf{x}_k^T A\mathbf{x}_k = \mathbf{x}_k^T\mathbf{x}_{k+1} = \sum_{j=1}^n \alpha_j^2\lambda_j^{2k+1}$ and $\mathbf{x}_k^T\mathbf{x}_k = \sum_{j=1}^n \alpha_j^2\lambda_j^{2k}$ (Problem 5). Now we form the sequence of *Rayleigh quotients*, $\{\mu_k\}$ where

$$\mu_k = \frac{\mathbf{x}_k^T A\mathbf{x}_k}{\mathbf{x}_k^T\mathbf{x}_k}.$$

Then, precisely as in (3.12), we see that the sequence $\{\mu_k\}$ converges to λ_1 at essentially the same rate as $(\lambda_2/\lambda_1)^{2k}$ converges to zero. Thus the Rayleigh quotient method for symmetric matrices will converge to λ_1 approximately twice as fast as the power method in the general case. When A is not symmetric, the Rayleigh quotients, μ_k, will still converge to λ_1 (although establishing convergence is a bit tedious). In fact, subroutine POWERM (given in Fig. 3.1) employs Rayleigh quotients. However, convergence for nonsymmetric matrices is no faster than $(\lambda_2/\lambda_1)^k$ when Rayleigh quotients are used.

As additional motivation for the Rayleigh quotient method, let α be any scalar, let $\mathbf{x}$ be any nonzero $(n \times 1)$ vector, and define $\mathbf{r}(\alpha) = A\mathbf{x} - \alpha\mathbf{x}$ where A is symmetric. Thus for any α, $\mathbf{r}(\alpha)$ is an $(n \times 1)$ vector, and furthermore $\mathbf{r}(\alpha) = \theta$ if and only if α is an eigenvalue of A and $\mathbf{x}$ is a corresponding eigenvector. Thus the size of $\mathbf{r}(\alpha)$ somewhat measures how well α approximates an eigenvalue of A. We next show for a given vector $\mathbf{x}$ that $\|\mathbf{r}(\alpha)\|_2$ is minimized if we choose $\alpha = \mathbf{x}^TA\mathbf{x}/\mathbf{x}^T\mathbf{x}$. To see this, we let $(\mathbf{r}(\alpha))^T(\mathbf{r}(\alpha)) = \|\mathbf{r}(\alpha)\|_2^2 \equiv g(\alpha) \geq 0$. Now

$$g(\alpha) = (A\mathbf{x} - \alpha\mathbf{x})^T(A\mathbf{x} - \alpha\mathbf{x}) = \mathbf{x}^TA^2\mathbf{x} - 2\alpha\mathbf{x}^TA\mathbf{x} + \alpha^2\mathbf{x}^T\mathbf{x}$$

since A is symmetric. Then $g'(\alpha) = -2\mathbf{x}^TA\mathbf{x} + 2\alpha\mathbf{x}^T\mathbf{x}$; so $g'(\tilde{\alpha}) = 0$ when $\tilde{\alpha} = (\mathbf{x}^T A\mathbf{x})/(\mathbf{x}^T\mathbf{x})$. Also, $g''(\tilde{\alpha}) = 2\mathbf{x}^T\mathbf{x} > 0$; and thus $g(\alpha)$ (and $\|\mathbf{r}(\alpha)\|_2$) is *minimized* for $\tilde{\alpha} = (\mathbf{x}^TA\mathbf{x})/(\mathbf{x}^T\mathbf{x})$, the *Rayleigh quotient*. We will return to this observation in Section 3.2.4.

PROBLEMS, SECTION 3.2

1. Program the power method; use (3.11) with $\mathbf{v} = \mathbf{e}_r$ to generate an estimate to the dominant eigenvalue. Thus at each step, $\beta_k = x_{k+1}^{(r)}/x_k^{(r)}$ where $x_k^{(r)}$ is the dominant component of $\mathbf{x}_k$. Also at each step, "normalize" $\mathbf{x}_k$ by dividing each component of $\mathbf{x}_k$ by $x_k^{(r)}$ before forming $\mathbf{x}_{k+1}$. Thus, $\|\mathbf{x}_k\|_\infty = 1$ for $k = 0, 1, 2, \ldots$, and $\beta_k = x_{k+1}^{(r)}$. In addition at each step, print the vector $\mathbf{x}_k$.
   a) Test your program on the matrix A of Example 3.3 with $\mathbf{x}_0 = (1, 1, 1)^T$, and note that there is some improvement in convergence.
   b) Note that the normalized vectors $\mathbf{x}_k$ are converging to an eigenvector that corresponds to $\lambda_1 = 5$, as would be expected from (3.10).
   c) Defining $\varepsilon_i = \lambda_1 - \beta_i$, note that $\varepsilon_{i+1} \approx \delta\varepsilon_i$ where $\delta = \lambda_2/\lambda_1 = 3/5$. This result would be expected from (3.12).
2. Write a program that calls subroutine POWERM and test your program on the symmetric matrix A given by

$$A = \begin{bmatrix} 1 & 2 & 1 \\ 2 & 3 & 0 \\ 1 & 0 & 1 \end{bmatrix}.$$

   (Recall that POWERM uses Rayleigh quotients.)
   Use the power method program developed in Problem 1 on the matrix A, and note the improvement in convergence provided by the Rayleigh quotient method.
3. The power method generates a sequence of vectors $\{\mathbf{x}_i\}$ where (theoretically) $\mathbf{x}_k = A^k\mathbf{x}_0$. Suppose A is an $(n \times n)$ matrix and suppose that A^{i+1} is formed by multiplying

A times A^i. How many multiplications are necessary to find $\mathbf{x}_k$ if we use the formula $\mathbf{x}_k = A^k\mathbf{x}_0$? How many multiplications are necessary to find $\mathbf{x}_k$ if we generate $\mathbf{x}_k$ from $\mathbf{x}_{i+1} = A\mathbf{x}_i$, $i = 0, 1, \ldots, k-1$?

4. a) In the notation of (3.9), let $\mathbf{v}$ be any *fixed* vector such that $\mathbf{v}^T\mathbf{u}_1 \neq 0$. Let $\{\beta_k\}$ be formed as in (3.11) and show that $\lim_{k\to\infty} \beta_k = \lambda_1$. [Assume $\alpha_1 \neq 0$ in (3.7).]

b) In the three following cases, suppose that $|\lambda_2| > |\lambda_3|$, $\mathbf{v}^T\mathbf{u}_2 \neq 0$, and $\alpha_2 \neq 0$. Establish the *theoretical limits* in each case.

(1) If $\mathbf{v}^T\mathbf{u}_1 = 0$, then $\lim_{k\to\infty} \beta_k = \lambda_2$.

(2) If $\lambda_1 = \lambda_2$, $\alpha_1 \neq 0$, and $\mathbf{v}^T\mathbf{u}_1 \neq 0$, then $\lim_{k\to\infty} \beta_k = \lambda_1$.

(3) If $\lambda_1 = -\lambda_2$, $\alpha_1 \neq 0$, and $\mathbf{v}^T\mathbf{u}_1 \neq 0$, then $\lim_{k\to\infty} (\mathbf{v}^T\mathbf{x}_{k+2}/\mathbf{v}^T\mathbf{x}_k) = (\lambda_1)^2$.

5. Let A be an $(n \times n)$ symmetric matrix with eigenvectors $\mathbf{u}_i$ chosen to be orthonormal. Using the notation of (3.9), show that

$$\mathbf{x}_k^T\mathbf{x}_k = \sum_{j=1}^{n} \alpha_j^2\lambda_j^{2k} \qquad \text{and} \qquad \mathbf{x}_k^T\mathbf{x}_{k+1} = \sum_{j=1}^{n} \alpha_j^2\lambda_j^{2k+1}.$$

***6.** Let A be a symmetric matrix and suppose λ_i and λ_j are two eigenvalues of A with corresponding eigenvectors $\mathbf{u}_i$ and $\mathbf{u}_j$. Assume that $\lambda_i \neq \lambda_j$, and show $\mathbf{u}_i^T\mathbf{u}_j = 0$. (To be precise, $\mathbf{u}_i$ and $\mathbf{u}_j$ are called *orthogonal*. Hint: Consider $\mathbf{u}_i^T A\mathbf{u}_j = \lambda_j\mathbf{u}_i^T\mathbf{u}_j$.)

***7.** In (3.12), assume that $|\lambda_1| > |\lambda_2| > |\lambda_3| \geq \cdots \geq |\lambda_n|$ and that $\alpha_1\gamma_1 \neq 0$. Let the estimates β_i defined by (3.11) be generated with a fixed vector $\mathbf{v}$.

a) Show that $\lim_{i\to\infty} (\beta_i - \lambda_1)(\lambda_1/\lambda_2)^i = C$ where C is some constant. [This statement means that for i sufficiently large, $\beta_i \approx \lambda_1 + C(\lambda_2/\lambda_1)^i$.]

b) Using the notation of Problem 1(c), show from part (a) of this problem that

$$\lim_{i\to\infty} \frac{\varepsilon_{i+1}}{\varepsilon_i} = \frac{\lambda_2}{\lambda_1}.$$

This limit substantiates mathematically the empirical evidence suggested by Problem 1(c). Also note that this result shows that Aitkens Δ^2-process can be applied to accelerate convergence. (Again see Section 4.3.1.)

8. Show that scaling does not affect convergence of the power method. To show this, suppose that $\mathbf{x}_0$ is given by (3.7) where $\|\mathbf{x}_0\| = 1$. If scaling is used, the power method takes this form: $\mathbf{y}_i = A\mathbf{x}_{i-1}$, $\beta_i = \mathbf{v}^T\mathbf{y}_i/\mathbf{v}^T\mathbf{x}_{i-1}$, $\mathbf{x}_i = \mathbf{y}_i/\|\mathbf{y}_i\|$, for $i = 1, 2, \ldots$. [Hint: Show that $\mathbf{v}^T\mathbf{y}_i/\mathbf{v}^T\mathbf{x}_{i-1} = \mathbf{v}^T\mathbf{w}_i/\mathbf{v}^T\mathbf{w}_{i-1}$ where $\mathbf{w}_0 = \mathbf{x}_0$ and $\mathbf{w}_i = A\mathbf{w}_{i-1}$, $i = 1, 2, \ldots$.]

***9.** Show that the Rayleigh quotient method will converge for matrices that are not symmetric. Assume, as usual, that $|\lambda_1| > |\lambda_2| \geq \cdots \geq |\lambda_n|$ and $\mathbf{x}_0^T\mathbf{u}_1 \neq 0$.

10. An important problem that arises when solving partial differential equations is the general eigenvalue problem $A\mathbf{x} = \lambda B\mathbf{x}$ where A and B are $(n \times n)$. If B is nonsingular, show how the power method (or the inverse power method, see Section 3.2.1) can be used for this problem. What can be done if B is singular, but A is nonsingular?

11. Modify subroutine POWERM to ''recognize'' when the matrix has dominant eigenvalues of opposite sign. Quantities of the form $\mathbf{v}^T\mathbf{x}_{i+2}/\mathbf{v}^T\mathbf{x}_i$ must be calculated in each step, but they must be modified because of the scaling in POWERM. Explain this modification.

12. Test the program of Problem 11 on the matrices

$$A = \begin{bmatrix} 1 & 1 & 1 & 1 \\ 2 & 0 & 1 & 1 \\ 0 & -1 & -2 & -2 \\ 0 & 0 & 2 & 2 \end{bmatrix}, \quad B = \begin{bmatrix} 1 & -1 & -1 & -1 \\ -1 & 1 & -1 & -1 \\ -1 & -1 & 1 & -1 \\ -1 & -1 & -1 & 1 \end{bmatrix}.$$

***13.** Suppose that A is a (3×3) matrix with eigenvalues λ_1 and λ_2 where $|\lambda_1| > |\lambda_2|$ and λ_1 is of multiplicity 2. Suppose also that A has only two linearly independent eigenvectors $\mathbf{u}_1$ and $\mathbf{u}_2$ where $A\mathbf{u}_1 = \lambda_1\mathbf{u}_1$ and $A\mathbf{u}_2 = \lambda_2\mathbf{u}_2$. It can be shown that there is a nonzero vector $\mathbf{q}$ such that $(A - \lambda_1 I)\mathbf{q} = \mathbf{u}_1$. Show that $A^k\mathbf{q} = \lambda_1^k\mathbf{q} + k\lambda_1^{k-1}\mathbf{u}_1$ for $k = 1, 2, 3, \ldots$, and that the set $\{\mathbf{u}_1, \mathbf{u}_2, \mathbf{q}\}$ is linearly independent. Hence any (3×1) vector $\mathbf{x}_0$ can be expressed as $\mathbf{x}_0 = a_1\mathbf{u}_1 + a_2\mathbf{u}_2 + a_3\mathbf{q}$. Let $\mathbf{w}$ be any vector such that $\mathbf{w}^T\mathbf{q} \neq 0$ and $\mathbf{w}^T\mathbf{u}_1 \neq 0$. From the power method sequence $\mathbf{x}_k = A\mathbf{x}_{k-1}$, $k = 1, 2, \ldots$, show that $\lim_{k\to\infty}(\mathbf{w}^T\mathbf{x}_{k+1}/\mathbf{w}^T\mathbf{x}_k) = \lambda_1$. (The vector $\mathbf{q}$ above is called a "generalized eigenvector," and this problem gives an indication of why the power method can still work even though the matrix does not have a full set of eigenvectors.) With the situation described above, convergence is slow. To see this point, let $e_k = \lambda_1 - \beta_k$ and show that $\lim_{k\to\infty} \dfrac{e_{k+1}}{e_k} = 1$.

3.2.1. The Inverse Power Method

The inverse power method gives us a way to approximate eigenvalues other than the dominant one. The method is based on the observation that if $A\mathbf{u}_i = \lambda_i\mathbf{u}_i$, $i = 1, 2, \ldots, n$, then for a scalar α, not equal to an eigenvalue, the nonsingular matrix $A - \alpha I$ has eigenvalues $(\lambda_i - \alpha)$ [since if $A\mathbf{u}_i = \lambda_i\mathbf{u}_i$, then $(A - \alpha I)\mathbf{u}_i = (\lambda_i - \alpha)\mathbf{u}_i$]. Furthermore, the matrix $(A - \alpha I)^{-1}$ has eigenvalues $(\lambda_i - \alpha)^{-1}$ for $i = 1, 2, \ldots, n$ [for if $(A - \alpha I)\mathbf{u}_i = (\lambda_i - \alpha)\mathbf{u}_i$, then $(\lambda_i - \alpha)^{-1}\mathbf{u}_i = (A - \alpha I)^{-1}\mathbf{u}_i$]. If we were to apply the power method to the matrix $(A - \alpha I)^{-1}$, we would converge to the eigenvalue of $(A - \alpha I)^{-1}$ that is largest in magnitude. Since the eigenvalues of $(A - \alpha I)^{-1}$ are $1/(\lambda_1 - \alpha)$, $1/(\lambda_2 - \alpha), \ldots, 1/(\lambda_n - \alpha)$, it follows that we would obtain estimates to $1/(\lambda_j - \alpha)$ where $|\lambda_j - \alpha| < |\lambda_i - \alpha|$, $i = 1, 2, \ldots, n$.

To be a bit more precise, suppose that α is any scalar and that we form the sequence of vectors

$$\mathbf{y}_{k+1} = (A - \alpha I)^{-1}\mathbf{y}_k, \qquad k = 0, 1, \ldots. \tag{3.13}$$

Then under assumptions similar to those of the last section, the sequence $\{\beta_k\}$ defined by

$$\beta_k = \frac{\mathbf{y}_{k+1}^T\mathbf{y}_k}{\mathbf{y}_k^T\mathbf{y}_k}, \qquad k = 0, 1, \ldots \tag{3.14}$$

converges to $1/(\lambda_j - \alpha)$ where λ_j is the eigenvalue of A nearest to α. In particular, the numbers $(1/\beta_k) + \alpha$ converge to the eigenvalue of A that is nearest α. As an important special case, we note that the choice $\alpha = 0$ would lead to estimates of the eigenvalue of A that is smallest in magnitude. We also observe that the analysis of the last section shows that convergence will be quite rapid when α is close to an eigenvalue. Specifically, suppose α is a good approximation to λ_j and that $\lambda_i - \alpha$ is of moderate size when $i \neq j$. The rate of convergence of the power method is governed by the ratios of the nondominant eigenvalues to the dominant one; and in the context of the matrix $(A - \alpha I)^{-1}$ with $\alpha \simeq \lambda_j$, these ratios are

$$\frac{1/(\lambda_i - \alpha)}{1/(\lambda_j - \alpha)} = \frac{\lambda_j - \alpha}{\lambda_i - \alpha}, \qquad i \neq j.$$

If C denotes the ratio $(\lambda_j - \alpha)/(\lambda_i - \alpha)$ that is largest in magnitude, then from Problem 7, Section 3.2, we have $\varepsilon_{k+1} \simeq C\varepsilon_k$ where ε_k is the kth error, $\varepsilon_k = \beta_k - 1/(\lambda_j - \alpha)$. If C is small, then the errors are decreasing dramatically; for example, if $\lambda_j - \alpha$ is on the order of 10^{-2} or so, and if the eigenvalues are separated sufficiently so that $\lambda_i - \alpha$ is on the order of 1 or larger for $i \neq j$, then we expect that C is on the order of 10^{-2}. To the extent that $\varepsilon_{k+1} \simeq C\varepsilon_k$, we see that each successive error is cut down by a factor of about 100, and thus each iteration would give us about two additional decimal places of accuracy.

To see how we can exploit this rapid convergence and to see another positive feature of the inverse power method, suppose we have estimates α_j to the distinct eigenvalues λ_j of A. For each estimate α_j, the inverse power method [that is, the power method applied to $(A - \alpha_j I)^{-1}$] will converge quickly and give us a very accurate estimate to λ_j and to a corresponding eigenvector [recall that $(A - \alpha_j I)^{-1}$ and A have the same eigenvectors]. Besides being rapidly convergent, the inverse power method is effective even when A has multiple dominant eigenvalues such as, for instance, when the dominant eigenvalue of A is complex. Rather than discussing this idea abstractly, we consider the following example. Suppose A is a real (6×6) matrix with eigenvalues $\lambda_1, \lambda_2, \ldots, \lambda_6$. Suppose also that λ_1 is complex, $\lambda_1 = a + ib$, $\lambda_2 = \lambda_1$, and $\lambda_3 = \lambda_4 = \bar{\lambda}_1$. Finally suppose that λ_5 and λ_6 are real where the ordering is

$$|\lambda_1| = |\lambda_2| = |\lambda_3| = |\lambda_4| > |\lambda_5| = |\lambda_6|.$$

This hypothetical situation is diagramed in Fig. 3.2 and represents a case with which the power method in its primary form cannot deal. Given the situation in Fig. 3.2, suppose we have estimates α_1 to λ_1, α_3 to λ_3, α_5 to λ_5, and α_6 to λ_6 where α_j is in the circle centered at λ_j for $j = 1, 3, 5, 6$. Clearly the power method applied to $(A - \alpha_j I)^{-1}$ will converge to $1/(\lambda_j - \alpha_j)$, $j = 1, 3, 5, 6$. [Recall that a multiple eigenvalue, such as $1/(\lambda_1 - \alpha_1)$ or $1/(\lambda_3 - \alpha_3)$, does not affect the convergence of the power method.] In summary, the inverse power method is a very good means of finding accurate approximations to all the distinct eigenvalues (and corresponding eigenvectors) of a matrix A, provided that we can produce reasonable initial guesses to the eigenvalues. We will discuss later the ways of obtaining these initial guesses.

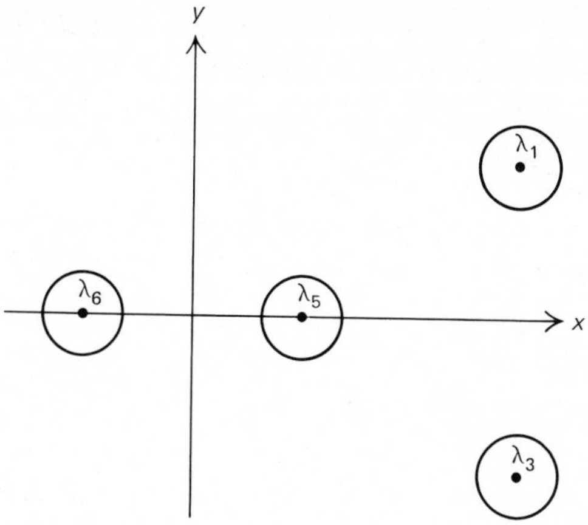

Figure 3.2 A distribution of eigenvalues in the complex plane.

In order to implement the inverse power method efficiently, we observe that the vectors $\mathbf{y}_{k+1}$ in (3.13) should be computed from the equation

$$(A - \alpha I)\mathbf{y}_{k+1} = \mathbf{y}_k, \qquad k = 0, 1, \ldots . \tag{3.15}$$

Gauss elimination would be a suitable procedure to solve (3.15); and the remarks of Section 2.2.4 are applicable since we are solving a series of problems with the same coefficient matrix as in (2.11). Given $A - \alpha I$, we can use a FACTOR routine (see Problem 11, Section 2.2.4) to reduce $A - \alpha I$ to triangular form. Then, given a starting vector $\mathbf{y}_0$, a SOLVE routine can be called repeatedly to generate $\mathbf{y}_1, \mathbf{y}_2, \ldots$. Note that the solve step requires essentially n^2 operations; and this number is the same as that required to generate $\mathbf{x}_k$ from $\mathbf{x}_{k+1}$ in the power method. Another complication that may occur to the reader is that if α is a good approximation to an eigenvalue, then the matrix $A - \alpha I$ is nearly singular and thus may be ill-conditioned. This situation turns out to be not as much of a problem as we might suspect in part because we are looking for a constant of proportionality, not an accurate solution of $(A - \alpha I)\mathbf{y}_k = \mathbf{y}_{k-1}$ [see, for example, Fox (1965)].

In Fig. 3.3 we include a subroutine, Subroutine INVPOW, that implements the inverse power method. This subroutine calls Subroutine GAUSS (see Fig. 2.1) and so is clearly not efficient as it is coded. In Problem 5, we ask the reader to modify INVPOW to incorporate the FACTOR and SOLVE routines of Section 2.2.4.

EXAMPLE 3.4. This example illustrates the use of the inverse power method as applied to the matrix A of Example 3.3. Setting $\alpha = 0$ in (3.13) so as to find the

```
      SUBROUTINE INVPOW(A,EIGNVC,ALPHA,EIGNVL,N,TOL,MAXITR,MAINDM,IFLAG)
      DIMENSION A(MAINDM,MAINDM),EIGNVC(MAINDM)
      DIMENSION AWORK(50,50),OLDVCT(50),REFVCT(50)
C
C  INITIALIZE THE ITERATION COUNTER AND SET UP THE WORKING
C  MATRIX AWORK=A-ALPHA*I.
C
      ITR=0
        DO 2 I=1,N
          DO 1 J=1,N
          AWORK(I,J)=A(I,J)
    1     CONTINUE
        AWORK(I,I)=AWORK(I,I)-ALPHA
    2   CONTINUE
C
C  CHOOSE A STARTING VECTOR OF NORM 1.
C
      AN=N
      Q=1./SQRT(AN)
        DO 3 I=1,N
        OLDVCT(I)=Q
    3   CONTINUE
C
C  CARRY OUT THE INVERSE POWER METHOD ITERATION.  REFVAL AND REFVCT
C  ARE REFINED ESTIMATES TO AN EIGENVALUE OF (A-ALPHA*I)**(-1)
C  AND A CORRESPONDING EIGENVECTOR.
C
      OLDVAL=ALPHA
    4 ITR=ITR+1
      CALL GAUSS(AWORK,OLDVCT,REFVCT,N,50,IERROR,RNORM)
      IF(IERROR.EQ.2)  GO TO 10
C
C  CALCULATE THE RAYLEIGH QUOTIENT AND THE NORM OF REFVCT.
C
      RAYQTN=0.
      SCALSQ=0.
        DO 5 I=1,N
        RAYQTN=RAYQTN+OLDVCT(I)*REFVCT(I)
        SCALSQ=SCALSQ+REFVCT(I)**2
    5   CONTINUE
C
C  CALCULATE A REFINED ESTIMATE OF AN EIGENVALUE OF A AND
C  SCALE REFVCT TO HAVE NORM 1.
C
      REFVAL=(1./RAYQTN)+ALPHA
      SCALE=SQRT(SCALSQ)
        DO 6 I=1,N
        REFVCT(I)=REFVCT(I)/SCALE
    6   CONTINUE
C
C  TEST FOR CONVERGENCE AND EXCESSIVE ITERATIONS.
C
      IF(ABS(OLDVAL-REFVAL).LE.TOL)  GO TO 8
      IF(ITR.GE.MAXITR)  GO TO 11
      OLDVAL=REFVAL
        DO 7 I=1,N
        OLDVCT(I)=REFVCT(I)
    7   CONTINUE
      GO TO 4
C
```

Figure 3.3 Calling program for INVPOW.

```
C   SUCCESSFUL RETURN.
C
      8 IFLAG=1
        EIGNVL=REFVAL
          DO 9 I=1,N
          EIGNVC(I)=REFVCT(I)
      9   CONTINUE
        RETURN
C
C   ABNORMAL RETURN --- AWORK MAY BE SINGULAR.
C
     10 IFLAG=2
        RETURN
C
C   ABNORMAL RETURN --- LIMIT ON NUMBER OF ITERATIONS IS EXCEEDED.
C
     11 IFLAG=3
        RETURN
        END
```

```
        DIMENSION A(20,20),EIGNVC(20)
        MAINDM=20
      1 READ 100,N
        IF(N.LT.2)   STOP
          DO 2 I=1,N
          READ 101,(A(I,J),J=1,N)
      2   CONTINUE
        READ 101,ALPHA
        TOL=.1E-06
        MAXITR=10
        CALL INVPOW(A,EIGNVC,ALPHA,EIGNVL,N,TOL,MAXITR,MAINDM,IFLAG)
        PRINT 102,IFLAG,EIGNVL
        PRINT 103,(EIGNVC(K),K=1,N)
        GO TO 1
    100 FORMAT(I2)
    101 FORMAT(10F4.0)
    102 FORMAT(1H0,6H FLAG=,I3,5X,36HREFINED ESTIMATE OF AN EIGENVALUE IS,
       1E15.5)
    103 FORMAT(1H0,30HA CORRESPONDING EIGENVECTOR IS,/,6E18.5)
        END
```

Figure 3.3 (continued)

eigenvalue of smallest magnitude, and choosing $\mathbf{y}_0 = \gamma[1, 1, 1]^T$ where $\gamma = 1/\sqrt{3}$, we obtained the results in Table 3.3 from subroutine INVPOW (where μ_k is the eigenvalue estimate).

TABLE 3.3

k	μ_k	k	μ_k	k	μ_k
1	−0.1184212E 01	5	−0.9719747E 00	9	−0.9996110E 00
2	−0.1527612E 01	6	−0.1010056E 01	10	−0.1000133E 01
3	−0.8223332E 00	7	−0.9965851E 00	11	−0.9999580E 00
4	−0.1082315E 01	8	−0.1001162E 01	12	−0.1000016E 01

Since we have the dominant eigenvalue of A from Example 3.3 ($\lambda_1 = 5$) and the smallest eigenvalue of A above ($\lambda_3 = -1$), we set $\alpha = (\lambda_1 + \lambda_3)/2 = 2$ and use subroutine INVPOW. We find after 11 iterations that $\lambda_2 \approx 2.999995$. To demonstrate the rapid convergence that results when α is a good estimate of an eigenvalue, we let $\alpha = 5.245$ (from Example 3.3, we have that this value of α approximates that obtained by the power method after five iterations). The inverse power method gave the results in Table 3.4. These results show a striking improvement over those in Example 3.3.

TABLE 3.4

k	μ_k
1	0.5122054E 01
2	0.5018278E 01
3	0.5002143E 01
4	0.5000240E 01
5	0.5000027E 01
6	0.5000003E 01

3.2.2. Localization of Eigenvalues

In order to make efficient use of the inverse power method we need some way to estimate the eigenvalues of a matrix A. In later sections we will develop some "transformation methods" that are quite effective; but for now, we want to present some results that describe regions in the complex plane that contain the eigenvalues. Theorems such as these, which give regions that contain the zeros of polynomials or the eigenvalues of matrices, are usually called localization theorems. In this section we state and prove a number of these theorems and indicate how to apply them in specific instances. In all these theorems, we assume A is a real, $(n \times n)$ matrix.

Theorem 3.1

If A is symmetric, then all the eigenvalues of A are real.

Proof. Suppose $A\mathbf{x} = \lambda\mathbf{x}$ where $\mathbf{x} \neq \theta$. Let $\bar{\mathbf{x}}$ denote the vector whose entries are the complex conjugates of those of $\mathbf{x}$. It is easy to show that $\overline{(A\mathbf{x})} = A\bar{\mathbf{x}}$ (since A is real) and that $\overline{(\lambda\mathbf{x})} = \bar{\lambda}\bar{\mathbf{x}}$. Therefore, if we have

$$\mathbf{x} = \begin{bmatrix} x_1 \\ x_2 \\ \vdots \\ x_n \end{bmatrix}, \qquad \text{then} \qquad \bar{\mathbf{x}} = \begin{bmatrix} \bar{x}_1 \\ \bar{x}_2 \\ \vdots \\ \bar{x}_n \end{bmatrix};$$

and since $A\bar{\mathbf{x}} = \bar{\lambda}\bar{\mathbf{x}}$, $\bar{\mathbf{x}}$ is an eigenvector of A corresponding to the eigenvalue $\bar{\lambda}$ (the eigenvalues and eigenvectors of a real matrix occur in conjugate pairs). Note also that

$$\bar{\mathbf{x}}^T\mathbf{x} = \mathbf{x}^T\bar{\mathbf{x}} = x_1\bar{x}_1 + x_2\bar{x}_2 + \cdots + x_n\bar{x}_n. \tag{3.16}$$

Recall that if $\beta = r + is$, then $\beta\bar{\beta} = (r + is)(r - is) = r^2 + s^2$. Thus since $\mathbf{x} \neq \theta$, $\mathbf{x}^T\bar{\mathbf{x}}$ is a real number; moreover, $\mathbf{x}^T\bar{\mathbf{x}} > 0$. In summary, from $A\bar{\mathbf{x}} = \bar{\lambda}\bar{\mathbf{x}}$ we obtain $\mathbf{x}^T A\bar{\mathbf{x}} = \bar{\lambda}\mathbf{x}^T\bar{\mathbf{x}}$, and from $A\mathbf{x} = \lambda\mathbf{x}$ we obtain $\bar{\mathbf{x}}^T A\mathbf{x} = \lambda\bar{\mathbf{x}}^T\mathbf{x}$. Now $A\bar{\mathbf{x}}$ is an $(n \times 1)$ vector and because $A^T = A$,

$$\mathbf{x}^T A\bar{\mathbf{x}} = (A\bar{\mathbf{x}})^T\mathbf{x} = \bar{\mathbf{x}}^T A^T\mathbf{x} = \bar{\mathbf{x}}^T A\mathbf{x}.$$

Thus, $\mathbf{x}^T A\bar{\mathbf{x}} = \bar{\mathbf{x}}^T A\mathbf{x}$ and therefore $\mathbf{x}^T(\overline{\lambda\mathbf{x}}) = \bar{\mathbf{x}}^T(\lambda\mathbf{x})$ or

$$\bar{\lambda}\mathbf{x}^T\bar{\mathbf{x}} = \lambda\bar{\mathbf{x}}^T\mathbf{x}.$$

Since $\mathbf{x} \neq \theta$, we have from (3.16) that $\bar{\lambda} = \lambda$, which shows that λ is real if λ is an eigenvalue of A. ∎

This result is significant since it shows that the search for eigenvalues of a symmetric matrix can be restricted to the real axis (as opposed to the entire complex plane). The following two theorems yield bounds on the magnitude of the eigenvalues of an arbitrary $(n \times n)$ matrix.

Theorem 3.2

Let λ be an eigenvalue of A, then $|\lambda| \leq \|A\|_1$, $|\lambda| \leq \|A\|_E$, and $|\lambda| \leq \|A\|_\infty$.

Proof. The proof is trivial and we do it only for the first inequality, $|\lambda| \leq \|A\|_1$. Note that the proof is valid for any compatible pair of vector and matrix norms. Suppose $A\mathbf{x} = \lambda\mathbf{x}$ where $\mathbf{x} \neq \theta$. Then $|\lambda|\,\|\mathbf{x}\|_1 = \|\lambda\mathbf{x}\|_1 = \|A\mathbf{x}\|_1 \leq \|A\|_1\|\mathbf{x}\|_1$; and since $\|\mathbf{x}\|_1 > 0$, the theorem follows. ∎

The applications of this theorem are obvious. Given a matrix A, we can compute easily a number $\|A\|$, which gives us the radius of a circular disk in the complex plane (centered at the origin) in which all the eigenvalues of A must lie. Returning to Example 3.1, we find $\|A\|_1 = 5$, $\|A\|_\infty = 4$, and $\|A\|_E = \sqrt{22}$. Our best bound from the theorem above is then that all the eigenvalues are in a disk of radius 4, centered at the origin. For this particular matrix, it happens that one of these three norms provides the best possible disk, but such is not usually the case.

Theorem 3.3 Gerschgorin Circle Theorem

For $A = (a_{ij})$, define the absolute off-diagonal row and column sums by

$$r_k = \sum_{\substack{j=1 \\ j\neq k}}^{n} |a_{kj}| \quad \text{and} \quad c_k = \sum_{\substack{j=1 \\ j\neq k}}^{n} |a_{jk}|, \quad \text{respectively.}$$

For $k = 1, 2, \ldots, n$, the sets $R_k = \{z: |z - a_{kk}| \le r_k\}$ and $C_k = \{z: |z - a_{kk}| \le c_k\}$ are circular disks in the complex plane, centered at a_{kk}, of radius r_k and c_k, respectively. If λ is any eigenvalue of A, then $\lambda \in R_k$ for some k, $1 \le k \le n$; and $\lambda \in C_m$ for some m, $1 \le m \le n$.

Proof. Suppose $A\mathbf{x} = \lambda\mathbf{x}$ where $\mathbf{x} \neq \theta$, and

$$\mathbf{x} = (x_1, x_2, \ldots, x_n)^{\mathrm{T}}. \tag{3.17}$$

Choose i such that $|x_i| \ge |x_j|, j = 1, 2, \ldots, n$, $(\|\mathbf{x}\|_\infty = |x_i|)$. Then, the ith row of the equation $A\mathbf{x} = \lambda\mathbf{x}$ is

$$a_{i1}x_1 + a_{i2}x_2 + \cdots + a_{ii}x_i + \cdots + a_{in}x_n = \lambda x_i, \tag{3.18}$$

and therefore

$$(a_{ii} - \lambda)x_i = -\sum_{\substack{j=1 \\ j \neq i}}^{n} a_{ij}x_j. \tag{3.19}$$

Dividing by x_i, taking absolute values, and using the triangle inequality on the right, we obtain the fact that $|a_{ii} - \lambda| \le r_i$. To show that λ is also in a disk C_m for some m, we note that $\det(A) = \det(A^{\mathrm{T}})$ and thus A and A^{T} have the same eigenvalues (but not necessarily the same eigenvectors). It can further be shown that if exactly p of these disks intersect, then there are exactly p eigenvalues in the union of the p disks. ■

EXAMPLE 3.5. Consider the (3×3) matrix

$$A = \begin{bmatrix} 1 & 2 & -1 \\ 2 & 7 & 0 \\ -1 & 0 & -5 \end{bmatrix}$$

As A is symmetric, all the eigenvalues of A are real. By Gerschgorin's theorem, any eigenvalues must be in one of the three intervals $[-2, 4]$, $[5, 9]$, and $[-6, -4]$.

3.2.3. Deflation

A technique called deflation is sometimes of use in the eigenvalue problem. Suppose A is $(n \times n)$ and has eigenvalues $\lambda_1, \lambda_2, \ldots, \lambda_n$; and suppose we somehow find one of these, say λ_1. The basic idea of deflation is to construct a new $((n-1) \times (n-1))$ matrix B from A where B has eigenvalues $\lambda_2, \lambda_3, \ldots, \lambda_n$. Deflation could be used (for example) in conjunction with the power method to find the first few dominant eigenvalues or in conjunction with the inverse power method to find the first few eigenvalues near some value α.

There are a number of deflation procedures, and we present just one here. Suppose A is $(n \times n)$; and let us denote the first row of A by $\mathbf{a}_1$ so that $\mathbf{a}_1$ is the $(1 \times n)$ vector

$$\mathbf{a}_1 = (a_{11}, a_{12}, \ldots, a_{1n}).$$

Also suppose $A\mathbf{u}_j = \lambda_j\mathbf{u}_j$ for $j = 1, 2, \ldots, n$ where

1. the set $\{\mathbf{u}_j\}_{j=1}^{n}$ is linearly independent,
2. no $\mathbf{u}_j$ is the zero vector, and
3. the eigenvector $\mathbf{u}_1$ is

$$\mathbf{u}_1 = \begin{bmatrix} 1 \\ \alpha_2 \\ \alpha_3 \\ \vdots \\ \alpha_n \end{bmatrix}. \tag{3.20}$$

Recall that any scalar multiple of an eigenvector is again an eigenvector. Thus if the first component of an eigenvector $\mathbf{u}_1$ belonging to λ_1 is nonzero, we can always assume that $\mathbf{u}_1$ has been normalized to make the first component 1. The case in which the first entry is zero can be easily handled by premultiplication and postmultiplication of A by a *permutation matrix*. If we know λ_1 and $\mathbf{u}_1$ where $\mathbf{u}_1$ is normalized as in Eq. (3.20), we next form A_1 where $A_1 = A - \mathbf{u}_1\mathbf{a}_1$. [Note that $\mathbf{u}_1$ is $(n \times 1)$ and $\mathbf{a}_1$ is $(1 \times n)$; so $\mathbf{u}_1\mathbf{a}_1$ is an $(n \times n)$ matrix.] It follows immediately that the first row of A_1 consists entirely of zeros; so let A_1 be given by

$$A_1 = \begin{bmatrix} 0 & 0 & \cdots & 0 \\ b_{21} & b_{22} & & b_{2n} \\ \vdots & & & \\ b_{n1} & b_{n2} & \cdots & b_{nn} \end{bmatrix}. \tag{3.21}$$

We next claim that A_1 has eigenvalues $\lambda_2, \lambda_3, \ldots, \lambda_n, 0$. Clearly, 0 is an eigenvalue since A_1 is singular. To verify the other eigenvalues, we proceed by noting (as above) that any eigenvector $\mathbf{u}_k$ of A may be assumed to have either form (3.22a) or form (3.22b).

$$\mathbf{u}_k = \begin{bmatrix} 1 \\ \beta_2 \\ \beta_3 \\ \vdots \\ \beta_n \end{bmatrix} \tag{3.22a}$$

$$\mathbf{u}_k = \begin{bmatrix} 0 \\ \gamma_2 \\ \gamma_3 \\ \vdots \\ \gamma_n \end{bmatrix}. \tag{3.22b}$$

If $\mathbf{u}_k$ has form (3.22a), then $A_1(\mathbf{u}_1 - \mathbf{u}_k) = A(\mathbf{u}_1 - \mathbf{u}_k) - \mathbf{u}_1\mathbf{a}_1(\mathbf{u}_1 - \mathbf{u}_k)$. Now $A(\mathbf{u}_1 - \mathbf{u}_k) = \lambda_1\mathbf{u}_1 - \lambda_k\mathbf{u}_k$; and we note that while $\mathbf{u}_1\mathbf{a}_1$ is an $(n \times n)$ matrix, the product $\mathbf{a}_1(\mathbf{u}_1 - \mathbf{u}_k)$ is a scalar. By the associative property, we write the $(n \times 1)$ vector $\mathbf{u}_1\mathbf{a}_1(\mathbf{u}_1 - \mathbf{u}_k)$ either as $(\mathbf{u}_1\mathbf{a}_1)(\mathbf{u}_1 - \mathbf{u}_k)$ or as $\mathbf{u}_1(\mathbf{a}_1(\mathbf{u}_1 - \mathbf{u}_k))$. Taking the latter course, we note that $\mathbf{a}_1\mathbf{u}_j$ is the first component of the vector $A\mathbf{u}_j$, $1 \leq j \leq n$; and hence $\mathbf{a}_1(\mathbf{u}_1 - \mathbf{u}_k) = \lambda_1 - \lambda_k$ by (3.22a). If one puts all this together, $A_1(\mathbf{u}_1 - \mathbf{u}_k) = \lambda_1\mathbf{u}_1 - \lambda_k\mathbf{u}_k - (\lambda_1 - \lambda_k)\mathbf{u}_1 = \lambda_k(\mathbf{u}_1 - \mathbf{u}_k)$. Thus λ_k is an eigenvalue of A_1 with corresponding eigenvector $(\mathbf{u}_1 - \mathbf{u}_k)$, and the first component of $\mathbf{u}_1 - \mathbf{u}_k$ is zero by (3.22a).

In the case $\mathbf{u}_k$ has the form (3.22b), we see immediately that $A_1\mathbf{u}_k = \lambda_k\mathbf{u}_k$ since in this case $\mathbf{a}_1\mathbf{u}_k = 0$ and so $\mathbf{u}_1\mathbf{a}_1\mathbf{u}_k = \mathbf{u}_1(\mathbf{a}_1\mathbf{u}_k) = \theta$. The next thing to note is that in either case (3.22a) or (3.22b), the eigenvectors belonging to the eigenvalues $\lambda_2, \ldots, \lambda_n$ of A_1 all have a zero in the first component. Thus the first column of A_1 is irrelevant as far as the Eq. (3.1) for $\lambda_2, \ldots, \lambda_n$ is concerned. In particular, let us define B from (3.21) by

$$B = \begin{bmatrix} b_{22} & \cdots & b_{2n} \\ \vdots & & \vdots \\ b_{n2} & \cdots & b_{nn} \end{bmatrix} \tag{3.23}$$

so that B is the lower $(n - 1) \times (n - 1)$ block of A_1. Then B has eigenvalues $\lambda_2, \ldots, \lambda_n$.

EXAMPLE 3.6. We use the matrix of Example 3.1 to illustrate the process. One eigenvalue of A is $\lambda = 4$, with corresponding eigenvector of the form

$$\begin{bmatrix} \alpha \\ \alpha \\ \alpha \end{bmatrix}, \qquad \alpha \neq 0.$$

Thus we set

$$\mathbf{u}_1 = \begin{bmatrix} 1 \\ 1 \\ 1 \end{bmatrix},$$

and observe that $A\mathbf{u}_1 = 4\mathbf{u}_1$ where

$$A = \begin{bmatrix} 1 & 2 & 1 \\ 0 & 1 & 3 \\ 2 & 1 & 1 \end{bmatrix}.$$

Then

$$A_1 = A - \mathbf{u}_1\mathbf{a}_1 = \begin{bmatrix} 1 & 2 & 1 \\ 0 & 1 & 3 \\ 2 & 1 & 1 \end{bmatrix} - \begin{bmatrix} 1 \\ 1 \\ 1 \end{bmatrix}(1 \quad 2 \quad 1) = \begin{bmatrix} 0 & 0 & 0 \\ -1 & -1 & 2 \\ 1 & -1 & 0 \end{bmatrix},$$

and the deflated matrix B is

$$B = \begin{bmatrix} -1 & 2 \\ -1 & 0 \end{bmatrix}.$$

Note that the characteristic equation for B is $\lambda^2 + \lambda + 2 = 0$, and thus B has all the eigenvalues of A, except for $\lambda = 4$, which was "divided out."

The process of deflation is subject to substantial numerical difficulties in practice. In most cases, we do not know λ_1 precisely (perhaps we have a very good estimate of λ_1 from the power method, but we still do not know λ_1 exactly). Consequently, we do not have $\mathbf{u}_1$ precisely, and so the deflated matrix B will not have precisely the eigenvalues $\lambda_2, \ldots, \lambda_n$. If we in turn deflate B, we introduce additional errors. We cannot be sure that the matrix that results after several deflations has eigenvalues that are very near those of the original matrix A. Because of all the uncertainties, deflation must be used with caution. When deflation is used, any eigenvalues we find (perhaps by using the power method on the sequence of deflated matrices) should be regarded as only approximations to the eigenvalues of the original matrix A. As such, these approximations should be refined or corrected with the inverse power method, using the original matrix A, which has not been contaminated by the deflation process.

skip

*3.2.4. The Rayleigh Quotient Iteration

In Section 3.2.1, we saw that the inverse power iteration, defined by

$$(A - \alpha I)\mathbf{y}_{k+1} = \mathbf{y}_k, \qquad k = 0, 1, \ldots,$$

can be used to estimate the eigenvalue of A that is nearest α. We also observed that convergence is rapid when α is close to an eigenvalue of A. This observation suggests that we might use a few steps of the power method to provide an estimate α to the dominant eigenvalue, λ_1, of A and then use inverse power iteration to converge rapidly to λ_1.

When A is a symmetric matrix, this idea can be exploited even further. Recall (from the concluding remarks of Section 3.2) that given a vector $\mathbf{y}$, the "best" estimate of λ_1 is found from the Rayleigh quotient, $\mathbf{y}^T A\mathbf{y}/\mathbf{y}^T\mathbf{y}$. Thus we can choose an initial vector $\mathbf{y}_0$, form $\alpha_0 = \mathbf{y}_0^T A\mathbf{y}_0/\mathbf{y}_0^T\mathbf{y}_0$, and then find $\mathbf{y}_1$ from $(A - \alpha_0 I)\mathbf{y}_1 = \mathbf{y}_0$. We then form $\alpha_1 = \mathbf{y}_1^T A\mathbf{y}_1/\mathbf{y}_1^T\mathbf{y}_1$ and find $\mathbf{y}_2$ from $(A - \alpha_1 I)\mathbf{y}_2 = \mathbf{y}_1$, etc. This procedure is called the *Rayleigh quotient iteration* and is usually carried out as follows.

Choose an initial vector $\mathbf{y}_0$ such that $\mathbf{y}_0^T\mathbf{y}_0 = 1$. Then for $i = 0, 1, 2, \ldots$

1. form the scalar $\alpha_i = \mathbf{y}_i^T A\mathbf{y}_i$,
2. solve the equation $(A - \alpha_i I)\mathbf{w}_{i+1} = \mathbf{y}_i$,
3. set $\eta_{i+1} = \sqrt{\mathbf{w}_{i+1}^T\mathbf{w}_{i+1}}$, and
4. let $\mathbf{y}_{i+1} = \mathbf{w}_{i+1}/\eta_{i+1}$ and return to step (1).

While a rigorous analysis of the Rayleigh quotient iteration is beyond the scope of this text, we can make some remarks of an intuitive nature. In the normal course of events, the vectors $\mathbf{y}_i$ converge to $\mathbf{u}_1$ where $\mathbf{u}_1$ is an eigenvector corresponding to the dominant eigenvalue λ_1 of A. To see this, suppose that $\mathbf{y}_i$ is a good approximation to $\mathbf{u}_1$, and let

$$\mathbf{y}_i = \gamma_1\mathbf{u}_1 + \gamma_2\mathbf{u}_2 + \cdots + \gamma_n\mathbf{u}_n$$

where $\gamma_2, \gamma_3, \ldots, \gamma_n$ are small with respect to γ_1. If α_i is a good approximation to λ_1, then we see that

$$\begin{aligned}\mathbf{w}_{i+1} &= \gamma_1 \frac{1}{\lambda_1 - \alpha_i}\mathbf{u}_1 + \gamma_2 \frac{1}{\lambda_2 - \alpha_i}\mathbf{u}_2 + \cdots + \gamma_n \frac{1}{\lambda_n - \alpha_i}\mathbf{u}_n \qquad (3.24)\\ &= \mu_1\mathbf{u}_1 + \mu_2\mathbf{u}_2 + \cdots + \mu_n\mathbf{u}_n.\end{aligned}$$

Clearly, if $\gamma_2, \gamma_3, \ldots, \gamma_n$ are small with respect to γ_1, then $\mu_2, \mu_3, \ldots, \mu_n$ are yet smaller with respect to μ_1. Thus $\mathbf{w}_{i+1}$ (and hence $\mathbf{y}_{i+1}$) is a better approximation to $\mathbf{u}_1$. It can be shown that when $\{\mathbf{y}_i\} \to \mathbf{u}_1$, that convergence is "cubic" (that is, there exists a constant C such that $\|\mathbf{y}_{i+1} - \mathbf{u}_1\| \le C\|\mathbf{y}_i - \mathbf{u}_1\|^3$ where the vector norm is the ℓ_2 norm). The price we pay for this cubic convergence is that each iteration requires significantly more operations then the inverse power method requires since the coefficient matrix changes with each step.

PROBLEMS, SECTION 3.2.4

1. Deflate the matrix A of Example 3.3, and find a (2×2) matrix A_1 with eigenvalues 3 and -1.

*2. As we noted in the discussion of the Rayleigh quotient method in Section 3.2, an $(n \times n)$ symmetric matrix A has a set of n orthonormal eigenvectors $\{\mathbf{u}_1, \mathbf{u}_2, \ldots, \mathbf{u}_n\}$. This fact forms a basis for a deflation technique that applies to *symmetric* matrices. In particular, if $A\mathbf{u}_i = \lambda_i\mathbf{u}_i$ for $i = 1, 2, \ldots, n$, show that the matrix $A_1 = A - \lambda_1\mathbf{u}_1\mathbf{u}_1^T$ has eigenvalues $\lambda_2, \lambda_3, \ldots, \lambda_n, 0$.

3. Write a program that utilizes subroutine INVPOW. Use $\alpha = 4.5$ and test your program on the matrix A of Example 3.3. You should note a dramatic improvement in the rate of convergence, compared to that of Example 3.3. Estimate the rate of convergence as in (3.10), and check this estimate with your computed results.

4. If a matrix A is deflated and an estimate α_2 to λ_2 is found, then the inverse power method should be applied to the original matrix rather than to the deflated one. Why?

5. Modify subroutine INVPOW to incorporate the FACTOR and SOLVE routines of Section 2.2.4. Test your routine using the matrix in Problem 3. Also print out an estimated eigenvector.

6. Write a program using the subroutine in Problem 5 to find the smallest eigenvalue λ_3 of the matrix A in Problem 2, Section 3.2.

7. Write a program that uses subroutine POWERM, deflation, and the subroutines of Problem 5 to find and refine all four eigenvalues and eigenvectors of A.

$$A = \begin{bmatrix} 6 & 4 & 4 & 1 \\ 4 & 6 & 1 & 4 \\ 4 & 1 & 6 & 4 \\ 1 & 4 & 4 & 6 \end{bmatrix}$$

8. Program the Rayleigh quotient iteration of Section 3.2.4 to estimate the smallest eigenvalue λ_3 of the matrix A in Problem 2, Section 3.2, with the same initial vector as that in Problem 6. Contrast the number of iterations in each attempt.

9. Adapt either subroutine INVPOW or the subroutine of Problem 5 to perform complex arithmetic. (This adaptation means using a complex Gauss elimination routine as well.)

10. Write a calling program to test the subroutine in Problem 9 on the matrix of Example 3.1, Section 3.1; use $\alpha = (-1.2 + 2.6i)/2$.

11. Use the program of Problem 10 to find eigenvalues and eigenvectors of A; use $\alpha_1 = 11$, $\alpha_2 = 4.5 + 6i$, $\alpha_3 = \bar{\alpha}_2$, $\alpha_4 = 2.3$. ($\lambda = 12,\ 1 \pm 5i,\ 2$)

$$A = \begin{bmatrix} 4 & -5 & 0 & 3 \\ 0 & 4 & -3 & -5 \\ 5 & -3 & 4 & 0 \\ 3 & 0 & 5 & 4 \end{bmatrix}$$

12. Suppose $\alpha = 0$ is used in the inverse power method where A has eigenvalues $\lambda_1, \lambda_2, \ldots, \lambda_n$ and where $|\lambda_n| \le |\lambda_{n-1}| \le \cdots \le |\lambda_1|$. If A is a real matrix and if λ_n is complex, explain why the inverse power method will fail. Explain also why the method will fail to find λ_n for any choice of real α. If we know $|\lambda_n| \ge \varepsilon > 0$, and if we choose $\alpha = \rho e^{i2\pi\gamma}$ where $0 < \rho \le \varepsilon$, $0 < \gamma < 1$ and where ρ and γ are selected randomly, convince yourself that we have a good chance of finding λ_n or $\bar{\lambda}_n$. (In what circumstances will we still fail to find λ_n or $\bar{\lambda}_n$?)

13. An example of a permutation matrix P is given by the $(n \times n)$ matrix P where for a particular choice r and s

$$p_{rs} = p_{sr} = 1; \qquad p_{rk} = 0, \quad k \ne s; \qquad p_{sk} = 0, \quad k \ne r$$

and for $i \ne r, s$; $p_{ii} = 1$ and $p_{ij} = 0$ for $j \ne i$. That is, P looks like the $(n \times n)$ identity matrix except for the rth and sth rows. Show that

a) P is symmetric and $PP = I$;

b) the matrix AP is the same as A, except with the rth and sth columns interchanged;

c) the matrix PA is the same as A, except with the rth and sth rows interchanged.

14. Referring to the deflation technique of Section 3.2.3 as demonstrated in Example 3.6, suppose that $A\mathbf{u}_1 = \lambda_1\mathbf{u}_1$ and that the first component of $\mathbf{u}_1$ is zero. Since $\mathbf{u}_1$ is an eigenvector, $\mathbf{u}_1$ has at least one nonzero component, say the second component. Let P be a permutation matrix as in Problem 13 with $r = 1$ and $s = 2$. Let $\mathbf{x}_1 = P\mathbf{u}_1$ and show that

a) the first component of the vector $\mathbf{x}_1$ is nonzero,

b) the matrix PAP is similar to A,

c) $\mathbf{x}_1$ is an eigenvector of PAP corresponding to λ_1.

15. Recall from the discussion of the Rayleigh quotient method that an $(n \times n)$ symmetric matrix A has a set of n orthonormal eigenvectors $\mathbf{u}_1, \mathbf{u}_2, \ldots, \mathbf{u}_n$ (see Section 3.4 for a proof of this important fact). Also recall that each eigenvector $\mathbf{u}_i$ is real (that is, each $\mathbf{u}_i$ has real components), and any $(n \times 1)$ vector $\mathbf{x}$ can be written as $\mathbf{x} = \alpha_1\mathbf{u}_1 +$

$\alpha_2\mathbf{u}_2 + \cdots + \alpha_n\mathbf{u}_n$. From Theorem 3.1, every eigenvalue of A is real, but more can be said. Show that if $\mathbf{x} \neq \theta$, then

$$\lambda_n \leq \frac{\mathbf{x}^T A\mathbf{x}}{\mathbf{x}^T\mathbf{x}} \leq \lambda_1$$

where we have ordered the eigenvalues of A so that $\lambda_n \leq \lambda_{n-1} \leq \cdots \leq \lambda_1$.

16. Consider the matrix A given by

$$A = \begin{bmatrix} 5 & 4 & 1 & 1 \\ 4 & 5 & 1 & 1 \\ 1 & 1 & 4 & 2 \\ 1 & 1 & 2 & 4 \end{bmatrix}.$$

This matrix has eigenvalues $\lambda_4 = 1$, $\lambda_3 = 2$, $\lambda_2 = 5$ and $\lambda_1 = 10$. Use Theorem 3.3 to find an interval (a, b) that contains all the eigenvalues of A. Use Problem 15 to obtain an upper bound for λ_4 and a lower bound for λ_1; try various vectors $\mathbf{x}$.

17. Repeat the calculations of Problem 16 for these matrices:

$$A = \begin{bmatrix} 6 & 4 & 4 & 1 \\ 4 & 6 & 1 & 4 \\ 4 & 1 & 6 & 4 \\ 1 & 4 & 4 & 6 \end{bmatrix}, \qquad B = \begin{bmatrix} 2 & 1 & 3 & 4 \\ 1 & -3 & 1 & 5 \\ 3 & 1 & 6 & -2 \\ 4 & 5 & -2 & -1 \end{bmatrix}$$

$(\lambda = 15, 5, 5, -1)$ $\qquad$ $(\lambda = -8.0286, 7.9329, 5.6689, 1.5732)$

3.3 TRANSFORMATION METHODS

In this section we begin a study of some methods, the transformation methods, that are quite effective for the eigenvalue problem. In broad terms, these methods seek to transform a matrix A into another matrix B, which has the same eigenvalues as A, but in which the eigenvalue problem for B is (relatively) easy. To understand how these methods work for the algebraic eigenvalue problem requires some elementary theoretical tools. The first is the idea of a *similarity transformation*. Let A be $(n \times n)$ and let S be any nonsingular $(n \times n)$ matrix. We say that $B = S^{-1}AS$ is *similar* to A. We call the process of transforming A into B a *similarity transformation*. The most significant fact about similar matrices is that they have the same eigenvalues. To see this, note that if $B\mathbf{x} = \lambda\mathbf{x}$ where $\mathbf{x}$ is a nonzero vector, then $S^{-1}AS\mathbf{x} = \lambda\mathbf{x}$ or $A(S\mathbf{x}) = \lambda(S\mathbf{x})$. Since S is nonsingular and $\mathbf{x} \neq \theta$, then the vector $S\mathbf{x}$ is nonzero; and hence λ is an eigenvalue of A. Conversely, suppose $A\mathbf{y} = \lambda\mathbf{y}$ where $\mathbf{y}$ is a nonzero vector (so that $\mathbf{x} = S^{-1}\mathbf{y}$ is nonzero). But then $A(S\mathbf{x}) = \lambda(S\mathbf{x})$ or $S^{-1}AS\mathbf{x} = \lambda\mathbf{x}$, and we have that λ is an eigenvalue of B also.

Similarity transformations are very important in both theoretical and applied problems and provide the foundation for many numerical procedures for determining eigenvalues. One example of the theoretical value of similarity transformations is that they can be used to provide a very nice proof of the famous Cayley-Hamilton theorem stated below. [See Fröberg (1969).]

Theorem 3.4

Let the $(n \times n)$ matrix A have characteristic equation $\lambda^n + a_1\lambda^{n-1} + \cdots + a_{n-1}\lambda + a_n = 0$. Then the matrix defined by $A^n + a_1A^{n-1} + \cdots + a_{n-1}A + a_nI$ is the zero matrix.

This theorem is usually stated as "A matrix satisfies its own characteristic equation." The theorem is a good example of a theoretical result that can be used in a practical way. For example, the theorem often provides an efficient means of calculating the coefficients of the characteristic polynomial, $p(\lambda) = \det(A - \lambda I)$. Note that a direct cofactor expansion to find $p(\lambda)$ is impractical even for seemingly modest-sized determinants. For instance, it would require more than 10^{18} multiplications to perform a cofactor expansion of a (20×20) determinant. Furthermore, the presence of the variable λ in $\det(A - \lambda I)$ severely limits the amount of simplification by row or column operations that can be used to reduce the number of multiplications.

The Cayley-Hamilton theorem provides a more practical procedure. Let $\mathbf{y}_0$ be any $(n \times 1)$ vector. Then although we do not know the coefficients $a_1, \ldots, a_n$, we do know by Theorem 3.4 that $(A^n + a_1A^{n-1} + \cdots + a_{n-1}A + a_nI)\mathbf{y}_0 = \theta$, or

$$A^n\mathbf{y}_0 + a_1A^{n-1}\mathbf{y}_0 + \cdots + a_{n-1}A\mathbf{y}_0 + a_nI\mathbf{y}_0 = \theta. \tag{3.25}$$

Now if we write out the known vectors, $A^{n-i}\mathbf{y}_0$, in Eq. (3.25), we have a system of n linear equations in the n unknowns $a_1, a_2, \ldots, a_n$

$$a_1A^{n-1}\mathbf{y}_0 + \cdots + a_{n-1}A\mathbf{y}_0 + a_nI\mathbf{y}_0 = -A^n\mathbf{y}_0, \tag{3.26}$$

which we can solve to find the coefficients of $p(\lambda)$. Again we note the obvious: there is no need to form the powers $A^2, A^3, \ldots, A^{n-1}, A^n$ since all we want is $A^2\mathbf{y}_0 = A(A\mathbf{y}_0)$, $A^3\mathbf{y}_0 = A(A(A\mathbf{y}_0))$, etc. We can illustrate this method (called *Krylov's method*) by reference again to the matrix of Example 3.1.

EXAMPLE 3.7. Choose

$$\mathbf{y}_0 = \begin{bmatrix} 1 \\ 0 \\ 0 \end{bmatrix};$$

then form

$$\mathbf{y}_1 = A\mathbf{y}_0, \qquad \mathbf{y}_2 = A\mathbf{y}_1 = A^2\mathbf{y}_0, \qquad \text{and} \qquad \mathbf{y}_3 = A\mathbf{y}_2 = A^3\mathbf{y}_0.$$

Hence as

$$A = \begin{bmatrix} 1 & 2 & 1 \\ 0 & 1 & 3 \\ 2 & 1 & 1 \end{bmatrix},$$

we have

$$\mathbf{y}_0 = \begin{bmatrix} 1 \\ 0 \\ 0 \end{bmatrix}, \quad \mathbf{y}_1 = \begin{bmatrix} 1 \\ 0 \\ 2 \end{bmatrix}, \quad \mathbf{y}_2 = \begin{bmatrix} 3 \\ 6 \\ 4 \end{bmatrix}, \quad \mathbf{y}_3 = \begin{bmatrix} 19 \\ 18 \\ 16 \end{bmatrix}.$$

Using (3.26), we obtain the system

$$\begin{aligned} 3a_1 + a_2 + a_3 &= -19 \\ 6a_1 &= -18 \\ 4a_1 + 2a_2 &= -16. \end{aligned} \tag{3.27}$$

Thus $a_1 = -3$, $a_2 = -2$, and $a_3 = -8$; and hence $p(\lambda) = \lambda^3 - 3\lambda^2 - 2\lambda - 8$. [The difference in signs between $p(\lambda)$ and $\det(A - \lambda I)$ comes from the form in which the Cayley-Hamilton theorem is usually stated. This difference is clearly unimportant for us since the roots of $p(\lambda) = 0$ and $-p(\lambda) = 0$ are the same.]

Krylov's method is conceptually simple; moreover since it requires only the first n vectors of a power method iteration and then the solution of a linear system of equations, Krylov's method is relatively easy to use as compared with a cofactor expansion to find $p(\lambda)$. Krylov's method, however, has two serious drawbacks. For (3.26) to have a unique solution, the vectors $\{\mathbf{y}_0, A\mathbf{y}_0, \cdots, A^{n-1}\mathbf{y}_0\}$ must be linearly independent. The solution we desire [the coefficients of $p(\lambda)$] is always a solution of (3.26); but if the coefficient matrix of (3.26) is singular, there will be an infinite number of other solutions, and none of these other solutions yields the characteristic polynomial. This drawback can be countered as we do in Section 3.3.1, but the second drawback is more serious computationally. Even if we succeed in choosing $\mathbf{y}_0$ so that (3.26) has a unique solution, we must be cautious in using Krylov's method. The reason is that the coefficients $a_1, a_2, \ldots, a_n$ cannot usually be calculated without some error caused by rounding. Even if quite small, these rounding perturbations in the coefficients can cause large perturbations in the roots of the resulting polynomial. (See Problem 10 for a numerical example of this phenomenon.) For this latter reason, any method that calculates the coefficients of the characteristic polynomial is usually not recommended, except possibly to obtain first estimates of eigenvalues to use in another technique such as the inverse power method.

PROBLEMS, SECTION 3.3

1. Use Krylov's method, as in Example 3.7, to determine the characteristic equation $p(\lambda) = 0$ for the matrix of Example 3.3. Use $\mathbf{y}_0 = (1, 1, 1)^T$.
2. Repeat Problem 1 for the matrix H using $\mathbf{y}_0 = [1, 0, 0, 0]^T$.

$$H = \begin{bmatrix} 1 & 1 & 1 & 1 \\ 2 & 0 & 1 & 1 \\ 0 & -1 & -2 & -2 \\ 0 & 0 & 2 & 2 \end{bmatrix}$$

3. Suppose that (3.26) has an infinite number of solutions. Argue that only one of them is the coefficients of the characteristic equation. [Hint: If $\{\lambda_k\}_{k=1}^n$ are roots of the nth degree polynomial $p(\lambda)$, then $p(\lambda)$ is unique up to a constant multiple.]
4. a) Let the $(n \times n)$ matrix A have eigenvalues $\{\lambda_k\}_{k=1}^n$ and eigenvectors $\{\mathbf{u}_k\}_{k=1}^n$ where $A\mathbf{u}_k = \lambda_k\mathbf{u}_k$, $1 \le k \le n$. Suppose $\mathbf{y}_0$ in (3.26) is given by $\mathbf{y}_0 = c_2\mathbf{u}_2 + c_3\mathbf{u}_3 + \cdots +$

$c_n\mathbf{u}_n$. Show that the set of vectors $\{\mathbf{y}_0, A\mathbf{y}_0, \ldots, A^{n-1}\mathbf{y}_0\}$ is linearly dependent. [Hint: Show that selecting $\{\alpha_k\}_{k=0}^{n-1}$ in $\alpha_0\mathbf{y}_0 + \alpha_1 A\mathbf{y}_0 + \cdots + \alpha_{n-1}A^{n-1}\mathbf{y}_0 = \theta$ is equivalent to solving $(n - 1)$ homogeneous equations in n unknowns.]

b) Verify the result in part (a) by attempting to use Krylov's method on the (3×3) matrix A of Problem 1 with $\mathbf{y}_0 = \mathbf{u}_2 + \mathbf{u}_3$ where $\mathbf{u}_2$ and $\mathbf{u}_3$ are given in Example 3.3.

5. Let $A = \begin{bmatrix} 1 & 1 \\ 0 & 2 \end{bmatrix}$ and $\mathbf{y}_0 = \begin{bmatrix} \alpha \\ \beta \end{bmatrix}$. Find all values of α and β for which Krylov's method fails on A.

6. Let A be any $(n \times n)$ matrix, and let P be an $(n \times n)$ matrix derived from the identity matrix I by interchanging the rth and sth columns.

a) Describe the product AP. [Hint: If $P = [\mathbf{P}_1, \mathbf{P}_2, \ldots, \mathbf{P}_n]$, then $AP = [A\mathbf{P}_1, A\mathbf{P}_2, \ldots, A\mathbf{P}_n]$. What is $\mathbf{P}_k$?]

b) Use part (a) to show that $PP = I$ (or $P^{-1} = P$). Also argue that $P^{\mathrm{T}} = P$.

c) Describe the product PA. [Hint: From part (b), $(PA)^{\mathrm{T}} = A^{\mathrm{T}}P$. Use part (a) on $A^{\mathrm{T}}P$.]

7. Write a program that uses Krylov's method to find the eigenvalues of an $(n \times n)$ matrix A. Have the program solve (3.26), call a library polynomial root-finder, and finally use the computed roots as initial estimates for the inverse power method.

8. Determine the number of operations necessary to use Krylov's method to find the characteristic polynomial of an $(n \times n)$ matrix. For $n = 20$, contrast this number with 10^{18}, the approximate operations count for a determinant cofactor expansion.

9. Test the program in Problem 7 on the matrices in Problems 16 and 17, Section 3.2.4.

10. To illustrate the effect that small variations can have on eigenvalues, consider the example of $(n \times n)$ matrices A and $B = A + E$ (from Forsythe) where $A = (a_{ij})$ and $a_{ij} = 1$ when $j = i$ and $j = i - 1$, and $a_{ij} = 0$ otherwise, and where the $(1, n)$ entry of E is a constant ε and the other entries are zero.

a) Verify that $\det(A - \lambda I) = (1 - \lambda)^n$ and $\det(B - \lambda I) = (1 - \lambda)^n + (-1)^{n+1}\varepsilon$.

b) Let $n = 10$ and $\varepsilon = 2^{-10}$; and verify the statement that a change of 2^{-10} in one entry of A produces a 50 percent change in one eigenvalue of A.

3.3.1. Reduction to Hessenberg Form

Similarity transformations are often used as a basis for numerical eigenvalue techniques. Given a matrix A, we would like to find a matrix S so that the problem of finding the eigenvalues of $B = SAS^{-1}$ is somehow simplified. For example if we could find S so that B is triangular, then the eigenvalues of A are just the diagonal entries of B. Unfortunately we can argue that such a triangularization cannot be done in a simple fashion. To see why, let $p(t)$ be any given monic nth degree polynomial. We can easily construct a matrix A for which $p(t)$ is the characteristic polynomial. [This matrix is called the "companion matrix" of $p(t)$. See Problem 1.] If we could easily find S so that $B = SAS^{-1}$ is triangular, then we would have a simple method for finding the zeros of $p(t)$. However [see Maxfield and Maxfield (1971)], Abel showed that for $n \geq 5$ there are no simple formulas to express the zeros of $p(t)$ in terms of its coefficients.

Hence we abandon the goal of making B triangular, and instead we consider the problem of finding S so that $B = SAS^{-1}$ is "almost triangular." Thus we are led to the following definition. A matrix H is called a *Hessenberg matrix* if H has the form

$$H = \begin{bmatrix} h_{11} & h_{12} & h_{13} & \cdots & h_{1,n-1} & h_{1n} \\ h_{21} & h_{22} & h_{23} & \cdots & h_{2,n-1} & h_{2n} \\ 0 & h_{32} & h_{33} & \cdots & h_{3,n-1} & h_{3n} \\ 0 & 0 & h_{43} & \cdots & h_{4,n-1} & h_{4n} \\ \vdots & & & \ddots & & \vdots \\ 0 & 0 & 0 & \cdots & h_{n,n-1} & h_{nn} \end{bmatrix}. \tag{3.28}$$

Thus $h_{ij} = 0$ when $i > j + 1$, and H would be upper triangular except for the presence of the subdiagonal terms $h_{21}, h_{32}, \ldots, h_{n,n-1}$. For example, a (5×5) Hessenberg matrix has the form

$$H = \begin{bmatrix} \times & \times & \times & \times & \times \\ \times & \times & \times & \times & \times \\ 0 & \times & \times & \times & \times \\ 0 & 0 & \times & \times & \times \\ 0 & 0 & 0 & \times & \times \end{bmatrix}.$$

We do not insist that the subdiagonal entries be nonzero; so (for example) an upper-triangular matrix is in Hessenberg form. Any (2×2) matrix is trivially in Hessenberg form. We will discuss two ways to transform an arbitrary matrix A into Hessenberg form. The first method is simpler for hand calculations and easier to illustrate. Once these ideas are firmly in mind, we will present the second technique (Householder transformations), which has several computational advantages over the first. The object of either approach is to find a matrix H that is similar to A and such that H is in the form (3.28). We shall also discuss techniques for finding the eigenvalues of a Hessenberg matrix.

Our first approach to reducing A to Hessenberg form resembles Gauss elimination, in which we essentially use elementary row operations to create zeros below the subdiagonal. However, creating zeros is not enough; we must also take measures to keep the newly formed matrices similar to A. If $A = (a_{ij})$ and $a_{21} \neq 0$, then we can create a zero in the $(i, 1)$ position, $3 \le i \le n$, by replacing the ith row by $-a_{i1}/a_{21}$ times the second row plus the ith row. These operations result in the matrix S_1A (see Problem 2) where S_1 is given by

$$S_1 = \begin{bmatrix} 1 & 0 & 0 & 0 & \cdots & 0 \\ 0 & 1 & 0 & 0 & \cdots & 0 \\ 0 & -\dfrac{a_{31}}{a_{21}} & 1 & 0 & \cdots & 0 \\ 0 & -\dfrac{a_{41}}{a_{21}} & 0 & 1 & \cdots & 0 \\ \vdots & \vdots & \vdots & \vdots & \ddots & \vdots \\ 0 & -\dfrac{a_{n1}}{a_{21}} & 0 & 0 & \cdots & 1 \end{bmatrix}. \tag{3.29a}$$

To preserve similarity, we next form $A_1 = S_1AS_1^{-1}$ where we see from Problem 2 that S_1^{-1} is given by

$$S_1^{-1} = \begin{bmatrix} 1 & 0 & 0 & 0 & \cdots & 0 \\ 0 & 1 & 0 & 0 & \cdots & 0 \\ 0 & \dfrac{a_{31}}{a_{21}} & 1 & 0 & \cdots & 0 \\ 0 & \dfrac{a_{41}}{a_{21}} & 0 & 1 & \cdots & 0 \\ \vdots & \vdots & \vdots & \vdots & \ddots & \vdots \\ 0 & \dfrac{a_{n1}}{a_{21}} & 0 & 0 & \cdots & 1 \end{bmatrix}, \tag{3.29b}$$

or $S_1^{-1} = 2I - S_1$. Problem 2 also shows that the zeros below the subdiagonal in the first column of S_1A are not disturbed by forming $A_1 = S_1AS_1^{-1}$.

In the case that $a_{21} = 0$, we modify the step above as follows. If $a_{i1} = 0$ for $3 \le i \le n$, we set $A_1 = A$ since the first column is already in the desired form. If not, we select any row k, $3 \le k \le n$, where $a_{k1} \ne 0$. We interchange row k and row 2, and then interchange column k and column 2, which interchange has the effect of placing the nonzero element a_{k1} in the (2, 1) position. Problem 4 shows that this row and column change operation is equivalent to forming the matrix product $\tilde{A} = PAP^{-1}$ where P is a permutation matrix. Since $\tilde{A}$ is similar to A and has a nonzero entry in the (2, 1) position, we use the process shown in (3.29a) and (3.29b) to create zeros below the subdiagonal in the first column of $\tilde{A}$.

The first step of the Hessenberg reduction is now completed. The remaining steps proceed as above where we successively create zeros below the subdiagonal in columns 2, 3, . . . , $n - 2$. For example, after the first step we have a matrix A_1 of the form

$$A_1 = \begin{bmatrix} a'_{11} & a'_{12} & \cdots & a'_{1n} \\ a'_{21} & a'_{22} & \cdots & a'_{2n} \\ 0 & a'_{32} & \cdots & a'_{3n} \\ 0 & a'_{42} & \cdots & a'_{4n} \\ \vdots & & & \\ 0 & a'_{n2} & \cdots & a'_{nn} \end{bmatrix}.$$

If $a'_{32} = 0$, we search for $a'_{k2} \ne 0$, $4 \le k \le n$; we interchange row k and row 3, and then interchange column k and column 3. With a nonzero element in the (3, 2) position (say, $a'_{32} \ne 0$ for simplicity), we form $A_2 = S_2A_1S_2^{-1}$ where

$$S_2 = \begin{bmatrix} 1 & 0 & 0 & 0 & \cdots & 0 \\ 0 & 1 & 0 & 0 & \cdots & 0 \\ 0 & 0 & 1 & 0 & \cdots & 0 \\ 0 & 0 & -\dfrac{a'_{42}}{a'_{32}} & 1 & \cdots & 0 \\ \vdots & \vdots & \vdots & \vdots & \ddots & \vdots \\ 0 & 0 & -\dfrac{a'_{n2}}{a'_{32}} & 0 & \cdots & 1 \end{bmatrix}, \quad S_2^{-1} = \begin{bmatrix} 1 & 0 & 0 & 0 & \cdots & 0 \\ 0 & 1 & 0 & 0 & \cdots & 0 \\ 0 & 0 & 1 & 0 & \cdots & 0 \\ 0 & 0 & \dfrac{a'_{42}}{a'_{32}} & 1 & \cdots & 0 \\ \vdots & \vdots & \vdots & \vdots & \ddots & \vdots \\ 0 & 0 & \dfrac{a'_{n2}}{a'_{32}} & 0 & \cdots & 1 \end{bmatrix} \tag{3.30}$$

or $S_2 + S_2^{-1} = 2I$. We also note that A_2 retains the zeros below the subdiagonal in the first column as well as the zeros newly created in the second column. Furthermore A_2 is similar to A since

$$A_2 = S_2 A_1 S_2^{-1} = S_2 S_1 A S_1^{-1} S_2^{-1} = (S_2 S_1) A (S_2 S_1)^{-1}$$

with obvious modifications with permutation matrices if interchanges are required. (An easy way to remember the form of the matrices $S_1, S_2, \ldots, S_{n-2}$ is to note that each S_k is the result of applying the desired row operations on the identity matrix I.)

The process concludes after $n - 2$ steps yielding a Hessenberg matrix H that is similar to A. There are certain computational modifications that can be included in this reduction. For example we could reduce round-off error by selecting the largest element in absolute value in the column below the subdiagonal and interchange that element by permutations into the pivot position. Also the products $S_k A_{k-1} S_k^{-1}$ are merely symbolic; the matrix $A_k = S_k A_{k-1} S_k^{-1}$ can be constructed by performing row and column operations on A_{k-1}. However since this method will not be our principal method of reduction, we will not pursue these items further. An example of Hessenberg reduction follows.

EXAMPLE 3.8 Consider the (4×4) matrix A given by

$$A = \begin{bmatrix} 1 & 1 & 8 & -2 \\ 0 & 3 & 5 & -1 \\ 1 & -1 & -3 & 2 \\ 3 & -1 & -4 & 9 \end{bmatrix}.$$

Since $a_{21} = 0$, we first interchange rows two and three, then interchange columns two and three, and obtain

$$A \to \begin{bmatrix} 1 & 1 & 8 & -2 \\ 1 & -1 & -3 & 2 \\ 0 & 3 & 5 & -1 \\ 3 & -1 & -4 & 9 \end{bmatrix} \to \begin{bmatrix} 1 & 8 & 1 & -2 \\ 1 & -3 & -1 & 2 \\ 0 & 5 & 3 & -1 \\ 3 & -4 & -1 & 9 \end{bmatrix} = \tilde{A}.$$

Since $\tilde{A} = PAP^{-1}$, $P = [\mathbf{e}_1, \mathbf{e}_3, \mathbf{e}_2, \mathbf{e}_4]$, we know that A and $\tilde{A}$ are similar and have the same eigenvalues. Now with S_1 and S_1^{-1} of (3.29a) and (3.29b), respectively, we have

$$A_1 = S_1 \tilde{A} S_1^{-1} = \begin{bmatrix} 1 & 0 & 0 & 0 \\ 0 & 1 & 0 & 0 \\ 0 & 0 & 1 & 0 \\ 0 & -3 & 0 & 1 \end{bmatrix} \begin{bmatrix} 1 & 8 & 1 & -2 \\ 1 & -3 & -1 & 2 \\ 0 & 5 & 3 & -1 \\ 3 & -4 & -1 & 9 \end{bmatrix} \begin{bmatrix} 1 & 0 & 0 & 0 \\ 0 & 1 & 0 & 0 \\ 0 & 0 & 1 & 0 \\ 0 & 3 & 0 & 1 \end{bmatrix}$$

$$= \begin{bmatrix} 1 & 2 & 1 & -2 \\ 1 & 3 & -1 & 2 \\ 0 & 2 & 3 & -1 \\ 0 & 14 & 2 & 3 \end{bmatrix}.$$

Finally, using S_2 and S_2^{-1} from (3.30), we achieve Hessenberg form with

$$H = S_2 A_1 S_2^{-1} = \begin{bmatrix} 1 & 0 & 0 & 0 \\ 0 & 1 & 0 & 0 \\ 0 & 0 & 1 & 0 \\ 0 & 0 & -7 & 1 \end{bmatrix} \begin{bmatrix} 1 & 2 & 1 & -2 \\ 1 & 3 & -1 & 2 \\ 0 & 2 & 3 & -1 \\ 0 & 14 & 2 & 3 \end{bmatrix} \begin{bmatrix} 1 & 0 & 0 & 0 \\ 0 & 1 & 0 & 0 \\ 0 & 0 & 1 & 0 \\ 0 & 0 & 7 & 1 \end{bmatrix}$$

$$= \begin{bmatrix} 1 & 2 & -13 & -2 \\ 1 & 3 & 13 & 2 \\ 0 & 2 & -4 & -1 \\ 0 & 0 & 51 & 10 \end{bmatrix}.$$

An elementary method for finding the characteristic polynomial of a Hessenberg matrix is provided by Krylov's method and the following basic theory. Let H be a Hessenberg matrix as in (3.28); and assume that the subdiagonal elements $h_{21}, h_{32}, \ldots, h_{n,n-1}$ are all nonzero. Let $\mathbf{w}_0 = \mathbf{e}_1 = [1, 0, 0, \ldots, 0]^T$; and define the vectors $\{\mathbf{w}_k\}_{k=1}^{n}$ by the algorithm

$$\mathbf{w}_{k+1} = H\mathbf{w}_k, \qquad k = 0, 1, \ldots, n-1. \tag{3.31}$$

Problem 6 shows that the $(k+1)$st component of $\mathbf{w}_k$ is $h_{21}h_{32} \ldots, h_{k+1,k}$ for $1 \le k \le n-1$, and the rth components of $\mathbf{w}_k$ for $r > k+1$ are all zero. Hence the matrix $W = [\mathbf{w}_0, \mathbf{w}_1, \ldots, \mathbf{w}_{n-1}]$ is upper triangular with nonzero diagonal elements, and is therefore nonsingular; that is, $\{\mathbf{w}_0, \mathbf{w}_1, \ldots, \mathbf{w}_{n-1}\}$ is linearly independent. Thus the system $W\mathbf{x} = -\mathbf{w}_n$ has a unique solution that can be found immediately by backsolving (note that the system is triangular). The solution $\mathbf{x}$ yields the coefficients of the characteristic polynomial. To show this result, we note that for $\mathbf{x} = [a_n, a_{n-1}, \ldots, a_1]^T$, the system $W\mathbf{x} = -\mathbf{w}_n$ can be written as

$$a_n\mathbf{w}_0 + a_{n-1}\mathbf{w}_1 + \cdots + a_1\mathbf{w}_{n-1} = -\mathbf{w}_n. \tag{3.32}$$

From (3.31) $\mathbf{w}_{k+1} = H\mathbf{w}_k = H^2\mathbf{w}_{k-1} = \cdots = H^{k+1}\mathbf{w}_0 = H^{k+1}\mathbf{e}_1$, and so (3.32) is the same system as (3.26) where we identify A with H and $\mathbf{y}_0$ with $\mathbf{w}_0 = \mathbf{e}_1$. Since one solution of (3.32) [or (3.26)] is the coefficients of the characteristic polynomial of H, and since (3.32) has a unique solution, then we have that $p(\lambda) = \lambda^n + a_1\lambda^{n-1} + \cdots + a_{n-1}\lambda + a_n$ is the monic characteristic polynomial of H. [Actually, $\det(H - \lambda I) = (-1)^n p(\lambda)$, but the constant factor $(-1)^n$ is irrelevant for our purposes.]

EXAMPLE 3.9 Let $\mathbf{w}_0 = \mathbf{e}_1$ and let H be given by

$$H = \begin{bmatrix} 1 & 1 & 1 & 1 \\ 2 & 0 & 1 & 1 \\ 0 & -1 & -2 & -2 \\ 0 & 0 & 2 & 2 \end{bmatrix}.$$

From (3.31) we obtain

$$\mathbf{w}_1 = \begin{bmatrix} 1 \\ 2 \\ 0 \\ 0 \end{bmatrix}, \quad \mathbf{w}_2 = \begin{bmatrix} 3 \\ 2 \\ -2 \\ 0 \end{bmatrix}, \quad \mathbf{w}_3 = \begin{bmatrix} 3 \\ 4 \\ 2 \\ -4 \end{bmatrix}, \quad \mathbf{w}_4 = \begin{bmatrix} 5 \\ 4 \\ 0 \\ -4 \end{bmatrix}.$$

The system $a_4\mathbf{w}_0 + a_3\mathbf{w}_1 + a_2\mathbf{w}_2 + a_1\mathbf{w}_3 = -\mathbf{w}_4$ is

$$\begin{aligned} a_4 + a_3 + 3a_2 + 3a_1 &= -5 \\ 2a_3 + 2a_2 + 4a_1 &= -4 \\ -2a_2 + 2a_1 &= 0 \\ -4a_1 &= 4; \end{aligned}$$

and the solution is $a_1 = -1$, $a_2 = -1$, $a_3 = 1$, and $a_4 = 0$. Hence $p(\lambda) = \lambda^4 - \lambda^3 - \lambda^2 + \lambda = \lambda(\lambda - 1)^2(\lambda + 1)$, and the eigenvalues of H are -1, 0, and 1.

In the analysis above we assumed that the subdiagonal elements h_{21}, $h_{32}, \ldots, h_{n,n-1}$ were all nonzero. If this is not the case, the problem of finding the eigenvalues of H actually simplifies by uncoupling into smaller problems of the same form.

EXAMPLE 3.10 Consider the (5×5) Hessenberg matrix (where $h_{43} = 0$):

$$H = \begin{bmatrix} 2 & 2 & -1 & 3 & -2 \\ -1 & -1 & 1 & 2 & 4 \\ 0 & 2 & 1 & 1 & -1 \\ 0 & 0 & 0 & 2 & 1 \\ 0 & 0 & 0 & 6 & 1 \end{bmatrix}.$$

We next write H in partitioned form as

$$H = \left[\begin{array}{c|c} H_{11} & H_{12} \\ \hline H_{21} & H_{22} \end{array}\right], \quad H_{11} = \begin{bmatrix} 2 & 2 & -1 \\ -1 & -1 & 1 \\ 0 & 2 & 1 \end{bmatrix}, \quad H_{22} = \begin{bmatrix} 2 & 1 \\ 6 & 1 \end{bmatrix}.$$

The equation $(H - \lambda I)\mathbf{x} = \theta$ can also be expressed in partitioned (or block) form as

$$\begin{aligned} (H_{11} - \lambda I)\mathbf{u} + H_{12}\mathbf{v} &= \theta \\ (H_{22} - \lambda I)\mathbf{v} &= \theta \end{aligned}$$

where we have partitioned $\mathbf{x}$ as

$$\mathbf{x} = \left[\begin{array}{c} \mathbf{u} \\ \hline \mathbf{v} \end{array}\right]$$

with $\mathbf{u}$ being (3×1) and $\mathbf{v}$ being (2×1). We will show that if λ is an eigenvalue of H_{11} or of H_{22}, then λ is an eigenvalue of H. (In Problem 8, the reader is asked to show that every eigenvalue of H is an eigenvalue of either H_{11} or H_{22}. Thus the eigenvalue problem for H has been uncoupled into two equivalent, but simpler problems.) Suppose that λ is an eigenvalue of H_{11}. Then there is a nonzero vector $\mathbf{u}_1$ such that $(H_{11} - \lambda I)\mathbf{u}_1 = \theta$. Since $\mathbf{u} = \mathbf{u}_1$ and $\mathbf{v} = \theta$ satisfy the (2×2) block system above, it follows that $\mathbf{x} = [\mathbf{u}_1, \theta]^T$

satisfies $(H - \lambda I)\mathbf{x} = \theta$ and therefore that λ is an eigenvalue of H. Next, suppose that λ is an eigenvalue of H_{22} but not an eigenvalue of H_{11}. There is a nonzero vector $\mathbf{v}_1$ such that $(H_{22} - \lambda I)\mathbf{v}_1 = \theta$; and to satisfy the first equation in the (2×2) block system, we need to solve $(H_{11} - \lambda I)\mathbf{u} = -H_{12}\mathbf{v}_1$. Since $H_{11} - \lambda I$ is nonsingular, this equation has a (unique) solution, say $\mathbf{u} = \mathbf{z}_1$. Thus, $\mathbf{x} = [\mathbf{z}_1, \mathbf{v}_1]^{\mathrm{T}}$ satisfies $(H - \lambda I)\mathbf{x} = \theta$, and again λ is an eigenvalue of H. With Problem 8 and the result above given, finding the eigenvalues of H is equivalent to finding the eigenvalues of H_{11} and H_{22}; we leave it as a problem to show that the eigenvalues are $-1, -1, 1, 2, 4$, and to compute the eigenvectors.

We note that the analysis in the example above does not require the subdiagonal elements of either H_{11} or H_{22} to be nonzero. If either block contains zero subdiagonal elements, that block can again be partitioned as above with the same theoretical results. Obviously the analysis is valid for an arbitrary $(n \times n)$ Hessenberg matrix H and shows that H can be partitioned a sufficient number of times so that its diagonal Hessenberg blocks either are (1×1) or have nonzero subdiagonal terms. Furthermore, the eigenvalues of these blocks are the eigenvalues of H.

We now have the basis for a rather crude numerical technique for finding eigenvalues and eigenvectors. Since this technique involves finding the characteristic polynomial, the process may be subject to severe errors caused by ill-conditioning. This technique does however illustrate the basics of a transformation method and can be used (cautiously) for hand calculations or small matrices. This procedure is summarized by the following steps.

1. Reduce the given matrix A to Hessenberg form H as illustrated in Example 3.8.
2. Partition H into block form where each of the square Hessenberg diagonal blocks has nonzero subdiagonal entries.
3. For each diagonal block H_{jj}, use Krylov's method to calculate the characteristic polynomial for H_{jj}.
4. Use a polynomial root-finder to find eigenvalue estimates.
5. Use the values in step (4) as approximate eigenvalues in the inverse power method on A (not H), and obtain refined estimates of eigenvalues and of corresponding eigenvectors.

We conclude this section with another technique for estimating eigenvalues of a Hessenberg matrix. This method, called Hyman's method, is less susceptible to rounding error. Again we assume that the subdiagonal elements are nonzero. For an arbitrary given value of α, consider the system of equations $(H - \alpha I)\mathbf{x} = \theta$

$$
\begin{array}{lllll}
(h_{11} - \alpha)x_1 + & h_{12}x_2 & + \cdots + h_{1n}x_n & = 0 \\
h_{21}x_1 & + (h_{22} - \alpha)x_2 & + \cdots + h_{2n}x_n & = 0 \\
 & h_{32}x_2 & + \cdots + h_{3n}x_n & = 0 \\
 & \ddots & \vdots & \vdots \\
 & & h_{n,n-1}x_{n-1} + (h_{nn} - \alpha)x_n & = 0
\end{array}
$$

Hyman's method consists of setting $x_n = 1$ and then successively backsolving the nth, $(n - 1)$st, . . . , 2nd equations of this system for $x_{n-1}, x_{n-2}, \ldots, x_1$. (Since the subdiagonal elements are nonzero, this backsolving can be done.) With these values computed, the first equation of the system has the form

$$g(\alpha) \equiv (h_{11} - \alpha)x_1 + h_{12}x_2 + h_{13}x_3 + \cdots + h_{1n}x_n$$

where each x_k is computed as above from the value of α being used. Problem 13 shows that $g(\alpha) = 0$ if and only if α is an eigenvalue of H. [Half of this result is immediate since if $g(\alpha) = 0$, then we have generated a nontrivial solution of $(H - \alpha I)\mathbf{x} = \theta$. In fact, $g(t)$ is a constant multiple of the characteristic polynomial $p(t)$.] Thus we essentially have a way of evaluating $g(\alpha)$ for any value α without calculating or using the values of its coefficients. We use this capability in the search for values of α to make $g(\alpha) = 0$. This approach eliminates one source of rounding-error contamination when we apply a root-finder to $g(t)$ to estimate eigenvalues. As a root-finder, we could use the secant method (Section 4.3.4) with the information generated thus far. It is also possible to differentiate the individual equations in the system $(H - tI)\mathbf{x} = \theta$, $x_n = 1$, with respect to t and to obtain an algorithm to evaluate $g'(\alpha)$ for any α. We then could use Newton's method (Section 4.3.3) to find eigenvalue estimates (Problem 13).

Finally, even if we could exactly calculate the characteristic polynomial $p(t)$ for H, we could still have problems because of round-off. That is, in transforming A to H, we will make some round-off errors and thus H is not truly similar to A; the eigenvalues of H are not exactly the same as the eigenvalues of A. In fact, suppose H_T denotes the result of transforming A to Hessenberg form with no rounding error and H_C denotes the calculated (machine) result. Even if H_C and H_T are nearly the same matrices, their eigenvalues might differ substantially. An example (from Forsythe) is

$$H_1 = \begin{bmatrix} 1 & 0 & 0 & \cdots & 0 & 0 \\ 1 & 1 & 0 & \cdots & 0 & 0 \\ 0 & 1 & 1 & \cdots & 0 & 0 \\ \vdots & & & & & \\ 0 & 0 & 0 & \cdots & 1 & 0 \\ 0 & 0 & 0 & \cdots & 1 & 1 \end{bmatrix} \qquad H_2 = \begin{bmatrix} 1 & 0 & 0 & \cdots & 0 & \varepsilon \\ 1 & 1 & 0 & \cdots & 0 & 0 \\ 0 & 1 & 1 & \cdots & 0 & 0 \\ \vdots & & & & & \\ 0 & 0 & 0 & \cdots & 1 & 0 \\ 0 & 0 & 0 & \cdots & 1 & 1 \end{bmatrix}.$$

If H_1 and H_2 are $(n \times n)$, then their respective characteristic polynomials are

$$p(t) = (1 - t)^n \qquad \text{and} \qquad q(t) = (1 - t)^n + (-1)^{n+1}\varepsilon.$$

For instance if $n = 10$ and $\varepsilon = 10^{-10}$, then $\lambda = .9$ is an eigenvalue of H_2, but the only eigenvalue of H_1 is $\lambda = 1$; the change in eigenvalues from H_1 to H_2 is nine orders of magnitude larger than the perturbation ϵ. Thus when we reduce A to Hessenberg form H and find the eigenvalues of H, we should regard these only as estimates to the eigenvalues of A. If necessary, these estimates can be refined using the inverse power method.

PROBLEMS, SECTION 3.3.1

1. Given the monic polynomial, $p(t) = t^n + \alpha_1 t^{n-1} + \cdots + \alpha_{n-1}t + \alpha_n$, construct the "companion matrix" $C = (c_{ij})$ from $p(t)$ by letting $c_{nj} = \alpha_{n-j+1}$, $1 \le j \le n$; $c_{j,j+1} = 1$, $1 \le j \le n-1$; and $c_{ij} = 0$ otherwise. Use induction and a cofactor expansion of the last column to show that $\det(C - tI) = (-1)^n p(t)$.

2. Given an $(n \times n)$ matrix $A = (a_{ij})$ with $a_{21} \neq 0$ and S_1 as in (3.29a), verify that the product S_1A has zeros in the (3, 1), (4, 1), . . . , (n, 1) positions, and verify that the product $S_1AS_1^{-1}$ does not disturb these zeros. Also verify that S_1 is formed by performing the desired row operations on the $(n \times n)$ identity I, and that $S_1^{-1} = 2I - S_1$.

3. Transform the matrix A to a Hessenberg matrix H that is similar to A where A is given by

$$\text{a)}\ \begin{bmatrix} -7 & 4 & -3 \\ 8 & -3 & 3 \\ 32 & -15 & 13 \end{bmatrix} \quad \text{b)}\ \begin{bmatrix} -6 & 3 & -14 \\ -1 & 2 & -2 \\ 2 & 0 & 5 \end{bmatrix} \quad \text{c)}\ \begin{bmatrix} 1 & 3 & 1 \\ 0 & 2 & 4 \\ 1 & 1 & 3 \end{bmatrix}$$

$$\text{d)}\ \begin{bmatrix} 6 & 1 & 4 & 4 \\ 1 & 6 & 4 & 4 \\ 4 & 4 & 6 & 1 \\ 4 & 4 & 1 & 6 \end{bmatrix} \quad \text{e)}\ \begin{bmatrix} 1 & 2 & 3 & -2 \\ 0 & -1 & 2 & 4 \\ 1 & -1 & 3 & 2 \\ 3 & 2 & 1 & 4 \end{bmatrix} \quad \text{f)}\ \begin{bmatrix} 1 & 2 & -1 & 3 & 4 \\ -3 & 1 & -3 & 2 & 1 \\ 3 & 2 & 1 & -1 & 2 \\ 0 & 2 & -1 & 1 & 4 \\ 6 & 8 & 7 & -5 & 2 \end{bmatrix}.$$

4. From the $(n \times n)$ matrix $A = (a_{ij})$ where $a_{21} = 0$ and $a_{31} \neq 0$, form the matrix $\tilde{A}$ by interchanging rows two and three and then interchanging columns two and three. Show that $\tilde{A} = PAP^{-1}$ where P is the (2, 3) permutation matrix. What is the (2, 1) entry of $\tilde{A}$?

5. Given an $(n \times n)$ matrix $A = (a_{ij})$ with $a_{11} \neq 0$, let L_1 be the matrix of the first step of Gauss elimination described in Section 2.2.4. What are the entries in the first column of L_1A? Using the results of this section, describe L_1^{-1}. Are the zeros in the first column of L_1A maintained in $L_1AL_1^{-1}$?

6. Given the Hessenberg matrix H in (3.28) with $h_{21}, h_{32}, \ldots, h_{n,n-1}$ all nonzero, consider the vectors $\{\mathbf{w}_k\}_{k=0}^{n-1}$. Show in the manner described in the text that the matrix $W = [\mathbf{w}_0, \mathbf{w}_1, \ldots, \mathbf{w}_{n-1}]$ is upper triangular and nonsingular.

7. Use Krylov's method, that is, (3.31) and (3.32), $\mathbf{w}_0 = \mathbf{e}_1$, to find the characteristic polynomial of H where H is given by

$$\text{a)}\ \begin{bmatrix} 2 & 2 & -1 \\ -1 & -1 & 1 \\ 0 & 2 & 1 \end{bmatrix} \quad \text{b)}\ \begin{bmatrix} 3 & -8 & 4 \\ -1 & -2 & -2 \\ 0 & 6 & 1 \end{bmatrix} \quad \text{c)}\ \begin{bmatrix} 2 & 3 & -21 & -3 \\ 2 & 7 & -41 & -5 \\ 0 & 1 & -5 & -1 \\ 0 & 0 & 4 & 4 \end{bmatrix}.$$

8. If $\mathbf{y} = [u_1, u_2, u_3, v_1, v_2]^T$ as given in Example 3.10, prove that $H\mathbf{y} = \lambda\mathbf{y}$. Also show that any eigenvalue of H is an eigenvalue of either H_{11} or H_{22}.

9. Show that $-1, -1, 1, 2, 4$ are the eigenvalues of the matrix H in Example 3.10; and use the technique of this example to find the corresponding eigenvectors.

10. Determine the number of multiplications necessary to reduce an $(n \times n)$ matrix A to Hessenberg form and to find the characteristic polynomial.

11. Find the eigenvalues and eigenvectors of the following matrices by partitioning as in Example 3.10.

a) $\begin{bmatrix} 1 & -1 & 1 & 4 \\ 1 & 3 & 2 & -1 \\ 0 & 0 & 2 & -1 \\ 0 & 0 & -1 & 2 \end{bmatrix}$ b) $\begin{bmatrix} -2 & 0 & -2 & 1 \\ -1 & 1 & -2 & 3 \\ 0 & 1 & -1 & -2 \\ 0 & 0 & 0 & 2 \end{bmatrix}$ c) $\begin{bmatrix} 1 & 1 & 2 & 1 & 3 \\ 1 & 1 & 1 & 3 & -1 \\ 0 & 0 & 1 & 2 & 1 \\ 0 & 0 & 0 & 3 & 0 \\ 0 & 0 & 0 & 1 & 4 \end{bmatrix}$

12. a) Write subroutines to transform an $(n \times n)$ matrix A into a Hessenberg matrix H, to partition H into block form with no zero subdiagonal entries in each diagonal block, and to apply Krylov's method to each diagonal block.

b) Write a program that finds eigenvalues and eigenvectors of A and that uses the subroutines in part (a), a machine library polynomial root-finder, and an inverse power method subroutine. Test the program on the matrices of Problem 3.

13. a) With $g(\alpha)$ given as in Hyman's method, show that if α is an eigenvalue of H, then $g(\alpha) = 0$.

b) With the matrix H in Example 3.9, find $g(\alpha)$ for the values $\alpha = -3, -2, -1, 0, 1, 2, 3$.

c) Note that each x_k is a polynomial function of α; and so $\dfrac{dx_k}{d\alpha} \equiv x_k'(\alpha)$ is defined. Thus we have

$$g'(\alpha) = -x_1(\alpha) - \alpha x_1'(\alpha) + h_{12}x_2'(\alpha) + \cdots + h_{1n}x_n'(\alpha).$$

For any H, describe how one could write a subroutine to evaluate $g'(\alpha)$ for any given value of α. [Hint: What is $x_n'(\alpha)$? How can this value be used to obtain $x_{n-1}'(\alpha)$? $x_{n-2}'(\alpha)$? and so on?]

d) Newton's method for finding zeros of a function $f(t)$ is to let t_0 be an initial guess for a zero and to generate values $t_1, t_2, t_3, \ldots$ to converge to a root by the formula $t_{k+1} = t_k - f(t_k)/f'(t_k)$. Explain how this method could be used with part (c) to search for eigenvalues of H.

3.3.2. Householder Transformations

In this section we consider a different technique for transforming an $(n \times n)$ matrix A to a similar Hessenberg matrix H. This technique is based on Householder transformations, which are explicitly defined transformations designed to zero blocks of entries in a given column vector. Householder transformations can be used to reduce a matrix to Hessenberg form and can be used also for singular-value decompositions (which we do not discuss) and as a part of the sophisticated QR algorithm (which we discuss briefly). An important advantage in using Householder transformations to reduce A to Hessenberg form H is that H is no more ill-conditioned with respect to the eigenvalue problem than was A [see Young and Gregory (1973)]. A second advantage is that Householder transformations preserve symmetry; if A is symmetric, then so is H.

Technically, a Householder transformation is an orthogonal matrix Q of the form (3.33) below. The Householder transformation Q is usually defined so that $Q\mathbf{v}$ has some desired form $\mathbf{w}$ where $\mathbf{v}$ and $\mathbf{w} = Q\mathbf{v}$ are given column vectors. Since we are concerned with the eigenvalue problem, we will use these matrices in pairs as similarity transformations to reduce A to Hessenberg form. As before, our objective is to find similarity transformations that create zeros below the subdiagonal. Again we proceed one column at a time. At each step our transformation matrices will have the form

$$Q = I - \frac{2}{\|\mathbf{u}\|^2}\mathbf{u}\mathbf{u}^T. \tag{3.33}$$

In (3.33) the norm is the ℓ_2 norm, I is the $(n \times n)$ identity, and $\mathbf{u}$ is an $(n \times 1)$ vector selected to create appropriate zeros. [Note that $\mathbf{u}\mathbf{u}^T$ is an $(n \times n)$ matrix.] Such transformations are called *Householder transformations*. We first note that Q in (3.33) is an orthogonal matrix; that is, $Q^TQ = I$. This statement is verified by noting that $Q^T = Q$ and using $\|\mathbf{u}\| \equiv \|\mathbf{u}\|_2 = \sqrt{\mathbf{u}^T\mathbf{u}}$, which yields

$$Q^TQ = QQ = I - \frac{4}{\|\mathbf{u}\|^2}\mathbf{u}\mathbf{u}^T + \frac{4}{\|\mathbf{u}\|^4}(\mathbf{u}\mathbf{u}^T)(\mathbf{u}\mathbf{u}^T).$$

Since $(\mathbf{u}\mathbf{u}^T)(\mathbf{u}\mathbf{u}^T) = \mathbf{u}(\mathbf{u}^T\mathbf{u})\mathbf{u}^T = \mathbf{u}(\|\mathbf{u}\|^2)\mathbf{u}^T = \|\mathbf{u}\|^2(\mathbf{u}\mathbf{u}^T)$, we have that $Q^TQ = I$. Hence the matrix $B = QAQ$ is similar to A; moreover if $A = A^T$, then $B = B^T$. [Note that Q in (3.33) is completely determined by the choice of $\mathbf{u}$.]

Now we consider how to choose $\mathbf{u}$ (and hence also Q) in (3.33) to perform a step of the Hessenberg reduction. In general let $\mathbf{v} = [v_1, \ldots, v_{k-1}, v_k, v_{k+1}, \ldots, v_n]^T$ be any $(n \times 1)$ vector; and suppose we wish $\mathbf{w} = Q\mathbf{v}$ to have the following properties.

1. The first $(k - 1)$ components of $\mathbf{w}$ are $v_1, v_2, \ldots, v_{k-1}$.
2. The last $(n - k)$ components of $\mathbf{w}$ are all zero.

(3.34a)

We shall see that $\mathbf{w}$ is of this desired form if we select $\mathbf{u} = [u_1, u_2, \ldots, u_n]^T$ to satisfy these conditions.

1. $u_1 = u_2 = \cdots u_{k-1} = 0$.
2. $u_{k+1} = v_{k+1}, u_{k+2} = v_{k+2}, \ldots, u_n = v_n$.
3. $u_k = v_k - s$ where $s = \pm\sqrt{v_k^2 + v_{k+1}^2 + v_{k+2}^2 + \cdots + v_n^2}$.

(3.34b)

Using $\mathbf{u}$ and $\mathbf{v}$ as given above, we have

$$\|\mathbf{u}\|^2 = 2s^2 - 2sv_k \quad \text{and} \quad \mathbf{u}^T\mathbf{v} = s^2 - sv_k = \frac{\|\mathbf{u}\|^2}{2}.$$

Thus when ordinary rules of matrix multiplication are used,

$$Q\mathbf{v} = \left(I - \frac{2}{\|\mathbf{u}\|^2}\mathbf{u}\mathbf{u}^T\right)\mathbf{v} = \mathbf{v} - \left(\frac{2\mathbf{u}^T\mathbf{v}}{\|\mathbf{u}\|^2}\right)\mathbf{u} = \mathbf{v} - \mathbf{u} = \mathbf{w}.$$

[Note that $\mathbf{v} - \mathbf{u}$ has the desired properties in (3.34a).]

If the matrix A is given in column form, $A = [\mathbf{A}_1, \mathbf{A}_2, \ldots, \mathbf{A}_n]$, then $QA = [Q\mathbf{A}_1, Q\mathbf{A}_2, \ldots, Q\mathbf{A}_n]$. Clearly the first step in our reduction is to choose Q so that $Q\mathbf{A}_1$ has zeros in the jth components for $j = 3, 4, \ldots, n$. We identify $\mathbf{v}$ with $\mathbf{A}_1$ above, set $k = 2$, use (3.34b), and choose $\mathbf{u}$ so that

$$u_1 = 0, \qquad u_2 = a_{21} \pm \sqrt{a_{21}^2 + a_{31}^2 + \cdots + a_{n1}^2}, \qquad u_3 = a_{31}, \ldots, u_n = a_{n1}.$$

(We will clarify the $\pm$ sign momentarily.) Then $H_1 = QAQ$ is similar to A and, as we shall see, has zeros below the subdiagonal in the first column.

Instead of forming $H_1 = QAQ$ by direct matrix multiplication, we pause to consider the following simplifications. The kth column of QA is $Q\mathbf{A}_k$, $1 \le k \le n$. If $\mathbf{y}$ is an arbitrary $(n \times 1)$ vector, then

$$Q\mathbf{y} = \left(I - \frac{2}{\|\mathbf{u}\|^2}\mathbf{u}\mathbf{u}^{\mathrm{T}}\right)\mathbf{y} = \mathbf{y} - \left(\frac{2\mathbf{u}^{\mathrm{T}}\mathbf{y}}{\|\mathbf{u}\|^2}\right)\mathbf{u} \equiv \mathbf{y} - c\mathbf{u}. \tag{3.35a}$$

Thus the multiplication $Q\mathbf{y}$ can be accomplished merely by computing the constant c in (3.35a) and performing the vector subtraction, $\mathbf{y} - c\mathbf{u}$; the elements of Q need never be computed. To form QA, we calculate and store $\beta = 2/\|\mathbf{u}\|^2$ (this calculation requires one scalar product, $\mathbf{u}^{\mathrm{T}}\mathbf{u}$). Next, for $2 \le k \le n$, we calculate $\alpha_k = \mathbf{u}^{\mathrm{T}}\mathbf{A}_k$ and form $Q\mathbf{A}_k = \mathbf{A}_k - (\beta\alpha_k)\mathbf{u}$ in order to produce the matrix product $QA = [Q\mathbf{A}_1, Q\mathbf{A}_2, \ldots, Q\mathbf{A}_n]$. We do not form $Q\mathbf{A}_1$ since we already know what $Q\mathbf{A}_1$ is supposed to be; a total of n scalar products is all that is required to form QA.

A similar simplification is possible in forming the product BQ where $B = QA$. Let $C = B^{\mathrm{T}}$ and write $C = [\mathbf{C}_1, \mathbf{C}_2, \ldots, \mathbf{C}_n]$ in column form; then

$$(BQ)^{\mathrm{T}} = Q^{\mathrm{T}}B^{\mathrm{T}} = QC = [Q\mathbf{C}_1, Q\mathbf{C}_2, \ldots, Q\mathbf{C}_n].$$

As above, we generate the columns of QC by

$$Q\mathbf{C}_k = \mathbf{C}_k - \left(\frac{2\mathbf{u}^{\mathrm{T}}\mathbf{C}_k}{\|\mathbf{u}\|^2}\right)\mathbf{u} = \mathbf{C}_k - d_k\mathbf{u} \tag{3.35b}$$

where $\beta = 2/\|\mathbf{u}\|^2$, $\gamma_k = \mathbf{u}^{\mathrm{T}}\mathbf{C}_k$, and $d_k = \beta\gamma_k$ for $1 \le k \le n$. The columns of C and QC are the rows of B and BQ, respectively; so γ_k is also the dot product of $\mathbf{u}$ with the kth row of B for $1 \le k \le n$. Moreover, the first component of $\mathbf{u}$ is zero; and so looking at the first components of the vectors in (3.35b), we see that the first column of BQ is the same as the first column of B. Since $B = AQ$ has zeros below the subdiagonal in the first column, these zeros are preserved in $H_1 = QAQ$.

Finally, we consider the $\pm$ sign in the construction of Q. The analysis above is valid using either $+$ or $-$; but to avoid the possibility of overflow, we choose the sign of s to be opposite to that of v_k in (3.34b).

The discussion above shows how to complete the first step of the Householder reduction to Hessenberg form. With $\mathbf{u}$ chosen as in (3.34b), the matrix $H_1 = QAQ$ is similar to A and has zeros below the subdiagonal in the first column. The remaining steps successively create zeros below the subdiagonal in columns 2, 3, . . . , $n-2$. (Of course, if the desired zeros are already present in any step, we go on to the next step.) In general, suppose H_1 has the form

$$H_1 = \begin{bmatrix} a'_{11} & a'_{12} & a'_{13} & \cdots & a'_{1n} \\ a'_{21} & a'_{22} & a'_{23} & \cdots & a'_{2n} \\ 0 & a'_{32} & a'_{33} & \cdots & a'_{3n} \\ 0 & a'_{42} & a'_{43} & \cdots & a'_{4n} \\ \vdots & \vdots & & & \\ 0 & a'_{n2} & a'_{n3} & \cdots & a'_{nn} \end{bmatrix}.$$

In the second step we wish to create zeros in the (4, 2), (5, 2), . . . , (n, 2) positions. Referring to (3.34b), we take $k = 3$ and choose $\mathbf{u} = [u_1, u_2, \ldots, u_n]^T$ such that

$$u_1 = u_2 = 0,$$

$$u_3 = a'_{32} \pm \sqrt{a'^2_{32} + a'^2_{42} + \cdots + a'^2_{n2}},$$

$$u_4 = a'_{42}, \ldots, u_n = a'_{n2}.$$

Using this $\mathbf{u}$ to determine Q in (3.33) and then forming QH_1 and QH_1Q in the manner prescribed by (3.35a) and (3.35b) yield $H_2 = QH_1Q$ where H_2 is similar to A. Since $u_1 = u_2 = 0$, the dot product of $\mathbf{u}$ with the first column of H_1 equals zero, and hence the zeros below the subdiagonal in the first column are maintained in QH_1. Also since $u_1 = u_2 = 0$, from (3.35b) we see that the first two columns of QH_1 are the same as the first two columns of QH_1Q. Thus $H_2 = QH_1Q$ has zeros below the subdiagonal in the first two columns.

After $(n-2)$ steps we obtain a Hessenberg matrix H that is similar to A. Moreover if $A = A^T$, then $H = H^T$. To verify this result, consider the first step, $H_1 = QAQ$. Then $H_1^T = Q^TA^TQ^T = QAQ = H_1$. The result $H = H^T$ follows immediately. We note that if H is in Hessenberg form and $H = H^T$, then H has zero entries above the superdiagonal as well. A matrix of this form is said to be *tridiagonal*; and nonzero entries can occur only in $(i, i-1)$, (i, i), and $(i, i+1)$ positions in a tridiagonal matrix.

Subroutine SIMTRN listed in Fig. 3.4 uses Householder transformations to produce a Hessenberg matrix H that is similar to A. Subroutine SIMTRN proceeds in stages in the obvious fashion. That is, given A, SIMTRN calculates a Householder transformation Q_1 such that $H_1 = Q_1AQ_1$ has zeros in the (3, 1), (4, 1), . . . , (N, 1) positions. Given H_1, SIMTRN finds Q_2 so that $H_2 = Q_2H_1Q_2$ has zeros in the (4, 2), (5, 2), . . . , (N, 2) positions, etc. For clarity, SIMTRN calls Subroutine PRODCT to calculate $Q_iH_{i-1}Q_i$. In addition, Subroutine PRODCT calls Subroutine SCPROD to calculate the scalar products $\mathbf{u}^T\mathbf{v}$ required by (3.35a) and (3.35b). The required inputs are

A	An $(N \times N)$ matrix
N	The size of A
MAINDM	The declared dimension of the array containing A in the calling program.

```
      SUBROUTINE SIMTRN(A,H,N,MAINDM)
      DIMENSION A(MAINDM,MAINDM),H(MAINDM,MAINDM)
      DIMENSION U(50)
      NM2=N-2
C
C  INITIALIZE H--H WILL CONTAIN THE MACHINE ESTIMATE TO A HESSENBERG
C  MATRIX SIMILAR TO A WHEN CONTROL IS RETURNED TO THE CALLING PROGRAM.
C
        DO 2 I=1,N
          DO 1 J=1,N
          H(I,J)=A(I,J)
    1     CONTINUE
    2   CONTINUE
C
C  USE HOUSEHOLDER TRANSFORMATIONS TO ZERO ENTRIES BELOW THE
C  SUBDIAGONAL IN COLUMN J, J=1,2,...,N-2.
C
        DO 6 J=1,NM2
        JP1=J+1
        JP2=J+2
C
C  CALCULATE THE VECTOR U IN EQUATION (6.22), WHERE K=J+1.
C
        SSQ=0.
          DO 3 I=JP1,N
          SSQ=SSQ+H(I,J)**2
    3     CONTINUE
        IF(SSQ.EQ.0.)  GO TO 6
        S=SQRT(SSQ)
        IF(H(JP1,J).GE.0.)  S=-S
        B=1./(SSQ-H(JP1,J)*S)
          DO 4 L=1,J
          U(L)=0.
    4     CONTINUE
        U(JP1)=H(JP1,J)-S
        H(JP1,J)=S
          DO 5 L=JP2,N
          U(L)=H(L,J)
          H(L,J)=0.
    5     CONTINUE
C
C  CALCULATE QHQ, WHERE Q IS THE HOUSEHOLDER TRANSFORMATION
C  DEFINED BY THE VECTOR U.
C
        CALL PRODCT(H,U,B,JP1,N,MAINDM)
    6   CONTINUE
      RETURN
      END
```

Figure 3.4 Subroutine SIMTRN.

The output from Subroutine SIMTRN is

H A machine estimate to an $(N \times N)$ Hessenberg matrix that is similar to A.

The two subroutines called by SIMTRN and a simple program that uses SIMTRN to transform A to Hessenberg form are listed in Figs. 3.5 and 3.6.

EXAMPLE 3.11. The (4 × 4) symmetric matrix

$$A = \begin{bmatrix} 6 & 4 & 4 & 1 \\ 4 & 6 & 1 & 4 \\ 4 & 1 & 6 & 4 \\ 1 & 4 & 4 & 6 \end{bmatrix}$$

has eigenvalues $\lambda_1 = 15$, $\lambda_2 = \lambda_3 = 5$, and $\lambda_4 = -1$. Corresponding eigenvectors are

$$\mathbf{u}_1 = \begin{bmatrix} 1 \\ 1 \\ 1 \\ 1 \end{bmatrix}, \quad \mathbf{u}_2 = \begin{bmatrix} -1 \\ -1 \\ 1 \\ 1 \end{bmatrix}, \quad \mathbf{u}_4 = \begin{bmatrix} 1 \\ -1 \\ -1 \\ 1 \end{bmatrix}.$$

This matrix was read in by the program in Fig. 3.5 and produced this output:

$$H = \begin{bmatrix} 0.6000\text{E } 01 & -0.5745\text{E } 01 & -0.2271\text{E-}17 & 0.2225\text{E-}15 \\ -0.5745\text{E } 01 & 0.8909\text{E } 01 & -0.5143\text{E } 01 & -0.6661\text{E-}15 \\ 0.0 & -0.5143\text{E } 01 & 0.4091\text{E } 01 & 0.1998\text{E-}14 \\ 0.0 & 0.0 & 0.4441\text{E-}15 & 0.5000\text{E } 01 \end{bmatrix}.$$

Although the input matrix A is symmetric, the output H is not quite symmetric and reflects the effect of roundoff error. (We expected H to be tridiagonal, but h_{13}, h_{14}, and h_{24} are not quite zero. The entries below the subdiagonal were set to zero by SIMTRN.) With the small entries interpreted as being zero, the machine approximation (to four places) to a Hessenberg matrix that is similar to A is

$$H = \begin{bmatrix} 6.000 & -5.745 & 0 & 0 \\ -5.745 & 8.909 & -5.143 & 0 \\ 0 & -5.143 & 4.091 & 0 \\ 0 & 0 & 0 & 5.000 \end{bmatrix}.$$

Note that H is block diagonal and that $\lambda = 5.000$ is probably one of the eigenvalues of A. We will return to this example later.

After the Householder reduction has been performed to find a Hessenberg matrix H similar to A, either of the two techniques (Krylov's method or Hyman's method) of the previous section can be used to obtain eigenvalue estimates. *If* A is symmetric however, then H is tridiagonal and symmetric; and the following technique, attributable to Givens, provides an attractive alternative. Let H be given by

$$H = \begin{bmatrix} d_1 & b_1 & 0 & 0 & \cdots & 0 & 0 \\ b_1 & d_2 & b_2 & 0 & \cdots & 0 & 0 \\ 0 & b_2 & d_3 & b_3 & \cdots & 0 & 0 \\ 0 & 0 & b_3 & d_4 & \cdots & 0 & 0 \\ \vdots & & & & & & \vdots \\ 0 & 0 & 0 & 0 & \cdots & b_{n-1} & d_n \end{bmatrix}.$$

```
      DIMENSION A(20,20),H(20,20)
      MAINDM=20
    1 READ 100,N
      IF(N.LT.3)  STOP
        DO 2 I=1,N
        READ 101,(A(I,J),J=1,N)
    2   CONTINUE
      CALL SIMTRN(A,H,N,MAINDM)
        DO 3 I=1,N
        PRINT 102,(H(I,J),J=1,N)
    3   CONTINUE
      GO TO 1
  100 FORMAT(I2)
  101 FORMAT(10F4.0)
  102 FORMAT(1H0,8E12.4)
      END
```

Figure 3.5 Calling program.

```
      SUBROUTINE PRODCT(H,U,B,JP1,N,MAINDM)
      DIMENSION H(MAINDM,MAINDM)
      DIMENSION U(50),V(50)
C
C  CALCULATE QH.
C
        DO 3 K=JP1,N
          DO 1 I=1,N
          V(I)=H(I,K)
    1     CONTINUE
        CALL SCPROD(U,V,UDOTV,N)
          DO 2 I=JP1,N
          H(I,K)=H(I,K)-UDOTV*B*U(I)
    2     CONTINUE
    3   CONTINUE
C
C  CALCULATE (QH)Q.
C
        DO 6 K=1,N
          DO 4 I=1,N
          V(I)=H(K,I)
    4     CONTINUE
        CALL SCPROD(U,V,UDOTV,N)
          DO 5 I=JP1,N
          H(K,I)=H(K,I)-UDOTV*B*U(I)
    5     CONTINUE
    6   CONTINUE
      RETURN
      END


      SUBROUTINE SCPROD(U,V,UDOTV,N)
      DIMENSION U(50),V(50)
C
C  THIS SUBPROGRAM CALCULATES UDOTV, WHERE UDOTV IS THE
C  SCALAR PRODUCT OF TWO N-DIMENSIONAL VECTORS, U AND V.
C
      UDOTV=0.
        DO 1 I=1,N
        UDOTV=UDOTV+U(I)*V(I)
    1   CONTINUE
      RETURN
      END
```

Figure 3.6 Subroutines PRODCT and SCPROD.

Suppose we define the sequence of polynomials

$$
\begin{aligned}
p_0(t) &= 1 \\
p_1(t) &= d_1 - t \\
p_2(t) &= (d_2 - t)p_1(t) - b_1^2 p_0(t) \\
&\vdots \\
p_i(t) &= (d_i - t)p_{i-1}(t) - b_{i-1}^2 p_{i-2}(t) \\
&\vdots \\
p_n(t) &= (d_n - t)p_{n-1}(t) - b_{n-1}^2 p_{n-2}(t).
\end{aligned}
\tag{3.36}
$$

It is an easy exercise to show (Problem 6) that $p_n(t)$ is the characteristic polynomial for H.

If the subdiagonal entries $b_1, b_2, \ldots, b_{n-1}$ are all nonzero, then the algorithm of Givens can be used to isolate the roots of $p_n(t) = 0$. The algorithm proceeds as follows.

1. Let c be some real number.
2. Calculate the numbers $p_0(c), p_1(c), \ldots, p_n(c)$
3. Let $N(c)$ be the number of agreements in sign of adjacent terms in the sequence $p_0(c), p_1(c), \ldots, p_n(c)$.
4. $N(c)$ is equal to the number of roots of $p_n(t) = 0$ that are in the interval $[c, \infty)$.

In the event that $p_k(c) = 0$ for some k, we take the sign of $p_k(c)$ to be that of $p_{k-1}(c)$. [Note that two successive terms in (3.36) cannot both vanish at $t = c$ unless $p_i(c) = 0$ for all i, $0 \le i \le n$. Thus if $p_k(c) = 0$, then $p_{k-1}(c) \ne 0$ and $p_{k-1}(c)$ has a well-defined sign.]

It is not too hard to show that assertion (4) in the algorithm is true, but it is inappropriate to include a demonstration here. As an example, if $n = 4$ and if $\{p_0(c), p_1(c), p_2(c), p_3(c), p_4(c)\}$ has the sign pattern

$$\{+, +, +, -, -\}$$

then $N(c) = 3$; there are 3 roots of $p_4(t) = 0$ in the interval $[c, \infty)$. Similarly, the pattern $\{+, -, +, -, -\}$ leads to $N(c) = 1$; the pattern $\{+, -, 0, +, +\}$ gives $N(c) = 2$.

To use the Givens algorithm for computational purposes, we would first determine an interval $[a, b]$ that contains all the roots of $p_n(t) = 0$; that is, $[a, b]$ contains all the eigenvalues of H. (Since H is symmetric, all the eigenvalues of H are real.) Next, let c be the midpoint of $[a, b]$. If $N(c) > N(b)$, then there is at least one eigenvalue in $[c, b]$. Let d be the midpoint of $[c, b]$. If $N(d) > N(b)$, there is at least one eigenvalue in $[d, b]$; on the other hand, if $N(d) = N(b)$, then any eigenvalue in $[c, b]$ must be in $[c, d]$. In this fashion by repeatedly halving and testing subintervals, we can determine a small subinterval $[r, s]$ that contains $N(r) - N(s)$ eigenvalues of H. The remaining eigenvalues of H must be in $[a, r]$, and the "bisection" procedure described above can be applied to the interval $[a, r]$.

This process can be terminated when we have determined k small subintervals, $I_1, I_2, \ldots, I_k$, whose union contains all the eigenvalues of H. The midpoint of an interval, I_j, will serve as an estimate, α_j, to an eigenvalue of H. This estimate, α_j, can then be passed to the inverse power method and corrected to be a good estimate of an eigenvalue of A.

Several points must be made about the practical implementation of the Givens algorithm. First of all, we need not demand that the intervals I_j be terribly small since the midpoint of I_j will be used only as an initial guess to an eigenvalue of A. Second, we note that the coefficients of the polynomials $p_i(t)$ in (3.36) are not actually calculated since all we need is the value of $p_i(c)$ for various numbers c, $i = 0, 1, \ldots, n$. To calculate $\{p_0(c), p_1(c), \ldots, p_n(c)\}$, we merely compute the numbers

$$\begin{aligned} P_0 &= 1 \\ P_1 &= d_1 - c \\ &\vdots \\ P_i &= (d_i - c)P_{i-1} - b_{i-1}^2 P_{i-2} \end{aligned}$$

for $i = 2, 3, \ldots, n$ and note that the number P_k is, in fact, $p_k(c)$. Finally, to start the algorithm, we need an initial interval $[a, b]$ that we know contains all the eigenvalues of H. Now if $C = (c_{ij})$ is an $(n \times n)$ real matrix, then every eigenvalue λ of C satisfies the inequality

$$|\lambda| \leq \|C\|_E$$

where

$$\|C\|_E = \left(\sum_{i=1}^{n} \sum_{j=1}^{n} c_{ij}^2 \right)^{\frac{1}{2}}.$$

Thus for the symmetric Hessenberg matrix H, we know that every eigenvalue of H is in the interval $[-\|H\|_E, \|H\|_E]$ where

$$\|H\|_E = \left(\sum_{i=1}^{n} d_i^2 + 2 \sum_{i=1}^{n-1} b_i^2 \right)^{\frac{1}{2}}.$$

Subroutine GIVENS listed in Fig. 3.7 applies the Givens algorithm to a symmetric Hessenberg matrix H. These are required inputs.

D	An $(N \times 1)$ array containing the diagonal entries of H
B	An $((N - 1) \times 1)$ array containing the subdiagonal entries of H. These entries are all presumed to be nonzero.
RADIUS	A scalar defining the desired radius of a subinterval containing an eigenvalue of H
N	The size of H
MAINDM	The declared dimension of the array containing H in the calling program

```
      SUBROUTINE GIVENS(D,B,ESTEIG,NEIG,RADIUS,N,MAINDM,LSTEIG)
C
C  INITIALIZATION PHASE--K IS THE INDEX OF A SUBINTERVAL OF LENGTH
C  2*RADIUS WHICH CONTAINS NEIG(K) EIGENVALUES OF H.  ALL EIGENVALUES OF
C  H ARE BETWEEN A1=-HNORM AND A2=HNORM.  NTOGO IS THE NUMBER OF
C  EIGENVALUES YET TO BE FOUND AT STAGE K.  SGNCTR IS A SUBROUTINE THAT
C  CALCULATES THE NUMBER OF SIGN AGREEMENTS, N(C).
C
      DIMENSION D(MAINDM),B(MAINDM),ESTEIG(MAINDM),NEIG(MAINDM)
      NM1=N-1
      K=1
      NTOGO=N
      HNRMSQ=0.
        DO 1 I=1,NM1
        HNRMSQ=HNRMSQ+2.*B(I)**2+D(I)**2
    1   CONTINUE
      HNRMSQ=HNRMSQ+D(N)**2
      HNORM=SQRT(HNRMSQ)
      A1=-HNORM
      A2=HNORM
      CALL SGNCTR(D,B,A2,N,NOFA2,MAINDM)
C
C  BASIC LOOP--BEGINS WITH AN INTERVAL _A1,A2' AND DETERMINES A
C  SUBINTERVAL OF LENGTH 2*RADIUS CONTAINING NEIG(K) EIGENVALUES OF
C  H, FOR K=1,2,...,LSTEIG.  NBISCT IS THE NUMBER OF BISECTIONS
C  NECESSARY TO PRODUCE THE K-TH SUBINTERVAL.
C
    2 NBISCT=ALOG(2.*(A2-A1)/RADIUS)/ALOG(2.)
      ICTR=0
    3 EVALPT=(A1+A2)/2.
      CALL SGNCTR(D,B,EVALPT,N,NOFEPT,MAINDM)
      IF(NOFEPT.EQ.NOFA2)  GO TO 4
      A1=EVALPT
      GO TO 5
    4 A2=EVALPT
    5 ICTR=ICTR+1
      IF(ICTR.LT.NBISCT)  GO TO 3
C
C  SET K-TH ESTIMATE OF AN EIGENVALUE EQUAL TO MID-POINT OF
C  K-TH SUBINTERVAL, CALCULATE NEIG(K) AND RETURN IF ALL THE
C  EIGENVALUES OF H HAVE BEEN ISOLATED.
C
      ESTEIG(K)=(A1+A2)/2.
      CALL SGNCTR(D,B,A1,N,NOFA1,MAINDM)
      CALL SGNCTR(D,B,A2,N,NOFA2,MAINDM)
      NEIG(K)=NOFA1-NOFA2
      LSTEIG=K
      NTOGO=NTOGO-NEIG(K)
      IF(NTOGO.LE.0)  RETURN
      K=K+1
      A2=A1
      NOFA2=NOFA1
      A1=-HNORM
      GO TO 2
      END
```

Figure 3.7a Subroutine GIVENS.

```
      SUBROUTINE SGNCTR(D,B,EVALPT,N,NOFEPT,MAINDM)
      DIMENSION D(MAINDM),B(MAINDM)
C
C  THIS SUBPROGRAM CALCULATES THE NUMBER OF SIGN AGREEMENTS, N(C),
C  AT C=EVALPT.  THE NUMBER N(EVALPT) IS RETURNED AS NOFEPT.
C
      NOFEPT=0
      PIM2=1.
      PIM1=D(1)-EVALPT
      IF(PIM2*PIM1.GE.0.)  NOFEPT=1
        DO 1 I=2,N
        PI=(D(I)-EVALPT)*PIM1-B(I-1)**2*PIM2
        IF(PI*PIM1.GT.0.)  NOFEPT=NOFEPT+1
        IF(PI.EQ.0.)  NOFEPT=NOFEPT+1
        PIM2=PIM1
        PIM1=PI
    1   CONTINUE
      RETURN
      END
```

Figure 3.7b Subroutine SGNCTR.

These are outputs from Subroutine GIVENS.

ESTEIG	An array containing machine estimates to the eigenvalues of H. If ESTEIG(K) $= \alpha_k$, then there is at least one eigenvalue of H in the interval $[\alpha_k - r, \alpha_k + r]$ where $r =$ RADIUS.
NEIG	An array of integers. If NEIG(K) $= j$, there are j eigenvalues of H in $[\alpha_k - r, \alpha_k + r]$.
LSTEIG	An integer set equal to the number of subintervals found

In summary,

$$\text{NEIG}(1) + \text{NEIG}(2) + \cdots + \text{NEIG}(\text{LSTEIG}) = N$$

and NEIG(K) eigenvalues of H are between ESTEIG(K) − RADIUS and ESTEIG(K) + RADIUS. Subroutine GIVENS calls the subprogram Subroutine SGNCTR, which evaluates the quantity $N(c)$ required by the Givens algorithm; this subprogram also is listed in Fig. 3.7.

Fig. 3.8 lists a simple program that employs Subroutine GIVENS. The program reads in a *symmetric* matrix A, uses SIMTRN to reduce A to a symmetric Hessenberg matrix, and then calls GIVENS to isolate the eigenvalues of H. When the program in Fig. 3.8 was given as input the (4×4) matrix A from Example 3.11, the program found that

Number of eigenvalues within .1 of 15.02 is 1,
Number of eigenvalues within .1 of 5.022 is 2,
Number of eigenvalues within .1 of −.9585 is 1.

As we noted in Example 3.11, the eigenvalues of A are $\lambda_1 = 15$, $\lambda_2 = \lambda_3 = 5$, and $\lambda_4 = 1$. The combination of Householder transformations and the Givens algorithm has worked well on this example.

```
      DIMENSION A(20,20),H(20,20),D(20),B(20),ESTEIG(20),NEIG(20)
      MAINDM=20
    1 READ 100,N
      IF(N.LT.3)  STOP
        DO 2 I=1,N
        READ 101,(A(I,J),J=1,N)
    2   CONTINUE
      CALL SIMTRN(A,H,N,MAINDM)
        DO 3 I=1,N
        PRINT 102,(H(I,J),J=1,N)
    3   CONTINUE
      NM1=N-1
        DO 4 I=1,NM1
        D(I)=H(I,I)
        B(I)=H(I+1,I)
        IF(B(I).EQ.0.)  GO TO 1
    4   CONTINUE
      D(N)=H(N,N)
      RADIUS=.1
      CALL GIVENS(D,B,ESTEIG,NEIG,RADIUS,N,MAINDM,LSTEIG)
        DO 5 I=1,LSTEIG
        PRINT 103,RADIUS,ESTEIG(I),NEIG(I)
    5   CONTINUE
      GO TO 1
  100 FORMAT(I2)
  101 FORMAT(10F4.0)
  102 FORMAT(1H0,8E12.4)
  103 FORMAT(1H0,28HNUMBER OF EIGENVALUES WITHIN,F8.4,2X,2HOF,
     1E16.4,4H  IS,I3)
      END
```

Figure 3.8 A program to isolate the eigenvalues of a symmetric matrix A.

The inverse power method can be incorporated into the program in Fig. 3.8 to produce a complete package for finding the eigenvalues and the eigenvectors of a symmetric matrix. Such a program is listed in Fig. 3.9.

EXAMPLE 3.12. As a simple test of the program in Fig. 3.9, we input the matrix

$$A = \begin{bmatrix} 2 & 4 & -6 \\ 4 & 2 & -6 \\ -6 & -6 & -15 \end{bmatrix},$$

which has eigenvalues $\lambda_1 = -18$, $\lambda_2 = 9$, $\lambda_3 = -2$, and corresponding eigenvectors

$$\mathbf{u}_1 = \begin{bmatrix} 1 \\ 1 \\ 4 \end{bmatrix}, \quad \mathbf{u}_2 = \begin{bmatrix} 2 \\ 2 \\ -1 \end{bmatrix}, \quad \mathbf{u}_3 = \begin{bmatrix} 1 \\ -1 \\ 0 \end{bmatrix}.$$

The program returned eigenvalues correct to the five places printed and eigenvectors (of length 1):

$$\mathbf{v}_1 = \begin{bmatrix} -.23570 \\ -.23570 \\ -.94281 \end{bmatrix}, \quad \mathbf{v}_2 = \begin{bmatrix} .66667 \\ .66667 \\ -.33333 \end{bmatrix}, \quad \mathbf{v}_3 = \begin{bmatrix} .70711 \\ -.70711 \\ -.14343\text{E} - 10 \end{bmatrix}.$$

```
      DIMENSION A(20,20),H(20,20),D(20),B(20),ESTEIG(20),NEIG(20)
      DIMENSION EIGNVC(20)
      MAINDM=20
    1 READ 100,N
      IF(N.LT.3) STOP
        DO 2 I=1,N
        READ 101,(A(I,J),J=1,N)
    2   CONTINUE
      CALL SIMTRN(A,H,N,MAINDM)
        DO 3 I=1,N
        PRINT 102,(H(I,J),J=1,N)
    3   CONTINUE
      NM1=N-1
        DO 4 I=1,NM1
        D(I)=H(I,I)
        B(I)=H(I+1,I)
        IF(B(I).EQ.0.)  GO TO 1
    4   CONTINUE
      D(N)=H(N,N)
      RADIUS=.1
      CALL GIVENS(D,B,ESTEIG,NEIG,RADIUS,N,MAINDM,LSTEIG)
        DO 5 I=1,LSTEIG
        PRINT 103,RADIUS,ESTEIG(I),NEIG(I)
    5   CONTINUE
      TOL=.1E-06
      MAXITR=10
        DO 6 I=1,LSTEIG
        ALPHA=ESTEIG(I)
        CALL INVPOW(A,EIGNVC,ALPHA,EIGNVL,N,TOL,MAXITR,MAINDM,IFLAG)
        PRINT 104,IFLAG,EIGNVL
        PRINT 105,(EIGNVC(K),K=1,N)
    6   CONTINUE
      GO TO 1
  100 FORMAT(I2)
  101 FORMAT(10F4.0)
  102 FORMAT(1H0,8E12.4)
  103 FORMAT(1H0,28HNUMBER OF EIGENVALUES WITHIN,F8.4,2X,2HOF,
     1E16.4,4H  IS,I3)
  104 FORMAT(1H0,6H FLAG=,I3,5X,36HREFINED ESTIMATE OF AN EIGENVALUE IS,
     1E15.5)
  105 FORMAT(1H0,30HA CORRESPONDING EIGENVECTOR IS,/,6E18.5)
      END
```

Figure 3.9 A program to find the eigenvalues and the eigenvectors of a symmetric matrix A.

EXAMPLE 3.13. It is possible (especially when running a contrived problem rather than a "real-world" problem) that an input symmetric matrix might have an eigenvector each of whose entries is 1. This sort of matrix might fool the inverse power method, and Subroutine INVPOW may have to be modified to accommodate this situation. For example, when we input the matrix A from Example 3.11 to the program in Fig. 3.9, the program found, as noted above, estimates $\lambda_1 = 15.02$, $\lambda_2 = \lambda_3 = 5.022$, and $\lambda_4 = -.9585$. When these estimates are passed to the inverse power method, convergence is to $\lambda = 15$ for each of these estimates because the eigenvector $\mathbf{u}_1$ corresponding to $\lambda_1 = 15$ is

$$\mathbf{u}_1 = \begin{bmatrix} 1 \\ 1 \\ 1 \\ 1 \end{bmatrix}.$$

To get the inverse power method to work properly, we changed the definition of the starting vector (OLDVCT in Subroutine INVPOW) to

$$\begin{bmatrix} 0 \\ 1/\sqrt{3} \\ 1/\sqrt{3} \\ 1/\sqrt{3} \end{bmatrix}.$$

With this modification, all three eigenvectors and eigenvalues were found correctly. (Note that the output from the Givens algorithm alerted us that the inverse power method had been fooled.)

In practice, the eigenvalues and the eigenvectors of a matrix A are usually found by transforming A to Hessenberg form (with Householder transformations) and then applying some version of the QR algorithm. For a symmetric matrix, the Givens algorithm is most suited for finding a set of selected eigenvalues [for instance, finding the $n/3$ smallest eigenvalues of a symmetric $(n \times n)$ matrix]. We cannot go into all the details here, but we do wish to include a brief discussion of the QR algorithm. Although quite sophisticated, it currently is the most widely accepted technique for eigenvalue estimation, and is a principal element in the popular library of eigenvalue/eigenvector programs called EISPACK, developed through the Argonne National Laboratory.

Given a matrix A, there is a factorization of A such that $A = QR$ where Q is orthogonal and R is upper triangular. This factorization can be achieved through Householder transformations, which are chosen to create zeros below the main diagonal rather than below the subdiagonal as in the Hessenberg reduction. For example if we choose $\mathbf{u} = [u_1, u_2, \ldots, u_n]^T$ in (3.34b) such that

$$u_1 = a_{11} \pm \sqrt{a_{11}^2 + a_{21}^2 + \cdots + a_{n1}^2}, \qquad u_2 = a_{21}, \qquad u_3 = a_{31}, \qquad \ldots,$$
$$u_n = a_{n1}.$$

Then the matrix $\tilde{Q}_1 A$ with $\tilde{Q}_1$ defined by (3.33) for this vector $\mathbf{u}$ has zeros in the first column below the diagonal, that is, zeros in the $(i, 1)$ positions for $2 \le i \le n$. As opposed to the Hessenberg reduction, we do not form the product $\tilde{Q}_1 A \tilde{Q}_1$; forming the product would destroy the zeros just created (Problem 13). Instead we proceed by selecting a second $(n \times 1)$ vector $\mathbf{u}$ from (3.34b) to determine a matrix $\tilde{Q}_2$ such that the product $\tilde{Q}_2 \tilde{Q}_1 A$ has zeros in the second column below the diagonal. (This procedure will also preserve the zeros in the first column.) We continue this process, and after $(n - 1)$ steps, obtain the product

$$R = \tilde{Q}_{n-1} \ldots \tilde{Q}_2 \tilde{Q}_1 A \tag{3.37}$$

where R is upper triangular. We define the matrix Q by

$$Q^T = \tilde{Q}_{n-1} \ldots \tilde{Q}_2 \tilde{Q}_1 \qquad \text{or} \qquad R = Q^T A.$$

Since each $\tilde{Q}_i$ is orthogonal, then $QQ^T = I$ and so Q is orthogonal. Thus multiplying both sides of $R = Q^T A$ by Q yields $QR = A$, the desired factorization.

An elementary version of the QR algorithm does such a factorization in each step. Letting $A^{(1)} = A$, we compute the sequences of matrices $\{A^{(m)}\}$, $\{Q^{(m)}\}$, $\{R^{(m)}\}$ as follows for $m = 1, 2, 3, \ldots$.

1. Given $A^{(m)}$, find its QR factorization, $A^{(m)} = Q^{(m)}R^{(m)}$ where $Q^{(m)}$ is orthogonal and $R^{(m)}$ is upper triangular.
2. Define $A^{(m+1)} = R^{(m)}Q^{(m)}$, and go back to Step 1.

We note that $R^{(m)} = Q^{(m)\mathrm{T}}A^{(m)}$, and so $A^{(m+1)} = Q^{(m)\mathrm{T}}A^{(m)}Q^{(m)}$. Thus each $A^{(m)}$ is similar to A. The sequence $\{A^{(m)}\}$ will converge to a matrix from which the eigenvalues of A can be readily calculated. For example if the eigenvalues of A satisfy $|\lambda_1| > |\lambda_2| > \cdots > |\lambda_n| > 0$, then $\{A^{(m)}\}$ will converge to an upper-triangular matrix with the eigenvalues of A on its diagonal; if A is also symmetric, this matrix will be diagonal.

There are many additional possibilities, such as $|\lambda_{i-1}| = |\lambda_i|$, that we cannot develop here. In any case, however, the cost of the QR factorization in each step would be considerably lessened if A were in Hessenberg form originally. Hence we perform the preliminary step of using the Householder reduction to find a Hessenberg matrix H similar to A and iterate on H instead of A. Finally, the QR algorithm is usually applied with a shift of the origin in order to speed convergence. In this modification a sequence of shift constants $\{\alpha_m\}$ are determined; the factorization $(A^{(m)} - \alpha_m I) = Q^{(m)}R^{(m)}$ is performed; and $A^{(m+1)} = \alpha_m I + R^{(m)}Q^{(m)}$. We leave to the reader that the matrices $A^{(m)}$ are similar to A. The reader who wants to consider a fully operational QR algorithm should consult Wilkinson (1965) for theoretical and computational foundations, and Wilkinson and Reinsch (1971) for programs.

PROBLEMS, SECTION 3.3.2

1. Show by example that if the $(n \times n)$ matrix A is symmetric, $A = A^{\mathrm{T}}$, and $A_1 = S_1 A S_1^{-1}$ with S_1 as in (3.29a), then A_1 need not be symmetric.
2. a) For each of the matrices A given below find a vector $\mathbf{u}$ and the matrix Q as given by (3.33) such that QA has zeros in the first column below the subdiagonal. (Choose the $\pm$ sign as described in the text.)

(i) $\begin{bmatrix} 3 & 1 & 1 \\ 0 & 3 & -1 \\ 4 & 8 & 3 \end{bmatrix}$ (ii) $\begin{bmatrix} 12 & -1 & -1 \\ 3 & 9.75 & 1 \\ 4 & 13 & -1 \end{bmatrix}$ (iii) $\begin{bmatrix} 2 & 0 & 3 \\ 12 & 1 & -1 \\ -5 & 2 & 1 \end{bmatrix}$

(iv) $\begin{bmatrix} 4 & 1 & 2 & 1 \\ 2 & 3 & 2 & -1 \\ 1 & 0 & 4 & 1 \\ -2 & 1 & 0 & -2 \end{bmatrix}$ (v) $\begin{bmatrix} 2 & 1 & 0 & 3 \\ 3 & 2 & -1 & 1 \\ 2 & 1 & 1 & -2 \\ -6 & 1 & -1 & 0 \end{bmatrix}$ (vi) $\begin{bmatrix} 3 & 3 & 2 & -2 & 1 \\ 4 & 1 & 1 & -3 & 0 \\ 2 & 1 & 1 & -2 & 1 \\ 1 & 2 & -1 & 0 & 2 \\ 2 & 1 & 3 & -1 & 1 \end{bmatrix}$

   b) Use (3.35a) and (3.35b) to form the product QAQ for each matrix in part (a).

3. Transform the following to Hessenberg form using the Householder reduction.

a) $\begin{bmatrix} 2 & 1 & 2 \\ -4 & -1 & 0 \\ 3 & -2 & 1 \end{bmatrix}$ b) $\begin{bmatrix} 1 & 3 & 1 & 3 \\ 2 & 2 & 1 & 1 \\ 0 & 0 & -1 & 2 \\ 0 & -1 & 2 & 0 \end{bmatrix}$ c) $\begin{bmatrix} 1 & 2 & 1 & 2 \\ 3 & 3 & -1 & 2 \\ 0 & 3 & 0 & 1 \\ 0 & -4 & 1 & -1 \end{bmatrix}$

4. How many multiplications and divisions and how many square root evaluations must be performed to transform an $(n \times n)$ matrix A to Hessenberg form using Householder transformations?

5. Modify the eigenvalue/eigenvector program of Problem 12, Section 3.3.2, to replace the Hessenberg reduction subroutine by subroutine SIMTRN. Test the program on these matrices.

$$A = \begin{bmatrix} 2 & 1 & 3 & 4 \\ 1 & -3 & 1 & 5 \\ 3 & 1 & 6 & -2 \\ 4 & 5 & -2 & -1 \end{bmatrix} \quad \text{and} \quad B = \begin{bmatrix} -31 & 96 & 84 & -96 & -240 & 384 \\ -30 & 74 & 45 & -60 & -150 & 240 \\ -12 & 6 & -25 & 24 & 60 & -96 \\ -12 & 18 & -6 & 43 & 85 & -145 \\ -36 & 54 & -18 & 66 & 102 & -174 \\ -24 & 36 & -12 & 50 & 80 & -137 \end{bmatrix}$$

(Caution: The eigenvalues of A are real, but those of B are $5 \pm 12i$, $8 \pm 6i$, and $\pm 9i$; be sure the program has the appropriate complex subroutines.)

6. a) If $\{p_k(t)\}_{k=0}^{n}$ is the Sturm sequence (3.36) for a symmetric, tridiagonal $(n \times n)$ matrix H, show that $p_n(t)$ is the characteristic polynomial for H.

   b) For a given value of c, show that if two successive terms of the Sturm sequence are zero $[p_{k-1}(c) = p_k(c) = 0]$, then c is an eigenvalue and $p_i(c) = 0$, $1 \le i \le n$.

7. For the given matrix H, calculate the Sturm sequence for the following values of c in the given order: $c = -2, 5, 0, 1, 1/2, 1/4$. At each stage, what conclusions can you draw about the location of the eigenvalues?

$$H = \begin{bmatrix} 3 & 2 & 0 & 0 \\ 2 & 1 & 1 & 0 \\ 0 & 1 & 1 & -2 \\ 0 & 0 & -2 & 2 \end{bmatrix}$$

8. Modify subroutine GIVENS to use Gerschgorin's theorem to provide an initial interval containing the eigenvalues. With this modification, test the program of Fig. 3.8 on the matrix

$$A = \begin{bmatrix} 1 & 2 & 1 & 2 & 0 \\ 2 & 1 & 2 & -1 & 3 \\ 1 & 2 & 0 & 3 & 1 \\ 2 & -1 & 3 & 1 & 0 \\ 0 & 3 & 1 & 0 & 4 \end{bmatrix}.$$

9. Perform the QR factorization for the matrices A in Problems 2a(i) and 2a(ii); that is, write A as $A = QR$ where Q is orthogonal and R is upper triangular.

10. a) If a QR factorization of an $(n \times n)$ matrix A is given, $A = QR$, then the system of equations $A\mathbf{x} = \mathbf{b}$ is equivalent to $QR\mathbf{x} = \mathbf{b}$ or $R\mathbf{x} = Q^{T}\mathbf{b}$, which can immediately

be backsolved. With the matrix A of Problem 2a(i) use the factorization in Problem 9 to solve $A\mathbf{x} = \mathbf{b}$ where $\mathbf{b} = [-2, 0, 9]^T$.

b) The column-by-column QR-factorization technique of this section yields a product $\tilde{Q}_{n-1} \ldots \tilde{Q}_2\tilde{Q}_1 A = R$ in (3.37) where each $\tilde{Q}_i$ is a Householder matrix. Hence $A\mathbf{x} = \mathbf{b}$ is equivalent to $R\mathbf{x} = \tilde{Q}_{n-1} \ldots \tilde{Q}_2\tilde{Q}_1\mathbf{b} \equiv \tilde{\mathbf{b}}$. Describe an algorithm to transform $\mathbf{b}$ into $\tilde{\mathbf{b}}$ without computing the entries of each $\tilde{Q}_k$.

c) Write a program to solve $A\mathbf{x} = \mathbf{b}$; use a subroutine to produce the QR factorization of an $(n \times n)$ matrix A and another subroutine to take this information, transform $\mathbf{b}$ into $\tilde{\mathbf{b}}$, and backsolve $R\mathbf{x} = \tilde{\mathbf{b}}$. Test this program on some systems you have solved using programs from Chapter 2.

11. Write a program that includes the factor subroutine of Problem 10c and that generates the sequences $\{A^{(m)}\}$, $\{Q^{(m)}\}$, and $\{R^{(m)}\}$ of the QR algorithm. Run the program on the matrix of Problem 8 and determine the behavior of $\{A^{(m)}\}$.

12. If $Q_1, Q_2, \ldots, Q_k$ are orthogonal and symmetric $(n \times n)$ matrices, show that the product $Q_1Q_2 \ldots Q_k$ is orthogonal. Show by example that the product need not be symmetric.

13. Let $\tilde{Q}_1$ be the Householder matrix that represents the first step of the QR factorization of A. Verify by example that forming $\tilde{Q}_1 A \tilde{Q}_1$ can destroy the zeros created in $\tilde{Q}_1 A$.

*3.4 SCHUR'S THEOREM AND RELATED TOPICS

In this section, we discuss a few of the more advanced aspects of matrix theory, particularly those that have practical applications. Most of the results of this section are presented with only a sketch of their proofs. The interested reader should attempt to work through these proofs since they lend a great deal of insight into how numerical methods are derived and how well they work. This section is not required for any of the following chapters and may be omitted should the reader desire.

The focus of this section will be Schur's theorem. Schur's theorem is a theoretical result that is of great value in the analysis of numerical methods. As a preliminary, we establish the following characterization of $(n \times n)$ matrices that have a *complete* set of eigenvectors (that is, a set of n linearly independent eigenvectors). Recall, that this matter was of concern in developing procedures to find eigenvalues.

Theorem 3.5

Let A be an $(n \times n)$ matrix. Then A has a set of n linearly independent eigenvectors if and only if A is similar to a diagonal matrix.

Proof. Suppose $A\mathbf{u}_i = \lambda_i\mathbf{u}_i$, $i = 1, 2, \ldots, n$ where $\mathbf{u}_1, \mathbf{u}_2, \ldots, \mathbf{u}_n$ are linearly independent. Form the $(n \times n)$ matrix $C = [\mathbf{u}_1, \mathbf{u}_2, \ldots, \mathbf{u}_n]$, and note (Problem 1) that for

$$D = \begin{bmatrix} \lambda_1 & 0 & 0 & \cdots & 0 \\ 0 & \lambda_2 & 0 & \cdots & 0 \\ 0 & 0 & \lambda_3 & \cdots & 0 \\ \vdots & & & & \\ 0 & 0 & 0 & \cdots & \lambda_n \end{bmatrix}$$

we have $AC = CD$. Since the columns of C are linearly independent, C is nonsingular and hence the inverse matrix, C^{-1}, exists. Therefore $C^{-1}AC = D$ where D is diagonal. The converse is left (Problem 2) to the reader. ∎

This theorem, while characterizing matrices with a complete set of eigenvectors, is of little immediate practical use since a knowledge of the eigenvectors is necessary in order to construct the similarity transformation. The theorem does, however, suggest a numerical procedure called Jacobi's method. The method proceeds by an infinite sequence of similarity transformations that produce matrices that converge to a diagonal matrix D and is applicable only to symmetric matrices.

To be more precise, let A be symmetric. Suppose a_{pq} is the largest off-diagonal element of A in absolute value; so $|a_{pq}| \geq |a_{ij}|$, $1 \leq i, j \leq n$, $p \neq q$, $i \neq j$. There is a matrix U such that

1. $U^{\mathrm{T}}U = I$.
2. If $U^{\mathrm{T}}AU = A_1$, then the (p,q)th entry of A_1 is zero (Problem 3).

If r_i denotes the sum of the squares of the off-diagonal elements of A_i, $i = 0, 1, 2, \ldots$ (where $A_0 = A$), then it can be shown that $r_0 > r_1$ and that A_1 is also symmetric. This fact leads to the sequence

$$A_{k+1} = U_{k+1}^{\mathrm{T}} A_k U_{k+1} \tag{3.38}$$

where $A_0 = A$, $r_{k+1} < r_k$, $U_{k+1}^{\mathrm{T}}U_{k+1} = I$; and thus inductively A_{k+1} is symmetric. It can further be shown that $\lim_{k\to\infty} r_k = 0$ so that the matrices A_k tend to a diagonal matrix D. Since A_k is similar to A, the diagonal entries of A_k are tending to the engenvalues of A.

The proof of the convergence of Jacobi's method would provide a proof that every symmetric matrix is similar to a diagonal matrix, and hence that symmetric matrices have a complete set of eigenvectors. We will see, however, that this important result is also a simple consequence of Schur's theorem. To establish Schur's theorem, we need a few preliminaries that are necessary because real matrices may have complex eigenvalues (and hence eigenvectors with complex components). Given an $(m \times n)$ matrix $B = (b_{ij})$, we define the matrix $\overline{B}$ to be the $(m \times n)$ matrix whose ijth entry is $\overline{b}_{ij}$ where the bar denotes the complex conjugate. We next set $B^* = (\overline{B})^{\mathrm{T}}$. Finally if U is an $(n \times n)$ matrix such that $U^*U = I$, we say U is *unitary* (if U is a real matrix, it is customary to call U *orthogonal*).

Theorem 3.6. Schur

If A is any $(n \times n)$ matrix, then there is a unitary matrix U such a that $U^* AU = T$ where T is upper triangular.

Proof. We use induction on n. Let $\mathbf{u}$ be an eigenvector of A such that $A\mathbf{u} = \lambda\mathbf{u}$ and such that $\mathbf{u}^*\mathbf{u} = 1$. By Problem 4, we can obtain a set of vectors $\{\mathbf{u}, \mathbf{w}_2, \mathbf{w}_3, \ldots, \mathbf{w}_n\}$ such that

1. $\mathbf{u}^*\mathbf{w}_i = 0, \quad i = 2, 3, \ldots, n$
2. $\mathbf{w}_i^*\mathbf{w}_j = \delta_{ij}, \quad 2 \le i, j \le n$

(such a set is called *orthonormal*). Let S be the matrix whose first column is $\mathbf{u}$ and whose kth column is $\mathbf{w}_k$, $2 \le k \le n$. Then S is unitary (Problem 5). Now the first column of S^*AS equals $(S^*A)\mathbf{u} = S^*(A\mathbf{u}) = \lambda S^*\mathbf{u}$. But since $S^*S = I$ and $\mathbf{u}$ is the first colunn of S, then $S^*\mathbf{u} = \mathbf{e}_1$ where $\mathbf{e}_1^T = (1, 0, 0, \ldots, 0)$. Thus

$$S^*AS = \left[\begin{array}{c|c} \lambda & \mathbf{w} \\ \hline 0 & A_1 \end{array}\right] \tag{3.39}$$

where $\mathbf{w}$ is a $1 \times (n-1)$ vector and A_1 is an $(n-1) \times (n-1)$ matrix. By the induction hypothesis there exists an $(n-1) \times (n-1)$ unitary matrix W such that $W^*A_1 W$ is upper triangular. Now let

$$R = \left[\begin{array}{c|c} 1 & 0 \\ \hline 0 & W \end{array}\right] \tag{3.40}$$

and let $U = SR$. Then $U^*AU = R^*S^*ASR$ (Problem 5); and

$$\begin{aligned} U^*AU &= \left[\begin{array}{c|c} 1 & 0 \\ \hline 0 & W^* \end{array}\right] \times \left[\begin{array}{c|c} \lambda & \mathbf{w} \\ \hline 0 & A_1 \end{array}\right] \times \left[\begin{array}{c|c} 1 & 0 \\ \hline 0 & W \end{array}\right] \\ &= \left[\begin{array}{c|c} \lambda & \mathbf{w} \\ \hline 0 & W^*A_1 \end{array}\right] \times \left[\begin{array}{c|c} 1 & 0 \\ \hline 0 & W \end{array}\right] = \left[\begin{array}{c|c} \lambda & \mathbf{z} \\ \hline 0 & W^*A_1W \end{array}\right], \end{aligned} \tag{3.41}$$

with $\mathbf{z} = \mathbf{w}W$. Thus, U^*AU is upper triangular, and the induction is complete. ■

We note here that if A is a real matrix with all real eigenvalues, then U is real and orthogonal. One important theorem involving the iterative solution of linear systems, whose proof utilizes Schur's theorem, is given below. We say an $(n \times n)$ matrix A is *convergent* if $\lim_{m\to\infty} A^m = \mathcal{O}$ (where $\mathcal{O}$ denotes the zero matrix). Finally if A has eigenvalues $\lambda_1, \lambda_2, \ldots, \lambda_n$ with $|\lambda_1| \ge |\lambda_2| \ge \cdots \ge |\lambda_n|$, then we define the *spectral radius of* A, $\rho(A)$, by $\rho(A) = |\lambda_1|$.

Theorem 3.7

An $(n \times n)$ matrix A is convergent if and only if $\rho(A) < 1$.

Proof. If A is convergent, then clearly $\lim_{m\to\infty} \|A^m\|_1 = 0$. By Theorem 3.3, and since $\rho(A^m) = \rho(A)^m$ (Problem 6),

$$\lim_{m\to\infty} \|A^m\|_1 \geq \lim_{m\to\infty} \rho(A)^m = \lim_{m\to\infty} \rho(A^m) = 0.$$

Thus it must be that $\rho(A) < 1$.

Suppose now that $\rho(A) < 1$. By Schur's theorem, there is a unitary matrix U such that $U^*AU = T$ where T is triangular. Because $T^m = U^*A^mU$ (and hence $UT^mU^* = A^m$ as in Problem 7), it follows that if T is convergent, then so is A. Now, an $(n \times n)$ matrix B has n linearly independent eigenvectors if and only if B is similar to a diagonal matrix. Thus if T has n linearly independent eigenvectors, then there is a matrix C such that $C^{-1}TC = D$ where D is diagonal. As before, if D is convergent, then so is T, and hence A. Since D is diagonal, D is convergent if and only if the diagonal entries of D are all less than 1 in magnitude. Therefore [since $\rho(D) = \rho(T) = \rho(A) < 1$] if T has n linearly independent eigenvectors, then A is convergent.

To handle the case in which T does not have n linearly independent eigenvectors, we construct a matrix T_1 that "majorizes" T. For $T = (t_{ij})$, choose T_1 to be an $(n \times n)$ triangular matrix of the form

$$T_1 = \begin{bmatrix} \alpha_1 & \beta & \beta & \beta & \cdots & \beta \\ 0 & \alpha_2 & \beta & \beta & \cdots & \beta \\ 0 & 0 & \alpha_3 & \beta & \cdots & \beta \\ \vdots & & & & & \\ 0 & 0 & 0 & & \cdots & \alpha_n \end{bmatrix} \tag{3.42}$$

where $1 > \alpha_i \geq |t_{ii}|$ and $\beta \geq |t_{ij}|$, $1 \leq i, j \leq n$. Clearly we can make choices for T_1 such that the eigenvalues $\alpha_1, \alpha_2, \ldots, \alpha_n$ of T_1 are distinct. From Problem 9, Section 3.1, we see that T_1 is similar to a diagonal matrix; and thus, arguing as we did above, it follows that T_1 is convergent since $\rho(T_1) < 1$. If the ijth entry of T_1^m is denoted by $w_{ij}^{(m)}$, it is an easy exercise to verify that $w_{ij}^{(m)} \geq |t_{ij}^{(m)}|$ where $t_{ij}^{(m)}$ is the ijth entry of T^m. In summary, if $\rho(A) < 1$ and if $U^*AU = T$, then T is majorized by a convergent matrix T_1 and hence A is convergent. ∎

As an application of Theorem 3.7 we return briefly to the iterative methods of Section 2.4. Thus if

$$\mathbf{x} = M\mathbf{x} + \mathbf{g} \tag{3.43}$$

is an equivalent fixed-point form of the equation $A\mathbf{x} = \mathbf{b}$ with solution $\mathbf{x}_t$, then the errors $\mathbf{e}^{(k)} = \mathbf{x}^{(k)} - \mathbf{x}_t$ are given by

$$\mathbf{e}_k^{(k)} = M^k\mathbf{e}_0, \qquad k = 1, 2, \ldots. \tag{3.44}$$

[See formula (2.57).] Thus, it follows that the sequence $\{\mathbf{x}^{(k)}\}$ converges to $\mathbf{x}_t$ for every initial guess $\mathbf{x}^{(0)}$ if and only if M is convergent (Problem 8). This observation provides the basis for the proof of Theorem 2.4.

We next extend the idea of symmetric matrices to matrices with complex entries, and say that A is *Hermitian* if $A^* = A$. (For example, a real symmetric matrix is a Hermitian matrix.) We can use Schur's theorem to prove Theorem 3.8.

Theorem 3.8

Let A be an $(n \times n)$ Hermitian matrix with eigenvalues $\lambda_1, \lambda_2, \ldots, \lambda_n$. Then we can choose eigenvectors $\mathbf{u}_1, \mathbf{u}_2, \ldots, \mathbf{u}_n$ for A such that

1. $A\mathbf{u}_i = \lambda_i\mathbf{u}_i$; $i = 1, 2, \ldots, n$; and $\{\mathbf{u}_i\}_{i=1}^n$ are linearly independent;
2. $\mathbf{u}_i^*\mathbf{u}_j = \delta_{ij}$, $1 \le i, j \le n$ (the eigenvectors can be chosen to be orthonormal).

The proof of this theorem is left (Problem 9) to the reader. The reader will recall that this theorem is needed to derive the improved version of the power method, the Rayleigh quotient method, which is applicable to Hermitian matrices. Note also that if A is real and symmetric, then conclusion (2) becomes $\mathbf{u}_i^T\mathbf{u}_j = \delta_{ij}$.

Another consequence of Theorem 3.8 is to find the matrix norm that is the natural extension of the vector norm $\|\mathbf{x}\|_2$. Recall from Section 2.3 that for any real matrix A, we would like to find the nonnegative real number $g(A) \equiv L_2$ such that $\|A\mathbf{x}\|_2 \le L_2\|\mathbf{x}\|_2$ for all x in R^n and where $L_2 < L$ for any number L that also satisfies $\|A\mathbf{x}\|_2 \le L\|\mathbf{x}\|_2$ for all $\mathbf{x}$ in R^n. [That is, $g(A) \equiv L_2 = \min\{L\colon \|A\mathbf{x}\|_2 \le L\|\mathbf{x}\|_2$, for all $\mathbf{x}$ in $R^n\}$.] It can be shown that $g(A)$ satisfies all the properties of a matrix norm, and thus we write $g(A) \equiv \|A\|_2$. [Note: From the definition above of $g(A)$, it follows that $g(A) \equiv \|A\|_2 \le \|A\|_E$ for all A in M_n, and that $\|A\|_2$ is compatible with the $\|\cdot\|_2$ vector norm.] It can easily be shown (Problem 10) that the matrix A^TA is Hermitian and its eigenvalues $\{v_i\}_{i=1}^n$ are real and nonnegative; that is, A^TA is *positive semidefinite*. Let $v_1 \ge v_2 \ge \cdots \ge v_n \ge 0$. By Theorem 3.8 the eigenvectors $\mathbf{u}_i$ of A^TA can be chosen to be orthonormal: $\mathbf{u}_i^T\mathbf{u}_j = \delta_{ij}$, $1 \le i, j \le n$. Then any vector $\mathbf{x} \ne \theta$ can be written as

$$\mathbf{x} = c_1\mathbf{u}_1 + c_2\mathbf{u}_2 + \cdots + c_n\mathbf{u}_n.$$

Thus we can verify that

$$\left(\frac{\|A\mathbf{x}\|_2}{\|\mathbf{x}\|_2}\right)^2 \equiv \frac{\mathbf{x}^TA^TA\mathbf{x}}{\mathbf{x}^T\mathbf{x}} = \frac{\sum_{i=1}^n |c_i|^2 v_i}{\sum_{i=1}^n |c_i|^2}, \tag{3.45}$$

which implies

$$0 \le v_n \le \left(\frac{\|A\mathbf{x}\|_2}{\|\mathbf{x}\|_2}\right)^2 \le v_1, \tag{3.46}$$

or

$$\|A\mathbf{x}\|_2 \le v_1^{1/2}\|\mathbf{x}\|_2 \tag{3.47}$$

for all vectors $\mathbf{x}$. Thus $\nu_1^{1/2} \equiv \rho(A^TA)^{1/2} \geq L_2 = \|A\|_2$. Now substitute $\mathbf{u}_1$ for $\mathbf{x}$ in (3.45), and (3.47) becomes $\|A\mathbf{u}_1\|_2 = \nu_1^{1/2}\|\mathbf{u}_1\|_2$. This equation implies that $\rho(A^TA)^{1/2} = L_2$, and we have proved the following theorem.

Theorem 3.9

Note that if A is real and symmetric, then $A^TA = A^2$ and so $\|A\|_2 = \rho(A^TA)^{1/2} = \rho(A^2)^{1/2} = \rho(A)$ (see Problem 6).

$$\|A\|_2 \equiv \min\{L: \|A\mathbf{x}\|_2 \leq L\|\mathbf{x}\|_2, \text{ for all } \mathbf{x} \text{ in } R^n\} = \rho(A^TA)^{1/2}.$$

PROBLEMS, SECTION 3.4

1. In the proof of Theorem 3.5, establish that $AC = CD$.
2. Complete the converse of Theorem 3.5 and show that if A is similar to a diagonal matrix, then A has n linearly independent eigenvectors. [Hint: Consider the column vectors of the matrix of the similarity transformation.]
3. Let $A = (a_{ij})$ be symmetric and suppose the (p, q)th entry of A, a_{pq}, is nonzero (where also, a_{pq} is not on the diagonal). Define U by
   a) $u_{pp} = u_{qq} = \cos(\theta)$, $u_{pq} = -u_{qp} = \sin(\theta)$;
   b) $u_{ij} = \delta_{ij}$, for all other elements.
      (1) Show $U^TU = I$ and U^TAU is symmetric.
      (2) Choose θ, $-\pi/4 \leq \theta \leq \pi/4$, such that $\tan(2\theta) = 2a_{pq}/(a_{qq} - a_{pp})$. Show the (p, q)th entry of U^TAU is 0. [Note: For $\alpha = \tan(2\theta)$, $\cos^2(\theta) = 1/2 + 1/(2\sqrt{\alpha^2 - 1})$.]
4. This problem describes the Gram-Schmidt process of constructing an orthonormal set of k vectors from a given set of k linearly independent vectors. Let $\{\mathbf{t}_1, \mathbf{t}_2, \ldots, \mathbf{t}_k\}$ be a set of k linearly independent, n-dimensional column vectors (which may have complex components). To derive a set $\{\mathbf{w}_1, \mathbf{w}_2, \ldots, \mathbf{w}_k)$ of orthonormal vectors from $\{\mathbf{t}_1, \mathbf{t}_2, \ldots, \mathbf{t}_k\}$, we proceed in a step-by-step fashion.
   a) Let $\mathbf{v}_1 = \mathbf{t}_1$ and $\beta_1 = 1/\sqrt{\mathbf{v}_1^*\mathbf{v}_1}$. Next, set $\mathbf{w}_1 = \beta_1\mathbf{v}_1$.
   b) For $2 \leq j \leq k$, let $\mathbf{v}_j = \mathbf{t}_j + \alpha_{j-1}\mathbf{w}_{j-1} + \cdots + \alpha_1\mathbf{w}_1$ where $\alpha_i = -\mathbf{w}_i^*\mathbf{t}_j$ and $\beta_j = 1/\sqrt{\mathbf{v}_j^*\mathbf{v}_j}$. Then set $\mathbf{w}_j = \beta_j\mathbf{v}_j$ for each j.

   Show that
      (1) each step of this process is always defined (that is, we can always form β_i, $i = 1, 2, \ldots, k$);
      (2) each $\mathbf{t}_r$ is expressible (uniquely) as a linear combination $\mathbf{t}_r = \gamma_1\mathbf{w}_1 + \gamma_2\mathbf{w}_2 + \cdots + \gamma_r\mathbf{w}_r$, $r \leq k$;
      (3) for $1 \leq i, j \leq k$, $\mathbf{w}_i^*\mathbf{w}_j = \delta_{ij}$;
      (4) to complete the proof of Theorem 3.8, if $\mathbf{u}$ is any eigenvector of A such that $\mathbf{u}^*\mathbf{u} = 1$, then we can find $\mathbf{w}_2, \ldots, \mathbf{w}_n$ such that $\{\mathbf{u}, \mathbf{w}_2, \mathbf{w}_3, \ldots, \mathbf{w}_n\}$ is an orthonormal set. To find this, we need only find $\mathbf{t}_2, \mathbf{t}_3, \ldots, \mathbf{t}_n$ so that $\{\mathbf{u}, \mathbf{t}_2, \ldots, \mathbf{t}_n\}$ is linearly independent. Prove that this procedure can always be done. [Hint: Consider the vectors $\{\mathbf{e}_1, \mathbf{e}_2, \ldots, \mathbf{e}_n\}$].

5. In the proof of Theorem 3.6, show that
   a) the matrix S is unitary;
   b) if B is $(m \times n)$ and C is $(n \times p)$, then $(BC)^* = C^*B^*$ (only outline the proof of this).
6. If A is $(n \times n)$, show $\rho(A^k) = [\rho(A)]^k$.
7. Show by induction that if $C^{-1}AC = B$, then $C^{-1}A^rC = B^r$.
8. In (3.44) show that $\{\mathbf{e}^{(k)}\} \to \theta$ for every initial guess $\mathbf{x}^{(0)}$ if and only if $\rho(M) < 1$.
9. Show that an $(n \times n)$ Hermitian matrix A has an orthonormal set of n eigenvectors. [Hint: Use Schur's theorem to show that A is "unitarily" similar to a diagonal matrix.]
10. Verify that the following steps in Theorem 3.9 are valid.
    a) Show, in analogy to Theorem 3.1, that a Hermitian matrix has real eigenvalues.
    b) Show the eigenvalues of A^*A are real and nonnegative.
    c) Verify formula (3.45) by direct computation.
    d) Show that (3.45) implies (3.46).
11. Consider $A\mathbf{x} = \mathbf{b}$ where

$$A = \begin{bmatrix} 2 & -1 \\ -1 & 2 \end{bmatrix}, \quad \mathbf{b} = \begin{bmatrix} 1 \\ 1 \end{bmatrix}, \quad \mathbf{x}_t = \begin{bmatrix} 1 \\ 1 \end{bmatrix}.$$

    a) Show that the Jacobi iteration method will converge for any initial guess, $\mathbf{x}^{(0)}$.
    b) Show that the Gauss-Seidel method will converge for any initial guess, $\mathbf{x}^{(0)}$.
    c) Show that $0 < \rho(M_G) < \rho(M_J)$. What does this inequality say about the relative rates of convergence of these two methods?
    d) Carry out a few iterations of these methods and compare the estimates to the true solution.
12. Let U be an $(n \times n)$ orthogonal matrix. Show that
    a) if $\mathbf{x} \in R^n$, then $\|U\mathbf{x}\|_2 = \|\mathbf{x}\|_2$;
    b) if λ is an eigenvalue of U, then $|\lambda| = 1$.
13. Use Theorem 3.7 and Problem 8 to show why the Jacobi method cannot converge for the (4×4) matrix A in Example 2.6. [Hint: Show $\det(M_J) > 1$. What does this inequality say about $\rho(M_J)$?]
14. Let $A = (a_{ij})$ be a symmetric, positive definite $(n \times n)$ matrix [recall formula (2.68)]. Show that $a_{ii} > 0$ for $i = 1, 2, \ldots, n$. [Hint: Consider $\mathbf{e}_i^T A \mathbf{e}_i$]. Show that the eigenvalues of A are all positive.
15. Let $D = (d_{ij})$ be an $(n \times n)$ diagonal matrix, ($d_{ij} = 0$ for $i \neq j$) such that $d_{ii} > 0$ for $i = 1, 2, \ldots, n$. Show that D is positive definite and symmetric.

*16. This problem outlines the proof of Theorem 2.4 and shows that the Gauss-Seidel iteration converges for any matrix A that is symmetric and positive definite. As in Section 2.4.1, let $A = L + D + U$, and note that $A = L + D + L^T$ since A is symmetric. The first step in the proof is to show that the matrix $P = A - M_G^T A M_G$ is symmetric and positive definite where $M_G = -(D + L)^{-1}L^T$.

a) Show that P is symmetric.

b) Show $M_G = I - Q$, $Q = (D + L)^{-1} A$. [Hint: $A = L + D + L^T$.]

c) Show that $P = Q^T(AQ^{-1} + (Q^T)^{-1}A - A)Q$. [Hint: Use part (b).] Now show that $P = Q^TDQ$; use the above and the definition of Q.

d) Using Problems 14 and 15 above, show that P is positive-definite symmetric. To finish the proof, suppose λ is an eigenvalue of M_G and $\mathbf{u}$ is a corresponding eigenvector. Show that $\mathbf{u}^T P\mathbf{u} > 0$ implies that $|\lambda| < 1$. This inequality shows that $\rho(M_G) < 1$ and completes the proof by virtue of Theorem 3.9.

4

Solution of Nonlinear Equations

4.1 INTRODUCTION

One of the oldest and most basic problems encountered in practice is that of finding the roots of an equation; i.e., given $y = f(x)$, find all values s such that $f(s) = 0$. In this chapter we shall study several techniques for the numerical approximation of these values of s and will demonstrate how one of these techniques (Newton's method) can be easily extended to the problem of simultaneously solving a system of n equations in n unknowns for $n > 1$. Before delving into this particular problem, however, we pause to give the following general illustration.

In previous chapters we considered some relatively simple physical examples that involved only the illustration of the particular numerical method we were studying at the moment. Most real-world examples are more complicated and involve different numerical procedures to solve the different problems that arise at various stages of the solution process of the problem as a whole. The following example provides an illustration of a realistic problem of this type. We have given this example in some detail since contained in it are problems that require procedures for numerically solving ordinary differential equations (Chapter 7), numerically evaluating integrals (Chapter 6), least-squares data fitting (Chapter 5), as well as numerically solving nonlinear equations (the topic of this chapter). We shall return to this example as we discuss these procedures in subsequent sections.

EXAMPLE 4.0. In this example, we consider the ballistic reentry of a body into the earth's atmosphere (for instance, the return of a spacecraft or the recovery of a package from an orbiting satellite). In a ballistic trajectory, the only forces acting on the body are gravity and drag (that is, lift can be assumed to be zero). The equations of motion governing a ballistic reentry trajectory can be expressed by the equation (4.0a)

$$\begin{aligned} \ddot{\mathbf{R}} &= \frac{-\overline{q}C_D A}{M}\frac{\dot{\mathbf{R}}}{|\dot{\mathbf{R}}|} + \mathbf{G}(\mathbf{R}), \\ \mathbf{R}(0) &= \mathbf{R}_0, \\ \dot{\mathbf{R}}(0) &= \mathbf{V}_0 \end{aligned} \tag{4.0a}$$

where the three-dimensional position vector $\mathbf{R} = \mathbf{R}(t)$ gives the x, y, and z coordinates of the body at any time t (in an earth-centered coordinate frame). In this equation, $\mathbf{G}(\mathbf{R})$ denotes the gravitational force, and the scalar quantity $\overline{q}C_DA/M$ represents the drag forces exerted on the vehicle because of the atmosphere. The term $\overline{q}$ is the *dynamic pressure*, which is defined by $\overline{q} = \rho(H)|\dot{\mathbf{R}}|^2/2$ where $\rho(H)$ is the atmospheric density at an altitude H above the earth's surface, and $|\dot{\mathbf{R}}|$ is the magnitude of the velocity vector, $\dot{\mathbf{R}}(t)$. The drag coefficient, C_D, is normally an experimentally determined function, usually depending on velocity, altitude, and the orientation of the body along the path of the trajectory. The evaluation of C_D in the equations of motion (4.0a) typically requires table look-ups and/or the evaluation of a series of empirical formulas. The term A represents a reference cross-section area of the body. Drag forces can be increased by orienting the vehicle so that a larger frontal area is exposed. This orientation serves as a drag-brake, slowing the vehicle down. (The velocity increases if a smaller frontal area is exposed.) The term M in (4.0a) is the mass of the body. Finally, to avoid undue complication in the form of (4.0a), we have assumed a nonrotating earth so that the trajectory will lie in the plane determined by the initial position and velocity vectors, $\mathbf{R}_0$ and $\mathbf{V}_0$.

During the initial phases of engineering design for a vehicle (or for its guidance system, etc.), simplifying approximations are often made in order to obtain a rough idea of the performance characteristics of a particular design. For example, if we assume that C_DA/M is constant, that gravitational forces are negligible with respect to drag, and that atmospheric density is given by the formula $\rho(H) = \rho_0 e^{-\lambda H}$, then we obtain a reentry model that approximates (4.0a):

$$\ddot{\mathbf{R}} = -Ke^{-\lambda H}|\dot{\mathbf{R}}|\dot{\mathbf{R}}, \qquad \mathbf{R}(0) = \mathbf{R}_0, \quad \dot{\mathbf{R}}(0) = \mathbf{V}_0 \tag{4.0b}$$

where the constant K is given by $K = \rho_0 C_D A/2M$. This approximate model, which is usually called the Allen and Eggers reentry model, can be solved in closed form by the elementary calculus technique of separation of variables [see Thomas (1972) and Problem 5, Section 7.1]. This closed-form solution gives an expression for velocity as a function of altitude:

$$\dot{\mathbf{R}}(H) = \mathbf{V}_0 e^{Q[e^{-\lambda H} - e^{-\lambda H_0}]}$$

where H_0 denotes the initial altitude at $t = 0$ and $Q = -(C_D A\rho_0/2\lambda M \sin(\alpha))$, with α being the angle the velocity vector makes with the local horizontal (see Figure 4.0).

The Allen and Eggers reentry model in Eq. (4.0b) gives a trajectory that is a straight-line path since the body is initially traveling along the path defined by the vector $\mathbf{V}_0$, and the only force acting on the body is drag (which is also along the vector $\dot{\mathbf{R}}$ but in the opposite direction). The inclusion of a force term involving gravity would, of course, cause the trajectory to deviate from a straight-line path, but including a gravity term in (4.0b) would yield a differential equation that could not be solved in closed form. [In particular, Eq. (4.0a) cannot be solved except by numerical means (such as those in Chapter 7); and the reason for the approximate model, (4.0b), is to get an indication of the solution of (4.0a) for design purposes.]

Some sort of control system is often required for vehicles during reentry in order to preserve the structural integrity of the vehicle, to keep heating within a tolerable range, etc. Such control systems could, for example, involve the deployment of parachutes and/or reorientation of the body along the path of the trajectory. A device called an accelerometer (which can be visualized roughly as a mass attached to a spring) is frequently used as a part of a control system. The accelerometer can be used to measure the magnitude of the drag forces; and, on the basis of the output of the accelerometer, a

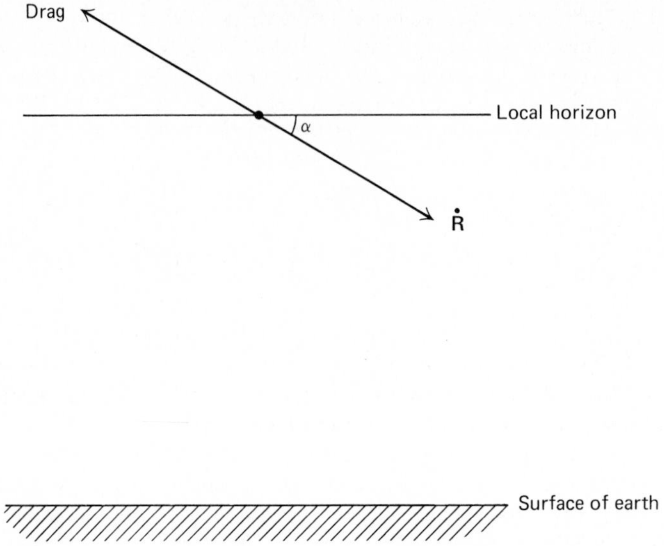

Figure 4.0 Atmospheric reentry with gravity neglected.

control maneuver can be initiated. In the reentry model (4.0b), the magnitude of the drag is proportional to

$$\bar{q} = \frac{|\dot{\mathbf{R}}(H)|^2 \rho_0 e^{-\lambda H}}{2};$$

so we can use the closed form expression for $\dot{\mathbf{R}}(H)$ to approximate the output of an accelerometer. In particular, to determine design parameters (or timer settings, etc.), we can estimate the altitude H at which the magnitude of drag is some specified value, say γ. This estimate would then amount to solving an equation of the form

$$\frac{\gamma}{K} = |\mathbf{V}_0|^2 e^{2Q[e^{-\lambda H} - e^{-\lambda H_0}]} e^{-\lambda H}$$

for H where γ/K, $|\mathbf{V}_0|$, H_0, λ, and Q are constants. We cannot normally expect to be able to solve such an equation by elementary means, and so we must use some numerical technique to approximate the solution. The construction of such root-finding procedures is the topic of this chapter.

As mentioned earlier, many of the problems that arise in practice require numerical approximations of one sort or another at various stages of the solution process. For instance, in this example an exponential atmosphere, $\rho(H) = \rho_0 e^{-\lambda H}$, was assumed in order to derive (4.0b). Since the standard atmospheric density models are not exponential, the constants ρ_0 and λ have to be selected so that $\rho_0 e^{-\lambda H}$ provides a good approximation to the standard atmospheric density model in the altitude range under consideration. Choosing ρ_0 and λ is then a problem in data-fitting (see Example 5.13). As another instance, (4.0a) must be solved to provide a benchmark against which to compare the approximate model (4.0b). Solving (4.0a) requires numerical techniques for the solution of ordinary differential equations; these techniques in turn require us to be able to evaluate the right-hand side of (4.0a) at various points $\mathbf{R}(t)$, $\dot{\mathbf{R}}(t)$ along the trajectory. This evaluation in turn will require interpolation in tables, developing a good approximate model for gravity, etc., (see Section 7.2.2).

4.2 BRACKETING METHODS

As is so often true in numerical analysis, the simplest methods are the ones that converge most slowly. However because of their inherent simplicity, one is often willing to use them even at the sacrifice of some speed of convergence.

One of the oldest and most geometrically intuitive root-finding methods is the bisection method. In this method, we assume $f(x)$ is a continuous function on the closed interval $[a, b]$ such that $f(a)f(b) < 0$. By the intermediate value theorem, there must exist at least one value $s \in [a, b]$ such that $f(s) = 0$. In the bisection method, we merely evaluate $f(x)$ at the midpoint $x = (a + b)/2$. If $f((a + b)/2) \neq 0$, then either $f(a)f((a + b)/2) < 0$ or $f((a + b)/2)f(b) < 0$. Without loss of generality let us assume that $f(a)f((a + b)/2) < 0$, and let $a = x_0$ and $(a + b)/2 = x_1$. Again by the intermediate value theorem, there must exist at least one value $s \in [x_0, x_1] \equiv I_1$ such that $f(s) = 0$. We next repeat this very same process on I_1, and find an interval, I_2, containing at least one root of $f(x) = 0$. We note here that the length of I_1 is $(b - a)/2$ and the length of I_2 is $(b - a)/4$. After this process is repeated n times we will obtain an interval I_n, of length $(b - a)/2^n$, which contains at least one root, s, of $f(x) = 0$. If x is any point of I_n, then $|x - s| \leq (b - a)/2^n$. Since every I_n must contain at least one such s, we say that each step "brackets" the root, and the error of the approximation at the nth step is bounded by $(b - a)/2^n$. We continue to repeat this process until n is sufficiently large so that we are satisfied with any approximation x^* for s that satisfies $|x^* - s| \leq (b - a)/2^n$. Thus once the initial a and b are found, the method is guaranteed to converge for any continuous function. The error bound for the nth step, $(b - a)/2^n$, however, is quite large compared to the errors of more sophisticated methods.

We next describe one further "bracketing" method, which usually converges faster than the bisection method. This technique is known as *Regula Falsi* or the method of false position, and its derivation is quite geometric as shown in Fig. 4.1. Again we must find an interval $[a, b]$ such that $f(a)f(b) < 0$. Next we find the point, x_1, where the secant line between the points $(a, f(a))$ and $(b, f(b))$ intersects the x axis. The equation of this secant line is

$$\frac{y - f(b)}{x - b} = \frac{f(a) - f(b)}{a - b}.$$

The secant line crosses the x-axis when $x = x_1$ where

$$x_1 = \frac{af(b) - bf(a)}{f(b) - f(a)},$$

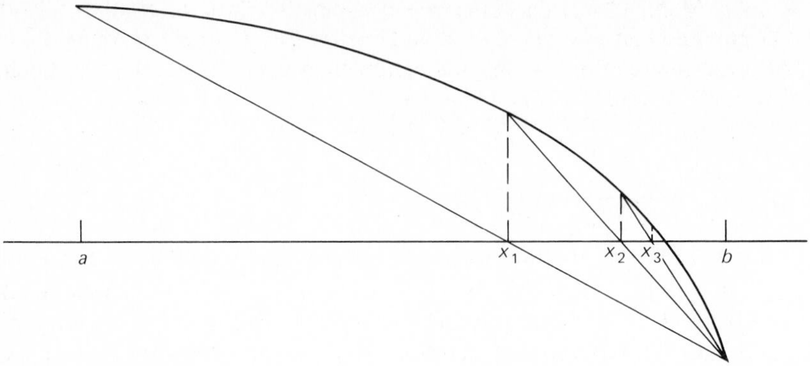

Figure 4.1 *Regula Falsi.*

and we take x_1 as our first approximation to the root s. As in the bisection method, if $f(x_1) \neq 0$, then either $f(a)f(x_1) < 0$ or $f(x_1)f(b) < 0$ [say, $f(a)f(x_1) < 0$]. Then again there must be at least one root of $f(x) = 0$ in the interval $I_1 = [x_0, x_1]$ (in this case, $x_0 = a$). This process is repeated again and again and generates intervals I_n, each of which contains at least one root of $f(x) = 0$, until we are satisfied with the approximation for s. It can happen that the lengths of I_n do not converge to zero as $n \to \infty$. This possibility is what happens for the particular function illustrated in Fig. 4.1. Furthermore, it can be shown that *Regula Falsi*

```
      SUBROUTINE BISECT(A,B,ROOT,TOL)
      REAL MIDPT
C
C  THIS SUBROUTINE USES THE BISECTION METHOD TO FIND A ROOT OF F(X)=0.
C  THE CALLING PROGRAM MUST SUPPLY NUMBERS A AND B SUCH THAT F(A)*F(B)<0
C  AND A STOPPING CRITERIA, TOL.  THE SUBROUTINE RETURNS AN
C  APPROXIMATION TO A ZERO OF F(X) (NAMED ROOT) WHEN AN INTERVAL OF
C  LENGTH TOL OR LESS IS FOUND IN WHICH F(X) CHANGES SIGN.  THE
C  SUBROUTINE ALSO REQUIRES A FUNCTION SUBPROGRAM, F(X)
C
      FA=F(A)
C
C  THE BISECTION ITERATION
C
    1 MIDPT=(A+B)/2.
      FM=F(MIDPT)
      IF(FA*FM.LE.0.)   GO TO 2
      A=MIDPT
      FA=FM
      GO TO 3
    2 B=MIDPT
C
C  TEST TO SEE IF EXIT CRITERION IS SATISFIED
C
    3 IF((B-A).LE.TOL)   GO TO 4
      GO TO 1
    4 ROOT=(A+B)/2.
      RETURN
      END
```

Figure 4.2 Subroutine BISECT.

will converge for any continuous function for which $f(a)f(b) < 0$ even though the lengths of the I_n may not go to zero. The *Regula Falsi* method is based on the premise that if $|f(a)| > |f(b)|$, then we expect $|s - b| < |s - a|$; that is, s is closer to b than a. Since in each successive step, x_k is computed using knowledge of the size of $|f(x)|$ at the endpoints of I_{k-1}, we would expect the *Regula Falsi* method usually to converge faster than the bisection method. It can be shown (see Section 4.3.4) that if $f''(x)$ is continuous, then $|x_k - s| \leq \lambda^{k-1}\varepsilon$ where $0 < \lambda < 1$ and ε is a fixed constant.

We have listed two subroutines that use the methods described in this section to find zeros of $f(x)$. Subroutine BISECT (Fig. 4.2) employes the bisection procedure; subroutine REGFAL (Fig. 4.3) uses *Regula Falsi*. Note that REGFAL requires two stopping criteria at which REGFAL terminates when

```
      SUBROUTINE REGFAL(A,B,ROOT,TOL,M,ITERM)
C
C  THIS SUBROUTINE USES REGULA FALSI TO FIND A ROOT OF F(X)=0.  THE
C  CALLING PROGRAM MUST SUPPLY NUMBERS A AND B SUCH THAT F(A)*F(B)<0
C  AND TWO STOPPING CRITERIA, TOL AND M.  THE SUBROUTINE RETURNS AN
C  APPROXIMATION TO A ZERO OF F(X) (NAMED ROOT) WHEN THE ABSOLUTE VALUE
C  OF F(ROOT) IS LESS THAN TOL OR WHEN M ITERATIONS HAVE BEEN EXECUTED.
C  IN THE FIRST CASE, A FLAG, ITERM, IS SET TO 1 AND IN THE SECOND CASE
C  ITERM IS SET TO 2.  THE SUBROUTINE ALSO REQUIRES A FUNCTION
C  SUBPROGRAM, F(X).
C
      ITR=1
      FA=F(A)
      FB=F(B)
C
C  REGULA FALSI ITERATION
C
    1 XN=(A*FB-B*FA)/(FB-FA)
      FN=F(XN)
C
C  TEST TO SEE IF MAXIMUM NUMBER OF ITERATIONS HAS BEEN EXECUTED
C
      IF(ITR.GE.M)    GO TO 4
C
C  TEST TO SEE IF A ROOT HAS BEEN FOUND
C
      IF(ABS(FN).LE.TOL)    GO TO 5
      IF(FA*FN.LE.0.)    GO TO 2
      A=XN
      FA=FN
      GO TO 3
    2 B=XN
      FB=FN
C
C  INCREMENT THE ITERATION COUNTER
C
    3 ITR=ITR+1
      GO TO 1
    4 ITERM=2
      ROOT=XN
      RETURN
    5 ITERM=1
      ROOT=XN
      RETURN
      END
```

Figure 4.3 Subroutine REGFAL.

$|f(x_n)| \le \text{TOL}$ or when $n \ge M$. An upper bound for the number of iterations is normally required for most root-finding programs since we cannot usually be certain that we will be able to find a number x_n such that $|f(x_n)| < \varepsilon$ for a prescribed ε (nor can we be certain that we can find two successive iterates, x_n and x_{n+1}, such that $|x_n - x_{n+1}| < \delta$ for a prescribed δ).

EXAMPLE 4.1. Let $f(x) = \cos(x) - x$, and note that $f(0) > 0 > f(1)$ so that $f(x) = 0$ has at least one root in [0, 1]. Subroutines BISECT and REGFAL were employed to estimate a root of $f(x) = 0$, with $a = 0$ and $b = 1$ in both cases. The results (Table 4.1) show the superiority of *Regular Falsi* for this particular function. The exact answer to eight places is $s = 0.73908513$. Round-off error causes both methods to give a final answer that is incorrect in the seventh place since we are near the limits of single precision for the machine used.

TABLE 4.1

Bisection Method		Regula Falsi	
x_i	$f(x_i)$	x_i	$f(x_i)$
0.5000000E 00	0.3775825E 00	0.6850733E 00	0.8929932E−01
0.7500000E 00	−0.1831114E−01	0.7362989E 00	0.4660129E−02
0.6250000E 00	0.1859630E 00	0.7389453E 00	0.2340078E−03
0.6875000E 00	0.8533489E−01	0.7390780E 00	0.1186132E−04
0.7187500E 00	0.3387933E−01	0.7390846E 00	0.8344650E−06
0.7343750E 00	0.7874667E−02	0.7390849E 00	0.2384185E−06
0.7421875E 00	−0.5195736E−02	0.7390850E 00	0.1788139E−06
0.7382812E 00	0.1345098E−02		
0.7402343E 00	−0.1923918E−02		
0.7392578E 00	−0.2890229E−03		
0.7387695E 00	0.5281567E−03		
0.7390136E 00	0.1195669E−03		
0.7391357E 00	−0.8475780E−04		
0.7390747E 00	0.1740455E−04		
0.7391052E 00	−0.3367662E−04		
0.7390899E 00	−0.8106231E−05		
0.7390823E 00	0.4649162E−05		
0.7390861E 00	−0.1728534E−05		
0.7390842E 00	0.1430511E−05		
0.7390851E 00	−0.1192092E−06		
0.7390847E 00	0.6556510E−06		
0.7390847E 00	0.6556510E−06		

PROBLEMS, SECTION 4.2

1. If we wish to compute the cube root of 5, we could proceed by finding the zeros of $f(x) = x^3 - 5$. Let $a = 0$ and $b = 3$ and calculate an approximation for $\sqrt[3]{5}$ using four

steps of the bisection and *Regula Falsi* methods. Which method is converging faster? ($\sqrt[3]{5} \simeq 1.70998$)

2. The function $f(x) = x^2 - 3x + 1$ has a sign change in [0, 1] and in [1, 3]. In each interval, execute six steps of the bisection method and the *Regula Falsi* method. Compare your estimates with the actual roots of $f(x) = 0$.

3. The equation $f(x) = 0$ has two real roots where $f(x) = x^4 + 2x^3 - 10018.01x^2 + 2x - 10019.01$. Since $f(0) < 0, f(x)$ must have a sign change in intervals of the form $[-(k+1), -k]$ and $[j, j+1]$ where k and j are positive integers. Find such intervals and use bisection to isolate each of the roots in an interval of length 10^{-4}. If α is your estimate to one of the roots s, evaluate $f(\alpha)$ and use the mean-value theorem to explain why $|f(\alpha)|$ is not small relative to $|s - \alpha| \simeq 10^{-4}$.

4. Isolating the roots of $f(x) = 0$ for a bracketing method can be a difficult problem. Consider $f(x) = \cos x - \cos 3.1x$, which has a sign change in the interval $[-1, 8]$. The equation has a number of roots in this interval, but suppose you want to find the smallest positive root. From a rough graph of $\cos x$ and $\cos 3.1x$, determine a suitable interval for the bisection method and find the root to within 10^{-6}.

5. By graphing the curves $y = x$ and $y = \tan x$, convince yourself that $f(x) = x - \tan x$ has zeros in $((2k-1)\pi/2, (2k+1)\pi/2)$, $k = 0, \pm 1, \pm 2, \ldots$. Use the bisection method to find all the zeros of $f(x)$ between 0 and $25\pi/2$, to within 10^{-4}.

6. For $0 < \alpha$, consider $f(x) = \sin nx - \alpha \sin (n-1)x$. Show that $f(x)$ has a sign change in each interval $[(k-1)\pi/n, k\pi/n]$, $1 < k < n$. [Hint: Consider the sign of $f(j\pi/n)$.] For $\alpha = .5$ and $n = 15$, use the bisection method to find 13 roots of $f(x) = 0$, one in each interval, to within 10^{-4}.

7. The equation $f(x) = 0$ in Problem 6 has two roots in $[0, \pi/n]$ and clearly $x = 0$ is one of them. By considering $f'(0)$ and the sign of $f(\pi/n)$, show that $f(x) = 0$ has a second root in $[0, \pi/n]$.

8. From Fig. 4.1, it appears that all the secant lines will pass through the point $(b, f(b))$. This situation suggests a modification to speed up *Regula Falsi*: after x_1 is found, if $f(a)f(x_1) > 0$, then use a line joining $(b, f(b)/2)$ and $(x_1, f(x_1))$ to find x_2. If $f(x_2)f(x_1) > 0$, find x_3 from the line joining $(b, f(b)/4)$ and $(x_2, f(x_2))$. If $f(x_2)f(x_1) < 0$, then x_2 and x_1 bracket the root; and the procedure can start anew. If $f(x)$ is concave up rather than concave down, then $f(b)f(x_1) > 0$ and the role of a and b can be interchanged. Program this modification of subroutine REGFAL and test your program with the function $f(x)$ in Example 4.1.

9. For the function $f(x)$ of Example 4.1, consider $f'(x)$ and show that $f(x) = 0$ has just one root in [0, 1]. Next, since $f(x) = \cos(x) - x$, consider the equation $x = \cos(x)$ and show that $f(x) = 0$ can have but one root for $-\infty < x < \infty$. [Hint: A solution of $x - \cos(x)$ occurs only when the graphs $y = x$ and $y = \cos(x)$ intersect.]

10. Using *Regula Falsi*, estimate the roots of the equations in Problems 5 and 6. Also test the program in Problem 8 on these equations and contrast the results from the three methods.

11. Subroutine REGFAL (Fig. 4.3) signals a success if the subprogram finds a point α such that $|f(\alpha)| \leq$ TOL. Clearly, α can be far from the root s if $f(x)$ is slowly varying near $x = s$; and we can expect this behavior at multiple roots. To illustrate, consider $f(x) = (x-2)^6$ and let TOL $= 10^{-3}$. Find all values α such that $|f(\alpha)| \leq$ TOL.

4.3 FIXED-POINT METHODS

Two methods that normally converge more rapidly than bisection and *Regula Falsi* are Newton's method and the secant method. Although the Newton and secant methods have a simple geometric interpretation for "nice" functions (see, for instance, Figs. 4.4 and 4.8), they do not have the root-bracketing property, and do not guarantee convergence for all continuous functions as do the bisection and *Regula Falsi* methods. When they do converge, however, the Newton and secant methods are generally much faster. In order to explain these convergence properties, we shall derive these methods via a concept known as the *fixed-point problem* and illustrate them geometrically after they are derived. Another important reason for taking this approach is that it can easily be extended to solving systems of equations in several variables. Furthermore, we shall see that other problems such as the iterative methods of Chapter 2 and the predictor-corrector methods of Chapter 7 are special cases of fixed-point iterations and can be analyzed by the procedures of the next section.

To illustrate a specific fixed-point problem, we consider the geometric rationale for Newton's method as given in most calculus texts. This geometric interpretation is quite simple (see Fig. 4.4). Given an initial estimate x_0 to a root s of $f(x) = 0$, we first construct the tangent line to the curve $y = f(x)$ through the point $(x_0, f(x_0))$ and find that the equation of the tangent line is $y = f'(x_0)(x - x_0) + f(x_0)$. We then find the point x_1 where the tangent line intersects the x-axis, and x_1 is taken as the next approximation to s. The process is then repeated with x_1 playing the role of x_0; and a new approximation to s, x_2, is found. Since $x_1 = x_0 - f(x_0)/f'(x_0)$, $x_2 = x_1 - f(x_1)/f'(x_1)$, etc., we are generating the sequence

$$x_{n+1} = x_n - \frac{f(x_n)}{f'(x_n)}, \qquad n = 0, 1, \ldots .$$

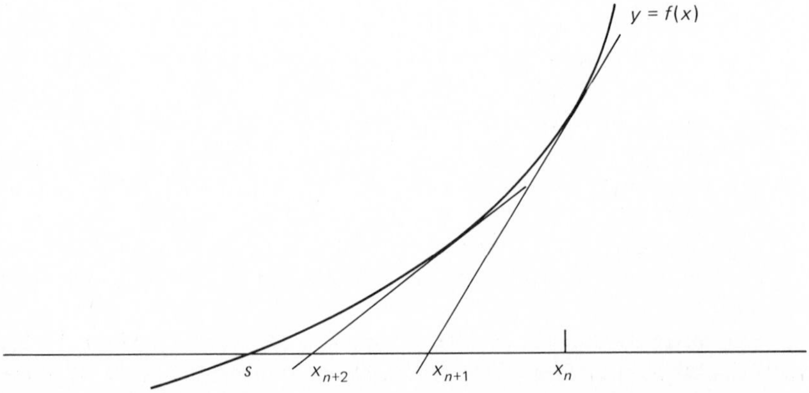

Figure 4.4 Newton's method.

Thus Newton's method is a special case of an iteration of the form $x_{n+1} = g(x_n)$ [where for the case of Newton's method, $g(x) = x - f(x)/f'(x)$]. The analysis of this general iteration, $x_{n+1} = g(x_n)$, is the topic of the next few sections.

4.3.1. The Fixed-Point Problem

Throughout this analysis let us be careful not to lose sight of our prime objective: given a function $f(x)$ where $a \leq x \leq b$, find values s such that $f(s) = 0$. Given such a function, $f(x)$, we now construct an auxiliary function, $g(x)$, such that $s = g(s)$ whenever $f(s) = 0$. The construction of $g(x)$ is not unique. For example, if $f(x) = x^3 - 13x + 18$, then possible choices for $g(x)$ might be (1) $g(x) = (x^3 + 18)/13$, (2) $g(x) = (13x - 18)^{1/3}$, (3) $g(x) = (13x - 18)/x^2$, and (4) $g(x) = x^3 - 12x + 18$, to list a few. In each of these cases, if $f(s) = 0$, then $s = g(s)$.

The problem of finding s such that $s = g(s)$ is known as the *fixed-point problem*, and s is said to be a *fixed point* of $g(x)$. Thus if we develop an efficient procedure for finding a fixed point for $g(x)$, $a \leq x \leq b$, then we automatically have an efficient procedure for finding a zero of $f(x)$, $a \leq x \leq b$. The fixed-point problem turns out to be quite simple, both theoretically and geometrically. It is immediate from Fig. 4.5 that $g(x)$ has a fixed point in the interval $[a, b]$ whenever the graph of $g(x)$ intersects the line $y = x$.

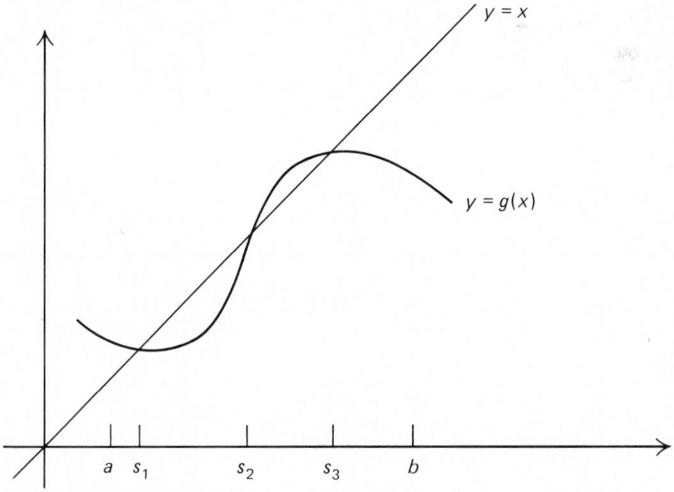

Figure 4.5 $s_i = g(s_i)$; $i = 1, 2, 3$.

It is also obvious that on any given interval, $I \equiv [a, b]$, $g(x)$ may have many fixed points or none at all. Thus, in order to ensure that $g(x)$ has a fixed point in I, certain restrictions must be placed on $g(x)$. First of all, let us assume for each $x \in I$, that $g(x) \in I$. This is plausible since, for $s \in I$, s cannot equal $g(s)$ if no

$g(x)$ belongs to I. [Hereafter this condition will be written as $g(I) \subseteq I$.] Even with this restriction, $g(x)$ may still not have a fixed point in I. For example if $g(x)$ is discontinuous, part of its graph may lie above $y = x$ and part may be below. However if we also require $g(x)$ to be continuous, we can prove that $g(x)$ must have a fixed point in I. To see this, suppose that $g(I) \subseteq I$ and $g(x)$ is continuous. Observe that $g(I) \subseteq I$ means $a \leq g(a) \leq b$ and $a \leq g(b) \leq b$. If either $a = g(a)$ or $b = g(b)$, then that endpoint is a fixed point. Let us assume that such is not the case so that $(g(a) - a) > 0$ and $(g(b) - b) < 0$. For $F(x) \equiv g(x) - x$, $F(x)$ is continuous and $F(a) > 0$ and $F(b) < 0$. Thus, by the intermediate value theorem, there exists at least one $s \in I$ such that $F(s) \equiv g(s) - s = 0$. Therefore we have established Theorem 4.1.

Theorem 4.1

If $g(I) \subseteq I$ and $g(x)$ is continuous, then $g(x)$ has at least one fixed point in I.

In order to ensure that $g(x)$ has a unique fixed point in I, we must not allow $g(x)$ to vary too rapidly. Thus we make the additional assumption that $g'(x)$ exists on I and that $|g'(x)| \leq L < 1$ for all $x \in I$. [Note that this condition implies that $g(x)$ is continuous on I.] Now let us assume that $s_1 \in I$, $s_2 \in I$, $s_1 \neq s_2$, and $s_1 = g(s_1)$ and $s_2 = g(s_2)$. Then by the mean-value theorem with ξ between s_1 and s_2,

$$|s_2 - s_1| = |g(s_2) - g(s_1)| = |g'(\xi)(s_2 - s_1)| \leq L|s_2 - s_1| < |s_2 - s_1|,$$

which is a contradiction. Thus we have proved Theorem 4.2.

Theorem 4.2

If $g(I) \subseteq I$ and $|g'(x)| \leq L < 1$ for all $x \in I$, then there exists exactly one $s \in I$ such that $g(s) = s$.

Now that we have established a condition for which $g(x)$ has a unique fixed point in I, there remains the problem of how to find it. The technique we shall employ is known as the fixed-point iteration, given by the following algorithm, and illustrated in Fig. 4.6.

Let x_0 be arbitrary in $I = [a, b]$, and let $x_{n+1} = g(x_n)$ for all $n \geq 0$.

Geometrically, this sequence can be pictured in the following way. Given any x_n of the above sequence, then $x_{n+1} = g(x_n)$ is the y-coordinate of the point $(x_n, g(x_n))$ on the graph of $y = g(x)$. Now consider the point $(x_{n+1}, x_{n+1}) = (x_{n+1}, g(x_n))$, which lies on the graph of $y = x$. The vertical projection from this point to the x-axis yields the point $(x_{n+1}, 0)$; so we know the position of x_{n+1} on the x-axis and are ready to repeat the process as illustrated in Fig. 4.6.

Note that if for any n, $x_n = s$, then $x_{n+1} = g(x_n) = g(s) = s = x_n$. Similarly, $x_m = s$, for all $m \geq n$, and the sequence stays "fixed" at s. We shall now show

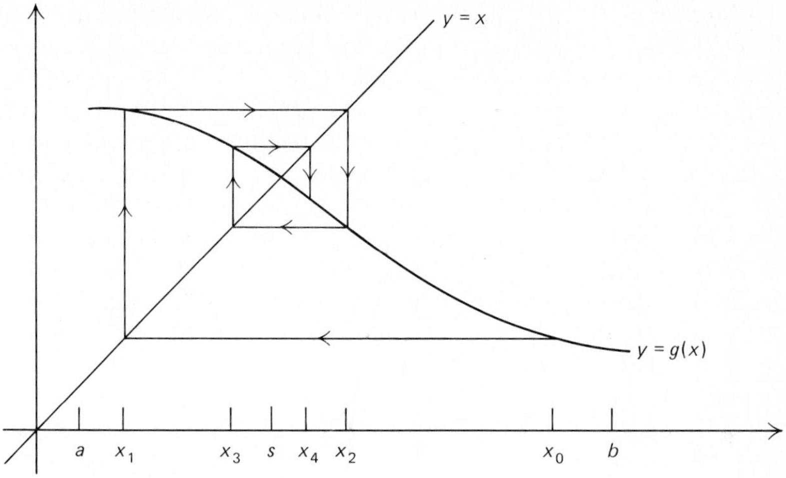

Figure 4.6 The fixed-point iteration.

that under the conditions of Theorem 4.2, the fixed-point iteration converges, and we shall give a bound on the error after n steps.

Theorem 4.3

Let $g(I) \subseteq I \equiv [a, b]$ and $|g'(x)| \leq L < 1$ for all $x \in I$. For $x_0 \in I$, the sequence $x_n = g(x_{n-1})$, $n = 1, 2, \ldots$ converges to the fixed-point s, and the nth error, $e_n \equiv x_n - s$, satisfies

$$|e_n| \leq \frac{L^n}{(1 - L)} |x_1 - x_0|.$$

Proof. By Theorem 4.2, we know there is precisely one fixed point s in I. Given any n, there exists a mean-value point ξ_n between x_{n-1} and s such that

$$|x_n - s| = |g(x_{n-1}) - g(s)| = |g'(\xi_n)|\,|x_{n-1} - s| \leq L|x_{n-1} - s|.$$

Successive repetition of this inequality yields $|x_n - s| \leq L^n|x_0 - s|$. Since $0 \leq L < 1$,

$$\lim_{n\to\infty} L^n = 0, \text{ and so } \lim_{n\to\infty} x_n = s.$$

Thus the method is convergent. To establish the error bound, note that

$$|x_0 - s| \leq |x_0 - x_1| + |x_1 - s| \leq |x_0 - x_1| + L|x_0 - s|.$$

Therefore, $(1 - L)|x_0 - s| \leq |x_1 - x_0|$; and since $|x_n - s| \leq L^n|x_0 - s|$, it follows that

$$|x_n - s| \leq L^n|x_1 - x_0|/(1 - L). \qquad \blacksquare$$

Theorem 4.3 bears a striking resemblance to Theorem 2.2 in Chapter 2, and the proofs are essentially the same. One important feature of Theorem 4.3 is that it provides an error estimate at each step and hence can be used as a test for terminating the iteration. Lacking an estimate such as the one provided by this theorem, we are left only with the possibly unattractive alternative of testing $|x_{n+1} - x_n|$ to determine whether or not to end the iteration.

Theorem 4.3 is called a "nonlocal" convergence theorem because it specifies a fixed, *known* interval, $I = [a, b]$, and displays convergence for any $x_0 \in I$. Often it is not possible to specify such an interval ahead of time, but we might still hope that the fixed-point iteration would converge if we could manage to make our initial guess, x_0, "sufficiently close" to the fixed point s. Any theorem that says, "If the initial guess is 'very close' to the solution, then the method will converge," is called a "local" convergence theorem because it does not specify beforehand precisely how close x_0 must be to s. The following is an example of a local convergence theorem.

Theorem 4.4

Let $g'(x)$ be continuous in some open interval containing s where s is a fixed point of $g(x)$. If $|g'(s)| < 1$, there exists an $\varepsilon > 0$ such that the fixed-point iteration is convergent whenever $|x_0 - s| < \varepsilon$.

Proof. Since $g'(x)$ is continuous in an open interval containing s and $|g'(s)| < 1$, then for any constant K satisfying $|g'(s)| < K < 1$, there exists an $\varepsilon > 0$ such that if $x \in [s - \varepsilon, s + \varepsilon] \equiv I_\varepsilon$, then $|g'(x)| \leq K$. By the mean-value theorem, given any $x \in I_\varepsilon$, there exists an η between x and s such that $|g(x) - s| = |g(x) - g(s)| = |g'(\eta)|\,|x - s| \leq K\varepsilon < \varepsilon$, and thus $g(I_\varepsilon) \subseteq I_\varepsilon$. Therefore by using I_ε in Theorem 4.3, our result is proved. ■

(Note that this theorem does not say what the value of ε is; the theorem merely assures us that there is an ε-neighborhood of s where the fixed-point iteration will converge.) Problem 7, gives a contrasting result, showing that if $|g'(s)| > 1$, then there is a neighborhood of s in which no initial guess (except $x_0 = s$) will work.

PROBLEMS, SECTION 4.3.1

1. The equation $x^3 - 13x + 18 = 0$ is equivalent to the fixed-point problem $x = g(x)$ for each of these choices:

   a) $g(x) = (x^3 + 18)/13$

   b) $g(x) = (13x - 18)^{\frac{1}{3}}$

   c) $g(x) = (13x - 18)/x^2$

   d) $g(x) = x^3 - 12x + 18.$

   Show by direct substitution that $s = 2$ is a fixed point in each case above. For which choices is $|g'(x)| < 1$?

2. The proof of Theorem 4.4 shows that if s is a fixed point of $g(x)$ and if $|g'(x)| \leq K < 1$ in $I = [s - \varepsilon, s + \varepsilon]$, then the fixed-point iteration will converge to s for any x_0 in I. For the appropriate choices in Problem 1, determine some value for ε. Having I, set $x_0 = s + \varepsilon$ and execute four steps of the fixed-point iteration.
3. Verify that the equation $x^2 - 5x + 6 = 0$ is equivalent to $x = g(x)$ for each of these choices:

   a) $g(x) = x^2 - 4x + 6$

   b) $g(x) = 5 - 6/x$

   c) $g(x) = (5x - 6)^{\frac{1}{2}}$

   d) $g(x) = (x^2 + 6)/5$.

   Find the roots of $x^2 - 5x + 6 = 0$ and repeat Problem 2 for each of the roots.
4. Verify that the equation $x^2 - c = 0$ $(c > 0)$ is equivalent to the fixed-point problem $x = (x^2 + c)/2x$. One fixed point is $s = \sqrt{c}$; verify that $0 < g'(x) < 1$ for $\sqrt{c} < x < \infty$. By Problem 5 below, the fixed-point iteration will converge for any x_0 in $(\sqrt{c}, \infty)$. Set $x_0 = c$ and execute six steps of the iteration for $c = 3, 5, 7$. Compare your estimates with the actual solution.
5. Suppose $g'(x)$ is continuous on $[s, b]$ where s is a fixed point of $g(x)$. Suppose also that $0 \leq g'(x) \leq K$ for x in $[s, b]$ where $K < 1$. Show that $[s, b]$ satisfies the hypotheses of Theorem 4.3. Also show that $s \leq \cdots \leq x_{n+1} \leq x_n \leq \cdots \leq x_1 \leq x_0$ when x_0 is in $[s, b]$.
6. Evaluate: $s = \sqrt[3]{6 + \sqrt[3]{6 + \sqrt[3]{6 + \cdots}}}$. [Hint: Let $x_0 = 0$ and consider $g(x) = \sqrt[3]{6 + x}$.]
7. Suppose $g(s) = s$, $g(x)$ is continuously differentiable in an interval containing s, and $|g'(s)| > 1$. Show there is $\varepsilon > 0$ such that if $0 < |x_0 - s| < \varepsilon$, then $|x_0 - s| < |x_1 - s|$ (thus, no matter how close x_0 is to s, the next iterate is farther away).
8. Figure 4.6 illustrates the case where $-1 < g'(s) < 0$. Draw similar graphs illustrating these cases:

   $$0 \leq g'(s) < 1, \qquad 1 < g'(s), \qquad g'(s) < -1.$$
9. Let $g(x) = x^2$. From a graph (as in Problem 8), deduce for what values x_0, $-\infty < x_0 < \infty$, the iteration $x_{i+1} = g(x_i)$ will converge to a fixed point of $g(x)$ and for what values x_0 will the iteration diverge.
10. For the equation $f(x) = 0$ with $f(x) = \cos(x) - x$ (as in Example 4.1), a natural associated fixed-point problem is $x = \cos(x)$. As in Problem 9, graph $g(x) = \cos(x)$ and convince yourself from the picture that the sequence $x_{i+1} = g(x_i)$ will converge for any $x_0 \in [0, 1]$. Program this iteration and test your program with $x_0 = 1$ and $x_0 = 0.1$. Use Theorem 4.3 to prove that the sequence converges for any $x_0 \in [0, 1]$, and print out both the theoretical error bounds and the actual errors at each step of the iteration.
11. Prove the following simple variation of Theorem 4.4. Suppose $g'(x)$ is continuous on $[a, b]$ and suppose $|g'(x)| \leq K < 1$ for all x in $[a, b]$. Suppose also that there is a fixed point s of $g(x)$ in $[a, b]$. Then for any x_0 in $[a, b]$, the fixed-point iteration converges to s. [Hint: as in the proof of Theorem 4.4, show that the hypotheses of Theorem 4.3 are satisfied.]

12. Show that uniqueness of the fixed point in $I = [a, b]$ can be established with the hypothesis that $g(x)$ is differentiable on I and $g'(x) \le L < 1$ (this hypothesis is weaker than the one employed in Theorem 4.2).

4.3.2. Rate of Convergence of the Fixed-point Algorithm

In this section, let s be a fixed point of $g(x)$ where $g(x)$ satisfies the hypotheses of Theorem 4.4 in an interval I. Let $x_0 \in I$; and for each k, let $e_k = x_k - s$. Further, let us suppose that the $(k + 1)$st derivative of $g(x)$ is continuous on I [hereafter written: $g \in C^{k+1}(I)$]. To determine how the errors decrease at each step of the iteration, we expand $g(x)$ in a Taylor's series about $x = s$. This expansion gives us

$$e_{n+1} = g(x_n) - g(s) = g'(s)e_n + \frac{g''(s)}{2!}e_n^2 + \cdots + \frac{g^{(k)}(s)}{k!}e_n^k + E_{k,n} \quad (4.1)$$

where

$$E_{k,n} = g^{(k+1)}(\alpha_n)e_n^{k+1}/(k + 1)!$$

(and where the mean-value point α_n is between x_n and s). Suppose first that $g'(x) \neq 0$ for all $x \in I$. From (4.1) with $k = 0$, we have

$$e_{n+1} = g'(\alpha_n)e_n \qquad \text{or} \qquad \frac{e_{n+1}}{e_n} = g'(\alpha_n).$$

By Theorem 4.4, we have $\lim_{n\to\infty} x_n = s$; so $\lim_{n\to\infty} \alpha_n = s$ as well. Using the continuity of $g'(x)$, we find

$$\lim_{n\to\infty} \frac{e_{n+1}}{e_n} = g'(s). \quad (4.1a)$$

The assumption that $g'(s) \neq 0$ means (for sufficiently large n) that $e_{n+1} \approx g'(s)e_n$. Such a rate of convergence is called *linear* or *first-order* convergence. By contrast, if $g'(s) = 0$ and $g''(x) \neq 0$ for all x in I, then [from (4.1) with $k = 1$] we obtain the stronger result

$$\lim_{n\to\infty} \frac{e_{n+1}}{e_n^2} = \frac{g''(s)}{2!}. \quad (4.1b)$$

This case, in which the $(n + 1)$st error is approximately proportional to the square of the nth error, is called *quadratic* or *second-order* convergence.

Similarly if $g'(s) = g''(s) = \cdots = g^{(k)}(s) = 0$ and $g^{(k+1)}(x)$ does not vanish on I, then we have "$(k + 1)$st order" convergence:

$$\lim_{n\to\infty} \frac{e_{n+1}}{e_n^{k+1}} = \frac{g^{(k+1)}(s)}{(k + 1)!}. \quad (4.1c)$$

Thus the more derivatives of $g(x)$ that vanish at $x = s$, the faster the rate of convergence of the fixed-point iteration. In the analysis above, we implicitly assumed that $e_n \neq 0$ for $n = 0, 1, 2, \ldots$. This assumption is always a valid one if $x_0 \neq s$ and if (as we required) $g^{(i)}(x) \neq 0$ for $x \in I$ when $g'(s) = \cdots = g^{(i-1)}(s) = 0$. For example, consider the linear case, in which $g'(s) \neq 0$. Let m be the first positive integer such that $e_m = 0$. Then

$$0 = e_m = g(x_{m-1}) - g(s) = g'(\alpha)(x_{m-1} - s) = g'(\alpha)e_{m-1}$$

where α is between x_{m-1} and s. As $g'(x) \neq 0$ for all x in I, then $e_{m-1} = 0$, which contradicts our choice of m. Thus if $e_0 \neq 0$, then $e_n \neq 0$ for $n = 1, 2, \ldots$.

We shall be particularly interested in quadratic convergence since this leads to the derivation of Newton's method. For now, however, we pause to consider the slowest case, $g'(x) \neq 0$ for all $x \in I$; and we shall develop a way to speed up the convergence of these sequences. The technique we shall employ is called the *Aitken's Δ^2-method.*

We shall emphasize that Aitken's method can be used on sequences other than those generated by fixed-point iteration and that we can employ the Δ^2-method in different procedures in other chapters. [For example, we note by Problem 14 that the power-method sequence (see Chapter 3) can be accelerated by Aitken's method by virtue of Theorem 4.5.] Thus we present Aitken's method in a context completely divorced from the fixed-point iteration and in its full generality. This method is given in the following algorithm.

Let $\{t_n\}_{n=1}^{\infty}$ be any sequence that converges to t^*. Form a new sequence $\{t_n'\}_{n=1}^{\infty}$ by the formula

$$t_n' = t_n - \frac{(\Delta t_n)^2}{\Delta^2 t_n} \tag{4.2}$$

where $\Delta t_n = t_{n+1} - t_n$ and $\Delta^2 t_n = t_{n+2} - 2t_{n+1} + t_n$. The symbols Δ and Δ^2 are *forward differences,* but we will defer the theory of differences until later. Equation (4.2) is an "extrapolation" procedure that is suggested by considering sequences $\{x_n\}$, which are converging in a "fairly regular" fashion, such as the sequences generated by a linearly converging fixed-point iteration where

$$\lim_{n\to\infty} \frac{x_{n+1} - s}{x_n - s} = g'(s)$$

[that is, $x_{n+1} - s \approx g'(s)(x_n - s)$ when n is large]. If we knew, for example, that $x_{n+1} - s = B(x_n - s)$ and $x_{n+2} - s = B(x_{n+1} - s)$, then we could use these two equations (in the unknowns B and s) to solve for s and obtain

$$s = \frac{x_n - (x_{n+1} - x_n)^2}{x_{n+2} - 2x_{n+1} + x_n},$$

which is exactly analogous to Eq. (4.2) (see Problem 13).

In practice, we use the fact that whenever we have regularity of convergence, we can take three successive terms of a sequence and "extrapolate" to the limit. The manner in which Aitken's Δ^2-method speeds convergence is given by the following theorem, which also explains more precisely what we mean by "regular convergence."

Theorem 4.5

Let $\{t_n\}_{n=1}^{\infty}$ and $\{t'_n\}_{n=1}^{\infty}$ be as in Eq. (4.2) where $\lim_{n\to\infty} t_n = t^*$. Further assume for all n that $\varepsilon_n = t_n - t^*$ satisfies $\varepsilon_{n+1} = (B + \beta_n)\varepsilon_n$ where $\varepsilon_n \neq 0$, $|B| < 1$ and $\lim_{n\to\infty} \beta_n = 0$. Then, for n sufficiently large, t'_n is well defined and the new sequence converges to t^* faster than the old sequence in the sense that

$$\lim_{n\to\infty} \frac{t'_n - t^*}{t_n - t^*} = 0.$$

Proof. Observe that $\varepsilon_{n+2} = (B + \beta_{n+1})\varepsilon_{n+1} = (B + \beta_{n+1})(B + \beta_n)\varepsilon_n$. Therefore

$$\begin{aligned}\Delta^2 t_n &= \varepsilon_{n+2} - 2\varepsilon_{n+1} + \varepsilon_n \\ &= (B + \beta_{n+1})(B + \beta_n)\varepsilon_n - 2(B + \beta_n)\varepsilon_n + \varepsilon_n \\ &= [(B - 1)^2 + \beta'_n]\varepsilon_n\end{aligned}$$

where

$$\beta'_n = B(\beta_n + \beta_{n+1}) - 2\beta_n + \beta_n\beta_{n+1}.$$

Moreover, $\lim_{n\to\infty} \beta'_n = 0$ since $\beta_n \to 0$ as $n \to \infty$. Thus for n sufficiently large, $\Delta^2 t_n \neq 0$; and t'_n is well defined. Now setting

$$\Delta\varepsilon_n = \varepsilon_{n+1} - \varepsilon_n$$

and

$$\Delta^2\varepsilon_n = \varepsilon_{n+2} - 2\varepsilon_{n+1} + \varepsilon_n,$$

we have

$$\begin{aligned}t'_n - t^* &= t_n - t^* - \frac{(\Delta t_n)^2}{\Delta^2 t_n} = t_n - t^* - \frac{(\Delta\varepsilon_n)^2}{\Delta^2\varepsilon_n} \\ &= t_n - t^* - \frac{(B - 1 + \beta_n)^2\varepsilon_n^2}{[(B - 1)^2 + \beta'_n]\varepsilon_n}.\end{aligned}$$

Therefore

$$\lim_{n\to\infty} \frac{t'_n - t^*}{t_n - t^*} = \lim_{n\to\infty} \left\{1 - \frac{(B - 1)^2 + 2\beta_n(B - 1) + \beta_n^2}{(B - 1)^2 + \beta'_n}\right\} = 0. \qquad \blacksquare$$

Corollary

Let $g(I) \subseteq I = [a, b]$ and $g'(x)$ be continuous on I where $0 < |g'(x)| \leq L < 1$. Then Theorem 4.5 may be applied to the fixed-point sequence, $\{x_n\}_{n=0}^{\infty}$, to speed its convergence.

Proof. In Theorem 4.5, identify $\{t_n\}_{n=0}^{\infty}$ with $\{x_n\}_{n=0}^{\infty}$ and t^* with s, the fixed point of $g(x)$. As noted previously, if $x_0 \neq s$, then $\varepsilon_n = t_n - t^* = x_n - s = e_n \neq 0$; and the theorem applies. $\blacksquare$

PROBLEMS, SECTION 4.3.2

1. Verify that $s = 1$ is a fixed point of $g(x) = (x^2 - 4x + 7)/4$. Using Problem 11, Section 4.3.1, show that the fixed-point iteration will converge to s for any x_0 satisfying $\varepsilon < x_0 < 4 - \varepsilon$ where $0 < \varepsilon < 1$. What is the rate of convergence?
2. If $x_{i+1} = g(x_i)$ is a fixed-point iteration converging linearly to s, then $e_{i+1} \simeq g'(s)e_i$, $i = 0, 1, \ldots$ For the iteration in Problem 1, print the values e_i and the ratios e_{i+1}/e_i for $i = 0, 1, \ldots, 10$; use $x_0 = 3.7$ and $x_0 = .1$ as starting values.
3. Code Aitken's Δ^2-method for the iteration in Problem 1; use $x_0 = 3.7$ and $x_0 = .1$ as starting values. Print the estimates x_i, x_i' and the ratio of the errors $(x_i' - s)/(x_i - s)$ for $i = 0, 1, \ldots, 10$.
4. Consider the fixed-point problem $x = g(x)$ where $g(x) = 5 - 6x^{-1}$. Verify that $s = 2$ and $s = 3$ are fixed points of $g(x)$. Using Problem 5, Section 4.3.1, show that the fixed point iteration will converge to $s = 3$ for any x_0 in $(3, \infty)$. Repeat Problems 2 and 3 for this function $g(x)$; choose $x_0 = 5$, $x_0 = 10$, and $x_0 = 1000$.
5. Let $g(x) = x^2 - 2x + 2$. What are the fixed points of $g(x)$? For each fixed point s, determine whether there is an $\varepsilon > 0$ such that if $|x_0 - s| < \varepsilon$, then the sequence $x_{i+1} = g(x_i)$ is convergent to s. What is the order of convergence of the iteration at those fixed points for which convergence occurs?
6. Let $\{t_n\}_{n=1}^{\infty} = \{(1/2)^n\}_{n=1}^{\infty}$ and $\{r_n\}_{n=1}^{\infty} = \{1/n^2\}_{n=1}^{\infty}$. Using Theorem 4.5, determine whether Aitken's Δ^2-method can be applied to either of these sequences.
7. Repeat Problem 6 for the sequence $\{t_n\}_{n=0}^{\infty}$ if

   a) $t_n = \dfrac{2n^2 + n - 3}{n^2 + 6}$

   b) $t_n = 1 - \left(\frac{7}{8}\right)^n$
8. The "geometric" series $\sum_{k=0}^{\infty} r^k$ converges for any value r in $(-1, 1)$, and the series converges to the value $1/(1 - r)$.

   a) Prove this statement by showing that if $S_n = 1 + r + r^2 + \cdots + r^n$, then $S_n = (1 - r^{n+1})/(1 - r)$. [Hint: Show that $(1 - r)(1 + r + r^2 + \cdots + r^n) = 1 - r^{n+1}$.]

   b) Complete the proof by calculating $\lim_{n\to\infty} S_n$.
9. Use Problem 8 and Theorem 4.5 to conclude that Aitken's Δ^2-method can be applied to the sequence of partial sums $\{S_n\}$ of the geometric series. For $r = .7$, calculate $S_0, S_1, \ldots, S_{20}$ and print these values. Next, apply the routine from Problem 3 to this sequence and note the improvement.
10. For $0 < x < 1$, the function $f(x) = 1/(1 - \sqrt{x})$ can be represented by a geometric series if we set $r = \sqrt{x}$. Repeat Problem 9 for this series representation for $f(x)$; use $x = .3$.
11. Using Problem 10, determine how you might estimate the integral

$$\int_{.04}^{.64} \frac{1}{1 - \sqrt{x}}\, dx.$$

12. In formula (4.1) assume that $x_0 \neq s$, $g'(s) = 0$, and that $g''(x)$ is continuous and nonzero on I. Show that $e_n \neq 0$ for all $n \geq 1$.

13. Suppose that $\{x_n\}$ is a sequence and $x_{i+1} - s = B(x_i - s)$ and $x_{i+2} - s = B(x_{i+1} - s)$ for some i. Under the assumption that $B \neq 1$, solve these two equations for s, and thus give an intuitive derivation for Aitken's Δ^2-method.

14. Consider the sequence $\{\beta_k\}$ generated by the power method as in formula (3.12). Use (3.12) to show that

$$\lim_{n\to\infty} \frac{\beta_{k+1} - \lambda_1}{\beta_k - \lambda_1} = r, \qquad |r| < 1.$$

Thus the sequence $\{\beta_k\}$ satisfies the conditions of Theorem 4.5, and can be accelerated by Aitken's method.

15. (Steffensen's method). Consider the fixed-point problem, $x = g(x)$, with an initial guess $x_0 = t_0$ and solution $s = g(s)$. Let $x_1 = g(x_0)$ and $x_2 = g(x_1) = g(g(x_0))$. Now apply the Aitken formula (4.2) to obtain

$$t_1 = x_0 - \frac{(\Delta x_0)^2}{\Delta^2 x_0} = x_0 - \frac{(x_1 - x_0)^2}{x_2 - 2x_1 + x_0}.$$

Now repeat the process; that is, let $x_0' = t_1$, $x_1' = g(t_1)$, $x_2' = g(x_1') = g(g(t_1))$, and $t_2 = t_1 - (g(t_1) - t_1)^2/[g(g(t_1)) - 2g(t_1) + t_1]$. Thus we see we are generating a sequence $\{t_k\}$ where each t_k is obtained by three steps of a fixed-point iteration and then an Aitken's method correction. We formalize this procedure by defining $R(x) = g(g(x)) - 2g(x) + x$; let $G(x) = x - (g(x) - x)^2/R(x)$ if $R(x) \neq 0$; and let $G(x) = x$ if $R(x) = 0$. Thus with t_0 given, the sequence $\{t_k\}$ is generated by the fixed-point formula: $t_{k+1} = G(t_k)$, $k \geq 0$.

a) Let $g(x) = \sqrt[3]{6 + x}$ and $x_0 = t_0 = 3$. Write a program that generates *both* the fixed-point sequence $x_{k+1} = g(x_k)$ and the Steffensen sequence $t_{k+1} = G(t_k)$. Compare the rates of convergence. [Obviously $s = 2$ satisfies $s = g(s)$.]

*b) Assume that an arbitrary $g(x)$ satisfies $g'(s) \neq 1$ and $g''(x)$ is continuous: Prove that $G'(s) = 0$, and thus Steffensen's method is quadratically convergent by (4.1) and (4.1b).

4.3.3. Newton's Method

We are now ready to apply the fixed-point analysis above to our principal problem of finding the zeros of a given function $f(x)$. One natural choice of the fixed-point function would be $g(x) = x + cf(x)$ where c is a nonzero constant. Then $f(s) = 0$ if and only if $s = g(s)$. From Section 4.3.2, we see that the fixed-point iteration is accelerated by making as many derivatives of $g(x)$ at $x = s$ equal to zero as possible. Now, $g'(s) = 1 + cf'(s)$; thus $g'(s) \neq 0$ unless $c = -(1/f'(s))$. Unfortunately since we do not know the value of s, it is impossible to make an *a priori* choice for c.

Thus we are led to make a different choice for $g(x)$. This time we let $g(x) = x + h(x)f(x)$, and try to select $h(x)$ such that $g'(s) = 0$. Now, $g'(s) = 1 + h'(s)f(s) + h(s)f'(s) = 1 + h(s)f'(s)$. Thus for $g'(s) = 0$, we must select $h(x)$ such that $h(s) = -(1/f'(s))$. Immediately we see that $h(x) \equiv -(1/f'(x))$ has this property, and furthermore we need not know the value of s to make this choice. Therefore, we select $g(x) \equiv x - f(x)/f'(x)$, and the fixed-point algorithm yields the following iteration, known as Newton's method. Given the function $f(x)$, let x_0 be an initial guess for s where $f(s) = 0$. Then let

$$x_{n+1} = x_n - f(x_n)/f'(x_n), \qquad n = 0, 1, \ldots . \tag{4.3}$$

The analysis above along with Theorem 4.4 yields the following local convergence theorem for Newton's method. (The proof is left to the reader.)

Theorem 4.6

Let $f''(x)$ be continuous and $f'(x) \neq 0$ in some open interval containing s where $f(s) = 0$. Then there exists an $\varepsilon > 0$ such that Newton's method is quadratically convergent whenever $|x_0 - s| < \varepsilon$.

Subroutine NEWTON (Fig. 4.7) employs Newton's method to find an approximate root of the equation $f(x) = 0$. Programming a root-finding procedure that does not possess the bracketing property of the bisection method or *Regula Falsi* presents one difficulty: selecting an appropriate test for terminating the iteration. When it is not practical to use a guaranteed error bound, such as that displayed in Theorem 4.3, we are left with the alternatives of prescribing either a tolerance $\varepsilon > 0$ or a tolerance $\delta > 0$, and stopping the iteration when $|x_{n+1} - x_n| < \delta$ or when $|f(x_n)| < \varepsilon$. Subroutine NEWTON uses both of these criteria and an upper bound on the number of iterations to be executed as well.

```
      SUBROUTINE NEWTON(XO,XTERM,FTERM,N,ITERM)
C
C     THIS SUBROUTINE USES NEWTON'S METHOD TO FIND A ROOT OF F(X)=0.  THE
C     CALLING PROGRAM MUST SUPPLY AN INITIAL GUESS, XO, AND 3 TERMINATION
C     CRITERIA, XTERM, FTERM AND N.  THE SUBROUTINE RETURNS AN APPROXIMATE
C     ROOT IN XO WHEN ONE OF THE TERMINATION REQUIREMENTS IS MET.  A FLAG,
C     ITERM, IS SET TO 1 WHEN THE ABSOLUTE VALUE OF F(XO) IS LESS THAN
C     FTERM; ITERM IS SET TO 2 WHEN 2 SUCESSIVE ITERATES DIFFER BY LESS
C     THAN XTERM AND ITERM IS SET TO 3 WHEN THE MAXIMUM NUMBER OF
C     ITERATIONS, N, IS REACHED.  TWO FUNCTION SUBPROGRAMS NAMED F(X)
C     AND FPRIM(X) MUST BE SUPPLIED TO CALCULATE F(X) AND THE DERIVATIVE
C     OF F(X).
C
      DO 1 I=1,N
      FO=F(XO)
      IF(ABS(FO).LT.FTERM)  GO TO 2
      FPO=FPRIM(XO)
      CORREC=FO/FPO
      IF(ABS(CORREC).LT.XTERM)   GO TO 3
    1 XO=XO-CORREC
      ITERM=3
      RETURN
    2 ITERM=1
      RETURN
    3 ITERM=2
      XO=XO-CORREC
      RETURN
      END
```

Figure 4.7 Subroutine NEWTON.

EXAMPLE 4.2. Let $f(x)$ be the simple function of Example 4.1, $f(x) = \cos(x) - x$. Starting with $x_0 = 0$, the following sequence of iterates was generated by Subroutine NEWTON. (See Table 4.2.) For this example, convergence is quite rapid.

TABLE 4.2 Newton's method.

x_i	$f(x_i)$
0.0000000E 00	0.1000000E 01
0.1000000E 01	−0.4596977E 00
0.7503638E 00	−0.1892304E−01
0.7391128E 00	−0.4643201E−04
0.7390850E 00	0.5960464E−07

We note that although quite simple, the geometric derivation illustrated in Figure 4.4 says nothing about quadratic convergence, sufficient conditions on $f(x)$ to ensure convergence, and the question of convergence when $f'(s) = 0$. Recall that we say the root s has *multiplicity* p if $f(s) = f'(s) = \cdots = f^{(p-1)}(s) = 0$, but $f^{(p)}(s) \neq 0$. We shall consider the case in which s has multiplicity 2, that is, s is a *double root* of $f(x)$, and consider higher multiplicities in the problems.

Let us assume that $f(s) = f'(s) = 0$ and $f^{(4)}(x)$ is continuous. Again let $g(x) = x - f(x)/f'(x)$; so $g'(x) = 1 - (f'(x)^2 - f(x)f''(x))/f'(x)^2$, or $g'(x) = f(x)f''(x)/f'(x)^2$. We easily verify by l'Hôpital's rule that $g'(s) = 1/2$. Since $|g'(s)| < 1$, and the rest of the hypotheses of Theorem 4.4 are satisfied, we see that Newton's method still converges locally to s. The convergence, however, is not quadratic. Instead we get only linear convergence where $\lim_{n\to\infty} (e_{n+1}/e_n) = g'(s) = 1/2$.

In the analysis above, however, if we choose $g(x) = x - 2f(x)/f'(x)$, then $g'(s) = 0$. Thus if we know *a priori* that the multiplicity of s is 2, then the sequence

$$x_{n+1} = x_n - 2f(x_n)/f'(x_n)$$

will converge to s quadratically for x_0 sufficiently close to s.

In general, if s has multiplicity p, then we can show (Problem 9) that if $g(x) = x - f(x)/f'(x)$, then $g'(s) = 1 - 1/p$; and if $g(x) = x - pf(x)/f'(x)$, then $g'(s) = 0$. Thus for x_0 sufficiently close to s, the sequence

$$x_{n+1} = x_n - pf(x_n)/f'(x_n) \tag{4.4}$$

is quadratically convergent to s. The formula (4.4) is of little practical use, however, since we rarely know the multiplicity of a root in advance.

EXAMPLE 4.3. In this example we illustrate the behavior of Newton's method near a point s where $f(s) = f'(s) = 0$. For $f(x) = x^3 + x^2 - 5x + 3$, we have $f(x) = (x-1)^2(x+3)$ so that $f(1) = f'(1) = 0$ and $f(-3) = 0$. Newton's method was run with an initial guess of $x_0 = 4$. Convergence to $s = 1$ is seen to be quite slow. As the iterates near 1, they exhibit an oscillatory behavior that is typical when the limits of machine accuracy are approached. The same program was run with an initial guess of $x_0 = -6$, with rapid convergence to $s = -3$. (See Table 4.3.) The same program run in double precision with $x_0 = 4$ yields results (Table 4.4) that are in agreement with geometric intuition (see Problem 9 of this section).

TABLE 4.3.

x_i	$f(x_i)$	x_i	$f(x_i)$
0.4000000E 01	0.6300000E 02	−0.6000000E 01	−0.1470000E 03
0.2764706E 01	0.1795237E 02	−0.4384615E 01	−0.4014566E 02
0.1999479E 01	0.4994273E 01	−0.3470246E 01	−0.9396978E 01
0.1545152E 01	0.1350779E 01	−0.3081737E 01	−0.1361792E 01
0.1287998E 01	0.3556594E 00	−0.3003147E 01	−0.5043125E−01
0.1148676E 01	0.9170532E−01	−0.3000004E 01	−0.7629394E−04
0.1075647E 01	0.2332305E−01	−0.3000000E 01	0.0000000E 00
0.1038170E 01	0.5883216E−02		
0.1019176E 01	0.1478195E−02		
0.1009609E 01	0.3700256E−03		
0.1004812E 01	0.9250640E−04		
0.1002412E 01	0.2384185E−04		
0.1001178E 01	0.5722045E−05		
0.1000572E 01	0.1907348E−05		
0.1000155E 01	0.9536743E−06		
0.9993886E 00	0.1907348E−05		
0.9997784E 00	0.9536743E−06		
0.1000315E 01	0.9536743E−06		
0.9999380E 00	0.0000000E 00		

TABLE 4.4

x_i	$f(x_i)$
0.40000000D 01	0.63000000D 02
0.27647059D 01	0.17952371D 02
0.19994794D 01	0.49942757D 01
0.15451534D 01	0.13507843D 01
0.12879985D 01	0.35566004D 00
0.11486779D 01	0.91707018D−01
0.10756476D 01	0.23323111D−01
0.10381716D 01	0.58838948D−02
0.10191756D 01	0.14778606D−02
0.10096106D 01	0.37034232D−03
0.10048111D 01	0.92696274D−04
0.10024070D 01	0.23187972D−04
0.10012038D 01	0.57987349D−05
0.10006020D 01	0.14499018D−05
0.10003010D 01	0.36250271D−06
0.10001505D 01	0.90629086D−07
0.10000753D 01	0.22657698D−07
0.10000376D 01	0.56644780D−08
0.10000188D 01	0.14161259D−08
0.10000094D 01	0.35403258D−09

Newton's method is very useful in problems in which $f'(x)$ is easily evaluated. [Such is the case where $f(x)$ is a polynomial as we shall see in Section 4.4.] We shall illustrate here a case of particular importance. Suppose we are given a constant $c > 0$ and wish to find the real, positive kth root of c. We then let $f(x) = x^k - c$, and the Newton iteration becomes

$$x_{n+1} = x_n - \frac{x_n^k - c}{kx_n^{k-1}} = \frac{(k-1)x_n + c/x_n^{k-1}}{k}. \tag{4.5}$$

EXAMPLE 4.4. In order to approximate $\sqrt{2}$, let $f(x) = x^2 - 2$ and $x_0 = 1$. Formula (4.5) reduces in this case to $x_{n+1} = (x_n + 2/x_n)/2$. Then to eight significant digits, $x_1 = 1.5000000$, $x_2 = 1.4166667$, $x_3 = 1.4142156$, and $x_4 = 1.4142135$. In fact, $\sqrt{2} = 1.4142136$ (to eight places).

For practical computational purposes, some modifications of Newton's method are desirable. These modifications are based on the observation that although Newton's method converges rapidly for a starting value near a simple root of $f(x) = 0$, the method may diverge rapidly (or exhibit other erratic behavior) in the presence of a zero of $f'(x)$ or for a starting value somewhat removed from the root. In order to account for this fact in a general root-finding program, we might include a test of whether $|f(x_{n+1})| < |f(x_n)|$ at each step (such a test would determine whether it is profitable to continue the iteration using Newton's method). A somewhat more comprehensive program might include a combination of bisection and Newton's method. For example, suppose $f(a)f(b) < 0$ and $m = (a + b)/2$. If $f(m)f(b) < 0$, then there is a zero of $f(x)$ in (m, b). If Newton's method starting with $x_0 = b$ (or $x_0 = a$) does not produce an estimate x_1 in (m, b), then x_1 can be rejected and several steps of the bisection method can be executed before Newton's method is tried again. Similar modifications may also be made to other "fast" procedures, such as the secant method, which do not possess the root-bracketing property of bisection.

PROBLEMS, SECTION 4.3.3

1. Apply Newton's method to $f(x) = x^3 - 2x^2 + 2x - 1$; use $x_0 = 0$ and $x_0 = 10$. Terminate the iteration if $|f(x_n)| \leq 10^{-6}$ or $|x_{n+1} - x_n| \leq 10^{-6}$ or $n \geq 25$. Print each iterate x_n and $f(x_n)$. The root is $s = 1$; so $e_n = x_n - 1$. For each n, also print e_n and e_n^2, and verify that $e_{n+1} \simeq e_n^2$.
2. In Problem 8 it is shown that the errors in Newton's method satisfy $e_{n+1} \simeq Ke_n^2$ where $K = f''(s)/2f'(s)$ [under the assumption that $f(s) = 0$ and $f'(s) \neq 0$]. For $f(x)$ in Problem 1, verify that $K = 1$.
3. Repeat Problem 1 for $f(x) = x^4 - 4x^3 + 6x^2 - 4x + 1$; this time, print x_n, $f(x_n)$, e_n, and $.75\, e_n$ ($s = 1$ is a root). Note that $e_{n+1} \simeq .75\, e_n$. By Problem 9, $e_{n+1} \simeq Ke_n$ when s is a root of multiplicity p where $p \geq 2$ and $K = 1 - 1/p$. Verify that $s = 1$ is a root of multiplicity 4.
4. To see that the estimates from Newton's method may have to be monitored, consider $f(x) = \sin 15x - .5 \sin 14x$. By Problem 6, Section 4.2, $f(x)$ has zeros in $[k\pi/15,$

$(k + 1)\pi/15]$ for $1 \le k \le 13$. Try to find each of these 13 zeros; use $x_0 = k\pi/15$, $1 \le k \le 13$, and the termination criteria in Problem 1. Observe that many of the zeros found are not in the desired range.

5. Design a subroutine that incorporates bisection and Newton's method. The input should include an interval where $f(x)$ changes sign and should use two bisections when a Newton iterate strays outside the current interval. Test your program on the functions in Problems 1 and 4.

6. Use Theorem 4.4 and formula (4.1) to prove Theorem 4.6, with the additional assumption that $f'''(x)$ is continuous.

7. Prove that the tangent line to the graph of $y = f(x)$ at the point $(x_n, f(x_n))$ intersects the x-axis when $x = x_n - f(x_n)/f'(x_n)$.

8. Theorem 4.6 can be proved without the assumption about $f'''(x)$ in Problem 6. First, Theorem 4.4 implies there is an $\varepsilon > 0$ such that $\{x_i\} \to s$ whenever $|x_0 - s| < \varepsilon$ (why?). Now, given that $\{e_n\} \to 0$, show that

$$\lim_{n\to\infty} \frac{e_{n+1}}{e_n^2} = K$$

where $K = f''(s)/2f'(s)$. [Hint: Consider $x_{n+1} - s = x_n - s - f(x_n)/f'(x_n)$, and use the expansion $f(s) = f(x_n) + f'(x_n)(s - x_n) + f''(\theta_n)(s - x_n)^2/2!$ where θ_n is between x_n and s.]

9. Assume that $f(s) = f'(s) = \cdots = f^{(p-1)}(s) = 0$, $f^{(p)}(x)$ is continuous, and $f^{(p)}(s) \ne 0$. Let $g(x) = x - f(x)/f'(x)$, and show that $g'(s) = 1 - 1/p$. [Hint: Let $h = x - s$, and expand both $f(x)$, and $f'(x)$ in a Taylor's expansion around s. Then let $h \to 0$.]

10. To illustrate the necessity of x_0 being "sufficiently close" to s for the convergence of the Newton iteration, we define

$$f(x) = \begin{cases} \sin x, & -\pi/2 \le x \le \pi/2 \\ 1, & x \ge \pi/2 \\ -1, & x \le -\pi/2. \end{cases}$$

Then $f'(x) = \begin{cases} \cos x, & -\pi/2 \le x \le \pi/2 \\ 0, & \text{otherwise.} \end{cases}$

Let t^*, $0 < t^* < \pi/2$, satisfy $\tan t^* = 2t^*$. (Since $\tan \pi/4 = 1 < \pi/2 = 2(\pi/4)$ and $\infty = \tan \pi/2 > \pi = 2(\pi/2)$, we know that such a t^* exists and $\pi/4 < t^* < \pi/2$.)

a) Assume that $x_0 = t^*$; then evaluate the rest of the Newton method iterates. Are they converging to $s = 0$? Are they diverging?

b) What happens to the Newton method iterates if $t^* < x_0 < \pi/2$?

c) What happens to the Newton method iterates if $0 < x_0 < t^*$?

11. Suppose $f(x) = (x - r_1)(x - r_2)\cdots(x - r_n)$ where $r_1 < r_2 < \cdots < r_n$. That is, $f(x)$ is an nth degree polynomial with n distinct real roots and with leading coefficient equal to 1. By a geometric argument, convince yourself that if $x_0 > r_n$, then the sequence $\{x_i\}$ generated by Newton's method satisfies

$$r_n < \cdots < x_{i+1} < x_i < \cdots < x_0 \qquad \text{for } i = 1, 2, \ldots .$$

Prove this mathematically, using Rolle's Theorem to observe that $f'(x) > 0$ for $x \ge r_n$.

12. As an extreme case of a function for which Newton's method is slowly convergent, consider $f(x) = (x - \alpha)^n$ for n some positive integer and α some real number. Show that Newton's method generates the sequence

$$x_{i+1} = (1 - 1/n)x_i + \alpha/n,$$

and then show that $x_{i+1} - \alpha = (1 - 1/n)(x_i - \alpha)$. This is a special case of Problem 9 above.

4.3.4. The Secant Method

In general Newton's method converges much faster than the bracketing methods, but it has serious disadvantages such as how close x_0 must be to s before convergence and the need to evaluate $f'(x_n)$, for each n. This derivative evaluation can be quite a cumbersome task. The secant method is, in a way, a compromise between Newton's method and the bracketing methods. The rate of convergence of the secant method is stronger than linear (called *superlinear*), but not quadratic. This fact, plus the fact that it does not require derivative evaluations, makes it a very attractive, practical method.

The secant method iteration comes directly from the Newton iteration by simply replacing $f'(x_n)$ by the difference quotient $(f(x_n) - f(x_{n-1}))/(x_n - x_{n-1})$. Note that when x_n and x_{n-1} are "close," then this difference quotient is an approximation to $f'(x_n)$. The secant method is represented by the following algorithm.

Given $f(x)$ such that $f(s) = 0$, let x_{-1} and x_0 be initial guesses for s. Then for $n \geq 0$,

$$x_{n+1} = x_n - \frac{f(x_n)(x_n - x_{n-1})}{f(x_n) - f(x_{n-1})} = \frac{f(x_n)x_{n-1} - f(x_{n-1})x_n}{f(x_n) - f(x_{n-1})}. \tag{4.6}$$

This algorithm should remind the reader of the *Regula Falsi* method. In fact if $x_{-1} = a$ and $x_0 = b$ bracket s, then x_1 is generated by exactly the same formula that generates the first iterate of *Regula Falsi*. (See Problem 1.) However, x_2

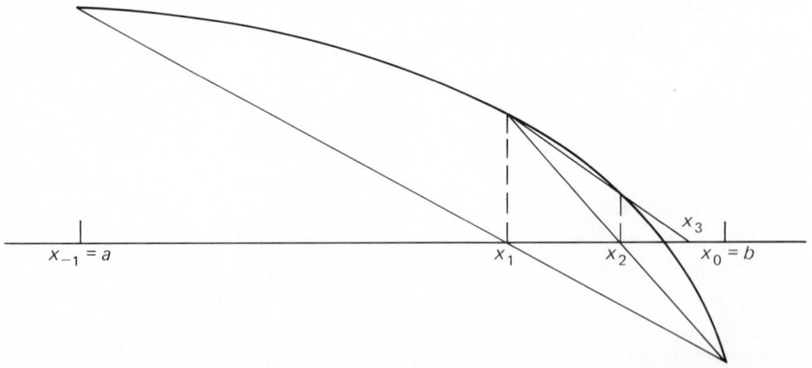

Figure 4.8 Secant method.

will not (in general) be the second iterate of *Regula Falsi* since x_0 and x_1 will not necessarily bracket the root. This point is illustrated in Fig. 4.8, which has the same object function as Fig. 4.1. We further note that since x_{n+1} depends explicitly on x_{n-1} and x_n, the secant method is not a fixed-point iteration although directly derived from the Newton iteration, which is. The secant method does have the property, similar to the fixed-point iteration, that if $x_{n-1} \neq s$ but $x_n = s$, then $x_{n+1} = s$. For the secant iteration to be well defined, we must assume that $f(x_n) - f(x_{n-1}) \neq 0$.

EXAMPLE 4.5. The secant method was programmed in single precision for $f(x) = \cos(x) - x$ with $x_0 = 0$ and $x_1 = 1$. Note from Eq. (4.6) that there are two mathematically equivalent ways to generate the sequence $\{x_n\}$:

Form 1
$$x_{n+1} = x_n - f(x_n)\frac{x_n - x_{n-1}}{f(x_n) - f(x_{n-1})};$$

Form 2
$$x_{n+1} = \frac{f(x_n)x_{n-1} - f(x_{n-1})x_n}{f(x_n) - f(x_{n-1})}.$$

Most numerical analysts feel that Form 1 is preferable to Form 2 and is less subject to rounding error. This particular example was computed using Form 1. (See Table 4.5.)

TABLE 4.5

x_i	$f(x_i)$
0.0000000E 00	0.1000000E 01
0.1000000E 01	−0.4596977E 00
0.6850733E 00	0.8929920E−01
0.7362989E 00	0.4660129E−02
0.7391192E 00	−0.5722045E−04
0.7390850E 00	0.5960464E−07

To analyze the rate of convergence for the secant method, let us suppose that $f''(x)$ is continuous, $s \neq x_{n-1} \neq x_n$; and let us define $e_n = s - x_n$ for all n. Then after algebraic reduction,

$$e_{n+1} = s - x_{n+1} = \frac{f(x_n)e_{n-1} - f(x_{n-1})e_n}{f(x_n) - f(x_{n-1})}$$
$$= \frac{f(x_n)e_{n-1} - f(x_{n-1})e_n}{x_{n-1} - x_n} \cdot \frac{x_{n-1} - x_n}{f(x_n) - f(x_{n-1})}. \tag{4.7}$$

Now,

$$\frac{f(x_n)e_{n-1} - f(x_{n-1})e_n}{x_{n-1} - x_n} = e_n e_{n-1}\frac{f(x_n)/e_n - f(x_{n-1})/e_{n-1}}{x_{n-1} - x_n}$$
$$= e_n e_{n-1}\left(\frac{\dfrac{f(x_n) - f(s)}{x_n - s} - \dfrac{f(x_{n-1}) - f(s)}{x_{n-1} - s}}{x_n - x_{n-1}}\right)$$
$$\equiv e_n e_{n-1} f[x_n, s, x_{n-1}].$$

Let $G(x) = (f(x) - f(s))/(x - s)$; then by the mean-value theorem, $f[x_n, s, x_{n-1}] = (G(x_n) - G(x_{n-1}))/(x_n - x_{n-1}) = G'(\zeta_n)$. Now

$$G'(x) = \frac{f'(x)(x - s) + f(s) - f(x)}{(x - s)^2}$$

and

$$f(s) = f(x) + f'(x)(s - x) + \frac{f''(\eta)}{2}(s - x)^2$$

by Taylor's Theorem and the continuity of $f''(x)$. Putting these two equations together, we get

$$G'(x) = \frac{f''(\eta)}{2}, \qquad \text{or} \qquad f[x_n, s, x_{n-1}] = G'(\zeta_n) \equiv \frac{f''(\eta_n)}{2}. \tag{4.8}$$

Thus from Eqs. (4.7) and (4.8) and the mean-value theorem,

$$\begin{aligned} e_{n+1} &= e_n e_{n-1} f[x_n, s, x_{n-1}](x_{n-1} - x_n)/(f(x_n) - f(x_{n-1})) \\ &= e_n e_{n-1}\left(\frac{f''(\eta_n)}{2}\right)\left(\frac{-1}{f'(\xi_n)}\right) \end{aligned} \tag{4.9}$$

where both η_n and ξ_n lie in the smallest interval containing x_{n-1}, x_n, and s. With the aid of (4.9) we are able to prove the following local convergence theorem for the secant method.

Theorem 4.7

Let $f(s) = 0, f'(s) \neq 0$; and let $f''(x)$ be continuous in a neighborhood of s. Then there exists an $\varepsilon > 0$ such that if $x_{-1}, x_0 \in I_\varepsilon \equiv [s - \varepsilon, s + \varepsilon]$, then the secant method converges to s.

Proof. In this proof, we will let M_α denote an upper bound for $|f''(x)/f'(x')|$ where x and x' are any points in $[s - \alpha, s + \alpha]$. Since $f''(x)$ and $f'(x)$ are continuous at s and since $f'(s) \neq 0$, we see, for $\beta > 0$ but sufficiently small, that the inequality $|f''(x)/f'(x')| \leq M_\beta$ is valid for x and x' in $[s - \beta, s + \beta]$. Choose $\varepsilon > 0$ such that $\varepsilon M_\beta \equiv K < 1$ where $\varepsilon < \beta$ so that $M_\varepsilon \leq M_\beta$. Let $|x_{-1} - s| \leq \varepsilon$ and $|x_0 - s| \leq \varepsilon$; then $|e_1| \leq |e_{-1}e_0|M_\beta \leq \varepsilon^2 M_\beta = \varepsilon K < \varepsilon$. Also, $|e_2| \leq |e_0 e_1|M_\beta = |e_0|(|e_0|M_\beta) < (\varepsilon K)(\varepsilon M_\beta) < \varepsilon K^2$.

Now with the induction hypotheses that $e_i < \varepsilon K^i$ and $e_{i-1} < \varepsilon K^{i-1} < \varepsilon$, we can easily see that $|e_{i+1}| < \varepsilon K^{i+1}$ and that $x_{i+1} \in [s - \varepsilon, s + \varepsilon]$ for all i. Therefore $\lim_{i\to\infty} e_i = 0$, and we have convergence on I_ε. [The reader can easily see that the inequality on e_{i+1} can be sharpened (see below), but this process was not necessary for the proof of simple convergence.] ■

It is beyond the scope of this text to establish rigorously the exact rate of convergence of the secant method [for a thorough coverage, see Ostrowski (1966)]. We can, however, present a convincing intuitive argument. Let M_β be given as in the proof above, and let $d_n \equiv M_\beta e_n$. Then

$$d_{n+1} = M_\beta e_{n+1} \leq e_{n-1} e_n M_\beta^2 = d_n d_{n-1}.$$

Let $d \equiv \max\{d_{-1}, d_0\}$; then

$$d_1 \leq d_{-1}d_0 \leq d^2, \qquad d_2 \leq d_0 d_1 \leq d^3, \qquad d_3 \leq d_1 d_2 \leq d^5;$$

and clearly by induction we see that $d_n \leq d^{\alpha_n}$ where $\alpha_{n+1} = \alpha_n + \alpha_{n-1}$, $\alpha_0 = 1$, $\alpha_1 = 2$. It is left (as Problem 3) for the reader to show that these integers, α_n, are given by

$$\alpha_n = \frac{3+\sqrt{5}}{2\sqrt{5}}\left(\frac{1+\sqrt{5}}{2}\right)^n - \frac{3-\sqrt{5}}{2\sqrt{5}}\left(\frac{1-\sqrt{5}}{2}\right)^n,$$

which is the famous *Fibonacci sequence* with the so-called *golden ratio,*

$$r^* = \left(\frac{1+\sqrt{5}}{2}\right).$$

The rate of convergence as shown in Ostrowski is superlinear; and under the conditions of Theorem 4.7, $\lim_{n\to\infty} (e_{n+1}/e_n^{r^*})$ exists and is usually not zero so the order of convergence is r^*. (Note: $1 < r^* < 2$.)

4.3.5. Newton's Method in Two Variables

In this section, we will derive Newton's method for solving a system of nonlinear equations, particularly the system of two equations in two unknowns:

$$\begin{aligned} f(x, y) &= 0 \\ g(x, y) &= 0. \end{aligned} \tag{4.10}$$

In (4.10), $f(x, y)$ and $g(x, y)$ are real-valued functions of two variables. We will consider this problem in more detail in Section 4.5; we are content in this section to present an intuitive derivation. As we shall see in Section 4.4.2, Newton's method for the special case of the system, (4.10), can be used to find zeros of polynomials where the zeros may be either real or complex.

When solving (4.10), we are looking for a simultaneous solution, that is, numbers s and t such that

$$\begin{aligned} f(s, t) &= 0 \\ g(s, t) &= 0 \end{aligned}$$

In order to derive Newton's method, let us suppose that we have approximations x_0 and y_0 to s and t, respectively. Expanding $f(x, y)$ and $g(x, y)$ in a Taylor's expansion (in two variables) about the point (x_0, y_0), we find

$$\begin{aligned} f(x, y) &= f(x_0, y_0) + f_x(x_0, y_0)(x - x_0) + f_y(x_0, y_0)(y - y_0) + R_f(x, y) \\ g(x, y) &= g(x_0, y_0) + g_x(x_0, y_0)(x - x_0) + g_y(x_0, y_0)(y - y_0) + R_g(x, y) \end{aligned} \tag{4.11a}$$

where $f_x(x_0, y_0)$ denotes the partial derivative of $f(x, y)$ with respect to x evaluated at the point (x_0, y_0), and where $R_f(x, y)$ is given by

$$R_f(x, y) = \frac{[f_{xx}(\alpha, \beta)(x - x_0)^2 + 2f_{xy}(\alpha, \beta)(x - x_0)(y - y_0) + f_{yy}(\alpha, \beta)(y - y_0)^2]}{2}$$

with the mean-value point, (α, β), being somewhere on the line segment joining the points (x, y) and (x_0, y_0). If we substitute (s, t) for (x, y) in (4.11a), we find [since (s, t) is a solution of (4.10)] that

$$\begin{aligned} 0 &= f(x_0, y_0) + f_x(x_0, y_0)(s - x_0) + f_y(x_0, y_0)(t - y_0) + R_f(s, t) \\ 0 &= g(x_0, y_0) + g_x(x_0, y_0)(s - x_0) + g_y(x_0, y_0)(t - y_0) + R_g(s, t). \end{aligned} \tag{4.11b}$$

If we now suppose that the point (x_0, y_0) is close to (s, t), then the factors $(s - x_0)^2$, $(s - x_0)(t - y_0)$, and $(t - y_0)^2$, which appear in $R_f(s, t)$ and $R_g(s, t)$, are small with respect to the first-order terms, $(s - x_0)$ and $(t - y_0)$. Thus, we consider the *linear* system of equations, (4.11c), obtained by neglecting $R_f(s, t)$ and $R_g(s, t)$:

$$\begin{aligned} 0 &= f(x_0, y_0) + f_x(x_0, y_0)(s - x_0) + f_y(x_0, y_0)(t - y_0) \\ 0 &= g(x_0, y_0) + g_x(x_0, y_0)(s - x_0) + g_y(x_0, y_0)(t - y_0). \end{aligned} \tag{4.11c}$$

If we solve (4.11c) for s and t, the resulting approximation to the solution of (4.10) should be better than (x_0, y_0). Note, however, that a solution of (4.11c) will not be precisely the pair (s, t) since Eq. (4.11c) is not the same as Eq. (4.11b). If we let (x_1, y_1) denote the solution of (4.11c), then

$$\begin{aligned} f_x(x_0, y_0)(x_1 - x_0) + f_y(x_0, y_0)(y_1 - y_0) &= -f(x_0, y_0) \\ g_x(x_0, y_0)(x_1 - x_0) + g_y(x_0, y_0)(y_1 - y_0) &= -g(x_0, y_0). \end{aligned} \tag{4.11d}$$

It is convenient to write (4.11d) in matrix form as $J_0(\mathbf{x}_1 - \mathbf{x}_0) = -\mathbf{F}(\mathbf{x}_0)$ where

$$\mathbf{x}_0 = \begin{bmatrix} x_0 \\ y_0 \end{bmatrix}, \quad \mathbf{x}_1 = \begin{bmatrix} x_1 \\ y_1 \end{bmatrix}, \quad \mathbf{F}(\mathbf{x}_0) = \begin{bmatrix} f(x_0, y_0) \\ g(x_0, y_0) \end{bmatrix},$$

$$J_0 = \begin{bmatrix} f_x(x_0, y_0) & f_y(x_0, y_0) \\ g_x(x_0, y_0) & g_y(x_0, y_0) \end{bmatrix}.$$

If $\mathbf{c}_0 = \left[\begin{smallmatrix} u_0 \\ v_0 \end{smallmatrix}\right]$ is the solution of $J_0\mathbf{x} = -\mathbf{F}(\mathbf{x}_0)$, then $\mathbf{x}_1 = \mathbf{x}_0 + \mathbf{c}_0$ is an updated (or corrected) estimate to (s, t).

We note that (4.11c) is normally written in an equivalent form (obtained by multiplying by J_0^{-1}): $\mathbf{x}_1 = \mathbf{x}_0 - J_0^{-1}\mathbf{F}(\mathbf{x}_0)$. Having the new estimate, $\mathbf{x}_1$, to the solution, we repeat this correcting process, which leads to the iteration

$$\mathbf{x}_{i+1} = \mathbf{x}_i - J_i^{-1}\mathbf{F}(\mathbf{x}_i), \qquad i = 0, 1, \ldots . \tag{4.11e}$$

The algorithm described by (4.11e) is Newton's method in several variables. The matrix, J_i, of partial derivatives evaluated at $\mathbf{x}_i$ is called the *Jacobian* matrix and plays the role of a derivative (see Section 4.5). Thus the form of Newton's method displayed in (4.11e) is similar to Newton's method for a single function of one variable. This algorithm and variations of it are the basis for Bairstow's method (Section 4.4.2) and also can be effectively used in optimization problems (Appendix) and collocation methods for differential equations (Chapter 7) as well as in the problem of finding solutions to systems of nonlinear equations.

PROBLEMS, SECTION 4.3.5

1. If $x_{-1} = a$ and $x_0 = b$ where $f(a)f(b) < 0$, verify that the first iteration of the secant method is the same as for *Regula Falsi*.
2. Note that Eq. (4.9) can be rewritten as

$$s - \alpha = -(s - a)(s - b)\frac{f''(\eta)}{2f'(\xi)} \tag{4.9a}$$

where α is the point where the line segment through $(a, f(a))$ and $(b, f(b))$ crosses the x-axis. Thus (4.9) can be used for an analysis of *Regula Falsi*. Suppose x_0 and x_1 are such that $f(x_0) > 0 > f(x_1)$, and suppose $f''(x)f'(x) > 0$ for $x_0 \le x \le x_1$ (see Fig. 4.1). If $x_2, x_3, \ldots, x_n, \ldots$ are the iterates of *Regula Falsi*, and if $f(s) = 0$, then

   a) show $s > x_i$ for $i = 2, 3, \ldots$;

   b) let $s - x_i = e_i$ and show (in the notation of Theorem 4.7) that $e_{i+1} \le M_\beta e_i e_0$;

   c) for $\varepsilon = \text{Max}\{|e_0|, |e_1|\}$, show $e_{i+1} \le \lambda^i \varepsilon$ where $\lambda = \varepsilon M_\beta$ (see Section 2).
3. Consider the sequence, $\{\alpha_n\}_{n=0}^{\infty}$, such that $\alpha_{n+1} = \alpha_n + \alpha_{n-1}$, $\alpha_0 = 1$, $\alpha_1 = 2$. Find the only two values of γ (γ_1 and γ_2) that satisfy $\gamma^{n+1} = \gamma^n + \gamma^{n-1}$ for *all* integers $n \ge 0$. Verify that $c_1\gamma_1^n + c_2\gamma_2^n \equiv \alpha_n$ satisfies the equation $\alpha_{n+1} = \alpha_n + \alpha_{n-1}$ for any two values of c_1 and c_2. Verify that for $\alpha_0 = 1$ and $\alpha_1 = 2$, $c_1 = (3 + \sqrt{5})/2\sqrt{5}$ and $c_2 = -(3 - \sqrt{5})/2\sqrt{5}$.
4. Let $f(x) = x^2 - 2$ and $[a, b] \equiv [1, 3]$. Compute the first four iterates of both the secant method and *Regula Falsi*; note how they differ.
5. Program the secant method and test your program on the function $f(x) = \cos(x) - x$.
6. Program Newton's method for two variables and test your program on the system

$$x^2 + y^2 - 9 = 0$$

$$x + y - 1 = 0.$$

(This sytem has two solutions where a solution represents the intersection of the line and the circle described by the equations.)

7. Let $p(x) = x^3 - x^2 - x - 2$ and let u and v be any real numbers.

   a) Verify that $p(x) = (x^2 - ux - v)(x + u - 1) + f_1(u, v)(x - u) + f_2(u, v)$ where

$$f_1(u, v) = u^2 - u + v - 1 \quad \text{and} \quad f_2(u, v) = u^3 - u^2 + 2uv - u - v - 2.$$

Obviously, $p(x) = (x^2 - u^*x - v^*)(x + u^* - 1)$ if and only if u^* and v^* are solutions of $f_1(u^*, v^*) = 0 = f_2(u^*, v^*)$ (see Section 4.4.2). Thus if we know u^* and v^*, we can find two zeros of $p(x)$ by merely applying the quadratic formula to find the two zeros of $x^2 - u^*x - v^*$. Note that even if these two zeros are complex, we need not use any complex arithmetic in finding them.

b) Let $\mathbf{x}_0 \equiv (u_0, v_0) \equiv (0, 0)$, and apply Newton's method for four iterations to find approximations u_4 and v_4 for u^* and v^*, respectively. Apply the quadratic formula to $x^2 - u_4x - v_4 = 0$ to obtain approximations for two zeros of $p(x)$. [The true answers are $(u^*, v^*) = (-1, -1)$, and the resulting zeros are $\frac{1}{2}(-1 \pm \sqrt{3}i)$. This problem presents the basic idea of Bairstow's method, which is presented in Section 4.4.2.]

4.4 ZEROS OF POLYNOMIALS

In Section 4.3, it was noted that a disadvantage of Newton's method was that the derivative of the object function, $f(x)$, had to be calculated at each iterate. For many functions, the calculation of $f'(x)$ is a formidable task. Such is not true, however, if the function is an nth degree polynomial,

$$p(x) = a_0x^n + a_1x^{n-1} + \cdots + a_{n-1}x + a_n, \qquad a_0 \neq 0. \tag{4.12}$$

Finding the zeros of an nth degree polynomial is one of the oldest and most studied problems in mathematics, and is also one of the more important problems in applied mathematics and has extensive practical applications. For $n = 1$, the only zero of $p(x)$ is given by $r = -a_1/a_0$. If $n = 2$, it is well known that the zeros are given by the quadratic formula,

$$r = \frac{-a_1 \pm \sqrt{a_1^2 - 4a_0a_2}}{2a_0}. \tag{4.13}$$

There are similar, but more cumbersome formulas for $n = 3$ and $n = 4$. What makes the problem so difficult in general is that for $n \geq 5$ there is *no* algebraic formula such as (4.13) that gives the zeros of $p(x)$ in terms of the coefficients. Thus we are forced to develop numerical procedures to approximate the zeros of $p(x)$. There are numerous such numerical procedures in the literature; but because of space and time considerations, we shall concentrate on just three—the Newton, Bairstow, and inverse power methods. (There is no reason, however, why the other methods we have discussed, such as the secant method, cannot be used for polynomials.) We shall also limit ourselves to the case in which all of the coefficients a_i in (4.12) are real.

Before proceeding further, we shall develop some basic and well-known theory about polynomials that is necessary for understanding and utilizing numerical root-finding techniques. The most basic result, which we recall from Chapter 1, is the *Fundamental Theorem of Algebra*.

Theorem 4.8

Given $p(x)$ as in (4.12) with $n \geq 1$, there exists at least one value r (possibly complex) such that $p(r) = 0$.

Given an r_1 such that $p(r_1) = 0$, it is then easily shown (Problem 1) that $p(x)$ can be written as $p(x) = (x - r_1)q_1(x)$ where $q_1(x)$ is an $(n - 1)$st degree polynomial. Applying Theorem 4.8 to $q(x)$, we obtain a value r_2 (not necessarily distinct from r_1) such that $q_1(r_2) = 0$. Thus there exists an $(n - 2)$nd degree polynomial $q_2(x)$ such that $q_1(x) = (x - r_2)q_2(x)$; so $p(x) = (x - r_1)(x - r_2)q_2(x)$. Repeating this process, we finally arrive at n values $r_1, r_2, \ldots, r_n$ such that $p(r_i) = 0, 1 \leq i \leq n$. Furthermore it can be shown that this set of roots is unique, and so $p(x)$ can be written as

$$p(x) = a_0(x - r_1)(x - r_2)\cdots(x - r_n). \tag{4.14}$$

This argument also shows that an nth degree polynomial can have no more than n zeros. As noted above, the r_j's need not all be distinct. If any r_j appears in (4.14) exactly m times, r_j is said to have "multiplicity m." The reader may easily verify that if r_j is a root of a multiplicity m, then $p(r_j) = p'(r_j) = \cdots = p^{(m-1)}(r_j) = 0$ and $p^{(m)}(r_j) \neq 0$. Note that this statement agrees with our earlier definition of "multiplicity m" for a zero of an arbitrary function.

The example, $p(x) = x^2 + 1 = (x + i)(x - i)$, demonstrates that even though the coefficients a_i are real, the zeros $\{r_j\}_{j=1}^n$ may be complex. It is well known that complex zeros of a polynomial with real coefficients occur in conjugate pairs. To be more precise, if $r = a + ib$, $b \neq 0$, is a complex number, then $\bar{r}$ (called the *conjugate* of r) is defined by $\bar{r} = a - ib$. If z_1 and z_2 are two complex numbers, then an elementary result is that $\overline{z_1 + z_2} = \bar{z}_1 + \bar{z}_2$ and $\overline{z_1 z_2} = \bar{z}_1\bar{z}_2$. By induction, it is easy to show that $\overline{z_1 + z_2 + \cdots + z_n} = \bar{z}_1 + \bar{z}_2 + \cdots + \bar{z}_n$ and $\overline{(r^k)} = (\bar{r})^k$. Finally, for any real number a, we have $\bar{a} = a$. Combining these properties for a polynomial $p(x)$ with real coefficients, we find that $\overline{p(r)} = p(\bar{r})$ for any complex number r. Thus if $p(r) = 0$, then

$$p(r) = 0 = \bar{0} = \overline{p(r)} = p(\bar{r}), \tag{4.15}$$

which shows that if r is a zero of $p(x)$, then so is $\bar{r}$. If r is a complex zero of $p(x)$ of multiplicity m, then $\bar{r}$ is also a zero of $p(x)$ of multiplicity m. This result follows since $p^{(i)}(x)$ is a polynomial with real coefficients for $i = 1, 2, \ldots, m - 1$; and therefore if $p^{(i)}(r) = 0$, then $p^{(i)}(\bar{r}) = 0$. Note that this equation implies that if n is odd, $p(x)$ must have at least one real zero. The last fundamental result we shall need here is the well-known *division algorithm*.

Theorem 4.9

Let $P(x)$ *and* $Q(x)$ be polynomials of degree n and m, respectively, where $1 \leq m \leq n$. Then there exists a unique polynomial $S(x)$ of degree $n - m$ and a unique polynomial $R(x)$ of degree $m - 1$ or less such that

$$P(x) = Q(x)S(x) + R(x). \tag{4.16}$$

In this formula, a polynomial of degree 0 is a constant. Thus if $Q(x)$ is a linear polynomial, the result of dividing $P(x)$ by $Q(x)$ is a quotient $S(x)$ and a remainder $R(x)$ where $R(x)$ is a constant.

PROBLEMS, SECTION 4.4

1. Let $p(x)$ be given by (4.12) and let r be a value such that $p(r) = 0$. Show that there exists a polynomial $q(x)$ of degree $n - 1$ such that $p(x) = (x - r)q(x)$. [Hint: Consider (4.16) with $Q(x) = (x - r)$.]
2. In (4.12), let $n = 4$ and $a_0 = 2$, and assume that $r_1 = 2 - 3i$ and $r_2 = -2 - 3i$ are zeros of $p(x)$. Find a_4, a_3, a_2, and a_1.
3. What are the relationships between the coefficients of (4.12) and the derivatives of p evaluated at zero. [Consider a Maclaurin expansion for $p(x)$.]
4. Using (4.16), show that if $P(r) = P'(r) = 0$, then $P(x) = (x - r)^2Q(x)$. Thus from the remarks following (4.14), r is a zero of multiplicity 2 of $P(x)$ if and only if $P(r) = P'(r) = 0$, $P''(r) \neq 0$. It should be clear that this statement extends to zeros of multiplicity $m > 2$.
5. Establish the special case of (4.16) when $Q(x)$ is the linear polynomial $Q(x) = x - \alpha$. To establish this case, let $P(x)$ be a given polynomial: $P(x) = a_0x^n + a_1x^{n-1} + \cdots + a_{n-1}x + a_n$. Set $S(x) = b_0x^{n-1} + b_1x^{n-2} + \cdots + b_{n-2}x + b_{n-1}$ where $b_0 = a_0$ and $b_j = \alpha b_{j-1} + a_j$ for $j = 1, 2, \ldots, n$. By comparing like powers in (4.16), establish the identity $P(x) = (x - \alpha)S(x) + b_n$.
6. Establish (4.16) for $Q(x) = (x - \alpha)(x - \beta)$ as follows. By Problem 5, there is $S(x)$ such that $P(x) = (x - \alpha)S(x) + R(x)$. Now by Problem 5 there is $T(x)$ such that $S(x) = (x - \beta)T(x) + R_1(x)$.
7. Apply the ideas in Problems 5 and 6 to find the quotient and remainder in (4.16) for $P(x) = x^5 - 3x^4 + 2x^2 + x - 1$ and for
   a) $Q(x) = x + 2$
   b) $Q(x) = x - 1$
   c) $Q(x) = x^2 - 5x + 6$
8. Using the ideas in Problems 5 and 6, prove Theorem 4.9.

4.4.1. Efficient Evaluation of a Polynomial and Its Derivatives

In order to use Newton's method on a polynomial $p(x)$ as given in Eq. (4.12), we must be able to evaluate $p(\alpha)$ and $p'(\alpha)$ for any α. Direct substitution of α into $p(x)$ and $p'(x)$, however, is far from the most efficient way to meet this objective. Furthermore as we have noted before, if the initial iterate x_0 of

Newton's method is not "close" to a zero, the Newton iteration may not converge. [This is obviously the case if all of the zeros of $p(x)$ are complex and x_0 is real.] Thus we need some means of "localizing" a zero, r, of $p(x)$ in order to choose x_0 "close" to r. Many of these "localization techniques" (see Section 4.4.3) require that we make a change of variable, $t = x - \alpha$, and write $p(x)$ as

$$p(x) = \beta_0(x - \alpha)^n + \beta_1(x - \alpha)^{n-1} + \cdots + \beta_{n-1}(x - \alpha) + \beta_n \qquad (4.17)$$

instead of in its original form, (4.12). By a Taylor's expansion, we can write

$$p(x) = p(\alpha) + p'(\alpha)(x - \alpha) + \frac{p''(\alpha)}{2!}(x - \alpha)^2 + \cdots + \frac{p^{(n)}(\alpha)}{n!}(x - \alpha)^n. \quad (4.18)$$

Equating (4.17) and (4.18), we see that

$$\beta_j = p^{(n-j)}(\alpha)/(n - j)! \qquad \text{for } 0 \leq j \leq n. \qquad (4.19)$$

This equation shows that it will be useful to have an efficient way of evaluating not only $p(\alpha)$ and $p'(\alpha)$, but $p^{(j)}(\alpha)$, $2 \leq j \leq n$, as well.

The technique we shall employ is known as *nested multiplication* or *synthetic division* and is based on the division algorithm (4.16). We first note that there is an alternative way of writing $p(x)$. For example with $n = 4$, we can write (4.12) as

$$p(x) = x(x(x(a_0x + a_1) + a_2) + a_3) + a_4. \qquad (4.20)$$

We see that $p(x)$ in (4.20) can be evaluated at $x = \alpha$ by only four multiplications and four additions whereas (Problem 2) direct substitution of α into (4.12) for $n = 4$ requires at least seven multiplications and four additions (if there is no $a_i = 0$). We can easily see that Eq. (4.20) can be extended for any value of n, and this extension is given by the following algorithm.

Let $p(x)$ be given by (4.12) and let α be any constant. Let $b_0 = a_0$, and generate $\{b_j\}_{j=1}^n$ by

$$b_j = \alpha b_{j-1} + a_j, \qquad 1 \leq j \leq n;$$

then $p(\alpha) = b_n$.

This algorithm is known as synthetic division and is exactly analogous to (4.20) where $b_n = p(\alpha)$. A less intuitive but more rigorous development of synthetic division is given in Problem 3. Note that the synthetic division algorithm requires only n multiplications to form $b_n = p(\alpha)$. This method is an efficient means of evaluating $p(\alpha)$, and moreover the iteration scheme can also be used to evaluate *all* derivatives, $p^{(m)}(\alpha)$, $0 \leq m \leq n$.

To see this point, we consider the division algorithm (4.16) and write $p(x)$ in the form

$$p(x) = (x - \alpha)q_{n-1}(x) + r_0(x). \qquad (4.21)$$

In (4.12), $a_0 \neq 0$; so $q_{n-1}(x)$ has degree $n - 1$ and its leading coefficient is a_0. Further, $r_0(x)$ is the constant $p(\alpha)$ as can be seen by setting $x = \alpha$ in (4.21).

Letting $q_{n-1}(x) = \gamma_0 x^{n-1} + \gamma_1 x^{n-2} + \cdots + \gamma_{n-1}$, substituting into Eq. (4.21), and equating like powers of x, we easily see (Problem 3) that

$$q_{n-1}(x) = b_0 x^{n-1} + b_1 x^{n-2} + \cdots + b_{n-1}$$

where $\{b_j\}_{j=0}^{n-1}$ are given by the synthetic division algorithm. Thus this algorithm generates the coefficients of $q_{n-1}(x)$ as well as $p(\alpha)$. By the same reasoning as above, the algorithm can be repeated and will yield $q_{n-1}(\alpha)$ and the coefficients of $q_{n-2}(x)$ (of degree $n - 2$) where

$$q_{n-1}(x) = (x - \alpha)q_{n-2}(x) + r_1, \qquad r_1 = q_{n-1}(\alpha). \tag{4.22}$$

Substituting (4.22) into (4.21), we obtain

$$p(x) = (x - \alpha)^2 q_{n-2}(x) + r_1(x - \alpha) + r_0. \tag{4.23}$$

Differentiating (4.23), we see that $p'(\alpha) = q_{n-1}(\alpha) = r_1$. Obviously this procedure may be continued, using synthetic division on each successive $q_k(x)$, $k = n - 2, n - 3, \ldots, 0$ to yield

$$p(x) = r_n(x - \alpha)^n + r_{n-1}(x - \alpha)^{n-1} + \cdots + r_1(x - \alpha) + r_0. \tag{4.24}$$

Equating (4.24) with the unique Taylor's expansion (4.18), we finally obtain our desired result,

$$r_m = p^{(m)}(\alpha)/m!, \qquad 0 \le m \le n.$$

EXAMPLE 4.6. Let $p(x) = x^6 + 5x^5 + 4x^4 + 3x^3 + 2x^2 + x + 1$ and let $\alpha = 2$. We illustrate synthetic division in Table 4.6 and find $p^{(j)}(\alpha)/j!$ for $j = 0, 1, \ldots, 6$. The entries in any row are the coefficients for $q_{6-j}(x)$, with the last entry in the row being $p^{(j)}(\alpha)/j!$. Thus, $p'(2) = 765$, $p''(2)/2! = 756$, etc. In this illustration, we see, for example, that

$$p(x) = (x - 2)(x^5 + 7x^4 + 18x^3 + 39x^2 + 80x + 161) + 323$$

and

$$p(x) = (x - 2)^6 + 17(x - 2)^5 + 114(x - 2)^4 + 395(x - 2)^3 + 756(x - 2)^2 + 765(x - 2) + 323.$$

TABLE 4.6

$p(x)$	1	5	4	3	2	1	1
$q_5(x)$	1	7	18	39	80	161	323
$q_4(x)$	1	9	36	111	302	765	
$q_3(x)$	1	11	58	227	756		
$q_2(x)$	1	13	84	395			
$q_1(x)$	1	15	114				
$q_0(x)$	1	17					
	1						

It is obvious how Newton's method should utilize synthetic division. In each iteration, let $x_m = \alpha$ and generate $p(x_m) \equiv r_0$ and $p'(x_m) \equiv r_1$ in exact analogy to the procedure in Example 4.6. Then proceed by setting $x_{m+1} = x_m - p(x_m)/p'(x_m)$. There are many other methods (such as the secant method) that involve evaluation of $p(x)$ or its derivatives at specific points, and synthetic division can be utilized by these methods as well. Once again we emphasize that the initial guess of Newton's method, x_0, should be "close" to r, a zero of $p(x)$, in order to ensure convergence. Section 4.4.3 will consider the problem of the approximate location of the zeros of $p(x)$, and will be of immense aid in making a good choice for x_0.

If our goal is to find *all* the zeros of an nth degree polynomial, $p(x)$, then it seems reasonable first to find one zero, say r_1, of $p(x)$. We can next say that $p(x) = (x - r_1)p_1(x)$, and hence any zero of the $(n - 1)$st degree polynomial $p_1(x)$ is a zero of $p(x)$. We now search for a zero of $p_1(x)$, say r_2, write $p_1(x) = (x - r_2)p_2(x)$, and then search for a zero of $p_2(x)$, etc. This process of finding a zero and dividing it out is called *deflation*. Polynomial deflation must be used judiciously, for large errors can result from our inability to find roots exactly. Subroutine POLRT, listed in Fig. 4.9, uses Newton's method and deflation to find all the real roots of a polynomial $p(x)$. Synthetic division is used for evaluation and for finding the coefficients of the deflated polynomials. An initial guess of $x_0 = 0$ is used for Newton's method at each stage of the deflation. Note that subroutine POLRT cannot determine complex roots, but modifications are easily made that enable the subroutine to find complex roots (see Problem 1).

EXAMPLE 4.7. Subroutine POLRT was used to find the zeros of $p(x) = x^5 + x^4 - 9x^3 - x^2 + 20x - 12$ where $p(x) = (x - 2)(x + 2)(x + 3)(x - 1)^2$. A tolerance TOL = 0.00001 was used in this example. The program found the following estimates to the roots, listed in the order found:

0.9998991E 01
0.2000000E 01
0.1000101E 01
−0.2000031E 01
−0.2999979E 01.

To illustrate the effect of not having these roots precisely, we listed the coefficients of the deflated polynomial of degree 4, found after the first root was divided out. These coefficients were

0.1000000E 01	(coefficient of x^4)
0.1999899E 01	(coefficient of x^3)
−0.7000303E 01	(coefficient of x^2)
−0.7999597E 01	(coefficient of x)
0.1200121E 02	(constant term).

These coefficients indicate that the deflated polynomial does not have precisely the remaining zeros of $p(x)$ as its zeros. For example, the constant term, 12.00121, is the product of the roots of the deflated polynomial whereas the product of 2, −2, −3, and 1 is 12.

```
      SUBROUTINE POLRT(A,ROOT,TOL,N,MTOL,NROOT)
      DIMENSION A(21),B(21),C(21),ROOT(20)
C
C  SUBROUTINE POLRT USES NEWTON'S METHOD AND DEFLATION TO FIND THE
C  REAL ROOTS OF A POLYNOMIAL.  THE CALLING PROGRAM MUST SUPPLY THE
C  DEGREE OF THE POLYNOMIAL, N, THE COEFFICIENTS A(I) (WHERE A(I) IS
C  THE COEFFICIENT OF X**(N+1-I)), A TOLERANCE TOL, AND AN INTEGER
C  MTOL.  THE SUBROUTINE RETURNS AN INTEGER NROOT=NUMBER OF REAL ROOTS
C  FOUND AND THE ROOTS (IN AN ARRAY ROOT).  AT EACH STAGE OF THE
C  DEFLATION, NEWTON'S METHOD TERMINATES WHEN MTOL ITERATIONS HAVE
C  BEEN EXECUTED OR WHEN TWO SUCCESSIVE ITERATES ARE LESS THAN TOL
C  IN ABSOLUTE VALUE.  N MUST BE GREATER THAN 2.
C
      NROOT=0
      NP1=N+1
    1 ITR=0
      X=0.
C
C  SYNTHETIC DIVISION ALGORITHM TO EVALUATE POLYNOMIAL AND ITS
C  DERIVITIVE
C
    2 B(1)=A(1)
      C(1)=B(1)
      DO 3 I=2,N
      B(I)=X*B(I-1)+A(I)
    3 C(I)=X*C(I-1)+B(I)
      B(NP1)=X*B(N)+A(NP1)
C
C  NEWTON'S METHOD UPDATE TO OLD ESTIMATE OF ROOT
C
      XCORR=B(NP1)/C(N)
      X=X-XCORR
      IF(ABS(XCORR).LT.TOL)   GO TO 4
      ITR=ITR+1
      IF(ITR.LT.MTOL)   GO TO 2
      RETURN
    4 NROOT=NROOT+1
      ROOT(NROOT)=X
C
C  SET UP COEFFICIENTS OF DEFLATED POLYNOMIAL
C
      NP1=N
      N=N-1
      DO 5 I=1,NP1
    5 A(I)=B(I)
C
C  USE QUADRATIC FORMULA WHEN DEFLATED POLYNOMIAL HAS DEGREE 2
C
      IF(N.GT.2)   GO TO 1
      DISCRM=A(2)*A(2)-4.*A(1)*A(3)
      IF(DISCRM.GE.0.)   GO TO 6
      RETURN
    6 ROOT(NROOT+1)=(-A(2)+SQRT(DISCRM))/2.
      ROOT(NROOT+2)=(-A(2)-SQRT(DISCRM))/2.
      NROOT=NROOT+2
      RETURN
      END
```

Figure 4.9 Subroutine POLRT.

To analyze the effects of not knowing r_1 exactly as in Example 4.7, suppose the Newton iterates $\{x_n\}_{n=0}^{\infty}$ are converging to a zero, r_1, of $p(x)$. No matter how large n may be, we can expect that $x_n \neq r_1$. Thus we must settle for some $x_n \equiv r'_1$ as an approximation for r_1. (In Section 4.4.3 we will say more about when to terminate an iteration.) By Theorem 4.9, there is a polynomial $p_1(x)$ of degree $n - 1$ such that $p(x) = (x - r_1')p_1(x) + p(r_1')$. Even if we had no round-off error in computing the coefficients of $p_1(x)$, $p(r_1')$ is probably not zero. Thus the zeros of $p_1(x)$ are not, in general, exactly the same as the remaining zeros, $\{r_j\}_{j=2}^{n}$, of $p(x)$. As a matter of fact, it can happen that even though r_1' is extremely close to r_1, the zeros of $p_1(x)$ can be quite different from $\{r_j\}_{j=2}^{n}$. If this phenomenon occurs, $p(x)$ is said, as before, to be ill-conditioned (once again we refer to Wilkinson's example of ill-conditioning given in Section 3.3.2). Therefore the most practical strategy to find r_2 is to use the Newton iteration on $p_1(x)$ until it seems to be converging to some approximate zero, r_2'', of $p_1(x)$. Then we let $x_0 = r_2''$, and use Newton's iteration on the *original* polynomial $p(x)$ to generate a corrected approximation, r_2', for the actual zero, r_2. This correction process should be repeated for all of the other approximates as well. This correction strategy is useful in any root-finding method that finds roots one by one.

When deflation is employed, the error (caused both by rounding errors and by truncation of the Newton iteration) incurred by accepting r_1' as an approximation to r_1 not only influences the approximation $r_2' \approx r_2$ but also influences the approximation $r_3' \approx r_3$ and all successive approximations. Furthermore, the error incurred in generating r_2' also influences $r_3' \approx r_3$ and all successive approximations. We can also see that the smaller roots (in magnitude) are more sensitive to error than the larger ones. For example, let r_1 and r_2 be two roots of $p(x)$ with $|r_1| < |r_2|$, and assume that r_1' and r_2' can be found such that $|r_1 - r_1'| = \varepsilon$ and $|r_2 - r_2'| = \varepsilon$. Then the relative errors (which are of more practical importance than the absolute errors) satisfy

$$\frac{|r_1 - r_1'|}{|r_1|} > \frac{|r_2 - r_2'|}{|r_2|}.$$

Thus we see that the accumulation of error with respect to the deflation process has less effect on our approximations if we are able to approximate the smaller roots first. Once again this approximation requires some *a priori* knowledge of the approximate location of the roots, which is the subject of Section 4.4.3. However, an intuitive approach to finding the smallest zero first would be to select the initial guess quite close to 0.

EXAMPLE 4.8. To illustrate some of the effects of deflation, the polynomial $p(x) = x^4 - 4x^3 + 6x^2 - 4x + 1$ was run using subroutine POLRT with TOL = 0.00001. [In this case, $p(x) = (x - 1)^4$.] After 19 iterations, Newton's method provided $r_1' = 1.021412$ as one root. The program deflated $p(x)$ using r_1' as an assumed root, and after 23 iterations provided $r_2'' = 0.9776155$ as a second root. The next deflated polynomial was $x^2 - 2.000972x + 1.001454$, which is seen to have no real zeros (in fact, the zeros of this quadratic are approximately 1.000486 ± 0.021949i). Newton's method applied to this quadratic gave a sequence of iterates that exhibited a characteristic oscillatory behavior. The first 20 iterates are shown in Table 4.7.

TABLE 4.7

x_i	$f(x_i)$
0.0000000E 00	0.1001454E 01
0.5004840E 00	0.2504845E 00
0.7509677E 00	0.6274182E−01
0.8766937E 00	0.1580685E−01
0.9405380E 00	0.4076481E−02
0.9745383E 00	0.1156807E−02
0.9968296E 00	0.4968643E−03
0.1064779E 01	0.4615963E−02
0.1028881E 01	0.1289368E−02
0.1006177E 01	0.5149841E−03
0.9609319E 00	0.2047658E−02
0.9868165E 00	0.6704330E−03
0.1011340E 01	0.6008148E−03
0.9836637E 00	0.7667542E−03
0.1006454E 01	0.5187988E−03
0.9629857E 00	0.1888931E−02
0.9881714E 00	0.6341934E−03
0.1013921E 01	0.6637573E−03
0.9892181E 00	0.6103516E−03
0.1016302E 01	0.7333755E−03

PROBLEMS, SECTION 4.4.1

1. A number of modifications of subroutine POLRT are possible and desirable. Either modify this program, or write your own root-finding program that incorporates two obvious improvements.

a) Use the root-refinement idea; take each zero of a deflated polynomial and use it as an initial guess for Newton's method applied to the *original* polynomial $p(x)$. This technique will refine both the root and the deflated polynomial.

b) If after a number of iterations, Newton's method does not appear to be converging, it is possible that $p(x)$ has a complex zero. Include a provision to start the iteration with a complex initial guess, (say $x_0 = i$), when this situation happens. The program will need a capability to perform complex arithmetic if a complex initial guess is used.

Test your program with $p(x) = (x - 1)^4$ and with $p(x) = x^6 - 2x^5 + 5x^4 - 6x^3 + 2x^2 + 8x - 8$ (which has zeros $1 + i$, $1 - i$, 1, -1, $2i$, $-2i$).

2. Let $p(x)$ be given by (4.12) and assume that none of its coefficients is zero. Show that evaluation of $p(\alpha)$ by direct substitution requires at least $2n - 1$ multiplications.
3. Establish that the number, b_n, generated by the synthetic division algorithm satisfies $b_n = p(\alpha)$. [Hint: In (4.16), let $p(x) = P(x)$ and $Q(x) = (x - \alpha)$. Let the coefficients of $Q(x)$ be $b_0, b_1, \ldots, b_{n-1}$ and equate like powers on both sides of (4.16).]
4. Start with Eq. (4.23) and show that $r_2 = p''(\alpha)/2$ where r_2 is obtained in the same manner as r_0 and r_1.
5. Let $p(x) = x^6 + 5x^5 + 4x^4 + 3x^3 + 2x^2 + x + 1$. Utilize synthetic division to find $\{\gamma_j\}_{j=0}^6$, and write $p(x)$ in the form

$$p(x) = \gamma_0(x + 2)^6 + \gamma_1(x + 2)^5 + \gamma_2(x + 2)^4 + \gamma_3(x + 2)^3 + \gamma_4(x + 2)^2 + \gamma_5(x + 2) + \gamma_6.$$

What does this value of γ_6 along with the last entry in the first row of Table 4.6 tell you?

4.4.2. Bairstow's Method

The basic purpose of Bairstow's method is to find a quadratic factor of a polynomial, $p(x)$. Let $p(x)$ be given by (4.12) and let u and v be any two real numbers. Then $p(x)$ can be written in the form

$$p(x) = (x^2 - ux - v)q(x) + b_{n-1}(x - u) + b_n \tag{4.25}$$

$$q(x) \equiv b_0x^{n-2} + b_1x^{n-3} + \cdots + b_{n-3}x + b_{n-2}. \tag{4.26}$$

We first note that $b_0 = a_0$ and the degree of $q(x)$ is $n - 2$. We also emphasize that each b_k is actually a function of u and v that we could find explicitly, should we desire (see Problem 1). Obviously for n very large, the explicit calculation of each b_k in terms of u and v could become quite cumbersome. Fortunately however, we can derive an algorithm that calculates each b_k quite efficiently.

Let $p(x)$ be given by (4.12), let u and v be arbitrary real numbers, and let $b_{-2} = b_{-1} = 0$. Then generate $\{b_k\}_{k=0}^n$ by

$$b_k = a_k + ub_{k-1} + vb_{k-2}, \qquad 0 \le k \le n. \tag{4.27}$$

We leave to the reader (Problem 2) to verify that the $\{b_k\}_{k=0}^n$ given by (4.27) satisfy (4.26) and (4.25). [This is merely a problem of comparing like powers of x in (4.25).] Before we lose sight of our principal problem of finding zeros of $p(x)$, we state the following theorem.

Theorem 4.10

Let u, v, and $p(x)$ be given as above, and let $r(x) = x^2 - ux - v = (x - s_1)(x - s_2)$. Then s_1 and s_2 are zeros of $p(x)$ if and only if $p(x) = r(x)q(x)$. [We see that finding a quadratic factor $r(x)$ of $p(x)$ is equivalent to finding u and v so that $b_{n-1} = b_n = 0$.]

The proof of Theorem 4.10 is quite straightforward and is also left to the reader (Problem 3). (We caution the reader to be careful to analyze the special case where $s_1 = s_2$, as well as the case where $s_1 \neq s_2$.) We emphasize the fact that $b_{n-1} \equiv b_{n-1}(u, v)$ and $b_n \equiv b_n(u, v)$; that is, they are functions of u and v. Because of Theorem 4.10 and Eq. (4.25), our problem reduces to finding u^* and v^* such that $b_{n-1}(u^*, v^*) = b_n(u^*, v^*) = 0$ and then using the quadratic formula to find the two roots t^* and s^* such that $r^*(x) \equiv x^2 - u^*x - v^* = 0$ for $x = t^*$ and $x = s^*$.

In general, $b_{n-1}(u, v)$ and $b_n(u, v)$ will be nonlinear functions in u and v as, for example, they are in Problem 1. We are then led to use Newton's method for two equations in two unknowns as given by (4.11e). We note the apparent difficulty here, however, that (4.11e) requires the evaluation of the functions $\partial b_{n-1}/\partial u$, $\partial b_{n-1}/\partial v$, $\partial b_n/\partial u$, and $\partial b_n/\partial v$ at each step of the iteration. This evaluation actually presents no real problem, however, as these quantities can be evaluated by differentiating formula (4.27). [Recall that *every* b_k in (4.27) is actually a function of u and v.]

Now $b_0 = a_0$; so $\partial b_0/\partial u = \partial b_0/\partial v = 0$. Also since $b_1 = a_1 + ub_0$, we see that $\partial b_1/\partial u = b_0$ and $\partial b_1/\partial v = 0$. Now we define $c_k \equiv \partial b_{k+1}/\partial u$, $0 \leq k \leq n - 1$, and for simplicity of notation let $c_{-2} = c_{-1} = 0$. From above, $c_0 = \partial b_1/\partial u = b_0$; and by (4.27)

$$c_1 = \frac{\partial b_2}{\partial u} = b_1 + u\frac{\partial b_1}{\partial u} + v\frac{\partial b_0}{\partial u} = b_1 + uc_0$$

$$c_2 = \frac{\partial b_3}{\partial u} = b_2 + u\frac{\partial b_2}{\partial u} + v\frac{\partial b_1}{\partial u} = b_2 + uc_1 + vc_0.$$

Continuing in this fashion, we see that for $0 \leq k \leq n - 1$

$$c_k = \frac{\partial b_{k+1}}{\partial u} = b_k + u\frac{\partial b_k}{\partial u} + v\frac{\partial b_{k-1}}{\partial u} = b_k + uc_{k-1} + vc_{k-2}. \qquad (4.28)$$

Thus from (4.28) with $k = n - 2$ and $n - 3$,

$$c_{n-2} = \frac{\partial b_{n-1}(u, v)}{\partial u} \quad \text{and} \quad c_{n-1} = \frac{\partial b_n(u, v)}{\partial u}. \qquad (4.29)$$

Furthermore, the iteration (4.28) is precisely the same as (4.27) with $\{a_k\}_{k=0}^{n-1}$ replaced by $\{b_k\}_{k=0}^{n-1}$ and $\{b_k\}_{k=-2}^{n-1}$ by $\{c_k\}_{k=-2}^{n-1}$.

Since (4.27) can be used to generate $\partial b_{n-1}/\partial u$ and $\partial b_n/\partial u$ for any u and v, it is not surprising that this algorithm can be used also to generate $\partial b_{n-1}/\partial v$ and $\partial b_n/\partial v$ for any u and v. This use can be seen once again by differentiating (4.27), this time with respect to v. First we define $d_k \equiv \partial b_{k+2}/\partial v$, $0 \leq k \leq n - 2$. We have already observed that

$$d_{-2} = \frac{\partial b_0}{\partial v} = d_{-1} = \frac{\partial b_1}{\partial v} = 0.$$

Now

$$d_0 = \frac{\partial b_2}{\partial v} = u\frac{\partial b_1}{\partial v} + b_0 + v\frac{\partial b_0}{\partial v} = b_0$$

$$d_1 = \frac{\partial b_3}{\partial v} = u\frac{\partial b_2}{\partial v} + v\frac{\partial b_1}{\partial v} + b_1 = ud_0 + b_1$$

$$d_2 = \frac{\partial b_4}{\partial v} = u\frac{\partial b_3}{\partial v} + v\frac{\partial b_2}{\partial v} + b_2 = ud_1 + vd_0 + b_2.$$

Therefore for $0 \le k \le n - 2$, we obtain

$$d_k = \frac{\partial b_{k+2}}{\partial v} = u\frac{\partial b_{k+1}}{\partial v} + v\frac{\partial b_k}{\partial v} + b_k = ud_{k-1} + vd_{k-2} + b_k. \tag{4.30}$$

So for $k = n - 3$ and $n - 2$ in (4.30),

$$d_{n-3} = \frac{\partial b_{n-1}(u, v)}{\partial v} \quad \text{and} \quad d_{n-2} = \frac{\partial b_n(u, v)}{\partial v} \tag{4.31}$$

where again (4.27) generates (4.30) and (4.31). Furthermore by comparing (4.30) with (4.31) (Problem 4) we see that $d_k = c_k$ for $-2 \le k \le n - 2$. Thus (4.27) need only be used once and yields

$$\begin{aligned} c_{n-3} &= \frac{\partial b_{n-1}(u, v)}{\partial v}, \\ c_{n-2} &= \frac{\partial b_{n-1}(u, v)}{\partial u} = \frac{\partial b_n(u, v)}{\partial v}, \\ c_{n-1} &= \frac{\partial b_n(u, v)}{\partial u}. \end{aligned} \tag{4.32}$$

Given the partials in (4.32), it is easy to write Newton's method in two variables for the system

$$b_{n-1}(u, v) = 0$$
$$b_n(u, v) = 0.$$

Using (4.11e), we see that Bairstow's method generates the sequence $\{u_i\}$ and $\{v_i\}$ where

$$u_{k+1} = u_k - [b_{n-1}c_{n-2} - b_n c_{n-3}]/\lambda$$
$$v_{k+1} = v_k - [b_n c_{n-2} - b_{n-1}c_{n-1}]/\lambda$$

and where $\lambda = c_{n-2}c_{n-2} - c_{n-1}c_{n-3}$ [that is, λ is the determinant of the Jacobian matrix in (4.11e)]. In the iteration above, of course, the terms b_{n-1}, b_n, c_{n-3}, c_{n-2}, and c_{n-1} are all obtained using $u = u_k$ and $v = v_k$.

EXAMPLE 4.9. Let $p(x) = x^3 - 6x^2 + 11x - 6 = (x - 1)(x - 2)(x - 3)$. [Note: $(x - 1)(x - 2) = x^2 - 3x + 2$, $(x - 1)(x - 3) = x^2 - 4x + 3$, and $(x - 2)(x - 3) = x^2 - 5x + 6$.] Bairstow's method was programmed in single precision with the following results.

Initial guess	(u, v) Convergents	Roots	Number of iterations
(8, 3)	(4, −3)	(1, 3)	12
(10, 5)	(5, −6)	(2, 3)	12
(50, 50)	(3, −2)	(1, 2)	9

We note here how sensitive Bairstow's method is with the three different initial guesses leading to the three different possible convergents. We ran this program with numerous initial guesses with magnitudes as large as 1000, and the program did not fail to converge for this simple polynomial for any iteration. Given a particular initial guess, (u_0, v_0), however, there is no readily discernible pattern as to which of the three (u, v) convergents the iteration finds. For example, we considered the three possible line segments connecting the points $(u, v) = (4, -3)$, $(5, -6)$, and $(3, -2)$. (See Fig. 4.10.) Each segment was divided into ten equal parts and each of the twenty-seven partition points was used as an initial point, (u_0, v_0), in the iteration. The points to which these iterations converged are indicated on the graph in Fig. 4.10.

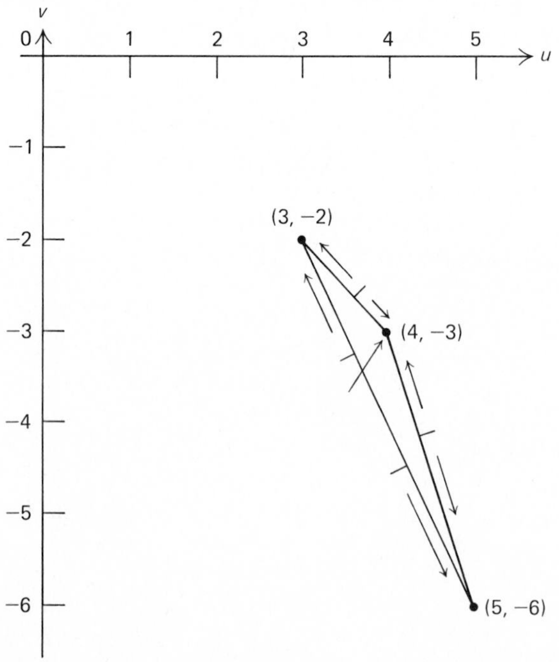

Figure 4.10 Successive applications of Bairstow's method.

Since it uses Newton's method in two variables, Bairstow's method is sensitive to the accuracy of the initial guess, and may fail to converge for that particular initial guess. Thus for an arbitrary polynomial, a reasonable procedure would be to use the localization methods of Section 4.4.3 and find the real roots by Newton's method first. Then use Bairstow's method to find the complex conjugate pairs.

PROBLEMS, SECTION 4.4.2

1. For the polynomial $p(x) = x^3 - 6x^2 + 11x - 6$ explicitly find the functions $b_1(u, v)$, $b_2(u, v)$, and $b_3(u, v)$ by equating like powers of x in (4.25).
2. Verify that $\{b_k\}_{k=0}^{n}$ given by (4.27) satisfy Eq. (4.25) via (4.26) with $p(x)$ given by (4.12).
3. Use Theorem 4.9 to prove Theorem 4.10.
4. From (4.28) and (4.30) verify that $d_k = c_k$ for $-2 \leq k \leq n - 2$.
5. Let $p(x) = x^3 - 4x^2 + 6x - 4$, and let $(u_0, v_0) = (0, 0)$. Program Bairstow's method and test your program on this polynomial.
6. Continue the numerical experiment conducted in Example 4.9 by refining the partitions of the three line segments and by considering various points within the triangle defined by these line segments in order to attempt to establish regions of convergence for three points.

4.4.3. Localization of Polynomial Zeros

There are numerous results in the literature on locating regions in the complex plane that contain zeros of a polynomial. We shall present only a few such results that are easy to implement and have a direct bearing on the numerical procedures discussed thus far. Since $a_0 \neq 0$ in (4.12), we may as well consider only polynomials of the form

$$p(x) = x^n + a_1 x^{n-1} + \cdots + a_{n-1}x + a_n. \tag{4.33}$$

The first localization technique has already been discussed in terms of eigenvalues of a matrix, but has a special modification for our particular problem. [This result can also be proved directly without the use of matrix theory (see Problem 1), but is most easily seen using the Gerschgorin circle theorem as a tool.] We note that the $(n \times n)$ *companion matrix* for $p(x)$,

$$A = \begin{bmatrix} 0 & 1 & 0 & \cdots & 0 \\ 0 & 0 & 1 & \cdots & 0 \\ \vdots & & & & \vdots \\ 0 & & & & 1 \\ -a_n & -a_{n-1} & & \cdots & -a_1 \end{bmatrix}, \tag{4.34}$$

is a matrix that has (4.33) as its characteristic polynomial. Using the Gerschgorin theorem in column form on (4.34), we have the following theorem and its corollary.

Theorem 4.11

Let $p(x)$ be given by (4.33); then all of the zeros of $p(x)$ lie in the union of the circular disks $C_1, C_2, \ldots, C_n$ in the complex plane:

$$\begin{aligned} C_n &= \{z: |z| \le |a_n|\} \\ C_k &= \{z: |z| \le 1 + |a_k|\} \qquad 2 \le k \le n-1 \\ C_1 &= \{z: |z + a_1| \le 1\}. \end{aligned}$$

Corollary

Let $p(x)$ be given by (4.33), and

$$r = 1 + \max_{1 \le j \le n} |a_j|.$$

Then every zero of $p(x)$ lies in the circular disk, $C = \{z: |z| \le r\}$.

Although it follows immediately from Theorem 4.11, this corollary can also be proved by induction using synthetic division (Problem 1), yielding the further information that if α is any number satisfying $|\alpha| \ge r$, then $|p(\alpha)| \ge 1$.

We note that all but one of the circular disks in Theorem 4.11 have centers at the origin; that is, at $z = 0$. Thus if our polynomial has some large coefficients, then Theorem 4.11 will yield large circular disks about the origin. Also, the theorem will not say anything about the location of the zeros within these disks. This is the point at which it becomes valuable to change the variable x to $t = x - \alpha$ (as we have seen can be done quite efficiently using synthetic division) in order to write $p(x)$ in the form of Eq. (4.18). Application of Theorem 4.11 to $p(x)$ rewritten in this form yields a set of circular disks whose center is $z = \alpha$. We illustrate this result with the following example.

EXAMPLE 4.10. Let $p(x) = x^4 + 8x^3 - 8x^2 - 200x - 425$. Then

$$p(x) = (x - 5)(x + 5)(x + 4 + i)(x + 4 - i).$$

From Theorem 4.11, $C_4 = \{z: |z| \le 425\}$ contains all of the other disks, and so the most we can say is that all of the zeros lie in C_4. If we let $x = -4$ and generate $p^{(m)}(-4)/m!$, $1 \le m \le 4$, we can write

$$\begin{aligned} p(x) &= (x + 4)^4 - 8(x + 4)^3 - 8(x + 4)^2 - 8(x + 4) - 9 \\ &= t^4 - 8t^3 - 8t^2 - 8t - 9. \end{aligned}$$

Using Theorem 4.11 with respect to the variable t, we see that $C_4 = \{z: |z| \le 9\}$ contains the other disks. By this equation we know that all of the zeros of $p(x)$ lie in a circular disk of radius 9 with center $\alpha = -4$, a considerable improvement.

We next state, without proof, a result that is helpful in determining the number of real zeros of a polynomial with real coefficients: *Descartes rule of signs*. The procedure of this rule is simply to write down the signs (in their respective order) of the nonzero coefficients of $p(x)$, and count the number of sign changes. (We call this number "v.") For example, let $p_1(x) = x^3 + x - 2$. Then the coefficient signs (in order) are $+, +, -$; and thus $v = 1$. For $p(x)$ given in Example 4.10, we get $+, +, -, -, -$; and so $v = 1$ again. The statement of the rule and how to use it is as follows.

Let k be the number of positive real roots of $p(x)$. Then $k \le v$, and $(v - k)$ is a nonnegative even integer.

For $p_1(x)$, we have $v = 1$. Since $v - k$ is nonnegative and even, we see that $v - k = 0$; so k must be 1. The same analysis holds for the second polynomial as well; so each of these polynomials has one positive real root.

It is easily verified (Problem 3) that if r is a zero of $p(x)$, then $-r$ is a zero of $p(-x)$. Thus we can obtain information on the number of negative real roots of $p(x)$ by using Descartes' rule of signs on $p(-x)$. For example, $p_1(-x) = -x^3 - x - 2$, and its coefficient signs are $-, -, -$; so $v = 0$. Therefore $p_1(x)$ has no negative real roots, and by combining these results we see that $p_1(x)$ has one positive real root and a pair of complex conjugate roots. From Example 4.10, $p(-x) = x^4 - 8x^3 - 8x^2 + 200x - 425$, and its sign pattern is $+, -, -, +, -$. Here $v = 3$, and so $v - k = 2$ or 0. Thus $p(x)$ has either 1 or 3 negative real roots, but it is not possible from this analysis to determine which.

The next root location result we shall derive is directly related to the Newton method iteration and provides some information as to when to terminate the iteration of Newton's method.

Theorem 4.12

Given any number α such that $p'(\alpha) \ne 0$ where $p(x)$ is given by (4.33), then there exists at least one zero of $p(x)$ in the circular disk centered at α given by

$$C = \left\{ z\colon |z - \alpha| \le n \left| \frac{p(\alpha)}{p'(\alpha)} \right| \right\}. \tag{4.35}$$

Proof. Let $p(x) = (x - r_1)(x - r_2) \ldots (x - r_n)$; then

$$p(x) = (x - \alpha)^n + \beta_1(x - \alpha)^{n-1} + \cdots + \beta_{n-1}(x - \alpha) + \beta_n$$

where $\beta_j = p^{(n-j)}(\alpha)/(n - j)!$, $1 \le j \le n$. Now we let $t \equiv (x - \alpha)$ and define $s(t) = t^n + \beta_1 t^{n-1} + \cdots + \beta_{n-1}t + \beta_n$. Obviously $s(t_j) = 0$ if and only if $t_j = r_j - \alpha$, $1 \le j \le n$. We next define the nth degree polynomial, $q(t) \equiv t^n s(1/t) = 1 + \beta_1 t + \cdots + \beta_{n-1}t^{n-1} + \beta_n t^n$. By Problem 3, the zeros of $q(t)$ are

$$u_j = \frac{1}{t_j} = \frac{1}{r_j - \alpha}, \qquad 1 \le j \le n.$$

Writing $q(t) = \beta_n(t - u_1)(t - u_2) \ldots (t - u_n)$, we can easily see (Problem 4) that

$$\frac{\beta_{n-1}}{\beta_n} = -(u_1 + u_2 + \cdots + u_n),$$

and so

$$\left|\frac{p'(\alpha)}{p(\alpha)}\right| = \left|\frac{\beta_{n-1}}{\beta_n}\right| \le n \max_{1\le j\le n} |u_j| = \frac{n}{\min_{1\le j\le n} |r_j - \alpha|}.$$

Therefore

$$\min_{1\le j\le n} |r_j - \alpha| \le n \left|\frac{p(\alpha)}{p'(\alpha)}\right|, \tag{4.36}$$

and so at least one r_j lies in C given by (4.35). ■

We again consider the polynomial, $p(x)$, of Example 4.10. If $\alpha = -4$, then $p(-4) = -9$, $p'(-4) = -8$, and $n = 4$. By Theorem 4.12, at least one zero lies in $C = \{z: |z + 4| \le 4(9/8) = 4.5\}$.

To use this theorem, we note that if $\{x_m\}_{m=0}^{\infty}$ are the iterates of the Newton method, then $p(x_m)/p'(x_m)$ is already calculated at each step. Therefore we can easily calculate $n|p(x_m)/p'(x_m)|$ and test to see if it is less than some prescribed tolerance. If so, then (4.36) guarantees that x_m is within that tolerance of some root, r, of $p(x)$.

The procedure above yields information as to when to terminate a Newton iteration in terms of the absolute error. As we have noted previously, a measure of the relative error is usually of more practical significance than a measure of the absolute error. We now give a procedure for estimating the relative error; this procedure is applicable to any iteration technique. In this analysis we return to formula (4.12) for $p(x)$ and write

$$p(x) = a_0x^n + a_1x^{n-1} + \cdots + a_n = a_0(x - r_1)(x - r_2)\ldots(x - r_n)$$

where $a_0 \ne 0$. It is easily seen (Problem 4) that

$$a_n/a_0 = (-1)^n r_1 r_2 \ldots r_n \tag{4.37}$$

where we assume that $a_n \ne 0$; that is, $r_j \ne 0$, $1 \le j \le n$. Let s (real or complex) be an approximation for a zero of $p(x)$ that satisfies $|p(s)| \le \varepsilon$, for some prescribed tolerance, $\varepsilon > 0$. Using (4.37), we see that

$$\frac{p(s)}{a_n} = \frac{a_0(s - r_1)(s - r_2)\ldots(s - r_n)}{a_0(-1)^n r_1 r_2 \ldots r_n} = (-1)^n\left(\frac{s}{r_1} - 1\right)\left(\frac{s}{r_2} - 1\right)\ldots\left(\frac{s}{r_n} - 1\right).$$

Since $|p(s)| \le \varepsilon$,

$$\left|\frac{p(s)}{a_n}\right| = \left|1 - \frac{s}{r_1}\right|\left|1 - \frac{s}{r_2}\right|\ldots\left|1 - \frac{s}{r_n}\right| \le \frac{\varepsilon}{|a_n|}.$$

Therefore

$$\left|\frac{p(s)}{a_n}\right| \ge \min_{1\le j\le n} \left|1 - \frac{s}{r_j}\right|^n,$$

and thus,

$$\min_{1 \le j \le n} \left| 1 - \frac{s}{r_j} \right| \le \left(\frac{\varepsilon}{|a_n|} \right)^{1/n}. \tag{4.38}$$

In (4.38) we have the answer to the question: "If $|p(s)| \le \varepsilon$, then what is an upper bound on the relative error of s with respect to some zero of $p(x)$?"

There are quite a few methods that we have not included in these sections. Some of the methods designed especially for polynomials include Bernoulli's method, Graeffe's root-squaring method, Lin's method, and Laguerre's method. Some more-sophisticated methods (some of which are based on localization procedures) include the use of Sturm sequences, the Lehmer-Schur method, the quotient-difference algorithm, and some recent procedures investigated by Henrici. In addition, there are a number of methods we have not discussed, such as Muller's method and those based on inverse interpolation, which are applicable to the problem $f(x) = 0$ even when $f(x)$ is not a polynomial. We refer the interested reader to Householder (1970), Henrici (1974), and Ralston and Rabinowitz (1978).

Considerable effort has been expended in developing sophisticated mathematical software for general use in computation libraries. These subroutines provide reliable and efficient estimates for the roots of an arbitrary polynomial, regardless of its particular characteristics. Much of this effort has centered around the work of Jenkins and Traub. Ralston and Rabinowitz (1978) give a detailed description of the Jenkins-Traub method along with references to their original papers. In the following section we present a related method based on the inverse power method.

PROBLEMS, SECTION 4.4.3

1. a) Prove the corollary of Theorem 4.11 by direct use of Theorem 4.11.
   b) Prove the corollary of Theorem 4.11 by using induction on the synthetic division algorithm. [Hint: For $|\alpha| \ge r$, show that $|b_j| \ge 1$ for all j where $b_0 = 1$, $b_j = \alpha b_{j-1} + a_j$, $j = 1, 2, \ldots, n$.]
2. Use the Gerschgorin Theorem in row form on the companion matrix (4.34) to derive other localization results.
3. Let $p(x)$ be given by (4.33) with zeros $\{r_j\}_{j=1}^n$. Find the zeros of $p(-x)$ and $x^n p(1/x)$.
4. Let $p(x) = a_0x^n + a_1x^{n-1} + \cdots + a_{n-1}x + a_n = a_0(x - r_1)(x - r_2) \ldots (x - r_n)$.
   a) Prove that $(a_1/a_0) = -(r_1 + r_2 + \cdots + r_n)$.
   b) Prove that $(a_n/a_0) = (-1)^n r_1 r_2 \ldots r_n$.
5. Program Newton's method and use (4.36) to include the provision that the iteration terminates when the absolute error is less than 10^{-5}. Apply this program to the polynomial of Example 4.10; use $x_0 = -4$.

6. In a Bairstow's method program, use (4.38) to include the provision that the iteration terminates when the relative error is less than 10^{-5}. Apply this program to the polynomial of Example 4.10; use $(u_0, v_0) = (6, 20)$.
7. Show that (4.33) is the characteristic polynomial of (4.34).

4.4.4 The Inverse Power Method for Polynomials

Some current methods for finding the roots of $p(x) = x^n + a_1x^{n-1} + \cdots + a_{n-1}x + a_n$ are based on eigenvalue procedures applied to the companion matrix A of (4.34). Here we present one such technique based on the inverse power method (Section 3.2.1) and the Rayleigh quotient iteration (Section 3.2.4). Our overall approach will be to iterate to find λ_1, the smallest eigenvalue in magnitude of A [the smallest root in magnitude of $p(x)$], eliminate the factor containing that root, iterate on the reduced polynomial to find λ_2, the next smallest eigenvalue, etc. (Finding the eigenvalues in this order can be shown to be less sensitive to rounding error.) The estimates $\tilde{\lambda}_2, \tilde{\lambda}_3, \ldots, \tilde{\lambda}_r$, $r \leq n$, derived from reduced polynomials can be refined by repeating the inverse power method using each as the initial guess α with the original companion matrix A. We choose to use the inverse power method since as we noted in Section 3.2.1, the inverse power method does not suffer most of the difficulties of the direct power method in the cases of multiple or complex eigenvalues.

Some comments are necessary concerning the use of the inverse power method on this problem. Since we are not using the method to refine estimates from another method, we must develop a way to find a value for α at each stage. Also, the eigenvectors (and generalized eigenvectors) of A have a particular form of which we may take advantage to choose the initial vector $\mathbf{x}_0$ for faster convergence. Finally, we can take advantage of the zero entries of A to solve the linear system in each iteration efficiently. We discuss these items below.

In each iteration we are solving a system of the form $(A - \alpha I)\mathbf{x} = \mathbf{y}$ with A given by (4.34). Gauss elimination can be modified to form an equivalent upper-triangular system $U\mathbf{x} = \tilde{\mathbf{y}}$ with $2n - 2$ multiplications. Each elimination multiple, however, has a division by α; and since we are initially searching for the smallest eigenvalue, we take a different approach.

We consider the augmented matrix $[A - \alpha I \vdots \mathbf{y}]$,

$$\begin{bmatrix} -\alpha & 1 & 0 & 0 & \cdots & 0 & 0 & y_1 \\ 0 & -\alpha & 1 & 0 & \cdots & 0 & 0 & y_2 \\ 0 & 0 & -\alpha & 1 & \cdots & 0 & 0 & y_3 \\ \vdots & & & & & & & \vdots \\ 0 & 0 & 0 & 0 & \cdots & -\alpha & 1 & y_{n-1} \\ -a_n & -a_{n-1} & -a_{n-2} & -a_{n-3} & \cdots & -a_2 & (-a_1-\alpha) & y_n \end{bmatrix}.$$

We eliminate $-\alpha$ on the diagonal in rows 2 through $n - 1$ as follows: add αRow(1) to Row(2), then αRow(2) to Row(3), . . . , and then αRow($n - 2$) to Row($n - 1$). This process yields

$$\begin{bmatrix} -\alpha & 1 & 0 & 0 & \cdots & 0 & 0 & r_1 \\ -\alpha^2 & 0 & 1 & 0 & \cdots & 0 & 0 & r_2 \\ -\alpha^3 & 0 & 0 & 1 & \cdots & 0 & 0 & r_3 \\ \vdots & & & & & & & \vdots \\ -\alpha^{n-1} & 0 & 0 & 0 & \cdots & 0 & 1 & r_{n-1} \\ -a_n & -a_{n-1} & -a_{n-2} & -a_{n-3} & \cdots & -a_2 & (-a_1-\alpha) & y_n \end{bmatrix},$$

where $r_1 = y_1$ and $r_i = y_i + \alpha r_{i-1}$ for $2 \leq i \leq n - 1$. We next create zeros in entries 2 through n of the last row in this fashion: add a_{n-1}Row(1) to Row(n), a_{n-2}Row(2) to Row(n), . . . , a_2Row($n - 2$) to Row(n), and $(a_1 + \alpha)$Row($n - 1$) to Row(n). This procedure changes only the last row, which now becomes

$$[R\,,0\,,0\,,0\,,\cdots,0\,,0\,,Q].$$

Hence the last equation becomes $x_1 = Q/R$. We leave it to the reader to verify that $R = -p(\alpha)$ and

$$Q = a_{n-1}r_1 + a_{n-2}r_2 + \cdots + a_2 r_{n-2} + (a_1 + \alpha)r_{n-1} + y_n.$$

Using $r_1 = y_1$ and $r_i = y_i + \alpha r_{i-1}$, $2 \leq i \leq n-1$, and collecting like coefficients of y_i, we can rewrite Q as

$$Q = s_0 y_n + s_1 y_{n-1} + s_2 y_{n-2} + \cdots + s_{n-1} y_1$$

where $s_0 = 1$ and $s_i = a_i + \alpha s_{i-1}$ for $1 \leq i \leq n - 1$. [Note that since $s_0 = 1$, it requires $(2n - 3)$ multiplications to compute the value of Q.]

To solve for $\mathbf{x}$, we use $Rx_1 = Q$ and the first $(n - 1)$ equations of the original system. This procedure yields the iteration

$$x_1 = -Q/p(\alpha), \qquad x_i = \alpha x_{i-1} + y_{i-1} \qquad \text{for } 2 \leq i \leq n.$$

[This iteration is defined when $p(\alpha) \neq 0$. If $p(\alpha) = 0$, we already have a root.] With nested multiplication used to compute $p(\alpha)$, the process requires $4(n - 1)$ multiplications to solve for $\mathbf{x}$. We note that the formulas above yield the mathematical results of this particular reduction of $(A - \alpha I)\mathbf{x} = \mathbf{y}$. To compute $\mathbf{x}$, all we need do is (1) generate $\{s_i\}_{i=0}^{n-1}$ and Q, (2) compute $p(\alpha)$, and (3) compute $\mathbf{x}$ by the iteration above.

If we use the basic form of the inverse power method and generate $\mathbf{x}_k$ by $(A - \alpha I)\mathbf{x}_k = \mathbf{x}_{k-1}$, $k = 1, 2, 3, \ldots$, then α is the same for each system and we need compute $p(\alpha)$ only once. We see, however, that it is probably feasible to let α vary with each iteration as in the Rayleigh quotient iteration of Section 3.2.4. Hence we could let $\alpha = \alpha_k = (\mathbf{x}_{k-1}^T A \mathbf{x}_{k-1})/(\mathbf{x}_{k-1}^T \mathbf{x}_{k-1})$ for each k, with the additional cost of the extra multiplications in each step necessary to evaluate

$p(\alpha_k)$. Young and Gregory (1972) state that this technique is often successful in practice even though it does not seem that its convergence has been established mathematically in the case of multiple eigenvalues. The usual inverse power method (α constant in each iteration) will always converge as long as α is closer to one eigenvalue than any other, but the convergence may not be so fast as in the Rayleigh quotient iteration. (Again see Section 3.2.4.)

Now that we have given an algorithm for solving $(A - \alpha I)\mathbf{x} = \mathbf{y}$, we turn to the choice of α in order to converge to one of the smallest eigenvalues in magnitude. Suppose the eigenvalues [roots of $p(x)$], $\{\lambda_i\}_{i=1}^{n}$, are labeled such that $0 < |\lambda_1| \le |\lambda_2| \le \cdots \le |\lambda_n|$; and we wish to have $|\lambda_1 - \alpha| < |\lambda_i - \alpha|$ whenever $\lambda_1 \ne \lambda_i$. We first note that if the coefficients of $p(x)$ are real, then we should choose α to be complex. This choice lessens the chance that a real value for α is equidistant from a pair of complex conjugate eigenvalues. We recall that the roots of the polynomial, $t^n p(1/t) = a_n t^n + a_{n-1}t^{n-1} + \cdots + a_1 t + 1$, are $\{1/\lambda_i\}_{i=1}^{n}$. Since $a_n \ne 0$ [otherwise $p(0) = 0$], we apply Theorem 3.3 (Gershgorin) to the companion matrix for $(t^n p(1/t)/a_n)$ and obtain either

$$|1/\lambda_i| \le 1 \qquad \text{or} \qquad \left|\frac{1}{\lambda_i} + \frac{a_{n-1}}{a_n}\right| < \left|\frac{1}{a_n}\right| + \sum_{i=1}^{n-2} \left|\frac{a_i}{a_n}\right|.$$

Hence, either

$$|\lambda_i| \ge 1 \qquad \text{or} \qquad |\lambda_i| \ge K = \left(\frac{1}{|a_n|} + \sum_{i=1}^{n-1} \left|\frac{a_i}{a_n}\right|\right)^{-1}.$$

Thus we know that no root of $p(x)$ lies inside the circle $C = \{z : |z| \le \rho\}$ where $\rho = \min\{1, K\}$. We can be fairly certain that $|\lambda_1 - \alpha| < |\lambda_j - \alpha|$ for $\lambda_j \ne \lambda_1$ if we select $\alpha = (\beta\rho)e^{2\pi i\gamma}$ where β and γ are numbers selected randomly in the interval $(0, 1)$ with $\gamma \ne 1/2$ so that $Im(\alpha) \ne 0$.

Last, we consider the choice of the initial vector $\mathbf{x}_0$. For simplicity of presentation, we assume that the eigenvalues are distinct although similar results can be obtained for multiple eigenvalues. We first note that each $\mathbf{u}_i = [1, \lambda_i, \lambda_i^2, \ldots, \lambda_i^{n-1}]^T$ for $1 \le i \le n$ is an eigenvector of A where A is given by (4.34) and λ_i is an eigenvalue of A. We consider the sequence $\{\gamma_j\}_{j=0}^{n-1}$ defined by

$$\begin{aligned}
\gamma_0 &= n,\\
\gamma_1 &= -a_{n-1}/a_n,\\
\gamma_2 &= -(a_{n-1}\gamma_1 + 2a_{n-2})/a_n,\\
\gamma_3 &= -(a_{n-1}\gamma_2 + a_{n-2}\gamma_1 + 3a_{n-3})/a_n,\\
\gamma_4 &= -(a_{n-1}\gamma_3 + a_{n-2}\gamma_2 + a_{n-3}\gamma_1 + 4a_{n-4})/a_n, \ldots,\\
\gamma_{n-1} &= -(a_{n-1}\gamma_{n-2} + a_{n-2}\gamma_{n-3} + \cdots + a_2\gamma_1 + (n-1)a_1)/a_n.
\end{aligned}$$

By rather extensive mathematical manipulation it can be shown [Young and

Gregory (1972)] that

$$\gamma_i = \sum_{j=1}^{n} \lambda_j^{-i}, \qquad \text{for } i = 0, 1, 2, \ldots, n-1.$$

Hence we have

$$\mathbf{w} = \begin{bmatrix} \gamma_{n-1} \\ \gamma_{n-2} \\ \vdots \\ \gamma_1 \\ \gamma_0 \end{bmatrix} = \begin{bmatrix} \lambda_1^{-n+1} \\ \lambda_1^{-n+2} \\ \vdots \\ \lambda_1^{-1} \\ 1 \end{bmatrix} + \begin{bmatrix} \lambda_2^{-n+1} \\ \lambda_2^{-n+2} \\ \vdots \\ \lambda_2^{-1} \\ 1 \end{bmatrix} + \cdots + \begin{bmatrix} \lambda_n^{-n+1} \\ \lambda_n^{-n+2} \\ \vdots \\ \lambda_n^{-1} \\ 1 \end{bmatrix},$$

or

$$\mathbf{w} = \lambda_1^{-n+1}\mathbf{u}_1 + \lambda_2^{-n+1}\mathbf{u}_2 + \cdots + \lambda_n^{-n+1}\mathbf{u}_n.$$

Since the sequence of approximations of the inverse power method is converging to $(\lambda_1 - \alpha)^{-1}$, and since $(A - \alpha I)^{-1}\mathbf{u}_1 = (\lambda_1 - \alpha)^{-1}\mathbf{u}_1$, we wish to choose the initial vector $\mathbf{x}_0$ to have a significant contribution from $\mathbf{u}_1$ in its expansion $\mathbf{x}_0 = c_1\mathbf{u}_1 + c_2\mathbf{u}_2 + \cdots + c_n\mathbf{u}_n$. For this reason, a usually good choice is $\mathbf{x}_0 = \mathbf{w}/\|\mathbf{w}\|_2$.

PROBLEMS, SECTION 4.4.4

1. In solving $(A - \alpha I)\mathbf{x} = \mathbf{y}$, two expressions were given for Q: $Q = (\sum_{i=1}^{n-1} a_{n-i}r_i) + \alpha r_{n-1} + a_n$ and $Q = \sum_{i=1}^{n} s_{n-i}y_i$. Show these expressions are equal when $n = 2$ and $n = 3$. Use this and show by induction that equality holds for all n.
2. a) For $p(x) = x^3 - 7x^2 + 7x + 15$, $\mathbf{x}_0 = [1,2,2]^T$, and A the companion matrix of $p(x)$, use $\alpha = 0$ and solve $(A - \alpha I)\mathbf{x}_1 = \mathbf{x}_0$ for $\mathbf{x}_1$ and $(A - \alpha I)\mathbf{x}_2 = \mathbf{x}_1$ for $\mathbf{x}_2$. Calculate $\beta_i = \mathbf{x}_{i-1}^T\mathbf{x}_i/\mathbf{x}_{i-1}^T\mathbf{x}_{i-1}$ for $i = 1,2$, and compute $p(\alpha + (1/\beta_i))$.

   b) Repeat all of these calculations for $\alpha = 2$.
3. Use $p(x) = x^3 - 5x^2 + 4x - 20$ and $\alpha = i \equiv \sqrt{-1}$, and repeat the calculations in Problem 2a. (Use conjugate transposes for the β_i's when the vectors are complex.)
4. For each of the polynomials in Problems 3 and 4 find the radius of a circle about the origin that contains no root of the polynomial.
5. In the use of the inverse power method to find the smallest root, describe pictorially how the method might converge to λ_2 rather than λ_1 even though $|\lambda_1| < |\lambda_2|$. Does this possibility adversely affect the method when trying to locate all roots? Explain.
6. Write a program to find the roots of an nth degree polynomial. Incorporate subroutines to choose α_0 and $\mathbf{x}_0$ as in the text. Have the program estimate an initial root, deflate, estimate a second root, improve the second estimate from the original polynomial, deflate using the estimate for the initial root and the improved estimate for the second, and continue in this manner. The estimating subroutine may use the same α in each iteration or may calculate different α_k's as in the text. One may also wish the program to extract complex conjugates as roots once a complex root is found. (Be sure to use conjugate transposes in complex vector scalar products.)

*4.5. NEWTON'S METHOD IN SEVERAL VARIABLES

We wish to consider the problem of simultaneously solving a system of n nonlinear equations in n unknowns. Let $x_1, x_2, \ldots, x_n$ be real variables, and let $f_i(x_1, x_2, \ldots, x_n)$ be a real-valued function for each i, $1 \le i \le n$. Then we wish to find values $s_1, s_2, \ldots, s_n$ such that $f_i(s_1, s_2, \ldots, s_n) = 0$ simultaneously for $1 \le i \le n$. We may simplify the statement of the problem by using vector notation. Let $\mathbf{x} = (x_1, x_2, \ldots, x_n)$ be a vector of n variables; then for each i, $f_i(x_1, x_2, \ldots, x_n)$ can be written as $f_i(\mathbf{x})$. Furthermore we let

$$\mathbf{F}(\mathbf{x}) = \begin{bmatrix} f_1(\mathbf{x}) \\ f_2(\mathbf{x}) \\ \vdots \\ f_n(\mathbf{x}) \end{bmatrix}. \tag{4.39}$$

Now our problem can be restated in vector form: find a vector $\mathbf{s} = (s_1, s_2, \ldots, s_n)$ such that $\mathbf{F}(\mathbf{s}) = \theta$ where θ is the $(n \times 1)$ vector with all components equal to zero. For a thorough understanding of the material in this section, the reader should have some familiarity with vector and matrix norms (see Chapter 2).

EXAMPLE 4.11. Let $n = 3$, and $f_1(x_1, x_2, x_3) = x_1 \cos(x_2) - x_3$, $f_2(x_1, x_2, x_3) = x_1^2 + x_3$, and $f_3(x_1, x_2, x_3) = e^{x_1 + x_3} \sin(x_2/2) + x_3$. Then

$$\mathbf{F}(\mathbf{x}) = \begin{bmatrix} x_1 \cos(x_2) - x_3 \\ x_1^2 + x_3 \\ e^{x_1 + x_3} \sin(x_2/2) + x_3 \end{bmatrix}.$$

Obviously, $\mathbf{s} = (1, \pi, -1)$ satisfies $\mathbf{F}(\mathbf{s}) = \theta$. Since the equations are nonlinear, however, we can probably expect other solutions to exist; for example, $\mathbf{s} = (0, 0, 0)$.

In problems of this form, we lose the geometric simplicity that led us to the derivation of simple methods for one variable. At this point the theory of the fixed-point problem becomes invaluable. As we shall see, almost all of the theorems about the fixed-point iteration are true in this setting, and their proofs are almost identical to the proofs in the case of a single variable. Using the same vector notation as above, we construct a vector-valued function

$$\mathbf{g}(\mathbf{x}) = \begin{bmatrix} g_1(x_1, x_2, \ldots, x_n) \\ g_2(x_1, x_2, \ldots, x_n) \\ \vdots \\ g_n(x_1, x_2, \ldots, x_n) \end{bmatrix} \tag{4.40}$$

such that if $\mathbf{g}(\mathbf{s}) = \mathbf{s}$, then $\mathbf{F}(\mathbf{s}) = \theta$. For example, given any such $\mathbf{F}(\mathbf{x})$, one choice for $\mathbf{g}(\mathbf{x})$ would be $\mathbf{g}(\mathbf{x}) = \mathbf{x} + A\mathbf{F}(\mathbf{x})$ where A is a nonsingular $(n \times n)$ matrix.

If $\mathbf{g}(\mathbf{x})$ has a fixed-point $\mathbf{s}$ (actually a "fixed-vector"), then we define the fixed-point iteration analogously to be

$$\mathbf{x}_{k+1} = \mathbf{g}(\mathbf{x}_k). \tag{4.41}$$

Before saying anything about the convergence of the fixed-point iteration, we must pause to introduce some vector concepts. Let $\mathcal{R}$ be a region in R^n (the space of all real n-tuples) such that $\mathbf{x} = (x_1, x_2, \ldots, x_n)$ belongs to $\mathcal{R}$ if and only if $a_1 \le x_1 \le b_1$, $a_2 \le x_2 \le b_2, \ldots, a_n \le x_n \le b_n$ where $\{a_i\}_{i=1}^n$ and $\{b_i\}_{i=1}^n$ are specified real numbers. (Thus $\mathcal{R}$ is the generalization of a closed interval. For example, $\mathcal{R}$ is a rectangle in R^2.) A *line segment* from the vector $\mathbf{x}$ to the vector $\mathbf{y}$ is defined to be the set of all vectors of the form $\mathbf{w} = \lambda\mathbf{x} + (1 - \lambda)\mathbf{y}$ where λ is any real number satisfying $0 \le \lambda \le 1$. Note that if $\lambda = 0$, $\mathbf{w} = \mathbf{y}$; and if $\lambda = 1$, $\mathbf{w} = \mathbf{x}$. It is left to the reader to show that if $\mathbf{x}$ and $\mathbf{y}$ belong to $\mathcal{R}$, then $\mathbf{w} \in \mathcal{R}$, for all λ, $0 \le \lambda \le 1$.

In order to talk about the rate of convergence of a fixed-point iteration, we must introduce the concept of a *derivative of a vector-valued function.* For a function of a real variable, g, the derivative of g at x is simply

$$\lim_{h\to 0} \frac{g(x + h) - g(x)}{h} = g'(x). \tag{4.42}$$

Equivalently, Eq. (4.42) means that

For every $\varepsilon > 0$, there exists a $\delta > 0$, such that

$$|(g(x + h) - g(x))/h - g'(x)| \le \varepsilon \qquad \text{whenever} \qquad 0 < |h| \le \delta. \tag{4.43}$$

For $\mathbf{h} = (h_1, h_2, \ldots, h_n) \in R^n$, division by $\mathbf{h}$ is not defined; so we may not extend (4.42) directly to R^n. However, (4.43) does have a natural extension to R^n. Let $\|\cdot\|$ be any of the three norms introduced in Chapter 2. Then the derivative of $\mathbf{g}$ at $\mathbf{x}$ is defined to be the $(n \times n)$ matrix, $A_{\mathbf{x}}$, that satisfies this condition.

For every $\varepsilon > 0$, there exists a $\delta > 0$ such that

$$\|\mathbf{g}(\mathbf{x} + \mathbf{h}) - \mathbf{g}(\mathbf{x}) - A_{\mathbf{x}}\mathbf{h}\| < \varepsilon\|\mathbf{h}\| \qquad \text{whenever} \qquad 0 < \|\mathbf{h}\| \le \delta. \tag{4.44}$$

We shall write $A_{\mathbf{x}} \equiv g'(\mathbf{x})$, and note that the matrix $A_{\mathbf{x}}$ changes as $\mathbf{x}$ changes. This definition also applies if $\mathbf{g}\colon R^n \to R^p$, in which case $A_{\mathbf{x}}$ is $(p \times n)$.

It is cumbersome but not difficult (see Problem 2) to verify that $g'(\mathbf{x})$ is the familiar Jacobian matrix whose (i, j)th element is

$$\frac{\partial g_i(\mathbf{x})}{\partial x_j};$$

that is,

$$g'(\mathbf{x}) \equiv J(\mathbf{x}) = \left[\frac{\partial g_i(\mathbf{x})}{\partial x_j}\right]. \tag{4.45}$$

EXAMPLE 4.12. For the function $\mathbf{F}(\mathbf{x})$ of Example 4.11, and $\mathbf{x} = (1, 2, 3)$,

$$F'(\mathbf{x}) = \begin{bmatrix} \cos(2) & -\sin(2) & -1 \\ 2 & 0 & 1 \\ e^4 \sin(1) & (1/2)e^4 \cos(1) & e^4 \sin(1) + 1 \end{bmatrix}.$$

The concept of the second derivative can also be defined, but it is not necessary for us to do so now. It will appear later in a Taylor expansion, in which it will simply be denoted by $g''(\mathbf{x}, \mathbf{h}, \mathbf{h})$. The only thing we will need to know about the second derivative is that if it is continuous for all $\mathbf{x} \in \mathcal{R}$, then it satisfies

$$\|g''(\mathbf{x}, \mathbf{h}, \mathbf{h})\| \leq K\|\mathbf{h}\|^2 \qquad \text{for all } \mathbf{x} \tag{4.46}$$

where K is a constant independent of $\mathbf{x}$.

The last preliminary result we need is a generalization of the mean-value theorem, which we state without proof.

Mean-value theorem.

Given any two vectors $\mathbf{x}$ and $\mathbf{y}$ in $\mathcal{R}$, assume that for all $\mathbf{w}$ on the line segment between $\mathbf{x}$ and $\mathbf{y}$

$$\left|\frac{\partial g_i(\mathbf{w})}{\partial x_j}\right| \leq L_{ij}.$$

Let

$$L_1 = \max_{1 \leq j \leq n}\left\{\sum_{i=1}^{n} L_{ij}\right\}, \qquad L_\infty = \max_{1 \leq i \leq n}\left\{\sum_{j=1}^{n} L_{ij}\right\},$$

and

$$L_2 = \left[\sum_{i=1}^{n}\sum_{j=1}^{n} L_{ij}^2\right]^{1/2}.$$

Then

$$\|\mathbf{g}(\mathbf{x}) - \mathbf{g}(\mathbf{y})\|_p \leq L_p\|\mathbf{x} - \mathbf{y}\|_p, \qquad p = 1, 2, \text{ or } \infty. \tag{4.47}$$

As an example of the theorem above, assume that each $L_{ij} = \lambda/n$ where $0 \leq \lambda < 1$. Then $L_1 = L_2 = L_\infty = \lambda < 1$; and $\|\mathbf{g}(\mathbf{x}) - \mathbf{g}(\mathbf{y})\|_p \leq \lambda\|\mathbf{x} - \mathbf{y}\|_p$; $p = 1, 2$, or ∞. Now we are able to prove the following convergence theorem for the fixed-point iteration. (Compare its results and proof with Theorem 4.3.)

Theorem 4.13

Let $\mathcal{R}$ be a rectangle in R^n (as above), and assume that $\mathbf{g}(\mathcal{R}) \subseteq \mathcal{R}$. Also assume that for all $\mathbf{w} \in \mathcal{R}$, at least one of the constants L_1, L_2, L_∞ defined above (say L_p) satisfies $0 < L_p < 1$. Then,

1. $\mathbf{x} = \mathbf{g}(\mathbf{x})$ has exactly one solution, $\mathbf{s}$, in $\mathcal{R}$;
2. the sequence (4.41) converges to $\mathbf{s}$;
3. for all n, $\|\mathbf{x}_n - \mathbf{s}\|_p \leq L_p^n\|\mathbf{x}_1 - \mathbf{x}_0\|_p/(1 - L_p)$.

Proof. By repeated use of (4.47),

$$\begin{aligned}\|\mathbf{x}_{n+1} - \mathbf{x}_n\|_p &= \|\mathbf{g}(\mathbf{x}_n) - \mathbf{g}(\mathbf{x}_{n-1})\|_p \le L_p\|\mathbf{x}_n - \mathbf{x}_{n-1}\|_p \\ &\le L_p^2\|\mathbf{x}_{n-1} - \mathbf{x}_{n-2}\|_p \le \cdots \le L_p^n\|\mathbf{x}_1 - \mathbf{x}_0\|_p.\end{aligned} \tag{4.48}$$

Let $m > n$; then by the triangle inequality and (4.48),

$$\begin{aligned}\|\mathbf{x}_m - \mathbf{x}_n\|_p &\le \|\mathbf{x}_m - \mathbf{x}_{m-1}\|_p + \|\mathbf{x}_{m-1} - \mathbf{x}_{m-2}\|_p + \cdots + \|\mathbf{x}_{n+1} - \mathbf{x}_n\|_p \\ &\le (L_p^{m-1} + \cdots + L_p^n)\|\mathbf{x}_1 - \mathbf{x}_{0b}\|_p \\ &\le L_p^n\|\mathbf{x}_1 - \mathbf{x}_0\|_p/(1 - L_p).\end{aligned} \tag{4.49}$$

Since $0 \le L_p < 1$, given any $\varepsilon > 0$, there exists an N such that if m and $n > N$, then $\|\mathbf{x}_m - \mathbf{x}_n\|_p < \varepsilon$; that is, choose N such that $L_p^n\|\mathbf{x}_1 - \mathbf{x}_0\|/(1 - L_p) < \varepsilon$. This property is called the *Cauchy Criterion*. Just as in the case of a sequence, $\{x_n\}_{n=1}^{\infty}$, of real numbers in a closed interval $[a, b]$, it is also true that if a sequence, $\{\mathbf{x}_n\}_{n=1}^{\infty}$, in $\mathcal{R}$ satisfies the Cauchy Criterion, then the sequence must converge to some $\mathbf{s} \in \mathcal{R}$. Since $\mathbf{g}(\mathbf{x})$ is continuous,

$$\mathbf{s} = \lim_{n\to\infty} \mathbf{x}_{n+1} = \lim_{n\to\infty} \mathbf{g}(\mathbf{x}_n) = \mathbf{g}(\mathbf{s}).$$

Now assume that there is another fixed-point $\mathbf{q} \in \mathcal{R}$; that is, $\mathbf{q} = \mathbf{g}(\mathbf{q})$, $\mathbf{q} \ne \mathbf{s}$. Then $\|\mathbf{s} - \mathbf{q}\|_p = \|\mathbf{g}(\mathbf{s}) - \mathbf{g}(\mathbf{q})\|_p \le L_p\|\mathbf{s} - \mathbf{q}\|_p < \|\mathbf{s} - \mathbf{q}\|_p$, but this expression is a contradiction and completes the proof. ■

Our analogy to the fixed-point problem of a single variable continues here with the following local convergence theorem.

Theorem 4.14

Let $r > 0$ be given such that the set of vectors, $S = \{\mathbf{x}: \|\mathbf{x} - \mathbf{s}\| < r\}$, contains a fixed point, $\mathbf{s}$, of $\mathbf{g}(\mathbf{x})$. Further, let $S \subseteq \mathcal{R}$, g' be continuous on S, and $\|g'(\mathbf{s})\| < 1$. Then there exists an $\varepsilon > 0$ such that the fixed-point iteration is convergent whenever $\|\mathbf{x}_0 - \mathbf{s}\| < \varepsilon$.

The proof is exactly analogous to the proof of Theorem 4.4 and is omitted. We now define the error vector, $\mathbf{e}_n = \mathbf{x}_n - \mathbf{s}$. By formula (4.46) and the remarks preceding it, there is a Taylor's expansion of the form

$$\begin{aligned}\|\mathbf{e}_{n+1}\| &= \|\mathbf{x}_{n+1} - \mathbf{s}\| = \|\mathbf{g}(\mathbf{x}_n) - \mathbf{g}(\mathbf{s})\| \\ &= \|g'(\mathbf{s})\mathbf{e}_n + g''(\boldsymbol{\zeta}; \mathbf{e}_n, \mathbf{e}_n)\| \le \|g'(\mathbf{s})\mathbf{e}_n\| + K\|\mathbf{e}_n\|^2.\end{aligned} \tag{4.50}$$

If $g'(\mathbf{s}) \ne \mathcal{O}$ (the zero matrix), then, in general, $g'(\mathbf{s})\mathbf{e}_n \ne \theta$. In this case, the best possible rate of convergence we can obtain is linear; that is,

$$\lim_{n\to\infty} \frac{\|\mathbf{e}_{n+1}\|}{\|\mathbf{e}_n\|} \le \|g'(\mathbf{s})\|.$$

If $g'(\mathbf{s}) = \mathcal{O}$, however, we obtain quadratic convergence,

$$\lim_{n\to\infty} \frac{\|\mathbf{e}_{n+1}\|}{\|\mathbf{e}_n\|^2} \le K.$$

Here we again return to our principal problem of finding vectors, $\mathbf{s}$, such that $\mathbf{F}(\mathbf{s}) = \theta$. Again if we let $\mathbf{g}(\mathbf{x}) = \mathbf{x} + A\mathbf{F}(\mathbf{x})$ where A is a nonsingular $(n \times n)$ matrix, then $\mathbf{x} = \mathbf{g}(\mathbf{s})$ if and only if $\mathbf{F}(\mathbf{s}) = \theta$. By considering the Jacobian of $\mathbf{g}(\mathbf{x}) = \mathbf{x} + A\mathbf{F}(\mathbf{x})$ (Problem 2d), we see that $g'(\mathbf{x}) = I + AF'(\mathbf{x})$. Thus $g'(\mathbf{s}) = \mathcal{O}$ only if $A = -[F'(\mathbf{s})]^{-1}$, that is, the inverse of the Jacobian of $\mathbf{F}(\mathbf{x})$ evaluated at $\mathbf{s}$. Once again this choice for A cannot be made without *a priori* knowledge of $\mathbf{s}$. We are then forced to consider the choice $\mathbf{g}(\mathbf{x}) = \mathbf{x} + A(\mathbf{x})\mathbf{F}(\mathbf{x})$ where $A(\mathbf{x})$ is an $(n \times n)$ matrix whose elements are real-valued functions of $\mathbf{x}$. It is beyond the level of this text to take the derivative of $A(\mathbf{x})\mathbf{F}(\mathbf{x})$ (and very cumbersome, at best, to calculate its Jacobian and arrive at the desired result). In order that $g'(\mathbf{s}) = \mathcal{O}$, however, it can be shown that $A(\mathbf{x}) = -[\mathbf{F}'(\mathbf{x})]^{-1}$. Therefore with this choice of $A(\mathbf{x})$, the fixed-point iteration becomes Newton's method in R^n where $\mathbf{x}_0$ is an initial guess:

$$\mathbf{x}_{n+1} = \mathbf{x}_n - J(\mathbf{x}_n)^{-1}\mathbf{F}(\mathbf{x}_n), \qquad n \ge 0, \tag{4.51}$$

where $J(\mathbf{x}_n)$ is the Jacobian of $\mathbf{F}$ at $\mathbf{x}_n$.

Thus the Newton method in R^n is notationally the same as (4.11e), possesses a local convergence theorem analogous to Theorem 4.6 (via Theorem 4.14), and again has quadratic convergence whenever $J(\mathbf{s})$ is nonsingular.

We have noted previously that it is usually computationally inefficient to compute inverses of matrices. Thus (4.51) is usually performed in the following manner. Assume that $\mathbf{x}_n$ has just been calculated, and define $\mathbf{z}_{n+1} \equiv \mathbf{x}_{n+1} - \mathbf{x}_n$. Rewrite (4.51) as $J(\mathbf{x}_n)\mathbf{z}_{n+1} = -\mathbf{F}(\mathbf{x}_n)$. Solve this linear system for $\mathbf{z}_{n+1}$ by one of the Chapter 2 methods (usually Gauss elimination), and then find $\mathbf{x}_{n+1}$ by $\mathbf{x}_{n+1} = \mathbf{z}_{n+1} + \mathbf{x}_n$.

EXAMPLE 4.13. Let $f_1(x_1, x_2, x_3) = x_1^5 + x_2^3x_3^4 + 1$, $f_2(x_1, x_2, x_3) = x_1^2x_2x_3$, and $f_3(x_1, x_2, x_3) = x_3^4 - 1$. It can be easily shown algebraically that there are exactly four real solutions:

$$\mathbf{s}_1 = (0, -1, -1),$$

$$\mathbf{s}_2 = (0, -1, 1),$$

$$\mathbf{s}_3 = (-1, 0, 1),$$

and

$$\mathbf{s}_4 = (-1, 0, -1).$$

We list in Table 4.8 the results of several trials of Newton's iteration carried out to six-place accuracy where $\mathbf{x}_0$ is the initial iteration vector and k represents the kth iteration.

TABLE 4.8

(1) $\mathbf{x}_0 = (-100, 0, -100)$: (Yields $\mathbf{s}_4$)

k	$x_1(k)$	$x_2(k)$	$x_3(k)$
10	−10.73743	0	−5.63222
20	−1.21805	0	−1.00000
24	−1.000000	0	−1.00000

(2) $\mathbf{x}_0 = (-1000, -1000, -1000)$: (Yields $\mathbf{s}_1$)

k	$x_1(k)$	$x_2(k)$	$x_3(k)$
10	−9.08509	−10000.00000	−56.31350
20	−0.08265	−998.47800	−3.17604
30	−0.00110	−96.58270	−1.00000
40	−0.00002	−1.74844	−1.00000
45	−0.00000	−1.00000	−1.00000

(3) $\mathbf{x}_0 = (-0.01, -0.01, -0.01)$: (Yields $\mathbf{s}_1$)

k	$x_1(k)$	$x_2(k)$	$x_3(k)$
10	−149328	40254800.	−18771.2
20	154152	40255700.	−1057.07
30	42024.2	40255700.	−59.5274
40	438.705	40206600.	−3.35629
50	4.64666	4204370.	−1.00000
60	0.08058	72910.2	−1.00000
70	0.00140	1264.37	−1.00000
80	0.00002	21.9258	−1.00000
90	0	−1.00718	−1.00000
92	0	−1.00000	−1.00000

(4) $\mathbf{x}_0 = (0.1, 0.1, 0.1)$: (Yields $\mathbf{s}_2$)

k	$x_1(k)$	$x_2(k)$	$x_3(k)$
10	−47.1446	3799.41	25.0358
20	−0.43201	3657.72	1.46340
30	−0.00685	117.065	1.00000
40	−0.00012	1.97841	1.00000
48	−0	−1.00000	1.00000

(5) $\mathbf{x}_0 = (-100, 0, 100)$: (Yields $\mathbf{s}_3$)

k	$x_1(k)$	$x_2(k)$	$x_3(k)$
10	−10.7374	0	5.63222
20	−1.21805	0	1.00000
24	−1.00000	0	1.00000

PROBLEMS, SECTION 4.5

***1.** Let $\mathcal{R}$ be a rectangle in R^n. Let $\mathbf{x}$ and $\mathbf{y}$ be any two vectors in $\mathcal{R}$, and let $\mathbf{w}$ be any vector of the form $\mathbf{w} = \lambda\mathbf{x} + (1 - \lambda)\mathbf{y}$ where $0 \le \lambda \le 1$.

a) What is the geometrical interpretation of $\mathbf{w}$ if $\lambda = 1/2$? (Consider $\|\mathbf{w} - \mathbf{x}\|$ and $\|\mathbf{w} - \mathbf{y}\|$ with respect to $\|\mathbf{x} - \mathbf{y}\|$ for any norm.)

b) Show that if $\mathbf{x}$ and $\mathbf{y}$ are in $\mathcal{R}$, then $\mathbf{w} \in \mathcal{R}$ for any λ, $0 \le \lambda \le 1$.

c) The vector-valued function $\mathbf{F}$, as given by (4.39), is continuous at a vector $\mathbf{z} \in \mathcal{R}$ if for all $\varepsilon > 0$, there exists a $\delta > 0$ such that $\|\mathbf{F}(\mathbf{z} + \mathbf{h}) - \mathbf{F}(\mathbf{z})\| \le \varepsilon$ whenever $0 < \|\mathbf{h}\| \le \delta$. Let $\mathbf{x}$ and $\mathbf{y}$ be any two points in $\mathcal{R}$ that satisfy $f_i(\mathbf{x}) < 0$ and $f_i(\mathbf{y}) > 0$ for all i, $1 \le i \le \eta$. Assume that $\mathbf{F}$ is continuous at all $\mathbf{z} \in \mathcal{R}$. Does it necessarily follow that an $\mathbf{s}$ on the line segment between $\mathbf{x}$ and $\mathbf{y}$ exists satisfying $\mathbf{F}(\mathbf{s}) = \theta$? (Note that if this result were true, then the bisection method could be extended to this problem. Hint: Consider R^2.)

d) Use the definition in part (c) to prove that if $F'(\mathbf{x})$ exists, then $\mathbf{F}$ is continuous at $\mathbf{z}$. [Remember, $F'(\mathbf{z})$ is an $n \times n$ matrix.]

***2.** Again let $\mathbf{e}_j \in R^n$ be 1 in the jth component and zero elsewhere. Also recall from calculus that

$$\frac{\partial f_i(\mathbf{z})}{\partial x_j} = \lim_{t\to\infty} \left(\frac{f_i(\mathbf{z} + t\mathbf{e}_j) - f_i(\mathbf{z})}{t}\right). \tag{4.52}$$

a) Verify that $\|\mathbf{e}_j\|_1 = \|\mathbf{e}_j\|_2 = \|\mathbf{e}_j\|_\infty = 1$, $1 \le j \le n$, and let $\mathbf{h} = t\mathbf{e}_j$ where $|t| \le \delta$. Thus $0 < \|\mathbf{h}\| \le \delta$ for any of the three norms.

b) Verify that if A is any $(n \times n)$ matrix, then $A(t\mathbf{e}_j) = t\mathbf{A}_j$ where $\mathbf{A}_j$ is the jth column of A, $1 \le j \le n$.

c) Assume that $\mathbf{F}$ is given by (4.39) and that $F'(\mathbf{z})$ exists; that is, given any ε there exists a δ by (4.44) such that

$$\|\mathbf{F}(\mathbf{z} + \mathbf{h}) - \mathbf{F}(\mathbf{z}) - F'(\mathbf{z})\mathbf{h}\| < \varepsilon\|\mathbf{h}\| \tag{4.53}$$

whenever

$$0 < |t| \le \delta.$$

Let

$$\mathbf{h}_j = t\mathbf{e}_j$$

for a fixed j, $1 \le j \le n$.

Assuming that $\partial f_i(\mathbf{z})/\partial x_j$ exists, $1 \le i \le n$, prove by use of (4.52) that (4.53) can be true for this particular $\mathbf{h}_j$ only if the jth column of $F'(\mathbf{z})$ equals

$$\begin{bmatrix} \dfrac{\partial f_1(\mathbf{z})}{\partial x_j} \\ \dfrac{\partial f_2(\mathbf{z})}{\partial x_j} \\ \vdots \\ \dfrac{\partial f_n(\mathbf{z})}{\partial x_j} \end{bmatrix}$$

(Use any of the three norms that you wish.) Thus we argue that if (4.53) holds for all $\mathbf{h}$ satisfying $0 < \|\mathbf{h}\| \le \delta$, then it must hold for each $\mathbf{h}_j$, $1 \le j \le n$. But if all partials $\partial f_i(\mathbf{z})/\partial x_j$ exist, their existence implies that $F'(\mathbf{z}) = J(\mathbf{z})$, the Jacobian of $\mathbf{F}$ at $\mathbf{z}$.

d) Let $\mathbf{F}(\mathbf{x})$ be given by (4.39) and let A be a nonsingular $(n \times n)$ matrix whose elements are constants. If $\mathbf{g}(\mathbf{x}) = \mathbf{x} + A\mathbf{F}(\mathbf{x})$, show that $g'(\mathbf{x}) = I + AF'(\mathbf{x})$.

3. Verify that the equations of Example 4.13 have only the four real solutions that are given. Find one complex solution. [Recall that $e^{i\theta} = \cos\theta + i\sin\theta$, and that $(e^{i\theta})^m = e^{im\theta} = \cos m\theta + i\sin m\theta$.]

4. One modification of Newton's method is as follows. Let $\mathbf{x}_0$ be an initial guess for $\mathbf{s}$ where $\mathbf{F}(\mathbf{s}) = \theta$, and define $\{\mathbf{x}_n\}_{n=0}^{\infty}$ by

$$\mathbf{x}_{n+1} = \mathbf{x}_n - J(\mathbf{x}_0)^{-1}\mathbf{F}(\mathbf{x}_n) \tag{4.54}$$

where $J(\mathbf{x}_0)$ is the Jacobian of $\mathbf{F}$ at $\mathbf{x}_0$. This method is still of the fixed-point type, but its convergence is linear. Apply this algorithm to one of the initial guesses in Example 4.13 and compare their rates of convergence. This method is obviously easier to perform since $J(\mathbf{x}_0)$ remains stationary. Can a different algorithm along these lines be developed to be faster than (4.54), but still not as fast as (4.51)?

5. Let $\mathbf{F}(\mathbf{x})$ be given as in Example 4.11. Write a program for Newton's method to approximate solutions of $\mathbf{F}(\mathbf{x}) = \theta$ with $\mathbf{x}_0 = (1/4, \pi/2, -1/2)$. Continue the iteration until $\|\mathbf{F}(\mathbf{x}_n)\|_2 < 10^{-5}$.

5

Interpolation and Approximation

5.1 INTRODUCTION

An important problem often encountered in scientific work is that of approximating some very "complicated" function, $f(x)$, by a "simpler" function, $p(x)$. The reader has already seen one form of this problem in calculus: if $f(a)$, $f'(a), f''(a), \ldots, f^{(k)}(a)$ are known at some point a, then the truncated Taylor's expansion,

$$p(x) = f(a) + f'(a)(x - a) + \frac{f''(a)}{2!}(x - a)^2 + \cdots + \frac{f^{(k)}(a)}{k!}(x - a)^k,$$

is a kth degree polynomial that approximates $f(x)$ for x near a. If $f^{(k+1)}(x)$ is continuous, then the error of the approximation at x is given by

$$\frac{f^{(k+1)}(\xi)}{(k + 1)!}(x - a)^{k+1}$$

where ξ lies between x and a. There are, however, several obvious drawbacks to this type of an approximation: the derivatives may not exist, calculating them may be very difficult [or even impossible if, for instance, $f(x)$ is given in tabular form instead of in terms of a formula], $f^{(k+1)}(\xi)$ may become quite large and thus make the error large, etc. Even for a "nice" function, such as $f(x) = \cos(x)$ with $a = 0$ where the Taylor (Maclaurin) expansion converges to $f(x)$ for any x and $f^{(k)}(0)$ is known for any k, this type of approximation is not usually practical since many terms may be required to maintain accuracy for values of x somewhat removed from 0 (see Problem 2).

Thus we see that the problem of approximation of functions must be analyzed quite carefully to obtain practical computational procedures. In this chapter we shall be mainly concerned with polynomial interpolation, spline interpolation, and Fourier approximations. In the following chapter, we will study their use in such problems as numerical integration and numerical differentiation. For example if $f(x)$ is approximated by $p(x)$ and we wish to find a numerical approximation for $\int_a^b f(x)\,dx$ or for $f'(\alpha)$ for some value α, then we could use $\int_a^b p(x)\,dx$ or $p'(\alpha)$, respectively, as these approximations. However, it may be that while $\int_a^b p(x)\,dx$ is a good approximation for $\int_a^b f(x)\,dx$, $p'(\alpha)$ is a poor approximation for $f'(\alpha)$ (and vice versa). Therefore an integral part of designing an approximation for a function is the knowledge of how the approximation is to be used subsequently. Thus Chapters 5 and 6 are intrinsically related in that Chapter 5 develops some approximation techniques and the main part of Chapter 6 discusses their utilization in integration and differentiation. Then with this background, Chapter 6 concludes by discussing some further approximation techniques.

We shall be principally interested in approximating continuous real-valued functions, $f(x)$, where x belongs to the closed finite interval $[a, b]$. We use the standard notation $C[a, b]$ to denote the set of real-valued functions that are continuous on $[a, b]$. If $p(x)$ is an approximation to $f(x)$ where $f \in C[a, b]$, then we need some way of measuring how good this approximation is. That is, we must be able to answer this question: How "close" is $p(x)$ to $f(x)$? We will naturally say that $p(x)$ is a good approximation to $f(x)$ if the function $f(x) - p(x)$ is small in some sense. So to measure closeness we need some sort of a measure of size for functions in $C[a, b]$. In (5.1), we have listed three commonly used measures for the size of the function $(f - p)$. These measures are called *norms*, and the magnitude of the number $\|f - p\|$ provides us with a quantitative way of gauging how good an approximation $p(x)$ is to $f(x)$. [The reader who has covered the material on vector norms in Chapter 2 will recognize the norms defined in (5.1) as natural extensions of the ℓ_p vector norms where $(f - p)$ is used in place of the vector $(x - y)$ and integration is used in place of summation.]

$$\|f - p\|_1 \equiv \int_a^b |f(x) - p(x)|\,w(x)\,dx \tag{5.1a}$$

$$\|f - p\|_2 \equiv \left(\int_a^b (f(x) - p(x))^2 w(x)\,dx\right)^{1/2} \tag{5.1b}$$

$$\|f - p\|_\infty \equiv \max_{a \le x \le b} |f(x) - p(x)|. \tag{5.1c}$$

In (5.1a) and (5.1b) the function $w(x)$ is a fixed weighting function that provides us with some flexibility in measuring closeness. In all the cases we consider, $w(x)$ is continuous and nonnegative on (a, b), $\int_a^b w(x)\,dx$ exists, and $\int_a^b w(x)\,dx > 0$. Finally, by way of notation, we shall denote the "zero function" as $\theta(x)$ where $\theta(x) \equiv 0$ for all x in $[a, b]$; and we will let $\mathcal{P}_n$ denote the set of all polynomials of degree n or less. The following example illustrates that two functions can be "close" in one norm but *not* in another.

EXAMPLE 5.1. Let $f(x) = \theta(x)$ and let $w(x) \equiv 1$ for x in $[a, b]$ with $a = 0$ and $b = 3$. For any positive integer k, let $f_k(x)$ be given by (see Fig. 5.1)

$$f_k(x) = \begin{cases} k(k^2x - 1), & \text{for} \quad 1/k^2 \le x \le 2/k^2 \\ -k(k^2x - 3), & \text{for} \quad 2/k^2 \le x \le 3/k^2 \\ 0, & \text{otherwise.} \end{cases}$$

Using these three formulas, we obtain

$$\|f - f_k\|_1 = 1/k, \qquad \|f - f_k\|_2 = \sqrt{2}/\sqrt{3}, \qquad \|f - f_k\|_\infty = k.$$

Thus, in the sense of (5.1a), the distance between $f(x)$ and $f_k(x)$ becomes small for large k; for (5.1b) the distance is constant for any k; and for (5.1c) the distance is large for large k. Now we consider $\{\|f - f_k\|\}_{k=1}^{\infty}$ as a sequence of real numbers and see that

$$\lim_{k\to\infty} \|f - f_k\|_1 = 0,$$

$$\lim_{k\to\infty} \|f - f_k\|_2 = \sqrt{2}/\sqrt{3},$$

and

$$\lim_{k\to\infty} \|f - f_k\|_\infty = \infty.$$

In this context we therefore say that the sequence of functions, $\{f_k(x)\}_{k=1}^{\infty}$, *converges to* $f(x)$ only in terms of the $\|\cdot\|_1$ distance measurement. [A sequence of functions, $\{q_k(x)\}_{k=1}^{\infty}$, is said to converge to a function $g(x)$ with respect to a given norm, $\|\cdot\|$, if and only if

$$\lim_{k\to\infty} \|q_k - g\| = 0.]$$

The preceding example illustrates that the quality of an approximation is completely dependent on how we choose to measure distance between func-

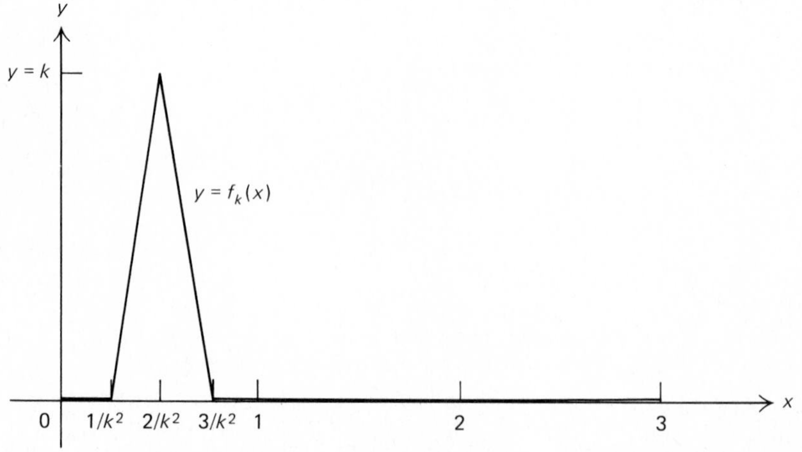

Figure 5.1 Graph of $y = f_k(x)$; $k \ge 1$.

tions. In most practical problems we would probably accept $f_k(x)$ for large k as a "good" approximation for $f(x) = \theta(x)$ since it is "bad" only in a small neighborhood of a single point. However, there are practical problems in which even moderate errors in a small neighborhood are unacceptable. For example, the cosine routine in a computer must provide *uniformly* good approximations for all x in $[0, \frac{\pi}{2}]$. The choice of a distance measurement (norm) is dependent on the underlying physical or mathematical problem. We note here that

$$\|f - p\|_1 = \int_a^b |f(x) - p(x)| \, w(x) \, dx \le \max_{a \le x \le b} |f(x) - p(x)| \int_a^b w(x) \, dx$$

$$= \|f - p\|_\infty \left(\int_a^b w(x) \, dx \right)$$

and similarly

$$\|f - p\|_2 \le \|f - p\|_\infty \left(\int_a^b w(x) \, dx \right)^{\frac{1}{2}}.$$

The number $\int_a^b w(x)\,dx$ is a constant; so if $\|f - p\|_\infty$ is "small," then both $\|f - p\|_1$ and $\|f - p\|_2$ are also "small." Thus the $\|\cdot\|_\infty$ norm is "stronger" than the other two norms, and we strive for goodness of approximation with respect to the $\|\cdot\|_\infty$ norm whenever possible. (The reader with some advanced calculus background will easily recognize that convergence of a sequence of functions in the $\|\cdot\|_\infty$ norm is equivalent to *uniform convergence*.)

As we shall see shortly, we will normally be approximating functions with polynomials. Two important reasons for this practice are that polynomials are easy to use (for example, in integration and differentiation as in Problem 1) and that polynomials can be used to provide very good approximations for functions in $C[a, b]$. The first reason is obvious to the reader and evidence for the second is provided by the following classical result (given here without proof) of Weierstrass.

Theorem 5.1 Weierstrass

Let $f \in C[a, b]$. For each $\varepsilon > 0$ there exists a polynomial $p(x)$ of degree N_ε (N_ε depends on ε) such that $\|f - p\|_\infty < \varepsilon$.

This theorem says that any continuous function on the finite interval $[a, b]$ may be uniformly approximated by some polynomial. We conclude this section with another classical result (again without proof).

Theorem 5.2

Let $f(x)$ be given in $C[a, b]$ and let n be a fixed positive integer. If $\|\cdot\|$ is any one of the three norms given above, then there exists a unique polynomial $p^*(x)$ of degree n or less such that $\|f - p^*\| \le \|f - p\|$ for all $p(x) \in \mathcal{P}_n$.

This theorem tells us that there is a unique "best" nth degree polynomial approximation to $f(x)$ with respect to any of the above norms. The reader should be warned that the polynomial that is best with respect to one norm is usually *not* the same polynomial that is best with respect to another. We shall see in Section 5.3.1 that the polynomial that minimizes $\|f - p\|_2$ for $p(x) \in \mathcal{P}_n$ can be explicitly constructed whereas this is not usually true with respect to the other two norms. However, Theorem 5.2 still has important theoretical implications with respect to these norms. Finally, let us define $E_n(f) \equiv \|f - p_n^*\|_\infty$ where $p_n^*(x) \in \mathcal{P}_n$ satisfies $\|f - p_n^*\|_\infty \leq \|f - p\|_\infty$ for all $p(x) \in \mathcal{P}_n$. [Usually, $E_n(f)$ is called the *degree of approximation for* $f(x)$ and $p_n^*(x)$ is called the *best nth degree uniform approximation to* $f(x)$.] It is easily seen that if $m > n$, then $E_m(f) \leq E_n(f)$. So, by Theorem 5.1, we obtain the result that $\lim_{n\to\infty} E_n(f) = 0$ for any $f(x)$ in $C[a, b]$.

PROBLEMS, SECTION 5.1

1. One argument for using polynomials $p(x)$ to approximate complicated functions $f(x)$ is that polynomials are easy to integrate, evaluate, and differentiate on the computer. Write a program that accepts the coefficients of any 20th-degree or less polynomial $p(x)$ as input, together with either an interval $[a, b]$ or a number c. Develop the program to have the capability of calculating $\int_a^b p(x)\,dx$ and the ith derivative of $p(x)$ evaluated at $x = c$, for any i, $0 \leq i \leq 20$.
2. Find the truncated kth degree Taylor's series expansion $p_k(x)$ for $f(x) = \cos(x)$ with $a = 0$ and k arbitrary. Show that a bound for the error $|f(x) - p_k(x)|$ is given by $|x^{k+1}|/(k+1)!$. Given this bound, how large must k be in order that $|p_k(x) - f(x)| \leq 10^{-6}$ for $x = 1$, for $x = 2$, and for $x = 3$? For $k = 6$, 8, and 10, write a short program that calculates and lists $p_k(x)$, $\cos(x)$, and $\cos(x) - p_k(x)$ for x varying between 0 and 3 in steps of 0.1.
3. Verify that the function $\|\cdot\|_\infty$ defined on $C[a, b]$ by (5.1c) satisfies the following three conditions (where f and g are in $C[a, b]$):

   a) $\|g\|_\infty \geq 0$ and $\|g\|_\infty = 0$ if and only if $g(x) \equiv \theta(x)$

   b) $\|\alpha g\|_\infty = |\alpha|\,\|g\|_\infty$ for any scalar α

   c) $\|f + g\|_\infty \leq \|f\|_\infty + \|g\|_\infty$.

   To do this problem, recall that a continuous function defined on $[a, b]$ attains its maximum at some point x_0 in $[a, b]$.
4. The *sine-integral*, Si(x), occurs frequently in certain applied problems where $x > 0$ and Si(x) is defined by

$$\text{Si}(x) = \int_0^x \frac{\sin(t)}{t}\,dt.$$

One way to estimate Si(x) is to take the truncated kth-degree Taylor's series expansion $p_k(t)$, for $f(t) = \sin(t)$ with $a = 0$, and use

$$\int_0^x \frac{p_k(t)}{t}\, dt$$

as an approximation to Si(x). How large must k be in order that

$$\left| \int_0^4 \frac{p_k(t)}{t}\, dt - \text{Si}(4) \right| \le 10^{-6}?$$

To bound the error, use the fact that if $|h(t)| \le |q(t)|$ for $0 \le t \le x$, then $\int_0^x |h(t)|\, dt \le \int_0^x |q(t)|\, dt$.

5.2 POLYNOMIAL INTERPOLATION

Perhaps the simplest and best known way to construct an nth-degree polynomial approximation $p(x)$ to a function $f(x)$ in $C[a, b]$ is by interpolation. Let x_0, $x_1, \ldots, x_n$ be $(n + 1)$ distinct points in the interval $[a, b]$. Then $p(x) \in \mathcal{P}_n$ is said to *interpolate* $f(x)$ at these points if $p(x_j) = f(x_j)$ for $0 \le j \le n$. For example, the second-degree polynomial $p(x) = -(4/\pi^2)x^2 + (4/\pi)x$ interpolates $f(x) = \sin(x)$ at the points $x_0 = 0$, $x_1 = \pi/2$ and $x_2 = \pi$. We must first show that such an interpolating polynomial always exists and is unique. In doing so we shall present two separate proofs since they both illustrate ways of constructing the interpolating polynomial.

Theorem 5.3

Let $\{x_j\}_{j=0}^n$ be $(n + 1)$ distinct points in the interval $[a, b]$, and let $\{y_j\}_{j=0}^n$ be any set of $(n + 1)$ real numbers. Then there exists a unique polynomial $p(x)$ in $\mathcal{P}_n$ such that $p(x_j) = y_j$ for $0 \le j \le n$. [Often the y_j's are determined by some function, $f(x)$; so we have the property that $p(x_j) = f(x_j)$ for $0 \le j \le n$.]

Proof 1. For each j, $0 \le j \le n$, let $\ell_j(x)$ be the nth-degee polynomial defined by

$$\ell_j(x) \equiv \frac{(x - x_0)(x - x_1)\cdots(x - x_{j-1})(x - x_{j+1})\cdots(x - x_n)}{(x_j - x_0)(x_j - x_1)\cdots(x_j - x_{j-1})(x_j - x_{j+1})\cdots(x_j - x_n)}$$

$$\equiv \prod_{\substack{i=0 \\ i \ne j}}^{n} \frac{(x - x_i)}{(x_j - x_i)}. \tag{5.2}$$

Then it follows for each i and j, that $\ell_i(x_j) = \delta_{ij}$. (The symbol δ_{ij} is defined to be 1 when $i = j$ and 0 otherwise. The symbol δ_{ij} is called the Kronecker delta.)

Since the sum of nth-degree polynomials is again a polynomial of at most nth degree, the polynomial $p(x)$ defined by (5.3) is in $\mathcal{P}_n$:

$$p(x) \equiv \sum_{j=0}^{n} y_j \ell_j(x). \tag{5.3}$$

Moreover, for $0 \le i \le n$, we have

$$p(x_i) = y_0\ell_0(x_i) + \cdots + y_n\ell_n(x_i) = y_i\ell_i(x_i) = y_i.$$

If $f(x)$ is a function such that $f(x_i) = y_i$, $0 \le i \le n$, then it has in $\mathcal{P}_n$ an interpolating polynomial of the form

$$p(x) = \sum_{j=0}^{n} f(x_j)\ell_j(x). \tag{5.4}$$

To establish uniqueness, let us assume that there are two different polynomials, $p(x)$ and $q(x)$ in $\mathcal{P}_n$, such that $p(x_j) = q(x_j) = y_j$ for $0 \le j \le n$. If $r(x) \equiv p(x) - q(x)$, then $r(x) \in \mathcal{P}_n$ and furthermore $r(x_j) = p(x_j) - q(x_j) = 0$ for $0 \le j \le n$. By the Fundamental Theorem of Algebra, $r(x) \equiv 0$; so $p(x) \equiv q(x)$, which contradicts our assumption.

Proof 2. Let $p(x) = a_0 + a_1x + a_2x^2 + \cdots + a_nx^n$ where the coefficients, a_j, are to be determined. Consider the $(n + 1)$ equations

$$p(x_j) = a_0 + a_1x_j + a_2x_j^2 + \cdots + a_nx_j^n = y_j, \qquad 0 \le j \le n. \tag{5.5}$$

In matrix form, Eqs. (5.5) become

$$\begin{bmatrix} 1 & x_0 & x_0^2 & \cdots & x_0^n \\ 1 & x_1 & x_1^2 & \cdots & x_1^n \\ \vdots & & & & \\ 1 & x_n & x_n^2 & \cdots & x_n^n \end{bmatrix} \begin{bmatrix} a_0 \\ a_1 \\ \vdots \\ a_n \end{bmatrix} = \begin{bmatrix} y_0 \\ y_1 \\ \vdots \\ y_n \end{bmatrix} \tag{5.6}$$

or $V\mathbf{a} = \mathbf{y}$ where V is the coefficient matrix in (5.6), $\mathbf{a} = [a_0, a_1, \ldots, a_n]^T$, and $\mathbf{y} = [y_0, y_1, \ldots, y_n]^T$. The matrix V is called a *Vandermonde matrix*, and it is easy to see that V is nonsingular when $x_0, x_1, \ldots, x_n$ are distinct. To see this, recall from Section 2.1 that V is nonsingular if and only if θ is the only solution of $V\mathbf{a} = \theta$. So if the vector $\mathbf{a}$ is any solution of $V\mathbf{a} = \theta$, then $p(x) = a_0 + a_1x + \cdots + a_nx^n$ is an nth-degree polynomial such that $p(x_j) = 0$ for $j = 0, 1, \ldots, n$. Since the only nth-degree polynomial with $(n + 1)$ zeros is the zero polynomial, it must be $\mathbf{a} = \theta$ and hence V is nonsingular. Thus Eqs. (5.5) have a unique solution for the a_i's, and the polynomial $p(x)$ found by solving (5.6) is the unique interpolating polynomial in $\mathcal{P}_n$. ■

From the uniqueness of the interpolating polynomial in $\mathcal{P}_n$, both (5.4) and (5.6) must yield the same polynomial even though it is written in different forms. The form given by (5.4) is called the *Lagrange form* whereas the construction of $p(x)$ by (5.6) is called the *method of undetermined coefficients*. We should note three obvious facts here. First, if $f(x)$ is itself in $\mathcal{P}_n$, then $f(x) \equiv p(x)$ for all x by the uniqueness property. Second, it is possible that the unique solution of (5.6) may yield $a_n = 0$ (and possibly other coefficients may be zero also); so $p(x)$ may be a polynomial of degree strictly less than n. Third, if $m > n$, there is an infinite number of polynomials, $q(x)$, in $\mathcal{P}_m$ which satisfy $q(x_j) = y_j$, $0 \le j \le n$. [Note that we can arbitrarily specify another point, x_{n+1}, in $[a, b]$ and any other value y_{n+1}, and then construct $q(x) \in \mathcal{P}_{n+1}$ such that $q(x_j) = y_j$, $0 \le j \le n + 1$. Thus the uniqueness holds only for $\mathcal{P}_n$.]

EXAMPLE 5.2 As a simple example of a problem involving data fitting, suppose we want a second-degree polynomial, $p(x)$, such that $p(0) = -1$, $p(1) = 2$, and $p(2) = 7$. Using the Lagrange form, we have

$$\ell_0(x) = \frac{(x-1)(x-2)}{2}, \qquad \ell_1(x) = -x(x-2), \qquad \text{and} \qquad \ell_2(x) = \frac{x(x-1)}{2}.$$

Thus $p(x)$ is given by the formula $p(x) = -\ell_0(x) + 2\ell_1(x) + 7\ell_2(x)$, or upon simplification $p(x) = x^2 + 2x - 1$. Alternatively, using the method of undetermined coefficients with $x_0 = 0$, $x_1 = 1$ and $x_2 = 2$ in (5.5), we obtain

$$\begin{aligned} a_0 \qquad\qquad\qquad &= -1 \\ a_0 + \ a_1 + \ a_2 &= 2 \\ a_0 + 2a_1 + 4a_2 &= 7. \end{aligned}$$

Solving this system gives $p(x) = x^2 + 2x - 1$.

The method of undetermined coefficients is a procedure that has wide application to other types of interpolation problems. For example, let $x_0 = 0$, $x_1 = \pi/6$, $x_2 = \pi/4$, and $x_3 = \pi/3$ with $y_0 = 0$, $y_1 = 2 - \sqrt{3}/2$, $y_2 = 1 + \sqrt{2}/2$, and $y_3 = 3/2$. Suppose we wish to find a *trigonometric* polynomial, $p(x)$, of the form $p(x) = C_0 + C_1\cos(x) + C_2\cos(2x) + C_3\cos(3x)$ such that $p(x_j) = y_j$ for $0 \le j \le 3$. These four equations yield the matrix equation [as in the derivation of (5.6)]

$$\begin{bmatrix} 1 & 1 & 1 & 1 \\ 1 & \sqrt{3}/2 & 1/2 & 0 \\ 1 & \sqrt{2}/2 & 0 & -\sqrt{2}/2 \\ 1 & 1/2 & -1/2 & -1 \end{bmatrix} \begin{bmatrix} C_0 \\ C_1 \\ C_2 \\ C_3 \end{bmatrix} = \begin{bmatrix} 0 \\ 2 - \sqrt{3}/2 \\ 1 + \sqrt{2}/2 \\ 3/2 \end{bmatrix}. \tag{5.7}$$

The solution of this system by Gauss elimination yields $C_0 = 1$, $C_1 = -1$, $C_2 = 2$, $C_3 = -2$. Therefore $p(x) = 1 - \cos(x) + 2\cos(2x) - 2\cos(3x)$ is a trigonometric polynomial satisfying $p(x_i) = y_i$, $i = 0, 1, 2, 3$.

5.2.1. Divided Differences and the Newton Form of the Interpolating Polynomial

If $f(x)$ is a function defined on $[a, b]$ and if $x_0, x_1, \ldots, x_n$ are distinct points in $[a, b]$, then there is a unique polynomial $p(x)$ in $\mathcal{P}_n$ that satisfies $p(x_i) = f(x_i)$, $0 \le i \le n$. There are many ways to represent this interpolating polynomial $p(x)$, but some representations are more useful for computation than others (in mathematical terms, we would like to use a basis for $\mathcal{P}_n$ that is convenient with respect to computation). For example, although we can immediately express $p(x)$ in the Lagrange form

$$p(x) = c_0\ell_0(x) + c_1\ell_1(x) + \cdots + c_n\ell_n(x)$$

simply by choosing $c_i = f(x_i)$, $0 \le i \le n$, this form is cumbersome to use in common operations such as differentiation and integration. The most convenient form for these operations is usually $p(x) = b_0 + b_1x + \cdots + b_nx^n$; but here, to find the values of the b_i's we must go through the effort of solving the system in (5.6) or collecting coefficients of like powers of x from another form of $p(x)$ such as the Lagrange form. Hence, the manner in which $p(x)$ is to be used is an important factor in the choice of the form of $p(x)$. In this section we consider yet another representation for $p(x)$, the *Newton form*, which facilitates computations with $p(x)$ when the interpolation points are equally spaced or when we wish to add further interpolation points to get a higher-degree approximation.

The Newton form for $p(x)$ arises when we express the interpolating polynomial $p(x)$ in the form

$$p(x) = a_0 + a_1(x - x_0) + a_2(x - x_0)(x - x_1) + a_3(x - x_0)(x - x_1)(x - x_2) + \cdots + a_n(x - x_0)(x - x_1)(x - x_2)\cdots(x - x_{n-1}). \tag{5.8}$$

If we impose the interpolatory constraints $p(x_0) = f(x_0)$, $p(x_1) = f(x_1)$, . . . , $p(x_n) = f(x_n)$, then it is clear that we can solve for the coefficients $a_0, a_1, \ldots, a_n$ in (5.8). For example, setting $x = x_0$ in (5.8), we have $f(x_0) = a_0$. Next setting $x = x_1$, we have $f(x_1) = a_0 + a_1(x_1 - x_0)$ or $a_1 = [f(x_1) - f(x_0)]/(x_1 - x_0)$. For $x = x_2$, we have $f(x_2) = a_0 + a_1(x_2 - x_0) + a_2(x_2 - x_0)(x_2 - x_1)$; and it is not hard to see that this leads to

$$a_2 = \frac{\dfrac{f(x_2) - f(x_1)}{x_2 - x_1} - \dfrac{f(x_1) - f(x_0)}{x_1 - x_0}}{x_2 - x_0}.$$

Clearly, determining the coefficients a_i in (5.8) is recursive in nature and we can mechanize the procedure (in a way that is suited for computation) by introducing the related ideas of divided differences and the divided difference table.

Given $x_0, x_1, \ldots, x_n$ in $[a, b]$, we define the *first divided difference*, $f[x_i, x_{i+1}]$, by

$$f[x_i, x_{i+1}] = \frac{f(x_{i+1}) - f(x_i)}{x_{i+1} - x_i}, \qquad 0 \le i \le n - 1.$$

We define the kth divided difference inductively by

$$f[x_i, x_{i+1}, \ldots, x_{i+k}] = \frac{f[x_{i+1}, x_{i+2}, \ldots, x_{i+k}] - f[x_i, x_{i+1}, \ldots, x_{i+k-1}]}{x_{i+k} - x_i} \tag{5.9}$$

where $0 \le i \le n - k$. For example,

$$f[x_2, x_3] = \frac{f(x_3) - f(x_2)}{x_3 - x_2}, \quad f[x_3, x_4, x_5] = \frac{f[x_4, x_5] - f[x_3, x_4]}{x_5 - x_3},$$

$$f[x_1, x_2, x_3, x_4] = \frac{f[x_2, x_3, x_4] - f[x_1, x_2, x_3]}{x_4 - x_1}.$$

We will show shortly that the coefficients a_i in (5.8) are given precisely by $a_i = f[x_0, x_1, \ldots, x_i]$. (Note that as found above, $a_1 = f[x_0, x_1]$ and $a_2 = f[x_0, x_1, x_2]$.)

The calculation of divided differences can be organized by constructing a *divided difference table*, (see Table 5.1). Table 5.1 presents the definition (5.9) in a rather schematic form; entries in the kth column are obtained by "differencing" successive entries in the $(k-1)$st column and then dividing by the appropriate difference $x_{i+k} - x_i$

TABLE 5.1 A divided difference table.

x_0	$f(x_0)$				
		$f[x_0, x_1]$			
x_1	$f(x_1)$		$f[x_0, x_1, x_2]$		
		$f[x_1, x_2]$		$f[x_0, x_1, x_2, x_3]$	
x_2	$f(x_2)$		$f[x_1, x_2, x_3]$		$f[x_0, x_1, x_2, x_3, x_4]$
		$f[x_2, x_3]$		$f[x_1, x_2, x_3, x_4]$	
x_3	$f(x_3)$		$f[x_2, x_3, x_4]$		.
		$f[x_3, x_4]$		.	.
x_4	$f(x_4)$		.	.	.
.	.	.	.	.	
.	.	.	.		
.	.	.			
			$f[x_{n-2}, x_{n-1}, x_n]$		
		$f[x_{n-1}, x_n]$			
x_n	$f(x_n)$				

EXAMPLE 5.3. For $f(x) = x^5 - x^4 + 2x^2 + 1$, we construct the divided difference table for the data

x	$f(x)$
−2	−39
−1	1
0	1
1	3
2	25
3	181
4	801

The divided difference table is

−2	−39						
		40					
−1	1		−20				
		0		7			
0	1		1		−1		
		2		3		1	
1	3		10		4		0
		22		19		1	
2	25		67		9		
		156		55			
3	181		232				
		620					
4	801						

For instance, $x_3 = 1$, $x_4 = 2$, $x_5 = 3$; so $f[x_3, x_4] = 22$, $f[x_4, x_5] = 156$, and $f[x_3, x_4, x_5] = 67$. To display the recursive nature of (5.9), we consider the entries along the upper diagonal, beginning with $f(x_0) = -39$. The next term along the diagonal is $f[x_0, x_1] = (1 - (-39))/1 = 40$, followed by $f[x_0, x_1, x_2] = (0 - 40)/2 = -20$; then $f[x_0, x_1, x_2, x_3] = (1 - (-20))/3 = 7$; then $f[x_0, x_1, x_2, x_3, x_4] = (3 - 7)/4 = -1$, etc.

A divided difference table contains a good deal of information that is relevant to interpolation. The downward slanting diagonals contain the coefficients of various interpolating polynomials, and we will see that the columns contain information that can be used to estimate interpolation errors. Our first objective is to find the coefficients a_i in (5.8), and to that end we state Theorem 5.4 below.

Theorem 5.4

Suppose that $f(x)$ is defined on $[a, b]$ and suppose that $x_0, x_1, \ldots, x_n$ are distinct points in $[a, b]$. The kth degree polynomial $p(x)$ interpolating $f(x)$ at $x_i, x_{i+1}, \ldots, x_{i+k}$ is given by

$$p(x) = f(x_i) + f[x_i, x_{i+1}](x - x_i) + f[x_i, x_{i+1}, x_{i+2}](x - x_i)(x - x_{i+1}) + \cdots + f[x_i, x_{i+1}, \ldots, x_{i+k}](x - x_i)(x - x_{i+1}) \ldots (x - x_{i+k-1}).$$

Since the notation is somewhat involved, we give several illustrations of the theorem before sketching the proof. For example, according to Theorem 5.4, the quadratic polynomial interpolating $f(x)$ at x_2, x_3, x_4 is given by $p(x) = f(x_2) + f[x_2, x_3](x - x_2) + f[x_2, x_3, x_4](x - x_2)(x - x_3)$; and we observe that the coefficients $f(x_2)$, $f[x_2, x_3]$, $f[x_2, x_3, x_4]$ are found in Table 5.1 on the downward diagonal that begins at $f(x_2)$. As another special case, the cubic polynomial interpolating $f(x)$ at x_0, x_1, x_2, x_3 is (according to Theorem 5.4) given by

$$p(x) = f(x_0) + f[x_0, x_1](x - x_0) + f[x_0, x_1, x_2](x - x_0)(x - x_1) + f[x_0, x_1, x_2, x_3](x - x_0)(x - x_1)(x - x_2).$$

Again, we note that the coefficients of $p(x)$ are found on the downward diagonal that starts at $f(x_0)$ (see Table 5.1).

Proof. The proof of Theorem 5.4 is by induction on k. The proof is simple and we sketch only a part of it here; the rest is left to the exercises. For $k = 1$, Theorem 5.4 asserts that $p(x) = f(x_i) + f[x_i, x_{i+1}](x - x_i)$ is the first-degree polynomial interpolating $f(x)$ at x_i and x_{i+1}. Clearly $p(x)$ is a linear polynomial, and it is also obvious that $p(x_i) = f(x_i)$. At $x = x_{i+1}$, we have $p(x_{i+1}) = f(x_i) + f[x_i, x_{i+1}](x_{i+1} - x_i) = f(x_i) + (f(x_{i+1}) - f(x_i))$; so $p(x_{i+1}) = f(x_{i+1})$. This argument shows that Theorem 5.4 is valid for $k = 1$.

Rather than doing the general induction step, we show that Theorem 5.4 is valid for $k = 2$, given that the theorem holds for $k = 1$ (virtually the same proof will work for general k and is left to the exercises). Let $p(x)$ be the polynomial of degree two interpolating $f(x)$ at x_i, x_{i+1}, x_{i+2}. We know we can represent $p(x)$ in the form

$$p(x) = a_0 + a_1(x - x_i) + a_2(x - x_i)(x - x_{i+1}) = g(x) + a_2(x - x_i)(x - x_{i+1}).$$

Now since $p(x_i) = f(x_i)$ and $p(x_{i+1}) = f(x_{i+1})$, it follows that $g(x_i) = f(x_i)$ and $g(x_{i+1}) = f(x_{i+1})$. Since $g(x)$ is linear and since Theorem 5.4 is valid for $k = 1$, we know that $g(x) = f(x_i) + f[x_i, x_{i+1}](x - x_i)$. From this, $p(x)$ has the form

$$p(x) = f(x_i) + f[x_i, x_{i+1}](x - x_i) + a_2(x - x_i)(x - x_{i+1}),$$

and all that remains is to show that $a_2 = f[x_i, x_{i+1}, x_{i+2}]$. To show this, let $h(x)$ be the linear polynomial that interpolates $f(x)$ at x_{i+1} and x_{i+2} so that $h(x) = f(x_{i+1}) + f[x_{i+1}, x_{i+2}](x - x_{i+1})$. Define $Q(x)$ by

$$Q(x) = \frac{(x - x_i)h(x) - (x - x_{i+2})g(x)}{x_{i+2} - x_i}.$$

Note that $Q(x)$ has degree two, and by direct substitution, $Q(x)$ interpolates $f(x)$ at x_i, x_{i+1}, x_{i+2}. Since polynomial interpolation is unique, $Q(x) \equiv p(x)$ and hence the coefficient of x^2 must be the same in both $Q(x)$ and $p(x)$. However, the coefficient of x^2 in $Q(x)$ is

$$\frac{f[x_{i+1}, x_{i+2}] - f[x_i, x_{i+1}]}{x_{i+2} - x_i};$$

so we have shown that $a_2 = f[x_i, x_{i+1}, x_{i+2}]$. ∎

The Newton form of the interpolating polynomial is given explicitly in a corollary to Theorem 5.4.

Corollary

The kth degree polynomial interpolating $f(x)$ at $x_0, x_1, \ldots, x_k$ is given by

$$p(x) = f(x_0) + f[x_0, x_1](x - x_0) + f[x_0, x_1, x_2](x - x_0)(x - x_1) + \cdots$$
$$+ f[x_0, x_1, \ldots, x_k](x - x_0)(x - x_1) \ldots (x - x_{k-1}).$$

As an example, consider the function $f(x)$ from Example 5.3. According to Theorem 5.4, the cubic polynomial interpolating $f(x)$ at $-2, -1, 0, 1$ is

$$q(x) = -39 + 40(x + 2) - 20(x + 2)(x + 1) + 7(x + 2)(x + 1)x \quad (5.10)$$

and this fact can be verified directly. An observation that may not be obvious is that when we use the Newton form, we can add another interpolation constraint without sacrificing our previous effort. To be explicit, suppose $p_k(x)$ interpolates $f(x)$ at $(k + 1)$ points $x_0, x_1, \ldots, x_k$. To construct a polynomial $p(x)$ that interpolates $f(x)$ at $(k + 2)$ points $x_0, x_1, \ldots, x_k, x_{k+1}$, we merely add another term to $p_k(x)$:

$$p(x) = p_k(x) + f[x_0, x_1, \ldots, x_k, x_{k+1}](x - x_0)(x - x_1) \ldots (x - x_k). \quad (5.11)$$

The observation above is one computational feature that recommends the Newton form of the interpolating polynomial.

EXAMPLE 5.4. As an illustration of (5.11), we can build the polynomial $p(x)$ that interpolates $f(x)$ at $-2, -1, 0, 1, 2$ from $q(x)$ in (5.10):

$$p(x) = q(x) - (x + 2)(x + 1)x(x - 1).$$

A variety of different interpolating polynomials can be found from a divided difference table such as the one in Example 5.3. For instance, the cubic polynomial interpolating $f(x)$ at 0, 1, 2, 3 is given by

$$p(x) = 1 + 2x + 10x(x - 1) + 19x(x - 1)(x - 2). \quad (5.12)$$

The coefficients of $p(x)$ in (5.12) are found (using Theorem 5.4) from the downward diagonal that starts at $f(0) = 1$. Finally, it should be clear that additional data can easily be added to either end of a divided difference table. To illustrate, if we add the point $x = -3$ and $f(x) = -305$ to the head of the table in Example 5.3, we can quickly construct a new diagonal that incorporates this additional information into the existing table.

PROBLEMS, SECTION 5.2.1

1. For each set of tabulated data below, let $p(x)$ be the cubic interpolating polynomial. Using the Lagrange form of $p(x)$, evaluate $p(x)$ at $x = -2$. [Note that you need not find the coefficients of x^j in $p(x)$; you need only evaluate $\ell_j(-2)$ for $0 \le j \le 3$.]

a)

x	y
−1	−1
0	3
2	11
3	27

b)

x	y
-1	-3
0	1
1	1
1	3

2. Repeat Problem 1, but use the method of undetermined coefficients to find $p(x)$.

3. For each set of data in Problem 1, set up the divided difference table and use it to construct $p(x)$ as in the corollary to Theorem 5.4. For each interpolating polynomial $p(x)$, evaluate $p(-2)$. [Again, note that you need not write $p(x)$ in the form $a_3x^3 + a_2x^2 + a_1x + a_0$ to calculate $p(-2)$.]

4. Add the data point $x = 4$, $y = 10$ to each of the tables in Problem 1. Form the divided difference table for these data by adding an upward slanting diagonal to the table constructed in Problem 3. Form the quartic polynomial $q(x)$ that interpolates these five data points, and evaluate $q(x)$ at $x = -2$.

5. Using the divided difference table in Example 5.3, construct the polynomial interpolating $f(x)$ at

a) $x = -1, 0, 1$ b) $x = -1, 0, 1, 2$ c) $x = -1, 0, 1, 2, 3$

d) $x = 0, 1$ e) $x = 0, 1, 2, 3, 4$

6. Use the method of undetermined coefficients to find a quadratic polynomial, $p(x)$ such that $p(1) = 0$, $p'(1) = 7$, and $p(2) = 10$. [Note the derivative evaluation. You must modify (5.6) slightly.]

7. Suppose we know that $f(t)$ has the form $f(t) = ae^{2t} + be^{-t}$ where a and b are unknown. Use the method of undetermined coefficients to find a and b given that $f(t_0) = 1$ and $f(t_1) = 7.5$ where $t_0 = 0$ and $t_1 = \ln(2)$. Suppose next that $f(t)$ is to be determined by some experimental observations in which $f(t_0) = 1$, $f(t_1) = 7$, and $f(t_2) = 9$ with $t_2 = \ln(3)$. Since we have three data points and only two parameters (a and b), we do not expect to be able to find a and b by the method of undetermined coefficients. Use the procedure of Section 2.5 to find the best least-squares solution for a and b.

8. Write a subroutine for polynomial interpolation. Your subprogram should accept an integer N ($N \leq 20$), N-dimensional arrays of data X(I) and Y(I), and an evaluation point ALPHA. The subroutine should calculate the divided difference table, construct the interpolating polynomial (of degree $N - 1$), and return the value of the polynomial at $x =$ ALPHA. Test your routine with the input data (x_i, y_i) where $x_i = i$ and $y_i = p(x_i)$, $0 \leq i \leq 4$, $p(x) = x^4 + 1$. Evaluate at $\alpha = -1 + k/2$, $k = 0, 1, \ldots, 12$; and note that $p(\alpha)$ should equal $\alpha^4 + 1$ for any α.

9. To see the sorts of results that can be expected, use the program in Problem 8 to interpolate $f(x) = \sin(x)$ at the knots $x_k = .1 + k/10$, $0 \leq k \leq 9$. If $p(x)$ denotes the ninth degree interpolating polynomial for $f(x)$, print $f(\alpha)$, $p(\alpha)$, and the relative error $(f(\alpha) - p(\alpha))/f(\alpha)$ for $\alpha = .05 + k/10$, $0 \leq k \leq 11$.

10. Repeat Problem 9 for $f(x) = 1/(1 + x^2)$ and the knots $x_k = -5 + k$, $0 \leq k \leq 10$ and $\alpha = -5.5 + k$, $0 \leq k \leq 12$. Also call a plotting routine and plot $f(x)$ and $p(x)$ over

$-5 \le x \le 5$ in steps of .1; note that $p(x)$ does not reproduce $f(x)$ well between the knots. Repeat this problem for the same function $f(x)$, but use knots $x_k = k$, $0 \le k \le 10$; note that interpolation over $[0, 10]$ is better behaved than over $[-5, 5]$. (Interpolation errors are discussed in Section 5.2.3.)

11. Suppose that P and Q are both sets of $k + 1$ points where $P \cap Q$ has k points. Suppose that $p(x)$ in $\mathcal{P}_k$ interpolates $f(x)$ at the points of P, and $q(x)$ in $\mathcal{P}_k$ interpolates $f(x)$ at the points of Q. Let α denote the only point in $P - (P \cap Q)$; and β, the only point in $Q - (P \cap Q)$. Show that $r(x)$ in $\mathcal{P}_{k+1}$ interpolates $f(x)$ on $P \cup Q$ where

$$r(x) = \frac{(x - \beta)p(x) - (x - \alpha)q(x)}{\alpha - \beta}.$$

5.2.2. Interpolation at Equally Spaced Points

In applications in which functions are given in tabular form, the tabular points are frequently equally spaced (for example, in tables of experimental data or in trigonometric and logarithmic tables). For problems of interpolating functions or data at equally spaced points there are several simplifications in the Newton form of the interpolating polynomial. As clarification, suppose that the interpolation points $x_0, x_1, \ldots, x_n$ are equally spaced in $[a, b]$ with $x_0 = a$ and $x_n = b$. In this case, we have

$$x_i = x_0 + ih, \qquad h = (b - a)/n, \qquad 0 \le i \le n \tag{5.13}$$

(equivalently, $x_i = x_{i-1} + h$, $1 \le i \le n$).

For equally spaced data, divided differences simplify somewhat. For example since $x_{i+1} - x_i = h$, $x_{i+2} - x_i = 2h$, and $x_{i+3} - x_i = 3h$, after some calculation we have

$$\begin{aligned} f[x_i, x_{i+1}] &= \frac{f(x_{i+1}) - f(x_i)}{h}, \\ f[x_i, x_{i+1}, x_{i+2}] &= \frac{f(x_{i+2}) - 2f(x_{i+1}) + f(x_i)}{2h^2}, \\ f[x_i, x_{i+1}, x_{i+2}, x_{i+3}] &= \frac{f(x_{i+3}) - 3f(x_{i+2}) + 3f(x_{i+1}) - f(x_i)}{6h^3}. \end{aligned} \tag{5.14}$$

The idea of forward differences can be used to exploit the simplications afforded by equally spaced data. For any function $f(x)$ and for a fixed step-size h, we define the forward difference operator Δ by

$$\Delta f(x) = f(x + h) - f(x). \tag{5.15}$$

In (5.15) the operator Δ associates a new function $f(x + h) - f(x)$ with a given function $f(x)$. Thus for example if $f(x) = \cos(x)$ and $h = 0.1$, then $\Delta f(x)$ is the function $\Delta f(x) = \cos(x + 0.1) - \cos(x)$. Higher-order forward differences are defined recursively by

$$\Delta^k f(x) \equiv \Delta(\Delta^{k-1} f(x)), \qquad k = 1, 2, \ldots \tag{5.16}$$

where we set $\Delta^0 f(x) \equiv f(x)$. For example,

$$\begin{aligned}\Delta^2 f(x) &= \Delta(\Delta f(x)) = \Delta(f(x + h) - f(x)) \\ &= (f(x + 2h) - f(x + h)) - (f(x + h) - f(x)) \\ &= f(x + 2h) - 2f(x + h) + f(x).\end{aligned}$$

Similarly we can easily verify that

$$\Delta^3 f(x) = f(x + 3h) - 3f(x + 2h) + 3f(x + h) - f(x).$$

A pattern soon emerges for higher-order forward differences, and we can easily prove that

$$\Delta^k f(x) = \sum_{i=0}^{k} \binom{k}{i} (-1)^i f(x + (k - i)h) \tag{5.17}$$

where $\binom{k}{i}$ is the usual symbol for binomial coefficients,

$$\binom{k}{i} = \frac{k!}{i!(k - i)!}$$

and where $\binom{k}{0} \equiv \binom{k}{k} \equiv 1$.

The connection between forward differences and divided differences is fairly direct. We observe that since $\Delta f(x) = f(x + h) - f(x)$, then $\Delta f(x_i) = f(x_{i+1}) - f(x_i)$ or

$$f[x_i, x_{i+1}] = \frac{\Delta f(x_i)}{h}. \tag{5.18}$$

It is easy to establish the general formula

$$f[x_i, x_{i+1}, \ldots, x_{i+k}] = \frac{\Delta^k f(x_i)}{k!h^k} \tag{5.19}$$

by induction. Specifically, (5.19) is true for $k = 1$ by (5.18). Assuming (5.19) is true for $k = j - 1$, consider

$$f[x_i, x_{i+1}, \ldots, x_{i+j}] = \frac{f[x_{i+1}, x_{i+2}, \ldots, x_{i+j}] - f[x_i, x_{i+1}, \ldots, x_{i+j-1}]}{x_{i+j} - x_i}. \tag{5.20}$$

But $x_{i+j} - x_i = jh$ and, given the validity of (5.19) for $k = j - 1$, (5.20) reduces to

$$f[x_i, x_{i+1}, \ldots, x_{i+j}] = \frac{\dfrac{\Delta^{j-1} f(x_{i+1})}{(j - 1)!h^{j-1}} - \dfrac{\Delta^{j-1} f(x_i)}{(j - 1)!h^{j-1}}}{jh}. \tag{5.21}$$

From (5.21), it follows that

$$f[x_i, x_{i+1}, \ldots, x_{i+j}] = \frac{\Delta^{j-1}f(x_{i+1}) - \Delta^{j-1}f(x_i)}{j!h^j} = \frac{\Delta^{j-1}f(x_i + h) - \Delta^{j-1}f(x_i)}{j!h^j};$$

and the definition (5.16) shows that (5.19) is valid for $k = j$.

Having shown how divided differences simplify when the interpolation points are equally spaced, we now turn our attention to the Newton form of the polynomial interpolating $f(x)$ at $x_0, x_1, \ldots, x_n$. From the last section, this polynomial $p(x)$ is given by

$$\begin{aligned} p(x) &= f(x_0) + f[x_0, x_1](x - x_0) + f[x_0, x_1, x_2](x - x_0)(x - x_1) + \cdots \\ &\quad + f[x_0, x_1, \ldots, x_n](x - x_0)(x - x_1) \ldots (x - x_{n-1}). \end{aligned} \tag{5.22}$$

In order to incorporate equally spaced knots into (5.22) we make the change of variable

$$x = x_0 + rh.$$

Here r is any number (not necessarily an integer), and in this regard r is a dimensionless variable and measures the directed distance from x_0 to x in units of the step size h. (For example with $r = 1.5$, $x = x_0 + rh$ is halfway between x_1 and x_2.) If we consider a typical term in the Newton form for $p(x)$ in (5.22),

$$f[x_0, x_1, \ldots, x_j](x - x_0)(x - x_1) \ldots (x - x_{j-1}),$$

we see that we can use (5.19) and the substitution $x = x_0 + rh$ to simplify this term. In particular, $x - x_0 = rh$, $x - x_1 = (r - 1)h$, $x - x_2 = (r - 2)h, \ldots$, and $x - x_{j-1} = (r - (j - 1))h$; and therefore we can write

$$f[x_0, x_1, \ldots, x_j](x - x_0)(x - x_1) \ldots (x - x_{j-1}) = \frac{\Delta^j f(x_0)}{j!} r(r - 1)(r - 2) \ldots (r - (j - 1)).$$

In this expression the factor $r(r - 1)(r - 2) \ldots (r - (j - 1))/j!$ is to be regarded as a polynomial of degree j in the variable r. It is conventional to denote this polynomial by the following binomial coefficient notation:

$$\binom{r}{j} = \frac{r(r - 1)(r - 2) \ldots (r - (j - 1))}{j!}, \qquad j \geq 1, \binom{r}{0} \equiv 1.$$

For example,

$$\binom{r}{1} = r, \quad \binom{r}{2} = \frac{r(r - 1)}{2}, \quad \binom{r}{3} = \frac{r(r - 1)(r - 2)}{6}.$$

Using this notation in the expression above, we can write the Newton form of $p(x)$ in (5.22) as

$$p(x) = p(x_0 + rh) = f(x_0) + \Delta f(x_0)\binom{r}{1} + \Delta^2 f(x_0)\binom{r}{2} + \cdots + \Delta^n f(x_0)\binom{r}{n},$$

or more briefly as

$$p(x_0 + rh) = \sum_{i=0}^{n} \Delta^i f(x_0)\binom{r}{i}. \tag{5.23}$$

The representation (5.23) is known as *Newton's forward formula* for the interpolating polynomial.

There are efficient ways to evaluate the Newton forward formula, given the numbers $\Delta^i f(x_0)$ (see Problem 10). In addition, the required differences, $\Delta^i f(x_0)$, are easily generated from a table (see Table 5.2).

Note that the table of forward differences is generated one column at a time, from left to right. For example, $\Delta f(x_1) = f(x_2) - f(x_1)$ and $\Delta^2 f(x_0) = \Delta f(x_1) - \Delta f(x_0)$. Note also that in order to compute $\Delta^{n+1} f(x_0)$, we need not start over. Instead, we merely compute one more entry in each column, that is, x_{n+1}, $f(x_{n+1})$, $\Delta f(x_n)$, $\Delta^2 f(x_{n-1})$, . . . , $\Delta^n f(x_1)$ and $\Delta^{n+1} f(x_0)$. The subroutine DIFINT shown in Fig. 5.2 implements the ideas outlined above. This subroutine first calculates the forward difference table, as in Table 5.2, for the interpolation points $x_1, x_2, \ldots, x_n$. When these differences are obtained, the interpolating polynomial is evaluated at $x = x_1 + rh$; and the nested multiplication algorithm given in Problem 10 is used. For ease of programming, the range on the subscript i is $1 \le i \le n$ instead of $0 \le i \le n$.

TABLE 5.2 The forward-difference table for $f(x_0), f(x_1), \ldots, f(x_n)$.

x	$f(x)$	$\Delta f(x)$	$\Delta^2 f(x)$	$\Delta^3 f(x)$	$\cdots$	$\Delta^n f(x)$
x_0	$f(x_0)$					
		$\Delta f(x_0)$				
x_1	$f(x_1)$		$\Delta^2 f(x_0)$			
		$\Delta f(x_1)$		$\Delta^3 f(x_0)$		
x_2	$f(x_2)$		$\Delta^2 f(x_1)$			
		$\Delta f(x_2)$			$\ddots$	
x_3	$f(x_3)$					$\Delta^n f(x_0)$
$\vdots$	$\vdots$					
x_{n-1}	$f(x_{n-1})$			$\cdot^{\cdot^{\cdot}}$		
		$\Delta f(x_{n-1})$				
x_n	$f(x_n)$					

```
      SUBROUTINE DIFINT(X,Y,R,PXPRH,N)
      DIMENSION X(20),Y(20),P(20),Q(20)
C
C   SUBROUTINE DIFINT CONSTRUCTS AND EVALUATES THE NEWTON FORWARD FORM OF
C   THE INTERPOLATING POLYNOMIAL, P(X), AT X=X(1)+R*H, WHERE P(X(I))=Y(I)
C   AND WHERE X(I)=X(1)+(I-1)*H, I=1,2,...,N.  THE CALLING PROGRAM MUST
C   SUPPLY THE ARRAYS X(I) AND Y(I), THE NUMBER R AND THE INTEGER N.
C   THE SUBROUTINE RETURNS P(X(1)+R*H) AS PXPRH.
C
      DO 1 I=1,N
    1 P(I)=Y(I)
C
C   SET UP THE FORWARD DIFFERENCE TABLE
C
      DO 3 J=2,N
      DO 2 K=J,N
    2 Q(K)=P(K)-P(K-1)
      DO 3 K=J,N
    3 P(K)=Q(K)
C
C   EVALUATE THE FORWARD FORM OF THE INTERPOLATING
C   POLYNOMIAL BY NESTED MULTIPLICATION
C
      C=P(N)
      DO 4 J=2,N
      I=N-J
    4 C=P(I+1)+(R-I)*C/(I+1)
      PXPRH=C
      RETURN
      END
```

Figure 5.2

EXAMPLE 5.5. We consider the function $f(x) = \cos(x)$. Following Table 5.2, we construct this table.

x	$f(x)$	$\Delta f(x)$	$\Delta^2 f(x)$	$\Delta^3 f(x)$
0.3	0.955336			
		−0.034275		
0.4	0.921061		−0.009203	
		−0.043478		0.000434
0.5	0.877583		−0.008769	
		−0.052247		
0.6	0.825336			

Therefore $p(x) = 0.955336 - 0.034275\binom{r}{1} - 0.009203\binom{r}{2} + 0.000434\binom{r}{3}$. To find $p(0.44)$, since $0.44 = 0.3 + (1.4)h$, we set $r = 1.4$ in the expression above and obtain $p(0.44) = 0.904750$.

In many applications, such as those involving numerical solutions of differential equations, we are interested in interpolating at points labeled x_m, x_{m+1}, . . . , x_{m+n} instead of at points labeled x_0, x_1, . . . , x_n. Although we could rename the points $x_m, x_{m+1}, \ldots, x_{m+n}$ and then use (5.23) to find the interpolating polynomial, this procedure is both inconvenient and unnecessary. Instead, we can use the formula

$$p(x) = p(x_m + rh) = \sum_{j=0}^{n} \binom{r}{j} \Delta^j f(x_m) \tag{5.24}$$

where now $r = (x - x_m)/h$. Clearly, (5.24) is a valid formula for the interpolating polynomial since the formula can be gotten from (5.23) by relabeling points. [For example, set $z_0 = x_m$, $z_j = z_0 + jh$ and (5.24) reduces directly to (5.23).]

A somewhat different labeling problem occurs when, as often happens, we wish to interpolate at the points $x_m, x_{m-1}, \ldots, x_{m-n}$. This is a problem of "backwards interpolation" where we again want a formula in which we can add an additional point x_{m-n-1} without losing the effort we expended to interpolate at $x_m, x_{m-1}, \ldots, x_{m-n}$. The formula needed is known as *Newton's backward formula* for the interpolating polynomial:

$$p(x) = p(x_m - rh) = \sum_{j=0}^{n} \binom{r}{j} (-1)^j \Delta^j f(x_{m-j}). \tag{5.25}$$

Note that each difference $\Delta^j f(x_{m-j})$ used in (5.25) can be obtained from a difference table like that displayed in Table 5.2. In fact, when $m = n$, each difference $\Delta^j f(x_{n-j})$ is found as the *last entry* of the jth column of Table 5.2.

To show how formula (5.25) is derived, let us rename the points $\{x_m, x_{m-1}, \ldots, x_{m-n}\}$ as $\{t_m, t_{m+1}, \ldots, t_{m+n}\}$. Then $t_{m+i} = t_m + ih'$ where $h' = -h$ and where therefore $t_{m+i} = x_{m-i}$, $i = 0, 1, \ldots, n$. Using (5.24) with $x_m = t_m$, we obtain

$$p(x) = p(t_m + rh') = \sum_{j=0}^{n} \binom{r}{j} \Delta^j f(t_m). \tag{5.26}$$

To translate $\Delta^j f(t_m)$ in terms of our original labeling, note by (5.17)

$$\begin{aligned} \Delta^j f(t_m) &= \sum_{i=0}^{j} \binom{j}{i} (-1)^i f(t_{m+j-i}) \\ &= \sum_{i=0}^{j} \binom{j}{i} (-1)^i f(x_{m+i-j}). \end{aligned}$$

Next we change the index of the summation by setting $k = j - i$, and observe that $\binom{j}{i} = \binom{j}{j-i}$ and that $(-1)^i = (-1)^j(-1)^k$. With these changes, we find

$$\Delta^j f(t_m) = (-1)^j \sum_{k=0}^{j} \binom{j}{k} (-1)^k f(x_{m-k}) = (-1)^j \Delta^j f(x_{m-j})$$

where the last equality comes from using (5.17) again. When inserted into (5.26), this identity gives the Newton backward formula of (5.25).

EXAMPLE 5.6. As an illustration of how the backward and forward formulas can be set up, we consider interpolating $f(x) = e^{3x}$ at five points with $x_0 = -1$ and $h = 0.5$. As in Table 5.2, we obtain the following difference table:

x	$f(x)$	$\Delta f(x)$	$\Delta^2 f(x)$	$\Delta^3 f(x)$	$\Delta^4 f(x)$
−1	0.049787				
		0.173343			
−0.5	0.223130		0.603527		
		0.776870		2.10129	
0.	1.		2.70482		7.31599
		3.48169		9.41728	
0.5	4.48169		12.1221		
		15.6038			
1.	20.0855				

Using the backward formula (5.25) with $m = 4$ and $n = 2$, the second-degree polynomial interpolating $f(x)$ at x_4, x_3, amd x_2 is

$$p(x) = p(x_4 - rh) = f(x_4) - \binom{r}{1} \Delta f(x_3) + \binom{r}{2} \Delta^2 f(x_2).$$

For example for $x = 0.8$, we have $r = 0.4$ and $p(0.8) = 20.0855 - (0.4)(15.6038) + (-0.12)(12.1221) = 12.3893$. The forward formula (5.24), using the same three points, has $m = 2$ and $n = 2$, and gives

$$p(x) = p(x_2 + sh) = f(x_2) + \binom{s}{1} \Delta f(x_2) + \binom{s}{2} \Delta^2 f(x_2).$$

In this case, for $x = 0.8$ we have $s = 1.6$ and $p(0.8) = 1.0 + (1.6)(3.48169) + (0.48)(12.1221) = 12.3893$.

To include the tabulated information at x_1 and x_0 in our approximation to $f(0.8)$, we merely have to add the appropriate terms to the backward form of the interpolating polynomial. Thus the third-degree polynomial $p(x_4 - rh) - \binom{r}{3} \Delta^3 f(x_1)$ and the fourth-degree polynomial $p(x_4 - rh) - \binom{r}{3} \Delta^3 f(x_1) + \binom{r}{4} \Delta^4 f(x_0)$ interpolate $f(x)$ at $\{x_4, x_3, x_2, x_1\}$ and $\{x_4, x_3, x_2, x_1, x_0\}$, respectively. The respective estimates for $f(0.8)$ are $12.3893 - (0.064)(9.41728) = 11.7866$ and $11.7866 + (-0.0416)(7.31599) = 11.4823$ whereas the correct value of $f(0.8)$ is 11.0232.

PROBLEMS, SECTION 5.2.2

1. For $f(x) = x^2$ and h arbitrary, calculate $\Delta f(x)$, $\Delta^2 f(x)$, $\Delta^3 f(x)$, and $\Delta^4 f(x)$. Repeat this calculation for the function $f(x) = x^3$.

2. For $f(x) = x^3 + 1$ and $x_k = 2k$, $0 \le k \le 3$, construct the forward difference table and the interpolating polynomial $p(x)$ as in (5.23). Evaluate $p(x)$ at $x = -1, 1, 3$. [Note: Since the four data points come from a cubic polynomial, $p(x)$ is just a different representation for $f(x)$.]

3. Use the functions $f(x) = \sin(x)$ and $f(x) = \ell n(x)$, and rework Example 5.5.

4. For $x = x_0 + rh$ and $p(x)$ given by (5.23), we have

$$\frac{dp}{dx} = \frac{dp}{dr}\frac{dr}{dx} = \frac{1}{h}\frac{dp}{dr}.$$

Thus (5.23) can be used to approximate the derivative of $f(x)$; $f'(x) \simeq p'(x)$. For $n = 3$, derive such a "numerical differentiation" formula for use at the point $\alpha = x_0 + 1.5h$ by differentiating the right-hand side of (5.23). Check calculations by testing the formula on $f(x) = 1, f(x) = x, f(x) = x^2$, and $f(x) = x^3$ with $x_0 = 0$ and $h = 1$. [The formula should give the correct value for $f'(1.5)$]. Use the data in Example 5.5 to estimate $f'(.45)$ when $f(x) = \cos(x)$, and compare your estimate with the actual value.

5. If $p(x)$ interpolates $f(x)$, then we can use $\int_a^b p(x)dx$ as an approximation for $\int_a^b f(x)dx$. Use (5.23) to construct a "numerical integration" formula by choosing $n = 2$, $x_0 = a$, and $h = (b - a)/2$. That is, verify that

$$\int_a^b p(x)dx = h\int_0^2 p(x_0 + rh)dr;$$

and integrate the right-hand side of (5.23). Check your calculations by testing the formula on $f(x) = 1$, $f(x) = x$, and $f(x) = x^2$ with $a = 0$ and $b = 2$. [The formula should give the correct value for these integrals.) Use the data in Example 5.5 to estimate $\int_{.3}^{.5}\cos(x)dx$.

6. Use $n = 3$ and repeat Problem 5. Use the formula derived to estimate $\int_{.3}^{.6}\cos(x)dx$.

7. For $f(x) = x^k$, $1 \le k \le 3$, calculate $\Delta^i f(x)$ for $1 \le i \le 4$; and note that $\Delta^{i+1}f(x) \equiv 0$ for $f(x) = x^i$.

8. Show that $\Delta^m p(x) \equiv 0$ when $p(x) \in \mathcal{P}_n$ and $m \ge n + 1$. [Hint: If $p(x) \in \mathcal{P}_n$, show that $\Delta p(x)$ is a polynomial in $\mathcal{P}_{n-1}$; thus conclude that $\Delta^{n-1}p(x)$ is a linear polynomial, $\Delta^n p(x)$ is a constant, and $\Delta^{n+1}p(x) \equiv 0$.]

9. Continue Example 5.6, by adding the interpolation point $x_5 = 1.5$ and forming $\Delta^1 f(x_4), \Delta^2 f(x_3), \ldots, \Delta^5 f(x_0)$. Knowing the value of the fourth-degree interpolating polynomial at $x = 0.8$ (see Example 5.6), use the forward form to evaluate the fifth-degree polynomial interpolating $f(x)$ at $x_0, x_1, \ldots, x_5$ for $x = 0.8$. Next add the point $x_{-1} = -1.5$ and use the backwards form to evaluate (at $x = 0.8$) the sixth-degree polynomial interpolating $f(x)$ at $x_{-1}, x_0, \ldots, x_5$. Compare all these results with the estimates provided by the truncated Taylor's series of degrees 4, 5, and 6.

10. For $n = 1, 2$, and 3, verify that the following version of nested multiplication can be used to evaluate Newton's forward formula for the interpolating polynomial (5.23):

Given any number r, form the numbers $C_n, C_{n-1}, \ldots, C_1$ and C_0 by

$$C_n = \Delta^n f(x_0)$$

$$C_i = \Delta^i f(x_0) + (r - i)C_{i+1}/(i + 1), \qquad i = n - 1, n - 2, \ldots, 1, 0.$$

Then $C_0 = p(x_0 + rh)$. Use this iteration to evaluate $p(0.44)$ in Example 5.5. (Note: This form of nested multiplication is valid for all n; but, in general, verification is a somewhat difficult exercise.)

5.2.3. Error of Polynomial Interpolation

We know that given a function, $f(x)$, and $(n + 1)$ distinct points $\{x_j\}_{j=0}^n$ in $[a, b]$, we can construct in $\mathcal{P}_n$ a polynomial, $p(x)$, which passes through the points $(x_j, f(x_j))$, $0 \le j \le n$. The question still remains as to how well $p(x)$ approximates $f(x)$ for other values of x in $[a, b]$. (See Fig. 5.3.) For example, as in Fig. 5.3, $f(x)$ may be very ill-behaved at one or more points, α; and $p(x)$ may not be a good approximation to $f(x)$ in a neighborhood of α. Thus if $e(x) \equiv f(x) - p(x)$ is defined to be the error function, a natural question is this: How large can $e(x)$ be for any x in $[a, b]$? If $f^{(n+1)}(x)$ is continuous on $[a, b]$, then the following theorem provides an answer to this question. Before this theorem is presented, it is convenient to give some notation. We let $C^{n+1}[a, b]$ denote the set of functions defined on $[a, b]$ that have a continuous $(n + 1)$st derivative on $[a, b]$, and use $\text{Spr}\{y_1, y_2, \ldots, y_n\}$ to denote the smallest interval containing $y_1, y_2, \ldots, y_n$ (thus $\text{Spr}\{1, -3, 2\} = [-3, 2]$).

Theorem 5.5

If $p(x) \in \mathcal{P}_n$ interpolates $f(x) \in C^{n+1}[a, b]$ at $(n + 1)$ distinct points $\{x_j\}_{j=0}^n$ in $[a, b]$, then

$$e(x) = f(x) - p(x) = \frac{f^{(n+1)}(\xi)}{(n+1)!} W(x) \tag{5.27}$$

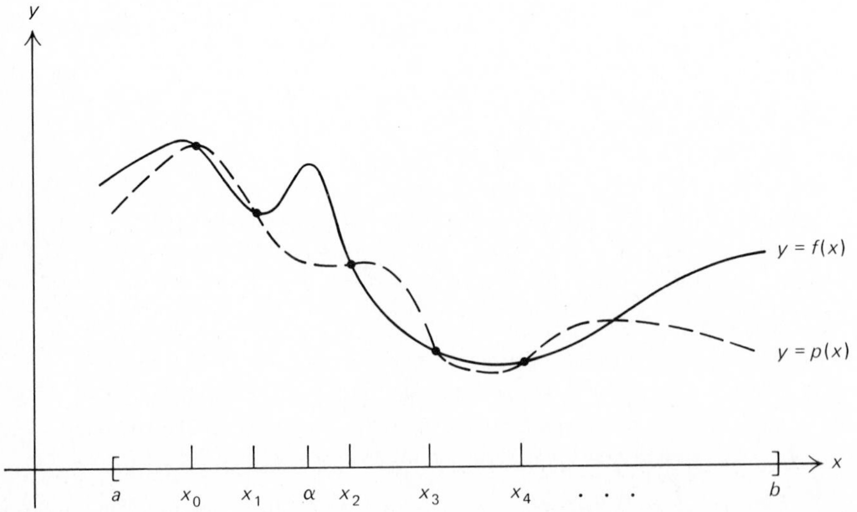

Figure 5.3 Graphs of $f(x)$ and its interpolating polynomial, $p(x)$.

where $W(x) \equiv (x - x_0)(x - x_1) \ldots (x - x_n)$ and where ξ is some point lying in $\text{Spr}\{x, x_0, x_1, \ldots, x_n\}$.

Proof. Let x be arbitrary but *fixed* in $[a, b]$. If $x = x_j$ for $0 \le j \le n$, then the theorem is trivially true. Thus we assume that $x \neq x_j$ for any j, and define the function $F(t) = f(t) - p(t) - CW(t)$ where the constant C is given by $C = (f(x) - p(x))/W(x)$. Now $F^{(n+1)}(t)$ is continuous on $[a, b]$, and furthermore $F(x) = F(x_0) = F(x_1) = \cdots = F(x_n) = 0$. By Rolle's theorem, $F'(t)$ has at least one zero between each distinct zero of $F(t)$, and thus $F'(t)$ has at least $(n + 1)$ distinct zeros in $\text{Spr}\{x, x_0, x_1, \ldots, x_n\}$. Using Rolle's theorem for $F'(t)$, we can reason that $F''(t)$ has at least n distinct zeros. Rolle's theorem applied to $F''(t)$ shows that $F'''(t)$ has at least $(n - 1)$ distinct zeros, etc. Finally we conclude that $F^{(n+1)}(t)$ has at least one zero, ξ, and $\xi \in \text{Spr}\{x, x_0, x_1, \ldots, x_n\}$. From the definition of $F(t)$ we see that $F^{(n+1)}(t) = f^{(n+1)}(t) - 0 - C(n + 1)!$. Thus, $0 = F^{(n+1)}(\xi) = f^{(n+1)}(\xi) - C(n + 1)!$. Substituting $(f(x) - p(x))/W(x)$ for C in this expression, we obtain (5.27). ■

The reader should note that the value ξ in (5.27) changes as x changes; so ξ can be regarded as a function of x. Here is one of the drawbacks in using (5.27); since ξ is an unknown function of x, it is impossible to evalute $f^{(n+1)}(\xi)$ exactly.

However, we have assumed in Theorem 5.5 that $f^{(n+1)}(x)$ is continuous; so there is a number K_n such that $|f^{(n+1)}(\xi)| \le K_n$ for any ξ in $[a, b]$. Consequently, a more useful form of (5.27), which we can use to bound the error for any x in $[a, b]$, is

$$|e(x)| = |f(x) - p(x)| \le \frac{K_n}{(n + 1)!} |W(x)|. \tag{5.28}$$

For any particular value of x, the number $|W(x)|$ can be calculated and thus (5.28) yields an upper bound on the error. If we wish to bound $|e(x)|$ for all possible x in $[a, b]$, however, we must undertake the more formidable task of calculating

$$\max_{a \le x \le b} |W(x)| \equiv \|W\|_\infty.$$

If n and the interpolating points, $\{x_j\}_{j=0}^n$, are prescribed beforehand, then this task can be performed numerically by the methods of Chapter 4, simply by finding the zeros of $W'(x)$ and checking them (and the endpoints a and b) to find the maxima of $|W(x)|$. Most often, however, one would like the choice of n and the x_j's to be flexible in using (5.28) so that they can be selected to satisfy

$$\frac{K_n}{(n + 1)!} \|W\|_\infty < \varepsilon$$

for some predetermined tolerance, $\varepsilon > 0$. This is a fairly difficult problem and we shall examine one approach to it after the following simple example of the use of (5.28).

EXAMPLE 5.7. We shall use two previous examples involving interpolation that will show that sometimes (5.28) is a very good bound, but on other occasions it may be overly pessimistic. In Example 5.5, $f(x) = \cos(x)$ was interpolated at four points, with $W(x) = (x - 0.3)(x - 0.4)(x - 0.5)(x - 0.6)$. In this case $n = 3$ and $f^{(4)}(x) = \cos(x)$. In the notation of (5.28), we have

$$K_3 = \max_{0.3 \le x \le 0.6} |\cos(x)| = |\cos(0.3)| \le 0.955336$$

and

$$|W(0.44)| = 0.5376 \times 10^{-4}.$$

Thus

$$|e(0.44)| \le K_3 |W(0.44)|/4! \le 0.214 \times 10^{-5}$$

whereas, in fact, $|e(0.44)| = 0.2 \times 10^{-5}$.

In Example 5.6, $f(x) = e^{3x}$ was interpolated at five points, with

$$W(x) = (x + 1)(x + 0.5)x(x - 0.5)(x - 1).$$

Thus $n = 4$ and $f^{(5)}(x) = 243e^{3x}$. Since the interpolation points are in $[-1, 1]$, $K_4 = 243e^3 \le 4880.79$ and $|W(0.8)| = 0.11232$. Thus,

$$|e(0.8)| \le K_4 |W(0.8)|/5! \le 4.57$$

whereas, in fact, $|e(0.8)| = 0.4591$. In this second case, the error bound has overestimated the true error by a factor of about 10.

When the error bound given by (5.28) is used, it is apparent that the function $|W(x)|$ plays an important role in determining the size of the bound. Clearly, $|W(x)|$ will be small when x is near an interpolation point x_j, but more can be said about the location of values x that make $|W(x)|$ small (or large). For instance, in the second case of Example 5.7 above, $W(x) = (x + 1)(x + 0.5)x(x - 0.5)(x - 1)$ and $|W(0.8)| = 0.11232$. Note that $x = 0.2$ is no nearer an interpolation point than is $x = 0.8$, but a simple calculation shows that $|W(0.2)| = 0.04032$. Intuitively, this statement makes sense since $W(0.2)$ has more "small" factors than $W(0.8)$ has. For equally spaced interpolation points in an interval $[a, b]$, it can be shown that the relative maxima of $|W(x)|$ decrease as x approaches the midpoint $(a + b)/2$ of $[a, b]$ (see Problems 5, 6, and 8). As a general rule, (5.28) shows that interpolation at equally spaced points is more accurate near the center of the interval than near the endpoints. [However, the error is also influenced by $f^{(n+1)}(\xi)$; and while $|W(x)|$ is small, $|f^{(n+1)}(\xi)|\,|W(x)|$ can be large.] Figure 5.4 shows the graph of $W(x)$ as given in the second case of Example 5.7.

Examination of the derivation of (5.28) reveals that we can do nothing to improve the value of K_n since we do not know the location of the mean-value point ξ. However, for $\|W\|_\infty = \max_{a \le x \le b} |W(x)|$, (5.28) tells us that the inequality $|e(x)| \le K_n \|W\|_\infty/(n + 1)!$ holds for all x in $[a, b]$. Thus if we wish to make the error bound as small as possible *for all* x in $[a, b]$, we should concentrate on the

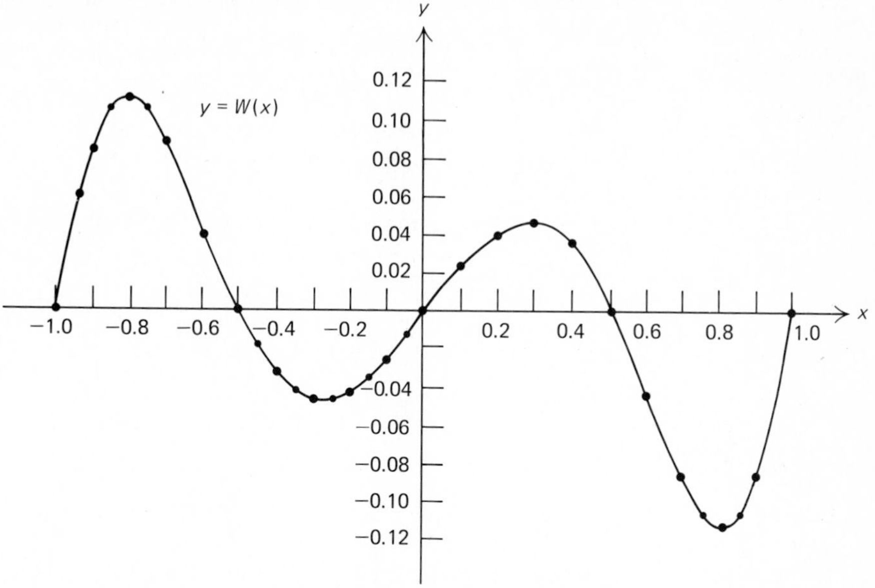

Figure 5.4 Graph of $W(x) = (x + 1)(x + 0.5)x(x - 0.5)(x - 1)$.

problem of choosing the interpolating points, $\{x_j\}_{j=0}^n$, so that $\|W\|_\infty$ is minimized. On the surface this choice seems to be a very difficult task. This problem was investigated by P. L. Chebyshev in the 19th century for the interval $[-1, 1]$, and the solution involves polynomials that were named after him. The solution is not too hard to obtain, and by a simple change of variable can even be extended to an arbitrary interval, $[a, b]$.

Before giving the result, we must introduce and investigate the *Chebyshev polynomials of the first kind*, $T_k(x) \equiv \cos(k \cos^{-1}(x))$, $k = 0, 1, 2, \ldots$. At first glance it probably seems difficult to believe that each $T_k(x)$ is actually an algebraic polynomial. However, we easily see that $T_0(x) \equiv 1$ and $T_1(x) \equiv x$. Now we make the variable substitution, $x = \cos(\theta)$, $0 \le \theta \le \pi$. Then $T_k(x) = T_k(\cos(\theta)) = \cos(k\theta)$. By elementary trigonometric identities, we can easily verify that

$$\begin{aligned} T_{k+1}(x) &= \cos((k+1)\theta) = 2\cos(\theta)\cos(k\theta) - \cos((k-1)\theta) \\ &= 2xT_k(x) - T_{k-1}(x), \qquad \text{for } k \ge 1. \end{aligned} \tag{5.29}$$

Therefore $T_2(x) = 2xT_1(x) - T_0(x) = 2x^2 - 1$, and $T_3(x) = 2xT_2(x) - T_1(x) = 4x^3 - 3x$, etc. We can easily verify inductively that $T_k(x)$ is a kth-degree polynomial for $k = 0, 1, 2, \ldots$. Furthermore, we see that the leading coefficient of $T_k(x)$ equals 2^{k-1}, and that $T_k(x_i) = 0$, $0 \le i \le k - 1$, when

$$x_i = \cos\frac{(2i+1)\pi}{2k}.$$

[Note: $\cos(kt) = 0$ whenever t is an odd multiple of $\pi/(2k)$.] Last we see that $|T_k(x)|$ is never larger than 1 for x in $[-1, 1]$; that is, $\|T_k\|_\infty \le 1$. Moreover, for $y_i = \cos(i\pi/k)$, we have $T_k(y_i) = \cos(i\pi) = (-1)^i$, so $\|T_k\|_\infty = |T_k(y_i)| = 1$ at each of the $(k + 1)$ distinct points $y_0, y_1, \ldots, y_k$. Chebyshev polynomials are very important in many types of numerical approximations as we shall see in subsequent sections. The graphs of the first five Chebyshev polynomials are given in Fig. 5.5.

Now we are ready to prove the famous result for minimizing

$$\max_{-1 \le x \le 1} |W(x)| \equiv \|W\|_\infty.$$

Theorem 5.6

Let $W(x) = (x - x_0)(x - x_1) \ldots (x - x_n) \in \mathcal{P}_{n+1}$. Among all possible choices for distinct $\{x_j\}_{j=0}^n$ in $[-1, 1]$, $\|W\|_\infty \equiv \max_{-1\le x\le 1} |W(x)|$ is minimized if $W(x) = (1/2^n)T_{n+1}(x)$ [i.e., the x_j's are the zeros of $T_{n+1}(x)$].

Proof. By the remarks above on Chebyshev polynomials, the leading coefficient of the $(n + 1)$st-degree polynomial, $(1/2^n)T_{n+1}(x)$, is 1. Therefore, $(1/2^n)T_{n+1}(x) = (x - x_0)(x - x_1) \ldots (x - x_n)$, $x_i = \cos[(2i + 1)\pi/2(n + 1)]$; and hence $W(x) = (1/2^n)T_{n+1}(x)$ is a candidate for the minimum W. Also from the

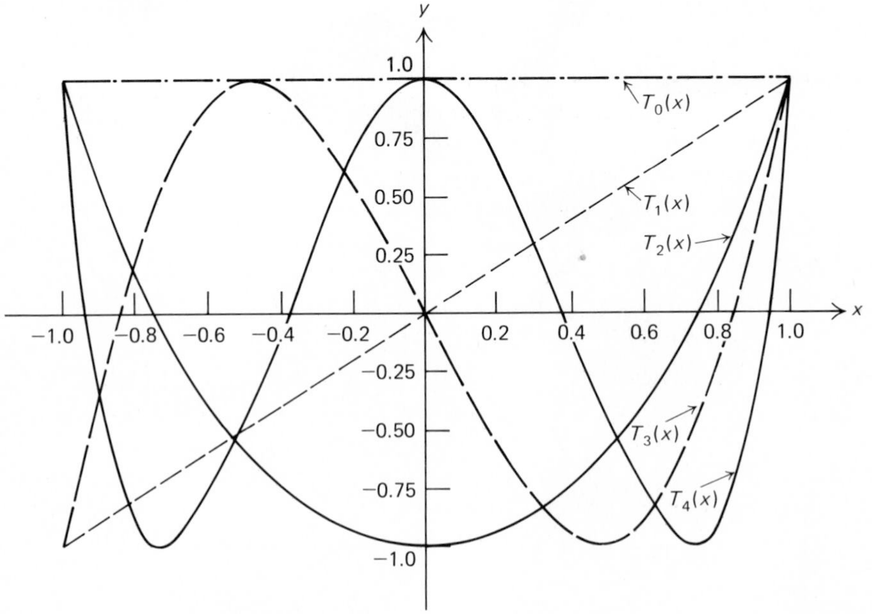

Figure 5.5 The Chebyshev polynomials, $\{T_k(x)\}_{k=0}^4$.

remarks above, $\|W\|_\infty = \|(1/2^n)T_{n+1}\|_\infty = 1/2^n$, $W(y_i) = (1/2^n)(-1)^i$, $0 \le i \le n+1$, and $W(y_i) = -W(y_{i+1})$, $0 \le i \le n$, where

$$y_i = \cos\frac{i\pi}{n+1}, \qquad 0 \le i \le n+1.$$

Now assume there exists a polynomial, $V(x) \in \mathcal{P}_{n+1}$, where $V(x)$ is monic (leading coefficient equals 1) and $\|V\|_\infty < \|W\|_\infty$. If i is even, then $V(y_i) < W(y_i) = 1/2^n$, or else $\|V\|_\infty \ge \|W\|_\infty$. If i is odd, then $V(y_i) > W(y_i) = -1/2^n$, or again $\|V\|_\infty \ge \|W\|_\infty$. Thus $V(y_0) < W(y_0)$, $V(y_1) > W(y_1)$, $V(y_2) < W(y_2)$, $V(y_3) > W(y_3)$, etc.; and so $H(x) \equiv V(x) - W(x)$ has a zero in each interval (y_i, y_{i+1}), $0 \le i \le n$. Moreover, $H(x)$ is in $\mathcal{P}_n$ since both $V(x)$ and $W(x)$ are monic. Therefore $H(x) \in \mathcal{P}_n$ and has at least $(n+1)$ zeros; so we may conclude that no such $V(x)$ exists. ■

Thus if $[a, b] = [-1, 1]$ with $f \in C^{n+1}[-1, 1]$, and if the interpolating points are the zeros of $T_{n+1}(x)$, the error bound on the interpolation given by (5.28) yields

$$|e(x)| = |f(x) - p(x)| \le \frac{K_n}{(n+1)!}\|W\|_\infty = \frac{K_n}{(n+1)!2^n}. \tag{5.30}$$

EXAMPLE 5.8. Again consider the function $f(x) = e^{3x}$ for x in $[-1, 1]$. In Example 5.7, we computed the error bound at $x = 0.8$ for interpolation at five equally spaced points ($n = 4$), and obtained a bound of 4.57. Finding this bound required us to compute $W(0.8)$ where $W(x) = (x+1)(x+0.5)x(x-0.5)(x-1)$. For interpolation at the zeros of $T_5(x)$, we can bound the error at any point x (including $x = 0.8$) by the constant $K_4/(2^4 5!)$ $= 4880.79/(2^4 5!) = 2.54$. In fact, the polynomial interpolating at the zeros of $T_5(x)$ gives a value of 11.2776 as an estimate to $f(0.8) = 11.0232$, which is an error of 0.2544 (roughly half the error of equally spaced interpolation).

To determine how many interpolation points are required to obtain a desired accuracy when interpolating $f(x)$ at the zeros of $T_{n+1}(x)$, we note that $f^{(n+1)}(x) = 3^{n+1}e^{3x}$; so $K_n = 3^{n+1}e^3$. Using (5.30), we have

$$|e(x)| \le \left(\frac{3}{2}\right)^n \frac{3e^3}{(n+1)!},$$

which can be shown to go to zero as $n \to \infty$. For example with $n = 10$, $|e(x)| \le 0.000087$; and with $n = 20$, $|e(x)| \le 0.39 \times 10^{-16}$. Using 11 equally spaced interpolation points ($n = 10$) and evaluating the error bound of formula (5.28) at $x = 0.9$ (a noninterpolation point), we find $|W(0.9)| = 0.0065$; and hence the error bound for the equally spaced case is 0.00058, which is more than six times as large as the Chebyshev interpolation error bound with $n = 10$.

Obviously we do not expect the error of Chebyshev interpolation, $|e_T(x)|$, to be smaller than the error for equally spaced interpolation, $|e_h(x)|$, for every x

in $[a, b]$. (A trivial example of this is given when $x = x_j = a + jh$, but x_j is not a Chebyshev zero. For then, $0 = |e_h(x_j)| \leq |e_T(x_j)|$.) Usually, however, if we are free to select the interpolating points at our own discretion, we prefer to use Chebyshev interpolation since we can determine an error bound *a priori* for all values of x.

There are other ways of using (5.27) to estimate the error of interpolation. For example, let $w(x)$ be a nonnegative weighting function as given in (5.1b). Multiplying the squares of both sides of (5.27) by $w(x)$ and integrating from a to b yield

$$\begin{aligned} \|e(x)\|_2 &= \left(\int_a^b (f(x) - p(x))^2 w(x)\,dx\right)^{1/2} \\ &= \left(\int_a^b \left(\frac{f^{(n+1)}(\xi)}{(n+1)!} W(x)\right)^2 w(x)\,dx\right)^{1/2} \\ &\leq \frac{\max\limits_{a \leq x \leq b} |f^{(n+1)}(x)|}{(n+1)!} \left(\int_a^b (W(x))^2 w(x)\,dx\right)^{1/2} \\ &= \frac{K_n}{(n+1)!}\|W\|_2. \end{aligned} \tag{5.31}$$

After we investigate orthogonal polynomials in a later section, we will show how to choose the interpolating points, $\{x_j\}_{j=0}^n$, in order to minimize $\|W\|_2$. We merely state here that they are not, in most cases, the zeros of a Chebyshev polynomial.

5.2.4. Translating the Interval

Often it is necessary for practical purposes to take a problem stated on an interval $[c, d]$ and reformulate the same problem on a different interval $[a, b]$. This necessity is especially true in problems that require the numerical approximation of integrals or in problems concerning orthogonal polynomials. We have already come across this problem in Chebyshev interpolation, in which Theorem 5.6 is applicable to the specific interval $[-1, 1]$ and says nothing about the optimal choice of interpolation points in an arbitrary interval $[a, b]$. The change of variable that usually achieves our purposes of reformulating a problem on $[c, d]$ to one on $[a, b]$ is the "straight line" transformation $x = mt + \beta$ for $t \in [c, d]$ and $x \in [a, b]$. Here m and β are chosen such that $x = a$ when $t = c$ and $x = b$ when $t = d$. A simple algebraic computation will show that this transformation is given by

$$x = \left(\frac{b - a}{d - c}\right)t + \left(\frac{ad - bc}{d - c}\right). \tag{5.32}$$

To illustrate this procedure, let us suppose that we wish to perform Chebyshev interpolation on an arbitrary interval $[a, b]$ instead of the interval $[-1, 1]$. Then $t \in [-1, 1] \equiv [c, d]$ and $x \in [a, b]$; so by (5.32)

$$x = \left(\frac{b-a}{2}\right)t + \left(\frac{a+b}{2}\right)$$

or (5.33)

$$t = \frac{2}{b-a}\left(x - \frac{a+b}{2}\right) = \frac{2(x-a)}{(b-a)} - 1.$$

We now define the "shifted" Chebyshev polynomial of degree k for $x \in [a, b]$ as

$$\tilde{T}_k(x) \equiv T_k(t) = T_k\left(\frac{2(x-a)}{(b-a)} - 1\right) = \cos\left[k \cos^{-1}\left(\frac{2(x-a)}{(b-a)} - 1\right)\right]. \tag{5.34}$$

Since $T_k(t_i) = 0$ for $t_i = \cos((2i+1)\pi)/2k$, $0 \le i \le k-1$, then $x_i = ((b-a)/2)t_i + (b+a)/2$, $0 \le i \le k-1$, are the zeros of $\tilde{T}_k(x)$. Furthermore, $\tilde{T}_0(x) = 1$, $\tilde{T}_1(x) = 2(x-a)/(b-a) - 1$; and so from (5.29)

$$\begin{aligned}\tilde{T}_{k+1}(x) &= T_{k+1}(t) = 2tT_k(t) - T_{k-1}(t)\\ &= 2\left(\frac{2(x-a)}{(b-a)} - 1\right)\tilde{T}_k(x) - \tilde{T}_{k-1}(x), \qquad k \ge 1. \end{aligned} \tag{5.35}$$

We leave it to the reader to verify that the leading coefficient of $\tilde{T}_k(x)$ is $2^{k-1}(2/(b-a))^k$, $k \ge 1$. Therefore, if we interpolate in $[a, b]$ at the zeros $\{x_i\}_{i=0}^n$ of $\tilde{T}_{n+1}(x)$, we have $W(x) = 2^{-n}(2/(b-a))^{-n-1}\,\tilde{T}_{n+1}(x)$. Then

$$\max_{a \le x \le b} |W(x)| = 2^{-n}\left(\frac{2}{b-a}\right)^{-n-1}$$

since

$$\max_{a \le x \le b} |\tilde{T}_{n+1}(x)| = \max_{-1 \le t \le 1} |T_{n+1}(t)| = 1.$$

Thus the error bound, (5.30), becomes

$$|e(x)| = |f(x) - p(x)| \le \frac{K_n}{(n+1)!}\max_{a \le x \le b}|W(x)| = \frac{K_n}{2^n(n+1)!}\left(\frac{b-a}{2}\right)^{n+1}. \tag{5.36}$$

It is easily seen that Theorem 5.6 is valid on $[a, b]$ with respect to this $W(x)$, and so $W(x)$ is the smallest monic polynomial of degree $(n+1)$ with respect to

$$\|W\|_\infty \equiv \max_{a \le x \le b} |W(x)|.$$

EXAMPLE 5.9. Suppose $f(x) = \cos(x)$ and $[a, b] = [0, \pi/2]$, and suppose we wish to find a value of n and interpolating points $\{x_j\}_{j=0}^n$ such that the error of the nth-degree interpolation is less than 10^{-6} for all x in $[0, \pi/2]$. Now $K_n = 1$ for all n; so with Eq. (5.36), the smallest value of n for which

$$\frac{1}{2^n(n+1)!}\left(\frac{b-a}{2}\right)^{n+1} = \frac{1}{2^n(n+1)!}(\pi/4)^{n+1} < 10^{-6}$$

is $n = 6$. Thus the interpolating points are the zeros of $\tilde{T}_7(x)$, which are the points

$$x_j = \frac{\pi}{4}(\cos[(2j+1)\pi/14] + 1), \qquad 0 \le j \le 6.$$

From (5.36)

$$|e(x)| \le \frac{1}{2^6 7!}(\pi/4)^7 = 5.71 \times 10^{-7}, \qquad x \in [0, \pi/2].$$

The transformation technique of this section is quite simple but is used often and will be referred to in later sections.

PROBLEMS, SECTION 5.2.4

1. As in Example 5.7, use (5.28) to bound the interpolation error at $x = .15$ and $x = .35$ when $p(x)$ interpolates $f(x)$ at $x_i = .1 + i/10$, $0 \le i \le 5$, for

   a) $f(x) = \cos(x)$ b) $f(x) = \ell n(x)$ c) $f(x) = \dfrac{1}{1+x}$.

   Compare the bounds with the actual errors; use the program in Problem 8, Section 5.2.1.

2. Use (5.36) to bound the interpolation error at any point in $[.1, .6]$ for the functions in Problem 1 when the interpolation points are the zeros of $T_k(x)$ shifted to $[.1, .6]$. Compare your results with the results of Problem 1.

3. How large should k be if we wish to obtain an interpolation error of 10^{-6} or less throughout $[a, b]$ by interpolating $f(x)$ at the zeros of $\tilde{T}_k(x)$ for

   a) $f(x) = \cos(x)$, $0.3 \le x \le 0.6$

   b) $f(x) = 1/(2 + x)$, $-1 \le x \le 1$

   c) $f(x) = \ln(x)$, $0.1 \le x \le 1$

   d) $f(x) = e^{3x}$, $-1 \le x \le 1$.

4. Let $x_0, x_1, \ldots, x_n$ be equally spaced points in $[-2, 2]$, $x_0 = -2$, and $x_n = 2$. Call a plotting routine to sketch $W(x)$ in increments of .05 for $n = 10, 15, 20$. Note the characteristic shapes and that $|W(x)|$ is largest near the endpoints ± 2.

5. To use the error estimate (5.28), we need some way of estimating $|W(x)|$. For equally spaced interpolation points, there are a number of ways of estimating $|W(x)|$. For simplicity, suppose we are interpolating in $[-1, 1]$ with an odd number of equally spaced points. Thus suppose that n is even, $W(x) = (x - x_0)(x - x_1)\ldots$

$(x - x_n)$ where $h = 2/n$, $x_j = x_0 + jh$ for $j = 0, 1, \ldots, n$ and where $x_0 = -1$, $x_n = 1$. Let $N = n/2$ and show that

$$W(x) = (x + Nh)(x + (N - 1)h) \cdots (x + h)x(x - h) \cdots (x - (N - 1)h)(x - Nh).$$

For $x \in [-1, 1]$, write x as $x = rh$ where r is a number such that $-N \le r \le N$. Suppose next that $x_{n-1} < x < x_n$ so that $N - 1 < r < N$. Show that

$$(n - 1)!h^{n-1}|(x - x_{n-1})(x - x_n)| \le |W(x)| \le n!h^{n-1}|(x - x_{n-1})(x - x_n)|.$$

6. Show that the maximum value of the factor $|(x - x_{n-1})(x - x_n)|$ in Problem 5, for $x_{n-1} \le x \le x_n$, is $h^2/4$. Thus conclude that $|W(x)| \le n!2^{n-1}/n^{n+1}$. Also conclude, for $x = (x_n + x_{n-1})/2$, that $(n - 1)!2^{n-1}/n^{n+1} \le |W(x)|$.
7. In Problem 6, for $n = 5$, 10, and 20, evaluate the upper and lower bounds for $|W(x)|$. Contrast the upper bounds with $1/2^n$, which is the bound on $|(x - t_0)(x - t_1) \ldots (x - t_n)|$ where $t_0, t_1, \ldots, t_n$ are the zeros of $T_{n+1}(x)$.
8. Given $(n + 1)$ equally spaced interpolation points in $[-1, 1]$ where n is even as in Problem 5, show that $W(x)$ is an odd function [i.e., $W(x) = -W(-x)$]. Thus if we can find the maximum of $|W(x)|$ for $0 \le x \le 1$, we have a maximum for $|W(x)|$ for $-1 \le x \le 1$. Show that the maximum of $|W(x)|$ occurs somewhere in (x_{n-1}, x_n). [Hint: Show that

$$\left|\frac{W(x + h)}{W(x)}\right| > 1$$

for $0 < x < x_{n-1}$ and for x not an interpolation point.] Use the representation of $W(x)$ in Problem 5 to establish this inequality. This result means that the upper and lower bounds established in Problem 6 are upper and lower bounds for $\|W\|_\infty$. The inequality in the hint also shows why interpolation at equally spaced points is usually more reliable near the center of the entire interval.
9. Let $f(x) = 1/(x + 2)$. Using (5.28) and the results of Problems 6 and 8, give an error estimate for interpolating $f(x)$ at n equally spaced points in $[-1, 1]$ where n is even. What is the corresponding error estimate if $f(x)$ is interpolated at the zeros of $T_{n+1}(x)$?
10. Let n be odd and let $x_0, x_1, \ldots, x_n$ be equally spaced in $[-1, 1]$, $x_0 = -1$, $x_n = 1$. The length of each subinterval is $2h$ where $h = 1/n$. Show that

$$W(x) = (x + nh)(x + (n - 2)h) \ldots (x + h)(x - h) \ldots (x - (n - 2)h)(x - nh).$$

Show that the maximum value of $W(x)$ for $-h \le x \le h$ occurs at $x = 0$. [Hint: Group the factors of $W(x)$ as $(x^2 - n^2h^2) \ldots (x^2 - h^2)$ and show that $W'(x)$ and $W'(-x)$ have opposite signs for any x in $(-h, h)$.] Evaluate $W(0)$ and determine the maximum interpolation error for $f(x) = 1/(x + 2)$ with $x \in (-h, h)$.
11. Verify (5.29) by showing that

$$\cos((k + 1)\theta) = 2\cos(\theta)\cos(k\theta) - \cos((k - 1)\theta).$$

12. Using Eq. (5.29), show by induction that

a) $T_k(x)$ is a kth-degree polynomial with leading coefficient 2^{k-1};

b) $T_k(x)$ is an even (odd) function if k is even (odd) [recall that $f(x)$ is an even function if $f(x) = f(-x)$ and that $f(x)$ is an odd function if $f(x) = -f(-x)$].

13. Show that $\tilde{T}_k(x)$ defined by (5.35) has leading coefficients equal to

$$2^{k-1}\left(\frac{2}{b-a}\right)^k, \qquad k \geq 1.$$

What are the polynomials $\tilde{T}_k(x)$ in the special case in which $[a, b] = [0, 1]$?

*5.2.5. Polynomial Interpolation with Derivative Data

Examination of (5.5) and (5.6) reveals that the type of nth-degree polynomial interpolation that we have studied up to this point is based on the fact that there exists a unique solution to the $(n+1)$ linear equations, $p(x_j) = f(x_j)$, $0 \leq j \leq n$, in the $(n+1)$ unknowns, a_j [the coefficients of $p(x)$]. Thus $p(x)$ "matches" $f(x)$ at the interpolating points. It is natural to ask if we might not get a better polynomial approximation to $f(x)$ if we forced some derivatives of $p(x)$ at various points to match the respective derivatives of $f(x)$ at these points [for then we would be using more information about $f(x)$ and would expect a better approximation]. The extreme example of this procedure is to take only one point, x_0, and choose the coefficients of $p(x)$ such that $p^{(j)}(x_0) = f^{(j)}(x_0)$ for $0 \leq j \leq n$. This method yields the following $(n+1)$ equations in $\{a_j\}_{j=0}^n$:

$$\begin{aligned}
a_0 + a_1x_0 + a_2x_0^2 + \cdots + \qquad\qquad a_nx_0^n &= f(x_0)\\
a_1 + 2a_2x_0 + \cdots + \qquad na_nx_0^{n-1} &= f'(x_0)\\
2a_2 + \cdots + n(n-1)a_nx_0^{n-2} &= f''(x_0)\\
\ddots \qquad\qquad \vdots \quad &\quad \vdots\\
n!a_n &= f^{(n)}(x_0).
\end{aligned}$$

We note that this triangular system is nonsingular (the coefficient matrix has positive diagonal elements), and so $p(x)$ exists and is unique. The reader will recall from calculus that the Taylor polynomial,

$$q(x) = \sum_{j=0}^{n} \frac{f^{(j)}(x_0)}{j!}(x - x_0)^j,$$

also has the property that $q^{(j)}(x_0) = f^{(j)}(x_0)$, $0 \leq j \leq n$; and so, by the uniqueness of $p(x)$, $q(x) = p(x)$. Thus truncated Taylor's expansions are actually special cases of this generalized interpolation problem. Since the error of the Taylor approximation at x is

$$\frac{f^{(n+1)}(\xi)}{(n+1)!}(x - x_0)^{n+1},$$

this approximation is quite good if x is "close" to x_0 and usually worsens as x moves away from x_0. Note the similarity of this error formula to that of (5.27).

There are obviously many ways of matching derivatives of a polynomial to respective derivatives of $f(x)$ at various points in order to obtain a linear system of $(n + 1)$ equations in $(n + 1)$ unknowns [the coefficients of $p(x)$]. This problem in its most general form is known as *Hermite-Birkoff interpolation*. For example, we might ask for a third-degree polynomial, $p(x)$, such that $p(x_0) = f(x_0)$, $p'''(x_0) = f'''(x_0)$, $p'(x_1) = f'(x_1)$ and $p''(x_1) = f''(x_1)$. In this general setting, the problem need not always have a solution, and the question of finding conditions under which solutions exist remains unanswered and the subject of extensive research.

We wish now to consider a special form of this problem known as *Hermite* or *osculatory* interpolation. This form of interpolation not only is important as an approximation technique, but is useful in the derivation of the cubic spline approximations of the next section. For simplicity we assume that n is even and define $N = n/2$. We consider $N + 1$ distinct points $\{x_j\}_{j=0}^{N}$, and impose the conditions that $p(x_j) = f(x_j)$ *and* $p'(x_j) = f'(x_j)$, $0 \le j \le N$. Thus $p(x)$ matches $f(x)$ and $p'(x)$ matches $f'(x)$ at the interpolating points. Since we have $2(N + 1) = n + 2$ equations; so we will be searching for $p(x)$ in $\mathcal{P}_{2N+1} = \mathcal{P}_{n+1}$ in order that we have $2N + 2$ unknowns, $\{a_j\}_{j=0}^{2N+1}$. As an illustration of the problem, consider the following example.

EXAMPLE 5.10. Let $x_0 = \alpha$ and $x_1 = \beta$; and assume that $f(x)$ is given such that $f(\alpha) = y_0$, $f'(\alpha) = y_0'$, $f(\beta) = y_1$, and $f'(\beta) = y_1'$. Then $N = 1$ and so $2N + 1 = 3$. Find $p(x) \in \mathcal{P}_3$ such that $p(x_j) = f(x_j) = y_j$ and $p'(x_j) = f'(x_j) = y_j'$ for $j = 0$ and 1. If $p(x) = a_0 + a_1x + a_2x^2 + a_3x^3$, then the equations above become (in matrix form)

$$\begin{bmatrix} 1 & \alpha & \alpha^2 & \alpha^3 \\ 0 & 1 & 2\alpha & 3\alpha^2 \\ 1 & \beta & \beta^2 & \beta^3 \\ 0 & 1 & 2\beta & 3\beta^2 \end{bmatrix} \begin{bmatrix} a_0 \\ a_1 \\ a_2 \\ a_3 \end{bmatrix} = \begin{bmatrix} y_0 \\ y_0' \\ y_1 \\ y_1' \end{bmatrix}.$$

The determinant of the coefficient matrix for this system equals $(\beta - \alpha)^4$. Then when $\alpha \neq \beta$, $p(x)$ exists and is unique. We could solve this system for the a_i's to obtain $p(x)$, but we shall show for $\alpha < \beta$ that $p(x)$ is given by

$$p(x) = y_0\left[\frac{(x-\beta)^2}{(\beta-\alpha)^2} + 2\frac{(x-\alpha)(x-\beta)^2}{(\beta-\alpha)^3}\right] + y_1\left[\frac{(x-\alpha)^2}{(\beta-\alpha)^2} - 2\frac{(x-\beta)(x-\alpha)^2}{(\beta-\alpha)^3}\right]$$
$$+ y_0'\left[\frac{(x-\alpha)(x-\beta)^2}{(\beta-\alpha)^2}\right] + y_1'\left[\frac{(x-\alpha)^2(x-\beta)}{(\beta-\alpha)^2}\right]. \tag{5.37}$$

We could approach the Hermite interpolation problem in the manner of undetermined coefficients as in Example 5.10, but we choose a Lagrangian-type approach instead. For each j, $0 \le j \le N$, let $\ell_j(x)$ be the Nth-degree polynomial, as defined by (5.2), with N substituted for n. Now for each j, $0 \le j \le N$, define the $(2N + 1)$st-degree polynomials

$$A_j(x) \equiv [1 - 2(x - x_j)\ell_j'(x_j)]\ell_j^2(x)$$
$$B_j(x) \equiv (x - x_j)\ell_j^2(x).$$

We leave to the reader to verify that $A_j(x_i) = \delta_{ij}$, $B_j(x_i) = 0$, $A_j'(x_i) = 0$, and $B_j'(x_i) = \delta_{ij}$ where $0 \le i, j \le N$, and δ_{ij} is again the Kronecker delta. Let $\{y_0, y_1, \ldots, y_N, y_0', y_1', \ldots, y_N'\}$ be any set of $(2N + 2)$ values, and define $p(x) \in \mathcal{P}_{2N+1}$ by

$$p(x) = \sum_{j=0}^{N} (y_j A_j(x) + y_j' B_j(x)). \tag{5.38a}$$

Because of the properties of $A_j(x)$ and $B_j(x)$, it is easily seen that

$$p(x_i) = \sum_{j=0}^{N} (y_j \cdot \delta_{ij} + y_j' \cdot 0) = y_i, \qquad 0 \le i \le N,$$

and

$$p'(x_i) = \sum_{j=0}^{N} (y_j \cdot 0 + y_j' \cdot \delta_{ij}) = y_i', \qquad 0 \le i \le N.$$

We can also show that the Hermite interpolating polynomial, $p(x)$, in (5.38a) is unique. Assume that $q(x) \in \mathcal{P}_{2N+1}$, $q(x) \ne p(x)$, and $q(x)$ also has the property that $q(x_j) = y_j$ and $q'(x_j) = y_j'$, $0 \le j \le N$. Let $r(x) \equiv p(x) - q(x)$. Then $r(x) \in \mathcal{P}_{2N+1}$ and $r(x_j) = r'(x_j) = 0$, $0 \le j \le N$. Thus, counting multiplicities, $r(x)$ has $2N + 2$ zeros, a result which contradicts the Fundamental Theorem of Algebra. Therefore no such $q(x)$ exists and $p(x)$ is unique. Hence we have just proved Theorem 5.7.

Theorem 5.7.

Let $\{x_j\}_{j=0}^N$ be a set of $(N + 1)$ distinct points and $\{y_j\}_{j=0}^N$ and $\{y_j'\}_{j=0}^N$ be any two sets of $N + 1$ values. Then there exists a unique $p(x) \in \mathcal{P}_{2N+1}$ such that $p(x_j) = y_j$ and $p'(x_j) = y_j'$, $0 \le j \le N$, and $p(x)$ is given by (5.38a).

Usually the y_j and y_j' are given by some function $f(x)$ where $f(x_j) = y_j$ and $f'(x_j) = y_j'$, $0 \le j \le N$. In this case, (5.38a) becomes

$$p(x) = \sum_{j=0}^{N} (f(x_j)A_j(x) + f'(x_j)B_j(x)). \tag{5.38b}$$

We leave as a problem for the reader that (5.38b) reduces to (5.37) in the case in which $N = 1$.

In the case in which $f'(x)$ does not exist or is unknown, Hermite interpolation is still sometimes used as in (5.38a) where we set $y_j = f(x_j)$ and $y_j' = 0$, $0 \le$

$j \leq n$. This is known as *Hermite-Fejér interpolation,* and the following result was proved by Fejér.

For any N, let $\{x_j\}_{j=0}^{N}$ be the zeros of the Chebyshev polynomial $T_{N+1}(x)$. Let $f(x)$ be any function in $C[-1, 1]$ and let $p_{2N+1}(x)$ be the Hermite-Fejér interpolating polynomial in $\mathcal{P}_{2N+1}$ for $f(x)$. Then

$$\lim_{N\to\infty}\left\{\max_{-1\leq x\leq 1}\left|f(x) - p_{2N+1}(x)\right|\right\} = 0.$$

This theorem of Fejér is a curious result since we ask for the condition $p'_{2N+1}(x_j) = 0$ [which is unrelated to $f(x)$] and still obtain uniform convergence of the Hermite interpolating polynomials. The result is doubly curious in that no matter what set of interpolation points are chosen, there are functions $f \in C[-1, 1]$ such that $\lim_{N\to\infty} \|p_N - f\|_\infty = +\infty$ where $p_N(x)$ denotes the Lagrange interpolating polynomial as in (5.4) (c.f., Natanson 1965).

The error analysis for Hermite interpolation, (5.38b), can be carried out in much the same way as in Theorem 5.5. This time, however, we define the auxiliary function as $F(t) = f(t) - p(t) - CW(t)^2$ with $C = (f(x) - p(x))/W(x)^2$, $W(t) = \Pi_{j=0}^{N}(t - x_j)$, and $p(t)$ as in (5.38b). Using Rolle's theorem as before, we obtain an expression for the error:

$$f(x) - p(x) = \frac{f^{(2N+2)}(\xi)W(x)^2}{(2N+2)!}. \tag{5.39}$$

5.2.6. Interpolation by Cubic Splines

In some practical problems, interpolating polynomials are not suitable for use as an approximation. For example, in order to obtain a "good" approximation to a function $f(x)$ by an nth-degree interpolating polynomial, it may be necessary to use a fairly large value of n. Unfortunately, polynomials of high degree often have a very oscillatory behavior, which is not desirable in approximating functions that are reasonably smooth. This disadvantage of polynomial interpolation becomes particularly apparent when interpolating tabular data as is shown in the example below. Also, computational problems arise when the number of data points is large. For instance, given 100 data points (x_0, y_0), (x_1, y_1), . . . , (x_{99}, y_{99}), it would usually be foolhardy to find the 99th-degree polynomial $p(x)$ such that $p(x_i) = y_i$, $i = 0, 1, \ldots, 99$. One alternative to interpolation, as we shall see in Section 5.3, would be to find a polynomial of a low degree that "best fits" the data. Unfortunately, the polynomial that best fits the data will not generally interpolate the data as well. When it is desirable to have an approximation $q(x)$ such that $q(x_i) = y_i$, $i = 0, 1, \ldots, n$, then *piecewise* polynomial interpolation is an attractive alternative and the one that we focus on in this section. In piecewise polynomial interpolation, several lower-degree polynomials are joined together in a continuous fashion so that the resulting piecewise polynomial, $q(x)$, interpolates the data. The extreme case of piecewise polynomial interpolation is the "broken line" as in Example 5.11, in which the approximating function $q(x)$ is obtained by joining (x_i, y_i) and (x_{i+1}, y_{i+1}) by a straight line for $i = 0, 1, \ldots, n-1$. In this case, $q(x_i) = y_i$, and $q(x)$ is a linear polynomial in each interval $[x_i, x_{i+1}]$.

EXAMPLE 5.11. The table below represents data taken from a hypothetical experiment.

x_i	0	1	2	3	4	5	6	7	8	9	10
y_i	3	2	3	5	3	4	3	2	2	3	2

In Fig. 5.6, we show the piecewise linear polynomial $q(x)$ (the broken line) and the tenth-degree interpolating polynomial, $p(x)$. Clearly, unless the function from which the data were taken is strange indeed, the broken line is a more acceptable approximation than is the interpolating polynomial. (Note that as mentioned before, the interpolating polynomial appears to be more reasonable near the center of the interval.)

Although polynomial interpolation may not give an acceptable approximation, as indicated by Fig. 5.6, the broken line approximation also has its disad-

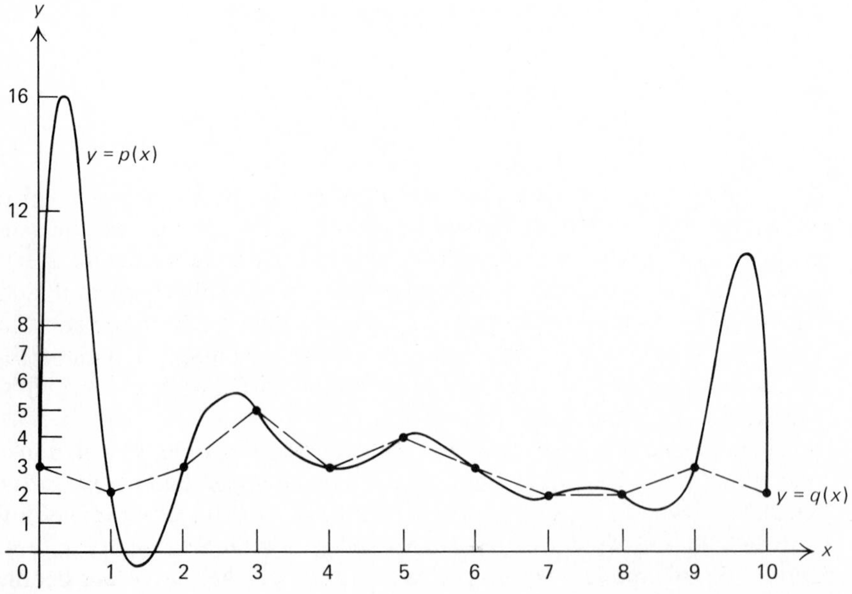

Figure 5.6 Piecewise linear and interpolating approximations of Example 5.11.

vantages. In particular, the broken line suffers from a lack of smoothness and has a discontinuous derivative. Thus, except to give a rough idea of the shape of the graph, the broken line is not well suited to approximate most of the functions that arise in physical problems since such functions are usually fairly smooth. To overcome the oscillatory behavior of polynomials and still provide a smooth approximation, an approximation technique using *splines* was presented in a paper by Schoenberg (1946), and has been the subject of extensive research ever since. This approximation technique resembles a physical process that has been used by drafters for many years. Given a set of points, $X_n = \{x_j\}_{j=0}^n$ where $a = x_0 < x_1 < \cdots < x_n = b$ and a set of functional values, $\{f(x_j)\}_{j=0}^n$, drafters will plot the data points $P_j = (x_j, f(x_j))$, $0 \le j \le n$. They will then take a thin elastic rod (called a *spline*) and a set of weights and will place the weights on the rod so that the rod must pass over each point P_j, $0 \le j \le n$. The resulting curve traced out by the spline then interpolates $f(x)$ at each x_j, and furthermore smooths out as much as possible between the points because of the elasticity of the rod. Thus the oscillatory behavior of the approximation is minimized as much as possible, but the approximation still retains the interpolation property.

We shall now consider a mathematical procedure that models the drafters' technique. First consider the set of all functions $Sp(X_n)$ such that if $S(x) \in Sp(X_n)$ then $S(x)$ satisfies the following three properties.

$S(x) \in C^2[a, b]$; that is, $S(x)$, $S'(x)$, and $S''(x)$ are continuous on $[a, b]$. (5.40a)

$S(x_j) = f(x_j) \equiv f_j$, $0 \le j \le n$; that is, $S(x)$ interpolates $f(x)$ on $[a, b]$. (5.40b)

$S(x)$ is a cubic polynomial on each subinterval $[x_j, x_{j+1}]$, $0 \le j \le n - 1$. (5.40c)

The function $S(x)$ defined by the three conditions of (5.40) is a piecewise cubic polynomial and is called a *cubic spline*. Higher-order splines are defined similarly. For example, a quartic spline would be a function $S(x)$ where $S(x) \in C^3[a, b]$ and $S(x)$ is a fourth-degree polynomial on each subinterval $[x_j, x_{j+1}]$. Note the dependence of the set of functions, $Sp(X_n)$, on the particular function, $f(x)$, being interpolated. Also note that $S(x)$ may be a *different* cubic on each subinterval; so we let $S_j(x)$ denote the cubic polynomial such that $S(x) = S_j(x)$ for x in $[x_j, x_{j+1}]$, $0 \le j \le n - 1$.

Before beginning the mathematical derivation needed to construct cubic splines, we pause to consider intuitively why we expect that there is any such function, $S(x)$, satisfying the three conditions of (5.40). For simplicity we let $n = 3$ so that $S(x)$ is represented by the cubic polynomials $S_0(x)$ on $[x_0, x_1]$, $S_1(x)$ on $[x_1, x_2]$, and $S_2(x)$ on $[x_2, x_3]$. Since each $S_j(x) \in \mathcal{P}_3$, it has four coefficients that we would like to choose to satisfy the conditions of (5.40). Therefore, in this case, we have twelve coefficients or unknowns at our disposal. Since each $S_j(x) \in \mathcal{P}_3$, all of its derivatives are continuous for any $x \in (x_j, x_{j+1})$—the *open interval*. Thus we need worry about the continuity conditions only at the interior interpolating points, x_1 and x_2, where the cubics must "patch together" with second-order continuity. With this point in mind, let us work from left to right on $[a, b] \equiv [x_0, x_3]$ and determine the number of equations that must be satisfied. Since $S(x)$ must interpolate $f(x)$ at each x_j, and since $S(x)$ must be continuous at x_1 and x_2, we have the following six equations to be satisfied:

$$S_0(x_0) = f_0; \qquad S_1(x_1) = f_1; \qquad S_0(x_1) = S_1(x_1); \qquad S_2(x_2) = f_2;$$
$$S_1(x_2) = S_2(x_2); \qquad \text{and} \qquad S_2(x_3) = f_3.$$

Finally since $S'(x)$ and $S''(x)$ must also be continuous at x_1 and x_2, we must have $S_0'(x_1) = S_1'(x_1)$, $S_0''(x_1) = S_1''(x_1)$, $S_1'(x_2) = S_2'(x_2)$, and $S_1''(x_2) = S_2''(x_2)$—four more equations. Therefore we have a total of ten equations in twelve unknowns. Thus we not only expect just one solution, but we hope that we can specify two of the unknowns to be designated arbitrarily and still have a solution for the remaining ten equations in ten unknowns. So, intuitively we expect that $Sp(X_n)$ is an infinite two-parameter family of functions. Two logical ways to obtain a good approximation for $f(x)$ would be to choose the two free parameters by specifying either $S'(x_0)$ and $S'(x_n)$ or $S''(x_0)$ and $S''(x_n)$. We choose the second alternative in the following mathematical derivation, but will also show how the first alternative may be implemented.

To construct a function $S(x)$ that satisfies the three conditions of (5.40), we first introduce some notation and then use the three conditions to obtain a linear system of equations that will enable us to determine the coefficients of each cubic polynomial, $S_j(x)$. We define $h_j \equiv \Delta x_j \equiv x_{j+1} - x_j$ and $\Delta f_j \equiv f(x_{j+1}) - f(x_j) \equiv f_{j+1} - f_j$. We next define $S''(x_j) \equiv y_j''$ and note that the quantities $y_0'', y_1'', \ldots, y_n''$ will appear as *unknowns* in a linear system of equations that will define the cubic spline $S(x)$. [The reader should note that usually $S''(x_j) \neq f''(x_j)$ and even $S'(x_j) \neq f'(x_j)$. All that (5.40) requires is that $S(x_j) = f(x_j)$, $0 \le j \le n$.]

The construction of the cubic spline proceeds roughly as follows. Suppose we choose any $n + 1$ values $y_0'', y_1'', \ldots, y_n''$ and then let $q(x)$ be the broken line such that $q(x_j) = y_j''$, $0 \le j \le n$. If we integrate $q(x)$ twice, we obtain a function $S(x)$ in $C^2[a, b]$, which is a piecewise cubic polynomial. The question is whether we can choose $y_0'', y_1'', \ldots, y_n''$ so that $S(x_j) = f_j$, $0 \le j \le n$. If we can, then $S(x)$ interpolates $f(x)$; so by (5.40), $S(x) \in Sp(X_n)$. Following this idea, we define the first-degree polynomials, $S_j''(x)$, on $[x_j, x_{j+1}]$ by

$$S_j''(x) = y_j'' \frac{x_{j+1} - x}{h_j} + y_{j+1}'' \frac{x - x_j}{h_j}, \qquad 0 \le j \le n - 1. \tag{5.41}$$

[In (5.41) we have not yet specified y_0'', y_1'', . . . , y_n''; they can be any set of values.] Note that $S_j''(x_j) = y_j''$, $S_j''(x_{j+1}) = y_{j+1}''$, $0 \le j \le n - 1$. Note also that $S_{j+1}''(x_{j+1}) = S_j''(x_{j+1})$ for $0 \le j \le n - 2$; so the function $S''(x)$, which has the value $S_j''(x)$ for $x \in [x_j, x_{j+1}]$, is a continuous function defined on $[a, b]$ [$S''(x)$ is a broken line.] Integrating (5.41) twice, we obtain for $x \in [x_j, x_{j+1}]$

$$S_j(x) = \frac{y_j''}{6h_j}(x_{j+1} - x)^3 + \frac{y_{j+1}''}{6h_j}(x - x_j)^3 + c_j(x - x_j) + d_j(x_{j+1} - x) \quad (5.42)$$

where c_j and d_j are constants of integration. To make $S(x)$ continuous and to satisfy the interpolation constraints, we need to choose the constants of integration so that $S_j(x_j) = f_j$, $S_j(x_{j+1}) = f_{j+1}$ for $0 \le j \le n - 1$. Substituting these two conditions into (5.42), we obtain

$$c_j = \frac{f_{j+1}}{h_j} - \frac{y_{j+1}''h_j}{6} \qquad \text{and} \qquad d_j = \frac{f_j}{h_j} - \frac{y_j''h_j}{6};$$

so

$$\begin{aligned} S_j(x) &= \frac{y_j''}{6h_j}(x_{j+1} - x)^3 + \frac{y_{j+1}''}{6h_j}(x - x_j)^3 + \left(\frac{f_{j+1}}{h_j} - \frac{y_{j+1}''h_j}{6}\right)(x - x_j) \\ &+ \left(\frac{f_j}{h_j} - \frac{y_j''h_j}{6}\right)(x_{j+1} - x), \qquad 0 \le j \le n - 1. \end{aligned} \quad (5.43)$$

Now, $S_j(x)$ as defined in (5.43) was chosen to match $f(x)$ at x_j and x_{j+1} by adjusting the constants of integration, c_j and d_j. Since we do this adjustment independently in each interval $[x_j, x_{j+1}]$, we have no guarantee that $S_j'(x_j) = S_{j-1}'(x_j)$ for $j = 1, 2, \ldots, n - 1$ [that is, that $S'(x)$ is continuous]. Differentiating (5.43) yields

$$S_j'(x) = -\frac{y_j''}{2h_j}(x_{j+1} - x)^2 + \frac{y_{j+1}''}{2h_j}(x - x_j)^2 + \frac{\Delta f_j}{h_j} - \frac{h_j}{6}(y_{j+1}'' - y_j''). \quad (5.44)$$

So the last condition that we must meet to satisfy all of (5.40) is to make $S'(x)$ continuous at the interior interpolating points. We do this by using (5.44) and choosing the values y_j'', $1 \le j \le n - 1$, so that $S_j'(x_j) = S_{j-1}'(x_j)$ for $1 \le j \le n - 1$. This procedure yields the following linear system of $(n - 1)$ equations in the unknowns $\{y_j''\}_{j=0}^n$:

$$\begin{aligned} &h_{j-1}y_{j-1}'' + 2(h_j + h_{j-1})y_j'' + h_jy_{j+1}'' = b_j \\ &b_j \equiv 6\left(\frac{\Delta f_j}{h_j} - \frac{\Delta f_{j-1}}{h_{j-1}}\right), \qquad 1 \le j \le n - 1. \end{aligned} \quad (5.45)$$

Since we have $(n + 1)$ unknowns and $(n - 1)$ equations in (5.45), we can specify fixed values for y_0'' and y_n'' and transfer the two terms involving these values to the right-hand side. We then have the following $(n-1) \times (n-1)$ linear system (written in matrix form and denoted by $A\mathbf{y} = \mathbf{b}$):

$$\begin{bmatrix} \gamma_1 & h_1 & 0 & 0 & \cdots & 0 & 0 & 0 \\ h_1 & \gamma_2 & h_2 & 0 & \cdots & 0 & 0 & 0 \\ \vdots & & & & & & & \\ 0 & 0 & 0 & 0 & \cdots & h_{n-3} & \gamma_{n-2} & h_{n-2} \\ 0 & 0 & 0 & 0 & \cdots & 0 & h_{n-2} & \gamma_{n-1} \end{bmatrix} \begin{bmatrix} y_1'' \\ y_2'' \\ \vdots \\ y_{n-2}'' \\ y_{n-1}'' \end{bmatrix} = \begin{bmatrix} b_1 - h_0 y_0'' \\ b_2 \\ \vdots \\ b_{n-2} \\ b_{n-1} - h_{n-1} y_n'' \end{bmatrix} \tag{5.46}$$

where $\gamma_i = 2(h_i + h_{i-1})$.

It is clear that the coefficient matrix A is tridiagonal and diagonally dominant and hence nonsingular (see Chapter 2). Thus, no matter what values we select for y_0'' and y_n'', (5.46) has a unique solution for $\{y_j''\}_{j=1}^{n-1}$ (the solution does depend on our choice of y_0'' and y_n'', of course). Furthermore the $\{y_j''\}_{j=1}^{n-1}$ can be easily found by the finite recurrence given for the LU-decomposition of a tridiagonal matrix. These values are then substituted into (5.43) and thus yield $S(x)$. It is common, as we shall see shortly, to set $y_0'' = y_n'' = 0$; and the unique cubic spline that results from this choice is called the *natural cubic spline*. A final point that should be made is that once the values $y_1'', y_2'', \ldots, y_{n-1}''$ are determined from (5.46), we then use (5.43) to evaluate $S(x)$ at $x = \alpha$. That is, if $\alpha \in [x_j, x_{j+1}]$, then $S(\alpha) = S_j(\alpha)$.

EXAMPLE 5.12. In this example, we consider approximating $f(x) = |x|$ by a cubic spline and contrast the spline approximation with an interpolating polynomial. Let $x_0 = -2$, $x_1 = -1$, $x_2 = 0$, $x_3 = 1$, and $x_4 = 2$. In this case, $h_j = 1$ for $j = 0, 1, 2, 3$ and $b_j = 6(f_{j+1} - 2f_j + f_{j-1}) = 6\Delta^2 f_{j-1}$ for $j = 1, 2, 3$. Thus we are led to the linear system [as in Eq. (5.46)]

$$\begin{aligned} 4y_1'' + y_2'' \qquad &= 0 \\ y_1'' + 4y_2'' + y_3'' &= 12 \\ y_2'' + 4y_3'' &= 0. \end{aligned}$$

Solving this system, we find $y_1'' = -6/7$, $y_2'' = 24/7$, and $y_3'' = -6/7$. Using these values in (5.43), we can evaluate the cubic spline in any subinterval $[x_j, x_{j+1}]$.

The fourth-degree interpolating polynomial for $f(x)$ at the points x_j above is easily seen to be given by $p(x) = \frac{7}{6}x^2 - \frac{1}{6}x^4$. We will compare these two approximations in the interval $[1, 2]$, where to evaluate the cubic spline in $[1, 2]$ we use [from (5.43)]

$$S_3(x) = -(2-x)^3/7 + 2(x-1) + 8(2-x)/7.$$

Since both of these approximations and $f(x)$ as well have such simple forms, it is easy to verify that the maximum value of $p(x) - |x|$ occurs at approximately $x = 1.6$. The maximum value of $(S_3(x) - |x|)$ occurs approximately at $x = 1.423$ and $|S_3(x) - |x|| \le 0.055$ for all x in $[1, 2]$. In Table 5.3, we have listed values of $S_3'(x)$ and $p'(x)$ as well. [As we shall see, the derivative of the cubic spline provides a fairly good approximation to

TABLE 5.3

x	$f(x)$	$S_3(x)$	$S_3'(x)$	$p(x)$	$p'(x)$
1.0	1.0	1.000	1.286	1.000	1.667
1.2	1.2	1.241	1.131	1.334	1.648
1.4	1.4	1.455	1.011	1.646	1.437
1.6	1.6	1.648	0.926	1.894	1.003
1.8	1.8	1.827	0.874	2.030	0.312
2.0	2.0	2.000	0.857	2.000	−0.667

$f'(x)$. On the other hand, derivatives of the interpolating polynomial are not generally good approximations to the derivatives of $f(x)$.] Since $f'(x) \equiv 1$ for $1 \le x \le 2$, this example shows how well the cubic spline can duplicate the general shape of $f(x)$. We shall return to this point when we discuss numerical differentiation in the next chapter.

If the derivatives of $f(x)$ at the endpoints are known, that is, if $f'(x_0) \equiv y_0'$ and $f'(x_n) \equiv y_n'$ are given, then we may suspect that we will get a better cubic spline approximation to $f(x)$ if we choose the two free parameters, y_0'' and y_n'', in a manner such that the resulting cubic spline, $S(x)$, satisfies $S'(x_0) = f'(x_0) = y_0'$ and $S'(x_n) = f'(x_n) = y_n'$ as well as the interpolation property, $S(x_j) = f_j$, $0 \le j \le n$. From Eq. (5.44) we see for $j = n - 1$ that

$$f'(x_n) = S_{n-1}'(x_n) = \frac{y_n''}{2h_{n-1}}(x_n - x_{n-1})^2 + \frac{\Delta f_{n-1}}{h_{n-1}} - \frac{h_{n-1}}{6}(y_n'' - y_{n-1}''),$$

or

$$\frac{h_{n-1}}{3} y_n'' + \frac{h_{n-1}}{6} y_{n-1}'' = y_n' - \frac{\Delta f_{n-1}}{h_{n-1}}. \tag{5.47a}$$

Similarly, from (5.44) with $j = 0$, we obtain

$$\frac{h_0}{6} y_1'' + \frac{h_0}{3} y_0'' = \frac{\Delta f_0}{h_0} - y_0'. \tag{5.47b}$$

The continuity of $S'(x)$ must be maintained so that all of the equations of (5.45) must still hold. If Eqs. (5.45) are written together with (5.47), we get an $(n + 1) \times (n + 1)$ tridiagonal diagonally dominant linear system of equations. These may be easily solved as before and again substituted into (5.43) to obtain $S(x)$. (There are other ways of deriving this particular cubic spline, but they involve a different derivation. See Rivlin, 1969.)

For brevity of notation we shall denote the cubic spline above as $S^{(1)}(x)$ and the natural cubic spline as $S^{(2)}(x)$. These two particular cubic splines are the ones that are most often used in practice because of the so-called *extremal properties* that we shall now develop. First we note by a simple algebraic reduction that for any $g(x)$ and $S(x) \in C^2[a, b]$

$$\int_a^b [g''(x) - S''(x)]^2\,dx = \int_a^b g''(x)^2\,dx - \int_a^b S''(x)^2\,dx - 2\int_a^b S''(x)[g''(x) - S''(x)]\,dx. \tag{5.48}$$

We concentrate on the last integral in (5.48) and see that integration by parts yields

$$\int_a^b S''(x)[g''(x) - S''(x)]\,dx = S''(x)[g'(x) - S'(x)]\Big|_a^b - \int_a^b S'''(x)[g'(x) - S'(x)]\,dx. \tag{5.49}$$

For the first extremal property we let $g(x)$ be any function in $C^2[a, b]$ that interpolates $f(x)$ at $\{x_j\}_{j=0}^n$. If $S(x) \in Sp(X_n)$, then $S'''(x)$ is a constant, say α_j, on each subinterval (x_j, x_{j+1}). Therefore

$$\int_a^b S'''(x)[g'(x) - S'(x)]\,dx = \sum_{j=0}^{n-1} \alpha_j \int_{x_j}^{x_{j+1}} [g'(x) - S'(x)]\,dx = \sum_{j=0}^{n-1} \alpha_j[g(x) - S(x)]\Big|_{x_j}^{x_{j+1}} = 0$$

since $g(x_j) = f(x_j) = S(x_j)$, $0 \le j \le n$.

Now if $S(x) = S^{(2)}(x)$ (the natural cubic spline), then $S''(b) = S''(a) = 0$; and so the integral on the left-hand side of (5.49) is zero. Using this result in (5.48) yields

$$\int_a^b g''(x)^2\,dx = \int_a^b S''(x)^2\,dx + \int_a^b [g''(x) - S''(x)]^2\,dx \ge \int_a^b S''(x)^2\,dx. \tag{5.50}$$

We leave it to the reader to show that equality holds in (5.50) if and only if $g(x) \equiv S^{(2)}(x)$. Hence, by (5.50), among all possible functions in $C^2[a, b]$ that interpolate $f(x)$ [including all of $Sp(X_n)$, all interpolating polynomials, and even $f(x)$ itself if $f''(x)$ is continuous], the integral $\int_a^b g''(x)^2\,dx$ is minimized if and only if $g(x) = S^2(x)$. This is the first extremal property and it explains the origin of the name "spline" for the mathematical approximation since the "strain energy" of the drafter's elastic rod is essentially proportional to the integral of the square of the second derivative, which we see is minimized by $S^{(2)}(x)$. The property is also called the "minimum curvature" property since the curvature of any approximation is essentially the integral of the square of the second derivative. Thus the oscillatory behavior of the approximation is minimized by $S^{(2)}(x)$. Yet another interpretation of (5.50) is that among all functions in $C^2[a, b]$ that interpolate $f(x)$, the natural cubic spline is closest to being a broken line since the broken line $q(x)$ has zero curvature; that is, $\int_a^b q''(x)^2\,dx = 0$.

For the second extremal property we now restrict the function $g(x)$ in (5.49) to be in the smaller set of functions that interpolate $f(x)$ *and* also satisfy $g'(a) = f'(a)$ and $g'(b) = f'(b)$ (such as the Hermite interpolating polynomial, for instance). This time we let $\hat{S}(x) = S^{(1)}(x)$; so $\hat{S}'(a) = f'(a) = g'(a)$ and $\hat{S}'(b) = f'(b) = g'(b)$. Using this, we again see that the integral in (5.49) is zero. Therefore, by (5.48), for all $g(x)$ of this particular form.

$$\int_a^b g''(x)^2\,dx = \int_a^b \hat{S}''(x)^2\,dx + \int_a^b [g''(x) - \hat{S}''(x)]^2\,dx. \tag{5.51}$$

[Equation (5.51) should not be confused with (5.50) since $g(x)$ is now more restricted, but note that $S^{(1)}(x)$ has minimum curvature in this smaller class of functions.] Now we let $u(x)$ be any cubic spline on the points $\{x_j\}_{j=0}^n$, whether it interpolates $f(x)$ or not. We then let $g(x) \equiv f(x) - u(x)$ and $S(x) \equiv S^{(1)}(x) - u(x)$. Note that $S(x)$ is still a cubic spline and has the properties that $S(x_j) = g(x_j)$, $0 \le j \le n$, and $S'(a) = g'(a)$ and $S'(b) = g'(b)$. Thus this particular choice of $g(x)$ and $S(x)$ can be used in (5.51) and yields $\int_a^b g''(x)^2\,dx \ge \int_a^b [g''(x) - S''(x)]^2\,dx$, or

$$\int_a^b \left[\frac{d^2f(x)}{dx^2} - \frac{d^2u(x)}{dx^2}\right]^2 dx \ge \int_a^b \left[\frac{d^2f(x)}{dx^2} - \frac{d^2S^{(1)}(x)}{dx^2}\right]^2 dx. \tag{5.52}$$

[We leave it to the reader to verify that equality holds in (5.52) if and only if $u(x) = S^{(1)}(x) + \alpha x + \beta$.] Formula (5.52) is the second extremal property. It says that if we measure the distance between $f(x)$ and any cubic spline $u(x)$ by the formula $\int_a^b [f''(x) - u''(x)]^2\,dx$, then this distance is minimized when $u(x) = S^{(1)}(x)$ [so $S^{(1)}(x)$ is a "best approximation" to $f(x)$ in this sense]. In summary, among all cubic splines $u(x)$, the *error* $f(x) - u(x)$ has *minimum curvature* when $u(x) = S^{(1)}(x)$. Among all functions $g(x)$ in $C^2[a, b]$ that interpolate $f(x)$, the *function* with minimum curvature is $g(x) = S^{(2)}(x)$. Among all functions $g(x)$ that interpolate $f(x)$ and satisfy $g'(a) = f'(a)$, $g'(b) = f'(b)$, the function with minimum curvature is $g(x) = S^{(1)}(x)$.

We conclude this section by giving error bounds on the approximation of $f(x)$ by $S(x)$ *and* the approximation of $f'(x)$ by $S'(x)$ where $S(x)$ can be taken to be either $S^{(1)}(x)$ or $S^{(2)}(x)$. For the sake of notation we let the error function be given by $E(x) = f(x) - S(x)$ for $x \in [a, b]$. Then $E'(x) = f'(x) - S'(x)$ is the error of the derivative approximation. We also let

$$h \equiv \max_{0 \le j \le n-1} (x_{j+1} - x_j),$$

the maximum step size. Let x be any arbitrary fixed point in $[a, b]$; then there exists some j, $0 \le j \le n - 1$, such that $x \in [x_j, x_{j+1}]$. Since $E(x_j) = E(x_{j+1}) = 0$, by Rolle's theorem there exists a point $c \in [x_j, x_{j+1}]$ such that $E'(c) = 0$. Thus, $\int_c^x E''(t)\,dt = E'(x) - E'(c) = E'(x)$. By the Cauchy-Schwarz inequality (see Problem 9, Section 2.6, or Theorem 5.8, Section 5.3), we have

$$\begin{aligned} |E'(x)|^2 &= \left|\int_c^x E''(t) \cdot 1\,dt\right|^2 \le \left(\int_c^x E''(t)^2\,dt\right)\left(\int_c^x 1^2\,dt\right) \\ &\le \left(\int_c^x E''(t)^2\,dt\right)|x - c| \le h\int_c^x E''(t)^2\,dt. \end{aligned} \tag{5.53}$$

From either (5.50) or (5.51) we find

$$\begin{aligned} \int_c^x E''(t)^2\,dt &\le \int_a^b E''(t)^2\,dt = \int_a^b [f''(t) - S''(t)]^2\,dt \\ &= \int_a^b f''(t)^2\,dt - \int_a^b S''(t)^2\,dt \le \int_a^b f''(t)^2\,dt. \end{aligned}$$

Substituting this expression into (5.53) and taking square roots yield

$$|E'(x)| \equiv |f'(x) - S'(x)| \le h^{1/2}\left(\int_a^b f''(t)^2\,dt\right)^{1/2} \qquad \text{for all } x \in [a, b]. \tag{5.54}$$

Thus for $x \in [a, b]$, $|f'(x) - S'(x)|$ is bounded by an expression proportional to $h^{1/2}$.

Now, as above, let x be fixed in $[a, b]$ and thus $x \in [x_j, x_{j+1}]$ for some j. Since $\int_{x_j}^x E'(t)\,dt = E(x) - E(x_j) = E(x) - 0 = E(x)$, we have

$$|E(x)| = \left|\int_{x_j}^x E'(t)\,dt\right| \le \int_{x_j}^x \left(\max_{a \le z \le b} |E'(z)|\right) dt \le h \max_{a \le z \le b} |E'(z)|.$$

Then by (5.54) we have

$$|E(x)| \le h^{3/2}\left(\int_a^b f''(t)^2\,dt\right)^{1/2}, \tag{5.55}$$

giving a bound for $|f(x) - S(x)|$ that is proportional to $h^{3/2}$.

Formulas (5.54) and (5.55) are important as error bounds, but also since the bounds are independent of x, they tell us that if we increase the number of interpolating points in a manner such that $h \to 0$ as $n \to \infty$, then $S(x)$ and $S'(x)$ converge uniformly to $f(x)$ and $f'(x)$, respectively. The inequality (5.54) is also significant in that it tells us that $S'(\alpha)$ is a good approximation to $f'(\alpha)$ for $\alpha \in [a, b]$. Thus cubic splines can be used as a method to find a numerical approximation for not only $f(x)$ but $f'(x)$ as well (see Example 5.12). This property is not shared by interpolating polynomials, $p_n(x)$, as their oscillatory behavior tends to exaggerate the difference between $f'(x)$ and $p_n'(x)$, and one must be extremely cautious in using interpolating polynomials for the purpose of numerical differentiation.

With a more careful mathematical analysis, sharper error bounds can be derived. Typical results are the following from Hall (1968). If $f^{(iv)}(x)$ is continuous for $a \le x \le b$, then

$$|E(x)| \le (5/384)\|f^{(iv)}\|_\infty h^4 \quad \text{and} \quad |E'(x)| \le \left(\frac{9+\sqrt{3}}{216}\right)\|f^{(iv)}\|_\infty h^3.$$

PROBLEMS, SECTION 5.2.6

1. To illustrate what can happen in even the simplest Hermite-Birkhoff interpolation problems, consider the following three problems.
   a) Find $p \in \mathcal{P}_3$ such that $p(0) = 1$, $p'(0) = 1$, $p'(1) = 2$, $p(2) = 1$.
   b) Find $p \in \mathcal{P}_3$ such that $p(-1) = 1$, $p'(-1) = 1$, $p'(1) = 2$, $p(2) = 1$.
   c) Find $p \in \mathcal{P}_3$ such that $p(-1) = 1$, $p'(-1) = -6$, $p'(1) = 2$, $p(2) = 1$.

   Using the method of undetermined coefficients, show that problem (a) has a unique solution, problem (b) has no solution, and problem (c) has infinitely many solutions.
2. Find the third-degree Hermite interpolating polynomial for $f(x) = \cos(x)$ on $[0.3, 0.6]$, and compare the results with those of Example 5.3 [$\sin(0.3) \approx 0.295520$ and $\sin(0.6) \approx 0.564642$].
3. Find in $\mathcal{P}_4$ the polynomial $p(x)$ that interpolates $f(x) = |x|$ as follows: $p(-2) = f(-2)$, $p'(-2) = f'(-2)$, $p(0) = f(0)$, $p(2) = f(2)$, and $p'(2) = f'(2)$. Compare your results with those of Example 5.12 to see that this polynomial is generally better than the interpolating polynomial but not as good as the cubic spline.
4. Write a set of subroutines that can be used to generate the natural cubic spline interpolator on equally spaced knots and that can be used to evaluate, differentiate, and integrate the resulting cubic spline. These subroutines could take the form outlined below.

   As a start, assume the main program has the first knot X0, the knot spacing H, and a value for N; the knots x_j are given by $x_j = \text{X0} + \text{J} * \text{H}$, $0 \le \text{J} \le \text{N}$. Also assume the main program has arrays DATA and YPP where DATA(J) $= f(x_j)$, and where

YPP(J) will hold the value y_j'' found by solving (5.46) with $y_0'' = y_n'' = 0$. For simplicity, assume all arrays have dimension 100. Write the following subroutines.

a) Subroutine SPLINE(DATA, H, YPP, N) should set up and solve the tridiagonal system (5.46).

b) Subroutine LOCATE (ALPHA, X0, H, N, JINT) is used in order to decide the index j that should be used in (5.42) to evaluate $S(x)$ at $x = \alpha$. For $x_j \le \alpha < x_{j+1}$, $S(\alpha) = S_j(\alpha)$ and JINT $= j$.

c) Subroutine EVALSP(DATA, YPP, X0, H, ALPHA, SPLVAL, N, JINT) evaluates $S(x)$ at $x = \alpha$ and returns the value $S(\alpha)$ as SPLVAL.

d) Also write subroutines similar to EVALSP that calculate $S'(x)$ and $S''(x)$ at $x = \alpha$ and that calculate $\int_a^b S(x)dx$. Note that $S'(x)$ is given by (5.44) and $S''(x)$ by (5.41); the integral can be calculated from (5.43).

As an initial check, consider $f(x) = 2x + 3$ on $[0, 2]$ with knots $x_j = j/10$, $0 \le j \le 20$. The spline $S(x)$ should be identical to $f(x)$. Next, for $f(x) = 1/(1 + x^2)$ on $[-5, 5]$, calculate the cubic spline interpolator with N $= 11, 21, 41$ and print $S(x), f(x), S'(x), f'(x), S''(x), f''(x)$ for $x = -5 + i/10$, $0 \le i \le 100$. Also, for each N, calculate $\int_{-5}^{5} S(x)dx$ and compare it with the exact value.

5. Test the program in Problem 4 on $f(x) = \cos(2x)$, $0 \le x \le 6$; use N $= 21, 31, 41$ and call a plotting routine to sketch $f(x)$ and $S(x)$.

6. Graph the derivative of the broken line $q(x)$ in Example 5.11.

7. Verify that the polynomials $A_j(x)$ and $B_j(x)$ in (5.38a) satisfy these conditions:

$$A_j(x_i) = \delta_{ij} \qquad B_j(x_i) = 0$$
$$A_j'(x_i) = 0 \qquad B_j'(x_i) = \delta_{ij}.$$

8. For $N = 1$ verify that (5.38a) reduces to (5.37).

9. In some cases, the error bound (5.55) for cubic spline approximation may be pessimistic. How large must n be in order that $|E(x)| \le 10^{-3}$ for $f(x) = e^{3x}$, $[a, b] = [-1, 1]$, and $h = 2/n$? Contrast this result with results from the program in Problem 4 and with Hall's bounds.

***10.** Verify that equality holds in (5.50) if and only if $g(x) = S^{(2)}(x)$. Verify that equality holds in (5.52) if and only if $u(x) = S^{(1)}(x) + \alpha x + \beta$. [Recall that $\int_a^b h(x)^2\,dx > 0$ for any function $h(x)$ that is continuous on $[a, b]$ unless $h(x) \equiv 0$ on $[a, b]$.]

5.2.7 Interpolation in Several Variables and Bicubic Splines

The problem of interpolating in a two-way table is fairly common in many engineering and scientific applications. Specifically if $f(x, y)$ is a function of two variables that has been tabulated at the points (x_i, y_j), $0 \le i \le n$, $0 \le j \le m$, then we can construct a two-way table that represents this information:

	x_0	x_1	$\cdots$	x_n
y_0	$f(x_0, y_0)$	$f(x_1, y_0)$	$\cdots$	$f(x_n, y_0)$
y_1	$f(x_0, y_1)$	$f(x_1, y_1)$	$\cdots$	$f(x_n, y_1)$
$\vdots$	$\vdots$			$\vdots$
y_m	$f(x_0, y_m)$	$f(x_1, y_m)$	$\cdots$	$f(x_n, y_m)$

If we want to estimate $f(x, y)$ at a point $(\hat{x}, \hat{y})$, which is not in the table, we are faced with an interpolation problem. We will see momentarily that we can construct a polynomial in x and y of the form

$$p(x, y) = \sum_{r=0}^{n} \sum_{s=0}^{m} a_{rs} x^r y^s,$$

which satisfies $p(x_i, y_j) = f(x_i, y_j)$, $0 \le i \le n$, $0 \le j \le m$. Given this result, we can estimate $f(x, y)$ at $(\hat{x}, \hat{y})$ by $f(\hat{x}, \hat{y}) \approx p(\hat{x}, \hat{y})$.

A variation of the problem of interpolating in a two-way table is the problem of approximating a function $f(x, y)$ of two variables; this problem is sometimes called "surface fitting." The graph of the function $z = f(x, y)$ is a surface in three-space, and we can ask for a simple function $p(x, y)$ such that $p(x, y) \simeq f(x, y)$. As an application, if we want the normal to the surface $z = f(x, y)$ at the point $(\hat{x}, \hat{y}, f(\hat{x}, \hat{y}))$, we can estimate the normal by finding the normal to $z = p(x, y)$.

If we restrict our approximation problem and consider how we might approximate only functions $f(x, y)$ that are defined over a *rectangular* region in the xy-plane, then we can very easily extend our previous results for polynomial and spline approximation. So for simplicity we consider how we can approximate $f(x, y)$ only for (x, y) in R where

$$R = \{(x, y): a \le x \le b, c \le y \le d\}.$$

We first treat the problem of interpolating $f(x, y)$ by polynomials in two variables and then consider the case of bicubic spline approximation. To begin, suppose $a \le x_0 < x_1 < \cdots < x_n \le b$ and $c \le y_0 < y_1 < \cdots < y_m \le d$; and let $\pi_{nm} = \{(x_i, y_j) : 0 \le i \le n, 0 \le j \le m\}$; that is, π_{nm} is a rectangular grid of $(n + 1)(m + 1)$ points in R. Given a function $f(x, y)$, we would like to find a polynomial in two variables, $p(x, y)$, such that $p(x_i, y_j) = f(x_i, y_j)$ for all (x_i, y_j) in the grid π_{nm}. Now if we define $\mathcal{P}_{nm}$ by

$$\mathcal{P}_{nm} = \{p(x, y)\ 1{:}p(x, y) = \sum_{i=0}^{n} \sum_{j=0}^{m} a_{ij} x^i y^j, \text{ for all real } a_{ij}\},$$

then each $p(x, y)$ has $(n + 1)(m + 1)$ coefficients. We then hope that we could choose these coefficients to satisfy the $(n + 1)(m + 1)$ interpolation constraints $p(x_i, y_j) = f(x_i, y_j)$, $0 \le i \le n$, $0 \le j \le m$.

Following the lines of one-variable polynomial interpolation in Section 5.2, we define the Lagrange polynomials of two variables by

$$\ell_{ij}(x, y) = \ell_i(x)\tilde{\ell}_j(y), \qquad 0 \le i \le n,\ 0 \le j \le m$$

where $\ell_i(x)$ is given in (5.2) and $\tilde{\ell}_j(y)$ is defined similarly for the points $c = y_0 < y_1 < \cdots < y_m = d$. Thus we have $\ell_i(x_k) = \delta_{ik}$ and $\tilde{\ell}_j(y_k) = \delta_{jk}$; and so $\ell_{ij}(x_r, y_s) = 1$ if $i = r$ and $j = s$; and $\ell_{ij}(x_r, y_s) = 0$ otherwise. From this we see that

$$p(x, y) = \sum_{i=0}^{n} \sum_{j=0}^{m} f(x_i, y_j)\ell_{ij}(x, y)$$

is a polynomial in $\mathcal{P}_{nm}$ that interpolates $f(x, y)$ on the set of points π_{nm}, and we will call this form of $p(x, y)$ the Lagrange form of the interpolating polynomial.

To see that $p(x, y)$ is unique in $\mathcal{P}_{nm}$, suppose that $q(x, y)$ in $\mathcal{P}_{nm}$ satisfies $q(x_i, y_j) = f(x_i, y_j)$, $0 \le i \le n$, $0 \le j \le m$, where $q(x, y) = \sum_{i=0}^{n} \sum_{j=0}^{m} b_{ij}x^i y^j$. Let us rewrite $q(x, y)$ as

$$q(x, y) = \sum_{i=0}^{n} x^i \sum_{j=0}^{m} b_{ij}y^j = \sum_{i=0}^{n} x^i q_i(y)$$

where

$$q_i(y) = \sum_{j=0}^{m} b_{ij}y^j.$$

If we set $y = y_s$ for a fixed value of s, $0 \le s \le m$, then $q(x, y_s)$ interpolates $f(x, y_s)$ at $x = x_0, x_1, \ldots, x_n$. Thus $q(x, y_s)$ is a polynomial only in x, and its coefficients $q_i(y_s)$ are uniquely determined (see Theorem 5.3) by the data $f(x_0, y_s), f(x_1, y_s)$ $\ldots, f(x_n, y_s)$.

Knowing that each $q_i(y_s)$ is determined uniquely by the data $f(x_i, y_s)$, $0 \le i \le n$, $0 \le s \le m$, we can show that the b_{ij} are also uniquely determined. If we fix i and consider the system of $(m + 1)$ equations

$$q_i(y_s) = b_{i0} + b_{i1}y_s + \cdots + b_{im}y_s^m, \qquad 0 \le s \le m,$$

then it is clear that $b_{i0}, b_{i1}, \ldots, b_{im}$ are uniquely determined by $q_i(y_0)$, $q_i(y_1), \ldots, q_i(y_m)$ since the coefficient matrix for the system is a Vandermonde matrix. Thus we have shown that interpolation by $p(x, y)$ is unique.

For computational purposes, it is frequently useful to express the interpolating polynomial $p(x, y)$ in the form

$$p(x, y) = \sum_{i=0}^{n} \ell_i(x) \sum_{j=0}^{m} f(x_i, y_j)\tilde{\ell}_j(y) = \sum_{i=0}^{n} \ell_i(x)p_i(y)$$

where

$$p_i(y) = \sum_{j=0}^{m} f(x_i, y_j)\tilde{\ell}_j(y).$$

As we noted before, $p_i(y)$ is a one-variable polynomial (in y) interpolating $f(x, y)$ along the line $x = x_i$, at the points (x_i, y_0), (x_i, y_1), . . . , (x_i, y_m). This means that $p(x, y)$ can be built up using one-variable polynomial interpolation. For example, $p(\hat{x}, \hat{y})$ is given by

$$p(\hat{x}, \hat{y}) = \sum_{i=0}^{n} \ell_i(\hat{x})p_i(\hat{y}).$$

From the expression above, $p(\hat{x}, \hat{y}) = q(\hat{x})$ where $q(x)$ is the polynomial interpolating the data $p_0(\hat{y})$, $p_1(\hat{y})$, . . . , $p_n(\hat{y})$, at the points $x = x_i$, $0 \le i \le n$. To get the values $p_i(\hat{y})$, we interpolate the data $f(x_i, y_j)$, $0 \le j \le m$, by $p_i(y)$ and evaluate $p_i(y)$ at $y = \hat{y}$.

This two-dimensional interpolation scheme can be rephrased thus.

1. For each fixed grid line $x = x_i$, interpolate the data $f(x_i, y_j)$ in the y-direction at the knots y_0, y_1, . . . , y_m and evaluate the (one-variable) interpolating polynomial $p_i(y)$ at $y = \hat{y}$.
2. Interpolate the values $p_0(\hat{y})$, $p_1(\hat{y})$, . . . , $p_n(\hat{y})$ in the x-direction at the knots x_0, x_1, . . . , x_n and evaluate the (one-variable) interpolating polynomial $q(x)$ at $x = \hat{x}$. The result, $q(\hat{x})$, is equal to $p(\hat{x}, \hat{y})$.

This algorithm can be illustrated as a "wire-frame" model as in Fig. 5.7. We also emphasize what is probably obvious: an algorithm to implement two-dimensional interpolation can be constructed from any routine for one-dimensional interpolation. For step (1) above, we call a polynomial interpolator $(n + 1)$ times to obtain the values $p_0(\hat{y})$, $p_1(\hat{y})$, . . . , $p_n(\hat{y})$ and then call it once more to get $q(\hat{x}) = p(\hat{x}, \hat{y})$.

Writing $p(x, y)$ in the form

$$p(x, y) = \sum_{i=0}^{n} \sum_{j=0}^{m} \ell_i(x)\tilde{\ell}_j(y)f(x_i, y_i)$$

makes it clear (see Fig. 5.7) that we could equally well construct one-dimensional interpolators in the x-direction (at the grid lines $y = y_j$), evaluate these at $x = \hat{x}$, and then pass a one-dimensional interpolator through these data along the line $x = \hat{x}$ in the y-direction. In addition, from the form of $p(x, y)$ above, we see how to calculate quantities such as $p_{xy}(\hat{x}, \hat{y})$. In particular, the mixed partial p_{xy} is clearly given by

$$p_{xy}(x, y) = \sum_{i=0}^{n} \ell_i'(x) \sum_{j=0}^{m} \tilde{\ell}_j'(y)f(x_i, y_j).$$

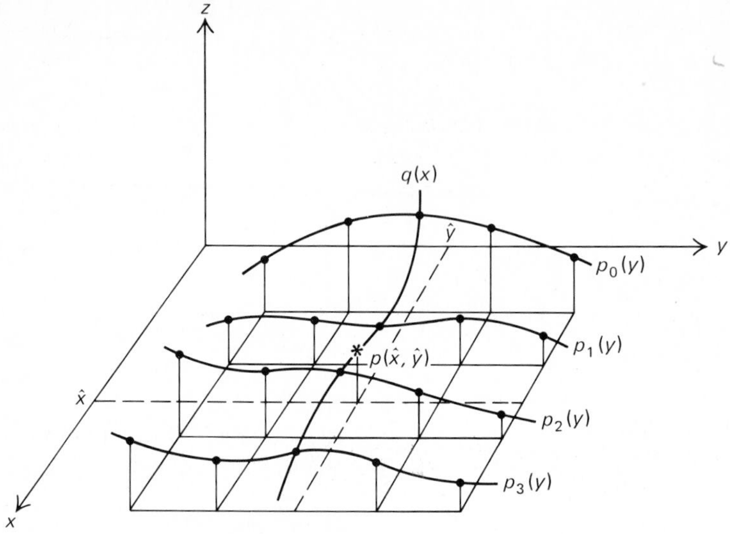

Figure 5.7 A wire-frame model illustrating the calculation $p(\hat{x}, \hat{y})$.

Therefore,

$$p_{xy}(\hat{x}, \hat{y}) = \sum_{i=0}^{n} \ell_i'(\hat{x})p_i'(\hat{y})$$

where

$$p_i'(\hat{y}) = \sum_{j=0}^{m} \tilde{\ell}_j'(\hat{y})f(x_i, y_j).$$

By way of interpretation, we pass $p_i(y)$ through the data $f(x_i, y_j)$, $0 \le j \le m$ (along the grid line $x = x_i$), differentiate $p_i(y)$, and evaluate $p_i'(\hat{y})$. Given the values $p_i'(\hat{y})$, $0 \le i \le n$, we pass $q(x)$ through these values at the knots $x_0, x_1, \ldots, x_n$, differentiate $q(x)$, and evaluate $q'(\hat{x})$. The net result is $q'(\hat{x}) = p_{xy}(\hat{x}, \hat{y}) \simeq f_{xy}(\hat{x}, \hat{y})$.

An exactly analogous development can be given for spline interpolation. In particular, a function $S(x, y)$ is called a *bicubic spline* if $S(x, y)$ is a cubic spline in x for fixed y and a cubic spline in y for fixed x. The natural bicubic spline is easiest to treat, and so we restrict our attention to it. As with two-variable polynomial interpolation, we will see that a bicubic spline interpolator can be built up from one-dimensional cubic spline interpolators. The wire-frame representation in Fig. 5.7 is also valid for bicubic splines. We can run cubic spline interpolators along the grid lines $x = x_i$, through the data $f(x_i, y_j)$ for $0 \le j \le m$, and then evaluate them at $y = \hat{y}$. We then run a cubic spline interpolator along $y = \hat{y}$, through the values obtained above, and evaluate this cubic spline at $x = \hat{x}$.

To see that the construction described above is valid, we can use the idea of a *cardinal* natural spline. Given the knots $y_0, y_1, \ldots, y_m$, we say that $S_j(y)$ is a cardinal natural spline if $S_j(y)$ is a natural cubic spline on knots $y_0, y_1, \ldots, y_m$ and if $S_j(y_k) = \delta_{jk}$, $0 \le k \le m$. Clearly, these cardinal natural splines always exist; and if $g(y)$ is a function defined on $[y_0, y_m]$, then the natural cubic spline interpolator for $g(y)$ can be represented as

$$S(y) = \sum_{j=0}^{m} S_j(y)g(y_j).$$

[Note that the sum of natural cubic splines is a natural cubic spline; so $S(y)$ is obviously the cubic spline interpolator for $g(y)$; the form of the representation is similar to the Lagrange form of the interpolating polynomial.]

Given the concept of a cardinal natural spline, it is easy to see how to construct a natural bicubic spline $S(x, y)$. In particular, if $S_i(x)$, $0 \le i \le n$, denote the cardinal natural splines for $x_0, x_1, \ldots, x_n$, and $\tilde{S}_j(y)$ denote the cardinal natural splines for $y_0, y_1, \ldots, y_m$, then

$$S(x, y) = \sum_{i=0}^{n} \sum_{j=0}^{m} S_i(x)\tilde{S}_j(y)f(x_i, y_j)$$

is a bicubic spline that interpolates $f(x, y)$ on the grid π_{nm}. As before, we can write $S(x, y)$ as

$$S(x, y) = \sum_{i=0}^{n} S_i(x)\phi_i(y)$$

where

$$\phi_i(y) = \sum_{j=0}^{m} \tilde{S}_j(y)f(x_i, y_j).$$

From the representation of $S(x, y)$ above, we see that we can evaluate $S(x, y)$ and various partial derivatives of $S(x, y)$ by repeatedly calling a one-dimensional spline routine, $(n + 1)$ times in the y-direction and then once in the x-direction. [Note, we do not need to calculate $S_i(x)$ or $\tilde{S}_j(y)$.]

Bicubic spline approximation to functions $f(x, y)$ are quite effective. A number of applications should be evident; we can use these approximations to estimate mixed partials, estimate normals to a surface, calculate approximations to surface area, etc. Finally, cardinal splines can be used to obtain higher-dimensional analogs of cubic splines. Given $f(x, y, z)$ to approximate over a solid rectangular region that is gridded by (x_i, y_j, z_k), $0 \le i \le n$, $0 \le j \le m$, $0 \le k \le \ell$, we can form

$$S(x, y, z) = \sum_{i=0}^{n} \sum_{j=0}^{m} \sum_{k=0}^{\ell} S_i(x)\tilde{S}_j(y)S_k^*(z)f(x_i, y_j, z_k).$$

As before,

$$S(x, y, z) = \sum_{k=0}^{\ell} S_k^*(z)B_k(x, y)$$

where $B_k(x, y)$ is the bicubic spline interpolator to $f(x, y, z)$ at the level $z = z_k$. Again, the three-dimensional spline approximation to $f(x, y, z)$ can be built up by calling a one-dimensional spline routine repeatedly. To evaluate $f(\hat{x}, \hat{y}, \hat{z})$, we call the routine $(n + 2)$ times to get $B_k(\hat{x}, \hat{y})$, a total of $(\ell + 1)(n + 2)$ calls. A final call in the z-direction interpolating $B_k(\hat{x}, \hat{y})$, $0 \le k \le \ell$ will produce $S(\hat{x}, \hat{y}, \hat{z})$.

PROBLEMS, SECTION 5.2.7

1. Using the ideas illustrated in Fig. 5.7, write a subroutine that evaluates the bivariate interpolating polynomial $p(x, y)$ at a point $(\hat{x}, \hat{y})$. In particular, suppose the interpolation points are (x_i, y_j) for $0 \le i \le n$, $0 \le j \le m$. For each fixed i, call the routine in Problem 8, Section 5.2.1, to evaluate $p_i(\hat{y})$ where $p_i(y)$ interpolates the data $f(x_i, y)$ for $y = y_j$, $0 \le j \le m$. Call the routine again to interpolate the data $p_i(\hat{y})$, $0 \le i \le n$ (the routine is working in the x-direction here). Test your program on $f(x, y) = x^2y + 3y^2x$ on the rectangular grid (x_i, y_j) where $x_i = i$, $0 \le i \le 3$ and $y_j = 2j$, $0 \le j \le 2$. Evaluate $p(x, y)$ at $(x, y) = (.5, .7)$ and $(x, y) = (2.1, 3.5)$; note that $p(x, y)$ should agree with $f(x, y)$ everywhere.
2. Test the program in Problem 1 on the data $f(x_i, y_j)$ where $f(x, y) = 1/(1 + 5x^2y^2)$ and where the grid is defined by $x_i = -1 + .2i$, $y_j = -1 + .2j$, $0 \le i, j \le 10$. Evaluate and print $p(x, y)$ and $f(x, y)$ at the points $(x, y) = (r/10, 1 - s/10)$, $r = 0, 1, \ldots, 10$, $s = 0, 1, \ldots, 10$.
3. What are the four Lagrange basis functions $\ell_{ij}(x, y)$ for interpolation at (0, 0), (0, 1), (1, 1), (1, 0)?
4. Use the functions in Problem 3 to construct $p(x, y)$ for an arbitrary $f(x, y)$, and use Fig. 5.7 to interpret $p(.5, .5)$ as an "average of averages."
5. Construct a routine for bicubic spline interpolation using the ideas of Problem 1 and the routines from Problem 4, Section 5.2.6. Test your program as in Problem 2.
6. Expand the program in Problem 5 so that the expanded program has the capability to calculate the mixed partial $S_{xy}(x, y)$ and to estimate $f_{xy}(.3, .7)$ where $f(x, y)$ is as in Problem 2.

7. The integral $\int_R\int f(x, y)dA$ where R is a rectangle, $a \le x \le b$, $c \le y \le d$, can be calculated as an iterated integral:

$$\int_R\int f(x, y)dA = \int_a^b\int_c^d f(x, y)dy\,dx.$$

Expand the program in Problem 5 so that the expanded program can calculate $\int_R\int S(x, y)dA$ where $S(x, y)$ is the bicubic spline interpolator for $f(x, y)$. Test your routine by estimating the integral of $f(x, y) = 1/(1 + xy)$ where R is defined by $0 \le x \le 1$, $0 \le y \le 1$. Use a grid that is equally spaced in x and y and that has 21 knots in each direction. Compare your result with the exact value of the integral.

5.3 ORTHOGONAL POLYNOMIALS AND LEAST-SQUARES APPROXIMATIONS

Given a function $f(x)$, all of the approximations for $f(x)$ that we have discussed thus far have been interpolatory [in that they matched $f(x)$ and/or its derivatives at a predetermined set of points]. We now consider another type of method to approximate $f(x)$: the least-squares or Fourier approach. We shall restrict ourselves to approximation by polynomials (algebraic or trigonometric) although the theory extends to more general approximations. Two simple but important problems serve to illustrate our objective.

Problem A: Let $w_0, w_1, \ldots, w_m$ be a set of positive constants (weights). Given data points $(x_0, y_0), (x_1, y_1), \ldots, (x_m, y_m)$, $m > n$, find $p^*(x) \in \mathcal{P}_n$ such that $\sum_{i=0}^m w_i[p^*(x_i) - y_i]^2$ is minimized.

Problem B: Let $w(x)$ be a function that is continuous and positive on (a, b); find $p^*(x) \in \mathcal{P}_n$ such that $\int_a^b w(x)[p^*(x) - f(x)]^2\,dx$ is minimized.

Perhaps the most important feature about these two problems is that they are "easy" to solve. In both cases, a unique solution exists, and, moreover, the solution can be computed in a finite number of steps by a sequence of formulas, once the underlying mathematical theory is understood. By contrast, the problem of finding $p^*(x) \in \mathcal{P}_n$ to minimize $\|p^* - f\|_\infty$ [see (5.1c)] is not usually solvable explicitly in terms of formulas. [Although $p^*(x)$ can be approximated by various iterative procedures, we cannot normally hope to find $p^*(x)$ precisely.] To give an idea of the sorts of approximations the method of least-squares generates, we present an example.

EXAMPLE 5.13. In this example, we consider Problem A with $m = 10$ and $w_i = 1$, $i = 0, 1, \ldots, 10$. Using the data of Example 5.11, we can find polynomials $p(x) \in \mathcal{P}_3$ and $q(x) \in \mathcal{P}_5$ to minimize the summation for Problem A. The procedure we use in this example is the procedure (outlined in Section 2.5) using the matrix equation (2.74).

Solving (2.74) [in a least-squares sense, using (2.73)], we find that

$$p(x) = 0.010685x^3 - 0.202239x^2 + 0.939874x + 2.34959$$

$$q(x) = -0.002162x^5 + 0.056869x^4 - 0.517548x^3 + 1.82337x^2 - 1.79051x + 2.80705$$

and

$$\sum_{i=0}^{10} [p(x_i) - y_i]^2 = 5.78, \qquad \sum_{i=0}^{10} [q(x_i) - y_i]^2 = 3.08.$$

The graphs of these two approximations are given in Fig. 5.8.

A more realistic example of data fitting is provided in Example 4.0, in which the setting is that of finding an approximate aerodynamic model for a ballistic reentry trajectory. In the derivation of the Allen and Eggers reentry model in Example 4.0, one assumption made was that atmospheric density, $\rho(H)$, was given by $\rho(H) = \rho_o e^{-\lambda H}$. In reality, atmospheric density is only approximately exponential; so we ask for constants ρ_0 and λ that minimize

$$\sum_{i=1}^{N} [\rho(H_i) - \rho_0 e^{-\lambda H_i}]^2$$

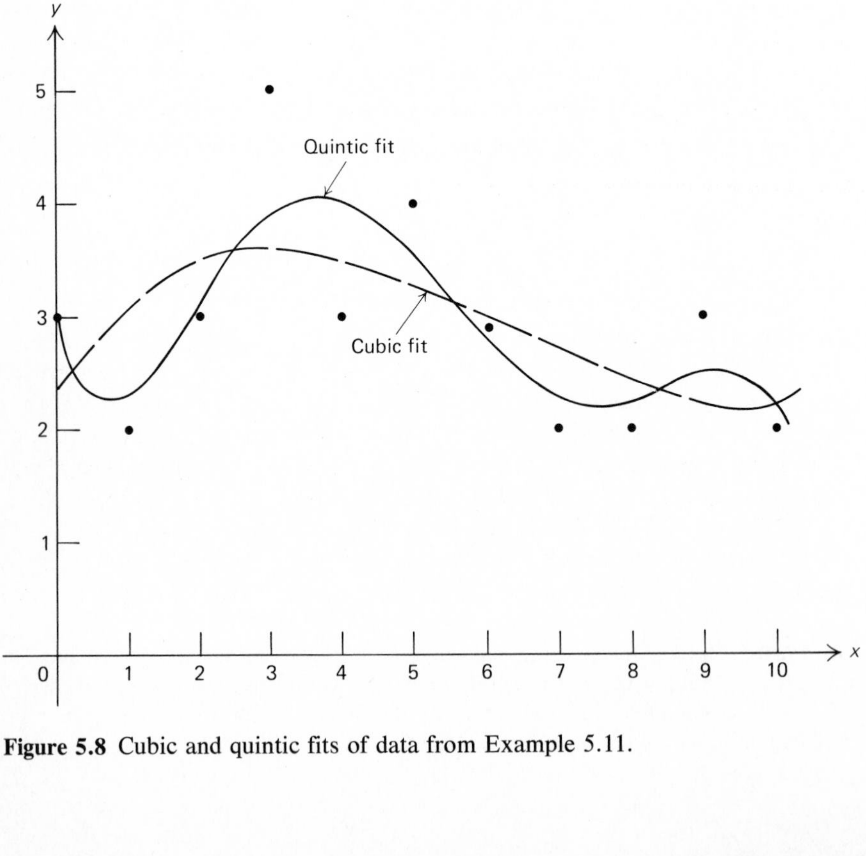

Figure 5.8 Cubic and quintic fits of data from Example 5.11.

where the values $\rho(H_i)$ denote the actual atmospheric density at selected altitudes, H_i. Solving the problem above gives a reasonable density model for the Allen and Eggers reentry model. The least-squares problem above is an example of a nonlinear least-squares problem since the parameters ρ_0 and λ occur in a nonlinear fashion [unlike the coefficients a_i of the polynomial $p^*(x)$ in Problem A or Problem B]. Usually such nonlinear problems are quite difficult, but in this case we can convert the problem to one in which the parameters occur linearly by fitting $\ln(\rho_0 e^{-\lambda H_i})$ to the values $\ln(\rho(H_i))$. This conversion gives a minimization problem of the form

$$\sum_{i=1}^{N} [y_i - (a - \lambda H_i)]^2$$

where $y_i = \ln(\rho(H_i))$ and $a = \ln(\rho_0)$. Thus we have reduced the problem to that of fitting a straight line to N data values $y_1, \ldots, y_N$. The resulting problem can be solved; and the approximation to $\rho(H)$ is $\rho_0 e^{-\lambda H}$, which has a relative error of 25 percent or less over an altitude range of 0 to 200,000 feet.

Although it serves to solve the problem posed in Example 5.13, the method of Section 2.5 is not the most efficient method computationally. In addition, the matrix equation (2.73), which was used to generate $p(x)$ and $q(x)$, can frequently be ill-conditioned. Finally, the method of Chapter 2 cannot be used to solve Problem B. For all of these reasons, many modern computational procedures for finding the best least-squares approximation are based on the ideas of *inner products* and *orthogonal polynomials*. Furthermore these concepts are also fundamental to other numerical procedures. These powerful tools are relatively easy to understand and we present the basics in this section.

For any $f(x)$ and $g(x)$ in $C[a, b]$, we define

$$\langle f, g \rangle_1 = \sum_{j=0}^{m} w_j f(x_j) g(x_j) \tag{5.56a}$$

$$\langle f, g \rangle_2 = \int_a^b w(x) f(x) g(x)\, dx. \tag{5.56b}$$

The reader can easily verify that for any $f(x)$, $g(x)$, and $h(x)$ in $C[a, b]$ and for any real number α, the following properties for (5.56b) are valid:

$$\langle f, f \rangle_2 \geq 0 \text{ and } \langle f, f \rangle_2 = 0 \text{ if and only if}$$
$$f(x) = \theta(x) \text{ where } \theta(x) \text{ is the zero function} \tag{5.57a}$$

$$\langle f, g \rangle_2 = \langle g, f \rangle_2 \tag{5.57b}$$

$$\langle \alpha f, g \rangle_2 = \alpha \langle f, g \rangle_2 \tag{5.57c}$$

$$\langle f, g + h \rangle_2 = \langle f, g \rangle_2 + \langle f, h \rangle_2. \tag{5.57d}$$

The reader can also easily see that properties (b), (c), and (d) are valid also for $\langle f, g\rangle_1$ given by (5.56a) and that $\langle f, f\rangle_1 \geq 0$. However, there are innumerable nonzero continuous functions, $f(x)$, such that $f(x_j) = 0, 0 \leq j \leq m$. For any such function, we have $\langle f, f\rangle_1 = 0$ but $f(x) \neq \theta(x)$. However, if we consider only those functions $f(x)$ that are polynomials of degree m or less, then it is not possible to have $f(x_j) = 0, 0 \leq j \leq m$. Thus for all $f(x) \in \mathcal{P}_m$, $\langle f, f\rangle_1 = 0$ if and only if $f(x) = \theta(x)$. So the four properties of (5.57) are satisfied by (5.56a) if the functions are restricted to be in $\mathcal{P}_m$. We note that the four properties of (5.57) are satisfied also for all vectors in R^n (that is, all n-tuples) where $\langle \mathbf{x}, \mathbf{y}\rangle$ is defined to be the usual inner product: $\langle \mathbf{x}, \mathbf{y}\rangle \equiv \mathbf{x}^T\mathbf{y}$ for $\mathbf{x}$ and $\mathbf{y}$ in R^n (recall 2.75, Chapter 2). We call both (5.56a and b) *inner products*. [Equation (5.56a) is often called the *discrete inner product*.] In the material to follow we will use $\langle f, g\rangle$ to denote both $\langle f, g\rangle_1$ and $\langle f, g\rangle_2$ when we make statements that are valid for both.

We first notice that

$$\langle f - p, f - p\rangle_2^{1/2} = \left(\int_a^b (f(x) - p(x))^2 w(x)\, dx\right)^{1/2} \equiv \|f - p\|_2, \quad (5.58a)$$

which is the norm given by (5.1b). This norm can be used as a way of measuring "distance" between $f(x)$ and $p(x)$ as was discussed in Section 1, and suggests that we can get a different distance measurement between $f(x)$ and $p(x)$ if we define $\|f - p\|_d$ by

$$\langle f - p, f - p\rangle_1^{1/2} = \left(\sum_{j=0}^{m} (f(x_j) - p(x_j))^2 w_j\right)^{1/2} \equiv \|f - p\|_d. \quad (5.58b)$$

(We use the subscript "d" to denote the discrete norm.) Now we can show that $\|f\|_d \equiv \langle f, f\rangle_1^{1/2}$ satisfies the properties of a norm [formula (2.37), Section 2.3.1] *except* that it is possible to have a continuous function $f(x)$ such that $\|f\|_d = 0$ but $f(x) \neq \theta(x)$. [Again this possibility is eliminated if we restrict $f(x)$ to be in $\mathcal{P}_m$.] Both of these norms, or distance measurements, are important in practical computational approximation problems because we are able to develop an efficient technique for finding polynomials $p^*(x)$ and $p_d^*(x)$ in $\mathcal{P}_n$ such that $\|f - p^*\|_2 \leq \|f - p\|_2$ and $\|f - p_d^*\|_d \leq \|f - p\|_d$ for all $p(x) \in \mathcal{P}_n$; that is, we can find the *best* nth-degree polynomial approximation to $f(x)$. [We note that it is usually the case that $p^*(x) \neq p_d^*(x)$.]

Both of these inner products satisfy the Cauchy-Schwarz inequality given by Theorem 5.8.

Theorem 5.8. Cauchy-Schwarz

Let $f(x)$ and $g(x) \in C[a, b]$; then

$$|\langle f, g\rangle| \leq (\langle f, f\rangle)^{1/2}(\langle g, g\rangle)^{1/2}. \quad (5.59)$$

Proof. For either inner product, if $\langle g, g\rangle = 0$, then $\langle f, g\rangle = 0$, and the theorem is trivially true. Let us assume, then, that $\langle g, g\rangle \neq 0$. Then by (5.57), for *any* scalar α,

$$0 \leq \langle f - \alpha g, f - \alpha g\rangle = \langle f, f\rangle - 2\alpha\langle f, g\rangle + \alpha^2\langle g, g\rangle.$$

We use the particular value $\alpha = \langle f, g\rangle/\langle g, g\rangle$, and this expression yields

$$0 \leq \langle f, f\rangle - \langle f, g\rangle^2/\langle g, g\rangle.$$

Transferring $\langle f, g\rangle^2/\langle g, g\rangle$ to the left-hand side and taking square roots yield the desired result. ■

Letting $\|\cdot\|$ denote either $\|\cdot\|_2$ or $\|\cdot\|_d$, we can use the Cauchy-Schwarz inequality to prove that the triangle inequality, $\|f + g\| \leq \|f\| + \|g\|$, holds for both.

Corollary.

Given $f(x)$ and $g(x) \in C[a, b]$ and $\|\cdot\|$ as above, then

$$\|f + g\| \leq \|f\| + \|g\|.$$

Proof.
$$\begin{aligned}\|f + g\|^2 &= \langle f + g, f + g\rangle = \langle f, f\rangle + 2\langle f, g\rangle + \langle g, g\rangle \\ &\leq \|f\|^2 + 2\|f\|\,\|g\| + \|g\|^2 = (\|f\| + \|g\|)^2.\end{aligned}$$
■

To develop our least-squares approximation technique, we must introduce the concepts of orthogonality and orthonormality. Since we are primarily interested in polynomial approximation, we shall give these definitions only for polynomials in $\mathcal{P}_m$ although the generalizations to a broader class of functions are obvious. If $p(x)$ and $q(x)$ are in $\mathcal{P}_m$, we say that $p(x)$ and $q(x)$ are *orthogonal* if $\langle p, q\rangle = 0$. Furthermore, if $\{p_j(x)\}_{j=0}^n$ is a set of polynomials in $\mathcal{P}_m$, we say that this set is *orthogonal* if $\langle p_i, p_j\rangle = 0$ for all $i \neq j$. (Thus we see that the concept of orthogonality in $\mathcal{P}_m$ is a direct generalization of the familiar concept of perpendicularity in R^n.) We say that the set is *orthonormal* if $\langle p_i, p_j\rangle = \delta_{ij}$. Note that if $p(x) \in \mathcal{P}_m$ and if $q(x) = p(x)/\|p\|$, then $\langle q, q\rangle = \langle p, p\rangle/\|p\|^2 = 1$. Thus to make an orthogonal set orthonormal, all we need do is to divide each element in the set by its norm. Note also that a set can be orthogonal with respect to one inner product and not to another.

We see by Problem 4 that the set of polynomials, $S = \{1, x, x^2, \ldots, x^n\}$, $n \geq 2$, can never be orthogonal with respect to either of the inner products (5.56a or b), no matter how the weighting function or the weights are chosen. Yet any polynomial, $p(x) \in \mathcal{P}_n$, can be written as a linear combination of the elements of S; that is, there exist constants, $\{a_i\}_{i=0}^n$ such that $p(x) = a_0 + a_1x + \cdots + a_nx^n$. Given an inner product we would like to be able to express any $p(x) \in \mathcal{P}_n$ as a linear combination of orthogonal polynomials. The first step is to transform S into a different set of polynomials that is orthogonal. This transformation can be done via the Gram-Schmidt algorithm given by the following theorem. (See also Problem 4, Section 3.4.)

Theorem 5.9. Gram-Schmidt

Given $S = \{1, x, x^2, \ldots, x^n\}$, let

$$q_0(x) \equiv 1 \qquad \text{and} \qquad p_0(x) \equiv q_0(x)/\|q_0\|;$$

and let

$$q_k(x) = x^k - \sum_{j=0}^{k-1} \langle x^k, p_j\rangle p_j(x)$$

and

$$p_k(x) = q_k(x)/\|q_k\|, \qquad \text{for} \qquad 1 \le k \le n.$$

Then $\|q_k\| > 0$ for each k, the set $\{q_k(x)\}_{k=0}^n$ is orthogonal, and the set $\{p_k(x)\}_{k=0}^n$ is orthonormal.

Proof. We use induction and assume that the set $\{q_k(x)\}_{k=0}^{r-1}$ is orthogonal and that $\|q_k\| > 0$ where $r \le n$. Note that for $j \le r-1, j \neq k, \langle q_k, p_j\rangle = \langle q_k, q_j/\|q_j\|\rangle = \langle q_k, q_j\rangle/\|q_j\| = 0$ by the induction hypotheses. Now, with $k \le r-1$, and with $p_k(x) = q_k(x)/\langle q_k, q_k\rangle^{1/2}$

$$\begin{aligned}\langle q_k, q_r\rangle &= \langle q_k, x^r\rangle - \sum_{j=0}^{r-1} \langle x^r, p_j\rangle\langle p_j, q_k\rangle \\ &= \langle q_k, x^r\rangle - \langle x^r, q_k\rangle\langle q_k, q_k\rangle/\langle q_k, q_k\rangle = 0.\end{aligned}$$

Since $q_r(x)$ is a monic polynomial of degree r, $\|q_r\| > 0$, completing the proof. ■

EXAMPLE 5.14. In this example, we illustrate the Gram-Schmidt process for each of the inner products of (5.56). Let $x_0 = -1, x_1 = -1/2, x_2 = 0, x_3 = 1/2, x_4 = 1$; and find $p_0(x), p_1(x), p_2(x)$ for the inner product (5.56a) with $w_i = 1, 0 \le i \le 4$. First, we set $q_0(x) = 1$ and find

$$\langle q_0, q_0\rangle = \sum_{i=0}^{4} q_0(x_i)q_0(x_i) = 5;$$

so $p_0(x) = 1/\sqrt{5}$. Next, $q_1(x) = x - \langle x, p_0\rangle p_0(x)$; but

$$\langle x, p_0\rangle = \frac{1}{\sqrt{5}}\sum_{i=0}^{4} x_i = 0;$$

so $q_1(x) = x$. Since $\langle q_1, q_1\rangle = 5/2$, we have $p_1(x) = \sqrt{2/5}x$. Finally, $q_2(x) = x^2 - \langle x^2, p_0\rangle p_0(x) - \langle x^2, p_1\rangle p_1(x)$ where we see that $\langle x^2, p_0\rangle = 5/(2\sqrt{5})$ and $\langle x^2, p_1\rangle = 0$. Thus $q_2(x) = x^2 - 1/2$ and $p_2(x) = \sqrt{8/7}(x^2 - 1/2)$.

Next, let us consider the inner product $\langle f, g\rangle = \int_{-1}^{1} f(x)g(x)\,dx$; and let $S = \{1, x, x^2\}$. In this case, the Gram-Schmidt process of Theorem 5.9 generates the orthonormal polynomials $p_0(x) = 1/\sqrt{2}$, $p_1(x) = \sqrt{3/2}x$, and $p_2(x) = \sqrt{45/8}(x^2 - 1/3)$. For any integer n, we could continue the Gram-Schmidt process and find the orthonormal set $\{p_0(x), p_1(x), \ldots, p_n(x)\}$. The polynomials in this set are known as the *Legendre* polynomials and are widely used for approximation and for numerical integration. The Gram-Schmidt process becomes cumbersome for large values of n, and in the next section we will derive a more efficient procedure for generating orthonormal polynomials.

Theorem 5.9 presents a simplified version of the Gram-Schmidt theorem, but it goes beyond the purposes of our text to introduce the concepts necessary to generalize it. [The reader using the Gram-Schmidt process should be careful not to make the mistake of assuming that $\|1\| = \|q_0\| = 1$. This is usually *not* true. For instance, in Example 5.14,

$$\langle f, g\rangle = \int_{-1}^{1} f(x)g(x)\,dx; \qquad \text{so} \qquad \|1\| = \left(\int_{-1}^{1} 1^2\,dx\right)^{1/2} = \sqrt{2}.$$

Thus in this case, $p_0(x) = q_0(x)/\|q_0\| = 1/\sqrt{2}$.] In this version of Theorem 5.9 we note that each $q_k(x)$ [and thus each $p_k(x)$] has degree exactly equal to k and that $q_k(x)$ has leading coefficient equal to 1; that is, it is monic. We see further that for each $k \geq 1$, x^k can be written as

$$x^k = q_k(x) + \sum_{j=0}^{k-1} \langle x^k, p_j\rangle p_j(x) = \|q_k\| p_k(x) + \sum_{j=0}^{k-1} \langle x^k, p_j\rangle p_j(x).$$

Thus each x^k can be written as a linear combination of the polynomials $\{p_j(x)\}_{j=0}^{k}$. Hence if $p(x) = a_0 + a_1x + a_2x^2 + \cdots + a_rx^r = \sum_{k=0}^{r} a_kx^k$ where $r \leq n$, then each x^k may be replaced by its expression in terms of $\{p_j(x)\}_{j=0}^{k}$; and by collecting like coefficients of each $p_j(x)$, we may write $p(x) = \sum_{j=0}^{r} \beta_j p_j(x)$. Now that we know that any $p(x) \in \mathcal{P}_n$ can be expressed in this manner, it is a simple task to determine the value of each β_j. Let k be fixed, $0 \leq k \leq n$. Then by the orthonormality of the $p_j(x)$'s

$$\langle p_k, p\rangle = \sum_{j=0}^{n} \beta_j\langle p_j, p_k\rangle = \beta_k\langle p_k, p_k\rangle = \beta_k.$$

Therefore for any $p(x) \in \mathcal{P}_n$, we can represent $p(x)$ in terms of $p_0(x), p_1(x), \ldots, p_n(x)$ by

$$p(x) = \sum_{j=0}^{n} \langle p, p_j\rangle p_j(x). \tag{5.60}$$

Now if $p(x) \in \mathcal{P}_r$ where $r < n$, by the argument above we still have $p(x) = \sum_{j=0}^{r} \langle p, p_j\rangle p_j(x)$. Therefore $\langle p_n, p\rangle = \sum_{j=0}^{r} \langle p, p_j\rangle\langle p_n, p_j\rangle = 0$ (since $r < n$), and we have proved Corollary 1.

Corollary 1

If the set $\{p_j(x)\}_{j=0}^n$ is orthonormal as in Theorem 5.9, and if $p(x)$ is *any* polynomial of degree less than k, then $\langle p_i, p\rangle = 0$ for $i \geq k$. ■

We recall that in R^3 there are infinitely many orthonormal sets besides the canonical one, $\{\mathbf{e}_1, \mathbf{e}_2, \mathbf{e}_3\} \equiv \{(1, 0, 0)^T, (0, 1, 0)^T, (0, 0, 1)^T\}$ (for example, any constant rotation of $\{\mathbf{e}_1, \mathbf{e}_2, \mathbf{e}_3\}$). The same is true for $\mathcal{P}_n$. (For example, using the Gram-Schmidt process on S in the order $x^n, x^{n-1}, \ldots, x, 1$ results in an orthonormal set of polynomials, *each* of which has degree n.) However, the orthonormal set $\{p_j(x)\}_{j=0}^n$ of Theorem 5.9 *is* unique in the following respect.

Corollary 2

Let $\{q_j(x)\}_{j=0}^n$ be the orthogonal polynomials generated by the Gram-Schmidt algorithm of Theorem 5.9. Let $\{r_j(x)\}_{j=0}^n$ be any other set of orthogonal polynomials with respect to the given inner product that also satisfies degree $(r_j(x)) = j$ $(=$ degree $(q_j(x)))$, $0 \leq j \leq n$. Then for each j, $r_j(x)$ is a constant multiple of $q_j(x)$. Furthermore, $\pm p_j(x) = \pm q_j(x)/\|q_j\|$, $0 \leq j \leq n$, is the only orthonormal set that satisfies degree $(p_j(x)) = j$.

Proof. Let $k \leq n$ be fixed, and consider the equation

$$q_k(x) = \sum_{j=0}^{k} \alpha_j r_j(x).$$

Then for any i where $0 \leq i \leq k$, the orthogonality of the $r_j(x)$'s yields

$$\langle q_k, r_i\rangle = \sum_{j=0}^{k} \alpha_j \langle r_i, r_j\rangle = \alpha_i \langle r_i, r_i\rangle.$$

By Corollary 1 we have $\langle q_k, r_i\rangle = 0$ for $0 \leq i \leq k-1$, and so the equation above yields $\alpha_0 = \alpha_1 = \cdots = \alpha_{k-1} = 0$ and $\alpha_k = \langle q_k, r_k\rangle / \langle r_k, r_k\rangle$. Thus $q_k(x) = \alpha_k r_k(x)$ for $0 \leq k \leq n$.

Now suppose that $\{\tilde{p}_j(x)\}_{j=0}^n$ is any orthonormal set of polynomials where degree $(\tilde{p}_j(x)) = j$, $0 \leq j \leq n$, and $c_j \neq 0$ is the leading coefficient of each $\tilde{p}_j(x)$. Let $r_j(x) = \tilde{p}_j(x)/c_j$, $0 \leq j \leq n$; then $\{r_j(x)\}_{j=0}^n$ is a monic orthogonal set with degree $(r_j(x)) = j$. By the argument above, $q_j(x) = \alpha_j r_j(x)$, $0 \leq j \leq n$. Thus for each j,

$$p_j(x) = q_j(x)/\|q_j\| = \alpha_j r_j(x)/\|\alpha_j r_j\| = \pm r_j(x)/\|r_j\| = \pm\, \tilde{p}_j(x).$$ ■

Now that we have developed some basic concepts of orthogonal polynomials and how to generate them, we are ready to prove the fundamental theorem of least-squares or Fourier approximation. Before proving the theorem, however, we give a few more examples to illustrate orthogonality in several different situations.

EXAMPLE 5.15.

a) Let $\langle f, g\rangle = \int_{-\pi}^{\pi} f(x)g(x)\,dx$; and consider the set of trigonometric functions, $S_n = \{1, \cos(jx), \sin(jx)\}_{j=1}^{n}$, for any n. Using identities such as

$$\cos(mx)\cos(kx) = 1/2(\cos((m+k)x) + \cos((m-k)x)),$$

the reader can easily verify that S_n is an orthogonal set. Furthermore, $\|1\|^2 = \int_{-\pi}^{\pi} 1^2\,dx = 2\pi$ and $\|\cos(jx)\|^2 = \int_{-\pi}^{\pi} \cos^2(jx)\,dx = \|\sin(jx)\|^2 = \int_{-\pi}^{\pi} \sin^2(jx)\,dx = \pi$, $1 \le j \le n$. Thus $S_n^* = \{1/\sqrt{2\pi}, (\cos(jx))/\sqrt{\pi}, (\sin(jx))/\sqrt{\pi}\}_{j=1}^{n}$ is an orthonormal set. This example illustrates our previous statement that the concept of orthogonality goes far beyond the simple results we have presented here. This example is also historically important in that it was the orthogonal set first used by Fourier when he introduced the concept of orthogonal expansions in 1807.

b) Let $\langle f, g\rangle = \int_{-1}^{1} f(x)g(x)(1-x^2)^{-1/2}\,dx$; and let $S_n = \{T_j(x)\}_{j=0}^{n}$ where $T_j(x) = \cos(j\cos^{-1}(x))$, the jth-degree Chebyshev polynomial. If we make the change of variable, $x = \cos(\theta)$, $-1 \le x \le 1$, and $0 \le \theta \le \pi$, then $T_j(x) = \cos(j\theta)$; and we have

$$\langle T_k(x), T_m(x)\rangle = \int_0^{\pi} \cos(k\theta)\cos(m\theta)\,d\theta = 0$$

for $k \ne m$. Moreover,

$$\|T_0(x)\|^2 = \int_{-1}^{1} 1^2(1-x^2)^{-1/2}\,dx = \pi$$

and

$$\|T_j(x)\|^2 = \int_{-1}^{1} T_j(x)^2(1-x^2)^{-1/2}\,dx = \int_0^{\pi} \cos^2(m\theta)\,d\theta = \frac{\pi}{2}.$$

Thus $S_n^* = \{T_0(x)/\sqrt{\pi}, \sqrt{2/\pi}\,T_j(x)\}_{j=1}^{n}$ is an orthonormal set with respect to this inner product.

c) Let $x_j = \cos(j\pi/n)$, $0 \le j \le n$, and let $\langle f, g\rangle = \sum_{j=0}^{n}{}'' f(x_j)g(x_j)$ where the double prime on the summation means to halve the first and last terms; that is, in Eq. (5.56a), $w_0 = w_n = 1/2$ and $w_j = 1$, $1 \le j \le n-1$. This example is important not only in terms of practical discrete least-squares approximation, but also in the analysis of certain numerical integration methods. We claim that the Chebyshev polynomials, $T_j(x) = \cos(j\cos^{-1}(x))$, $0 \le j \le n$, are orthogonal with respect to this inner product. To verify this claim we need the following lemma, which is useful also in other settings.

Lemma 5.1

Let $S_1(x) = \sum_{j=1}^{n} \cos(jx)$ and $S_2(x) = \sum_{j=0}^{n}{}'' \cos(jx)$. Then

$$S_1(x) = \frac{1}{2}\left(\frac{\sin((n+1/2)x)}{\sin(x/2)} - 1\right) \quad \text{and} \quad S_2(x) = \frac{\sin(nx)\cos(x/2)}{2\sin(x/2)} \tag{5.61}$$

where since $S_1(x)$ and $S_2(x)$ have continuous derivatives, we can use L'Hôpital's rule to evaluate them when $x = k\pi$ for an even integer k.

Proof. Assume $x \neq k\pi$; then by the trigonometric identity,

$$\sin(x/2)\cos(jx) = 1/2(\sin((j + 1/2)x) - \sin((j - 1/2)x)),$$

we can write $S_1(x)$ as the telescoping summation:

$$\begin{aligned} S_1(x) &= (2\sin(x/2))^{-1} \sum_{j=1}^{n} (\sin((j + 1/2)x) - \sin((j - 1/2)x)) \\ &= (2\sin(x/2))^{-1}(\sin((n + 1/2)x) - \sin(x/2)). \end{aligned}$$

To obtain a closed form for the sum $S_2(x)$, note that

$$\begin{aligned} S_2(x) &= \sum_{j=0}^{n}{}'' \cos(jx) = 1/2 + S_1(x) - (\cos(nx))/2 \\ &= 1/2 + \left(\frac{\sin((n + 1/2)x)}{2\sin(x/2)}\right) - 1/2 - (\cos(nx))/2 \\ &= \frac{\sin(nx)\cos(x/2) + \sin(x/2)\cos(nx) - \cos(nx)\sin(x/2)}{2\sin x/2} \\ &= \frac{\sin(nx)\cos(x/2)}{2\sin(x/2)}. \end{aligned}$$

■

Now, continuing Example 5.15(c) with $x = \cos(\theta)$, $T_j(x) = \cos(j\theta)$, $0 \le j \le n$. Let $0 \le k, m \le n$ (and without loss of generality let $m \ge k$); then

$$\begin{aligned} \langle T_m, T_k \rangle &= \sum_{j=0}^{n}{}'' T_m(x_j)T_k(x_j) = \sum_{j=0}^{n}{}'' \cos\frac{mj\pi}{n} \cos\frac{kj\pi}{n} \\ &= 1/2 \sum_{j=0}^{n}{}'' \left(\cos\frac{(m + k)j\pi}{n} + \cos\frac{(m - k)j\pi}{n}\right) \\ &= 1/2(S_2(y) + S_2(z)) \end{aligned}$$

where $y \equiv (m + k)\pi/n$ and $z \equiv (m - k)\pi/n$. Then by (5.61),

$$\begin{aligned} \langle T_m, T_k \rangle &= \frac{\sin[(m + k)\pi]\cos[(m + k)\pi/2n]}{4\sin[(m + k)\pi/2n]} \\ &\quad + \frac{\sin[(m - k)\pi]\cos[(m - k)\pi/2n]}{4\sin[(m - k)\pi/2n]}. \end{aligned}$$

Thus since $\sin[(m+k)\pi] = \sin[(m-k)\pi] = 0$, $\langle T_m, T_k\rangle = 0$ unless $\sin[(m+k)\pi/2n]$ and/or $\sin[(m-k)\pi/2n] = 0$. This situation occurs when (i) $m = k = n$, (ii) $m = k = 0$, or (iii) $m = k$, $1 \le m \le n-1$. In cases (i) and (ii), by the definition of $S_2(x)$ we see easily that $S_2(y) = S_2(z) = n$. In case (iii), $S_2(y) = 0$ and $S_2(z) = n$. Putting these equations all together, we have

$$\langle T_m, T_k\rangle = \begin{cases} n, & \text{if } m = k = n \quad \text{or} \quad m = k = 0 \\ n/2, & \text{if } m = k, \quad 1 \le m \le n-1 \\ 0, & \text{if } m \ne k. \end{cases} \tag{5.62}$$

Hence the orthogonality of the Chebyshev polynomials for this discrete inner product is established.

We now prove a classical theorem that will show us how to construct the best least-squares nth-degree polynomial approximation for any continuous function, $f(x)$.

Theorem 5.10. Best Least-squares Approximation.

Let $\langle f, g\rangle$ be the inner product given by either (5.56a) or (b), and let $\|f\| \equiv \langle f, f\rangle^{1/2}$. If $f(x) \in C[a, b]$, then the polynomial, $p_n^*(x) \in \mathcal{P}_n$, which satisfies $\|f - p_n^*\| \le \|f - p\|$ for all $p(x) \in \mathcal{P}_n$, is given by

$$p_n^*(x) = \sum_{j=0}^{n} \langle f, p_j\rangle p_j(x) \tag{5.63}$$

where $\{p_j(x)\}_{j=0}^n$ is the *orthonormal* set of polynomials generated by the Gram-Schmidt theorem. [Among all polynomials in $\mathcal{P}_n$, $p_n^*(x)$ is "closest" to $f(x)$ with respect to the given least-squares norm; that is, formula (5.63) provides the solution to Problem A *and* Problem B.]

Proof. Let $p(x)$ be any polynomial in $\mathcal{P}_n$. Then by (5.60) $p(x)$ may be written as $p(x) = \sum_{j=0}^n \alpha_j p_j(x)$. Since $\langle p_i, p_j\rangle = \delta_{ij}$, we may write

$$\begin{aligned} 0 \le \|f - p\|^2 &= \left\langle f - \sum_{j=0}^n \alpha_j p_j, \quad f - \sum_{i=0}^n \alpha_i p_i \right\rangle \\ &= \langle f, f\rangle - 2\sum_{j=0}^n \alpha_j \langle f, p_j\rangle + \sum_{j=0}^n \alpha_j^2 \\ &= \langle f, f\rangle - \sum_{j=0}^n \langle f, p_j\rangle^2 + \sum_{j=0}^n (\alpha_j^2 - 2\alpha_j\langle f, p_j\rangle + \langle f, p_j\rangle^2) \\ &= \langle f, f\rangle - \sum_{j=0}^n \langle f, p_j\rangle^2 + \sum_{j=0}^n (\alpha_j - \langle f, p_j\rangle)^2. \end{aligned} \tag{5.64}$$

Since the right-hand sum is the only term containing the α_j's, the expression above is minimized when we choose $\alpha_j = \langle f, p_j \rangle$, $0 \le j \le n$. Thus $\|f - p\|$ is minimum when

$$p(x) = \sum_{j=0}^{n} \langle f, p_j \rangle p_j(x) = p_n^*(x).$$

[The coefficients, $\langle f, p_j \rangle$, of (5.63) are called *generalized Fourier coefficients*.] ∎

Therefore we can find a best least-squares polynomial approximation to $f(x)$ with respect to either the discrete norm or the integral norm by the following steps. (1) Generate the orthonormal set $\{p_j(x)\}_{j=0}^{n}$ by the Gram-Schmidt algorithm. (2) Calculate $\langle f, p_j \rangle$ for $0 \le j \le n$. (3) Construct $p_n^*(x)$ by formula (5.63). We could give examples of this procedure here, but we defer them to the next section in which we develop a more efficient technique. We reiterate here that $p_n^*(x)$ will be different for different inner products. It is up to the reader to decide how best to choose the weighting function of (5.56b) or the weights of (5.56a) to get a best fit for the particular problem. For example, suppose that $\{f(x_j)\}_{j=0}^{m}$ represent the results of some physical experiment, and suppose that the reader has reason to believe that $f(x_0)$, $f(x_1)$, $f(x_{m-1})$, and $f(x_m)$ are less reliable than the others. Then in using (5.56a) choose w_0, w_1, w_{m-1}, and w_m to be smaller than the other weights in order to place more emphasis on the more reliable results.

Approximation in a least-squares sense also has many "geometric" properties. For instance, suppose that Π is a plane in Euclidean three-dimensional space and P is a point that does not lie in Π. If P^* is the point in Π closest to P, then the vector from P^* to P is perpendicular to all vectors in Π. The following corollary shows that this geometric concept is valid for the Fourier approximation, $p_n^*(x)$ in (5.63), as well.

Corollary 1

Under the hypotheses of Theorem 5.10, the remainder function, $(f(x) - p_n^*(x))$, is orthogonal to every polynomial in $\mathcal{P}_n$.

Proof. Since any $p(x)$ in $\mathcal{P}_n$ can be written as $p(x) = \sum_{j=0}^{n} \alpha_j p_j(x)$ where $\{p_j(x)\}_{j=0}^{n}$ is orthonormal as above, then our result is valid if $(f(x) - p_n^*(x))$ is orthogonal to each $p_k(x)$, $0 \le k \le n$. Now,

$$\langle f - p_n^*, p_k \rangle = \langle f, p_k \rangle - \sum_{j=0}^{n} \langle f, p_j \rangle \langle p_j, p_k \rangle = \langle f, p_k \rangle - \langle f, p_k \rangle = 0. \qquad ∎$$

Another corollary of Theorem 5.10 is the following classical result, which is immediate from formula (5.64).

Corollary 2 Bessel's Inequality

Under the hypotheses of Theorem 5.10, for any $f \in C[a, b]$,

$$\|f\|^2 \geq \sum_{j=0}^{n} \langle f, p_j \rangle^2.$$

We shall see in Section 5.3.2 that Theorem 5.10 and its two corollaries lead to some important practical results that can be used to estimate errors and to accelerate convergence. We conclude this section by again emphasizing that the results presented here are only very special cases of the theory of orthogonal expansions, which is fundamental to many areas of mathematical sciences. For example, discrete least-squares estimation is a primary tool in the statistical analysis of data. As another instance, the integral least-squares expansion is often used in the solution of differential equations that model a physical phenomenon. For example, let $L(y)$ be the differential operator $L(y) = (1 - x^2)y''(x) - xy'(x)$, $-1 \leq x \leq 1$; and suppose we wish to solve the differential equation $L(y) = h(x)$ for some function $h(x)$. The reader can easily verify that $L(T_n) = -n^2T_n(x)$ where $T_n(x)$ is the nth degree Chebyshev polynomial, $n = 0, 1, 2, \ldots$; that is, $T_n(x)$ satisfies $(1 - x^2)T_n''(x) - xT_n'(x) + n^2T_n(x) = 0$. Suppose that $h(x)$ may be expressed as $h(x) = \Sigma_{j=0}^{\infty} c_jT_j(x)$ where $c_j = \langle h, T_j \rangle / \langle T_j, T_j \rangle$ and where the inner product is as in Example 5.15(b). If the solution $y(x)$ can be written $y(x) = \Sigma_{j=0}^{\infty} \alpha_jT_j(x)$ and if the operation $L(y) = \Sigma_{j=0}^{\infty} \alpha_jL(T_j)$ is valid, then $L(y) = h$ becomes

$$L(y) = \sum_{j=0}^{\infty} \alpha_jL(T_j) = \sum_{j=0}^{\infty} \alpha_j(-j^2T_j(x)) = h(x) = \sum_{j=0}^{\infty} c_jT_j(x).$$

Therefore, $\alpha_j = -c_j/j^2$, $j = 1, 2, 3, \ldots$, which determines the solution $y(x)$ in terms of the Fourier coefficients of $h(x)$ [note that $L(T_0) = 0$]. This type of problem is very common in the solution of partial differential equations by a technique known as separation of variables. The solution $y(x)$ above is not the complete solution of $L(y) = h$, but we shall not finish the solution since our intention here is only to illustrate the utility and importance of integral least-squares approximations.

5.3.1. Efficient Computation of Least-squares Approximations

We have seen by Theorem 5.10 and formula (5.63) that in order to find the best least-squares approximation in $\mathcal{P}_n$ for $f(x)$, we must calculate an orthonormal set, $\{p_j(x)\}_{j=0}^{n}$, and then calculate the Fourier coefficients, $\{\langle f, p_j \rangle\}_{j=0}^{n}$. For $p_n^*(x)$ to be a useful approximation, we should also be able to evaluate $p_n^*(x)$ easily for any value of x. We shall now develop efficient and practical computational procedures for each of these steps.

We have seen in Example 5.14 that the Gram-Schmidt process can become quite cumbersome even for relatively small values of n. For example, if we wish to find $p_{10}(x)$, not only must we first find $p_j(x)$, $0 \le j \le 9$, but each $p_j(x)$ is actually explicitly used in the computation of $p_{10}(x)$. Thus we first seek an efficient alternative to the Gram-Schmidt process. The key to finding this alternative is to note that the orthogonal polynomials $\{q_j(x)\}_{j=0}^{n}$ of Theorem 5.9 are monic (leading coefficient is 1) and that for either inner product, (5.56a or b), it is true that $\langle xf(x), g(x)\rangle = \langle f(x), xg(x)\rangle$ for any functions $f(x)$ and $g(x)$. This observation will enable us to write $q_k(x)$ in terms of $q_{k-1}(x)$ and $q_{k-2}(x)$. Let k be any positive integer such that $2 \le k \le n$. Since $q_k(x)$ is monic,

$$\begin{aligned} q_k(x) &= x^k + \alpha_{k-1}x^{k-1} + \alpha_{k-2}x^{k-2} + \cdots + \alpha_1 x + \alpha_0 \\ &= x(x^{k-1} + \alpha_{k-1}x^{k-2} + \cdots + \alpha_1) + \alpha_0 \equiv xr_{k-1}(x) + \alpha_0. \end{aligned}$$

Since $r_{k-1}(x)$ and $q_{k-1}(x)$ are monic in $\mathcal{P}_{k-1}$, we see $r_{k-1}(x) - q_{k-1}(x) \equiv u_{k-2}(x)$ is a polynomial of degree $k-2$. By (5.60), $(xu_{k-2}(x) + \alpha_0)$ can be written as

$$\sum_{j=0}^{k-1} \beta_j q_j(x);$$

so

$$\begin{aligned} q_k(x) &= xr_{k-1}(x) + \alpha_0 = x(q_{k-1}(x) + u_{k-2}(x)) + \alpha_0 \\ &= xq_{k-1}(x) + \sum_{j=0}^{k-1} \beta_j q_j(x) = (x - \beta_{k-1})q_{k-1}(x) + \sum_{j=0}^{k-2} \beta_j q_j(x). \end{aligned}$$

For $i < k-2$,

$$\begin{aligned} 0 = \langle q_k, q_i\rangle &= \left\langle (x - \beta_{k-1})q_{k-1}(x) + \sum_{j=0}^{k-2} \beta_j q_j(x), q_i(x)\right\rangle \\ &= \langle xq_{k-1}, q_i\rangle + \beta_i\langle q_i, q_i\rangle. \end{aligned}$$

But $\langle xq_{k-1}, q_i\rangle = \langle q_{k-1}, xq_i\rangle = 0$ by Corollary 1, Theorem 5.9; and hence $\beta_i = 0$ for $i < k-2$. Thus we have just demonstrated for $k \ge 2$ that $q_k(x)$ can be written in the form

$$q_k(x) = (x - a_k)q_{k-1}(x) - b_k q_{k-2}(x), \qquad k \ge 2. \tag{5.65}$$

Therefore, in order to generate $q_k(x)$, all we need use is $q_{k-1}(x)$ and $q_{k-2}(x)$, a considerable savings of computation over the Gram-Schmidt algorithm. Formula (5.65) is called a *three-term recurrence relation* for orthogonal polynomials. Now all we need to do is find the formulas for a_k and b_k in (5.65). They are quite easily found as follows:

$$0 = \langle q_k, q_{k-1}\rangle = \langle xq_{k-1}, q_{k-1}\rangle - a_k\langle q_{k-1}, q_{k-1}\rangle + 0$$

and

$$\begin{aligned} 0 = \langle q_k, q_{k-2}\rangle &= \langle xq_{k-1}, q_{k-2}\rangle - a_k\langle q_{k-1}, q_{k-2}\rangle - b_k\langle q_{k-2}, q_{k-2}\rangle \\ &= \langle q_{k-1}, xq_{k-2}\rangle - 0 - b_k\langle q_{k-2}, q_{k-2}\rangle. \end{aligned}$$

Thus for $k \geq 2$, we have that $a_k = \langle xq_{k-1}, q_{k-1}\rangle/\langle q_{k-1}, q_{k-1}\rangle$ and $b_k = \langle q_{k-1}, xq_{k-2}\rangle/\langle q_{k-2}, q_{k-2}\rangle$ in Eq. (5.65). (We leave to the reader to verify that $\langle q_{k-1}, xq_{k-2}\rangle = \langle q_{k-1}, q_{k-1}\rangle$, and so $b_k = \|q_{k-1}\|^2/\|q_{k-2}\|^2$.) The reader should also verify from the Gram-Schmidt algorithm that (5.65) is valid for $k = 1$ if we set $a_1 = \langle xq_0, q_0\rangle/\langle q_0, q_0\rangle$, and $b_1 = 0$. We wish to make clear that to obtain $q_{10}(x)$, for example, we still need to know $q_j(x)$, $0 \leq j \leq 9$. But in the computation of $q_{10}(x)$, needing to use only $q_8(x)$ and $q_9(x)$ is the advantage of (5.65) over the Gram-Schmidt process. For many of the widely used sets of orthogonal polynomials, the constants a_k and b_k appearing in (5.65) have been tabulated. These tables further facilitate the use of (5.65) in least-squares approximation.

Efficient computation of $\langle f, p_j\rangle$, $0 \leq j \leq n$, we will defer to the next chapter. (See especially the sections on the Fast Fourier Transform and Gaussian quadrature and the subsequent development of interpolation at the zeros of orthogonal polynomials.) We merely note here that the computation of $\langle f, p_k\rangle$ can often be facilitated by the use of (5.65). Because $p_k(x) = q_k(x)/\|q_k\|$, $\langle f, p_k\rangle = \langle f, q_k\rangle/\|q_k\|$; and by (5.65) the term $\langle f, q_k\rangle$ is given by

$$\langle f, q_k\rangle = \langle f, xq_{k-1}\rangle - a_k\langle f, q_{k-1}\rangle - b_k\langle f, q_{k-2}\rangle.$$

The last two terms of this expression can be easily calculated from the previously calculated values of $\langle f, p_{k-1}\rangle$ and $\langle f, p_{k-2}\rangle$, $k \geq 2$. This procedure leaves only the computation of $\langle f, xq_{k-1}\rangle$, which can often be simplified by the use of integration or summation by parts.

We pause to note here that

$$p_n^*(x) = \left(\sum_{j=0}^{n-1} \langle f, p_j\rangle p_j(x)\right) + \langle f, p_n\rangle p_n(x) = p_{n-1}^*(x) + \langle f, p_n\rangle p_n(x).$$

Thus for any $n \geq 1$, the best nth degree approximation may be obtained by adding one term to the best $(n - 1)$st degree approximation. So if we wish to increase our accuracy by going from $n - 1$ to n, the previous work in computing $p_{n-1}^*(x)$ is not wasted. When $p_n^*(x)$ is actually calculated on a computer, rounding errors will occur in the computation of the Fourier coefficients. Hence, if we compute $p_n^*(x)$ from $p_n^*(x) = p_{n-1}^*(x) + \langle f, p_n\rangle p_n(x)$, then $p_n^*(x)$ will inherit the accumulated errors of $p_{n-1}^*(x)$ as well as the error introduced by calculating $\langle f, p_n \rangle$. Computational experience and some theoretical results have led many analysts to suggest [particularly for the inner product (5.56a)] that $p_n^*(x)$ be computed from the formula

$$p_n^*(x) = \sum_{j=0}^{n} \langle f - p_{j-1}^*, p_j \rangle p_j(x)$$

where $p_{j-1}^*(x)$ is the *computed* $(j-1)$st degree approximation to $f(x)$ with $p_{-1}^* = 0$ in the formula above. Since (theoretically) $\langle p_{j-1}^*, p_j \rangle = 0$ by Corollary 1 of Theorem 5.9, this method of finding $p_n^*(x)$ is equivalent to (5.63) under the assumption that no errors are made in calculating $\langle f - p_{j-1}^*, p_j \rangle$.

From (5.63) we know that the best least-squares approximation is given by

$$p_n^*(x) = \sum_{j=0}^{n} \langle f, p_j \rangle p_j(x) = \sum_{j=0}^{n} \frac{\langle f, q_j \rangle}{\langle q_j, q_j \rangle} q_j(x) \equiv \sum_{j=0}^{n} c_j q_j(x).$$

In practice we usually wish to evaluate $p_n^*(x)$ at one or more values, $x = \alpha$. Assuming that c_j and the values a_j and b_j of the three-term recurrence (5.65) are known quantities, we may take advantage of (5.65) and calculate $p_n^*(\alpha)$ by using only $(2n - 1)$ multiplications. The evaluation algorithm proceeds as follows. Set $d_{n+2} = d_{n+1} = 0$; and compute

$$d_k = c_k + (\alpha - a_{k+1})d_{k+1} - b_{k+2}d_{k+2}, \qquad k = n, n-1, \ldots, 1, 0. \qquad (5.66)$$

[Note that although a_{n+1}, b_{n+1}, and b_{n+2} may not be known, they are always multiplied by zero when they occur in (5.66). Thus each d_k in this algorithm is well-defined.] We now claim that $d_0 = p_n^*(\alpha)$. To see that this is true, consider the following, using (5.66) for each c_k:

$$p_n^*(\alpha) = \sum_{k=0}^{n} c_k q_k(\alpha) = \sum_{k=0}^{n} [d_k - (\alpha - a_{k+1})d_{k+1} + b_{k+2}d_{k+2}]q_k(\alpha).$$

Now, collecting like coefficients of each d_k, we obtain,

$$p_n^*(\alpha) = d_0 q_0(\alpha) + d_1[q_1(\alpha) - (\alpha - a_1)q_0(\alpha)]$$
$$+ \sum_{k=2}^{n} d_k[q_k(\alpha) - (\alpha - a_k)q_{k-1}(\alpha) + b_k q_{k-2}(\alpha)].$$

But by (5.65) all the coefficients of d_k, $1 \le k \le n$, are $[q_k(\alpha) - q_k(\alpha)] = 0$. So since $q_0(\alpha) = 1$ for any α, we have $p_n^*(\alpha) = d_0$.

EXAMPLE 5.16. To show how the three-term recurrence relation in (5.65) simplifies the computation of orthogonal polynomials, we calculate the first few Legendre polynomials [i.e., we use the inner product $\langle f, g \rangle = \int_{-1}^{1} f(x)g(x)\,dx$]. As suggested by

(5.65), we let $q_0(x) \equiv 1$ and find $\langle q_0, q_0 \rangle = \int_{-1}^{1} dx = 2$. Next $q_1(x) = (x - a_1)q_0(x)$ where $a_1 = \langle xq_0, q_0 \rangle / \langle q_0, q_0 \rangle$; but $\langle xq_0, q_0 \rangle = \int_{-1}^{1} x\, dx = 0$; so $q_1(x) = x$ and $\langle q_1, q_1 \rangle = \int_{-1}^{1} x^2\, dx = 2/3$. The next step is to find $q_2(x)$ from the formula $q_2(x) = (x - a_2)q_1(x) - b_2 q_0(x)$ where $a_2 = \langle xq_1, q_1 \rangle / \langle q_1, q_1 \rangle$ and $b_2 = \langle q_1, xq_0 \rangle / \langle q_0, q_0 \rangle$. Again $a_2 = 0$ since $\langle xq_1, q_1 \rangle = \int_{-1}^{1} x^3\, dx = 0$; so $q_2(x) = xq_1(x) - b_2 q_0(x)$.

The pattern soon emerges and we suspect that $a_i = 0$ for all i. This pattern is easy to show, for $a_i = \langle xq_{i-1}, q_{i-1} \rangle / \langle q_{i-1}, q_{i-1} \rangle$ where $\langle xq_{i-1}, q_{i-1} \rangle = \int_{-1}^{1} x[q_{i-1}(x)]^2\, dx$. Recalling the elementary result that $\int_{-a}^{a} h(x)\, dx = 0$ whenever $h(x)$ is an odd function, we see $a_i = 0$ since $h(x) = x[q_{i-1}(x)]^2$ is an odd function. Thus, for all i, $q_i(x) = xq_{i-1}(x) - b_i q_{i-2}(x)$. By Problem 8, we note that $b_i = \langle q_{i-1}, q_{i-1} \rangle / \langle q_{i-2}, q_{i-2} \rangle$ for $i \geq 2$; so the computation of the Legendre polynomials is easily systematized as follows:

$$
\begin{array}{lll}
q_0(x) = 1 & \langle q_0, q_0 \rangle = 2 & \\
q_1(x) = x & \langle q_1, q_1 \rangle = \dfrac{2}{3}, & b_2 = \dfrac{1}{3} \\
q_2(x) = xq_1(x) - \dfrac{1}{3} q_0(x) & & \\
\quad = x^2 - \dfrac{1}{3} & \langle q_2, q_2 \rangle = \dfrac{8}{45}, & b_3 = \dfrac{4}{15} \\
q_3(x) = xq_2(x) - \dfrac{4}{15} q_1(x) & & \\
\quad = x^3 - \dfrac{3}{5} x & \langle q_3, q_3 \rangle = \dfrac{8}{175}, & b_4 = \dfrac{9}{35} \\
\vdots & &
\end{array}
$$

Clearly these calculations can be programmed to be carried out on the computer since we can program the calculations of $\int_{-1}^{1} p(x)\, dx$ for any polynomial, $p(x)$. Note that the existence of the three-term recurrence relation was crucial to our development of an efficient and systematic approach for finding the orthogonal polynomials. The simplification given by knowing $a_i = 0$, $i = 1, 2, \ldots$ is not peculiar to Legendre polynomials but holds for a large class of inner products (see Problem 10).

Finally, let us use Theorem 5.10 to find the best least-squares approximation of degree three or less to $f(x) = \cos(\pi x)$. Since the orthonormal polynomials $p_i(x)$ are given by $p_i(x) = q_i(x)/\|q_i\|$, we have

$$p_3^*(x) = \sum_{i=0}^{3} \langle f, p_i \rangle p_i(x) = \sum_{i=0}^{3} \left\langle f, \frac{q_i}{\|q_i\|} \right\rangle \frac{q_i(x)}{\|q_i(x)\|},$$

or

$$p_3^*(x) = \sum_{i=0}^{3} \frac{\langle f, q_i \rangle}{\langle q_i, q_i \rangle} q_i(x).$$

Since we calculated $\langle q_i, q_i \rangle$ in the process of finding the orthogonal polynomials $q_0(x)$, . . . , $q_3(x)$, we need only the numbers

$$\langle f, q_i \rangle = \int_{-1}^{1} \cos(\pi x) q_i(x)\, dx, \qquad i = 0, 1, 2, 3.$$

Since $\cos(\pi x)$ is an even function and $q_i(x)$ is an odd function for $i = 1$ and $i = 3$, we see immediately that $\langle f, q_1 \rangle = \langle f, q_3 \rangle = 0$. Next, we find $\langle f, q_0 \rangle = 0$; and integration by parts shows that

$$\langle f, q_2 \rangle = \int_{-1}^{1} \cos(\pi x) \left(x^2 - \frac{1}{3} \right) dx = \frac{-4}{\pi^2}.$$

Thus

$$p_3^*(x) = \frac{-45}{2\pi^2} \left(x^2 - \frac{1}{3} \right).$$

PROBLEMS

1. Write a computer program to solve Problem A, Section 3, where the input is the following: m; weights $w_0, \ldots, w_m$; data points $(x_0, y_0), \ldots, (x_m, y_m)$; and the desired degree of approximation, n. Generate the orthogonal polynomials $q_i(x)$ according to (5.65) and find $p_n^*(x)$ by

$$p_n^*(x) = \sum_{j=0}^{n} \frac{\langle f, q_j \rangle}{\langle q_j, q_j \rangle} q_j(x).$$

Evaluate $p_n^*(x)$ according to the recurrence (5.66). Test your program with the data in Example 5.13. (Also see Problem 10 for some programming simplifications.)

2. Reference to Problem A shows that for $n = m$, $p_n^*(x)$ is the interpolating polynomial. Thus Theorem 5.10 provides another way to generate interpolating polynomials. Verify this statement directly for the discrete inner product of Example 5.14; generate the orthogonal polynomials via (5.65); and find $p_4^*(x)$ that interpolates $f(x) = 1/(2 + x)$.
3. Verify (5.57), (b), (c), and (d) for both inner products $\langle f, g \rangle_1$ and $\langle f, g \rangle_2$.
4. Let $\langle f, g \rangle$ be either of the inner products of (5.56). Show that $S = \{1, x, x^2, \ldots, x^n\}$, $n \geq 2$, cannot be an orthogonal set. [Hint: Consider $\langle 1, x^k \rangle$ for $k \geq 2$.]
5. For $p(x) \in \mathcal{P}_n$, we can write $p(x) = \sum_{j=0}^{n} \alpha_j p_j(x)$ by (5.60). Show that the coefficients α_j are unique by supposing that $p(x) = \sum_{j=0}^{n} \beta_j p_j(x)$ is another expansion. [Hint: What is $\langle \theta, p_i \rangle$?]
6. Consider the inner product $\langle f, g \rangle = \int_0^\infty e^{-x} f(x) g(x)\, dx$. Generate the first four monic polynomials that are orthogonal with respect to this inner product (these are called the *Laguerre polynomials*).

7. For $F(x) = \sqrt{1 - x^2}$, find $p_n^*(x)$ when

$$\langle f, g \rangle = \int_{-1}^{1} \frac{f(x)g(x)}{\sqrt{1 - x^2}}\,dx$$

[recall Example 5.15(b)].

8. Show that b_k in (5.65) is given by $b_k = \langle q_{k-1}, q_{k-1}\rangle / \langle q_{k-2}, q_{k-2}\rangle$. [Hint: Show by (5.60) that

$$xq_{k-2}(x) = \sum_{i=0}^{k-1} \alpha_i q_i(x).]$$

9. Write a computer program to do an nth degree discrete least-squares fit for $f(x) = 1/(2 + x)$ at $x_0, x_1, \ldots, x_m$; use the inner product of Example 5.15(c) and (5.62). For $m = 20$ and $n = 10$, evaluate the approximation and print the errors for $x = -1 + ih$, $h = 0.01$, $0 \le i \le 200$.

*10. Suppose $w(x)$ is an even function on $[-a, a]$ and $\langle f, g\rangle = \int_{-a}^{a} w(x)f(x)g(x)\,dx$. Show that the three-term recurrence of (5.65) becomes $q_i(x) = xq_{i-1}(x) - b_i q_{i-2}(x)$. Next, let $\langle f, g\rangle = \int_a^b w(x)f(x)g(x)\,dx$, $\alpha = (a + b)/2$ and suppose $w(\alpha - h) = w(\alpha + h)$ for all h such that $0 \le h \le (b - a)/2$. Use the change of variable $x = y + \alpha$ to show that $q_i(x) = (x - \alpha)q_{i-1}(x) - b_i q_{i-2}(x)$. Show that this simplification is also valid for the discrete inner product $\langle f, g\rangle = \sum_{i=0}^{m} w_i f(x_i)g(x_i)$ whenever the x_i are symmetrically spaced about α and the weights are symmetric [that is, $(x_i + x_{m-i})/2 = \alpha$ and $w_i = w_{m-i}$]. Use this simplification in the program in Problem 1 with the data of Example 5.13.

*11. Do an operations count to get an estimate of the relative efficiency of solving Problem A (Section 5.3) by using orthogonal polynomials (as in Problem 1) as against solving Problem A by the matrix method of Chapter 2.

*5.3.2. Error Estimates for Least-squares Approximations

Suppose $\{p_0(x), p_1(x), \ldots, p_k(x), \ldots\}$ are the orthonormal polynomials generated by the Gram-Schmidt theorem with respect to the inner product $\langle f, g\rangle = \int_a^b f(x)g(x)w(x)\,dx$; and for each k let

$$p_k^*(x) = \sum_{j=0}^{k} \langle f, p_j\rangle p_j(x)$$

as in (5.63). Then by Theorem 5.10 if $m > k$, $\|f - p_m^*\| \le \|f - p_k^*\|$ since $\mathcal{P}_k \subseteq \mathcal{P}_m$. By the Weierstrass theorem stated in Section 5.1, for any $\varepsilon > 0$ there exists a polynomial of degree N, $q_N(x)$, (where N depends on ε) such that

$$\max_{a \le x \le b} |f(x) - q_N(x)| \equiv \|f - q_N\|_\infty < \varepsilon.$$

By the best least-squares approximation property of Theorem 5.10,

$$\|f - p_N^*\|^2 = \int_a^b (f(x) - p_N^*(x))^2 w(x)\, dx \le \|f - q_N\|^2$$

$$= \int_a^b (f(x) - q_N(x))^2 w(x)\, dx \le \max_{a \le x \le b} |f(x) - q_N(x)|^2 \int_a^b w(x)\, dx$$

$$< \varepsilon^2 \int_a^b w(x)\, dx \equiv \varepsilon'$$

(since $\int_a^b w(x)\, dx$ is a constant). Thus we have $\lim_{k\to\infty} \|f - p_k^*\| = 0$; or the polynomials, $p_k^*(x)$, converge to $f(x)$ in the least-squares norm. By placing relatively mild restrictions on $f(x)$, we can even show uniform convergence; that is,

$$\lim_{k\to\infty} \|f - p_k^*\|_\infty = 0$$

(see Problem 3).

From Bessel's inequality we have that

$$\|f\|^2 = \int_a^b (f(x))^2 w(x)\, dx \ge \sum_{j=0}^{n} \langle f, p_j\rangle^2$$

holds for all n. Since $\|f\|^2$ is a finite number for any $f(x) \in C[a, b]$, the partial sums of the infinite series, $\sum_{j=0}^{\infty} \langle f, p_j\rangle^2$, form a nondecreasing sequence that is bounded from above; and hence the series is convergent. Thus $\lim_{j\to\infty} \langle f, p_j\rangle^2 = 0$ and thus $\lim_{j\to\infty} |\langle f, p_j\rangle| = 0$. From Corollary 1, Theorem 5.10, we see that $\langle f - p_n^*, p_n^*\rangle = 0$ for each n. Using formula (5.63), we see that $\langle f, p_n^*\rangle = \langle f(x), \sum_{j=0}^{n} \langle f, p_j\rangle p_j(x)\rangle = \sum_{j=0}^{n} \langle f, p_j\rangle^2$. From this information we can calculate the least-squares error, $\|f - p_n^*\|$, since

$$\|f - p_n^*\|^2 = \langle f - p_n^*, f - p_n^*\rangle = \langle f - p_n^*, f\rangle - \langle f - p_n^*, p_n^*\rangle$$

$$= \langle f - p_n^*, f\rangle = \langle f, f\rangle - \langle f, p_n^*\rangle = \|f\|^2 - \sum_{j=0}^{n} \langle f, p_j\rangle^2. \quad (5.67)$$

In addition since $\lim_{n\to\infty} \|f - p_n^*\| = 0$, we have by (5.67) that

$$\|f\|^2 = \sum_{j=0}^{\infty} \langle f, p_j\rangle^2 \quad \text{or} \quad \|f - p_n^*\|^2 = \sum_{j=n+1}^{\infty} \langle f, p_j\rangle^2. \quad (5.68)$$

This formula is known as Parseval's Equality and can be used in a very practical way to estimate the error of the nth degree least-squares approximation. We note that the error is explicitly dependent on the size of the Fourier coefficients, $\langle f, p_j\rangle$, $j = n + 1, n + 2, \ldots$.

We now examine one way to estimate the error in (5.68) without having to calculate each $\langle f, p_j\rangle$ for $j \geq n + 1$. Theorem 5.2 tells us that for each k there is a unique polynomial, $\tilde{p}_k(x)$, such that $\max_{a \leq x \leq b} |f(x) - \tilde{p}_k(x)| \equiv \|f - \tilde{p}_k\|_\infty \leq \|f - p\|_\infty$ for all $p(x) \in \mathcal{P}_k$; and again we let $E_k(f) \equiv \|f - \tilde{p}_k\|_\infty$. Many results in the literature give upper bounds for $E_k(f)$. Any such result is usually called a Jackson theorem in honor of Dunham Jackson, who first presented broad significant results of this form. For instance, we already have a simple example of a Jackson theorem by way of formula (5.30). That is to say, for $f(x) \in C^{k+1}[a, b]$ let $q_k(x) \in \mathcal{P}_k$ be the polynomial that interpolates $f(x)$ at the zeros of the shifted Chebyshev polynomial, $\tilde{T}_{k+1}(x)$. Then (5.36) yields

$$E_k(f) \equiv \|f - \tilde{p}_k\|_\infty \leq \|f - q_k\|_\infty \leq \frac{\max_{a \leq x \leq b} |f^{(k+1)}(x)|}{2^k(k+1)!}\left(\frac{b-a}{2}\right)^{k+1}. \quad (5.69a)$$

A stronger Jackson theorem, which is beyond the scope of this text to prove, is (5.69b) [see Cheney (1966), p. 147]. In (5.69b), we let $[a, b] = [-1, 1]$ and suppose that $f^{(m)}(x) \in C[-1, 1]$ with $k \geq m$. Then

$$E_k(f) \leq (\pi/2)^m \|f^{(m)}\|_\infty / [(k+1)(k) \ldots (k - m + 2)]. \quad (5.69b)$$

We now use these bounds for $E_k(f)$ in estimating the error in (5.68). Since $\langle p_k, \tilde{p}_{k-1}\rangle = 0$ by Corollary 1, Theorem 5.10, then for $k \geq n + 1$,

$$\langle f, p_k\rangle = \langle f - \tilde{p}_{k-1}, p_k\rangle = \int_a^b (f(x) - \tilde{p}_{k-1}(x))p_k(x)w(x)\,dx. \quad (5.70)$$

Two ways to bound (5.70) immediately come to mind. In the first we use the familiar results for the absolute values of integrals and obtain

$$\begin{aligned} |\langle f, p_k\rangle| &\leq \int_a^b |f(x) - \tilde{p}_{k-1}(x)|\,|p_k(x)|w(x)\,dx \\ &\leq \max_{a \leq x \leq b} |f(x) - \tilde{p}_{k-1}(x)| \int_a^b |p_k(x)|w(x)\,dx \\ &= E_{k-1}(f) \int_a^b |p_k(x)|w(x)\,dx. \end{aligned} \quad (5.70a)$$

In the second we use Cauchy-Schwarz and the orthonormality of the $p_k(x)$'s, and obtain

$$\begin{aligned} |\langle f, p_k\rangle| &\leq \left(\int_a^b (f(x) - \tilde{p}_{k-1}(x))^2 w(x)\,dx\right)^{1/2} \left(\int_a^b p_k(x)^2 w(x)\,dx\right)^{1/2} \\ &\leq \max_{a \leq x \leq b} |f(x) - \tilde{p}_{k-1}(x)| \left(\int_a^b w(x)\,dx\right)^{1/2} (\|p_k\|^2) \\ &= E_{k-1}(f) \left(\int_a^b w(x)\,dx\right)^{1/2}. \end{aligned} \quad (5.70b)$$

We now give an example of how to use the bounds on the Fourier coefficients and the Jackson theorems to bound the least-squares error, $\|f - p_n^*\|$, in (5.68).

EXAMPLE 5.17. Let $[a, b] = [-1, 1]$ and $\langle f, g\rangle = \int_{-1}^{1} f(x)g(x)(1 - x^2)^{-1/2}\,dx$. Then as in Example 5.15(b), the orthonormal polynomials are $p_k(x) = \sqrt{2/\pi}T_k(x)$ for $k \geq 1$. For (5.70a) we have

$$\int_a^b |p_k(x)|w(x)\,dx = \sqrt{\frac{2}{\pi}}\int_{-1}^{1} |T_k(x)|(1 - x^2)^{-1/2}\,dx = \sqrt{\frac{2}{\pi}}\int_0^{\pi} |\cos(k\theta)|\,d\theta,$$

using $x = \cos(\theta)$. We leave to the reader to verify that $\int_0^{\pi} |\cos(k\theta)|\,d\theta = 2$, and so (5.70a) yields $|\langle f, p_k\rangle| \leq (2\sqrt{2}/\sqrt{\pi})E_{k-1}(f)$. In (5.70b), $\int_a^b w(x)\,dx = \int_{-1}^{1}(1 - x^2)^{-1/2}\,dx = \pi$, and so (5.70b) yields $|\langle f, p_k\rangle| \leq /\sqrt{\pi}E_{k-1}(f)$. In this example $\sqrt{\pi} \geq 2\sqrt{2}/\sqrt{\pi}$; thus (5.70a) yields a better (smaller) bound on $|\langle f, p_k\rangle|$. [Such is usually true for other examples as well. However, $\int_a^b |p_k(x)|w(x)\,dx$ in (5.70a) can be a formidable task to compute, especially when the zeros of $p_k(x)$ are not known precisely, or when they vary irregularly for different values of k.] If we assume that $f(x) \in C^2[-1, 1]$, then by (5.69b) and (5.70a)

$$\langle f, p_k\rangle^2 \leq \frac{8}{\pi}E_{k-1}^2(f) \leq \pi^3\|f''\|_\infty^2/2k^2(k - 1)^2.$$

Thus, by (5.68)

$$\|f - p_n^*\|^2 \leq \frac{\pi^3}{2}\|f''\|_\infty^2 \sum_{k=n+1}^{\infty}(1/k^2(k - 1)^2) \leq \frac{\pi^2\|f''\|_\infty^2}{6(n - 1)^3}$$

where the integral test for infinite series from calculus was used to provide a crude bound for the series above. We note that we would obtain a smaller error bound if $f(x) \in C^m[-1, 1]$ for $m > 2$ by virtue of (5.69b) if we assume that $\|f^{(m)}\|_\infty$ is not "too much larger" than $\|f''\|_\infty$.

PROBLEMS, SECTION 5.3.2

1. Using (5.67) and the expansion found in Problem 7 in the last section, determine $\|f - p_n^*\|^2$ for $n = 2, 4, 6$.
2. Find the Chebyshev series expansion for $f(x) = |x|$, and determine $\|f - p_n^*\|^2$ and $|f(1) - p_n^*(1)|$ for $n = 2, 4, 6$.

*3. a) Suppose that $f \in C^2[-1, 1]$ and suppose that the Chebyshev series expansion for $f(x)$ is

$$\sum_{i=0}^{\infty} \alpha_i p_i(x)$$

where $p_i(x) = \sqrt{2/\pi}T_i(x)$, $i \geq 1$; $p_0(x) = \sqrt{1/\pi}$; and

$$\langle f, g\rangle = \int_{-1}^{1} \frac{f(x)g(x)}{\sqrt{1 - x^2}}\,dx.$$

Show that the Chebyshev series converges uniformly to $f(x)$ for $-1 \le x \le 1$. [Hint: By integrating by parts twice, show that each α_i satisfies an inequality of the form $|\alpha_i| \le M/i^2$. Thus conclude that the series $\Sigma_{i=0}^{\infty} \alpha_i p_i(x)$ converges uniformly to a continuous function $g(x)$. Show next that $\|f - g\| = 0$ by considering $\|f - g\| \le \|f - p_n^*\| + \|p_n^* - g\|$. Thus since f and g are continuous, $f(x) \equiv g(x)$.]

b) For any positive integer n, let $Q_{n-1}(x) \in \mathcal{P}_{n-1}$; then interpolate $f(x)$ at $x_j = \cos[(2j - 1)\pi/2n]$, $1 \le j \le n$ [the zeros of $T_n(x)$]. For any $x \in [-1, 1]$ define $R_{n-1}(f, x) \equiv f(x) - Q_{n-1}(x)$. Let k be any positive integer where $k \ge n$ and write k as $k = 2rn + \alpha$, $r \ge 1$ and $\beta \equiv |\alpha| \le n$. Show that $T_k(x_j) = (-1)^r T_\beta(x_j)$, $1 \le j \le n$, and therefore that $(-1)^r T_\beta(x)$ for $\beta < n$ or $\theta(x)$ for $\beta = n$ is the interpolating polynomial in $\mathcal{P}_{n-1}$ for $T_k(x)$. Show that $R_{n-1}(T_k; x) = 0$ for $k \le n - 1$ and that $|R_{n-1}(T_k; x)| \le 2$ for $k \ge n$. From part (a) we see that $f(x)$ can be expressed in a uniformly convergent Fourier expansion of the form

$$f(x) = \sum_{k=0}^{\infty} a_k T_k(x).$$

Because of the uniform convergence we may write

$$R_{n-1}(f; x) = \sum_{k=0}^{\infty} a_k R_{n-1}(T_k; x) = \sum_{k=n}^{\infty} a_k R_{n-1}(T_k; x).$$

Use part (a) in order to argue that $\lim_{n\to\infty} \|f - Q_{n-1}\|_\infty = 0$, and use a Jackson theorem to bound $\|R_{n-1}(f; x)\|_\infty$ for any n. [Note: Dropping the hypothesis that $f \in C^2[-1, 1]$, we may still argue that $Q_{n-1}(x)$ is uniformly convergent to $f(x)$ whenever $\Sigma_{k=0}^{\infty} |a_k|$ is convergent.]

4. Let $f(x) = \cos(x)$ and let

$$\sum_{i=0}^{\infty} \alpha_i p_i(x)$$

be the Chebyshev series for $\cos(x)$. By Problem 3, this series converges uniformly and absolutely to $\cos(x)$ for $-1 \le x \le 1$. Use the error bound for Chebyshev interpolation to show that $|\langle f, p_k\rangle| \le 4\sqrt{2}/(k!2^k\sqrt{\pi})$ as in Example 5.17. Since the series converges uniformly to $f(x)$, we have

$$f(x) - p_n^*(x) = \sum_{k=n+1}^{\infty} \alpha_k p_k(x).$$

Use this equation to show that $|f(x) - p_n^*(x)| \le 8/\pi \, \Sigma_{k=n+1}^{\infty} |T_k(x)|/2^k k!$. Recalling that $\|T_k\|_\infty = 1$ and using the Maclaurin's series expansion for e^x with $x = \frac{1}{2}$, find an upper bound for $\|f - p_n^*\|_\infty$.

5. Verify that $\int_0^\pi |\cos(k\theta)| \, d\theta = 2$.

6. Determine the Chebyshev expansion for $f(x) = \arccos(x)$. Give both a lower and an upper bound for $|f(1) - p_n^*(1)|$, using the integral test on

$$\sum_{i=n+1}^{\infty} \langle f, p_i\rangle p_i(x)$$

for $x = 1$. How many terms of the series must be taken in order that $\|f - p_n^*\|_\infty \le 10^{-8}$?

5.3.3. B Splines and the Rayleigh-Ritz Procedure

In this section we consider a particular type of least-squares approximation called the Rayleigh-Ritz procedure, and we illustrate its usage with splines in the approximation of the solution of certain ordinary differential equations. (See also Section 8.5.1.)

To begin, let $X = \{x_j\}_{j=0}^n$ be a set of distinct points in the interval $[a, b]$, $a = x_0 < x_1 < \cdots < x_{n-1} < x_n = b$; and let $\mathcal{S}(X)$ be the set of all cubic splines on these points; that is, $S(x) \in \mathcal{S}(X)$ if and only if $S(x)$ is a cubic polynomial on each subinterval $[x_j, x_{j+1}]$, $0 \le j \le n-1$, and $S^{(j)}(x)$ is continuous on $[a, b]$ for $j = 0, 1, 2$. Just as each polynomial $p(x)$ in $\mathcal{P}_n$ can be expressed uniquely as a linear combination of the "basis" functions $\{1, x, x^2, \ldots, x^n\}$, we shall show that there is a set of functions $\{B_{-1}(x), B_0(x), B_1(x), \ldots, B_n(x), B_{n+1}(x)\}$ such that each spline function $S(x)$ in $\mathcal{S}(X)$ can be uniquely written in the form

$$S(x) = c_{-1}B_{-1}(x) + c_0B_0(x) + \cdots + c_nB_n(x) + c_{n+1}B_{n+1}(x). \tag{5.71}$$

[In vector space terminology, $\mathcal{S}(X)$ is a vector space of dimension $n + 3$; this fact is established below and in the exercises.]

We now assume that the points $\{x_j\}_{j=0}^n$ are equally spaced, $x_j = x_0 + jh$, $0 \le j \le n$, $h = (b - a)/n$; and for purposes of construction we shall also employ similarly defined points $x_{-3}, x_{-2}, x_{-1}, x_{n+1}, x_{n+2}, x_{n+3}$ lying outside $[a, b]$. For each j, $-1 \le j \le n + 1$, we have from Section 5.2.6 that a unique cubic spline $c_j(x)$ defined on the interval $[x_{j-2}, x_{j+2}]$ exists that satisfies the conditions $c_j(x_{j-2}) = 0$, $c_j(x_{j-1}) = 1$, $c_j(x_j) = 4$, $c_j(x_{j+1}) = 1$, $c_j(x_{j+2}) = 0$, and $c_j''(x_{j-2}) = c_j''(x_{j+2}) = 0$. From (5.46) we can obtain the formula

$$c_j(x) = \frac{1}{h^3}\begin{cases} (x - x_{j-2})^3 & \text{for } x_{j-2} \le x \le x_{j-1} \\ h^3 + 3h^2(x - x_{j-1}) + 3h(x - x_{j-1})^2 - 3(x - x_{j-1})^3 & \\ & \text{for } x_{j-1} \le x \le x_j \\ h^3 + 3h^2(x_{j+1} - x) + 3h(x_{j+1} - x)^2 - 3(x_{j+1} - x)^3 & \\ & \text{for } x_j \le x \le x_{j+1} \\ (x_{j+2} - x)^3 & \text{for } x_{j+1} \le x \le x_{j+2}. \end{cases} \tag{5.72a}$$

From this formula we see also that $c_j'(x_{j-2}) = c_j'(x_{j+2}) = 0$. Thus for each j, $-1 \le j \le n + 1$, we can define $B_j(x)$ for all x by

$$B_j(x) = \begin{cases} c_j(x) & \text{for } x \in [x_{j-2}, x_{j+2}] \\ 0 & \text{otherwise.} \end{cases} \tag{5.72b}$$

Since $B_j^{(k)}(x_{j-2}) = B_j^{(k)}(x_{j+2}) = 0$ for $k = 0, 1, 2$, then each $B_j(x)$ is a cubic spline on $[a, b]$ with respect to the points $\{x_j\}_{j=0}^n$; that is, $B_j(x) \in \mathcal{S}(X)$, $-1 \le j \le n + 1$.

The functions $\{B_j(x)\}_{j=-1}^{n+1}$ are called B splines. For $-1 \le j \le n+1$, the graph of $B_j(x)$ is given in Fig. 5.9.

Theorem 5.11

Let $\{B_j(x)\}_{j=-1}^{n+1}$ be defined as in (5.72). For any cubic spline $S(x)$ in $\mathcal{S}(X)$ there exist unique constants $\{c_j\}_{j=-1}^{n+1}$ such that

$$S(x) = c_{-1}B_{-1}(x) + c_0 B_0(x) + c_1 B_1(x) + \cdots + c_n B_n(x) + c_{n+1}B_{n+1}(x). \quad (5.73)$$

Proof. For any $S(x)$ in $\mathcal{S}(X)$, let $S(x_j) = y_j$, $0 \le j \le n$, $S'(x_0) = y_0'$, and $S'(x_n) = y_n'$. Consider the linear system of $(n+3)$ equations:

$$\begin{aligned} c_{-1}B'_{-1}(x_0) + c_0 B'_0(x_0) + \cdots + c_{n+1}B'_{n+1}(x_0) &= y'_0 \\ c_{-1}B_{-1}(x_j) + c_0 B_0(x_j) + \cdots + c_{n+1}B_{n+1}(x_j) &= y_j, \qquad 0 \le j \le n \qquad (5.74) \\ c_{-1}B'_{-1}(x_n) + c_0 B'_0(x_n) + \cdots + c_{n+1}B'_{n+1}(x_n) &= y'_n. \end{aligned}$$

From (5.72) the coefficient matrix of this system is

$$A = \begin{bmatrix} -\frac{3}{h} & 0 & \frac{3}{h} & 0 & 0 & \cdots & & 0 \\ 1 & 4 & 1 & 0 & 0 & & & 0 \\ 0 & 1 & 4 & 1 & 0 & & & 0 \\ 0 & 0 & 1 & 4 & 1 & \cdots & & 0 \\ \vdots & & & & \vdots & & & \vdots \\ 0 & 0 & 0 & 0 & 0 & 1 & 4 & 1 \\ 0 & 0 & 0 & 0 & 0 & -\frac{3}{h} & 0 & \frac{3}{h} \end{bmatrix}. \quad (5.75)$$

The matrix A is strictly diagonally dominant except for the first and the last rows. We leave to the reader to modify the proof of Theorem 3.4 to show that zero is not an eigenvalue of A. Hence A is nonsingular and there is a unique solution $\{c_j\}_{j=-1}^{n+1}$ of (5.74). From Section 5.2.6, $S(x)$ is the unique cubic spline satisfying $S(x_j) = y_j$, $0 \le j \le n$, $S'(x_0) = y'_0$, and $S'(x_n) = y'_n$; and so (5.73) is valid for all x in $[a, b]$. ■

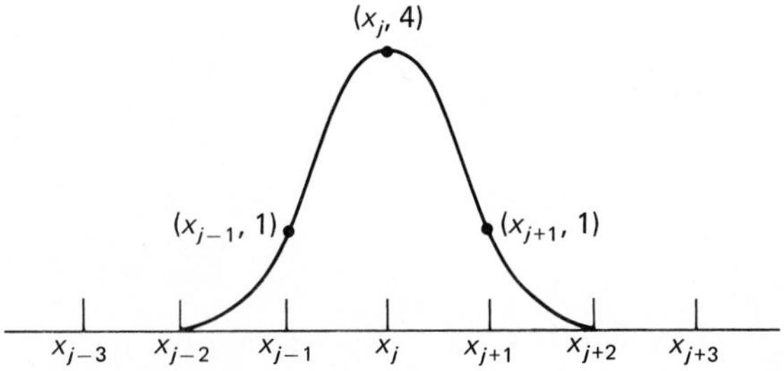

Figure 5.9 Graph of $B_j(x)$.

There are many other bases for $\mathcal{S}(X)$, but we have chosen to use B splines because each $B_j(x)$ is nonzero on at most four subintervals of $[a, b]$; that is, $B_j(x) = 0$ unless $x \in [x_{j-2}, x_{j+2}]$. The computational advantages of this property will become apparent momentarily. Furthermore, if one wishes to use (5.73) to express the cubic spline $S(x)$ that interpolates a given function $f(x)$ by $S'(x_0) = f'(x_0)$, $S'(x_n) = f'(x_n)$, $S(x_j) = f(x_j)$, $0 \le j \le n$, then one merely uses $y_0' = f'(x_0)$, $y_n' = f'(x_n)$, $y_j = f(x_j)$, $0 \le j \le n$, in the equations (5.74).

We now turn to an example of the Rayleigh-Ritz procedure that makes use of B splines. Consider the differential equation $L[y] = F(x)$ with boundary conditions $y(0) = y(1) = 0$ where $L[y] = -y''(x) + w(x)y(x)$ and $w(x) > 0$. [The solution of this problem can be used to solve $L[y] = F(x)$, $y(0) = \alpha$, $y(1) = \beta$ (see Problem 3).] Let the inner product of two arbitrary functions $f(x)$ and $g(x)$ be given by

$$\langle f, g\rangle = \int_0^1 f(x)g(x)dx. \tag{5.76}$$

Then the associated norm is given by

$$\|f\| = \sqrt{\langle f,f\rangle} = \sqrt{\int_0^1 f(x)^2dx.} \tag{5.77}$$

With $a = 0$ and $b = 1$, and $\mathcal{S}(X)$ being the space of all cubic splines for the distinct points $X = \{x_j\}_{j=0}^n$, suppose we wish to find in $\mathcal{S}(X)$ the cubic spline $S^*(x)$ that is the best approximation to a given function $f(x)$ with respect to the norm above; that is, find $S^*(x)$ such that $\|f - S^*\| \le \|f - S\|$ for all $S(x)$ in $\mathcal{S}(X)$. The analysis proceeds precisely as in Section 5.3, and the analogy to Corollary 1 of Theorem 5.10 yields that $S^*(x)$ is the unique cubic spline in $\mathcal{S}(X)$ such that

$$\langle S^* - f, S\rangle = 0 \qquad \text{or} \qquad \langle S^*, S\rangle = \langle f, S\rangle, \qquad \text{for all } S(x) \text{ in } \mathcal{S}(X). \tag{5.78}$$

By Theorem 5.11 every $S(x)$ in $\mathcal{S}(X)$ [including $S^*(x)$] can be written as a linear combination of the B splines, $\{B_j(x)\}_{j=-1}^{n+1}$. Thus (5.78) is equivalent to

$$\langle S^*, B_j\rangle = \langle f, B_j\rangle \qquad \text{for } -1 \le j \le n+1. \tag{5.79}$$

Since $S^*(x)$ can be written as $S^*(x) = c_{-1}^*B_{-1}(x) + c_0^*B_0(x) + \cdots + c_{n+1}^*B_{n+1}(x)$, we can substitute this expression in (5.79) and get the system of $(n + 3)$ linear equations:

$$\sum_{i=-1}^{n+1} c_i^*\langle B_i, B_j\rangle = \langle f, B_j\rangle \qquad \text{for } -1 \le j \le n+1. \tag{5.80}$$

Solving this system for $\{c_j^*\}_{j=-1}^{n+1}$ identifies $S^*(x)$.

In solving the boundary value problem,

$$L[y] = F(x), \qquad y(0) = y(1) = 0,$$

we desire a cubic spline approximation for $y(x)$. We cannot directly apply the equations in (5.80) since we do not know $y(x)$. The Rayleigh-Ritz procedure consists of determining a new inner product and norm so that the computations in (5.80) can be performed. The new inner product, denoted by $\langle f, g\rangle_L$ for two given functions $f(x)$ and $g(x)$, is defined by

$$\langle f, g\rangle_L = \langle L[f], g\rangle = \int_0^1 (-f''(x) + w(x)f(x))g(x)dx. \tag{5.81}$$

The associated norm is then given by $\|f\|_L = \sqrt{\langle f, f\rangle_L}$ and is called the "energy norm." To be mathematically consistent and meaningful, the energy norm must satisfy the three norm properties given in (2.37). It can be shown that these three properties follow if the differential operator L satisfies the conditions that $\langle L[f], g\rangle = \langle f, L[g]\rangle$ (or $\langle f, g\rangle_L = \langle g, f\rangle_L$) and $\langle L[f], f\rangle > 0$ (or $\langle f, f\rangle_L > 0$) for $f(x) \neq \theta(x)$. If these two conditions are satisfied, then L is said to be a symmetric, positive-definite operator. [To be precise, we should place certain restrictions on $f(x)$ and $g(x)$, but we shall not pursue this further.]

If we define the set of functions V by

$$V = \{f(x) : f''(x) \text{ is continuous and } f(0) = f(1) = 0\},$$

then the solution $y(x)$ of $L[y] = F(x)$, $y(0) = y(1) = 0$, is in V. Furthermore if $f(x)$ and $g(x)$ are any two functions in V, we can integrate twice by parts and obtain

$$\begin{aligned}
\langle f, g\rangle_L = \langle L[f], g\rangle &= \int_0^1 [-f''(x) + w(x)f(x)]g(x)dx \\
&= -f'(x)g(x)\Big|_0^1 + \int_0^1 f'(x)g'(x)dx + \int_0^1 w(x)f(x)g(x)dx \\
&= 0 + f(x)g'(x)\Big|_0^1 - \int_0^1 f(x)g''(x)dx + \int_0^1 w(x)f(x)g(x)dx \\
&= 0 + \int_0^1 f(x)[-g''(x) + w(x)g(x)]dx = \langle f, L[g]\rangle = \langle g, f\rangle_L.
\end{aligned}$$

A similar integration by parts shows that

$$\begin{aligned}
\langle f, f\rangle_L = \langle L[f], f\rangle &= \int_0^1 [-f''(x) + w(x)f(x)]dx \\
&= \int_0^1 w(x)[f(x)]^2 dx + \int_0^1 [-f'(x)]^2 dx > 0
\end{aligned}$$

for any $f(x)$ in V, $f(x) \neq 0$. Hence L is a symmetric, positive-definite operator.

With the obvious results that $\langle \alpha f, g\rangle_L = \alpha\langle f, g\rangle_L$ and $\langle f, g + h\rangle_L = \langle f, g\rangle_L + \langle f, h\rangle_L$, we have that $\langle f, g\rangle_L$ in (5.81) satisfies the properties (5.57).

We now use this new inner product to find a best least-squares approximation to the solution $y(x)$ of $L[y] = F(x)$, $y(0) = y(1) = 0$. We first note that the B splines $\{B_j(x)\}_{j=2}^{n-2}$ satisfy the boundary conditions $B_j(0) = B_j(1) = 0$, $2 \le j \le n$. The remaining B splines do not, but obviously we can form combinations of them that do. For example, let $\tilde{B}_0(x) = 4B_{-1}(x) - B_0(x)$, $\tilde{B}_1 = 4B_1(x) - B_0(x)$, $\tilde{B}_{n-1}(x) = 4B_{n-1}(x) - B_n(x)$, and $\tilde{B}_n(x) = 4B_{n+1}(x) - B_n(x)$. If one defines $\tilde{B}_j(x) = B_j(x)$ for $2 \le j \le n - 2$, then $\tilde{B}_j(0) = \tilde{B}_j(1) = 0$ for $0 \le j \le n$. We let V_n be the subset of cubic splines of $\mathcal{S}(X)$ defined by

$$V_n = \{S(x) | S(x) = c_0\tilde{B}_0(x) + c_1\tilde{B}_1(x) + \cdots + c_n\tilde{B}_n(x)\}.$$

Then V_n is also a subspace of V; and we wish to find in V_n the cubic spline $S^*(x)$ that is the best least-squares approximation to the solution $y(x)$; that is, we wish to find constants $\{c_j^*\}_{j=0}^n$ where $S^*(x) = c_0^*\tilde{B}_0(x) + c_1^*\tilde{B}_1(x) + \cdots + c_n^*\tilde{B}_n(x)$ and $\|y - S^*\|_L \le \|y - S\|_L$ for all $S(x)$ in V_n.

In a like manner to (5.79) and (5.80), we have

$$\begin{aligned} &\langle S^*, \tilde{B}_j\rangle_L = \langle y, \tilde{B}_j\rangle_L \qquad \text{for } 0 \le j \le n, \text{ and} \\ &\sum_{i=0}^{n} c_i^*\langle \tilde{B}_i, \tilde{B}_j\rangle_L = \langle y, \tilde{B}_j\rangle_L \qquad \text{for } 0 \le j \le n. \end{aligned} \tag{5.81}$$

Unlike (5.80), for (5.81) we can compute the constants on the right-hand side of the equations. This computation is possible since

$$\langle y, \tilde{B}_j\rangle_L = \langle L[y], \tilde{B}_j\rangle = \langle F, \tilde{B}_j\rangle.$$

[Recall that $L[y] = F(x)$ was the given differential equation.] The solution $\{c_j^*\}_{j=0}^n$ identifies the cubic spline approximation $S^*(x)$ to the solution $y(x)$.

In (5.81) we see the computational advantage of using B splines. Since each $B_k(x)$ is nonzero only for $x \in [x_{k-2}, x_{k+2}]$, then $\langle \tilde{B}_i, \tilde{B}_j\rangle_L = 0$ whenever $|i - j| \ge 4$. Thus the coefficient matrix of the linear system of equations in (5.81) is a "7-banded" matrix; that is, the main diagonal and the three successive subdiagonals and superdiagonals contain all of the nonzero entries. Furthermore, the property above of the $\tilde{B}_j(x)$'s simplifies the computations in calculating or approximating the integrals $\langle F, \tilde{B}_j\rangle$. (For the solution technique using linear splines, see Problem 7.)

We close this section by emphasizing that we have only touched on the applications of the Rayleigh-Ritz procedure. For example, the homogeneous boundary conditions $y(0) = y(1) = 0$ are not necessary; the procedure can be applied to partial differential equations (as in Chapter 8); the procedure can be adapted to problems in which the differential operator is not symmetric, positive definite (called the Galerkin procedure); etc. These studies constitute what is known as the "finite element" technique, and the reader is directed to Prenter (1975) as an initial reference.

PROBLEMS, SECTION 5.3.3

1. a) In (5.72a) let $a = 0$, $b = 1$, $h = .1$, and $x_j = (.1)j$ for $-3 \le j \le 13$. Give the specific formula for $c_0(x)$, and graph $B_0(x)$. Compare the graph with Fig. 5.9.

b) Verify that $B_2''(0) = B_2''(1) = 0$ from (5.72a) and (5.72b). In the system (5.46) use this information along with $B_2(0) = 0$, $B_2(.1) = 1$, $B_2(.2) = 4$, $B_2(.3) = 1$, and $B_2(x_j) = 0$ for $4 \le j \le 10$. Show that the solution is the same as that prescribed by (5.72a) and (5.72b); that is, show that $y_j'' = B_2''(x_j)$, $0 \le j \le 10$, satisfy (5.46).

c) Repeat part (b) for $B_0(x)$; use $y_j'' = B_0''(x_j)$ in (5.46) for $0 \le j \le 10$.

2. Let $n = 5$, $a = 0$, $b = 1$, $h = .2$, and $x_j = (.2)j$, $0 \le j \le 5$. For the function $f(x)$ find the interpolating cubic spline $S(x)$ in the form of (5.73), $S'(0) = f'(0)$ and $S'(1) = f'(1)$ where

a) $f(x) = x^4$ b) $f(x) = \cos 5\pi x$ c) $f(x) = \sin(5\pi x/2)$.

3. With $L[y] = -y''(x) + w(x)y(x)$, consider the nonhomogeneous problem $L[y] = F(x)$ where $y(0) = \alpha$, $y(1) = \beta$. Let $u(x)$ be any twice continuously differentiable function such that $u(0) = \alpha$ and $u(1) = \beta$; for example, $u(x) = (\beta - \alpha)x + \alpha$. If $v(x) = y(x) - u(x)$, show that $L[v] = F(x) - L[u]$ and $v(0) = v(1) = 0$. How can one use this result with the techniques in this section to approximate $y(x)$? For $u(x) = (\beta - \alpha)x + \alpha$, what is the natural cubic interpolating spline for $u(x)$ with $x_j = jh$, $0 \le j \le n$, $h = 1/n$?

4. If $p(x)$ is given where $p'(x)$ is continuous and there exists a constant c such that $p(x) \ge c > 0$ for $0 \le x \le 1$, let $\tilde{L}[y] = -\dfrac{d}{dx}(p(x)y'(x)) + w(x)$, $w(x) \ge 0$. Show that $\tilde{L}$ is symmetric and positive definite; that is, $\langle L[v], u\rangle = \langle v, L[u]\rangle$ and $\langle L[v], v\rangle > 0$ for all functions $u(x)$ and $v(x)$ where $u''(x)$ and $v''(x)$ are continuous and $u(0) = u(1) = v(0) = v(1) = 0$. Argue that the Rayleigh-Ritz technique applies also to $\tilde{L}[y] = F(x)$, $y(0) = y(1) = 0$.

5. Using the composite Simpson's rule (6.13a) to approximate the necessary integrals and using the results of Problems 3 and 4, write a program to find the Rayleigh-Ritz approximation for $\tilde{L}[y] = F(x)$, $y(a) = \alpha$ and $y(b) = \beta$ where $\tilde{L}$ is given in Problem 4. Test the program on

a) $-\dfrac{d}{dx}(e^{-x}y') + 2e^{-x}y = -(1 + e^{-2x})/2, \qquad y(0) = y(1) = 0$

[Answer: $y(x) = c_1e^{-x} + c_2e^{2x} + (1/6)xe^{2x} + (1/8)e^{-2x}$ where c_1 and c_2 are chosen so that $y(0) = y(1) = 0$.]

b) $-\dfrac{d}{dx}(2e^{-2x}y') + 6e^{-2x}y = -3, \qquad y(0) = y(1) = 0$

[Answer: $y(x) = c_1e^{3x} + c_2e^{-x} - (1/2)e^{2x}$ where c_1 and c_2 are chosen so that $y(0) = y(1) = 0$.]

c) $-\dfrac{d}{dx}(x^2y') + 2y = -4x^2, \qquad y(1) = y(2) = 0$

[Answer: $y(x) = (8/7)x^{-2} - (15/7)x + x^2$.]

6. Given the equation $a_2(x)y'' + a_1(x)y' + a_0(x)y = G(x)$, let $p(x) = \exp(\int [a_1(x)/a_2(x)]dx)$. Show that the equation can be written in the form $\dfrac{d}{dx}(p(x)y'(x)) +$

$w(x)y = F(x)$. Transform the following equations to the latter form, determine intervals where the transformation is valid, and determine intervals where the technique of this section is applicable.

a) $(1 - x^2)y'' - 2xy' + 6y = 0$

b) $(x^3 - 2)y'' - y' + 2xy = 0$

7. Often first-degree splines are used instead of cubic splines in (5.81). Consider the points $0 = x_0 < x_1 < \cdots < x_N < x_{N+1} = 1$ where $h_i = x_{i+1} - x_i$. For each x_i, $1 \le i \le N$, we construct the first-degree spline $A_i(x)$ such that $A_i(x_i) = 1$, $A_i(x) = 0$ for $x \le x_{i-1}$ and $x \ge x_{i+1}$, $A_i(x) = (x - x_{i-1})/h_{i-1}$ for $x_{i-1} \le x \le x_i$, and $A_i(x) = (x_{i+1} - x)/h_i$ for $x_i \le x \le x_{i+1}$. Graph a typical $A_i(x)$. Argue that (5.81) yields the best least-squares energy-norm approximation to y from the subspace $W_N = \{w(x) | w(x) = c_1A_1(x) + c_2A_2(x) + \cdots + c_NA_N(x)\}$. Argue that the resultant linear system is tridiagonal. Program this Rayleigh-Ritz technique and compare accuracies with the cubic-spline program for various values of n and N.

8. Show that the $((n + 3) \times (n + 3))$ matrix in (5.75) is nonsingular by using a variation of the proof of the Gershgorin Circle theorem. If A were singular, then there would be a nonzero vector $\mathbf{z}$ such that $A\mathbf{z} = \boldsymbol{\theta}$. Let $\mathbf{z} = [z_1, z_2, \ldots, z_{n+3}]^T$ and let k be the index for which $|z_i|$ is maximum; so $|z_k| \ge |z_i|$, $1 \le i \le n + 3$. By considering the kth equation of $A\mathbf{z} = \boldsymbol{\theta}$, show that k cannot fall in the range $1 < k < n + 3$; that is, k must be either 1 or $n + 3$. Now deduce a contradiction to $A\mathbf{z} = \boldsymbol{\theta}$, $\mathbf{z} \ne \boldsymbol{\theta}$ by looking at the first and $(n + 3)$rd equations of $A\mathbf{z} = \boldsymbol{\theta}$.

*5.4 APPROXIMATION BY RATIONAL FUNCTIONS

In this section, we will briefly describe some methods for approximating functions $f(x)$ by *rational functions*. Recall that a rational function $r(x)$ is defined to be a quotient of polynomials. Thus

$$r(x) = \frac{p(x)}{q(x)} = \frac{p_0 + p_1x + \cdots + p_nx^n}{q_0 + q_1x + \cdots + q_mx^m} \tag{5.82}$$

where we assume that $p(x)$ and $q(x)$ are reduced to lowest terms [that is, $p(x)$ and $q(x)$ have no common factors]. Further, if we wish to approximate $f(x)$ by $r(x)$ where $f(x)$ is in $C[a, b]$, then we will naturally insist that $q(x)$ does not vanish in $[a, b]$.

One reason for introducing rational functions as a means to approximate a function $f(x)$ is that ordinary polynomial approximation may not be practical. For example, in order to approximate $f(x) = \arccos(x)$ for x in $[-1, 1]$ with an accuracy of 10^{-8}, it can be shown that a polynomial of degree 10^4 would be necessary. Clearly, since graphs of rational functions can assume shapes that the graph of a polynomial cannot, rational functions should provide a useful class of approximations. The theory of rational function approximation is still under active investigation and what has been developed is fairly complex; so we will not present it in detail here. Instead, we will illustrate some of the techniques and give some examples. (We emphasize that polynomial approximation is a special case of rational approximation with $m = 0$. So in theory, rational approximation must always yield better results although the computation may become more cumbersome.)

The approach most widely used to choose a rational function approximation is that of *Padé approximation*. Suppose $f(x)$ has a Maclaurin's series expansion $f(x) = \sum_{i=0}^{\infty} a_i x^i$. In Padé approximation, we seek a rational function, $r(x)$, such that the Maclaurin's series expansion for $r(x)$ agrees with that of $f(x)$ to as many terms as possible. Thus, we seek $p(x)$ and $q(x)$ such that

$$f(x) - \frac{p(x)}{q(x)} = \sum_{i=s}^{\infty} c_i x^i \tag{5.83}$$

where s is as large as possible. If $r(x)$ is to have a Maclaurin's series expansion, then clearly $q_0 \neq 0$ in (5.82). By dividing $p(x)$ and $q(x)$ by q_0, we may as well assume that $q(x)$ has the form $q(x) = 1 + q_1 x + \cdots + q_m x^m$. Hence, in (5.82), $r(x)$ has $(n + 1) + m$ parameters and we might hope that the series for $r(x)$ can be made to match that of $f(x)$ for at least the first $(n + 1) + m$ terms (that is, $s = n + m + 1$). As we shall see (Problem 1), it is not always possible to obtain $s \geq n + m + 1$. However, by choosing $m = 0$ and $p(x)$ to be the first $n + 1$ terms of the series for $f(x)$, we see that we can always obtain $s \geq n + 1$.

To determine the coefficients of $r(x)$, we write (5.83) as

$$f(x)q(x) - p(x) = q(x) \sum_{i=s}^{\infty} c_i x^i.$$

Now, $q(x) \sum_{i=s}^{\infty} c_i x^i$ has the form $q(x) \sum_{i=s}^{\infty} c_i x^i = \sum_{i=s}^{\infty} d_i x^i$. Therefore, we want $p(x)$ and $q(x)$ so that the expression $f(x)q(x) - p(x)$ has the form

$$f(x)q(x) - p(x) = \sum_{i=s}^{\infty} d_i x^i$$

where s is as large as possible. Therefore we want to collect like powers of x in the expression $f(x)q(x) - p(x)$ and choose the coefficients p_j and q_j so that as many powers of x vanish as possible.

This program leads us to consider

$$\begin{aligned} f(x)q(x) - p(x) &= \left(\sum_{i=0}^{\infty} a_i x^i\right)\left(\sum_{j=0}^{m} q_j x^j\right) - \sum_{j=0}^{n} p_j x^j \\ &= (a_0 q_0 - p_0) + (a_0 q_1 + a_1 q_0 - p_1)x \\ &\quad + (a_0 q_2 + a_1 q_1 + a_2 q_0 - p_2)x^2 + \cdots. \end{aligned}$$

In general, to get as many powers of x to vanish as possible, we want

$$\sum_{i=0}^{k} a_i q_{k-i} = p_k, \qquad k = 0, 1, \ldots, n$$

$$\sum_{i=0}^{k} a_i q_{k-i} = 0, \qquad k = n+1, n+2, \ldots, s-1$$

where $q_{k-i} = 0$ when $k - i > m$.

EXAMPLE 5.18. Let

$$f(x) = \cos(x) = 1 - \frac{x^2}{2!} + \frac{x^4}{4!} - \frac{x^6}{6!} + \frac{x^8}{8!} - \cdots.$$

Since $f(x)$ is even, we consider an approximation $r(x)$ of the form

$$r(x) = \frac{p_0 + p_2x^2 + p_4x^4 + \cdots + p_{2n}x^{2n}}{1 + q_2x^2 + q_4x^4 + \cdots + q_{2m}x^{2m}}.$$

Since $\cos(x)$ looks roughly like a quadratic polynomial on $[-1, 1]$, it is plausible to try $n = m + 1$. So, for example, we might try

$$r(x) = \frac{p_0 + p_2x^2 + p_4x^4}{1 + q_2x^2}.$$

Since $r(x)$ has four parameters, we try to choose them so that $f(x) - r(x) = c_8x^8 + c_{10}x_{10} + \cdots$. As above, we find that we want to match coefficients in as many powers of x as possible in

$$(1 + q_2x^2)\left(1 - \frac{x^2}{2!} + \frac{x^4}{4!} - \frac{x^6}{6!} + \cdots\right) = p_0 + p_2x^2 + p_4x^4.$$

Multiplying this out and equating like powers of x lead to

$$1 = p_0$$

$$q_2 - \frac{1}{2!} = p_2$$

$$\frac{-q_2}{2!} + \frac{1}{4!} = p_4.$$

$$\frac{q_2}{4!} - \frac{1}{6!} = 0.$$

The last equation gives us $q_2 = \frac{1}{30}$, and hence we find $p_0 = 1$, $p_2 = -\frac{7}{15}$ and $p_4 = \frac{1}{40}$. Therefore,

$$r(x) = \frac{1 - \frac{7}{15}x^2 + \frac{1}{40}x^4}{1 + \frac{1}{30}x^2}.$$

It is not hard to show (Problem 3) that $r(x)$ is a better approximation to $\cos(x)$ than is

$$h(x) = 1 - \frac{x^2}{2!} + \frac{x^4}{4!} - \frac{x^6}{6!},$$

which also has four parameters. (See Table 5.4.)

TABLE 5.4

x	$\cos(x)$	$r(x)$
0	1.0	1.0
0.25	0.96891242	0.96891242
0.5	0.87758256	0.87758264
0.75	0.73168887	0.73169095
1.0	0.54030231	0.54032258

Other frequently used methods for determining the coefficients of a rational function approximation are to select $r(x)$ so that

$$r(x_i) = f(x_i), \qquad i = 0, 1, \ldots, s \tag{5.84a}$$

or

$$\sum_{i=0}^{N} [f(x_i) - r(x_i)]^2 = \text{minimum}. \tag{5.84b}$$

A least-squares criterion similar to (5.84b) is to write $p(x) = \sum_{i=0}^{n} p_i\phi_i(x)$ and $q(x) = \sum_{i=0}^{m} q_i\phi_i(x)$ where $\phi_0(x), \phi_1(x), \ldots$ is an orthonormal set of polynomials. By analogy with Padé approximation, we ask that $p(x)$ and $q(x)$ be chosen so that

$$f(x)q(x) - p(x) = \sum_{i=s}^{\infty} \gamma_i\phi_i(x) \tag{5.84c}$$

where s is as large as possible. There is empirical evidence to suggest that choosing the Chebyshev polynomials $T_i(x)$ as the orthonormal set is a good choice. Using Chebyshev polynomials in (5.84c) is particularly nice from a computational point of view because of the identity $2T_i(x)T_j(x) = T_{i+j}(x) + T_{i-j}(x)$ when $i \geq j$ (Problem 4).

PROBLEMS

1. To see that $s = n + m + 1$ in (5.83) is not always possible, let

$$r(x) = \frac{p_0 + p_1x}{1 + q_1x} \quad \text{and} \quad f(x) = a_0 + a_1x + a_2x^2 + \cdots.$$

If $a_1 = 0$ and $a_2 \neq 0$, show that $s = 2$ is the best that can be done. If $a_1 \neq 0$, show that $s = 3$ is possible.

2. To see another limitation of Padé approximation, suppose we wish to approximate $f(x) = e^{3x}$ on $[-1, 1]$ by a quotient of linear polynomials as in Problem 1. By Problem 1, a Padé approximation exists with $s = 3$. Find the approximation and determine why it is not suitable for the interval $[-1, 1]$.

***3.** Using the standard error estimates for alternating series, show in Example 5.18 that

$$|\cos(x) - r(x)| \leq \left(\frac{26}{x^2 + 30}\frac{x^8}{8!}\right).$$

Compare this estimate with the table in Example 5.18.

4. Using the change of variable $x = \cos(\theta)$, establish the identity $2T_i(x)T_j(x) = T_{i+j}(x) + T_{i-j}(x)$, for $i \geq j$.

5. The most straight-forward way to determine $p(x)$ and $q(x)$ in (5.84c) is by taking inner products and obtaining

$$\langle f(x)q(x), \phi_k(x)\rangle = \langle p(x), \phi_k(x)\rangle, \qquad k = 0, 1, \ldots, n$$

$$\langle f(x)g(x), \phi_k(x)\rangle = 0, \qquad k = n + 1, \ldots, s - 1.$$

For $\phi_i(x) = T_i(x)$ and $m = n = 3$, write the resulting equations, using the identity of Problem 4. Using these equations and Problem 7 of Section 5.3.1, find a rational function approximation to $f(x) = \sqrt{1 - x^2}$. Use the inner product of Example 5.17.

6

Numerical Integration and Differentiation

6.1 INTRODUCTION

In Chapter 5 we developed numerous ways for efficiently approximating a function, $f(x) \in C[a, b]$. If $g(x)$ is any approximation whatsoever to $f(x)$, and we wish to approximate either the integral or the derivative of $f(x)$, then our first inclination is to use either the integral or the derivative of $g(x)$, respectively, as our approximation. In the case of approximating derivatives, we have already pointed out that the derivative of the interpolatory cubic spline is fairly reliable whereas the derivative of the nth degree interpolating polynomial usually is not because of its oscillatory behavior. One other fairly reliable means of derivative approximation is to use the derivative of the discrete least-squares approximation of Section 5.3.1. [This method is especially reliable when the values of $f(x)$ are given in tabular form since this approximation tends to "smooth" the data when the degree of the approximation is fairly small.] In general, however, it remains true that numerical differentiation is a particularly unstable process and quite difficult to analyze carefully. We shall present in detail only one more numerical-differentiation method (based on Richardson extrapolation—Section 6.4). The rest of this chapter will be primarily devoted to numerical integration procedures.

6.2 INTERPOLATORY NUMERICAL INTEGRATION

For any function $f(x)$ that is integrable on the interval $[a, b]$, we define

$$I(f) \equiv \int_a^b f(x)w(x)\,dx \tag{6.1}$$

where $w(x)$ is a fixed nonnegative weight function as defined in (5.1a) or (5.1b). [Often we shall be using $w(x) \equiv 1$, but we shall see some important cases later where we shall desire more flexibility in choosing $w(x)$.] Any formula that approximates $I(f)$ is called a numerical integration or quadrature formula. As mentioned above, if $g(x)$ approximates $f(x)$ on $[a, b]$, then we expect $I(g) \approx I(f)$. For example if $g(x)$ is a cubic interpolating spline, then it is usually true that $I(g)$ approximates $I(f)$ quite well. However, because of the complex formula for the cubic spline we shall not pursue this point further here. We shall instead concentrate on the use of interpolating polynomials, and we shall find that they yield quite favorable results and lead to efficient, yet simple formulas. Formulas of this type are called *interpolatory quadratures*.

Interpolatory quadratures all have one standard form. From (5.4), the interpolating polynomial in $\mathcal{P}_n$ at the points $\{x_j\}_{j=0}^n$ can be written as

$$p_n(x) = \sum_{j=0}^{n} f(x_j)\ell_j(x);$$

and we define $Q_n(f) \equiv I(p_n)$ as the quadrature formula to approximate $I(f)$. Thus

$$Q_n(f) \equiv I(p_n) = \int_a^b p_n(x)w(x)\,dx = \int_a^b \left(\sum_{j=0}^{n} f(x_j)\ell_j(x)\right)w(x)\,dx$$

$$= \sum_{j=0}^{n} f(x_j)\int_a^b \ell_j(x)w(x)\,dx \equiv \sum_{j=0}^{n} A_j f(x_j) \tag{6.2}$$

where

$$A_j \equiv \int_a^b \ell_j(x)w(x)\,dx, \qquad 0 \le j \le n.$$

The values $\{A_j\}_{j=0}^n$ are called the *weights* of the quadrature and the points $\{x_j\}_{j=0}^n$ are called the *nodes*. Thus an interpolatory quadrature formula,

$$Q_n(f) = \sum_{j=0}^{n} A_j f(x_j),$$

is nothing more than a *weighted* sum of the function values $f(x_0), f(x_1), \ldots, f(x_n)$. The weights, A_j, are computed once and for all and depend only on the nodes $x_0, x_1, \ldots, x_n$, the weight function $w(x)$, and the interval $[a, b]$. That is, the same weights are used no matter what function $f(x)$ appears in the integral $\int_a^b w(x)f(x)\,dx$ that we are trying to estimate. This simple form makes interpolatory quadrature formulas easy both to use and to analyze.

We note again that if $f(x)$ is a polynomial of degree n or less [that is, $f(x) \in \mathcal{P}_n$], then $f(x) = p_n(x)$ by Theorem 5.3 [$f(x)$ is its own interpolating polynomial]. Hence for any $f(x) \in \mathcal{P}_n$, $Q_n(f) = I(f)$; and so the quadrature of (6.2) gives the exact value for the integral. If for some integer m, $Q_n(f) = I(f)$ for all $f(x) \in \mathcal{P}_m$, then we say that the quadrature has *precision m*. From our remarks above, we see that any $(n + 1)$-point interpolatory quadrature, $Q_n(f)$, always has precision at least n. Later we shall develop quadratures with precision strictly greater than n.

Since from (6.2) it has precision n, $Q_n(f)$ must give the exact values of the integrals of $1, x, x^2, \ldots, x^n$. This fact gives us an alternative way to determine the quadrature weights, $\{A_j\}_{j=0}^n$, once the nodes, $\{x_j\}_{j=0}^n$, are given and fixed. This procedure is similar to the method of undetermined coefficients in that we have a linear system of $(n + 1)$ equations in $(n + 1)$ unknowns (the weights). To be specific, for $f_k(x) = x^k$, $0 \le k \le n$, we get the $(n + 1)$ equations

$$I(f_k) = \int_a^b x^k w(x)\,dx = Q_n(f_k) = \sum_{j=0}^{n} A_j x_j^k, \qquad 0 \le k \le n. \tag{6.3}$$

In (6.3) the nodes, $\{x_j\}_{j=0}^n$, were fixed beforehand and the weights are determined by solving the linear system. If, however, we treat the nodes, as well as the weights, as unknowns, then we have $2(n + 1)$ unknowns. Considering Eq. (6.3) we might reason, as Gauss did in 1814, that it could be possible to let k range from 0 to $2n + 1$ so that (6.3) represents a system (nonlinear this time) of $(2n + 2)$ equations in $(2n + 2)$ unknowns. If (6.3) has a solution, it will yield a quadrature of precision $(2n + 1)$. This procedure is indeed possible and we will study Gaussian quadrature in Section 6.5. For now we will be content to illustrate the construction of a quadrature with the following simple example.

EXAMPLE 6.1. Let $[a, b] = [-h, h]$ be the interval of integration where h is a positive constant, and let $w(x) \equiv 1$. We shall derive two simple quadrature formulas: the first by direct integration of the interpolating polynomial, and the second by undetermined coefficients as in (6.3).

1. Let $x_0 = -h$ and $x_1 = h$; then the first-degree interpolating polynomial is

$$p_1(x) = f(-h)\,\frac{(h - x)}{2h} + f(h)\,\frac{(h + x)}{2h}.$$

Therefore

$$Q_1(f) = I(p_1) = \int_{-h}^{h} p_1(x)\,dx$$

$$= \frac{-f(-h)}{2h}\,\frac{(h - x)^2}{2}\bigg|_{-h}^{h} + \frac{f(h)}{2h}\,\frac{(h + x)^2}{2}\bigg|_{-h}^{h}.$$

So, $Q_1(f) = hf(-h) + hf(h)$, which is the familiar *trapezoidal rule* estimate for $\int_{-h}^{h} f(x)\,dx$.

2. Let $x_0 = -h$, $x_1 = 0$ and $x_2 = h$. By forcing the quadrature formula $Q_2(f) = A_0 f(x_0) + A_1 f(x_1) + A_2 f(x_2)$ to equal $\int_{-h}^{h} f(x)\,dx$ for $f(x) = 1, x, x^2$, we obtain this system:

$$I(1) = \int_{-h}^{h} 1\,dx = 2h = A_0 + A_1 + A_2$$

$$I(x) = \int_{-h}^{h} x\,dx = 0 = -A_0 h + A_2 h$$

$$I(x^2) = \int_{-h}^{h} x^2\,dx = \frac{2h^3}{3} = A_0 h^2 + A_2 h^2.$$

Solving these equations yields

$$A_0 = A_2 = \frac{h}{3} \quad \text{and} \quad A_1 = \frac{4h}{3}.$$

Thus

$$Q_2(f) = \frac{h}{3}\,[f(-h) + 4f(0) + f(h)],$$

which is the well-known *Simpson's rule* for estimating $\int_{-h}^{h} f(x)\,dx$. Note in this case that $I(x^3) = 0 = Q_2(x^3)$ but $I(x^4) \neq Q_2(x^4)$. Thus, Simpson's rule has precision 3 and integrates all cubic polynomials exactly. For small values of h, these formulas often provide quite good estimates. For example, for $h = 0.2$ and $f(x) = \cos(x)$, we have

$$I(f) = 0.397339, \; Q_2(f) = 0.397342, \; Q_1(f) = 0.392027;$$

and for $f(x) = e^x$ and $h = 0.2$, we obtain

$$I(f) = 0.402672, \qquad Q_2(f) = 0.402676, \qquad Q_1(f) = 0.408027.$$

It is natural at this point to ask if the interpolatory quadratures of formula (6.2) will converge to $\int_a^b w(x)f(x)\,dx$ as we increase the number of interpolating points, that is, as $n \to \infty$. The following theorem provides a condition for which this result is true.

Theorem 6.1

For any positive integer n and any $f \in C[a, b]$, let $Q_n(f) = \sum_{j=0}^{n} A_j^{(n)} f(x_j^{(n)})$ be the interpolatory quadrature given by (6.2). If there exists a constant $K > 0$ such that $\sum_{j=0}^{n} |A_j^{(n)}| \leq K$ for all n, then $\lim_{n\to\infty} Q_n(f) = I(f)$ for all $f \in C[a, b]$.

Proof. By the Weierstrass theorem, for any $\varepsilon > 0$ there exists a polynomial $q_N(x) \in \mathcal{P}_N$ (where N depends on ε) such that

$$\max_{a \leq x \leq b} |f(x) - q_N(x)| = \|f - q_N\|_\infty \leq \varepsilon.$$

We note, since the quadrature formula is interpolatory, that $Q_n(q_N) = I(q_N)$ when $n \geq N$. Choosing $n \geq N$ and letting $\int_a^b w(x)\,dx = c$, we obtain

$$\begin{aligned} |I(f) - Q_n(f)| &= |I(f) - I(q_N) + Q_n(q_N) - Q_n(f)| \\ &\leq |I(f) - I(q_N)| + |Q_n(q_N) - Q_n(f)|. \end{aligned}$$

Now,

$$\begin{aligned} |I(f) - I(q_N)| &= \left| \int_a^b w(x)[f(x) - q_N(x)]\,dx \right| \\ &\leq \|f - q_N\|_\infty \int_a^b w(x)\,dx \leq c\varepsilon \end{aligned}$$

and

$$\begin{aligned} |Q_n(q_N) - Q_n(f)| &= \left| \sum_{j=0}^{n} A_j^{(n)}[q_N(x_j^{(n)}) - f(x_j^{(n)})] \right| \\ &\leq \|f - q_N\|_\infty \sum_{j=0}^{n} |A_j^{(n)}| \leq K\varepsilon. \end{aligned}$$

Thus, for any $\varepsilon > 0$, there is an integer N such that $|I(f) - Q_n(f)| \leq (K + c)\varepsilon$ whenever $n \geq N$. Therefore, $\lim_{n\to\infty} Q_n(f) = I(f)$ for every f in $C[a, b]$. ■

It can also be shown that the converse of this theorem is true. That is, if $\lim_{n\to\infty} Q_n(f) = I(f)$ for each $f \in C[a, b]$ [where $Q_n(f)$ is an interpolatory quadrature], then there must be a number K such that $\sum_{j=0}^{n} |A_j^{(n)}| \leq K$ for all n. However, it goes beyond the scope of this text to prove the converse. Theorem 6.1 (and its converse) do have some practical applications in terms of selecting quadrature formulas. We note first that any interpolatory formula has the property that $Q_n(1) = \int_a^b w(x)\,dx$ since the constant polynomial 1 is integrated exactly. Since $w(x)$ is a nonnegative weight function, we must have

$$0 < I(1) = \int_a^b w(x)\,dx = Q_n(1) = \sum_{j=0}^{n} A_j^{(n)}.$$

Therefore, if the weights $A_j^{(n)}$ are all positive for $0 \leq j \leq n$, $\int_a^b w(x)\,dx = \sum_{j=0}^{n} |A_j^{(n)}|$. If for each n we can choose the nodes $x_0^{(n)}, x_1^{(n)}, \ldots, x_n^{(n)}$ so that the corresponding weights $A_0^{(n)}, A_1^{(n)}, \ldots, A_n^{(n)}$ are positive, then the hypotheses of Theorem 6.1 are satisfied and convergence is guaranteed. A further advantage of positive-weight quadrature formulas is that they have good rounding-error properties. For example, if $Q_N(f) = \sum_{j=0}^{N} A_j f(x_j)$, we can usually expect that the round-off error for the evaluation of the $f(x_j)$'s will be on the high side approximately as often as on the low side. Thus, if the A_j's are all positive, the errors to the high side will tend to cancel the errors to the low side when forming the sum, $Q_n(f)$. Furthermore, the expected value of the total rounding error will be minimized if the A_j's are nearly equal. (We shall see two examples of equal weight formulas when we investigate the composite trapezoidal rule and Gauss-Chebyshev quadrature.)

6.2.1. Transforming Quadrature Formulas to Other Intervals

Frequently we have a quadrature formula, $Q_n(g) = \sum_{j=0}^{n} A_j g(t_j)$, that is derived for a specific interval, say $[-1, 1]$, and is designed to approximate $\int_{-1}^{1} g(t)\,dt$. If the problem is to estimate $\int_a^b f(x)\,dx$, we can use the results of Section 5.2.4 to transform the formula from $[-1, 1]$ to $[a, b]$. To be specific, suppose $Q_n(g)$ is a given quadrature formula for $[-1, 1]$ and suppose $f(x)$ is defined on $[a, b]$. With the change of variable $x = \alpha t + \beta$ where $\alpha = (b - a)/2$ and $\beta = (a + b)/2$, we have

$$\int_a^b f(x)\,dx = \int_{-1}^{1} f(\alpha t + \beta)\alpha\,dt.$$

Letting $g(t) = \alpha f(\alpha t + \beta)$, we find $Q_n(g) = \sum_{j=0}^{n} \alpha A_j f(\alpha t_j + \beta)$ as an approximation to $\int_{-1}^{1} g(t)\,dt$. Thus we can define the corresponding transformed quadrature formula, $Q_n^*(f)$, for the interval $[a, b]$ by

$$Q_n^*(f) = \sum_{j=0}^{n} A_j^* f(x_j) = \frac{b-a}{2} \sum_{j=0}^{n} A_j f(x_j)$$
$$x_j = \frac{b-a}{2} t_j + \frac{a+b}{2}. \tag{6.4}$$

We leave as a problem for the reader to show that if Q_n has precision m on $[-1, 1]$, then Q_n^* has precision m on $[a, b]$.

6.2.2. Newton-Cotes Formulas

Given $\int_a^b f(x)\,dx$ to approximate, probably the most natural choice of nodes x_i to use in a quadrature formula are nodes that are equally spaced in $[a, b]$. Let $h = (b - a)/n$ and let $x_i = x_0 + ih$ where $x_0 = a$. An interpolatory quadrature formula $Q_n(f) = \sum_{j=0}^{n} A_j f(x_j)$ constructed using these equally spaced nodes is called a *closed* Newton-Cotes formula. The interpolatory quadrature formula constructed using the equally spaced nodes $y_i = a + ih$, $i = 1, 2, \ldots, n + 1$; $h = (b - a)/(n + 2)$ is called an *open* Newton-Cotes formula. (The closed formula uses the end-points a and b; the open formula has $a < y_1$, and $y_{n+1} < b$.)

From Example 6.1 with $h = 1$, we see that

$$Q_1(f) = f(-1) + f(1)$$

$$Q_2(f) = \frac{1}{3}[f(-1) + 4f(0) + f(1)]$$

are the two- and three-point closed Newton-Cotes rules for $[-1, 1]$. Using the results of Section 6.2.1 on these formulas, we have for an arbitrary interval $[a, b]$, that

$$T(f) = \frac{(b-a)}{2}[f(a) + f(b)] \tag{6.5}$$

$$S(f) = \frac{(b-a)}{6}\left[f(a) + 4f\left(\frac{a+b}{2}\right) + f(b)\right] \tag{6.6}$$

are respectively the closed two-point and the three-point Newton-Cotes formulas. [Formula (6.5) is the trapezoidal rule for $[a, b]$ and (6.6) is Simpson's rule for $[a, b]$.] As two more examples of Newton-Cotes formulas, we have the three-point open formula for $[-1, 1]$ and the four-point closed formula for $[-1, 1]$ (see Problem 9):

$$\int_{-1}^{1} f(x)\,dx \approx \frac{2}{3}\left[2f\left(-\frac{1}{2}\right) - f(0) + 2f\left(\frac{1}{2}\right)\right]$$

$$\int_{-1}^{1} f(x)\,dx \approx \frac{1}{4}\left[f(-1) + 3f\left(-\frac{1}{3}\right) + 3f\left(\frac{1}{3}\right) + f(1)\right].$$

Note in the three-point open formula above that not all the weights are positive. In fact, it is known for $n \geq 10$ that the weights of any Newton-Cotes formula are of mixed sign. As we noted, this is a bad feature in terms of rounding error. Moreover, as n increases, the weights themselves grow without bound; thus there are continuous functions for which the quadratures do not converge to the integral. For these reasons, higher order Newton-Cotes formulas are rarely used in practice. However, lower order Newton-Cotes formulas such as the trapezoidal rule and Simpson's rule are *extremely* useful, and the time invested in analyzing their properties in the next section will be well spent.

We now give one further type of quadrature that involves derivative evaluations of $f(x)$ as well as values of $f(x)$ itself. The cubic polynomial, $p_3(x)$, which satisfies $p_3(a) = f(a)$, $p_3'(a) = f'(a)$, $p_3(b) = f(b)$, and $p_3'(b) = f'(b)$ is given by formula (5.37); and we can approximate $I(f)$ by $\tilde{Q}_3(f) \equiv I(p_3)$. We could laboriously compute $I(p_3) \equiv \int_a^b p_3(x)\,dx$ by direct integration, but we can simplify this task by noting, since Simpson's rule has precision 3, that $S(p_3) = I(p_3)$. From

Eq. (5.37) we can easily verify that $p_3(a) = f(a)$, $p_3((a+b)/2) = 1/2(f(a)+f(b)) + ((b-a)/8)(f'(a) - f'(b))$, and $p_3(b) = f(b)$. Then, using Eq. (6.6), we have

$$\tilde{Q}_3(f) = S(p_3) = \frac{(b-a)}{2}[f(a) + f(b)] + \frac{(b-a)^2}{12}[f'(a) - f'(b)],$$

or

$$\tilde{Q}_3(f) = T(f) + \frac{(b-a)^2}{12}[f'(a) - f'(b)]. \tag{6.7}$$

Because of the presence of $T(f)$ in this formula, $\tilde{Q}_3(f)$ is called the *corrected trapezoidal rule* and is often denoted by $CT(f) \equiv \tilde{Q}_3(f)$. By the uniqueness of the Hermite interpolating polynomial in Theorem 5.7, if $f(x)$ is any polynomial of degree 3 or less, then $f(x) = p_3(x)$; and so $I(f) = CT(f)$. We can easily check that $I(x^4) \neq CT(x^4)$, and therefore $CT(f)$ has precision 3.

EXAMPLE 6.2. Given $\int_{-1}^{1} \cos(x)\, dx$ to estimate, we find for $f(x) = \cos(x)$ that

$$I(f) = 1.683$$

$$T(f) = 1.081$$

$$CT(f) = T(f) + 0.561 = 1.642.$$

This example indicates that when derivative data are readily available, the corrected trapezoidal rule may be expected to provide better answers. The theoretical analysis in the next section bears this out.

PROBLEMS, SECTION 6.2.2

1. By integrating the Lagrange basis functions as in (6.2), derive an interpolatory quadrature formula of the form

$$\int_0^{2h} f(x)dx \simeq A_0 f(0) + A_1 f(h).$$

2. Use the method of undetermined coefficients to derive an interpolatory quadrature formula

$$\int_0^{3h} f(x)dx \simeq A_0 f(0) + A_1 f(h) + A_2 f(3h).$$

3. Use each of (6.5), (6.6), and (6.7) to estimate the integrals and compare your results with the exact result.

a) $\int_0^1 x^4 dx$ b) $\int_{.1}^{.2} \ell n(x) dx$ c) $\int_0^{.3} \frac{1}{1+x} dx$

4. Determine the precision of the formulas for $\int_{-1}^{1} f(x)dx$; use the results of Problem 8.
   a) $\frac{2}{3}[2f(-\frac{1}{2}) - f(0) + 2f(\frac{1}{2})]$ b) $\frac{1}{4}[f(-1) + 3f(-\frac{1}{3}) + 3f(\frac{1}{3}) + f(1)]$
5. Apply each of the formulas in Problem 4 to the integrals in Problem 3; use (6.4) to translate the formulas to the proper interval.
6. Interpolatory numerical differentiation formulas can be constructed using techniques similar to (6.2) and (6.3). If $p_n(x)$ interpolates $f(x)$ at $x_0, x_1, \ldots, x_n$, and if α is some point, then $f'(\alpha) \simeq p_n'(\alpha)$. Furthermore, using the Lagrange form for $p_n(x)$, $p_n'(\alpha) = \sum_{j=0}^{n} A_j f(x_j)$ where $A_j = \ell_j'(\alpha)$. Derive a numerical differentiation formula of the form
$$f'(\alpha) \simeq A_0 f(\alpha - h) + A_1 f(\alpha) + A_2 f(\alpha + h),$$
and test the formula on $f(x) = \cos(x)$ with $\alpha = 2$ and $h = .05$.
7. Use undetermined coefficients, as in (6.3), to derive a differentiation formula of the form
$$f'(\alpha) \simeq A_0 f(\alpha - 2h) + A_1 f(\alpha - h) + A_2 f(\alpha).$$
[That is, choose A_0, A_1, A_2 so that the approximation above is exact for $f(x) = 1, f(x) = x$, and $f(x) = x^2$.] Test your formula on $f(x) = \cos(x), f(x) = e^x$, and $f(x) = \sqrt{x}$ with $\alpha = 1$ and $h = .05$.
8. Suppose $Q_n(f) = \sum_{j=0}^{n} A_j f(x_j)$. Show that $Q_n(f + g) = Q_n(f) + Q_n(g)$. Next, suppose that $Q_n(f) = \int_a^b w(x)f(x)dx$ for the functions $1, x, \ldots, x^m$. Use the fact that $Q_n(f + g) = Q_n(f) + Q_n(g)$ to show that Q_n has precision at least m.
9. Using (6.3), construct the open Newton-Cotes formula for $\int_{-1}^{1} f(x)\,dx$ with nodes $-1/2, 0, 1/2$ and the closed Newton-Cotes formula with nodes $-1, -1/3, 1/3, 1$.
10. Construct the interpolatory quadrature for $\int_{-1}^{1} f(x)\,dx$ with nodes $-1, -1/2, 1/2, 1$.
11. Prove that the quadrature formula Q_n^* given in (6.4) has precision m on $[a, b]$ whenever Q_n has precision m on $[-1, 1]$. [Hint: If $f(x)$ is a polynomial in x of degree m or less, what is $g(t)$?]
12. In Example 6.1 derive $Q_1(f)$ by undetermined coefficients and $Q_2(f)$ by integration of the interpolating polynomial.
13. Suppose that $\int_c^d g(t)dt$ is approximated by the quadrature $Q(g) = \sum_{i=0}^{n} A_i g(t_i)$. Find the corresponding quadrature $\tilde{Q}(f) = \sum_{i=0}^{n} B_i f(x_i)$ to approximate $\int_a^b f(x)dx$. [Hint: Let $x = mt + \beta$ where $x = a$ when $t = c$ and $x = b$ when $t = d$.]
14. a) Let $p(x) = b_3(x - 1)^2(x + 1) + b_2(x - 1)(x + 1)^2 + b_1(x - 1)^2 + b_0(x + 1)^2$. For a given $f(x)$ find b_0, b_1, b_2, and b_3 such that $p(-1) = f(-1), p(1) = f(1), p'(-1) = f'(-1)$, and $p'(1) = f'(1)$.
    b) Show that $p(x)$ in part (a) is unique in $\mathcal{P}_3$. [Hint: Let $q(x)$ in $\mathcal{P}_3$ satisfy $q(\pm 1) = f(\pm 1)$ and $q'(\pm 1) = f'(\pm 1)$, and consider $s(x) = p(x) - q(x)$.]

c) Let $Q(f) = \int_{-1}^{1} p(x)dx$ and express $Q(f)$ in the form $B_0 f(-1) + B_1 f(1) + B_2 f'(-1) + B_3 f'(1)$.

d) Transform $Q(f)$ in part (c) to approximate $\int_a^b f(x)dx$ and obtain the corrected trapezoidal rule (6.7).

e) Use part (b) and Exercise 11 to argue that the precision of Q is at least 3.

15. a) With $\tilde{Q}_3(f)$ as in (6.7) show that $\tilde{Q}_3(\alpha f + \beta g) = \alpha \tilde{Q}_3(f) + \beta \tilde{Q}_3(g)$, α and β constants, f and g functions.

b) Show that $\tilde{Q}_3(x^4) \neq \int_a^b x^4 dx$. If $q(x) = a_4 x^4 + a_3 x^3 + a_2 x^2 + a_1 x + a_0$ with $a_4 \neq 0$, show that $\tilde{Q}_3(q) \neq \int_a^b q(x)dx$.

16. For the interpolatory quadrature formula (6.2), show that at least one of the weights is positive. [Hint: Consider $I(1)$.]

6.2.3. Errors of Quadrature Formulas

In this section, we consider ways of estimating the quadrature error $I(f) - Q_n(f)$ and derive specific estimates for the important cases of the trapezoidal rule, Simpson's rule, and the corrected trapezoidal rule. By formula (5.27), the error of regular polynomial interpolation for $f(x) \in C^{n+1}[a, b]$ is given by

$$e_n(x) \equiv f(x) - p_n(x) = \frac{f^{(n+1)}(\xi)}{(n+1)!} W(x).$$

We can therefore express the error of interpolatory quadrature as

$$e_n = \int_a^b e_n(x) w(x)\, dx,$$

or

$$e_n \equiv \int_a^b f(x) w(x)\, dx - \int_a^b p_n(x) w(x)\, dx = \int_a^b \frac{f^{(n+1)}(\xi)}{(n+1)!} W(x) w(x)\, dx. \tag{6.8a}$$

Recalling that $f^{(n+1)}(\xi)$ is actually an unknown function of x, we see that the right-hand integral in (6.8a) cannot usually be evaluated explicitly. We can, however, obtain a computable bound on the error, e_n, from (6.8a) by using

$$|e_n| \leq \max_{a \leq x \leq b} |f^{(n+1)}(x)| \int_a^b \frac{|W(x)|}{(n+1)!} w(x)\, dx. \tag{6.8b}$$

We will use an error bound similar to this in a later section, but for now we consider two other ways of simplifying (6.8a). Recall that a simple form of the Second Mean-Value Theorem for integrals states that if $g(x)$ and $h(x)$ are continuous and if $g(x)$ does not change sign in the interval (a, b), then there exists a mean value point, $\eta \in (a, b)$, such that $\int_a^b g(x)h(x)\, dx = h(\eta) \int_a^b g(x)\, dx$. We can use this theorem immediately in (6.5) and (6.7). For (6.5), (6.8a) gives the error for the trapezoidal rule:

$$I(f) - T(f) \equiv e^T = \frac{1}{2}\int_a^b f''(\xi)(x - a)(x - b)\,dx.$$

Now $[(x - a)(x - b)]$ is of constant sign on (a, b); so the mean-value theorem above yields

$$e^{\mathrm{T}} = \frac{f''(\eta)}{2}\int_a^b (x - a)(x - b)\,dx = -\frac{f''(\eta)}{12}(b - a)^3. \tag{6.9}$$

Since (6.7) was the result of Hermite interpolation, we must use (5.39) instead of (5.27) to obtain the error. From (5.39), where $p_3(x)$ is given as in (6.7), we have

$$f(x) - p_3(x) = \frac{f^{(iv)}(\xi)}{4!}(x - a)^2(x - b)^2.$$

Now $[(x - a)^2(x - b)^2] > 0$ for $x \in (a, b)$; so the mean-value theorem yields

$$I(f) - CT(f) \equiv e^{CT} = \int_a^b \frac{f^{(iv)}(\xi)}{4!}(x - a)^2(x - b)^2\,dx$$

or

$$e^{\mathrm{CT}} = \frac{f^{(iv)}(\eta)}{4!}\int_a^b (x - a)^2(x - b)^2\,dx = \frac{f^{(iv)}(\eta)}{720}(b - a)^5. \tag{6.10}$$

The error for Simpson's rule is a bit more difficult to analyze since in (6.8a) the function $W(x)$ changes sign in (a, b). First, let $x_0 = a$, $x_1 = (a + b)/2$, and $x_2 = b$. We can get at the error in Simpson's rule by defining an auxiliary cubic polynomial, $p_3(x)$, such that $p_3(x_0) = f(x_0)$, $p_3(x_1) = f(x_1)$, $p_3(x_2) = f(x_2)$, and $p_3'(x_1) = f'(x_1)$ (we leave as Problem 7 to show that such a polynomial exists). Next, note that $S(f) = S(p_3)$ since $p_3(x)$ interpolates $f(x)$ at x_0, x_1, and x_2. Further, since Simpson's rule has precision 3, $S(p_3) = I(p_3)$. Therefore we obtain an alternative expression for the error:

$$I(f) - S(f) = I(f) - S(p_3) = I(f) - I(p_3).$$

In order to use this expression, we need an error formula for $f(x) - p_3(x)$. This formula is easily obtained, as below, using a method similar to that in the proof of Theorem 5.5.

For a *fixed* x, different from x_0, x_1, and x_2, define $\phi(t)$ for $t \in [a, b]$ by

$$\phi(t) = [f(t) - p_3(t)] - [f(x) - p_3(x)]\frac{(t - x_0)(t - x_1)^2(t - x_2)}{(x - x_0)(x - x_1)^2(x - x_2)}.$$

Then $\phi(t)$ has at least 4 zeros in $[a, b]$; namely, x_0, x_1, x_2, and x. By Rolle's theorem, $\phi'(t)$ has at least 3 zeros in (a, b) that are between the 4 zeros of $\phi(t)$. By construction, we see also that $\phi'(x_1) = 0$, and therefore $\phi'(t)$ has at least 4 *distinct* zeros in (a, b). If $f \in C^4[a, b]$, we find (as in the proof of Theorem 5.5) that there is a point ξ in (a, b) (which depends on x) such that $\phi^{(iv)}(\xi) = 0$. Therefore

$$f(x) - p_3(x) = \frac{f^{(iv)}(\xi)}{4!}(x - x_0)(x - x_1)^2(x - x_2).$$

Since the function $[(x - x_0)(x - x_1)^2(x - x_2)]$ does not change sign on (a, b), we can now use the Second Mean-Value Theorem to deduce

$$I(f) - I(p_3) = \int_a^b \frac{f^{(iv)}(\xi)}{4!}(x - x_0)(x - x_1)^2(x - x_2)\, dx,$$

or

$$I(f) - S(f) = \frac{f^{(iv)}(\eta)}{4!}\int_a^b (x - x_0)(x - x_1)^2(x - x_2)\, dx. \tag{6.11a}$$

We can obviously evaluate the integral above directly to obtain the error. To illustrate another idea, however, we use a standard trick to evaluate

$$\int_a^b (x - x_0)(x - x_1)^2(x - x_2)\, dx$$

by choosing $\tilde{f}(x) = (x - x_1)^4$. From (6.11a), we have

$$I(\tilde{f}) - S(\tilde{f}) = \int_a^b (x - x_0)(x - x_1)^2(x - x_2)\, dx$$

since the fourth derivative of $\tilde{f}(x) = (x - x_1)^4$ is the constant 4!. Letting $h = (b - a)/2 = x_1 - x_0 = x_2 - x_1$, we find

$$\int_{x_0}^{x_2} (x - x_1)^4\, dx = \frac{(x - x_1)^5}{5}\bigg|_{x_0}^{x_2} = \frac{2h^5}{5}$$

and

$$S((x - x_1)^4) = \frac{2h}{6}[h^4 + h^4] = \frac{2h^5}{3}.$$

Thus for $\tilde{f}(x) = (x - x_1)^4$,

$$\int_a^b (x - x_0)(x - x_1)^2(x - x_2)\, dx = I(\tilde{f}) - S(\tilde{f}) = \frac{-4h^5}{15}.$$

Using this equation in (6.11a) with $e^S = I(f) - S(f)$, we have

$$e^S = \frac{-f^{(iv)}(\eta)}{90}\left(\frac{b-a}{2}\right)^5. \tag{6.11b}$$

EXAMPLE 6.3. The sine-integral, $Si(x)$, is defined by

$$Si(x) = \int_0^x \frac{\sin(t)}{t}\,dt.$$

To illustrate the error analysis of this section, we estimate $Si(1)$ by the trapezoidal rule, Simpson's rule, and the corrected trapezoidal rule. In order to use the error bounds derived above on the estimates for $Si(1)$, we need in (6.9) to bound $|f''(\eta)|$ for $\eta \in [0, 1]$; and in (6.10) and (6.11b) we need to bound $|f^{(iv)}(\eta)|$ for $\eta \in [0, 1]$.

To obtain these bounds, we use the expansion

$$\sin(t) = t - \frac{t^3}{3!} + \frac{t^5}{5!} - \frac{t^7}{7!} + \cdots$$

to obtain for $f(t) = (\sin(t)/t)$

$$f(t) = 1 - \frac{1}{3}\frac{t^2}{2!} + \frac{1}{5}\frac{t^4}{4!} - \frac{1}{7}\frac{t^6}{6!} + \cdots$$

$$f''(t) = -\frac{1}{3} + \frac{1}{5}\frac{t^2}{2!} - \frac{1}{7}\frac{t^4}{4!} + \cdots$$

$$f^{(iv)}(t) = \frac{1}{5} - \frac{1}{7}\frac{t^2}{2!} + \cdots.$$

Since all these series are alternating series, we see that $|f''(\eta)| \le 1/3$ and $|f^{(iv)}(\eta)| \le 1/5$ for $\eta \in [0, 1]$. Hence, the error bounds are $|e^{\mathrm{T}}| \le 1/36 = 0.027778$, $|e^{CT}| \le 1/3600 = 0.000278$, and $|e^S| \le 1/14400 = 0.000069$. From a table, we obtain $Si(1) = 0.946083$. Noting that $f(0) = 1$ and $f'(0) = 0$, we have

$$T(f) = 0.920735; \qquad e^{\mathrm{T}} = 0.025348$$

$$CT(f) = 0.945832; \qquad e^{\mathrm{CT}} = 0.000251$$

$$S(f) = 0.946146; \qquad e^{\mathrm{S}} = -0.000063.$$

6.2.4. Composite Rules for Numerical Integration

If we wish to derive a highly accurate quadrature formula for the interval $[a, b]$, we immediately see two obvious choices.

1. Take the number of nodes, $(n + 1)$, to be large so that the quadrature,

$$Q_n(f) = \sum_{j=0}^{n} A_j f(x_j),$$

is the integral of a high-degree interpolating polynomial.

2. Divide $[a, b]$ into several subintervals of small length, use a simple quadrature of fairly low precision on each, and add the results to obtain an approximation for $I(f)$.

The methods generated by (2) above are called *composite* quadratures, and we shall study them in this section. First we consider $(N + 1)$ points $\{x_j\}_{j=0}^N$ such that $a = x_0 < x_1 < \cdots < x_N = b$. From calculus we know that

$$I(f) = \int_a^b f(x)w(x)\,dx = \sum_{j=0}^{N-1} \int_{x_j}^{x_{j+1}} f(x)w(x)\,dx.$$

Thus approximating $I(f)$ by composite quadrature amounts to approximating each $\int_{x_j}^{x_{j+1}} f(x)w(x)\,dx$ by a low order quadrature, $Q_k^{(j)}(f)$, and adding the resulting approximations. The three rules (trapezoidal, Simpson, and corrected trapezoidal) specifically studied in Section 6.2.2 are very well suited for our purposes here, and we shall concentrate on their use. We shall see that their form is especially simple if we take the points $\{x_j\}_{j=0}^N$ to be equally spaced; that is, $x_j = a + jh$ for $h = (b - a)/N$ and $0 \le j \le N$. [Using these rules, of course, we take $w(x) \equiv 1$ in $I(f)$.]

First we consider the composite trapezoidal rule. From (6.5) and the error formula (6.9), we see that for any j

$$\int_{x_j}^{x_{j+1}} f(x)\,dx = \frac{h}{2}\,(f(x_j) + f(x_{j+1})) - \frac{f''(\eta_j)}{12}\,h^3, \qquad x_j < \eta_j < x_{j+1}.$$

Thus we define the composite trapezoidal rule, $T_N(f)$, as

$$T_N(f) = \sum_{j=0}^{N-1} \frac{h}{2}\,(f(x_j) + f(x_{j+1})) = h \sum_{j=1}^{N-1} f(x_j) + \frac{h}{2}\,(f(x_0) + f(x_N)). \tag{6.12a}$$

The error, $I(f) - T_N(f) \equiv e_N^{\mathrm{T}}$, is the sum of the errors on each interval and so

$$e_N^{\mathrm{T}} = \sum_{j=0}^{N-1} f''(\eta_j)(-h^3/12).$$

We pause here to give a lemma that will significantly simplify e_N^{T} above, and can also be used when investigating the errors of other composite rules.

Lemma 6.1

Let $g(x) \in C[a, b]$; and let $\{a_j\}_{j=0}^{N-1}$ be any set of constants, all of which have the same sign. If $t_j \in [a, b]$ for $0 \le j \le N - 1$, then for some $\eta \in [a, b]$

$$\sum_{j=0}^{N-1} a_j g(t_j) = g(\eta) \sum_{j=0}^{N-1} a_j.$$

Proof. Let $m = \min_{a \le x \le b} (g(x)) = g(y_m)$ and $M = \max_{a \le x \le b} (g(x)) = g(y_M)$. Then if the a_j's are nonnegative,

$$m \sum_{j=0}^{N-1} a_j \le \sum_{j=0}^{N-1} a_j g(t_j) \le M \sum_{j=0}^{N-1} a_j.$$

Define r by $r = \sum_{j=0}^{N-1} a_j g(t_j)$ and let $G(x) \equiv g(x) \sum_{j=0}^{N-1} a_j$. Then $G(x) \in C[a, b]$; and by the above inequalities, $G(y_m) \le r \le G(y_M)$. By the intermediate-value theorem, there exists an $\eta \in [a, b]$ such that $G(\eta) = r$, or

$$g(\eta) \sum_{j=0}^{N-1} a_j = \sum_{j=0}^{N-1} a_j g(t_j).$$

The case in which the a_j's are all negative is similar and left to the reader. ■

If $f''(x) = g(x)$ and $a_j = -h^3/12$ are used in Lemma 6.1, the expression for e_N^T becomes

$$e_N^T = f''(\eta) \sum_{j=0}^{N-1} \left(\frac{-h^3}{12}\right) = -f''(\eta) \frac{Nh^3}{12} = -\frac{f''(\eta)h^2(b-a)}{12}. \tag{6.12b}$$

Continuing in the same fashion, from Simpson's rule, (6.6), and its error, we see that for any j

$$\int_{x_j}^{x_{j+1}} f(x)\,dx = \frac{h}{6}\left(f(x_j) + 4f((x_j + x_{j+1})/2) + f(x_{j+1})\right) - \frac{f^{(iv)}(\eta_j)}{90}\left(\frac{h}{2}\right)^5,$$

$$x_j < \eta_j < x_{j+1}.$$

Summing the approximations for each interval, we have the composite Simpson's rule:

$$S_N(f) = \frac{h}{6}\left\{f(x_0) + f(x_N) + 2\sum_{j=1}^{N-1} f(x_j) + 4\sum_{j=0}^{N-1} f((x_j + x_{j+1})/2)\right\}. \tag{6.13a}$$

Summing the individual errors for $0 \le j \le N-1$ and using Lemma 6.1, we obtain the error, $e_N^S \equiv I(f) - S_N(f)$:

$$\begin{aligned} e_N^S &= -\sum_{j=0}^{N-1} \frac{f^{(iv)}(\eta_j)}{90}\left(\frac{h}{2}\right)^5 = -\frac{f^{(iv)}(\eta)}{90} N \left(\frac{h}{2}\right)^5 \\ &= -\frac{f^{(iv)}(\eta)}{180}\left(\frac{h}{2}\right)^4 (b-a). \end{aligned} \tag{6.13b}$$

We note that Simpson's rule in the form (6.13a) actually requires the evaluation of $f(x)$ at $(2N+1)$ equally spaced points, $\{x_j\}_{j=0}^{N}$ and $\{(x_j + x_{j+1})/2\}_{j=0}^{N-1}$. For a fair comparison with the accuracy of the composite trapezoidal rule, we should actually display a form of the composite Simpson's rule using only $(N + 1)$ points, $\{x_j\}_{j=0}^{N}$. From above we see immediately that we must restrict N to be even and investigate the formulas

$$\int_{x_j}^{x_{j+2}} f(x)\,dx = \frac{2h}{6}\,(f(x_j) + 4f(x_{j+1}) + f(x_{j+2})) - \frac{f^{(iv)}(\eta_j)}{90}\,h^5, \qquad x_j < \eta_j < x_{j+2},$$

for $j = 0, 2, 4, \ldots, N-2$. Doing this and summing, we obtain

$$S_N'(f) = \frac{h}{3}\left(f(x_0) + f(x_N) + 2\sum_{j=1}^{(N-2)/2} f(x_{2j}) + 4\sum_{j=0}^{(N-2)/2} f(x_{2j+1})\right) \tag{6.14a}$$

and

$$e_N^{S'} = -\frac{f^{(iv)}(\eta)}{180}\,h^4(b-a). \tag{6.14b}$$

We leave to the reader to verify in the manner above that from the corrected trapezoidal rule, (6.7), and its error, (6.10), the composite corrected trapezoidal rule and its error are given by

$$CT_N(f) = h\sum_{j=1}^{N-1} f(x_j) + \frac{h}{2}\,(f(x_0) + f(x_N)) + \frac{h^2}{12}\,(f'(a) - f'(b))$$

$$= T_N(f) + \frac{h^2}{12}\,(f'(a) - f'(b)) \tag{6.15a}$$

$$e_N^{CT} = \frac{f^{(iv)}(\eta)}{720}\,h^4(b-a). \tag{6.15b}$$

The reader will note in the derivation of (6.15a) that the derivative evaluations of $f(x)$ at the interior nodes cancel out, and the only two required derivative evaluations are $f'(a)$ and $f'(b)$. This observation, plus the fact that the method has precision 3 and has an h^4 term in the error, makes this rule computationally attractive.

Since the precision of a composite quadrature formula is not increased by taking N larger and larger, we cannot use Theorem 6.1 to guarantee convergence. However, it is fairly easy to show that $\lim_{N\to\infty} T_N(f) = \lim_{N\to\infty} S_N(f) = I(f)$ for any $f \in C[a, b]$. To see this, we merely recall that $T_N(f) = \sum_{j=0}^{N-1} [f(x_j) + f(x_{j+1})](h/2)$. Since $[f(x_j) + f(x_{j+1})]/2$ is just an average, we have by the intermediate-value theorem that there is a point $z_j \in [x_j, x_{j+1}]$ such that $f(z_j) = [f(x_j) + f(x_{j+1})]/2$. Thus $T_N(f) = \sum_{j=0}^{N-1} f(z_j)h$ is a Riemann sum; so from the definition of the definite integral it follows that $T_N(f) \to I(f)$ as $N \to \infty$. A similar analysis can be carried out for Simpson's rule. (See Problem 11 for a general result on the convergence of composite quadratures.)

Because Simpson's rule is derived from quadratic interpolation on the subintervals whereas the trapezoidal rule comes from linear interpolation, we might jump to the conclusion that Simpson's rule is always preferable. This is often true, but in certain instances the trapezoidal rule yields surprisingly good results. For example if $f(x)$ has the property that $f'(a) = f'(b)$, then the trapezoidal rule is equivalent to the corrected trapezoidal rule. [See (6.15a) and compare the error (6.15b) to the Simpson error (6.14b) for application of both rules to $(N + 1)$ points.] Davis and Rabinowitz (1975) is rich with comparative numerical examples from the literature using both rules.

EXAMPLE 6.4. As an example showing how the error bounds of this section can be used, we consider again (as in Example 6.3), the problem of computing $Si(1)$. Suppose we want accuracy on the order of 10^{-8}. How large must N be in order that $|I(f) - T_N(f)| \leq 10^{-8}$ and how large must N be in order that $|I(f) - S'_N(f)| \leq 10^{-8}$ where

$$I(f) = Si(1) = \int_0^1 \frac{\sin(t)}{t}\, dt?$$

Using the information derived in Example 6.3, we know that $|f''(x)| \leq \frac{1}{3}$ and $|f^{(iv)}(x)| \leq \frac{1}{5}$ for $0 \leq x \leq 1$. From (6.12b) and (6.14b) with $b - a = 1$ and $h = 1/N$, we have

$$|e_N^T| \leq \frac{1}{36N^2}, \qquad |e_N^{S'}| \leq \frac{1}{900N^4}.$$

A quick computation then shows that $N \geq 1667$ is required to make $|e_N^T| \leq 10^{-8}$, and $N \geq 19$ is necessary to ensure that $|e_N^{S'}| \leq 10^{-8}$.

PROBLEMS, SECTION 6.2.4

1. Write a subroutine that uses the composite trapezoid rule T_N. Test your routine on

   a) $\displaystyle\int_1^2 \frac{1}{x}\, dx$ b) $\displaystyle\int_{-1}^1 \frac{1}{1 + x^2}\, dx$ c) $\displaystyle\int_0^2 \cos(x) dx$

   with $N = 2^k$, $k = 3, 4, \ldots, 10$. Print all your estimates and compare them with the exact result.
2. Apply (6.12b) to each of the integrals in Problem 1 and determine a value of N so that $|e_N^T| \leq 10^{-3}$. Compare this value with the actual errors.
3. Repeat Problem 1 with the Simpson's rule S'_N and the corrected trapezoid rule CT_N. Also repeat the analysis in Problem 2 for the integrals (1a) and (1c).
4. This problem and Problem 5 describe how one might construct an elementary "adaptive quadrature", using T_N. The idea is to design a subroutine that accepts an interval $[a, b]$, a function $f(x)$, and an input tolerance TOL. The routine should

return an estimate EST to the integral, and there should be a reasonable expectation that

$$\left|\int_a^b f(x)dx - \text{EST}\right| \leq \text{TOL}.$$

The estimate EST is given by $T_N(f)$; so the subroutine somehow has to select an appropriate value for N so that $T_N(f)$ meets the input accuracy request. The subroutine will select N by using an error monitor that is based on the theoretical form of the error in (6.12b); however, the subroutine will *not* calculate $f''(x)$.

To begin, consider the form of the error for (6.12a):

$$I(f) - T_N(f) = \frac{-h^3}{12} \sum_{j=0}^{N-1} f''(n_j) \tag{P.1}$$

where $x_1 \leq \eta_j \leq x_{i+1}$. Now, $T_{2N}(f)$ uses knots $y_0, y_1, \ldots, y_{2N}$ where $y_{2i} = x_i$ and $y_{2i+1} = (x_i + x_{i+1})/2$. Assume that h is small enough so that $f''(x)$ is approximately constant in $[x_i, x_{i+1}]$. Write the expression for $I(f) - T_{2N}(f)$ and show that

$$I(f) - T_{2N}(f) \simeq \frac{1}{4}(I(f) - T_N(f)). \tag{P.2}$$

[Note: In (P.1), h is replaced by $h/2$ to obtain (P.2).] Next, show that the approximation (P.2) is the same as

$$I(f) - T_{2N}(f) \simeq \frac{1}{3}\,[T_{2N}(f) - T_N(f)]. \tag{P.3}$$

5. By Problem 4 (to the extent that h is small enough so that $f''(x)$ is nearly constant over intervals of length h) we see that the error $I(f) - T_{2N}(f)$ can be estimated by a quantity we can monitor; $[T_{2N}(f) - T_N(f)]/3$. Write a subroutine that calculates $T_K(f)$, $T_{2K}(f)$, $T_{4K}(f)$, . . . and that accepts $T_{2N}(f)$ as an estimate to the integral when

$$|T_{2N}(f) - T_N(f)| \leq \text{TOL} \tag{P.4}$$

where TOL is an input accuracy request. [Note that the termination test (P.4) is three times more severe than is required by (P.3); this excess severity is a safety factor.] To ensure that the smoothness assumption on $f''(x)$ has a chance of holding, also insist that K be chosen so that $K \geq 8$ and $h \leq .125$. If (P.4) cannot be satisfied with $N \leq 1024$, set an error flag to signal failure of the method, and return. Code the subroutine so that only M evaluations of $f(x)$ are necessary in going from $T_M(f)$ to $T_{2M}(f)$. (Note: This formulation of an adaptive quadrature based on T_N is quite inefficient but is simple enough to serve as an introduction to Section 6.3.) Test your routine on the integrals in Problem 1; use TOL $= 10^{-2}$, TOL $= 10^{-3}$, TOL $= 10^{-4}$. Also test your routine on $\int_0^3 f(x)\,dx$ where $f(x) = 1/(x - \sqrt{2})$; the integral does not exist and the routine should signal a failure.

6. Once the construction of any adaptive quadrature is known, it is easy to devise a function that will fool the routine. Try to estimate $\int_0^{\frac{1}{2}} \cos(64\pi x)dx$ with the routine in Problem 5; use TOL $= 10^{-2}$. Why did the routine fail to get a good estimate?

7. Given x_0 and $h > 0$, let $x_1 = x_0 + h$, $x_2 = x_0 + 2h$. Suppose $f \in C^1[x_0, x_2]$. Using the method of undetermined coefficients, show that there is a cubic polynomial $p(x)$ such that $p(x_i) = f(x_i)$, $i = 0, 1, 2$ and $p'(x_1) = f'(x_1)$.
8. Verify the error formula (6.15b) for the corrected trapezoidal rule.
9. How small must h be in order that $|I(f) - S_N(f)| \leq 10^{-6}$ for $f(x) = \sin(10x)$ and $[a, b] = [0, \pi]$? How small must h be in order that $|I(f) - T_N(f)| \leq 10^{-6}$?
10. Show that $\lim_{N\to\infty} S_N(f) = I(f)$ for any $f \in C[a, b]$ where $S_N(f)$ denotes the composite Simpson's rule for $[a, b]$.
11. Suppose $Q(f) = \Sigma_{j=0}^{n} A_j f(x_j)$ is a quadrature formula to approximate $I(f) = \int_{-1}^{1} f(x)dx$ where $\Sigma_{i=0}^{n} A_j = 2$. Let Q_N be the composite formula corresponding to Q applied to $[a, b]$ and suppose that $g \in C[a, b]$. Using upper and lower Riemann sums, show that $Q_N(g) \to \int_a^b g(z)\, dz$.
12. Discuss how the composite Simpson's rule given in (6.13a) can be modified slightly so as to integrate cubic splines exactly.

6.2.5. Multiple Integrals

A fairly common problem is that of evaluating the multiple integral

$$\int_Q\int f(x, y)dA$$

where $f(x, y)$ is real valued and defined on a region Q of the xy-plane. An elementary but useful approach to this important problem is based on techniques familiar from calculus: if the region is simple enough and if the integrand satisfies mild continuity conditions, then we can express the multiple integral as an iterated integral. Specifically, suppose that Q is a region in the xy-plane and has the form pictured in Fig. 6.1. In Fig. 6.1, Q is bounded on the left and right

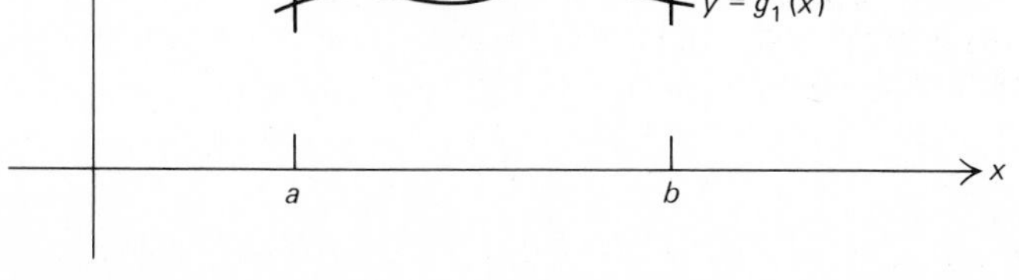

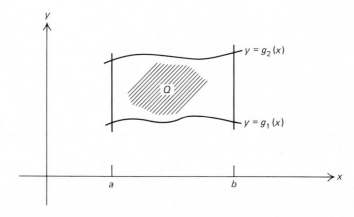

Figure 6.1 A region in the xy-plane

by the lines $x = a$ and $x = b$; Q is bounded above and below by the curves $y = g_2(x)$ and $y = g_1(x)$. With mild assumptions about Q and $f(x, y)$ we can express the multiple integral as an iterated integral:

$$\int_Q\int f(x, y)dA = \int_a^b \int_{g_1(x)}^{g_2(x)} f(x, y)dy\, dx.$$

If we recall the definition of the iterated integral,

$$\int_a^b \int_{g_1(x)}^{g_2(x)} f(x, y)dy\, dx = \int_a^b \phi(x)dx \qquad \text{where} \qquad \phi(x) = \int_{g_1(x)}^{g_2(x)} f(x, y)dy,$$

then it is easy to see how one-dimensional quadrature formulas can be used to evaluate multiple integrals.

Specifically, if we select nodes $a \le x_0 < x_1 < \cdots < x_n \le b$ and a quadrature formula

$$\int_a^b h(x)dx \simeq \sum_{i=0}^{n} A_i h(x_i),$$

then we can approximate the iterated integral as

$$\int_a^b \int_{g_1(x)}^{g_2(x)} f(x, y)dy\, dx \simeq \sum_{i=0}^{n} A_i\phi(x_i) \qquad \text{where} \qquad \phi(x_i) = \int_{g_1(x_i)}^{g_2(x_i)} f(x_i, y)dy.$$

To evaluate (approximately) the terms $\phi(x_i)$ in the expression $\sum_{i=0}^{n} A_i\phi(x_i)$, we select a quadrature formula for the interval $[g_1(x_i), g_2(x_i)]$ and use this formula to estimate $\phi(x_i)$:

$$\phi(x_i) \simeq \sum_{j=0}^{m} B_j f(x_i, y_j)$$

For example, suppose that we decide to use an "N-panel" compound Simpson's rule [such as (6.13a)] in the x-direction, and suppose that we use an "M-panel" compound Simpson's rule in the y-direction. We first set up the required $2N + 1$ nodes in the x-direction: $x_i = x_0 + ih/2$, $x_0 = a$, $h = (b - a)/N$. For each i, $0 \le i \le 2N$, we next approximate $\phi(x_i)$ by ψ_i where ψ_i is the M-panel Simpson's rule estimate to $\phi(x_i)$:

$$\psi_i = \frac{h_i}{6}\left[f(x_i, y_0) + f(x_i, y_M) + 2\sum_{j=1}^{M-1} f(x_i, y_i) + 4\sum_{j=0}^{M-1} f(x_i, \alpha_j)\right]$$

where [see (6.13a)] $h_i = [g_2(x_i) - g_1(x_i)]/M$, $y_j = y_0 + jh_i$, $y_0 = g_1(x_i)$, $\alpha_j = [y_j + y_{j+1}]/2$. Finally, we have an estimate to the multiple integral of the form

$$\int_Q\int f(x, y)\, dA \approx \frac{h}{6}\left[\psi_0 + \psi_{2N} + 2\sum_{j=1}^{N-1}\psi_{2j} + 4\sum_{j=0}^{N-1}\psi_{2j+1}\right].$$

In practice, "adaptive" quadrature rules (see Section 6.3) are frequently used for multiple integrals. In such an adaptive scheme, the nodes are selected "automatically" by the algorithm (the algorithm monitors the errors continually in order to decide the proper placement; that is, the algorithm selects points x_i at which to sample $\phi(x)$ in order to get an accurate numerical estimate of $\int_a^b \phi(x)dx$). Given a point x_i, the algorithm also adaptively selects nodes in the y-direction in $[g_1(x_i), g_2(x_i)]$ so that the quadrature approximation to

$$\phi(x_i) = \int_{g_1(x_i)}^{g_2(x_i)} f(x_i, y)dy$$

is sufficiently accurate (again, see Section 6.3).

It should be noted that numerical evaluations of multiple integrals (especially in dimensions larger than two) can be quite expensive in terms of function evaluations. For example if $f(x)$ and $g(x,y,z)$ are functions of comparable smoothness and if it requires approximately n evaluations of $f(x)$ to estimate $\int_a^b f(x)dx$ accurately, then we would expect to need on the order of n^3 evaluations of $g(x, y, z)$ to estimate

$$\iint_Q\int g(x, y, z)dV$$

[on the assumption that the volume of Q is about $(b - a)^3$]. This dramatic increase in the number of function evaluations required as the dimension grows is aptly termed the "curse of dimension" by Davis and Rabinowitz (1975). Finally, we mention that quadrature rules have been developed for certain standard regions such as m-dimensional cubes, m-dimensional spheres, etc.; and Stroud (1971) lists some of these rules.

6.2.6. The Euler-Maclaurin Summation Formula

In this section we show how the composite trapezoidal rule may be used for the problem of finding an approximate value for a series, $S = \sum_{j=0}^{\infty} a_j$. We attack this problem by observing that often we can find a function $f(x)$ such that $f(j) = a_j$ for all j and $f(x)$ can be integrated directly. Then the nth partial sum, $S_n =$

$\Sigma_{j=0}^{n} a_j$, is precisely equal to $(T_n(f) + 1/2(a_0 + a_n))$ by (6.12a) (where $[a, b] = [0, n]$ and $h = 1$). Thus we have the approximation

$$S_n - \frac{1}{2}(a_0 + a_n) = T_n(f) \approx \int_0^n f(x)\,dx \tag{6.16}$$

where, of course, $S \approx S_n$.

Formula (6.16) is called the Euler-Maclaurin summation formula and essentially reverses the problem of numerical integration since we are now approximating an unknown sum by a known integral. Since $b - a = n$ in this case, the error formula (6.12b) is of little practical use in analyzing this approximation. Instead we define the function $P_1(x) = x - [x] - 1/2$ where $[x] \equiv$ greatest integer in x. We can easily verify the following identity (by using integration by parts on the second integral):

$$\frac{1}{2}(f(j) + f(j+1)) = \int_j^{j+1} f(x)\,dx + \int_j^{j+1} P_1(x)f'(x)\,dx. \tag{6.17}$$

Now summing (6.17) from $j = 0$ to $j = n$, we obtain

$$T_n(f) = \int_0^n f(x)\,dx + \int_0^n P_1(x)f'(x)\,dx. \tag{6.18}$$

Through somewhat cumbersome, but straight-forward manipulations, $P_1(x)$ can be expanded in a trigonometric Fourier sine series and (6.18) can be successively integrated by parts to yield

$$\begin{aligned} T_n(f) &= (1/2)f(0) + f(1) + \cdots + f(n-1) + (1/2)f(n) \\ &= \int_0^n f(x)\,dx + \frac{B_2}{2!}(f'(n) - f'(0)) + \frac{B_4}{4!}(f'''(n) - f'''(0)) \\ &\quad + \cdots + \frac{B_{2j}}{(2j)!}(f^{(2j-1)}(n) - f^{(2j-1)}(0)) \\ &\quad + \int_0^n P_{2j+1}(x)f^{(2j+1)}(x)\,dx \end{aligned} \tag{6.19}$$

where

$$P_2'(x) = P_1(x), \qquad P_3'(x) = P_2(x), \text{ etc.,}$$

and where

$$B_{2j}/(2j)! \equiv (-1)^{j-1} \sum_{k=1}^{\infty} 2/(2k\pi)^{2j}, \qquad j > 1.$$

[Again see Davis and Rabinowitz (1975) for details.] The constants B_{2j} are called the *Bernoulli numbers,* and their values can be found in most mathematical tables. The first few Bernoulli numbers of even order are

$$B_0 = 1, \qquad B_2 = \frac{1}{6}, \qquad B_4 = -\frac{1}{30}, \qquad B_6 = \frac{1}{42}.$$

One important result that follows from formula (6.19) is the following alternative bound for the error of the composite trapezoidal rule.

Theorem 6.2

Let $g(x) \in C^{2j+1}[a, b]$ and let $h = (b - a)/N$; then

$$-e_N^T \equiv T_N(g) - \int_a^b g(x)\, dx = \frac{B_2}{2!} h^2(g'(b) - g'(a))$$
$$+ \frac{B_4}{4!} h^4(g'''(b) - g'''(a)) + \cdots + \frac{B_{2j}}{(2j)!} h^{2j}(g^{(2j-1)}(b) - g^{(2j-1)}(a))$$
$$+ h^{2j+1} \int_a^b P_{2j+1}\left(N\frac{x-a}{b-a}\right) g^{(2j+1)}(x)\, dx. \qquad (6.20a)$$

Proof. In formula (6.19) define $f(x) \equiv g(a + hx)$. ■

From Theorem 6.2 we immediately have the result stated in the corollary.

Corollary

Let $g(x) \in C^{(2j+1)}[a, b]$ with $g^{(2i-1)}(a) = g^{(2i-1)}(b)$ for $i = 1, 2, \ldots, j$. Then there exists a constant C such that

$$|e_N^T| = \left|\int_a^b g(x)\, dx - T_N(g)\right| \leq \frac{C}{N^{2j+1}}. \qquad (6.20b)$$

Formula (6.20b) emphasizes that the more periodic character $g(x)$ has, the stronger the composite trapezoidal rule becomes as a candidate for numerical integration.

Finally, the theory of Bernoulli polynomials can be used in deriving another form of the Euler-Maclaurin summation formula although developing the properties of Bernoulli polynomials is beyond our means in this brief introduction to summation. (The interested reader is referred to Ralston, 1965.) With the notation of Theorem 6.2, it can be shown for $g \in C^{2j+2}[a, b]$ that

$$T_N(g) - \int_a^b g(x)\, dx = \sum_{i=1}^{j} \frac{B_{2i}}{(2i)!} h^{2i}[g^{(2i-1)}(b) - g^{(2i-1)}(a)]$$
$$+ (b - a)h^{2j+2} \frac{B_{2j+2}}{(2j+2)!} g^{(2j+2)}(\eta). \qquad (6.20c)$$

Note that (6.20c) is just formula (6.20a) rewritten with a derivative evaluation replacing the integral that appears on the right-hand side of (6.20a).

EXAMPLE 6.5. There is a variety of information that can be obtained from the Euler-Maclaurin summation formula. For example, consider the problem of evaluating the sum:

$$S = 1 + \frac{1}{2} + \frac{1}{3} + \cdots + \frac{1}{20} = \sum_{k=0}^{19} \frac{1}{k+1}.$$

To estimate S, let us choose $h = 1$ and $[a, b] = [0, 19]$; and let $g(x) = 1/(x+1)$ so that $-e_{19}^{T} = T_{19}(g) - \int_0^{19} g(x)\,dx$. Since we have $S = T_{19}(g) + \frac{1}{2}[g(0) + g(19)] = T_{19}(g) + \frac{21}{40}$, we find

$$S = \int_0^{19} g(x)\,dx + \frac{21}{40} - e_{19}^{T}$$

where $-e_{19}^{T}$ is given by (6.20c). Now, $\int_0^{19} g(x)\,dx = \ln(x+1)\big|_0^{19} = \ln(20)$, so $S = \ln(20) + \frac{21}{40} - e_{19}^{T}$. By evaluating a few of the terms in the expression for e_{19}^{T}, we can obtain estimates for S (where the actual value of S is 3.597740). For example,

$$(j = 0) \qquad S \approx S_0 = \ln(20) + \frac{21}{40} = 3.520732$$

$$(j = 1) \qquad S \approx S_1 = S_0 + \frac{B_2}{2!}[g'(19) - g'(0)] = S_0 + \left(\frac{1}{12}\right)\left(\frac{399}{400}\right) = 3.603857$$

$$(j = 2) \qquad S \approx S_2 = S_1 + \frac{B^4}{4!}[g'''(19) - g'''(0)]$$

$$= S_1 - \frac{1}{720}[g'''(19) - g'''(0)] = 3.595524.$$

For $j = 2$, we find from (6.20c) that $S - S_2 = 19(B_6/6!)g^{(vi)}(\eta)$, which gives a bound of $|S - S_2| \leq 19/42 = 0.45$. The crudeness of this bound brings home the fact that caution must be exercised in using the Euler-Maclaurin summation formula. Part of the problem is the growth of the Bernoulli numbers, for it can be shown that $|B_{2i}| \to \infty$ since $|B_{2i}| \approx 2(2i)!/(2\pi)^{2i}$.

Another interesting case of (6.20c) is given when $g(x)$ is a polynomial. For $g \in \mathcal{P}_{2j+1}$, we see that $g^{(2j+2)}(\eta) = 0$. To see how this can be used, consider the sum $S = 1 + 2 + \cdots + n$. Again, choose $h = 1$ and $[a, b] = [0, n-1]$, and let $g(x) = x + 1$. Then $S = T_{n-1}(g) + \frac{1}{2}[g(0) + g(n-1)] = T_{n-1}(g) + (n+1)/2$. As before,

$$S = \int_0^{n-1} g(x)\,dx + \frac{n+1}{2} - e^{T}{}_{n-1};$$

but in this case since $g'''(x) \equiv 0$, we have $-e_{n-1}^{T} = (B_2/2)[g'(n-1) - g'(0)] = 0$. Therefore

$$S = \int_0^{n-1} g(x)\,dx + \frac{n+1}{2} = \frac{n^2-1}{2} + \frac{n+1}{2} = \frac{n(n+1)}{2},$$

which is the familiar formula for the sum of the first n integers. This technique will clearly work to evaluate $S = 1^k - 2^k + \cdots + n^k$ for any positive integer k.

Finally, choosing $j = 1$ in formula (6.20c), we obtain the corrected trapezoidal rule $CT(g)$ and the error of the corrected trapezoidal rule, e_N^{CT}. Choosing successively larger values of j, we can derive higher-order corrections to the trapezoidal rule. For example with $j = 2$, we obtain the quadrature formula

$$Q(g) = T_N(g) - \frac{h^2}{12}[g'(b) - g'(a)] + \frac{h^4}{720}[g'''(b) - g'''(a)]$$

and the error term

$$\int_a^b g(x) - Q(g) = -(b - a)h^6 \frac{B_6}{6!} g^{(vi)}(\eta).$$

PROBLEMS, SECTION 6.2.6

1. Estimate $\int_Q\int \frac{y}{1 + xy}\, dA$ where Q denotes the square $0 \le x \le 1$, $0 \le y \le 1$. Use T_N in both the x- and y-directions for $N = 2^k$, $k = 5, 6, 7$. Compare your estimate with the exact result.
2. Using S_N', repeat Problem 1.
3. Repeat Problem 2 for $f(x, y) = 1/\sqrt{x^2 + y^2}$ over the trapezoidal region Q bounded by the lines $y = 0$, $y = x$, $x = 1$, $x = 2$. Compare your estimate with the exact result, which can be obtained by changing to polar coordinates.
4. Write an adaptive routine for multiple integrals when Q is the rectangle $0 \le x \le a$, $0 \le y \le b$. This routine can use the program in Problem 5, Section 6.2.4, or a computer-center routine, such as CADRE in the IMSL library. Specifically,

$$\int_Q\int f(x, y)dA = \int_0^a \phi(x)dx \qquad \text{where} \qquad \phi(x) = \int_0^b f(x, y)dy;$$

so whenever $\phi(x_i)$ is required, it can be calculated by adaptively estimating $\int_0^b f(x_i, y)dy$. Test your program on the integral in Problem 1 with an accuracy request of 10^{-2}.
5. Following Example 6.5, find a formula for $1^4 + 2^4 + 3^4 + \cdots + n^4$.

6.3 ADAPTIVE QUADRATURE

Adaptive quadrature is a process designed to use a given set of quadrature rules to approximate $I(f)$ to within a given error tolerance, $\varepsilon > 0$, where the computer is programmed to make the decisions that produce the approximation with optimal efficiency and thus to relieve the user (as much as possible) of the necessity of analyzing the accuracy of the results. We shall concentrate here on describing an adaptive quadrature routine based on composite Simpson rules although extensions to other rules should become evident to the reader.

For the sake of intuition, let us suppose that $f(x)$ is "badly behaved" only on some small segment, $[\alpha, \beta]$, of the entire interval of integration $[a, b]$. Then composite Simpson's rules with relatively few points will produce fairly accurate approximations for $\int_a^\alpha f(x)\,dx$ and $\int_\beta^b f(x)\,dx$ although the entire estimate for $I(f) = \int_a^b f(x)\,dx$ may be badly in error. If we were to take the seemingly natural course of halving the step size over $[a, b]$, we would not appreciably increase the accuracy of our estimates for $\int_a^\alpha f(x)\,dx$ and $\int_\beta^b f(x)\,dx$, and we still might not have an acceptable estimate for $I(f)$. Thus the new work done in $[a, \alpha]$ and $[\beta, b]$ is essentially wasted whereas more refinement may still be necessary on $[\alpha, \beta]$.

In the area of mathematical software, an "adaptive" algorithm is one that automatically varies the problem-solving strategy as the problem being solved becomes easier or harder. For the problem above, an adaptive routine would use small steps in $[\alpha, \beta]$ and relatively large steps in $[a, \alpha]$ and $[\beta, b]$. However, an adaptive algorithm typically considers more than just computational efficiency. In particular, we would like to have a software package of an "automatic" nature; that is, a package that accepts an interval $[a, b]$, a function $f(x)$, and a tolerance ε, and that then returns an estimate EST where

$$\left| \int_a^b f(x)dx - \text{EST} \right| \le \varepsilon.$$

It is fairly easy to design an elementary version of an adaptive quadrature routine using Simpson's rule; but in order to construct such an automatic package, we will have to provide some sort of an error-sensing device so that the algorithm can control the quadrature errors in intermediate steps in order to meet the final error tolerance. To see how this procedure might be done, suppose that $[c, d]$ is some (typically small) subinterval of $[a, b]$ where we are trying to estimate $\int_a^b f(x)dx$. If $d - c = H$ and if we apply Simpson's rule on $[c, d]$, then by (6.11b) we have

$$\int_c^d f(x)dx - S(f) = \frac{-f^{(iv)}(\eta)}{90}\left(\frac{H}{2}\right)^5$$

where η is in $[c, d]$. If we next apply a two-panel composite Simpson's rule S_2 on $[c, d]$, we have by (6.13b) that

$$\int_c^d f(x)dx - S_2(f) = \frac{-f^{(iv)}(\gamma)}{90}\,2\left(\frac{H}{4}\right)^5$$

where γ is in $[c, d]$. We now make a critical assumption: that $f^{(iv)}(x)$ is reasonably constant over $[c, d]$ so that $f^{(iv)}(\eta) \simeq f^{(iv)}(\gamma)$; this assumption will be valid if $[c, d]$ is small compared with how rapidly $f^{(iv)}(x)$ varies in $[c, d]$.

If we set $I = \int_c^d f(x)dx$, the approximation $f^{(iv)}(\eta) \simeq f^{(iv)}(\gamma)$ leads us to

$$I - S_2(f) \simeq \frac{1}{16}[I - S(f)].$$

In words, we expect the error of a two-panel Simpson's rule to be about 16 times as small as the error of a one-panel Simpson's rule when $f^{(iv)}(x)$ is well behaved in $[c, d]$. Now, assuming that $16[I - S_2(f)] \simeq I - S(f)$, we subtract $(I - S_2(f))$ to get the equivalent approximation $15[I - S_2(f)] \simeq S_2(f) - S(f)$, or

$$I - S_2(f) \simeq \frac{1}{15}[S_2(f) - S(f)].$$

Since we can easily calculate $S_2(f)$ and $S(f)$, we have a practical and *computable* error estimate for using a two-panel Simpson's rule on $[c, d]$; the error is about $[S_2(f) - S(f)]/15$. In many applications, it is more important that the estimate to $\int_a^b f(x)dx$ be reliable than that the quadrature routine execute quickly. Assuming such to be the case here, we will employ a more conservative upper estimate for the error over $[c, d]$: $|I - S_2(f)| \leq |S_2(f) - S(f)|/2$.

By now, the outlines of an adaptive quadrature routine should be nearly apparent. If we have already calculated an accurate estimate to $\int_a^c f(x)dx$, we next try to estimate $\int_c^d f(x)dx$ by a two-panel Simpson's rule. If we can do this accurately, we can add the estimate for $\int_a^c f(x)dx$ and have an approximation to $\int_a^d f(x)dx$. Moreover, we have a practical way to assess the accuracy of the two-panel Simpson's rule for $\int_c^d f(x)dx$; and if the approximation is not sufficiently accurate on $[c, d]$, we can try for an accurate approximation on the smaller interval $[c, e]$ where $e = (c + d)/2$. In order to implement the algorithm sketched above, we need to be a bit more precise. In particular, since we want to estimate $\int_a^b f(x)dx$ to within ε, it seems reasonable to assign an appropriate proportion of the total error ε to the subinterval $[c, d]$. Thus, we will judge $S_2(f)$ to be an acceptable approximation to $\int_c^d f(x)dx$ if the (conservative) error estimate satisfies

$$\frac{|S_2(f) - S(f)|}{2} \leq \varepsilon \frac{(d - c)}{(b - a)}.$$

That is, we budget to the interval $[c, d]$ an appropriate fraction of the total error. This decision is also conservative, for it tacitly assumes that the errors are all of the same sign and thus accounts for the worst possible case. If $|S_2(f) - S(f)| > 2\varepsilon\,(d - c)/(b - a)$, then we attempt S_2 on a smaller interval $[c, e]$ where $e = (c + d)/2$; and we ask that $|S_2(f) - S(f)| \leq 2\varepsilon\,(e - c)/(b - a)$. By cutting the interval $[c, d]$ in half, we expect the error for S_2 to be reduced by a factor of $2^5 = 32$ [see (6.11b)]. Since the proportion of the error allocated to $[c,$

$d]$ is cut only in half when we reduce to $[c, e]$ while the quadrature error is cut by about 2^5, we expect that repeated halvings will eventually produce an interval $[c, q]$ where

$$\frac{|S_2(f) - S(f)|}{2} < \varepsilon \frac{(q - c)}{(b - a)}.$$

To see that this strategy of apportioning the errors is effective, let EST(r, s) denote our approximation to $\int_r^s f(x)dx$ and let err(r, s) denote the error

$$\text{err}(r, s) = \int_r^s f(x)dx - \text{EST}(r, s).$$

If $|\text{err}(a, z)| \le \varepsilon\,(z - a)/(b - a)$ with $z < t$ and if $|\text{err}(z, t)| \le \varepsilon\,(t - z)/(b - a)$, then clearly

$$|\text{err}(a, t)| \le |\text{err}(a, z)| + |\text{err}(z, t)| \le \varepsilon \frac{(z - a)}{(b - a)} + \varepsilon \frac{(t - z)}{(b - a)} = \varepsilon \frac{(t - a)}{(b - a)}.$$

In particular, when we finally reach $t = b$, we will have $|\text{err}(a, b)| \le \varepsilon$ as desired.

An important feature of the adaptive process is that only two additional evaluations of $f(x)$ are needed at each step of the algorithm. An example should serve to make this clear. Let us start with the interval $[a, b]$ and let $a = r_1 < s_1 < t_1 < u_1 < v_1 = b$ denote the five equally spaced points used in forming $S_2(f)$ and $S(f)$. We test $|S_2(f) - S(f)| \le 2\varepsilon(v_1 - r_1)/(b - a)$; and if the test succeeds, we use $S_2(f)$ as the estimate to the integral over $[r_1, v_1]$. On the other hand, if the test fails, we retreat to a smaller interval $[r_2, v_2]$ where $r_1 = r_2 < s_2 < t_2 < u_2 < v_2 = t_1$. In order to calculate $S_2(f)$ and $S(f)$ on $[r_2, v_2]$, we recall that we have already evaluated $f(x)$ at $r_2 = r_1$, $t_2 = s_1$, and $v_2 = t_1$. Thus, only two evaluations, those at s_2 and u_2, are necessary. We test $|S_2(f) - S(f)| \le 2\varepsilon(v_2 - r_2)/(b - a)$; and if the test fails again, we retreat to $[r_3, v_3]$ where $r_2 = r_3 < s_3 < t_3 < u_3 < v_3 = t_2$. [Again, we need calculate $f(x)$ only at s_3 and u_3.] If the test now succeeds on $[r_3, v_3]$, we are ready to advance to the next interval $[t_2, v_2]$. [Recall that $[r_2, v_2]$ was halved to produce $[r_3, v_3] = [r_2, t_2]$; so we have yet to calculate the contribution from $[t_2, v_2]$. Also, we need only two evaluations of $f(x)$ in $[t_2, v_2]$; and for purposes of programming the algorithm, it would be logical to relabel $[t_2, v_2]$ as $[\hat{r}_2, \hat{v}_2]$ where $t_2 = \hat{r}_2 < \hat{s}_2 < \hat{t}_2 < \hat{u}_2 < \hat{v}_2 = v_2$.]

As the example above indicates, we can save all past evaluations so that the algorithm proceeds according to the following steps.

1. Whether we advance or retreat, we define a new interval $[r_i, v_i]$.
2. Make the necessary evaluations of $f(x)$ at s_i and u_i.
3. Test $|S_2(f) - S(f)| \le 2\varepsilon(v_i - r_i)/(b - a)$.

4. If the test in (3) succeeds, add $S_2(f)$ to the running estimate for the integral over $[a, r_i]$.
5. If the test in (3) fails, prepare to halve the interval.

The bookkeeping details of this algorithm are not overly complicated, and we outline one possible record-keeping scheme in Problem 1. Although it is not included in steps (1) through (5) above, we also need some sort of a stopping criterion. Since we are working from left to right across $[a, b]$, the algorithm will have succeeded if for some interval $[r_i, v_i]$ we pass the test in step (3) and if $v_i = b$. On the other hand, the algorithm may fail. For example if we try to estimate

$$\int_1^3 \frac{1}{x - \sqrt{2}}\, dx,$$

we want the algorithm to fail since the integral does not exist. One simple exit criterion is to test $v_i - r_i \leq \text{HMIN}$ where HMIN is input to the algorithm. That is, we say the algorithm has failed if we cannot satisfy the error test (3) without using a subdivision smaller than some input value HMIN. For the integral displayed above, we would expect to be able to advance successfully from $x = 1$ until we get near $x = \sqrt{2}$. For x near $\sqrt{2}$, the integrand is changing rapidly; we would anticipate that very small subintervals would be required to maintain accuracy; and thus the algorithm should fail as it nears $x = \sqrt{2}$; however, the algorithm should have a reasonable approximation to the integral over $[a, \alpha]$ where $\alpha < \sqrt{2}$ was the last point reached before the failure.

A variety of adaptive quadrature routines have been published recently in the mathematical software literature. Many of these routines are quite sophisticated in their error-control strategy, but the underlying philosophy is similar to the elementary adaptive quadrature described in this section. In any adaptive routine, a fundamental problem is deciding whether or not an approximation is acceptable; this is the subject of much research; see for example Clenshaw and Curtis (1960), and Rowland and Varol (1972) for results concerning an adaptive Simpson's rule. Of course, once the error-control strategy of any routine is known, it is quite easy to construct a function $f(x)$ that will fool the algorithm. Thus, one thrust of current research is to determine ways to validate adaptive routines and to specify classes of problems for which the routines are reliable.

PROBLEMS, SECTION 6.3

1. Write a subroutine to implement an adaptive Simpson's rule. One form might be Subroutine ADSIMP (A, B, HMIN, TOL, EST, IERROR) where IERROR is a flag denoting either success or failure of the algorithm. To make the routine more

efficient, function values should be stored; and some sort of record-keeping scheme is required for this storage. One possible scheme is outlined below. This scheme can be visualized as creating a stack of intervals $[r_i, v_i]$, where $S_2(f)$ and $S(f)$ will be applied, and also creating stacks of function values $fr_i, fs_i, ft_i, fu_i, fv_i$ where $r_i < s_i < t_i < u_i < v_i$ are equally spaced in $[r_i, v_i]$. To store this information, arrays R, V, FR, FS, FT, FU, FV are required where the arrays can be dimensioned at 20. A level pointer can be used to keep track of the information. In a brief form, the algorithm proceeds as follows.

Initialization phase: Give an input tolerance ε and the interval $[a, b]$, set est = 0, $r_1 = a$, $v_1 = b$; calculate $fr_1, fs_1, ft_1, fu_1, fv_1$; and set the level pointer to 1.

General step for the interval $[r_i, v_i]$ with level pointer at i; Calculate $S_2(f)$, $S(f)$, err $= 2\varepsilon(v_i - r_i)/(b - a)$; and test $|S_2(f) - S(f)| \leq$ err.

If the test $|S_2(f) - S(f)| \leq$ err fails,

set $r_{i+1} = r_i$

set $v_{i+1} = (r_i + v_i)/2$

set $fr_{i+1} = fr_i$

set $ft_{i+1} = fs_i$

set $fv_{i+1} = ft_i$

calculate fs_{i+1} and fu_{i+1}

set the level pointer to $i + 1$

test to see that $v_{i+1} - r_{i+1} \geq$ HMIN; if $[r_{i+1}, v_{i+1}]$ is too small, return with r_{i+1} in b and a signal that the algorithm could not integrate $f(x)$ beyond $x = r_{i+1}$.

If the test $|S_2(f) - S(f)| \leq$ err succeeds,

add $S_2(f)$ to est

test to see if $v_i = b$ and return with a signal of success if $v_i = b$ (note that $v_i = b$ happens only when the level pointer has the value 1)

set $r_{i-1} = v_i$

set $fr_{i-1} = fv_i$

set $ft_{i-1} = fu_{i-1}$

calculate fs_{i-1} and fu_{i-1}

set the level pointer to $i - 1$.

2. Adaptive quadrature depends on an error monitor, and the routine in Problem 1 uses an error monitor derived from an estimate $|I - S_2(f)| \simeq |S_2(f) - S(f)|/15$. This estimate is valid if $[a, b]$ is small, but may not be good if $[a, b]$ is large. Write a main program that uses ADSIMP to evaluate $\int_c^d f(x)dx$. One way to partition $[c, d]$ is to define k to be the greatest integer in $10(d - c)$ and then subdivide $[c, d]$ into k equal subintervals. Test your program as in Problem 5, Section 6.2.4; use a tolerance of 10^{-8}.

3. Use the program in Problem 2 to estimate the Fresnel integral

$$C(x) = \int_0^x \cos\left(\frac{\pi}{2}t^2\right)dt$$

at $x = 5$. [to seven places, $C(5) = 0.5636312$.]

4. Estimate the following integrals with the program in Problem 2.

a) $\int_0^1 (1 - x^2)^{1.5}\,dx$ b) $\int_0^3 \sqrt{x}\,dx$ c) $\int_0^1 x \sin x\,dx$

d) $\int_0^2 \frac{1}{x - \sqrt{2}}\,dx$ e) $\int_0^5 e^{-x}dx$ f) $\int_{-1}^1 x^5 \cos x\,dx$

5. Modify and test the program in Problem 4, Section 6.2.6, so that the modified program uses ADSIMP.

6.4 RICHARDSON EXTRAPOLATION AND NUMERICAL DIFFERENTIATION

Richardson extrapolation is a procedure similar to Aitken's Δ^2-process. That is, we compute several estimates of some quantity and then combine these estimates in such a way as to (one hopes) provide a better estimate. To be more precise, let us assume that an *unknown* quantity, a_0, is approximated by a quantity $A(h)$ where $A(h)$ can be explicitly computed for each value of h, $h \neq 0$. Further, we assume that $a_o = \lim_{h\to 0} A(h)$ (typically, we consider only positive values of h). Finally, we assume that

$$a_0 = A(h) + \sum_{i=k}^{m} a_i h^i + C_m(h)h^{m+1} \tag{6.21}$$

where $C_m(h)$ is a function of h, the constants a_i are independent of h, and $a_k \neq 0$. We emphasize that to use the techniques of this section, we do not need to know explicitly what the coefficients a_i are, nor do we need to know the function $C_m(h)$.

EXAMPLE 6.6. As a simple example, suppose we wish to approximate $f'(\alpha)$ where $f(x)$ is differentiable at $x = \alpha$. From the definition of the derivative, we have that

$$f'(\alpha) = \lim_{h\to 0} \frac{f(\alpha + h) - f(\alpha)}{h}.$$

Although we will see this approximation is not optimal, we might think of approximating $a_0 = f'(\alpha)$ by $A(h) = [f(\alpha + h) - f(\alpha)]/h$ for small h. Doing this, we certainly have that

$\lim_{h \to 0} A(h) = a_0$. Using a Taylor's series expansion for $f(x)$ about $x = \alpha$, we easily see that a_0 and $A(h)$ are related by an expression of the form (6.21):

$$a_0 = A(h) - \left[\frac{f''(\alpha)}{2!} h + \cdots + \frac{f^{(m+1)}(\alpha)}{(m+1)!} h^m\right] - \frac{f^{(m+2)}(\eta_h)}{(m+2)!} h^{m+1}.$$

We have written the Taylor's series error term as

$$\frac{f^{(m+2)}(\eta_h)}{(m+2)!} h^{m+1}$$

to emphasize the dependence of the mean-value point, η_h, on the step size h. In the context of (6.21), we have that

$$C_m(h) = \frac{-f^{(m+2)}(\eta_h)}{(m+2)!}$$

and that

$$a_i = \frac{-f^{(i+1)}(\alpha)}{(i+1)!} \qquad \text{for} \qquad i = 1, 2, \ldots, m.$$

Note that the coefficients a_i are independent of h and that we probably do not know the values of a_i [since we do not even know $f'(\alpha)$].

Another example of the situation described in (6.21) is provided by the composite trapezoidal rule [see in particular (6.20c)] where we have $a_0 = \int_a^b g(x)\,dx$, $A(h) = T_N(g)$. [Here, $h = (b - a)/N$; so $T_N(g)$ is a function of h.] The constants a_i and the function $C_m(h)$ can be found from (6.20c).

Formula (6.21) provides the basis for a simple technique known as *Richardson extrapolation* or *extrapolation to the limit*. The basic idea is to compute two different approximations, $A(h_1)$ and $A(h_2)$, to a_0. We can then form a combination of $A(h_1)$ and $A(h_2)$ in such a way as to eliminate the h^k term in (6.21) and thus obtain (for small h) a better approximation to a_0. This combining is usually done by choosing $h_1 = h$ and $h_2 = rh$ where $0 < r < 1$ (normally, $r = \frac{1}{2}$). Carrying this idea out, we find from (6.21) that

$$\begin{aligned} a_0 &= A(h) + \sum_{i=k}^{m} a_i h^i + C_m(h) h^{m+1} \\ a_0 &= A(rh) + \sum_{i=k}^{m} a_i (rh)^i + C_m(rh)(rh)^{m+1}. \end{aligned} \tag{6.22}$$

Multiplying the top equation by r^k and subtracting from the bottom equation in (6.22), we find:

$$\begin{aligned} a_0 - r^k a_0 = A(rh) - r^k A(h) &+ \sum_{i=k+1}^{m} a_i [r^i - r^k] h^i \\ &+ [C_m(rh) r^{m+1} - C_m(h) r^k] h^{m+1}. \end{aligned} \tag{6.23}$$

Thus with $b_i = a_i[r^i - r^k]/[1 - r^k]$ and $\tilde{C}_m(h) = [C_m(rh)r^{m+1} - C_m(h)r^k]/[1 - r^k]$, we see from (6.23) that

$$a_0 = \frac{A(rh) - r^k A(h)}{1 - r^k} + \sum_{i=k+1}^{m} b_i h^i + \tilde{C}_m(h)h^{m+1}. \tag{6.24}$$

That is, $B(h) = [A(rh) - r^k A(h)]/[1 - r^k]$ is an approximation to a_0 that satisfies (6.21) except that the error begins with h^{k+1} instead of h^k. Therefore we would tend to feel that $B(h)$ is probably more accurate (for small h) than $A(h)$.

Note that in order to form a new approximation for $A(h)$, all we needed to know was the integer k. That is, it is not necessary to know the coefficients a_i in (6.21) or the function $C_m(h)$, but it is necessary to know k where $a_0 - A(h) = a_k h^k + \cdots$. If b_{k+1} in (6.24) is different from zero, we can carry out the extrapolation process again and find an approximation whose error term starts with h^{k+2}. This idea can be mechanized most naturally as follows. For $m = 0, 1, 2, \ldots$ define $A_{0,m} = A(r^m h)$. By (6.24), extrapolation with $A_{0,m}$ and $A_{0,m+1}$ gives

$$A_{1,m} = \frac{A_{0,m+1} - r^k A_{0,m}}{1 - r^k}; \tag{6.25}$$

and in general, we have

$$A_{i+1,m} = \frac{A_{i,m+1} - r^{k+i} A_{i,m}}{1 - r^{k+i}}, \qquad m = 0, 1, \ldots. \tag{6.26}$$

Using (6.26), we have an easily programmed algorithm to construct Table 6.1 one column at a time from left to right. As with all extrapolation procedures, we must be somewhat careful. For example, if in (6.24) it turns out that $b_{k+1} = 0$, then it is possible that a further extrapolation using (6.25) might cause a loss of accuracy in our approximation.

TABLE 6.1. Richardson extrapolation of *n*th order.

$A_{0,0}$					
$A_{0,1}$	$A_{1,0}$				
$A_{0,2}$	$A_{1,1}$	$A_{2,0}$			
$A_{0,3}$	$A_{1,2}$	$A_{2,1}$	$A_{3,0}$		
$\vdots$	$\vdots$	$\vdots$	$\vdots$	$\ddots$	
$A_{0,n}$	$A_{1,n-1}$	$A_{2,n-2}$	$A_{3,n-3}$	$\cdots$	$A_{n,0}$

EXAMPLE 6.7. An approximation that is sometimes used to estimate $f'(\alpha)$ is

$$A(h) = \frac{f(\alpha + h) - f(\alpha - h)}{2h}, \qquad h > 0. \tag{6.27}$$

Clearly, $\lim_{h \to 0} A(h) = f'(\alpha)$; moreover, using a Taylor's series expansion about $x = \alpha$, we see (Problem 3) that

$$A(h) = f'(\alpha) + \frac{f'''(\alpha)}{3!} h^2 + \frac{f^{(v)}(\alpha)}{5!} h^4 + \cdots. \tag{6.28}$$

As frequently happens, only even powers of h occur in (6.28). A moment's reflection shows that we need to use r^2, r^4, r^6, etc. in (6.26) in order to be successful in extrapolation. That is, since (Problem 4)

$$B(h) = \frac{A(rh) - r^2 A(h)}{1 - r^2} = f'(\alpha) + b_4 h^4 + \cdots, \tag{6.29}$$

we see that the next extrapolation must be of the form $[B(rh) - r^4 B(h)]/[1 - r^4]$. In general, we obtain as in (6.26)

$$A_{i+1,\, m} = \frac{A_{i,\, m+1} - r^{k+2i} A_{i,\, m}}{1 - r^{k+2i}}, \qquad i = 0, 1, \ldots. \tag{6.30}$$

Formula (6.30) should be used whenever we have

$$a_0 = A(h) + a_k h^k + a_{k+2} h^{k+2} + a_{k+4} h^{k+4} + \cdots.$$

To demonstrate this fact on a specific function, consider the Gamma function that is defined by

$$\Gamma(x) = \int_0^\infty t^{x-1} e^{-t}\, dt \qquad \text{for } x > 0.$$

Suppose we wish to estimate $\Gamma'(1.5)$ and use (6.27) and Richardson extrapolation. Choosing $h = 0.4$ and $r = \frac{1}{2}$, we set up a table similar to Table 6.1, in which the necessary values of $\Gamma(x)$ are obtained from a mathematical handbook. We find first

$$A_{0,\,0} = A(h) = \frac{\Gamma(1.9) - \Gamma(1.1)}{0.8} = 0.013019$$

$$A_{0,\,1} = A\left(\frac{h}{2}\right) = \frac{\Gamma(1.7) - \Gamma(1.3)}{0.4} = 0.027920$$

$$A_{0,\,2} = A\left(\frac{h}{4}\right) = \frac{\Gamma(1.6) - \Gamma(1.4)}{0.2} = 0.031258$$

$$A_{0,\,3} = A\left(\frac{h}{8}\right) = \frac{\Gamma(1.55) - \Gamma(1.45)}{0.1} = 0.032069.$$

Using (6.30), we then find:

0.013019			
0.027920	0.032887		
0.031258	0.032371	0.032337	
0.032069	0.032339	0.032337	0.032337.

If fact, $\Gamma'(1.5) = 0.032338$ to the places shown.

6.4.1. Romberg Integration

Classical Romberg integration is a numerical integration procedure that amounts to using Richardson extrapolation on the composite trapezoidal rule $T_N(g)$. It is customary to use $h = b - a$ and $r = \frac{1}{2}$ when using Romberg integration to approximate $I(g) = \int_a^b g(x)\,dx$. Thus we define

$$T_{0,m} = \frac{b-a}{2^m}\left[\frac{1}{2}g_0 + g_1 + \cdots + g_{s-1} + \frac{1}{2}g_s\right] \tag{6.31}$$

where $g_i = g(x_i)$, $x_i = a + (i(b-a)/2^m)$, and $s = 2^m$. This choice of h and r means that we can use $T_{0,m}$ in the computation of $T_{0,m+1}$, and save half the work in evaluating (or looking up) $g(x)$ when forming $T_{0,m+1}$.

By (6.20c), we see that the error term contains only even powers of h; so (6.30) is the appropriate form of Richardson extrapolation to use. Thus we have

$$T_{1,m} = \frac{T_{0,m+1} - \frac{1}{4}T_{0,m}}{1 - \frac{1}{4}}$$

or in general

$$T_{i,m} = \frac{T_{i-1,m+1} - (\frac{1}{4})^i T_{i-1,m}}{1 - (\frac{1}{4})^i}, \qquad i = 1, 2, \ldots .$$

This equation can be simplified to the more convenient form

$$T_{i,m} = T_{i-1,m+1} + \frac{T_{i-1,m+1} - T_{i-1,m}}{4^i - 1}, \qquad i = 1, 2, \ldots . \tag{6.32}$$

Since Romberg integration is an extrapolation procedure, some care must be exercised in the application of Romberg integration. As a check (when forming a table similar to Table 6.1), we can calculate the ratios

$$R_{i,m} = \frac{T_{i,m} - T_{i,m-1}}{T_{i,m+1} - T_{i,m}}. \tag{6.33}$$

As is shown in Problem 6, we expect the ratios $R_{i,m}$ to have the approximate value $R_{i,m} \approx 4^{i+1}$. If such is not the case, we should suspect that roundoff error is contaminating our estimates and we should stop the extrapolation at this point.

EXAMPLE 6.8. As in Example 6.4, we consider the problem of calculating

$$Si(1) = \int_0^1 \frac{\sin(t)}{t}\,dt.$$

The exact value of $Si(1)$ to eight places is $Si(1) = 0.94608307$. The result of Romberg integration is given in Table 6.2. Note that only nine evaluations of the integrand $\sin(t)/t$ are required to construct this table and that $T_{3,0}$ is correct to eight places.

TABLE 6.2

$T_{0,i}$	$T_{1,i}$	$T_{2,i}$	$T_{3,i}$
0.92073549			
0.93979328	0.94614588		
0.94451352	0.94608693	0.94608300	
0.94569086	0.94608331	0.94608307	0.94608307

PROBLEMS, SECTION 6.4.1

1. Repeat Example 6.7 for $f(x) = \sqrt{x}$; estimate $f'(2)$ with $h = .8$ and $r = 1/2$.
2. Use Richardson extrapolation to estimate $f'(x)$ at $x = 1$ where $f(x) = \ln(x)$. In Table 6.1, use $n = 4$ and $h = 0.4$ and $r =$. Try both forms of $A(h)$: the form given in Example 6.6 and that given in Example 6.7.
3. Show that formula (6.28) is valid by writing down the Taylor's series expansion for $f(x + h)$ and $f(x - h)$ and expanding about $x = \alpha$ in both cases.
4. Verify the expansion in (6.29).
5. Write a computer program to carry out Romberg integration; test your program on the Fresnel integral given in Problem 3, Section 6.3. To determine (in a rough fashion) the accuracy of your answers, print out the ratios defined in (6.33).
6. The fundamental assumption of Richardson extrapolation can be seen from (6.21). We are assuming that $a_0 - A(h) = a_k h^k + a_{k+1} h^{k+1} + \cdots$ and that eliminating $a_k h^k$ will lead to a better approximation. That is, we are assuming that $a_k h^k$ is the dominant term in the error $a_0 - A(h)$.

   a) Show that the assumption above means that the ratio

   $$[A_{0,m} - A_{0,m-1}]/[A_{0,m+1} - A_{0,m}]$$

   should be approximately r^{-k}.

   b) Use part (a) to establish that the ratios $R_{i,m}$ defined in (6.33) should be approximately 4^{i+1} if Romberg integration is proceeding without a substantial error.
7. Calculate the appropriate ratios, as defined in Problem 6, for the tables in Examples 6.7 and 6.8.
8. Show that $T_{1,m}$ defined in (6.32) corresponds to the composite Simpson's rule. (However, there is no relation between $T_{k,m}$ and Newton-Cotes rules for $k > 2$.)

6.5 GAUSSIAN QUADRATURE

As we previously remarked in Section 6.2, if we let the quadrature weights, $\{A_j\}_{j=0}^n$, *and* the quadrature nodes, $\{x_j\}_{j=0}^n$, be treated as unknown variables, then the equations of (6.3),

$$\int_a^b x^k w(x)\,dx = \sum_{j=0}^n A_j x_j^k, \qquad 0 \le k \le 2n + 1, \tag{6.34}$$

represent a nonlinear system of $(2n + 2)$ equations in $(2n + 2)$ unknowns. In 1814, Gauss was able to show that this system of equations has a unique solution for the unknowns $\{A_j\}_{j=0}^n$ and $\{x_j\}_{j=0}^n$ when $w(x) \equiv 1$ and $[a, b] = [-1, 1]$.

Once again we return to the notation of Section 5.3, in which we let $\langle f, g\rangle \equiv \int_a^b f(x)g(x)w(x)\, dx$, and $\|f\| \equiv \langle f, f\rangle^{1/2}$. We let $\{q_j(x)\}_{j=0}^\infty$ denote the *monic* orthogonal polynomials and $\{p_j(x)\}_{j=0}^\infty$ the orthonormal polynomials with respect to this inner product where degree $(q_j(x))$ = degree $(p_j(x)) = j$ for each j. We first prove the following theorem, which localizes the zeros of each $p_j(x)$. [Recall that for each j, $p_j(x)$ is a constant multiple of $q_j(x)$; and thus they have the same zeros.]

Theorem 6.3

Let $\{p_j(x)\}_{j=0}^\infty$ be given as above. Then for each $n \geq 1$, the zeros of $p_n(x)$ are real and distinct and lie in the interval (a, b).

Proof. Let $n \geq 1$ be fixed and suppose that none of the zeros of $p_n(x)$ are in (a, b) so that $p_n(x)$ is of constant sign on (a, b) [say $p(x) > 0$ for $x \in (a, b)$]. By the orthogonality of $p_n(x)$ and $q_0(x)$, $(q_0(x) \equiv 1)$, we would then have

$$0 = \langle 1, p_n\rangle = \int_a^b p_n(x)w(x)\, dx > 0$$

since we assume $w(x)$ is nonnegative and strictly positive for some subinterval of (a, b). Thus our initial assumption is contradicted, and $p_n(x)$ must have at least one zero, x_0, in (a, b).

If any zero, say x_0, is a multiple zero of $p_n(x)$, then $(x - x_0)^2$ factors $p_n(x)$ and so $r(x) \equiv p_n(x)/(x - x_0)^2$ is in $\mathcal{P}_{n-2}$. By Corollary 1 of Theorem 5.9, $\langle p_n, r\rangle = 0$; and so we can write

$$0 = \langle p_n, r\rangle = \int_a^b p_n(x)[p_n(x)/(x - x_o)^2]w(x)\, dx$$

$$= \int_a^b \frac{p_n(x)^2}{(x - x_0)^2} w(x)\, dx > 0,$$

which is again a contradiction. Hence we can infer that any zero of $p_n(x)$ lying in (a, b) is simple.

Now let $\{x_0, x_1, \ldots, x_j\}$ be the zeros of $p_n(x)$ lying in (a, b); and suppose that $j < n - 1$; that is, $p_n(x)$ has other zeros elsewhere. Since the zeros $\{x_i\}_{i=0}^j$ are simple, we can form the polynomial $p_n(x)[(x - x_0)(x - x_1) \cdots (x -$

$x_j)]$. We write this polynomial as $r(x) \cdot [(x - x_0)^2(x - x_1)^2 \cdots (x - x_j)^2]$ where $r(x) \in \mathcal{P}_{n-j-1}$, and we note that $r(x)$ is of constant sign (say >0) on (a, b). Again

$$\langle p_n(x), [(x - x_0)(x - x_1) \cdots (x - x_j)] \rangle = 0$$

by the corollary mentioned above. So

$$0 = \int_a^b p_n(x)[(x - x_0)(x - x_1) \cdots (x - x_j)]w(x)\, dx$$

$$= \int_a^b r(x)[(x - x_0)^2(x - x_1)^2 \cdots (x - x_j)^2]w(x)\, dx > 0,$$

which contradicts the assumption that $j < n - 1$. Thus all the zeros of $p_n(x)$ are in (a, b) and are simple. ■

Recall that the Eqs. (6.34) are satisfied for $0 \le k \le 2n + 1$ if and only if $Q_n(f) = \sum_{j=0}^n A_j f(x_j)$ has precision $2n + 1$; that is, if $f(x)$ is any polynomial of degree $(2n + 1)$ or less, then $I(f) = Q_n(f)$. (Any interpolatory quadrature formula with this property is called *Gaussian*.) With this point in mind we are able to prove the basic theorem of Gaussian quadrature.

Theorem 6.4

The formula $\int_a^b p(x)w(x)\, dx = \sum_{j=0}^n A_j p(x_j)$ holds for all $p(x)$ in $\mathcal{P}_{2n+1}$ if and only if $\{x_j\}_{j=0}^n$ are the zeros of $p_{n+1}(x)$ (as given in Theorem 6.3) and $\{A_j\}_{j=0}^n$ are given in (6.2).

Proof. 1. Let $p(x) \in \mathcal{P}_{2n+1}$ and let $p_{n+1}(x_j) = 0$, $0 \le j \le n$. By the Euclidean division algorithm, $p(x)$ can be written as $p(x) = p_{n+1}(x)S(x) + R(x)$ where $S(x)$ and $R(x)$ are in $\mathcal{P}_n$. [Note that since $p_{n+1}(x_j) = 0$, $0 \le j \le n$, then $p(x_j) = R(x_j)$, $0 \le j \le n$.] In addition, since the weights are given by (6.2), $I(R) = Q_n(R) = \sum_{j=0}^n A_j R(x_j)$ (since the quadrature formula has precision at least n). If one uses Corollary 1 of Theorem 5.9, $\langle p_{n+1}, S \rangle = 0$; so

$$I(p) \equiv \int_a^b p(x)w(x)\, dx = \int_a^b p_{n+1}(x)S(x)w(x)\, dx + \int_a^b R(x)w(x)\, dx$$

$$= \langle p_{n+1}, S \rangle + I(R) = 0 + \sum_{j=0}^n A_j R(x_j)$$

$$= \sum_{j=0}^n A_j p(x_j) = Q_n(p).$$

2. Now we assume that $\{x_j\}_{j=0}^n$ is any distinct set of points and $\int_a^b p(x)w(x)\, dx = \sum_{j=0}^n A_j p(x_j)$ for all $p(x) \in \mathcal{P}_{2n+1}$. Given any integer k, $0 \le k \le n$, let $r_k(x)$ be any polynomial of degree k or less. Let $W(x) = \prod_{j=0}^n (x - x_j)$ and define $p(x) \equiv$

$r_k(x)W(x)$. Then $p(x) \in \mathcal{P}_{2n+1}$; and by our hypothesis, $I(p) = Q_n(p)$. Using this equality, we have

$$\langle r_k, W\rangle = \int_a^b r_k(x)W(x)w(x)\,dx = \int_a^b p(x)w(x)\,dx$$

$$= \sum_{j=0}^{n} A_j p(x_j) = \sum_{j=0}^{n} A_j r_k(x_j)W(x_j) = 0$$

since $W(x_j) = 0$, $0 \le j \le n$. Hence we have just shown that $W(x)$, a monic polynomial of degree $(n + 1)$, is orthogonal to any polynomial of degree n or less. By Corollary 2, Theorem 5.9, $W(x) \equiv q_{n+1}(x)$; and thus x_j's are the zeros of $p_{n+1}(x)$. Now that the nodes x_j are known, we are nearly done. That the weights A_j are as in (6.2) follows immediately since the first $(n + 1)$ equations on the right-hand side of (6.34) constitute a linear system where the coefficient matrix is Vandermonde. Hence there is one and only one choice for the weights, and (6.2) gives the solution explicitly. ■

Gaussian quadratures are powerful numerical integration methods as the following corollary illustrates.

Corollary

Let $Q_n(f) = \sum_{j=0}^{n} A_j f(x_j)$ be Gaussian; then $\lim_{n\to\infty} Q_n(f) = I(f)$ for all $f(x) \in C[a, b]$.

Proof. By the definition of $\ell_k(x)$ [(5.2) in Chapter 5], for each k, $(\ell_k(x))^2 \in \mathcal{P}_{2n}$; and so $I(\ell_k^2) = Q_n(\ell_k^2)$. Thus

$$0 < \int_a^b (\ell_k(x))^2 w(x)\,dx = \sum_{j=0}^{n} A_j(\ell_k(x_j))^2 = A_k$$

since $(\ell_k(x_j))^2 = \delta_{jk}$. Thus the Gaussian quadrature weights are positive for each n; so by the remarks following Theorem 6.1 this corollary is proved. ■

We should also note that since this corollary shows that the weights are all positive, the formulas have nice rounding properties. In the literature there exist extensive tables for the weights and nodes of many common Gaussian quadratures. [For example, see Stroud and Secrest (1966).] We shall now consider efficient methods for calculating the weights and nodes of any Gaussian formula.

To compute the nodes we take the obvious course of generating $q_{n+1}(x)$ by the three-term recurrence formula, (5.65). Since the zeros of $q_{n+1}(x)$ are simple and lie in (a, b), Newton's method is ideally suited for computing these nodes.

Efficient computation of the weights is not so straightforward. We could use the computed values for the nodes and solve for the weights by formula (6.3) or we could use the formula $A_j = \int_a^b \ell_j(x)w(x)\,dx$, $0 \le j \le n$. There is a more efficient procedure, however, which uses the same recurrence relation as for $q_{n+1}(x)$, but has *different starting values* (otherwise we would be just generating the $q_j(x)$'s again).

We first define a new sequence of polynomials, $\{\phi_0, \phi_1, \phi_2, \ldots, \phi_{n+1}\}$, by $\phi_0(x) \equiv 0$, $\phi_1(x) \equiv \int_a^b w(x)\, dx$, and

$$\phi_k(x) = (x - a_k)\phi_{k-1}(x) - b_k\phi_{k-2}(x) \qquad \text{for } k \geq 2. \tag{6.35}$$

The constants $\{a_k\}_{k=2}^{n+1}$ and $\{b_k\}_{k=2}^{n+1}$ are the same as in (5.65), the three-term recurrence that we have just used to find $q_{n+1}(x)$ in order to compute the nodes. [The reader should note that $\phi_0(x)$ and $\phi_1(x)$ differ from $q_0(x)$ and $q_1(x)$, respectively. One should also note that the degree of $\phi_j(x)$ is $j - 1$.] We have introduced (6.35) so that we can calculate each A_j by

$$A_j = \phi_{n+1}(x_j)/q'_{n+1}(x_j), \qquad 0 \leq j \leq n. \tag{6.36}$$

The validity of (6.36) is an immediate consequence of the following theorem.

Theorem 6.5

Let $\phi_{n+1}(x)$ and $q_{n+1}(x)$ be generated by (6.35) and (5.65), respectively. Then $\phi_{n+1}(x)$ can alternatively be written as

$$\phi_{n+1}(x) = \int_a^b \frac{q_{n+1}(t) - q_{n+1}(x)}{(t - x)}\, w(t)\, dt. \tag{6.37}$$

Proof. The proof is by induction and we leave to the reader to verify (6.37) for $n = -1$ and $n = 0$. Then for $n \geq 1$, assume (6.37) is valid for all positive integers up to n. By the three-term recurrence, (5.65),

$$\int_a^b \frac{q_{n+1}(t) - q_{n+1}(x)}{(t - x)}\, w(t)\, dt$$

$$= \int_a^b \frac{[(t - a_{n+1})q_n(t) - b_{n+1}q_{n-1}(t) - (x - a_{n+1})q_n(x) + b_{n+1}q_{n-1}(x)]w(t)\, dt}{(t - x)}.$$

Adding and subtracting $xq_n(t)$ from the numerator of the integrand and recalling that $\int_a^b q_n(t)w(t)\, dt = 0$ reduce this expression to

$$(x - a_{n+1})\int_a^b \frac{q_n(t) - q_n(x)}{t - x}\, w(t)\, dt - b_{n+1}\int_a^b \frac{q_{n-1}(t) - q_{n-1}(x)}{t - x}\, w(t)\, dt$$

$$+ \int_a^b q_n(t)w(t)\, dt = (x - a_{n+1})\phi_n(x) - b_{n+1}\phi_{n-1}(x) = \phi_{n+1}(x). \quad \blacksquare$$

To see that (6.36) follows from this theorem, note that $\ell_j(x)$ can alternatively be written as $\ell_j(x) = q_{n+1}(x)/[(x - x_j)q'_{n+1}(x_j)]$, $0 \le j \le n$. Thus since $q_{n+1}(x_j) = 0$ and $q'_{n+1}(x_j) \ne 0$ (by Theorem 6.3), (6.37) yields

$$\frac{\phi_{n+1}(x_j)}{q'_{n+1}(x_j)} = \int_a^b \frac{q_{n+1}(t) - 0}{(t - x_j)q'_{n+1}(x_j)} w(t)\, dt = \int_a^b \ell_j(t)w(t)\, dt \equiv A_j.$$

EXAMPLE 6.9. We pause here to present a particularly nice type of Gaussian quadrature since we can derive closed-form formulas for its weights and nodes for any n. If we let $\langle f, g \rangle = \int_{-1}^{1} f(x)g(x)(1 - x^2)^{-1/2}\, dx$, then the orthogonal polynomials are the Chebyshev polynomials of the first kind, $T_k(x) = \cos[k \cos^{-1}(x)]$. Thus we immediately have the nodes since the zeros of $T_{n+1}(x)$ are $x_j = \cos((2j + 1)\pi/(2n + 2))$, $0 \le j \le n$. To find the weights, we must introduce the Chebyshev polynomials of the second kind, $U_k(x) \equiv \sin[(k + 1)\cos^{-1}(x)]/\sin[\cos^{-1}(x)]$, or $U_k(cos(\theta)) = \sin[(k + 1)\theta]/\sin(\theta)$ for $x \equiv \cos(\theta)$. Now obviously $U_0(x) \equiv 1$ and $U_1(x) = U_1(\cos(\theta)) = \sin(2\theta)/\sin(\theta) = 2\cos(\theta) = 2x$. By elementary trigonometric identities we can show that

$$\sin[(k + 2)\theta] = 2\cos(\theta)\sin[(k + 1)\theta] - \sin(k\theta);$$

and so for $k \ge 1$,

$$\begin{aligned} U_{k+1}(x) &\equiv U_{k+1}(\cos(\theta)) \equiv \sin[(k + 2)\theta]/\sin(\theta) \\ &= 2\cos(\theta)\sin[(k + 1)\theta]/\sin(\theta) - \sin(k\theta)/\sin(\theta) \\ &= 2\cos(\theta)U_k(\cos(\theta)) - U_{k-1}(\cos(\theta)) = 2xU_k(x) - U_{k-1}(x). \end{aligned} \tag{6.38a}$$

Formula (6.38a) shows that $U_k(x)$ is a polynomial of degree k with leading coefficient 2^k [since $U_0(x) \equiv 1$ and $U_1(x) = 2x$]. Thus $V_k(x) \equiv 2^{-k}U_k(x)$ is monic in $\mathcal{P}_k$ and by (6.38a) satisfies the recurrence formula $V_0(x) = 1$, $V_1(x) = x$,

$$V_k(x) = xV_{k-1}(x) - \frac{1}{4}V_{k-2}(x), \qquad k \ge 2. \tag{6.38b}$$

Similarly the monic Chebyshev polynomial of the first kind, $q_k(x) \equiv T_k(x)/2^{k-1}$, $k \ge 1$, satisfies (Problem 4)

$$q_k(x) = xq_{k-1}(x) - b_k q_{k-2}(x), \qquad k \ge 2,$$

where $b_2 = 1/2$ and $b_k = 1/4$ for $k \ge 3$. [This result follows from $T_0(x) = 1$, $T_1(x) = x$, and $T_k(x) = 2xT_{k-1}(x) - T_{k-2}(x)$, $k \ge 2$.]

Now to use (6.36), we must find $\phi_{n+1}(x)$ from (6.35). We have $\phi_0(x) \equiv 0$, $\phi_1(x) = \int_{-1}^{1} (1 - x^2)^{-1/2}\, dx = \pi$, and $\phi_k(x) = x\phi_{k-1}(x) - b_k\phi_{k-2}(x)$, $k \ge 2$. If $k = 2$, $\phi_2(x) = x(\pi) - b_2(0) = \pi x$. If $k = 3$, $\phi_3(x) = \pi x^2 - \pi/4 = \pi V_2(x)$. Since $\phi_2(x) = \pi V_1(x)$ and for $k > 3$ (6.35) is $\phi_k(x) = x\phi_{k-1}(x) - \phi_{k-2}(x)/4$, we see by (6.38b) that $\phi_{k+1}(x) = \pi V_k(x)$, $k \ge 0$. Hence $\phi_{n+1}(x) = \pi V_n(x) = \pi 2^{-n}U_n(x)$. Now

$$\frac{d}{dx}(T_{n+1}(x)) = \frac{d(\cos[(n + 1)\theta])}{d\theta}\frac{d\theta}{dx} = (n + 1)\frac{\sin[(n + 1)\theta]}{\sin(\theta)} = (n + 1)U_n(x).$$

Thus (6.36) becomes

$$A_j = \frac{\phi_{n+1}(x_j)}{q'_{n+1}(x_j)} = \frac{\pi 2^{-n}U_n(x_j)}{(n + 1)2^{-n}U_n(x_j)} = \frac{\pi}{(n + 1)}, \qquad 0 \le j \le n.$$

This equation yields the Gaussian formula

$$\int_{-1}^{1} f(x)(1 - x^2)^{-1/2}\, dx \approx \frac{\pi}{n+1}\sum_{j=0}^{n} f(x_j). \tag{6.39}$$

This particular Gaussian quadrature is called a *Gauss-Chebyshev* quadrature and has especially nice rounding characteristics since the weights are all equal.

Perhaps the most commonly used Gaussian quadrature is obtained from $w(x) \equiv 1$ and $\langle f, g\rangle = \int_{-1}^{1} f(x)g(x)\, dx$. The orthogonal polynomials for this case are the Legendre polynomials. There are no nice closed-form formulas for the nodes and the weights in this case. However, with the use of extensive previously tabulated results and our ability to translate easily the integral of integration from $[-1, 1]$ to $[a, b]$, this quadrature (called a *Gauss-Legendre* quadrature or sometimes even simply a *Gauss quadrature*) is a very practical tool.

If $f(x) \in C^{(2n+2)}[a, b]$, then it is possible to derive an error formula for Gaussian quadrature of this form (see Ralston, 1965):

$$R_n(f) \equiv I(f) - Q_n(f) = \frac{f^{(2n+2)}(\eta)}{(2n+2)!}\int_a^b (q_{n+1}(x))^2 w(x)\, dx. \tag{6.40a}$$

We shall take a slightly different approach that emphasizes the benefits of having precision $(2n + 1)$. Once again, we define the uniform degree of approximation as

$$E_m(f) = \min_{p(x)\in\mathcal{P}_m}\left\{\max_{a\le x\le b}|f(x) - p(x)|\right\} = \min_{p(x)\in\mathcal{P}_m}\{\|f - p\|_\infty\} \equiv \|f - p_m^*\|_\infty.$$

Since $I(1) = Q_n(1)$ for any Gaussian formula, we have $\int_a^b w(x)\, dx = \sum_{j=0}^{n} A_j$. Since the formula is Gaussian, $A_j \ge 0$, $0 \le j \le n$, and $I(p) = Q_n(p)$ for all $p(x) \in \mathcal{P}_{2n+1}$. Thus if $p_{2n+1}^*(x) \in \mathcal{P}_{2n+1}$ is the best uniform polynomial approximation to $f(x)$,

$$\begin{aligned} R_n(f) &\equiv I(f) - Q_n(f) = (I(f) - I(p_{2n+1}^*)) - (Q_n(f) - Q_n(p_{2n+1}^*)) \\ &= \int_a^b (f(x) - p_{2n+1}^*(x))w(x)\, dx - \sum_{j=0}^{n} A_j(f(x_j) - p_{2n+1}^*(x_j)). \end{aligned}$$

Therefore

$$|R_n(f)| \le E_{2n+1}(f)\left(\int_a^b w(x)\, dx + \sum_{j=0}^{n} A_j\right) = 2E_{2n+1}(f)\int_a^b w(x)\, dx. \tag{6.40b}$$

Obviously if any interpolatory quadrature, $Q_n(f) = \sum_{j=0}^{n} B_j f(x_j)$, has precision m and $B_j \ge 0$, $0 \le j \le n$, the argument above could be repeated to yield $|R_n(f)| \le 2E_m(f)\int_a^b w(x)\, dx$. Since the precision is maximized when the quadrature is Gaussian, that is, $m = 2n + 1$, and since $E_{2n+1}(f) \le E_m(f)$ for $m < 2n + 1$, then (6.40b) represents an optimal error bound of this type. For practical use of (6.40b) we can call on any Jackson theorem [for instance, (5.69a) or

(5.69b)] to yield a fairly simple bound on $|R_n(f)|$. We also note (Problem 6) that no formula of the type (6.34) can have precision equal to $2n + 2$. Thus Gaussian formulas are the outside limits in increasing precision.

EXAMPLE 6.10. As an illustration of the power of Gauss-Legendre quadrature, we consider again

$$Si(1) = \int_0^1 \frac{\sin(t)}{t}\, dt.$$

From a table, we obtain the weights and nodes for the five-point Gauss-Legendre formula for $[-1, 1]$:

$$\begin{aligned} x_0 &= -0.9061798459 & A_0 &= 0.2369268851 \\ x_1 &= -0.5384693101 & A_1 &= 0.4786286705 \\ x_2 &= 0.0 & A_2 &= 0.5688888889 \\ x_3 &= 0.5384693101 & A_3 &= 0.4786286705 \\ x_4 &= 0.9061798459 & A_4 &= 0.2369268851. \end{aligned}$$

[The symmetric character of the weights and the nodes should be expected since the nth degree Legendre polynomials is an even function when n is even and an odd function when n is odd.] Since our problem is to estimate an integral on $[0, 1]$ rather than on $[-1, 1]$, we must use (6.4) where $a = 0$ and $b = 1$. With this change, the Gauss-Legendre five-point formula provides the estimate of 0.94608307, which is correct to eight places.

6.5.1. Interpolation at the Zeros of Orthogonal Polynomials

Gaussian quadrature provides a valuable link between the integral inner product, $\langle f, g\rangle \equiv \int_a^b f(x)g(x)w(x)\,dx$, and the discrete inner product,

$$\langle f, g\rangle_d = \sum_{j=0}^{n} A_j f(x_j)g(x_j),$$

where A_j and x_j, $0 \le j \le n$, are the weights and the nodes respectively of the Gaussian quadrature $Q_n(f) \approx \int_a^b f(x)w(x)\,dx$. [Note that since the quadrature is Gaussian, then $A_j. > 0$ and $x_j \in (a, b)$, $0 \le j \le n$, so that $\langle f, g\rangle_d$ is a well-defined discrete inner product.]

Theorem 6.6

Given $\langle f, g\rangle$ and $\langle f, g\rangle_d$ as above, let $\{p_0(x), p_1(x), p_2(x), \ldots\}$ be orthonormal polynomials [with degree $(p_j(x)) = j$] with respect to $\langle f, g\rangle$. Then $\{p_0(x), p_1(x), \ldots, p_n(x)\}$ is an orthonormal set with respect to $\langle f, g\rangle_d$.

Proof. Let $p(x) \equiv p_k(x)p_m(x)$, $k + m \le 2n$. Then since $I(p) = Q_n(p)$, we have

$$\delta_{km} = \langle p_k, p_m\rangle = \int_a^b p_k(x)p_m(x)w(x)\,dx = I(p) = Q_n(p)$$

$$= \sum_{j=0}^{n} A_j p_k(x_j)p_m(x_j) = \langle p_k, p_m\rangle_d. \qquad \blacksquare$$

We can use Theorem 6.6 to give an easily computable formula for the polynomial, $P(x) \in \mathcal{P}_n$, that interpolates a function, $f(x)$, at the zeros of $p_{n+1}(x)$.

Theorem 6.7

Let $\{x_j\}_{j=0}^n$ be the zeros of $p_{n+1}(x)$ (as in Theorem 6.6), and let $P(x) = \Sigma_{k=0}^n \alpha_k p_k(x)$. If $\alpha_k = \Sigma_{j=0}^n A_j p_k(x_j) f(x_j)$, then $P(x_j) = f(x_j)$ for $0 \le j \le n$.

Proof. From Chapter 5 we know that the interpolating polynomial exists, is unique, and thus can be written in the form $P(x) = \Sigma_{k=0}^n \alpha_k p_k(x)$. If $P(x_j) = f(x_j)$, then $\Sigma_{k=0}^n \alpha_k p_k(x_j) = f(x_j)$, $0 \le j \le n$. Now for m fixed, multiplying both sides by $A_j p_m(x_j)$ and summing from $j = 0$ to $j = n$ yield

$$\sum_{j=0}^n A_j p_m(x_j) f(x_j) = \sum_{j=0}^n \left(A_j p_m(x_j) \sum_{k=0}^n \alpha_k p_k(x_j) \right)$$

$$= \sum_{k=0}^n \alpha_k \left(\sum_{j=0}^n A_j p_m(x_j) p_k(x_j) \right) = \sum_{k=0}^n \alpha_k \langle p_m, p_k \rangle_d = \alpha_m.$$

This equality holds for each m, $0 \le m \le n$, and so the proof is complete. ■

We next turn our attention to an estimate of the error that is made when interpolating at the zeros of orthogonal polynomials. Suppose $p(x) \in \mathcal{P}_n$ interpolates $f(x)$, $f^{(n+1)}(x) \in C[a, b]$, at *any* set of $(n + 1)$ distinct points, $\{z_j\}_{j=0}^n$, in $[a, b]$. Then we recall from formula (5.27) that for any $x \in [a, b]$

$$f(x) - p(x) = \frac{f^{(n+1)}(\xi)}{(n+1)!} W(x) \tag{6.41a}$$

where $\xi \in \text{Spr}\{x, z_0, z_1, \ldots, z_n\}$ and where $W(x) = \Pi_{j=0}^n (x - z_j)$. We have already seen the merits of interpolating at the zeros of the shifted Chebyshev polynomials, $\tilde{T}_{n+1}(x)$, in that this minimizes $\|W\|_\infty \equiv \max_{a \le x \le b} |W(x)|$ [see formula (5.35)]. There are similar advantages in the interpolation given in Theorem 6.7. Recall, as in (5.31), if we square both sides of (6.41a), multiply by $w(x)$, integrate from a to b, and take the square root, we have

$$\|f - p\| = \left(\int_a^b (f(x) - p(x))^2 w(x)\, dx \right)^{1/2}$$

$$\le \frac{\max_{a \le x \le b} |f^{(n+1)}(x)|}{(n+1)!} \left(\int_a^b (W(x))^2 w(x)\, dx \right)^{1/2} \tag{6.41b}$$

$$= \frac{\|f^{(n+1)}\|_\infty}{(n+1)!} \|W\|.$$

Now let $W(x)$ be any monic polynomial of degree $(n + 1)$ as above, and let

$q_{n+1}(x)$ be the monic orthogonal polynomial as before. Then $r(x) \equiv (W(x) - q_{n+1}(x)) \in \mathcal{P}_n$, so $\langle r, q_{n+1}\rangle = 0$. Now we have

$$\|W\|^2 = \int_a^b (W(x))^2 w(x)\,dx = \int_a^b (q_{n+1}(x) + r(x))^2 w(x)\,dx$$

$$= \int_a^b (q_{n+1}(x))^2 w(x)\,dx + 2\int_a^b q_{n+1}(x) r(x) w(x)\,dx + \int_a^b (r(x))^2 w(x)\,dx$$

$$= \|q_{n+1}\|^2 + 2\langle q_{n+1}, r\rangle + \|r\|^2 = \|q_{n+1}\|^2 + \|r\|^2.$$

Therefore $\|q_{n+1}\| \le \|W\|$ for all such $W(x)$; and so the error bound, (6.41b), is minimized for $W(x) = q_{n+1}(x)$, that is to say, for interpolation at the zeros of the orthogonal polynomial. In the special case in which $[a, b] = [-1, 1]$ and $w(x) = 1$, the minimum bound is achieved by interpolating at the zeros of the $(n + 1)$st-degree Legendre polynomial.

Another indication of the power of interpolating at the zeros of orthogonal polynomials is provided by the following theorem, which guarantees least-squares convergence of the interpolating polynomials.

Theorem 6.8 Erdös-Turán

Let $P_n(x) \in \mathcal{P}_n$ interpolate $f(x)$ at the zeros of $p_{n+1}(x)$ (as in Theorem 6.7). Then

$$\|f - P_n\| \equiv \left(\int_a^b (f(x) - P_n(x))^2 w(x)\,dx\right)^{1/2} \to 0, \qquad \text{as } n \to \infty.$$

Proof. Again for each n, let $p_n^*(x) \in \mathcal{P}_n$ be the best uniform approximation to $f(x)$; and let

$$E_n(f) \equiv \max_{a \le x \le b} |f(x) - p_n^*(x)| \equiv \|f - p_n^*\|_\infty.$$

Then by the triangle inequality, $\|f - P_n\| \le \|f - p_n^*\| + \|p_n^* - P_n\|$. Now

$$\|f - p_n^*\|^2 = \int_a^b (f(x) - p_n^*(x))^2 w(x)\,dx \le (E_n(f))^2 \int_a^b w(x)\,dx.$$

Also since $(p_n^*(x) - P_n(x))^2 \in \mathcal{P}_{2n}$ and the nth Gaussian quadrature is exact in $\mathcal{P}_{2n+1}$ and has positive weights with $\int_a^b w(x)\,dx = \sum_{j=0}^n A_j$, we see that

$$\|p_n^* - P_n\|^2 \equiv \int_a^b (p_n^*(x) - P_n(x))^2 w(x)\,dx$$

$$= \sum_{j=0}^n A_j (p_n^*(x_j) - P_n(x_j))^2 = \sum_{j=0}^n A_j (p_n^*(x_j) - f(x_j))^2$$

$$\le (E_n(f))^2 \sum_{j=0}^n A_j = (E_n(f))^2 \int_a^b w(x)\,dx.$$

Since $\int_a^b w(x)\,dx$ is a constant and $E_n(f) \to 0$ as $n \to \infty$, then both $\|f - p_n^*\|$ and $\|p_n^* - P_n\| \to 0$ as $n \to \infty$. ∎

Gaussian quadratures are particularly effective in approximating Fourier coefficients, $\langle f, p_k \rangle \equiv \int_a^b f(x)p_k(x)w(x)\,dx$, but provide the following result, which may be somewhat surprising at first reading. Let $f(x)$ be approximated by the truncated Fourier expansion $f(x) \approx F_n(x) \equiv \Sigma_{k=0}^n \langle f, p_k \rangle p_k(x)$. Using the Gaussian quadrature to approximate $\langle f, p_k \rangle$, we get $\langle f, p_k \rangle \approx \alpha_k \equiv \Sigma_{j=0}^n A_j f(x_j)p_k(x_j)$. Then $F_n(x) \approx \Sigma_{k=0}^n \alpha_k p_k(x)$, but we notice this expression is precisely the interpolating polynomial, $P_n(x)$, as in Theorem 6.7.

6.5.2. Interpolation Using Chebyshev Polynomials

In this section, we will consider some of the practical aspects of interpolation at the zeros of $T_{n+1}(x)$. We have already seen from (5.36) that there are advantages in using these interpolation points. The example given below establishes a useful discrete orthogonality property determined by the zeros of $T_{n+1}(x)$. This relation [and a companion result given in (6.43)] can be used in a variety of ways.

EXAMPLE 6.11. Let $\langle f, g \rangle = \int_{-1}^1 f(x)g(x)(1 - x^2)^{-1/2}\,dx$ and $x_j = \cos[(2j+1)\pi/2(n+1)]$, $0 \le j \le n$. For this inner product, the orthonormal polynomials are $p_0(x) = 1/\sqrt{\pi}$ and $p_k(x) = \sqrt{2/\pi}\,T_k(x)$ for $k \ge 1$ (see Example 5.15b, Chapter 5). By Example 6.9, $A_j = \pi/(n+1)$, $0 \le j \le n$. Thus by Theorem 6.6, the following relationship holds for $k + m \le 2n$:

$$\frac{2}{n+1} \sum_{j=0}^{n} T_k(x_j)T_m(x_j) = \begin{cases} 0, & k \ne m \\ 1, & k = m > 0 \\ 2, & k = m = 0. \end{cases} \tag{6.42}$$

Formula (6.42) very closely resembles another discrete orthogonality relation for the Chebyshev polynomials (displayed in Example 5.15c), which we can express for $k + m \le 2n$ as

$$\frac{2}{n} \sum_{j=0}^{n}{}'' \; T_k(t_j)T_m(t_j) = \begin{cases} 0, \; k \ne m \\ 1, \; k = m, & 1 \le k \le n-1 \\ 2, \; k = m, & k = 0 \quad \text{or} \quad k = n. \end{cases} \tag{6.43}$$

In (6.43), the points t_j are given by $t_j = \cos[j\pi/n]$, $0 \le j \le n$; and the double prime denotes halving the first and the last terms. Since the validity of (6.42) rests upon the fact that (6.39) is a Gaussian quadrature for $I(f) = \int_{-1}^1 f(x)(1 - x^2)^{-1/2}\,dx$, the similarity of (6.42) and (6.43) leads us to suspect that (6.43) also represents a quadrature for $I(f) = \int_{-1}^1 f(x)(1 - x^2)^{-1/2}\,dx$ of the form

$$Q_n'(f) = \frac{\pi}{n} \sum_{j=0}^{n}{}'' \; f(t_j). \tag{6.44}$$

In investigating the quadrature $Q_n'(f)$, we note first that $I(1) = Q_n'(1) = \pi$ and $I(T_k) = Q_n'(T_k) = 0$ for $1 \le k \le 2n - 1$. However, $I(T_{2n}) = 0$ but $Q_n'(T_{2n}) = \pi$, and so the precision of this quadrature is $(2n - 1)$. To verify these results, we note that $I(T_k) = \langle 1, T_k \rangle = 0$ for $k \ge 1$. To determine $Q_n'(T_k)$, we first suppose $1 \le k \le n$ and use (6.43) with $m = 0$ to find $Q_n'(T_k) = 0$. For $n < k \le 2n$, we let $k = n + m$ and observe that $T_k(t_j) = T_n(t_j)T_m(t_j)$. Thus, using (6.43), we see that $Q_n'(T_k) = 0$ for $n < k < 2n$ and that $Q_n'(T_{2n}) = \pi$. We note that $Q_n'(f)$ is not Gaussian since its precision is $(2n - 1)$ instead of $(2n + 1)$; but it is a powerful formula as we can see from the remarks following (6.40b). The nodes t_j are often called the "practical"Chebyshev nodes. Also observe that if we make the transformation $x = \cos\theta$, then $I(f) = \int_0^\pi f(\cos\theta)\,d\theta$. From (6.44), we see that $Q_n'(f)$ is the composite trapezoidal rule for appoximating $\int_0^\pi f(\cos\theta)\,d\theta$.

From Theorem 6.7 and formula (6.39) we easily find that the polynomial of degree n or less that interpolates $f(x)$ at $x_j = \cos[(2j + 1)\pi/2(n + 1)]$, $0 \le j \le n$, is given by

$$P(x) = \beta_0/2 + \sum_{k=1}^{n} \beta_k T_k(x), \qquad \beta_k = \frac{2}{n+1} \sum_{j=0}^{n} T_k(x_j) f(x_j). \tag{6.45a}$$

A similar formula can be developed for interpolation at the points $t_j = \cos(j\pi/n)$. However since (6.44) is not a Gaussian formula, we cannot directly apply Theorem 6.7. However, we can use (6.43) and mimic the proof of Theorem 6.7 to show that the polynomial of degree n or less that interpolates $f(x)$ at $t_j = \cos(j\pi/n)$, $0 \le j \le$, is (see Problem 8)

$$\tilde{P}(x) = \sum_{k=0}^{n}{}'' \gamma_k T_k(x) \qquad \text{where} \qquad \gamma_k = \frac{2}{n} \sum_{j=0}^{n}{}'' T_k(t_j) f(t_j). \tag{6.45b}$$

We pause to mention that the interpolating polynomials constructed via (6.45a) and (6.45b) can be easily evaluated at a point $x = \alpha$ since they have the form $p(x) = \sum_{k=0}^{n} b_k r_k(x)$ where $\{b_k\}_{k=0}^{n}$ are known constants and each $r_k(x)$ is a constant multiple of $p_k(x)$. We leave to the reader to verify that the efficient algorithm of formula (5.66) can be modified to accomplish this task with $(2n - 1)$ multiplications, once the three-term recursion is known for the $r_k(x)$'s. For example (see Problem 9), if $r_k(x) = T_k(x)$, then we have $T_0(x) = 1$, $T_1(x) = x$, and $T_k(x) = 2xT_{k-1}(x) - T_{k-2}(x)$ for $k \ge 2$. The modification of (5.66) yields in this case

$$\sum_{k=0}^{n} b_k T_k(\alpha) = s_0(\alpha) - \alpha s_1(\alpha) \tag{6.45c}$$

where $s_k(\alpha)$ is defined by $s_{n+1}(\alpha) = s_{n+2}(\alpha) = 0$; and for $k = n, n - 1, \ldots, 0$, by

$$s_k(\alpha) - 2\alpha s_{k+1}(\alpha) + s_{k+2}(\alpha) = b_k.$$

[Note in this particular example that since the coefficient of $s_{k+2}(\alpha)$ equals 1, only $(n + 1)$ multiplications are required.]

We turn now to the infinite Chebyshev expansion for $f(x)$. Recall that

$p_0(x) = 1/\sqrt{\pi}T_0(x)$ and $p_k(x) = \sqrt{2/\pi}T_k(x)$ for $k \geq 1$; so $\langle f, p_0 \rangle = 1/\sqrt{\pi}\langle f, T_0 \rangle$ and $\langle f, p_k \rangle = \sqrt{2/\pi}\langle f, T_k \rangle$, $k \geq 1$. Thus the Fourier-Chebyshev expansion for $f(x)$ for $-1 \leq x \leq 1$ is

$$f(x) \sim \langle f, p_0 \rangle p_0(x) + \sum_{k=1}^{\infty} \langle f, p_k \rangle p_k(x). \tag{6.46a}$$

Often, for the sake of convenience, the normalizing constants of the $p_k(x)$'s are multiplied together and (6.46a) is written in the equivalent form

$$f(x) \sim \sum_{k=0}^{\infty}{}' a_k T_k(x), \qquad a_k = \frac{2}{\pi}\int_{-1}^{1} f(x)T_k(x)(1 - x^2)^{-1/2}\,dx \tag{6.46b}$$

where the prime denotes that the first term is halved.

Using the Gauss-Chebyshev quadrature (6.39) to approximate a_k for $0 \leq k \leq n$, we obtain the interpolating polynomial $P(x)$ of formula (6.45a) as an approximation for $f(x)$ where each $\beta_k = (2/(n+1)) \sum_{j=0}^{n} T_k(x_j)f(x_j)$ is our approximation to a_k. An alternate approach is to use $\tilde{P}(x)$ of formula (6.45b) as our approximation to $f(x)$ where $\gamma_k = (2/n) \sum_{j=0}^{n}{}'' T_k(t_j)f(t_j)$ is our approximation to a_k, $0 \leq k \leq n$.

To assess further the accuracy of these interpolations, let us assume that the expansion (6.46b) is absolutely and uniformly convergent to $f(x)$ (which is the case, for example, if $f \in C^2[-1, 1]$). We first consider $\tilde{P}(x)$ in (6.45b); and for a fixed value of m, $0 \leq m \leq n$, we consider the approximation of γ_m for a_m. We can easily verify by trigonometric identities that if $k = 2rn + \alpha$, $r = 0, 1, 2, \ldots$, $|\alpha| \equiv \beta \leq n$; then $T_k(t_j) = T_\beta(t_j)$, $0 \leq j \leq n$. Making use of (6.43), we obtain for any m

$$\begin{aligned}\gamma_m &= \frac{2}{n}\sum_{j=0}^{n}{}'' \left(\sum_{k=0}^{\infty}{}' a_k T_k(t_j)\right) T_m(t_j) \\ &= \sum_{k=0}^{\infty}{}' a_k \left(\frac{2}{n}\sum_{j=0}^{n}{}'' T_k(t_j)T_m(t_j)\right) \\ &= a_m + (a_{2n-m} + a_{2n+m}) + (a_{4n-m} + a_{4n+m}) + \cdots,\end{aligned} \tag{6.47a}$$

and the resulting approximation of (6.45b)

$$f(x) \approx \sum_{m=0}^{n}{}'' \gamma_m T_m(x). \tag{6.47b}$$

Similarly if we write $k = 2r(n+1) + \alpha$, we can easily establish the identity $T_\alpha(x_j) = (-1)^r T_{2r(n+1)\pm\alpha}(x_j)$, $0 \leq j \leq n$. Using this and (6.42) in the same manner as above, we obtain for any m

$$\beta_m = a_m - (a_{2n+2-m} + a_{2n+2+m}) + (a_{4n+4-m} + a_{4n+4+m}) - \cdots, \tag{6.48a}$$

which yields the approximation

$$f(x) \approx \sum_{m=0}^{n}{}' \beta_m T_m(x). \tag{6.48b}$$

We leave (Problem 10) for the reader to verify that $\left|f(x) - \Sigma''^{\,n}_{m=0}\, \gamma_m T_m(x)\right|$ and $\left|f(x) - \Sigma'^{\,n}_{m=0}\, \beta_m T_m(x)\right|$ are both bounded by $2\,\Sigma^{\infty}_{m=n+1} |a_m|$. Thus, by Section 5.3.2, the error of (6.47b) and (6.48b) can never exceed twice the error of the truncated series, $\left|f(x) - \Sigma'^{\,n}_{m=0}\, a_m T_m(x)\right|$. We thus note that if the magnitudes of the Fourier coefficients are rapidly decreasing, then both (6.47b) and (6.48b) are good approximations. We also note from (6.47a) and (6.48a) that both γ_m and β_m are most likely to agree closely with a_m when m is small. We expect the most discrepancy in (6.47a) when $m = n - 1$ and in (6.48a) when $m = n$; in those cases we see that $\gamma_{n-1} \approx a_{n-1} + a_{n+1}$ and $\beta_n \approx a_n - a_{n+2}$, respectively. However if n is sufficiently large, then the coefficients a_{n-1}, a_{n+1}, and a_{n+2} should be relatively small and should not significantly affect the accuracy.

6.5.3. Clenshaw-Curtis Quadrature

We note that any time we have an approximation for $f(x)$ of the form $f(x) \approx p(x) \equiv \Sigma^n_{k=0}\, b_k T_k(x)$, then we can easily construct a numerical integration formula from the approximation. This fact follows from the simple observation that indefinite integration of $T_k(x)$ yields

$$\begin{aligned}\int T_k(x)\,dx &= -\int \cos(k\theta)\sin(\theta)\,d\theta \\ &= -\frac{1}{2}\int (\sin[(k+1)\theta] - \sin[(k-1)\theta])\,d\theta \qquad (6.49a)\\ &= \frac{1}{2}\left(\frac{T_{k+1}(x)}{k+1} - \frac{T_{k-1}(x)}{k-1}\right), \qquad k \ge 2,\end{aligned}$$

and

$$\int T_0(x)\,dx = T_1(x), \qquad \int T_1(x)\,dx = \frac{1}{4}(T_0(x) + T_2(x)). \qquad (6.49b)$$

Given $p(x)$ as above, then the result of the indefinite integration of $p(t)$ is an expression of the form

$$\int_{-1}^{x} p(t)\,dt = \sum_{k=0}^{n} b_k \int_{-1}^{x} T_k(t)\,dt \equiv \sum_{k=0}^{n+1} A_k T_k(x) \equiv P(x). \qquad (6.50)$$

Using the integral formulas (6.49a) and (6.49b) and equating like coefficients of each $T_k(x)$, we have the following equations for each A_k:

$$A_{n+1} = \frac{b_n}{2(n+1)}, \qquad A_n = \frac{b_{n-1}}{2n}, \qquad A_1 = b_0 - \frac{b_2}{2}, \qquad (6.51a)$$

and

$$A_k = \frac{1}{2k}(b_{k-1} - b_{k+1}), \qquad \text{for } 2 \le k \le n-1.$$

To obtain A_0, we notice that $P(-1) = 0$; and since $T_k(-1) = (-1)^k$, we find

$$A_0 = A_1 - A_2 + A_3 + \cdots + (-1)^n A_{n+1}. \tag{6.51b}$$

This technique is valid for all x in $[-1, 1]$; and moreover $P(x)$ can be efficiently evaluated at any x by (6.45c). If $x = 1$ and $p(x)$ in (6.50) is given by either the interpolating polynomials (6.45a) or (6.45b), then $P(1)$ is an interpolatory quadrature for $I(f) \equiv \int_{-1}^{1} f(x)\,dx$. Thus, (6.51) can be written in the form $Q_n(f) = \Sigma_{j=0}^{n} A_j f(z_j)$. Although this form is not computationally efficient, it can be shown in either case that the weights are positive; so the convergence result of Theorem 6.1 is applicable. Formulas of this type are called *Clenshaw-Curtis* quadratures.

Finally, we remark that we seem to have placed a great deal of emphasis on Chebyshev-type approximations in these sections. However, both from a theoretical background and from practical experience, Chebyshev methods have proven to yield excellent procedures in terms of truncation errors and round-off propagation.

EXAMPLE 6.12. As an illustration of some of these ideas, let $f(x) = \sin(x)/x$ for $-1 \le x \le 1$. First, using (6.45b), we construct the interpolating polynomial for $f(x)$ with $n = 4$. Since $f(x)$ is an even function on $[-1, 1]$ and $T_k(x)$ is odd when k is odd, we see that γ_1 and γ_3 in (6.45b) are zero. For even k, since $f(x)$ and $T_k(x)$ are even, we have (by symmetry between t_1 and t_3 and between t_0 and t_4)

$$\gamma_k = \frac{2}{4}\,[f(t_0)T_k(t_0) + 2f(t_1)T_k(t_1) + f(t_2)T_k(t_2)].$$

Now $T_k(t_0) = 1$ for $k = 0, 2, 4$; $T_k(t_1)$ has the value 1 for $k = 0$, 0 for $k = 2$, and -1 for $k = 4$; $T_k(t_2)$ has the value 1 for $k = 0$, -1 for $k = 2$, and 1 for $k = 4$. Thus with $f(0) = 1$,

$$\gamma_0 = \frac{1}{2}\left[f(1) + 2f\left(\cos\frac{\pi}{4}\right) + f(0)\right] = 1.839461$$

$$\gamma_2 = \frac{1}{2}[f(1) - f(0)] = -0.079265$$

$$\gamma_4 = \frac{1}{2}\left[f(1) - 2f\left(\cos\frac{\pi}{4}\right) + f(0)\right] = 0.002010.$$

Thus $p(x) = 0.919731 - 0.079265T_2(x) + 0.001005T_4(x)$ where the rapid decrease of the coefficients is typical of a well-behaved function $f(x)$.

To demonstrate Clenshaw-Curtis quadrature, we derive a polynomial $P(x)$ that approximates

$$Si(x) = \int_0^x \frac{\sin(t)}{t}\,dt.$$

Writing the interpolating polynomial $P(x)$ above in the form of (6.50), we have $b_0 = 0.919731$, $b_1 = 0$, $b_2 = -0.079265$, $b_3 = 0$, and $b_4 = 0.001005$. Thus from (6.51a) and (6.51b), we obtain $A_5 = 0.000101$, $A_4 = 0$, $A_3 = -0.013378$, $A_2 = 0$, $A_1 = 0.959364$, and $A_0 = 0.946087$. Thus

$$P(x) = 0.946087 + 0.959364T_1(x) - 0.013378T_3(x) + 0.000101T_5(x)$$

is an approximation for $\int_{-1}^{x} (\sin(t)/t)\, dt$. Since $\sin(t)/t$ is even, we have

$$\int_{-1}^{0} \frac{\sin(t)}{t}\, dt = \int_{0}^{1} \frac{\sin(t)}{t}\, dt = Si(1).$$

Thus $P(0) = 0.946807$ is an estimate to $Si(1) = 0.946083$. Moreover, we can use $P(x)$ to provide an estimate $Si(x), 0 < x < 1$, by observing (again since the integrand is even) that

$$\int_{-1}^{-1+x} \frac{\sin(t)}{t}\, dt = \int_{1-x}^{1} \frac{\sin(t)}{t}\, dt = Si(1) - Si(1 - x).$$

Therefore for $0 < x < 1$, we have the approximation

$$Si(1 - x) \approx P(0) - P(-1 + x),$$

which is an easily computed and accurate approximation. For example, with $x = 0.5$, $Si(0.5) = 0.493107$ and $P(0) - P(-0.5) = 0.493060$, which is in error by 0.000047.

PROBLEMS, SECTION 6.5.3.

1. Using the result of Theorem 6.4, find the weights and the nodes of the two- and three-point Gauss-Legendre quadrature formulas. Find the weights by undetermined coefficients. [The second- and third-degree monic Legendre polynomials are respectively $P_2(x) = x^2 - (1/3)$ and $P_3(x) = x^3 - (3/5)x$. The weights can be verified by checking precision in (6.34).]
2. Use the three-point Gauss-Legendre formula of Problem 1 and the five-point formula given in Example 6.10 to estimate

   a) $\int_{-1}^{1} \sin(3x)\, dx$ b) $\int_{1}^{3} \ln(x)\, dx$ c) $\int_{1}^{2} e^{x^2}\, dx$.
3. Write a computer program to generate the nth Legendre polynomial from the three-term recurrence relation and find the zeros by Newton's method. Next, use (6.36) to find the Gauss-Legendre quadrature weights. Check your results for various values of n against tabulated formulas.
4. Let $q_k(x) = (1/2^{k-1})T_k(x)$ denote the monic Chebyshev polynomial of the first kind. Verify that $q_k(x) = xq_{k-1}(x) - b_k q_{k-2}(x), k \geq 2$ where $b_2 = 1/2$ and $b_k = 1/4, k \geq 3$.
5. Use (6.40a) and (6.40b) to bound the error made in estimating the integral in Problem 2(a) by the three-point Gauss-Legendre formula [that is, $n = 2$ in (6.40a) and (6.40b)]. Use (6.40b) and an appropriate Jackson theorem to bound the error for five-point Gauss-Legendre formula applied to the integral in 2(a).
6. Show that no matter how the nodes and weights of a quadrature formula

   $$Q_n(f) = \sum_{j=0}^{n} A_j f(x_j)$$

   are chosen, the formula cannot have precision greater than $2n + 1$. [Hint: Assume that $Q_n(f)$ is designed to approximate $\int_a^b f(x)w(x)\, dx$. Use Theorem 6.4 and find a polynomial $p(x) \in \mathcal{P}_{2n+2}$ for which $Q_n(p) \neq \int_a^b p(x)w(x)\, dx$.]

7. The complete elliptic integral

$$K = \int_0^1 \frac{dx}{\sqrt{(1 - x^2)(1 - k^2x^2)}}$$

is usually written as

$$K = \int_0^{\pi/2} \frac{d\theta}{\sqrt{1 - k^2 \sin^2 \theta}}$$

where $0 \le k < 1$. For $k = 1/2$, the value of K is 1.6858. Using the fact that the integrand is even, estimate K by Gauss-Chebyshev quadrature.

8. Verify in (6.45b) that $\tilde{P}(x)$ interpolates $f(x)$ at t_j, $j = 0, 1, \ldots, n$.

9. Verify the recursion (6.45c) for evaluating a finite Chebyshev series.

***10.** Suppose that $f(x) = \Sigma'_{m=0}{}^{\infty}\, a_m T_m(x)$ where this Chebyshev expansion converges uniformly and absolutely to $f(x)$ for $-1 \le x \le 1$. For γ_m defined in (6.47a) and β_m defined in (6.48a), show that $|f(x) - \Sigma''_{m=0}{}^{n}\, \gamma_m T_m(x)|$ and $|f(x) - \Sigma'_{n=0}{}^{m}\, \beta_m T_m(x)|$ are each bounded by a $2\,\Sigma_{m=n+1}^{\infty} |a_m|$. [Note that in conjunction with (5.70a) or (5.70b), this bound gives a way to estimate interpolation errors.]

***11.** This rather long problem gives a sharper convergence result for the uniform convergence of best least-squares approximations. Suppose that $\langle f, g\rangle = \int_a^b f(x)g(x)w(x)\,dx$; and suppose that $p_i(x)$, $i = 0, 1, \ldots$ are the orthonormal polynomials associated with this inner product. Let $P_m(x) = \Sigma_{j=0}^{m} \langle f, p_j\rangle p_j(x)$ be the best least-squares approximation to $f(x)$.

a) Show that $P_m(x) = \int_a^b f(t)G_m(x, t)w(t)\,dt$ where $G_m(x, t) = \Sigma_{j=0}^{m} p_j(x)p_j(t)$. [The function $G_m(x, t)$ is usually called a "kernel."]

b) Let $q_i(x)$, $i = 0, 1, \ldots$ denote the monic orthogonal polynomials generated by the three-term recurrence relation (5.65). Then $p_i(x) = \lambda_i q_i(x)$ where

$$\lambda_i = \frac{1}{\|q_i\|}, \qquad \|q_i\| = \sqrt{\langle q_i, q_i\rangle}.$$

Thus (5.65) can be written so as to generate the p_i where

$$p_k(x) = \lambda_k^{-1}(x - a_k)\lambda_{k-1}p_{k-1}(x) - \lambda_k^{-1}b_k\lambda_{k-2}p_{k-2}(x), \qquad k \ge 2.$$

Rewrite this equation as $p_k(x) = (A_k x + B_k)p_{k-1}(x) - C_k p_{k-2}(x)$ and prove the *Christoffel-Darboux Identity*:

$$\frac{\lambda_m}{\lambda_{m+1}}\left[p_{m+1}(x)p_m(t) - p_{m+1}(t)p_m(x)\right] = (x - t)\sum_{j=0}^{m} p_j(x)p_j(t).$$

[Hint: Multiply the rewritten recurrence relation for $p_{j+1}(x)$ by $p_j(t)$ and the relation for $p_{j+1}(t)$ by $p_j(x)$. Thus obtain an expression for $(x - t)p_j(x)p_j(t)$ and sum these expressions.]

c) Show that $\int_a^b w(t)G_m(x, t)\,dt = 1$, and hence deduce that $f(x) = \int_a^b f(x)G_m(x, t)w(t)\,dt$.

d) Using the results above, show that $f(x) - P_m(x) = \int_a^b [f(x) - f(t)]G_m(x, t)w(t)\,dt$.

e) From the Christoffel-Darboux identity, we have a useful expression for the kernel. For example, suppose that $|\lambda_m/\lambda_{m+1}| \le C$ for $m = 0, 1, \ldots$ and suppose that $f(x)$ is differentiable at $x = x_0$, $a < x_0 < b$. Further suppose that $|f(x) - f(x_0)| \le K|x - x_0|$ for all $x \in [a, b]$ and that $|p_i(x_0)| \le M$ for $i = 0, 1, \ldots$. Show that $\lim_{m \to \infty} P_m(x_0) = f(x_0)$. [Hint: Let $g(t) = [f(x_0) - f(t)]/[x_0 - t]$ and use (b) and the fact that $\langle g, p_i \rangle \to 0$ as $i \to \infty$ (see Section 5.3.2).]

f) The conditions of (e) are not too hard to satisfy. What are the constants C and M in (e) when the $p_i(x)$ are the Chebyshev polynomials? For the Chebyshev polynomials, estimate $f(x_0) - P_m(x_0)$ when $f \in C^2[-1, 1]$.

6.6 TRIGONOMETRIC POLYNOMIALS AND THE FAST FOURIER TRANSFORM

As we have seen in Chapter 5, interpolating data or functions given at equally spaced points by algebraic polynomials $p(x) = a_n x^n + \cdots + a_1 x + a_0$ may lead to poor approximations. However, particularly in engineering or scientific problems, we may have no other practical alternative than to use information given at equally spaced points. In this case, we shall see that trigonometric polynomials (as opposed to algebraic polynomials) provide a good tool for approximation. The set of all nth-degree trigonometric polynomials, $\mathcal{T}_n$, is all functions that can be written in this form:

$$p(\theta) = a_0 + \sum_{j=1}^{n} a_j \cos(j\theta) + \sum_{j=1}^{n} b_j \sin(j\theta) \tag{6.52}$$

where θ is given in radians and a_j and b_j are constants. The most natural interval to consider for an approximation problem involving trigonometric polynomials is the interval $[0, 2\pi]$; and trigonometric polynomials are well suited to approximate functions that are continuous on $[0, 2\pi]$ and that satisfy $f(0) = f(2\pi)$. This latter condition allows us to extend each such $f(\theta)$ to the entire real axis such that this extension is continuous and periodic with period 2π. For notation, we define this set of functions as

$$C_{2\pi} = \{f(\theta) | f(\theta) \text{ continuous for } 0 \le \theta \le 2\pi, f(0) = f(2\pi)\}.$$

These approximations can be easily modified to suit any interval $[a, b]$ and functions that are not 2π-periodic, but we omit the details in the interest of brevity.

The power of trigonometric polynomial approximation is based on classical Fourier series [see Example 5.15(a)]. Classical Fourier series uses the inner product

$$\langle f, g \rangle = \int_0^{2\pi} f(\theta) g(\theta)\, d\theta \tag{6.53}$$

for which we have (see Problem 1)

$$\int_0^{2\pi} \cos(k\theta)\cos(j\theta)\,d\theta = \begin{cases} 0, & k \neq j \\ \pi, & k = j, \quad k \neq 0 \\ 2\pi, & k = j = 0 \end{cases}$$

$$\int_0^{2\pi} \cos(k\theta)\sin(j\theta)\,d\theta = 0, \qquad \text{for all } k \text{ and } j \tag{6.54}$$

$$\int_0^{2\pi} \sin(k\theta)\sin(j\theta)\,d\theta = \begin{cases} 0, & k \neq j \\ \pi, & k = j > 0. \end{cases}$$

Thus for each n, the set of functions

$$S_n = \left\{\frac{1}{\sqrt{2\pi}}, \frac{1}{\sqrt{\pi}}\cos(j\theta), \frac{1}{\sqrt{\pi}}\sin(j\theta)\right\}_{j=1}^{n} \tag{6.55}$$

is an orthonormal set in $C_{2\pi}$. Furthermore each trigonometric polynomial, $p(\theta)$ in $\mathcal{T}_n$, can be written as a linear combination of elements of S_n.

Hence, given any $f(\theta) \in C_{2\pi}$, we can mimic the proof of Theorem 5.10 and see that $P_n^*(\theta)$ given by

$$P_n^*(\theta) = \left\langle f, \frac{1}{\sqrt{2\pi}}\right\rangle \frac{1}{\sqrt{2\pi}} + \sum_{j=1}^{n} \left\langle f, \frac{\cos(j\theta)}{\sqrt{\pi}}\right\rangle \frac{\cos(j\theta)}{\sqrt{\pi}}$$
$$+ \sum_{j=1}^{n} \left\langle f, \frac{\sin(j\theta)}{\sqrt{\pi}}\right\rangle \frac{\sin(j\theta)}{\sqrt{\pi}} \tag{6.56a}$$

satisfies $\|f - P_n^*\| \leq \|f - p\|$ for all $p(\theta) \in \mathcal{T}_n$; that is,

$$\int_0^{2\pi} (f(\theta) - P_n^*(\theta))^2\,d\theta \leq \int_0^{2\pi} (f(\theta) - p(\theta))^2\,d\theta, \qquad p(\theta) \in \mathcal{T}_n.$$

[The reader should compare (6.56a) with (5.63) and see that the result is equally valid with trigonometric polynomials replacing the algebraic polynomials in Chapter 5.] We note that (6.56a) may be written in this equivalent, but slightly more simplified form:

$$P_n^*(\theta) = \frac{A_0}{2} + \sum_{j=1}^{n} A_j\cos(j\theta) + \sum_{j=1}^{n} B_j\sin(j\theta),$$

$$A_j = \frac{1}{\pi}\langle f, \cos(j\theta)\rangle = \frac{1}{\pi}\int_0^{2\pi} f(\theta)\cos(j\theta)\,d\theta, \tag{6.56b}$$

$$B_j = \frac{1}{\pi}\langle f, \sin(j\theta)\rangle = \frac{1}{\pi}\int_0^{2\pi} f(\theta)\sin(j\theta)\,d\theta.$$

As in the case of the closely related Chebyshev series expansions, we can guarantee that $P_n^*(\theta)$ converges uniformly to $f(\theta)$ for $0 \leq \theta \leq 2\pi$ by imposing mild smoothness conditions on $f(\theta)$ [for instance, $f''(\theta) \in C_{2\pi}$]. Thus trigonometric polynomials will give good approximations; but in order to provide a practical computational tool, it is necessary to develop algorithms for approximating the Fourier coefficients A_j and B_j in (6.56b). It turns out that the composite trapezoidal rule is a *Gaussian* quadrature formula with respect to trigonometric polynomials and hence will give good estimates of the integrals in (6.56b). Since the composite trapezoidal rule is based on equally spaced points, we see that trigonometric polynomials provide a practical and useful form of approximation that is well-suited for functions or data given at equally spaced points.

To be specific, let

$$\theta_j = j\frac{2\pi}{2m} = \frac{j\pi}{m}, \qquad j = 0, 1, \ldots, 2m, \tag{6.57}$$

be equally spaced in $[0, 2\pi]$; and let $T_{2m}(g)$ denote the composite trapezoidal rule applied to $g(\theta)$ on $[0, 2\pi]$. Thus

$$T_{2m}(g) = \frac{\pi}{m} \sum_{j=0}^{2m}{}'' \, g(\theta_j) \approx \int_0^{2\pi} g(\theta)\, d\theta.$$

Since $g \in C_{2\pi}$, we know that $g(0) = g(2\pi)$ [or $g(\theta_0) = g(\theta_{2m})$]. When we wish, then, we can write $T_{2m}(g)$ as

$$T_{2m}(g) = \frac{\pi}{m} \sum_{j=0}^{2m-1} g(\theta_j).$$

To show that this quadrature is Gaussian with respect to trigonometric polynomials, we note that

$$T_{2m}(\cos(k\theta)) = \frac{\pi}{m} \sum_{j=0}^{2m}{}'' \cos\frac{jk\pi}{m}$$

$$T_{2m}(\sin(k\theta)) = \frac{\pi}{m} \sum_{j=0}^{2m}{}'' \sin\frac{jk\pi}{m}.$$

Using Lemma 5.1 and (6.54), we see (Problem 3) that $T_{2m}(\cos(k\theta)) = 0 = \int_0^{2\pi} \cos(k\theta)\, d\theta$ for $1 \leq k \leq 2m - 1$; and $T_{2m}(1) = 2\pi = \int_0^{2\pi} d\theta$. Since $\sin(k\theta)$ is an odd function about the center of the interval of integration, $\theta = \pi$, it is easy to show (Problem 4) that $T_{2m}(\sin(k\theta)) = 0 = \int_0^{2\pi} \sin(k\theta)\, d\theta$ for $1 \leq k \leq 2m - 1$. Thus we have that $T_{2m}(p(\theta)) = \int_0^{2\pi} p(\theta)\, d\theta$ for any trigonometric polynomial of degree $2m - 1$ or less; and this equation is the Gaussian feature of the composite trapezoidal rule for $\mathcal{T}_{2m-1}$.

6.6.1. Least-squares Fits and Interpolation at Equally Spaced Points

Since the composite trapezoidal rule is Gaussian with respect to trigonometric polynomials, it is not surprising (as in Section 6.5.1) that with the composite trapezoidal rule weights and nodes there is a discrete inner product for which the first $2m$ functions in (6.55) are orthonormal. More precisely, the $2m$ functions

$$\frac{1}{\sqrt{2\pi}}, \quad \frac{1}{\sqrt{\pi}}\cos(\theta), \ldots, \quad \frac{1}{\sqrt{\pi}}\cos((m-1)\theta), \quad \frac{1}{\sqrt{2\pi}}\cos(m\theta),$$
$$\frac{1}{\sqrt{\pi}}\sin(\theta), \ldots, \quad \frac{1}{\sqrt{\pi}}\sin((m-1)\theta) \tag{6.58}$$

are orthonormal with respect to the discrete inner product

$$\langle f, g\rangle_d = \frac{\pi}{m} \sum_{j=0}^{2m-1} f\left(\frac{j\pi}{m}\right) g\left(\frac{j\pi}{m}\right) \tag{6.59}$$

(see Problem 5). Thus, given the problem of finding a kth-degree trigonometric polynomial to minimize the expression

$$\sum_{j=0}^{2m-1} [y_i - p(\theta_j)]^2$$

where $\{y_j\}_{j=0}^{2m-1}$ is a given set of values, we find (Problem 6) that the solution for $k \leq m - 1$ is given by

$$p_k(\theta) = \frac{a_0}{2} + \sum_{j=1}^{k} a_j \cos(j\theta) + \sum_{j=1}^{k} b_j \sin(j\theta),$$
$$a_j = \frac{1}{m}\langle f, \cos(j\theta)\rangle_d = \frac{1}{m}\sum_{r=0}^{2m-1} y_r \cos\frac{rj\pi}{m}, \tag{6.60}$$
$$b_j = \frac{1}{m}\langle f, \sin(j\theta)\rangle_d = \frac{1}{m}\sum_{r=0}^{2m-1} y_r \sin\frac{rj\pi}{m}$$

where $f(\theta)$ is any function in $C_{2\pi}$ such that $f(\theta_r) = y_r$, $0 \leq r \leq 2m - 1$.

EXAMPLE 6.13. In order to demonstrate the calculations in (6.60), we let the data y_j be given by

$$y_r = \frac{\sin\theta_j}{\theta_j}, \qquad \theta j = j\frac{2\pi}{10} = \frac{j\pi}{5},$$
$$j = 0, 1, \ldots, 9,$$

where we define $y_0 = 1$. In this case, $m = 5$ and we calculate as in (6.60)

$$a_0 = \frac{1}{5}\sum_{j=0}^{9} \frac{\sin(\theta_j)}{\theta_j} = 0.7061$$

$$a_1 = \frac{1}{5}\sum_{j=0}^{9} \frac{\sin(\theta_j)}{\theta_j} \cos(\theta_j) = 0.0823$$

$$b_1 = \frac{1}{5}\sum_{j=0}^{9} \frac{\sin(\theta_j)}{\theta_j} \sin(\theta_j) = 0.1501.$$

Similarly, we find $a_2 = 0.0918$ and $b_2 = 0.0759$ so that $p_2(\theta) = 0.3531 + 0.0823\cos(\theta) + 0.1501\sin(\theta) + 0.0918\cos(2\theta) + 0.0759\sin(2\theta)$ is the best least-squares fit of degree 2 to this data on the ten points θ_j.

When $k = m$ in (6.60), the situation is somewhat altered. As was the case in Section 6.5.1, we find that the trigonometric polynomial $p_m(\theta)$ given below in (6.61) is an *interpolating* polynomial. [Note in (6.61) that $p_m(\theta)$ is still the best least-squares approximation to the data $y_0, y_1, \ldots, y_{2m-1}$ with respect to the orthonormal set (6.58).] We have

$$p_m(\theta) = \sum_{j=0}^{m}{}'' \, a_j \cos(j\theta) + \sum_{j=1}^{m-1} b_j \sin(j\theta) \tag{6.61}$$

where the coefficients a_j and b_j are as given in (6.60) and the double prime means that the first and the last terms are halved. To see that $p_m(\theta_r) = y_r$ for $r = 0, 1, \ldots, 2m-1$, we write out $p_m(\theta_r)$ and obtain

$$\begin{aligned}
p_m(\theta_r) &= \sum_{s=0}^{m}{}'' \cos(s\theta_r) \frac{1}{m} \sum_{j=0}^{2m-1} y_j \cos(s\theta_j) \\
&\quad + \sum_{s=1}^{m-1} \sin(s\theta_r) \frac{1}{m} \sum_{j=0}^{2m-1} y_j \sin(s\theta_j) \\
&= \sum_{j=0}^{2m-1} y_j \frac{1}{m} \sum_{s=0}^{m}{}'' \cos(s\theta_r)\cos(s\theta_j) \\
&\quad + \sum_{j=0}^{2m-1} y_j \frac{1}{m} \sum_{s=1}^{m-1} \sin(s\theta_r)\sin(s\theta_j) \\
&= \sum_{j=0}^{2m-1} y_j \frac{1}{m} \sum_{s=0}^{m}{}'' \cos\left(s(r-j)\frac{\pi}{m}\right).
\end{aligned}$$

To obtain the last equality, we use the fact that $\sin(s\theta_r)\sin(s\theta_j) = 0$ for $s = 0$ and m, and then use the identity for $\cos(A - B)$. Using Lemma 5.1 (Problem 7), we see that the inner sum is zero unless $r = j$, in which case the inner sum is 1. Thus $p_m(\theta_r) = y_r$ for $r = 0, 1, \ldots, 2m-1$.

In considering (6.60) or (6.61), we see that the bulk of the computation involves calculating the coefficients a_j and b_j. In many engineering problems, the *power spectrum*, which is a plot of the numbers $a_j^2 + b_j^2$ as a function of j, has significance. Generally, we are interested in the behavior of the function $f(x)$ defined by the inverse Fourier transform

$$f(x) = \int_{-\infty}^{\infty} F(t)e^{2\pi ixt}\,dt = \int_{-\infty}^{\infty} F(t)[\cos(2\pi xt) + i\sin(2\pi xt)]\,dt \qquad (6.62a)$$

where $i = \sqrt{-1}$. Typically, $F(t)$ decays rapidly for $|t|$ large so that

$$\int_{-a}^{a} F(t)\cos(2\pi xt)\,dt + i\int_{-a}^{a} F(t)\sin(2\pi xt)\,dt \qquad (6.62b)$$

provides a good estimate for $f(x)$. The composite trapezoidal rule gives good approximations to the integrals in (6.62b) and again leads to sums of the form

$$\sum_{j=0}^{2m-1} F(t_j)\cos(\lambda t_j) \quad \text{and} \quad \sum_{j=0}^{2m-1} F(t_j)\sin(\lambda t_j)$$

where the points t_j are equally spaced in $[-a, a]$. Thus sums of this form occur frequently in important applications other than interpolation, and efficient techniques such as the *fast Fourier transform* have been developed to compute them.

*6.6.2. The Fast Fourier Transform

The fast Fourier transform (FFT) is an efficient procedure for organizing the computation of the a_j's and b_j's [and thus $p_k(\theta)$) in (6.60) or (6.61)]. As an example, suppose we wish to compute the $2m$ coefficients $a_0, a_1, \ldots, a_m, b_1, \ldots, b_{m-1}$ that are needed to calculate the interpolating polynomial in (6.61). We can verify that the direct calculation of the coefficients takes approximately $(2m)^2$ multiplications and $(2m)^2$ additions. By efficient programming, the fast Fourier transform can reduce the number of operations to approximately $2m \ln(2m)$ operations. For small values of m this change is perhaps not significant, but for large values of m (say, $m = 10^4$ or more) this amount is tremendously significant (see Problem 8) and reduces formerly impractical computations to ones that are fairly easily handled.

It is most convenient to define $C_j \equiv a_j + ib_j$, ($i \equiv \sqrt{-1}$). Using Euler's formula, $e^{ij\theta} = \cos(j\theta) + i\sin(j\theta)$, we have

$$C_j = a_j + ib_j = \frac{1}{m}\sum_{r=0}^{2m-1} f(\theta_r)e^{ij\theta_r}. \qquad (6.63)$$

Now let M and K be any two integers such that $2m = MK$ where K is even. Next, set

$$j = j_1 M + j_0, \qquad 0 \le j_0 < M,$$

and

$$r = r_1 K + r_0, \qquad 0 \le r_0 < K.$$

We leave to the reader (Problem 9) to verify that any sum of the form $\Sigma_{r=0}^{2m-1} v_r$ can be expressed as the double sum $\Sigma_{r_0=0}^{K-1} \Sigma_{r_1=0}^{M-1} v_{r_1K+r_0}$. Using this fact, we can write [with $\exp(x) \equiv e^x$]

$$\begin{aligned} C_j &= \frac{1}{m} \sum_{r_0=0}^{K-1} \sum_{r_1=0}^{M-1} f(\theta_{r_1K+r_0}) \exp[i(j_1M + j_0)(r_1K + r_0)2\pi/MK] \\ &= \frac{1}{m} \sum_{r_0=0}^{K-1} \sum_{r_1=0}^{M-1} f(\theta_{r_1K+r_0}) \exp[2\pi i(j_1r_1 + j_1r_0/K + j_0r_1/M + j_0r_0/MK)] \qquad (6.64) \\ &= \frac{1}{m} \sum_{r_0=0}^{K-1} \exp\left[2\pi i\left(j_1 + \frac{j_0}{M}\right)\frac{r_0}{K}\right] \sum_{r_1=0}^{M-1} f(\theta_{r_1K+r_0}) \exp\left[\frac{2\pi i j_0 r_1}{M}\right]. \end{aligned}$$

In the inner sum of the last expression we used the reduction

$$\begin{aligned} \exp\left[2\pi i\left(j_1r_1 + \frac{j_0r_0}{M}\right)\right] &= \exp[2\pi i j_1 r_1] \exp\left[\frac{2\pi i j_0 r_1}{M}\right] \\ &= \exp\left[\frac{2\pi i j_0 r_1}{M}\right], \end{aligned}$$

which is valid since j_1 and r_1 are integers (and so $\exp[2\pi i j_1 r_1] \equiv 1$). Thus the inner sum is independent of j_1. This independence is what makes the FFT "fast"; that is to say, the same inner sum is used to compute each C_j of the form $C_j = C_{j_1M+j_0}$ for any *fixed value* of j_0. For example, if $j_0 = 3$, then the same inner sum is used to compute $C_3, C_{M+3}, C_{2M+3}, \ldots$.

We now define

$$C(j_0, r_0) \equiv \frac{1}{m} \sum_{r_1=0}^{M-1} f(\theta_{r_1K+r_0}) \exp\left[\frac{2\pi i j_0 r_1}{M}\right], \qquad (6.65)$$

and so (6.64) becomes (for $j = j_1M + j_0$)

$$C_j = \sum_{r_0=0}^{K-1} C(j_0, r_0) \exp\left[2\pi i\left(j_1 + \frac{j_0}{M}\right)\frac{r_0}{K}\right]. \qquad (6.66)$$

Thus far we see that we must (1) form a table of $C(j_0, r_0)$ for $0 \le j_0 < M$, $0 \le r_0 < K$; and (2) compute each C_j according to formula (6.66). To be more precise, let us now investigate the number of multiplications required in the scheme above.

1. Each entry $C(j_0, r_0)$ of the table requires M multiplications. The table has MK entries; so step (1) requires $M(MK)$ multiplications.
2. With the table constructed, it takes K multiplications to compute each C_j. There are $m = MK/2$ coefficients C_j; so step (2) requires less than $K(MK/2)$ multiplications. [In fact, calculating C_0 and C_m requires no multiplications; so we need only $K(MK/2 - 1)$ multiplications. However, $K(MK/2)$ is an easier bound to deal with as we shall see. Additionally, the multiplications in (6.65) and (6.66) are "complex" multiplications because of the presence of the complex exponential.]

The total number of multiplications in (1) and (2) is then less than $M^2K + MK^2/2 = MK(M + K/2)$. The straightforward evaluation of C_j from (6.63) requires $2m$ multiplications for each j, or a total of $2m(m) = MK(MK/2) = M^2K^2/2$. For example if $m = 64$ and we select $M = 8$ and $K = 16$ ($2m = 128 = MK = 8(16)$), then FFT requires $MK(M + K/2) = 8(16)(8 + 16/2) = 128(16)$ multiplications; and straightforward computation requires $2m(m) = 128(64)$ multiplications, which is larger than FFT by a factor of 4.

Thus far, however, we have not utilized the FFT scheme to its full potential. For example we note that formula (6.65) for each $C(j_0, r_0)$ looks very much like formula (6.63) for each C_j where in (6.65) M and j_0 play the roles of $2m$ and j, respectively, in (6.63). We may then correctly reason that we can use the FFT scheme to construct the required table of $C(j_0, r_o)$'s. More precisely, fix r_0 and consider construction of the r_0 column of the table—$C(0, r_0)$, $C(1, r_0)$, . . . , $C(M - 1, r_0)$. Thus if $M = 2s$, then for $0 \le k \le M - 1$,

$$\begin{aligned} C(k, r_0) &= \frac{1}{m}\sum_{r_1=0}^{M-1} f(\theta_{r_1K+r_0}) \exp\left[\frac{2\pi i k r_1}{M}\right] \\ &= \frac{1}{s}\sum_{r_1=0}^{2s-1} \frac{1}{K} f(\theta_{r_1K+r_0}) \exp\left[\frac{\pi i k r_1}{s}\right]; \end{aligned} \tag{6.67}$$

but this is precisely the form of (6.63) with the role of $f(\theta_r)$ being played by $g(\theta_{r_1}) \equiv (1/K) f(\theta_{r_1K+r_0})$.

Once again we let u and v be two integers such that $2s = u \cdot v$; and we repeat the same process on each $C(k, r_0)$, $0 \le k \le M - 1$, instead of on each C_j, $1 \le j \le m$. We shall not write this out again in detail except to note that with the modification of indices above the same program may be used. Counting multiplications as outlined in (1) and (2) yields $u(uv) + v(uv) = u^2v + uv^2$ multiplications to compute $C(0, r_0)$, $C(1, r_o)$, . . . , $C(M - 1, r_0)$. [Note that we have in (2) a factor of $v(uv)$ instead of $K(MK/2)$ as in the previous calculation because there are $2s = w$ quantities in the latter versus $m = (MK/2)$ quantities in the former.] Thus each column of the table requires $u^2v + uv^2 = uv(u + v)$ multiplications or a total of $Kuv(u + v)$ for the entire table since there are K rows. Now having constructed the table, we know from (2) that we need about $K(MK/2)$ more multiplications or a total of $Kuv(u + v) + K(MK/2) = Kuv(u + v) + K^2uv/2 = Kuv(u + v + K/2)$.

If in the previous step, u can be written in the factored form $u = 2\sigma = \mu v$, we can again use the FFT scheme to economize our computation. In general we find that if $2m = h_1h_2 \ldots h_k$, then the multiplications necessary are $h_1h_2 \ldots h_k \times (h_1 + h_2 + \cdots + h_{k/2})$. A typical choice is $2m = 2^k$, which yields $2^k(2 + 2 + \cdots + 2 + 1) = 2^k(2k - 1)$ multiplications. Straightforward calculation requires $(2m)(m) = (2^k)(2^{k-1})$ multiplications. For example, let $m = 1024 = 2^{10}$ (which is not an exceptionally large number of data points). Then $k = 11$, $2^k(2k - 1) = 43{,}008$, and $2^k(2^{k-1}) = 2{,}097{,}152$. The FFT has a factor of approximately 49 times fewer multiplications.

PROBLEMS, SECTION 6.6.2

1. Verify the identities in (6.54).
2. Using (6.56b), find $P_n^*(\theta)$ for $f(\theta) = \theta(\theta - 2\pi)$ for $n = 2, 3$, and 4. What is $\|f - P_n^*\|$?
3. Verify that $T_{2m}(\cos(k\theta)) = \int_0^{2\pi} \cos(k\theta)\, d\theta$, $0 \le k \le 2m - 1$.
4. Verify that $T_{2m}(\sin(k\theta)) = \int_0^{2\pi} \sin(k\theta)\, d\theta$, $1 \le k \le 2m - 1$.
5. Verify that the functions in (6.58) are orthonormal with respect to the inner product in (6.59).
6. Show that $p_k(\theta)$ defined in (6.60) minimizes $\Sigma_{j=0}^{2m-1}[y_j - p(\theta_j)]^2$ among all $p(\theta) \in \mathcal{T}_n$.
7. Show that $1/m\, \Sigma_{s=0}^{\prime\prime\, m} \cos(s(r - j)\, \pi/m) = \delta_{rj}$.
8. Compare the numbers $10^4 10^4$ and $10^4 \ln(10^4)$.
9. Show that a sum of the form $\Sigma_{r=0}^{2m-1} v_r$ can be rewritten as $\Sigma_{r_0=0}^{K-1} \Sigma_{r_1=0}^{M-1} v_{r_1K+r_0}$ when $2m = MK$ and $r = r_1K + r_0$, $0 \le r_0 < K$.

*10. As an example, let $2m = 128 = MK = (8)(16)$ and sketch out the details of how the tables can be constructed for the FFT. Write a program to implement the FFT; test your routine by calculating numerically the coefficients A_j and B_j for $f(\theta) = \theta(\theta - 2\pi)$, $j \le 64$.

7 Numerical Solution of Ordinary Differential Equations

7.1 INTRODUCTION

As the reader is probably aware, differential equations serve as *mathematical descriptions* for many physical problems and phenomena. A few examples of commonly occurring differential equations follow.

1. $ay''(x) + by'(x) + cy(x) = F(x)$
2. $EIy^{(iv)}(x) = w(x)$
3. $y''(x) = \frac{w}{H}\sqrt{1 + (y'(x))^2}$
4. $\frac{d}{dx}\left(p(x)\frac{dy}{dx}\right) + q(x)y = r(x)$

Equation (1) occurs in the study of vibrating or oscillating mechanical systems or in electrical circuits, (2) arises in the study of beam deflections, (3) deals with problems in cable suspension, and (4) arises when a separation of variables technique can be applied to partial differential equations describing important problems in mathematical physics and engineering. Finding accurate and efficient procedures for solving differential equations has long been a problem of importance. However in many practical situations, an analytical solution (that is, a closed-form mathematical solution) is either impossible to find or extremely difficult to evaluate. In recent years, numerical procedures for approximating solutions have become increasingly popular, thanks in large part to the power of modern, high-speed computers. The basic ideas involved in deriving numerical methods for differential equations are relatively simple and do not require an extensive theoretical background (although a detailed analysis of the convergence of a method may be quite difficult). In this section, we will present most of the fundamentals and terminology necessary for an understanding of many of the more popular methods now in use.

The first problem we consider is the *initial-value problem*:

$$y'(x) = f(x, y(x)), \qquad a \le x \le b \tag{7.1a}$$

$$y(a) = y_0. \tag{7.1b}$$

In this problem, we are given a finite interval $[a, b]$ and a function $f(x, y)$. We seek a function $y(x)$ defined on $[a, b]$ such that $y'(x) \equiv f(x, y(x))$ for $a \le x \le b$ *and* such that $y(x)$ satisfies the initial condition $y(a) = y_0$ where y_0 is some prescribed value. Thus an initial-value problem consists of two parts: (7.1a), the differential equation $y' = f(x, y)$ that gives the desired relationship between $y(x)$ and $y'(x)$; and (7.1b), the initial condition $y(a) = y_0$. Equation $y' = f(x, y)$ is called a *first-order* differential equation because the highest derivative that appears is the first derivative. In a later section, we will show how many of the methods developed for solving the first-order equation (7.1) can be applied to solve *systems* of first-order differential equations simultaneously. Further, we will show how to convert an nth-order differential equation into a system of first-order equations. Thus methods for handling the apparently simple case considered in (7.1) are in fact applicable to systems of higher-order equations. Therefore we will concentrate almost exclusively on numerical procedures for the initial-value problem (7.1).

EXAMPLE 7.1. These are some examples of initial-value problems.

$$\begin{aligned} y' &= y, \qquad 0 \le x \le 1 \\ y(0) &= 1 \end{aligned} \tag{7.2a}$$

$$\begin{aligned} y' &= 2xy, \qquad 0 \le x \le 1 \\ y(0) &= 1 \end{aligned} \tag{7.2b}$$

$$\begin{aligned} y' &= y^2(x - 6)/x^3, \qquad 1 \le x \le 2 \\ y(1) &= -\tfrac{1}{2} \end{aligned} \tag{7.2c}$$

$$\begin{aligned} y'' &= (z^2 - 1)/y, \qquad 0 \le x \le 1 \\ z'' &= -z \\ y(0) = 1, \quad y'(0) &= 0, \quad z(0) = 0, \quad z'(0) = 1. \end{aligned} \tag{7.2d}$$

The differential equations in (7.2a) and (7.2b) are first-order *linear* equations (since y and y' occur linearly). The differential equation in (7.2c) is a first-order *nonlinear* equation because of the presence of y^2. Finally, (7.2d) provides an example of a second-order system of nonlinear differential equations. The solutions are given in Problem 1. A more realistic example of a nonlinear second-order system is provided by (4.0a), the equations of motion for a ballistic reentry trajectory. This equation cannot be solved in closed form and must be solved numerically.

As examples in this chapter, we will frequently use differential equations like (7.2a) and (7.2b). These equations are quite simple and not at all representative of the sorts of differential equations that occur in practice. Our purpose in using these equations as examples is merely to clarify the presentation. We return now to the simple initial-value problem $y' = y$ with $y(0) = 1$ as in Example 7.1. The differential equation $y' = y$ has many solutions, all of the form $y(x) = ce^x$ where c can be any constant (it is easy to verify that the relationship $y' = y$ holds when $y = ce^x$). So the set of functions $y = ce^x$ forms a one-parameter family of solutions for $y' = y$ with each member of the family determined by the value chosen for c. Although $y = ce^x$ solves $y' = y$ regardless of the value of c, there is only one choice of c for which $y(0) = 1$, namely $c = 1$. Thus the solution of the initial-value problem is $y = e^x$.

For the general case, $y' = f(x, y)$ with $y(a) = y_0$, we call each function $y(x)$ that satisfies $y' = f(x, y)$ an *integral curve*. Thus solving an initial-value problem is a two-stage process: (1) determine the integral curves for $y' = f(x, y)$; (2) find the member of this family of integral curves that satisfies $y(a) = y_0$. In the example above, we note that $y' = y$, $y(0) = y_0$ can always be solved no matter what value y_0 has (in fact, the solution is $y = y_0e^x$). Geometrically, this statement says that given any point in the xy-plane of the form $(0, y_0)$, there is a solution of $y' = y$ passing through this point (see Fig. 7.1).

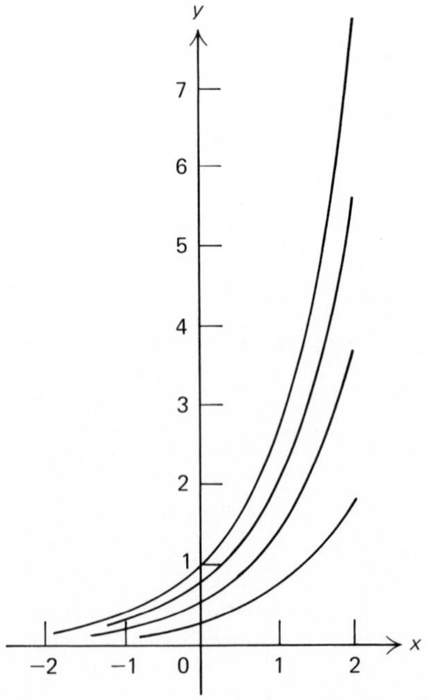

Figure 7.1 Integral curves for $y' = y$.

The idea of a *flow field* or *direction field* is related to Fig. 7.1 and can be used to visualize a family of integral curves for $y' = f(x, y)$. In addition, flow fields provide a useful framework for understanding how numerical methods behave and how the errors propagate. As an introduction to flow fields, suppose that (x^*, y^*) is any point in the xy-plane and suppose that a solution (an integral curve) of $y' = f(x, y)$ passes through the point (x^*, y^*). Given that $y(x)$ satisfies $y'(x) = f(x, y(x))$ and that $y(x^*) = y^*$, it follows that $y'(x^*) = f(x^*, y^*)$; that is, the tangent to the graph of $y(x)$ has slope $f(x^*, y^*)$ at the point (x^*, y^*). In general, the function $f(x, y)$ gives the slope of any solution of $y' = f(x, y)$ that passes through the point (x, y). We can represent this information pictorially as in Fig. 7.2, which shows the flow field for $y' = y$. A little imagination shows that we can use the flow field to get a rough idea of the shapes of the integral curves if we draw curves through the flow field with the slopes of these curves dictated by Fig. 7.2. If two integral curves $y_1(x)$ and $y_2(x)$ are near each other at $x = x_1$, their slopes $y_1'(x_1)$ and $y_2'(x_1)$ will determine whether the curves remain near each other as x moves away from x_1. Thus given nearby points (x_1, y_1) and (x_1, y_2) in the xy-plane, if $f(x_1, y_1)$ is substantially different from $f(x_1, y_2)$, then the integral curves through (x_1, y_1) and (x_1, y_2) are diverging from each other (at least for x near x_1).

As another example, the flow field for $y' = y(2 - y)$ is given in Fig. 7.3. The flow field in Fig. 7.3 suggests that any solution that starts at (x_0, y_0) where $y_0 > 2$ will fall quickly and become asymptotic to the line $y = 2$. A solution started at (x_0, y_0) where $0 < y_0 < 1$ will rise to the line $y = 2$; the solution will begin as concave upwards and then switch to being concave downwards. Finally, Fig.

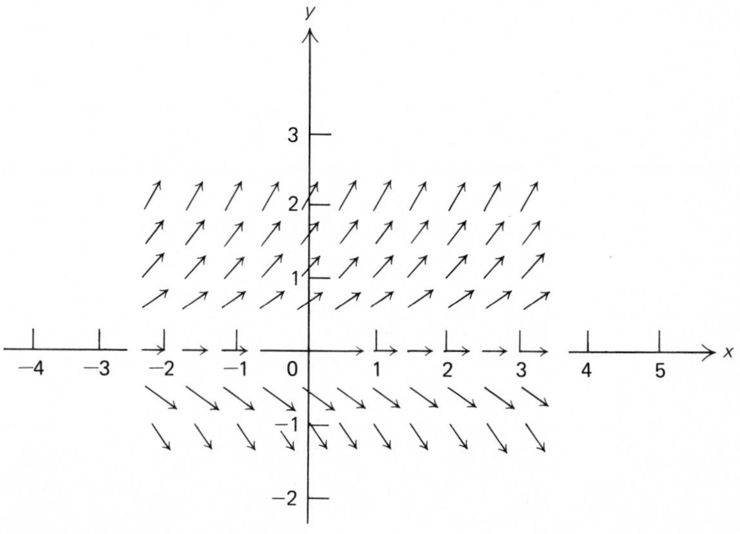

Figure 7.2 The flow field for $y' = y$.

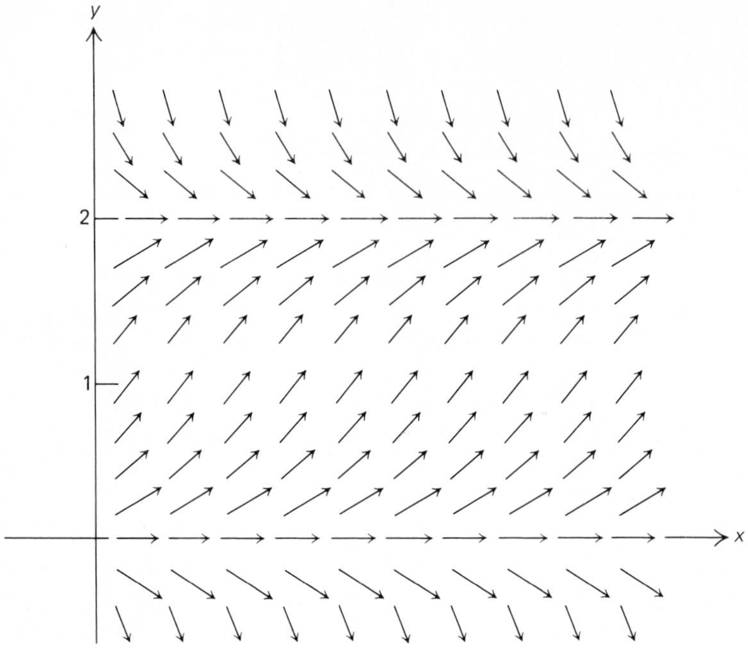

Figure 7.3 The flow field for $y = y\,(2 - y)$.

7.3 suggests that a solution started at (x_0, y_0) where $y_0 < 0$ will be unbounded negatively and will become asymptotic to some vertical line $x = c$ as x approaches c from the left. The equation $y' = y(2 - y)$ is easy to solve, and the reader may verify that solutions behave exactly as the flow field predicts.

Another important qualitative indicator for the equation $y' = f(x, y)$ is the sign of the partial derivative f_y at the point (x, y). Suppose that (x^*, y^*) is a point in the xy-plane and suppose that $f_y(x^*, y^*) > 0$. Then, as y ranges from $y^* - \varepsilon$ to $y^* + \varepsilon$ (where $\varepsilon > 0$), the function $f(x^*, y)$ is increasing. In terms of the flow field, $f_y(x^*, y^*) > 0$ means that slopes are increasing near (x^*, y^*) as we move up the line $x = x^*$ (see Fig. 7.4). A qualitative interpretation of this fact is that the integral curves are fanning out in a neighborhood of (x^*, y^*). For example in Fig. 7.2, the integral curves for $y' = y$ are apparently spreading out in a neighborhood of the point (1, 2). On the other hand, if $f_y(x^*, y^*) < 0$, then $f(x^*, y)$ is a decreasing function of y for y near y^*; and this fact can be interpreted as suggesting that the integral curves are funneling together in a neighborhood of (x^*, y^*). [See Dahlquist, Björk, and Anderson (1974) for a more detailed discussion.] The concept of a flow field and the related Fig. 7.3 are primarily heuristic tools, but they can be used as a rough aid for understanding the numerical methods we present later.

This section concludes with a well-known existence and uniqueness

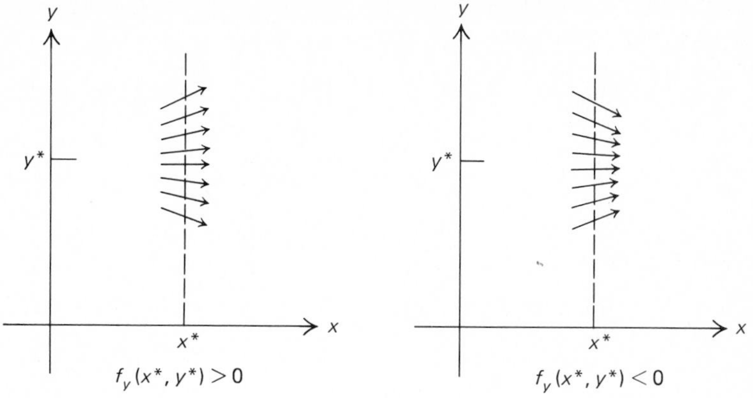

Figure 7.4

theorem for the initial-value problem (7.1). The hypotheses of this theorem are quite strong and there are a number of existence/uniqueness theorems with weaker hypotheses, but the theorem does indicate the sort of result that can be obtained. For a proof of Theorem 7.1, we refer the reader to Coddington and Levinson (1955). In all our discussions of numerical methods in later sections, we will assume whenever necessary that solutions exist and are unique.

Theorem 7.1

Let $f(x, y)$ be jointly continuous for $a \le x \le b$ and for all y. Suppose further that $f(x, y)$ satisfies the Lipschitz condition

$$|f(x, u) - f(x, v)| \le L|u - v| \tag{7.3}$$

for $a \le x \le b$ and all u and v where L is a constant. Then for any number y_0, there exists a unique differentiable function $y(x)$ defined on $[a, b]$ such that $y' = f(x, y)$ and $y(a) = y_0$.

From the mean-value theorem, it is easy to see that $f(x, y)$ satisfies the Lipschitz condition (7.3) if the partial derivative, $f_y(x, y)$, is bounded in absolute value for x in $[a, b]$ and all y. To see this, suppose that $|f_y(x, y)| \le L$; then

$$f(x, u) - f(x, v) = f_y(x, \theta)(u - v)$$

where θ is between u and v. Hence if $|f_y(x, y)| \le L$ for $a \le x \le b$, then $|f(x, u) - f(x, v)| \le L|u - v|$.

EXAMPLE 7.2. For the problem $y' = y$, $y(a) = y_0$, we have $f(x, y) = y$ and hence $|f(x, u) - f(x, v)| \le |u - v|$. Therefore solutions exist uniquely on any interval $[a, b]$. For the problem $y' = 2xy$, $y(0) = y_0$, we have $f_y(x, y) = 2x$. Thus on any interval $[0, b]$, $|f(x, u) - f(x, v)| \le 2b|u - v|$; so a unique solution exists. For problem (3) of Example

7.1, $f_y(x, y) = 2y(x - 6)/x^3$. This partial derivative is not bounded for all y when $1 \le x \le b$; so Theorem 7.1 is not applicable. In fact (see Problem 1), the solution is $y(x) = x^2/(x - 3)$, which becomes infinite as x nears 3. Similarly, the problem $y' = y/(x - 1)$, $y(0) = y_0$ has a unique solution on $[0, b]$ when $b < 1$, but Theorem 7.1 does not apply when $1 \le b$ (see Problem 2).

PROBLEMS, SECTION 7.1

1. Verify by substitution that the following functions are solutions of the initial-value problems in Example 7.1.

   b) $y(x) = e^{x^2}$ c) $y(x) = x^2/(x - 3)$ d) $y(x) = \cos(x)$, $z(x) = \sin(x)$

2. Rewrite $y' = y/(x - 1)$ as $y'/y = 1/(x - 1)$ and integrate both sides with respect to x. Draw the flow field for this equation and conclude that you can satisfy any initial conditions associated with this differential equation except those of the type $y(1) = y_0$, $y_0 \neq 0$.

3. Find all integral curves for $y' = x + 1$. Find the integral curve for $y' = x + 1$ that satisfies

   a) $y(0) = 0$, b) $y(0) = 1$, c) $y(1) = 5$.

4. Determine the Lipschitz constant for these equations.

   a) $f(x, y) = x^2 y$, $-1 \le x \le 1$

   b) $f(x, y) = xe^{-y^2}$, $-1 \le x \le 1$

   c) $f(x, y) = \dfrac{(x^4 - x^2 - 2)^{33}}{(x^2 - 5)^6}$, $-1 \le x \le 1$

5. Let λ be a given constant. Verify that the solution of $y' = \lambda y$, $y(0) = y_0$ is $y(x) = y_0 e^{\lambda x}$.

6. Solve $y' = \lambda y + c$, $y(0) = y_0$ by rewriting the equation as $y' = \lambda(y + c/\lambda)$, and making the change of variables $w(x) = y(x) + c/\lambda$, and then using Problem 5. What is $\lim_{x \to \infty} y(x)$ for the two cases $\lambda > 0$ and $\lambda < 0$? Sketch the flow field and the solution for $y' = -y + 1$, $y(0) = 2$ and for $y' = -100y + 100$, $y(0) = 2$ for $0 \le x \le 3$.

7. Solve $y' = y(2 - y)$, $y(0) = y_0$ and verify that the solution behaves as predicted in Fig. 7.3. For $y_0 < 0$, what is the value of c for which $y(x)$ becomes unbounded negatively as x approaches c from the left? (Note: The differential equation is a Bernoulli equation and the solution technique can be found in any elementary differential equations text.)

8. Show that the initial-value problem $y' = \sqrt{|y|}$, $y(0) = 0$ has infinitely many solutions. Show this by verifying (for any $\alpha \ge 0$) that the function $y(x)$ is a solution:

$$y(x) = \begin{cases} 0, & x \le \alpha \\ \dfrac{(x - \alpha)^2}{4}, & x > \alpha\,. \end{cases}$$

9. Sketch the flow field for $y' = x - y$ for $y \ge 0$ (this sketch is easily done by considering the slopes along lines $y = x + c$ for various c). Deduce the behavior of

solutions if $y(0) > 0$. Illustrate by solving the initial-value problem $y' = x - y$, $y(0) = 3$.

10. Repeat Problem 9 for $y' = y(2 - x)$; sketch slopes along lines $x = c$. Solve $y' = y(2 - x)$, $y(0) = 3$

11. For the equation $y' = y(2 - x)$ in Problem 10, find all initial conditions $y_0 > 0$ such that $|y(x)| \leq 10^{-3}$ at $x = 9$.

12. Derive the closed form solution to the Allen and Eggers reentry model (4.0b) as follows.

a) Suppose that the position vector $\mathbf{R}(t)$ is given by $\mathbf{R}(t) = [x_1(t), x_2(t), x_3(t)]^{\mathrm{T}}$. Write (4.0b) out componentwise and obtain the three equations

$$\ddot{x}_1(t) = -Ke^{-\lambda H}|\dot{\mathbf{R}}|\dot{x}_i(t), \qquad i = 1, 2, 3.$$

b) Since $\mathbf{R}(t)$ is a vector from the center of the earth to the point-mass, we see that $\mathbf{R}(t)$ is perpendicular to the local horizon. Show that the dot product $\mathbf{R}(t) \cdot \dot{\mathbf{R}}(t)$ is given by $\mathbf{R}(t) \cdot \dot{\mathbf{R}}(t) = -|\mathbf{R}(t)|\,|\dot{\mathbf{R}}(t)| \sin(\alpha)$. Note that α is a constant since the motion is in a straight line.

c) If R_e is the radius of the earth and if H denotes the altitude of the point-mass above the earth's surface, then $|\mathbf{R}(t)| = R_e + H(t)$. From this equation, show that

$$\frac{dH}{dt} = \frac{\mathbf{R}(t) \cdot \dot{\mathbf{R}}(t)}{|\mathbf{R}(t)|}.$$

d) Using (a), (b), and (c), derive the closed form expression for $\dot{\mathbf{R}}(t)$ given in Example 4.0.

13. One of the early iterative methods designed for solving $y' = f(x, y)$, $y(a) = y_0$ was Picard's method. In this method we let $y^{(0)}(x) \equiv y_0$; and since $\int_a^x y'(t)\,dt = y(x) - y(a) = \int_a^x f(t, y(t))\,dt$, we define a sequence of functions, $\{y^{(k)}(x)\}_{k=1}^{\infty}$, by

$$y^{(k+1)}(x) = y(a) + \int_a^x f(t, y^{(k)}(t))\,dt, \qquad k \geq 0.$$

a) Apply Picard's method to the first two parts of Example 7.1 and to the problem, $y' = \sin(x + y)$, $y(0) = 0$. (After this work, the reader should quickly see the drawbacks of this method: repeated indefinite integrations, slowness of convergence, etc. Hence Picard's method is seldom used for practical computation but is used extensively in developing the theory of differential equations.)

*b) The reader should note that Picard's method is actually a special case of *fixed-point* iteration as in Section 2.4, and Section 4.3. From this background the reader should be able to find sufficient conditions to ensure the convergence of Picard's method. [Hint: Consider Theorem 7.1.]

7.2 ONE-STEP METHODS

Our study of numerical procedures for differential equations begins with a class of methods known as one-step methods, which are both effective and easy to understand. Given the initial-value problem

$$y' = f(x, y), \ y(a) = y_0,$$

we seek an approximation to the solution $y(x)$. This approximation might take the form of a polynomial as in the method of collocation (see Section 7.7). Normally however, the approximation to $y(x)$ will be a discrete set of values y_1, $y_2, \ldots, y_n$ that approximate $y(x_1), y(x_2), \ldots, y(x_n)$ with $a < x_1 < x_2 < \cdots < x_n \le b$. Then since $y_i \approx y(x_i)$, the set of points (x_i, y_i) could be plotted to provide an approximation to $y(x)$. In some applications, only an approximation to $y(b)$ is wanted. In this case with $x_n = b$, the numbers $y_1, y_2, \ldots, y_{n-1}$ become just intermediate steps.

The general one-step method is an iteration that takes the form

$$y_{i+1} = y_i + h\phi(x_i, y_i; h), \qquad i = 0, 1, \ldots, n-1.$$

The function $\phi(x_i, y_i; h)$ is called the *increment function* and tells us how to proceed from an estimate y_i for $y(x_i)$ to an estimate y_{i+1} for $y(x_{i+1})$. Essentially, the only one-step methods we will consider are Taylor's series methods and the closely related Runge-Kutta methods. The interested reader is referred to Henrici (1962) for a more detailed treatment of one-step methods. One notable advantage of any one-step method is the ease of changing the size of the step size h as we proceed from a to b. For example, if an error analysis (see Section 7.2.4) indicates that y_{i+1} may not be a good approximation for $y(x_{i+1})$, then we can easily replace h in the formula above by a smaller value, say $h/2$, and repeat the calculation and the estimation of its accuracy.

7.2.1. Euler's Method

The simplest one-step method is Euler's method, which is suggested by Figs. 7.1 and 7.2. Given the problem $y' = f(x, y)$, $y(a) = y_0$, we know the value of the solution $y(x)$ at the point $x = a$. Moreover, we know the slope of the solution curve at $(a, y(a))$ since $y'(a) = f(a, y(a)) = f(a, y_0)$. It seems reasonable that we might proceed along a line of slope $y'(a)$ from the point $(a, y(a))$ and still be close to $y(x)$, at least for x near a. For small h, this expectation leads us to approximate $y(a + h)$ by

$$y(a + h) \simeq y(a) + hy'(a) = y_0 + hf(a, y_0).$$

Next, we choose a value n, let $h = (b - a)/n$, and set $x_i = x_0 + ih$ where $x_0 = a$. Since $y(a) = y(x_0) = y_0$, we have an approximation $y_1 = y_0 + hf(x_0, y_0)$ to $y(x_1)$. Unfortunately, the point (x_1, y_1) is probably not on the solution curve $y(x)$ (see Fig. 7.5). If $f(x, y)$ is well behaved and if y_1 is near $y(x_1)$, we can hope that $f(x_1, y_1) \approx f(x_1, y(x_1))$. Thus we next proceed along a line from (x_1, y_1) of slope $f(x_1, y_1)$ and find $y_2 = y_1 + hf(x_1, y_1)$ as an approximation to $y(x_2)$. Another way of looking at this problem is to realize that the point (x_1, y_1) is probably on an integral curve, $y_1(x)$, that is near the solution curve $y(x)$. In other words, we have a point (x_1, y_1) on the curve $y_1(x)$ where $y_1(x)$ is the solution of the initial-value problem $u' = f(x, u)$, $u(x_1) = y_1$. We have no information as to how to return from this integral curve to the solution curve $y(x)$ [the integral curve that also satisfies $y(a) = y_0$]. Therefore, the best we can hope to do is to try to remain on this new integral curve, $y_1(x)$. We thus repeat the same process and obtain $y_2 = y_1 + hf(x_1, y_1)$ as our approximation for $y(x_2)$ at $x = x_2$. Continuing, we have *Euler's method,* which is defined by the iteration

$$y_{i+1} = y_i + hf(x_i, y_i), \qquad i = 0, 1, \ldots, n-1. \tag{7.4}$$

Figure 7.5 shows the results of four steps of Euler's method applied to the problem $y' = y$, $y(0) = 1$ with $h = 0.25$. The behavior of Euler's method is governed by the flow field for $y' = f(x, y)$ as is illustrated in Figs. 7.1, 7.2, and 7.5, which show that the approximations y_i to $y(x_i)$ are becoming worse with increasing i (note that nearby integral curves are spreading out). On the other hand, consider the problem $y' = y(2 - y)$, $y(0) = 0.1$. The flow field sketched in Fig. 7.3 suggests (for small h) that the approximations y_i will move away from the solution curve initially but will move back towards the solution as the values y_i get above the line $y = 1$. In effect, Euler's method can be viewed as tracing a polygonal path through the flow field with changes in the direction of the path at (x_i, y_i) determined by $f(x_i, y_i)$. The example $y' = y(2 - y)$ also suggests that errors may propagate differently in different regions of the xy-plane. That is, in some regions the error $y(x_i) - y_i$ may be magnified so that $|y(x_{i+1}) - y_{i+1}|$ is larger than $|y(x_i) - y_i|$. In other regions, past errors may tend

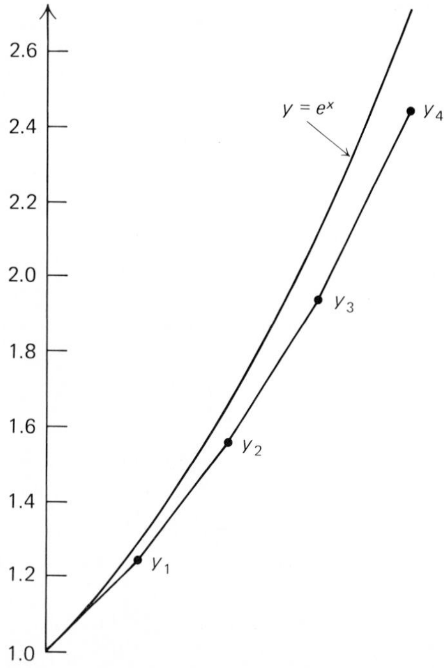

Figure 7.5 Euler's method with $h = 0.25$ for $y' = y$; $y(0) = 1$.

to damp out. We will discuss error propagation and the related problem of step-size control in Sections 7.2.4 and 7.3.

EXAMPLE 7.3. Consider the initial-value problem $y' = y$, $y(0) = 1$, which has $y(x) = e^x$ as a solution. In Table 7.1 we have tabulated the results of using Euler's method on $[0, 1]$, with $h = \frac{1}{4} = 0.25$ and $h = \frac{1}{8} = 0.125$. For this differential equation, the iteration, (7.4), reduces to simply $y_{i+1} = (1 + h)y_i$ with $0 \le i \le 3$ for $h = 0.25$ and $0 \le i \le 7$ for $h = 0.125$ (we start the iteration with $y_0 = 1$ in both cases).

TABLE 7.1

x	e^x	$h = 0.25$	$h = 0.125$
0.125	1.133148		$y_1 = 1.125000$
0.250	1.284025	$y_1 = 1.250000$	$y_2 = 1.265625$
0.375	1.454991		$y_3 = 1.423828$
0.500	1.648721	$y_2 = 1.562500$	$y_4 = 1.601807$
0.625	1.868246		$y_5 = 1.802032$
0.750	2.117000	$y_3 = 1.953125$	$y_6 = 2.027287$
0.875	2.398875		$y_7 = 2.280697$
1.000	2.718282	$y_4 = 2.441406$	$y_8 = 2.565784$

The subroutine in Fig. 7.6 was used to compute the results given in Table 7.1. The subroutine executes in a mode called *partial double precision,* which provides more accuracy than single precision but does not cost as much in

```
      SUBROUTINE EULER(Y,X,H,N)
      DOUBLE PRECISION Y
C
C  THIS SUBROUTINE USES EULER'S METHOD WITH PARTIAL DOUBLE PRECISION
C  TO SOLVE THE INITIAL-VALUE PROBLEM Y'(X)=F(X,Y), WITH INITIAL
C  CONDITION Y(XO)=YO.  A SINGLE PRECISION FUNCTION SUBPROGRAM F(X,Y)
C  MUST BE PROVIDED.  THE MAIN PROGRAM MUST SUPPLY A DOUBLE PRECISION
C  INITIAL-VALUE Y=YO, A STARTING VALUE X=XO AND STEP-SIZE H IN
C  SINGLE PRECISION, AND AN INTEGER N=NUMBER OF STEPS TO EXECUTE.
C
      ISTEP=0
C
C  PRINT INITIAL CONDITIONS
C
      PRINT 100,ISTEP,X,Y
C
C  EXECUTE EULER'S METHOD
C
      DO 1 ISTEP=1,N
      SY=Y
      Y=Y+H*F(X,SY)
      X=X+H
    1 PRINT 100,ISTEP,X,Y
  100 FORMAT(1H ,I3,2E20.7)
      RETURN
      END
```

Figure 7.6

terms of storage and execution time as double precision. To explain partial double precision, we note from the form of Euler's method

$$y_{i+1} = y_i + hf(x_i, y_i)$$

that we are typically adding a small number, $hf(x_i, y_i)$, to a number, y_i, which is usually large relative to $hf(x_i, y_i)$. In order to preserve accuracy, we make this *one* addition, $y_i + hf(x_i, y_i)$, in double precision and store the result, y_{i+1}, as a double precision number (y_i is also a double precision number of course). The function $f(x_i, y_i)$ is calculated in single precision; and in this subroutine, y_i is passed to f in single precision (by the statement $SY = Y$). Thus since the bulk of the computation is normally done in evaluating $f(x, y)$, we have not appreciably increased storage or execution, but we have lessened the effects of a potential source of rounding error. By way of example, we ran the subroutine with $N = 101$ and $H = 0.01$ in both single precision and partial double precision. In single precision we have $y_{100} = 2.704736$, and in partial double precision we had $y_{100} = 2.704813$ where $y_{100} \approx y(1)$. Using a smaller step size h would result in even more of a difference. Partial double precision is usually recommended for all the one-step methods.

7.2.2. Taylor's Series Methods of Order *k*

In the last section, we saw that Euler's method takes the form $y_{i+1} = y_i + hf(x_i, y_i)$ when applied to the initial-value problem $y' = f(x, y)\ y(x_o) = y_0$. For $i = 0$, we have $y_1 = y_0 + hf(x_o, y_0) = y(x_0) + y'(x_0)h$ since $y'(x_0) = f(x_0, y(x_o))$. Thus another way to view Euler's method is to think of y_1 as being the result of neglecting the higher-order terms in a Taylor's series expansion for $y(x_1)$. That is, since $x_1 = x_0 + h$, we have the expansion

$$y(x_1) = y(x_0) + y'(x_0)h + y''(x_0)\frac{h^2}{2!} + \cdots$$
$$+ y^{(k)}(x_0)\frac{h^k}{k!} + y^{(k+1)}(\theta)\frac{h^{k+1}}{(k+1)!} \tag{7.5}$$

when the solution $y(x)$ is sufficiently differentiable. Setting $k = 1$ in (7.5) and neglecting higher-order terms, we have $y_1 = y(x_0) + hy'(x_0)$, which is the Euler's method estimate to $y(x_1)$. If we knew some of the higher derivatives of $y(x)$ evaluated at $x = x_0$, then we suspect we could use (7.5) to find a better estimate to $y(x_1)$ by carrying more terms (i.e., by choosing $k > 1$).

In fact, the differential equation $y' = f(x, y)$ does provide as many derivatives, $y''(x_0), y'''(x_0), \ldots$, as we care to calculate. To see that such is the case, we note that if $y'(x) = f(x, y(x))$, then by taking the total derivative, we have

$$y''(x) = \frac{d}{dx}[f(x, y(x))] = f_x(x, y) + f_y(x, y)\frac{dy}{dx}. \tag{7.6}$$

Simplifying (7.6) by using $y' = f(x, y)$, we have $y''(x) = f_x(x, y) + f_y(x, y)f(x, y)$. Thus $y''(x_0) = f_x(x_0, y_0) + f_y(x_0, y_0)f(x_0, y_0)$, and we have a (presumably) more accurate estimate y_1 to $y(x_1)$:

$$y_1 = y_0 + hf(x_0, y_0) + \frac{h^2}{2}[f_x(x_0, y_0) + f_y(x_0, y_0)f(x_0, y_0)].$$

Continuing, we next find that

$$\begin{aligned} y'''(x) = f_{xx}(x, y) &+ 2f_{xy}(x, y)f(x, y) + f_{yy}(x, y)f^2(x, y) \\ &+ f_y(x, y)f_x(x, y) + f_y^2(x, y)f(x, y). \end{aligned}$$

Thus we see that in principle we can compute $y^{(i)}(x)$(if it exists) for as many values i as we choose; therefore (7.5) can be used to estimate $y(x_1)$. In general if we knew $y(x_i)$, we could use (7.5) with x_i replacing x_0 to estimate $y(x_{i+1})$.

In fact unless $y(x)$ is a polynomial, we can never hope that a truncated Taylor's series will provide more than just a good estimate, y_{i+1}, to $y(x_{i+1})$. Nevertheless if y_i is a good approximation to $y(x_i)$, then we would expect that

$$\begin{aligned} y'(x_i) &\approx f(x_i, y_i) \\ y''(x_i) &\approx f_x(x_i, y_i) + f_y(x_i, y_i)f(x_i, y_i) \\ y'''(x_i) &\approx f_{xx}(x_i, y_i) + 2f_{xy}(x_i, y_i)f(x_i, y_i) + f_{yy}(x_i, y_i)f^2(x_i, y_i) \\ &\quad + f_y(x_i, y_i)f_x(x_i, y_i) + f_y^2(x_i, y_i)f(x_i, y_i). \end{aligned}$$

To exploit these ideas, let $f^{(j)}(x, y)$ denote the jth total derivative of $f(x, y)$ with respect to x [that is, $f^{(j)}(x, y) = y^{(j+1)}(x)$] so that we have

$$y(x_{i+1}) = y(x_i) + hT_k(x_i, y(x_i); h) + y^{(k+1)}(\theta)\frac{h^{k+1}}{(k+1)!} \tag{7.7}$$

where

$$T_k(x, y; h) = f(x, y) + f'(x, y)\frac{h}{2!} + \cdots + f^{(k-1)}(x, y)\frac{h^{k-1}}{k!}.$$

We are then led to the iteration

$$y_{i+1} = y_i + hT_k(x_i, y_i; h) \qquad i = 0, 1, \ldots, n-1. \tag{7.8}$$

The method described by (7.8) is called a *kth-order Taylor's series method.* Note that (7.8) has the form of a one-step method as defined in Section 7.2. For $k = 1$, formula (7.8) reduces to Euler's method.

For values of k larger than 3, it is clear that a major effort is required to compute the necessary derivatives of $f(x, y)$. Moreover in many practical problems, it may be impossible or extremely costly to find and evaluate the necessary partial derivatives of $f(x, y)$. For this reason higher-order Taylor's series methods are not frequently used. As we shall see shortly, Runge-Kutta methods provide a good compromise, replacing suitably chosen combinations of evaluations of $f(x, y)$ for derivatives of $f(x, y)$. However as demonstrated in the next example, the problem of obtaining higher derivatives is not always so formidable as we might first think.

EXAMPLE 7.4. Consider the initial-value problem (b) in Example 7.1 for $0 \le x \le 1$:

$$y' = 2xy, \qquad y(0) = 1.$$

In this case, we quickly find

$$y'' = 2y + 2xy' = 2y(1 + 2x^2)$$

$$y''' = 2y'(1 + 2x^2) + 8xy = 4xy(3 + 2x^2).$$

Hence a third-order Taylor's series method is provided by the iteration

$$y_{i+1} = y_i + h\left[2x_iy_i + y_i(1 + 2x_i^2)h + 2x_iy_i(3 + 2x_i^2)\frac{h^2}{3}\right].$$

Most computer installations have symbol-manipulation packages, such as FORMAC, which can be used to automate the computation of higher derivatives for algebraic expressions such as $y' = 2xy$ or for even more cumbersome expressions such as

$$y' = \frac{x^2 \sin(y)}{1 + y^2}.$$

For certain types of differential equations, a Taylor's series might be appropriate if a symbol-manipulation package is available.

Before continuing, we wish to observe that the implementation of most numerical methods for solving $y' = f(x, y)$ will require the evaluation of $f(x, y)$ at various points (x_i, y_i). In fact, unlike the simple examples considered so far, the evaluation of $f(x, y)$ may be quite time consuming. To give an indication of what can be involved in these evaluations, let us return to the equations of motion for a ballistic reentry trajectory, (4.0a). For a given position vector $\mathbf{R}(t)$ and velocity vector $\dot{\mathbf{R}}(t)$, we see from (4.0a) that we need to evaluate the gravity vector $\mathbf{G}(\mathbf{R})$ and the scalar drag force $-\overline{q}C_DA/M$. In simple textbook problems, gravity is often assumed to be constant; however, to obtain an accurate solution of (4.0a), we cannot assume constant gravity. What is required for an accurate numerical solution is a good mathematical model for the earth's gravitational field. Thus, given a point $\mathbf{R}$ in space, we need to be able to determine a good approximation to the gravity vector, $\mathbf{G}(\mathbf{R})$. This approximation is frequently achieved in practice by converting $\mathbf{R}$ to spherical coordinates (r, θ, λ) and then evaluating various partial derivatives of expressions that have the form

$$\sum_{n=0}^{N}\sum_{m=0}^{n}\frac{P_{nm}(\theta)}{r^{n+1}}\left[a_{nm}\cos(m\lambda) + b_{nm}\sin(m\lambda)\right].$$

These expressions arise from the truncation of the theoretical spherical harmonic expansion for the earth's gravitational potential. The coefficients a_{nm} and b_{nm} are usually determined by using satellite data and least-squares fits.

The functions $P_{nm}(\theta)$ can be evaluated through a series of recurrence formulas. Even for a modest value of N, say $N = 10$ or $N = 20$, the evaluation of these expressions is quite time consuming.

When drag forces are present as in (4.0a), the term $-\overline{q}C_D A/M$ also needs to be evaluated. Recall that the function $\overline{q}$ is given by

$$\overline{q} = \frac{\rho(H)|\dot{\mathbf{R}}|^2}{2}$$

where $\rho(H)$ is atmospheric density at an altitude H. The drag coefficient, C_D, is usually an experimentally determined function of Mach and altitude where the Mach number, m, is given from $a_s = a_1\sqrt{T(H)}$ and $m = |\dot{\mathbf{R}}|/a_s$ with a_1 a constant and $T(H)$ denoting an atmospheric temperature function of altitude. The basic steps in evaluating the drag forces on a body moving with velocity $\dot{\mathbf{R}}$ at a point $\mathbf{R}$ are these.

1. Determine the altitude H.
2. Given H, determine $T(H)$. This determination is usually done by interpolation in a table of values for $T(H_0)$, $T(H_1)$, ...
3. Determine the density, $\rho(H)$; $\rho(H)$ is usually given as a set of empirical formulas depending on $T(H)$.
4. Determine the Mach number, m.
5. Evaluate C_D as a function of H and m.

In the scheme above, step (5) is normally the most costly. Determining C_D may depend on a long sequence of table look-ups, interpolation, and the evaluation of a series of empirical formulas wherein the tables and empirical formulas are derived from theoretical aerodynamic considerations, wind-tunnel data, and flight-test data.

In the discussion above, with a specific case we have tried to illustrate that evaluating $f(x, y)$ in the equation $y' = f(x, y)$ can be costly in terms of storage and execution time. The example above serves to illustrate also that many realistic differential equations cannot be solved except by numerical means. A further observation that can be drawn from this example is that the function $f(x, y)$ may not be very smooth. The roughness of $f(x, y)$ can have a drastic effect on the size of the steps that can be taken in a numerical method, and may require a large number of evaluations of $f(x, y)$ to maintain any accuracy in the solution.

PROBLEMS, SECTION 7.2.2

1. Apply Euler's method to the problems (7.2a), (7.2b), and (7.2c) in Example 7.1. Use step size $h = 1/n$ for $n = 4, 8, 16$, and 32.
2. Apply Euler's method to $y' = y(2 - y)$ with initial conditions $y(0) = .1$ and $y(0) = 5$. Use $h = 10^{-2}$; solve on the interval $[0, 10]$. Modify the program in Fig. 7.6 so that the

estimates are printed at intervals of .2, sketch the results, and compare with the flow field (see Fig. 7.3).

3. Repeat Problem 2 for the equations in Problems 9 and 10, Section 7.1. Use initial conditions $y(0) = 2$ and $y(0) = 5$.
4. By Problem 8, Section 7.1, $y' = \sqrt{|y|}$, $y(0) = 0$ has infinitely many solutions. Which solution will Euler's method follow?
5. Solve the initial-value problem $y' = 3x^2y^2$, $y(0) = -1$. Run this problem on $[0, 3]$; use Euler's method with step size $h = 2^{-r}$, $r = 2, 3, \ldots, 10$. Print your estimates and the actual errors at $x = 1, 2, 3$. The error in Euler's method at $x = x^*$ behaves (theoretically) like C^*h where C^* is a constant depending on x^* and the differential equation (see Section 7.3). Check this statement by plotting the error against h at $x = 1, 2, 3$.
6. Following the discussion in Example 7.4, write a program that uses a kth order Taylor's series method for $y' = 3x^2y^2$, $y(0) = -1$ for $k = 2, 3, 4$. Run this problem on $[0, 4]$, use a step size of .05, and print your results at $x = 1, 2, 3, 4$ for $k = 2, 3, 4$.
7. Repeat Problem 6 for $y' = y(2 - y)$, $y(0) = .1$.
8. The error in a kth order Taylor's series method at $x = x^*$ behaves (theoretically) like C^*h^k where C^* depends on x^* and the differential equation. Solve the initial-value problem $y' = x/y$, $y(0) = 1$ on $[0, 2]$; and run the second-order Taylor's series method on this equation with step size $h = 2^{-r}$, $r = 2, 3, \ldots, 10$. Print your estimates and the errors at $x = 2$, and plot the errors against h.
9. Apply Euler's method to $y' = -25y$, $y(0) = 1$ on $[0, 1]$. Use a step size $h = 1/10$ and $h = 1/20$, and print each step. Calculate the relative errors at $x = 1$.
10. Consider the initial-value problem $y' = \lambda y$, $y(0) = y_0$. Show that Euler's method with step size h applied to this problem generates the sequence $y_i = (1 + h\lambda)^i y_0$, $i = 0, 1, \ldots$. Observe that if $\lambda < 0$, then the solution $y(x)$ tends to zero as x increases. Given $\lambda < 0$, for what values of h will $\{y_i\} \to 0$? Apply the results of this problem to explain the output from Problem 9.
11. Show that the kth order Taylor's series method applied to $y' = \lambda y$, $y(0) = y_0$ generates the sequence $y_i = p(\bar{h})^i y_0$, $i = 0, 1, \ldots$ where $\bar{h} = h\lambda$ and

$$p(\bar{h}) = 1 + \bar{h} + \frac{\bar{h}^2}{2!} + \cdots + \frac{\bar{h}^k}{k!}$$

For $k = 2$, show that $|p(\bar{h})| < 1$ if and only if $-2 < \bar{h} < 0$. For $k = 3$, use Newton's method to find α such that $|p(\bar{h})| < 1$ if and only if $\alpha < \bar{h} < 0$. [Hint: Let $p(x) = 1 + x + x^2/2! + x^3/3!$ and show that $p'(x) > 0$ for all x by showing $p'(x) = 0$ has no real solutions. Use the fact that $p(x)$ is increasing to determine α.]
12. Consider $y' = -20y$, $y(0) = 1$ and note that $y(x) \to 0$ as x grows. Using the value of α found in Problem 11, choose h so that $\bar{h} \simeq \alpha - .2$ and run 100 steps of the third-order Taylor's series method on this equation; print each step. Next choose h so that $\bar{h} \simeq \alpha + .2$; run the problem again and note the improvement.
13. Given the differential equation $y' = y$, $y(0) = 1$ as in Example 7.3, suppose that Euler's method is employed first with a step size h and then with a step size $h_1 = h/2$. Suppose that y_i is the ith estimate of Euler's method using h, and u_i is the ith estimate using h_1. Show that $y_{i+1} = (1 + h)y_i$ and $u_{i+1} = (1 + h_1)u_i$. For a point $x_i = x_0 + ih = ih$, note that $y_i \approx y(x_i)$ and $u_{2i} \approx y(x_i)$. Show that $y_i < u_{2i} < y(x_i)$.

14. Again suppose that Euler's method is employed to estimate the solution of $y' = y$, $y(0) = 1$ and suppose that $\tilde{x}$ is fixed, $\tilde{x} > 0$. For $h_n = \tilde{x}/n$, let y_n denote the estimate of $y(\tilde{x})$ given by Euler's method with step size h_n. Show from a definition of e^x that

$$\lim_{n\to\infty} y_n = y(\tilde{x}).$$

[Hint: By Problem 13, there is a closed form expression for y_n.]

7.2.3. Runge-Kutta Methods

The general one-step method takes the form

$$y_{i+1} = y_i + h\phi(x_i, y_i; h);$$

so we can think of the number $h\phi(x_i, y_i; h)$ as representing an approximation to the amount by which the solution $y(x)$ increases (or decreases) as we move from x_i to x_{i+1}. Since the nature of the flow field in a neighborhood of $(x_i, y(x_i))$ governs the change in $y(x)$ between x_i and x_{i+1}, it seems plausible that we might "sample" the flow field at various points near $(x_i, y(x_i))$ in order to get some indication of changes to be expected in $y(x)$ for $x_i \le x \le x_{i+1}$. In the real situation, of course, we do not have $y(x_i)$; so we must think of sampling at points $(\theta_1, \gamma_1), \ldots, (\theta_k, \gamma_k)$ that are near (x_i, y_i). In particular, we could specify the increment function by

$$\phi(x_i, y_i; h) = A_1 f(\theta_1, \gamma_1) + A_2 f(\theta_2, \gamma_2) + \cdots + A_k f(\theta_k, \gamma_k) \tag{7.9}$$

so that $\phi(x_i, y_i; h)$ is a weighted average of the slopes near (x_i, y_i). For example, Euler's method uses just the one sample point (x_i, y_i) with $A_1 = 1$.

The problem with (7.9) is in choosing the weights A_j and the sample points (θ_j, γ_j). The Runge-Kutta methods that are developed in this section provide one logical means for choosing the weights and sample points. Runge-Kutta methods (named for the German mathematicians C. Runge and M. W. Kutta, who made important contributions around the turn of the century) are quite similar to Taylor's series methods but do not require the evaluation of derivatives of $f(x, y)$ (a requirement which is one major deficiency of a Taylor's series method). In order to demonstrate the essential ideas, we will develop a second-order Runge-Kutta method that uses the form of (7.9) with two sample points. As one sample point, let us choose (x_i, y_i); and for a sample point near (x_i, y_i), let us choose $(x_i + \alpha h, y_i + \alpha h f(x_i, y_i))$ where α is as yet undetermined. This choice of a nearby sample point is not as unnatural as it might seem at first glance since we want to sample near the integral curve passing through (x_i, y_i). [For example, with $\alpha = 1$, we would be using the sample point $(x_{i+1}, \tilde{y}_{i+1})$ where $\tilde{y}_{i+1}$ is the estimate of $y(x_{i+1})$ given by Euler's method.] In general, for two sample points we have $y_{i+1} = y_i + h\phi(x_i, y_i; h)$ where $\phi(x_i, y_i; h)$ is defined by

$$y_{i+1} = y_i + h[A_1 f(x_i, y_i) + A_2 f(x_i + \alpha h, y_i + \alpha h f(x_i, y_i))]. \tag{7.10}$$

The basis for a Runge-Kutta method is to choose the parameters A_1, A_2, and α so that $\phi(x_i, y_i; h)$, given by (7.10), agrees as well as is possible with the increment function $T_k(x_i, y_i; h)$ for a Taylor's series method.

To see how this is done, we first expand $f(x_i + \alpha h, y_i + \alpha hf(x_i, y_i))$ in a Taylor series to find

$$f(x_i + \alpha h, y_i + \alpha hf(x_i, y_i)) = f(x_i, y_i) + f_x(x_i, y_i)\alpha h + f_y(x_i, y_i)\alpha hf(x_i, y_i) + E$$

where E denotes the remainder in the expansion. If we let $K = \alpha hf(x_i, y_i)$, we see that E has the form $E = Ch^2$ since

$$2E = f_{xx}(\eta, \delta)(\alpha h)^2 + 2f_{xy}(\eta, \delta)\alpha hK + f_{yy}(\eta, \delta)(K)^2$$

where (η, δ) is the mean-value point for the Taylor's series expansion in two variables. Collecting like terms, we find that the increment function in (7.10) reduces to

$$\begin{aligned}\phi(x_i, y_i; h) = (A_1 + A_2)f(x_i, y_i) + A_2h[\alpha f_x(x_i, y_i)\\ + \alpha f_y(x_i, y_i)f(x_i, y_i) + Ch].\end{aligned} \tag{7.11a}$$

Recalling that $T_2(x_i, y_i; h)$ is given by

$$T_2(x_i, y_i; h) = f(x_i, y_i) + \frac{h}{2}[f_x(x_i, y_i) + f_y(x_i, y_i)f(x_i, y_i)], \tag{7.11b}$$

we see that (7.11a) can be made to agree fairly well with (7.11b) if we equate like coefficients in (7.11a) and (7.11b) and thereby choose the parameters so that

$$\begin{aligned}A_1 + A_2 &= 1\\ A_2\alpha &= \tfrac{1}{2}.\end{aligned} \tag{7.12}$$

With these choices, we will have

$$\phi(x_i, y_i; h) = T_2(x_i, y_i; h) + A_2Ch^2.$$

We observe that (7.12) reduces to $A_2 = 1/2\alpha$, and $A_1 = 1 - (1/2\alpha)$. That there are infinitely many solutions for the parameters is a peculiarity of Runge-Kutta methods. One natural choice is $\alpha = \frac{1}{2}$, giving the sequence

$$y_{i+1} = y_i + hf\left(x_i + \frac{h}{2}, y_i + \frac{h}{2}f(x_i, y_i)\right),$$

which is usually called the *modified Euler's method*. Another choice is $\alpha = 1$, which leads to the *improved Euler's method* (often called *Heun's method*):

$$y_{i+1} = y_i + \frac{h}{2}[f(x_i, y_i) + f(x_{i+1}, y_i + hf(x_i, y_i))].$$

We see that these Runge-Kutta methods do not require the evaluation of the derivative of $f(x, y)$ as does the second-order Taylor's series method.

EXAMPLE 7.5. The subroutine given in Fig. 7.7 uses Heun's method (the improved Euler's method). As an illustration of the sort of results that Heun's method might give, we ran the problem $y' = 2xy$, $y(0) = 1$ (with $h = 0.125$). The results are shown in Table 7.2. By contrast, Euler's method with $h = 0.125$ gives $y_8 = 2.820158$ where $y_8 \approx y(1)$. For $h = 0.01$, Heun's method provides $y_{100} = 2.718186$ where $y(1) = 2.718282$ (an error of 0.000096).

TABLE 7.2

x	y_i	$y(x_i)$
0.125	1.015625	1.015748
0.250	1.064224	1.064494
0.375	1.150485	1.150993
0.500	1.283060	1.284025
0.625	1.476020	1.477904
0.750	1.751332	1.755055
0.875	2.142987	2.150338
1.000	2.703847	2.718282

The second-order Runge-Kutta methods derived above used two terms in (7.9) in order to match closely a second-order Taylor's series method. By

```
      SUBROUTINE HEUN(Y,X,H,N)
      DOUBLE PRECISION Y
C
C  THIS SUBROUTINE USES HEUN'S METHOD WITH PARTIAL DOUBLE PRECISION
C  TO SOLVE THE INITIAL-VALUE PROBLEM Y'(X)=F(X,Y), WITH INITIAL
C  CONDITION Y(XO)=YO.   A SINGLE PRECISION FUNCTION SUBPROGRAM F(X,Y)
C  MUST BE PROVIDED.   THE MAIN PROGRAM MUST SUPPLY A DOUBLE PRECISION
C  INITIAL-VALUE Y=YO, A STARTING VALUE X=XO AND STEP-SIZE H IN
C  SINGLE PRECISION, AND AN INTEGER N=NUMBER OF STEPS TO EXECUTE.
C
      ISTEP=0
C
C  PRINT INITIAL CONDITIONS
C
      PRINT 100,ISTEP,X,Y
C
C  EXECUTE HEUN'S METHOD
C
      DO 1 ISTEP=1,N
      YS=Y
      YP1=F(X,YS)
      X=X+H
      YS=Y+H*YP1
      YP2=F(X,YS)
      Y=Y+.5*H*(YP1+YP2)
    1 PRINT 100,ISTEP,X,Y
  100 FORMAT(1H ,I3,2E20.7)
      RETURN
      END
```

Figure 7.7

carrying more terms in (7.9), we can match higher-order Taylor's series methods. The derivation of higher-order Runge-Kutta methods proceeds much as above [in (7.10) through (7.12)]; however the details are quite tedious and the interested reader may consult a text such as Lambert (1973). A special notation is usually used to describe the general form of a Runge-Kutta method. Specifically, an m-stage, kth order Runge-Kutta method has the form $y_{i+1} = y_i + h\phi(x_i, y_i; h)$ where

$$\phi(x_i, y_i; h) = T_k(x_i, y_i; h) + 0(h^k) \tag{7.13a}$$

and where

$$\phi(x_i, y_i; h) = \sum_{j=1}^{m} A_j K_j(x_i, y_i; h). \tag{7.13b}$$

We will make the term "kth order" a bit more precise in the next section; for now, we can think of kth order in the context of (7.13a). The term "m-stage" refers to the integer m in (7.13b), and we observe below that m is the number of evaluations of $f(x, y)$. The terms K_j on the right-hand side of (7.13b) are given by

$$K_1(x_i, y_i; h) = f(x_i, y_i);$$

and for $2 \le j \le m$, K_j is defined in terms of a weighted average of $K_1, K_2, \ldots, K_{j-1}$ by

$$K_j(x_i, y_i; h) = f(x_i + \alpha_j h, y_i + h \sum_{r=1}^{j-1} \beta_{jr} K_r(x_i, y_i; h)) \tag{7.13c}$$

where $0 < \alpha_j \le 1$ and $\alpha_j = \sum_{r=1}^{j-1} \beta_{jr}$.

From (7.13b) and (7.13c), it follows that using an m-stage Runge-Kutta method to solve $y' = f(x, y)$ has a cost of m evaluations of $f(x, y)$ in going from y_i to y_{i+1}. Higher-order Runge-Kutta methods are more accurate than lower-order ones, but it is known that m and k must satisfy $m \ge k$ for an m-stage, kth order method. In fact, k-stage, kth order methods exist only for $k = 1, 2, 3$, and 4. To achieve order 5, a six-stage method is required; for order 6, a seven-stage method is needed; and a method of order k with $k \ge 7$ requires at least $(k + 2)$ stages.

For a given value of m, there are many choices for the weights A_j in (7.13b) and for the parameters α_j and β_{jr} in (7.13c). For $m = 4$, a very popular four-stage, 4th order Runge-Kutta method is

$$y_{i+1} = y_i + \frac{h}{6}[K_1 + 2K_2 + 2K_3 + K_4]$$

where

$$K_1 = f(x_i, y_i)$$

$$K_2 = f\left(x_i + \frac{h}{2}, y_i + \frac{h}{2} K_1\right)$$

$$K_3 = f\left(x_i + \frac{h}{2}, y_i + \frac{h}{2} K_2\right)$$

$$K_4 = f(x_i + h, y_i + hK_3).$$

In Section 7.2.5 we will see a pair of Runge-Kutta methods, attributable to Fehlberg, which can be used together in order to give an estimate of the error made at each step. This pair forms the basis of a variable-step Runge-Kutta method in which the error-monitoring capability is used to adjust automatically the step size h in order to achieve a specified accuracy target.

Finally, we mention that recent research has considered another form of Runge-Kutta methods, an implicit form. An *implicit m-stage Runge-Kutta method* has the same form as (7.13b); but instead of using (7.13c) to define the K_j, we use

$$K_j(x_i, y_i; h) = f(x_i + \alpha_j h, y_i + h \sum_{r=0}^{m} \beta_{jr} K_r(x_i, y_i; h)). \tag{7.14}$$

Note that K_j appears on both sides of equation (7.14); so each step of an implicit method requires the solution of a system of m nonlinear equations in the m unknowns $K_1, K_2, \ldots, K_m$. The virtue of such implicit methods is that they have desirable stability properties and that for all m, there exist m-stage implicit methods having order $2m$. An example of a two-stage, fourth-order method is

$$y_{i+1} = y_i + \frac{h}{2}(K_1 + K_2)$$

$$K_1 = f(x_i + \alpha_1 h, y_i + \beta_{11} hK_1 + \beta_{12} hK_2)$$

$$K_2 = f(x_i + \alpha_2 h, y_i + \beta_{21} hK_1 + \beta_{22} hK_2)$$

where $\alpha_1 = (3 + \sqrt{3})/6$, $\alpha_2 = (3 - \sqrt{3})/6$, $\beta_{11} = \beta_{22} = 1/4$, $\beta_{12} = 1/4 + \sqrt{3}/6$, and $\beta_{21} = 1/4 - \sqrt{3}/6$. The interval of absolute stability (see Section 7.6.1) for this method is $(-\infty, 0)$. The effort needed to solve the system (7.14) can be reduced if $\beta_{jr} = 0$ for $r = j + 1, \ldots, m$; such methods are called *semiexplicit*. An example of a fourth-order, three-stage semiexplicit method attributable to Butcher is

$$y_{i+1} = y_i + \frac{h}{6}(K_1 + 4K_2 + K_3)$$

$$K_1 = f(x_i, y_i)$$

$$K_2 = f\left(x_i + \frac{h}{2}, y_i + \frac{h}{4}K_1 + \frac{h}{4}K_2\right)$$

$$K_3 = f(x_i + h, y_i + hK_2).$$

PROBLEMS, SECTION 7.2.3

1. Apply Heun's method to $y' = x/y$, $y(0) = 1$ on $[0, 2]$; use $h = 10^{-r}$, $r = 1, 2, 3$. Print the estimates and the exact solution at $x = 0., .1, .2, \ldots, 2.$
2. Repeat Problem 1 for $y' = 3x^2y^2$, $y(0) = -1$.
3. Repeat Problem 1 for $y' = -50y$, $y(0) = 1$; and note that the results for $h = 10^{-1}$ are terrible.
4. Code the modified Euler's method and apply it to Problems 1, 2, and 3.
5. Consider the three-stage method

$$y_{i+1} = y_i + \frac{h}{9}(2K_1 + 3K_2 + 4K_3)$$

$$K_1 = f(x_i, y_i), \qquad K_2 = f\left(x_i + \frac{h}{2}, y_i + \frac{h}{2}K_1\right),$$

$$K_3 = f\left(x_i + \frac{3h}{4}, y_i + \frac{3h}{4}K_2\right).$$

Verify that this is a third-order Runge-Kutta method by comparing $\phi(x_i, y_i; h)$ with $T_3(x_i, y_i; h)$.

6. The initial-value problem $y' = \lambda y$, $y(0) = y_0$ is often used as a test equation for a numerical method. Verify that if the second-order Taylor's series method is applied to $y' = \lambda y$, then the estimates satisfy $y_{i+1} = p(\bar{h})y_i$, $i = 0, 1, \ldots$ where $\bar{h} = h\lambda$ and $p(\bar{h}) = 1 + \bar{h} + \bar{h}^2/2!$
7. Show that Heun's method applied to $y' = \lambda y$, $y(0) = y_0$ generates estimates that satisfy $y_{i+1} = p(\bar{h})y_i$, $i = 0, 1, \ldots$ where $\bar{h}$ and $p(\bar{h})$ are exactly as in Problem 6. Use the result of this problem and Problem 11, Section 7.2.2, to explain the output in Problem 3.
8. Show that *any* two-stage, second-order Runge-Kutta method applied to $y' = \lambda y$, $y(0) = y_0$ generates estimates that satisfy $y_{i+1} = p(\bar{h})y_i$, $i = 0, 1, \ldots$ where $\bar{h}$ and $p(\bar{h})$ are as in Problem 6. [Hint: Use (7.10) and (7.12).]
9. Show that the three-stage method in Problem 5 applied to $y' = \lambda y$, $y(0) = y_0$ generates the estimates $y_{i+1} = p(\bar{h})y_i$ where $p(\bar{h}) = 1 + \bar{h} + \bar{h}^2/2! + \bar{h}^3/3!$
10. Code Butcher's fourth-order, three-stage semiexplicit method in a form suitable for equations of the type $y' = f(x)y + g(x)$. (Note: Semiexplicit methods are easy to use for *linear* differential equations.) Test your program on the equation in Problem 3; use $h = 10^{-1}$ on $[0, 5]$. Print your estimates and the relative errors at each step.

7.2.4. Variable-step Runge-Kutta Methods

Ideally, a differential-equation solver should be a software package that accepts an initial-value problem along with some sort of accuracy criterion and returns a numerical solution that meets the accuracy specification. In this section we will consider some of the elementary aspects of using Runge-Kutta methods to produce such an "automatic" differential-equation solver. Ignoring the effects of roundoff error for the moment (see Section 7.3), we note that we can increase the accuracy of a one-step method by decreasing the step size h (see for instance Examples 7.3 and 7.5). A variable-step Runge-Kutta method does exactly this, in an adaptive fashion: at each step, an estimate of the error is made; and if the error estimate is overly large, the step size is reduced. On the other hand, when the error estimate is small with respect to the accuracy desired, the step size can be increased and the cost of solving the problem thereby reduced without sacrificing accuracy.

Before describing how errors can be monitored in a one-step method $y_{i+1} = y_i + h\phi(x_i, y_i; h)$, we introduce a fundamental concept, the local truncation error. Let $y(x)$ be some solution of $y' = f(x, y)$ and let x^* and h be fixed. We define the *local truncation error*, τ, by the equation

$$y(x^* + h) = y(x^*) + h\phi(x^*, y(x^*); h) + h\tau. \tag{7.15}$$

Note that a single application of the one-step method beginning at $(x^*, y(x^*))$ would produce $y(x^*) + h\phi(x^*, y(x^*); h)$; therefore, $h\tau$ measures how much we miss $y(x^* + h)$ if we take a step of length h starting on the solution at $x = x^*$. Although the notation in (7.15) does not exhibit this dependence explicitly, τ depends on h, x^*, and the particular integral curve, $y(x)$.

A simple illustration of the local truncation error is provided by Euler's method, $y_{i+1} = y_i + hf(x_i, y_i)$. Let $y(x)$ be some solution of $y' = f(x, y)$. A single step of Euler's method starting with $y(x^*)$ at $x = x^*$ will produce an estimate of $y(x^*) + hf(x^*, y(x^*))$ to $y(x^* + h)$. According to (7.15), the local truncation error τ is given by

$$y(x^* + h) = y(x^*) + hf(x^*, y(x^*)) + h\tau. \tag{7.16a}$$

If the solution $y(x)$ is smooth enough in $[x^*, x^* + h]$, then we know also that

$$y(x^* + h) = y(x^*) + hy'(x^*) + \frac{h^2}{2} y''(\gamma) \tag{7.16b}$$

where $x^* < \gamma < x^* + h$. If one recalls that $y'(x^*) = f(x^*, y(x^*))$ and then equates (7.16a) with (7.16b), it follows that

$$\tau = \frac{h}{2} y''(\gamma). \tag{7.17}$$

In (7.17), we see that τ depends on the particular solution $y(x)$, the stepsize h, and the mean-value point γ. For the general kth order Taylor's series method, an argument like that in (7.16) above easily shows that the local truncation error

$$y(x^* + h) = y(x^*) + hT_k(x^*, y(x^*); h) + h\tau$$

is given by

$$\tau = \frac{h^k}{(k+1)!} y^{(k+1)}(\gamma) \tag{7.18}$$

where $x^* < \gamma < x^* + h$ (again on the assumption that $y(x)$ is sufficiently smooth in $[x^*, x^* + h]$).

Before continuing, we note that (7.18) suggests what we should mean by the order of a general one-step method. In particular, we say that the method $y_{i+1} = y_i + h\phi(x_i, y_i; h)$ is a *pth order method* if τ in (7.15) satisfies

$$\tau = O(h^p).$$

The definition of order for a one-step method is somewhat ambiguous since the definition depends on the smoothness of the solutions to $y' = f(x, y)$; a method that is classified as pth order might not be truly pth order for a particular equation $y' = g(x, y)$ if the function $g(x, y)$ is not smooth. [An analogous situation is found with Newton's method for solving $f(x) = 0$. Newton's method is classified as a quadratic method although it is not quadratically convergent to multiple roots.] Finally, our informal definition of a kth order Runge-Kutta method in the last section agrees with the classification scheme above since (7.13a) can be used to show that the local truncation error for such a Runge-Kutta method normally satisfies $\tau = O(h^k)$.

Having the concept of the local truncation error, we are in a position to discuss how errors can be monitored in a one-step method. Suppose that $y(x)$ is the solution to the initial-value problem $y' = f(x, y)$, $y(x_0) = y_0$, and suppose at some point $x = x_i$ we have an approximation y_i to $y(x_i)$. The quantity $y(x_i) - y_i$, called the *global error* at x_i, is the quantity we wish to keep within some specified bounds of accuracy. Unfortunately, it is fairly difficult to estimate the global error (see Section 7.3) and so instead we will monitor a quantity known as the local error. It can be shown that if we can control the local errors, then we will in fact control the global errors [see Shampine and Gordon (1975)]. The local error is pictured in Fig. 7.8, in which $y_{i+1} = y_i + h\phi(x_i, y_i; h)$. In Fig. 7.8, $u(x)$ is an integral curve for $y' = f(x, y)$ and passes through the point (x_i, y_i). Thus $u(x)$ satisfies $u' = f(x, u)$ with $u(x_i) = y_i$, and the *local error* at x_{i+1} is defined to be $u(x_{i+1}) - y_{i+1}$. Since $y_{i+1} = y_i + h\phi(x_i, y_i; h)$ and since $u(x_i) = y_i$, we are (in effect) trying to follow the curve $u(x)$ where we start on the curve at $x = x_i$. Therefore, the local error measures how well we can follow $u(x)$ for one step.

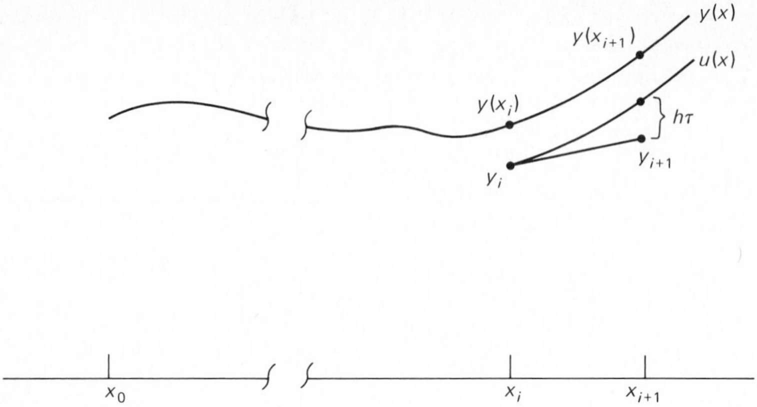

Figure 7.8 The local error.

The local error and the global error at x_{i+1} are related by

$$y(x_{i+1}) - y_{i+1} = [y(x_{i+1}) - u(x_{i+1})] + [u(x_{i+1}) - y_{i+1}], \tag{7.19}$$

and the simple relation (7.19) is one that is conceptually useful. In particular, we can use (7.19) to interpret the global error as having two components.

1. The component $y(x_{i+1}) - u(x_{i+1})$ measures how far apart the two integral curves $y(x)$ and $u(x)$ spread as x ranges from x_i to x_{i+1}.
2. The component $u(x_{i+1}) - y_{i+1}$ measures how well the method solves the problem $u' = f(x, u)$, $u(x_i) = y_i$.

Component (1) of the global error depends on the differential equation and is related to the "stability" of the differential equation in a neighborhood of $(x_i, y(x_i))$. Component (2) is linked to the method and can be made smaller by increasing the accuracy of the method (perhaps by decreasing h or by increasing the order of the method).

The local error can be expressed in terms of the local truncation error:

$$u(x_{i+1}) = u(x_i) + h\phi(x_i, u(x_i); h) + h\tau. \tag{7.20a}$$

As $u(x_i) = y_i$, it follows that $y_{i+1} = u(x_i) + h\phi(x_i, u(x_i); h)$ and therefore the local error is given by

$$u(x_{i+1}) - y_{i+1} = h\tau. \tag{7.20b}$$

If the method is a pth order method, then the local error satisfies $u(x_{i+1}) - y_{i+1} = 0(h^{p+1})$. From (7.20), it is fairly easy to see how to estimate the local error. Suppose that we use two different one-step methods, both starting at the same point (x_i, y_i):

$$y_{i+1} = y_i + h\phi_1(x_i, y_i; h) \tag{7.21a}$$

$$\tilde{y}_{i+1} = y_i + h\phi_2(x_i, y_i; h); \tag{7.21b}$$

and suppose (7.21a) is a pth order method and (7.21b) is a qth order method where $q > p$. As in (7.20), we have

$$\begin{aligned} u(x_{i+1}) &= u(x_i) + h\phi_1(x_i, u(x_i); h) + h\tau \\ u(x_{i+1}) &= u(x_i) + h\phi_2(x_i, u(x_i); h) + h\tilde{\tau}. \end{aligned} \tag{7.22}$$

Using (7.22) and the fact that (7.21b) is a qth order method, we have $\tilde{y}_{i+1} - y_{i+1} = h\tau - h\tilde{\tau}$, or

$$\tilde{y}_{i+1} - y_{i+1} = h\tau + 0(h^{q+1}).$$

Since $h\tau = 0(h^{p+1})$ and since $p < q$, we have an approximation to $h\tau$, the local error of (7.21a), which is very good for small h; namely, $h\tau \simeq \tilde{y}_{i+1} - y_{i+1}$. In view of (7.20b), we obtain the estimate

$$u(x_{i+1}) - y_{i+1} \simeq \tilde{y}_{i+1} - y_{i+1} \tag{7.23}$$

EXAMPLE 7.6. In order to illustrate (7.23) in a specific case, we consider the equation $y' = 2xy$ and calculate y_{i+1} from Euler's method (which is a first-order method) and $\tilde{y}_{i+1}$ from Heun's method (which is second order). At (x_i, y_i), Euler's method produces

$$y_{i+1} = y_i + 2hx_iy_i;$$

at (x_i, y_i) Heun's method (which is $\tilde{y}_{i+1} = y_i + h[f(x_i, y_i) + f(x_i + h, y_i + hf(x_i, y_i))]/2)$ produces

$$\begin{aligned} \tilde{y}_{i+1} &= y_i + \frac{h}{2}[2x_iy_i + 2(x_i + h)(y_i + h2x_1y_i)] \\ &= y_i + \frac{h}{2}[4x_iy_i + h(2y_i + 4x_i^2y_i) + h^2 4x_iy_i]. \end{aligned}$$

Now if $u(x)$ satisfies $u' = 2xu$, $u(x_i) = y_i$, then y_{i+1} above is the same as $y_{i+1} = u(x_i) + hu'(x_i)$. Next since $u'' = 2u + 2xu'$, it also follows that $\tilde{y}_{i+1}$ above is the same as

$$\tilde{y}_{i+1} = u(x_i) + hu'(x_i) + \frac{h^2}{2}u''(x_i) + h^3 2x_iu(x_i).$$

The estimate $\tilde{y}_{i+1} - y_{i+1}$ to the local error in Euler's method is

$$\tilde{y}_{i+1} - y_{i+1} = \frac{h^2}{2}u''(x_i) + h^3 2x_iu(x_i).$$

To obtain the actual local error, we expand $u(x)$ in a Taylor's series

$$u(x_{i+1}) = u(x_i) + hu'(x_i) + \frac{h^2}{2} u''(x_i) + \frac{h^3}{6} u'''(\gamma),$$

and thus we have [since $u(x_i) + hu'(x_i) = y_{i+1}$]

$$u(x_{i+1}) - y_{i+1} = \frac{h^2}{2} u''(x_i) + \frac{h^3}{6} u'''(\gamma).$$

Thus $u(x_{i+1}) - y_{i+1}$ and $\tilde{y}_{i+1} - y_{i+1}$ have the same leading term in an expansion in powers of h. Put another way,

$$\tilde{y}_{i+1} - y_{i+1} = u(x_{i+1}) - y_{i+1} + 0(h^3)$$

where $0(h^3)$ denotes the term $h^3[2x_iu(x_i) - u'''(\gamma)/6]$; and so the calculated quantity $\tilde{y}_{i+1} - y_{i+1}$ is a good estimate to the local error $u(x_{i+1}) - y_{i+1}$.

The approximation (7.23) gives us a practical way to estimate the local error; given (x_i, y_i), an application of (7.21a) yields y_{i+1} and then an application of (7.21b) provides $\tilde{y}_{i+1}$. The difference $\tilde{y}_{i+1} - y_{i+1}$ is very nearly the local error for the pth order method (7.21a). In the context of a variable-step method, the estimate $\tilde{y}_{i+1} - y_{i+1}$ is compared (see below) to an "error per unit step" criterion. If $\tilde{y}_{i+1} - y_{i+1}$ is too large, then y_{i+1} is rejected and new values of y_{i+1} and $\tilde{y}_{i+1}$ are calculated using a smaller step h. If $\tilde{y}_{i+1} - y_{i+1}$ is small with respect to the accuracy criterion, the value y_{i+1} is accepted but the next step from (x_{i+1}, y_{i+1}) is taken with a larger value of h. In such a variable-step method, the length of the step h may change each time; and the points $x_0, x_1, x_2, \ldots, x_i, \ldots$ are not necessarily equally spaced.

In order to give at least an intuitive justification for how controlling the local errors will control the global errors, we briefly consider a "linearized" version of the equation $y'(x) = f(x, y(x))$ in a neighborhood of $(x_i, y(x_i))$. Using a Taylor's series expansion for the dependent variable and truncating after two terms, we have

$$y'(x) \simeq f(x, y(x_i)) + f_y(x, y(x_i))(y(x) - y(x_i));$$

and this leads to a further approximation

$$y'(x) \simeq \lambda y(x) + c$$

where $\lambda = f_y(x_i, y(x_i))$, and $c = f(x_i, y(x_i)) - f_y(x_i, y(x_i))y(x_i)$. Thus, for x near x_i, solving $y' = f(x, y)$ is nearly the same as solving $y' = \lambda y + c$; note that the constants λ and c depend on the point $(x_i, y(x_i))$ and the behavior of $f(x, y)$ in a neighborhood of $(x_i, y(x_i))$. That is, the approximation $y' = \lambda y + c$ for $y' = f(x, y)$ is a local approximation; and the values of λ and c will change as we move away from $(x_i, y(x_i))$. To the extent that $y' = \lambda y + c$ is a good model for $y' = f(x,$

y), the behavior of a numerical method applied to $y' = f(x, y)$ can be predicted by considering how the method behaves on $y' = \lambda y + c$. Next, a further simplification can be obtained by change of variables. If we write $y' = \lambda y + c$ as $y' = \lambda(y + c/\lambda)$ and make the substitution $w(x) = y(x) + c/\lambda$, then $y' = \lambda y + c$ becomes the simple equation $w' = \lambda w$ [note that the curves $y(x)$ and $w(x)$ are parallel since $y(x) = w(x) - c/\lambda$]. Thus for (x_i, y_i) near $(x_i, y(x_i))$, we expect that the behavior of a method applied to $y' = f(x, y)$ can be well approximated by applying the method to $y' = \lambda y$ where $\lambda = f_y(x_i, y_i)$. Below and in later discussions as well, we will use the simple test equation $y' = \lambda y$ to analyze and compare methods.

With these preliminaries, we are ready to explain the mechanics of a variable-step method. Suppose that we wish to solve $y' = f(x, y)$ on an interval $[a, b]$ and suppose that we want to control the *relative* error so that at each step

$$\frac{|y(x_k) - y_k|}{|y(x_k)|} \le \varepsilon \frac{(x_k - x_0)}{(b - a)} \tag{7.24}$$

(in particular for $k = n$, we want the relative error at $x = b$ to be bounded by ε).

To give a heuristic derivation of a step-size-selection strategy that will control the relative errors in the sense of (7.24), we use (7.19) to express the relative error at x_{i+1}:

$$\frac{y(x_{i+1}) - y_{i+1}}{y(x_{i+1})} = \frac{y(x_{i+1}) - u(x_{i+1})}{y(x_{i+1})} + \frac{u(x_{i+1}) - y_{i+1}}{y(x_{i+1})}. \tag{7.25}$$

If we assume that we are solving the equation $y' = \lambda y$, then we have $y(x) = y(x_i)e^{\lambda(x - x_i)}$ or $y(x_{i+1}) = y(x_i)e^{\lambda h}$. Similarly since $u(x)$ satisfies $u' = \lambda u$, $u(x_i) = y_i$, it follows that $u(x_{i+1}) = y_i e^{\lambda h}$. Therefore, $y(x_{i+1}) - u(x_{i+1}) = [y(x_i) - y_i]e^{\lambda h}$; and for $y' = \lambda y$ we can write the first term on the right-hand side of (7.25) as

$$\frac{y(x_{i+1}) - u(x_{i+1})}{y(x_{i+1})} = \frac{y(x_i) - u(x_i)}{y(x_i)}. \tag{7.26a}$$

For the second term in (7.25), recall from (7.20b) that $u(x_{i+1}) - y_{i+1} = h\tau$ and that we can estimate $h\tau$ from $h\tau \simeq \tilde{y}_{i+1} - y_{i+1}$ [see (7.23)]. If h is reasonably small and if y_i is near $y(x_i)$, then we would expect that

$$\frac{u(x_{i+1}) - y_{i+1}}{y(x_{i+1})} = \frac{h\tau}{y(x_{i+1})} \simeq \frac{h\tau}{y_i} \simeq \frac{\tilde{y}_{i+1} - y_{i+1}}{y_i}. \tag{7.26b}$$

Combining (7.25) and (7.26), we arrive at a plausible inequality relating the relative error at x_{i+1} to the relative error at x_i and the relative local error:

$$\frac{|y(x_{i+1}) - y_{i+1}|}{|y(x_{i+1})|} \le \frac{|y(x_i) - y_i|}{|y(x_i)|} + \frac{|\tilde{y}_{i+1} - y_{i+1}|}{|y_i|}. \tag{7.27}$$

If in previous steps we have successfully controlled the relative error so that at $x = x_i$

$$\frac{|y(x_i) - y_i|}{|y(x_i)|} \le \varepsilon \frac{(x_i - a)}{(b - a)},$$

and if we use a step of length h at x_i where h is chosen so that

$$\frac{|\tilde{y}_{i+1} - y_{i+1}|}{|y_i|} \le \varepsilon \frac{h}{(b - a)},$$

then from (7.27) we would expect that

$$\frac{|y(x_{i+1}) - y_{i+1}|}{|y(x_{i+1})|} \le \varepsilon \frac{(x_i - a)}{(b - a)} + \varepsilon \frac{h}{(b - a)} = \varepsilon \frac{(x_{i+1} - a)}{(b - a)}.$$

A variable-step method proceeds in a fashion suggested by the argument above. At the point (x_i, y_i), a step h is selected so that an "error per unit step" criterion is satisfied; that is, h is chosen so that

$$|\tilde{y}_{i+1} - y_{i+1}| \le \varepsilon \frac{h|y_i|}{(b - a)}. \tag{7.28}$$

In the next section we see how this device can be implemented in practice. Finally we note that when we wish to control absolute errors rather than relative errors, we select h at (x_i, y_i) so that we control the absolute local error rather than the relative local error; that is, we choose h so that

$$|\tilde{y}_{i+1} - y_{i+1}| \le \varepsilon \frac{h}{(b - a)}.$$

7.2.5. Implementation of Variable-step Runge-Kutta Methods

One popular variable-step Runge-Kutta method using the concepts of Section 7.2.4 is based on a pair of fourth- and fifth-order methods attributable to Fehlberg. In these formulas, the parameters of the fourth- and fifth-order methods are selected so that only six evaluations per step are required. The formulas are

$$y_{i+1} = y_i + h\left(\frac{25}{216} K_1 + \frac{1408}{2565} K_3 + \frac{2197}{4104} K_4 - \frac{1}{5} K_5\right) \tag{7.29a}$$

$$\tilde{y}_{i+1} = y_i + h\left(\frac{16}{135} K_1 + \frac{6656}{12825} K_3 + \frac{28561}{56430} K_4 - \frac{9}{50} K_5 + \frac{2}{55} K_6\right) \tag{7.29b}$$

where (7.29a) is the fourth-order method and (7.29b) is the fifth-order method. The evaluations $K_1, \ldots, K_6$ are given by

$$K_1 = f(x_i, y_i).$$

$$K_2 = f\left(x_i + \frac{h}{4}, y_i + \frac{h}{4} K_1\right),$$

$$K_3 = f\left(x_i + \frac{3h}{8}, y_i + h\left(\frac{3}{32} K_1 + \frac{9}{32} K_2\right)\right),$$

$$K_4 = f\left(x_i + \frac{12h}{13}, y_i + h\left(\frac{1932}{2197} K_1 - \frac{7200}{2197} K_2 + \frac{7296}{2197} K_3\right)\right),$$

$$K_5 = f\left(x_i + h, y_i + h\left(\frac{439}{216} K_1 - 8K_2 + \frac{3680}{513} K_3 - \frac{845}{4104} K_4\right)\right),$$

$$K_6 = f\left(x_i + \frac{h}{2}, y_i + h\left(-\frac{8}{27} K_1 + 2K_2 - \frac{3544}{2565} K_3 + \frac{1859}{4104} K_4 - \frac{11}{40} K_5\right)\right).$$

From (7.29) it is clear that six evaluations will suffice to produce y_{i+1} and an estimate $\tilde{y}_{i+1} - y_{i+1}$ of the local error. By contrast, an unrelated pair of fourth-order and fifth-order methods would usually require four evaluations for y_{i+1} and six more evaluations for $\tilde{y}_{i+1}$. There are Runge-Kutta pairs that are similar to (7.29) but that have higher orders. For example, the subroutine DVERK in the IMSL library is a pair of fifth- and sixth-order methods requiring only eight evaluations per step to produce y_{i+1} and $\tilde{y}_{i+1}$.

Given that (7.29) is a suitable pair of formulas for a variable-step Runge-Kutta method, we now consider some of the practical aspects of designing a reasonably reliable differential-equation solver based on this pair. In order to solve $y' = f(x, y)$ on $[a, b]$ with a relative error of ε or less, we see from (7.28) that the step h used at (x_i, y_i) should satisfy

$$\frac{|\tilde{y}_{i+1} - y_{i+1}|}{h} \leq \varepsilon \frac{|y_i|}{(b - a)}.$$

Using (7.29a) and (7.29b), we can verify that

$$\frac{\tilde{y}_{i+1} - y_{i+1}}{h} = \frac{1}{360} K_1 - \frac{128}{4275} K_3 - \frac{2197}{75240} K_4 + \frac{1}{50} K_5 + \frac{2}{55} K_6. \quad (7.29\text{d})$$

For brevity, we let est $= (\tilde{y}_{i+1} - y_{i+1})/h$; and as a criterion for choosing the step size h at (x_i, y_i), we would like to use a value h for which

$$|\text{est}| \simeq \varepsilon \frac{|y_i|}{(b - a)}.$$

That is, we want to use the largest step for which we can safely assume that (7.28) is satisfied. To determine this "optimum" h, recall that (7.29a) is a fourth-order method so that the local error (estimated by $\tilde{y}_{i+1} - y_{i+1}$) is $O(h^5)$. Thus for small h, we can say that

$$\text{est} = \frac{\tilde{y}_{i+1} - y_{i+1}}{h} = Mh^4 + O(h^5).$$

If we were to use a step of length γh in (7.29), we would obtain a different value $\widehat{\text{est}}$ where

$$\widehat{\text{est}} = M(\gamma h)^4 + O(h^5) = \gamma^4 \text{est} + O(h^5). \tag{7.30}$$

If γh were the optimum step, then we would have $|\widehat{\text{est}}| = \varepsilon |y_i|/(b - a)$; or in view of (7.30)

$$\gamma^4 |\text{est}| \simeq \varepsilon \frac{|y_i|}{(b - a)}.$$

From the above, we see that

$$\gamma \simeq \left[\frac{\varepsilon |y_i|}{|\text{est}|(b - a)} \right]^{\frac{1}{4}}. \tag{7.31}$$

Put another way, if we take a step of length h from (x_i, y_i) and calculate est, then we can say (after the fact) that we should have taken a step of length γh. Given this hindsight, we will use a step of length γh at the next application of (7.29a) and (7.29b). If the step was successful [$|\text{est}| \le \varepsilon |y_i|/(b - a)$], then we accept y_{i+1} and use γh to step from (x_{i+1}, y_{i+1}) to (x_{i+2}, y_{i+2}); if the step was unsuccessful [$|\text{est}| > \varepsilon |y_i|/(b - a)$], then we use γh to step again from (x_i, y_i).

The information in (7.29) and (7.31) forms the core of a variable-step Runge-Kutta method. A design for a differential-equation solver can be formulated from this core with a few additional constraints based on experience. A simple but reasonable formulation is outlined below.

1. Suppose that we are at (x_i, y_i) and have an estimate h_0 of the step we should use to calculate y_{i+1}.
2. Use (7.29a) to calculate y_{i+1} and (7.29d) to calculate est with $h = h_0$.
3. Calculate γ from (7.31) and set $h_1 = .8\gamma h_0$. The value h_1 will be used for the next step and the reason for a factor of .8 is explained below.
4. Test est against $\varepsilon |y_i|/(b - a)$; if $|\text{est}| \le \varepsilon |y_i|/(b - a)$, accept y_{i+1} and use h_1 as the appropriate step with which to calculate y_{i+2}. If $|\text{est}| > \varepsilon |y_i|/(b - a)$, reject y_{i+1} and begin again from (x_i, y_i) using step size h_1.

In addition to the broad outline above, some additional constraints on h_1 are desirable [see, for instance, Shampine and Allen (1973)]. In particular, h_1 should be constrained so that

5. $.1\, h_0 \le h_1 \le 5\, h_0$.

6. $h_{\min} \le h_1 \le h_{\max}$ where $h_{\min}$ and $h_{\max}$ are input parameters. If (3) yields a value h_1 for which $h_1 > h_{\max}$, set $h_1 = h_{\max}$. If (3) yields a value of h_1 for which $h_1 < h_{\min}$, we conclude that the algorithm cannot safely step beyond (x_i, y_i).

In (3), we use a factor of .8 to define h_1 in order to keep down the number of rejections in step (4). If h_1 were defined by $h_1 = \gamma h_0$, it is likely that h_1 would frequently be just a little too large, causing us to reject y_{i+1} and then requiring another set of evaluations of $f(x, y)$ to step from (x_i, y_i). The factor .8 means (in effect) that we are aiming at an error per unit step of $\varepsilon/2$ in order to achieve a global error of ε or less [replacing ε by $\varepsilon/2$ in (7.31) has the effect of scaling γ by $2^{-\frac{1}{4}} \simeq .841$]. The constraint on h_1 in (5) is to prevent the steps from swinging wildly from small to large. In (6), it is natural to require that $h_1 \le h_{\max}$, in part because many of the approximations made in formulating the variable-step algorithm assume a "small" h. Finally, saying that the algorithm has failed to solve $y' = f(x, y)$ on $[a, b]$ if $h_1 < h_{\min}$ is also a sensible constraint; if the solution to $y' = f(x, y)$ is varying so rapidly that the algorithm cannot follow it without using very small steps, perhaps a person should look at the problem instead of having the machine calculate on. For example, the problem $y' = y^2$, $y(0) = 1$ has $y(x) = 1/(1 - x)$ as the solution; and attempting to solve this problem across the interval [0, 2] should require the algorithm to send a signal that all is not well. In addition, to execute (7.29), $f(x, y)$ has to be evaluated at points such as $(x + h/4, y + hK_1/4)$ and h must be large enough so that the machine numbers for x and $x + h/4$ are different. That is, we need $x + h/4 > x$ in the computer; so for $x > 0$ we need

$$1 + \frac{h}{4x} > 1.$$

Thus h must be large enough so that $h/(4x)$ is not smaller than machine epsilon.

In Fig. 7.9 we list a very simple version of a Runge-Kutta-Fehlberg program. The program was written in a modular form for ease of reading and consists of three subroutines: DESOLV, GAMCAL, and RKF. Basically, subroutine DESOLV accepts an initial-value problem on an interval $[a, b]$ and then controls the step-size selection as the problem is solved across $[a, b]$. In particular, DESOLV calls RKF to obtain a value for est [see (7.29d)] and then calls GAMCAL to obtain a value for γ [see (7.31)]. The specific differential equation to be solved must be formulated in another subroutine called YPRIME; and RKF calls YPRIME to evaluate $K_1, \ldots, K_6$ [see (7.29c)]. [Figure 7.10 (page 384) gives an example of YPRIME for the problem $y' = 2xy$. All that YPRIME does is to accept x and y and return the value $f(x, y)$.] The modular form of the program in Fig. 7.9 also makes it very easy to modify the program for systems of first-order equations, and we will consider this modification in the next section.

```
      SUBROUTINE DESOLV(A,B,YA,YB,TOL,HSTART,HMIN,HMAX,IFLAG)
C
C  INITIALIZATION PHASE.  START SOLUTION PROCESS AT (A,Y(A)) WITH
C  STEP SIZE HSTART.
C
      HOLD=HSTART
      X=A
      YOLD=YA
C
C  MAKE STEP OF LENGTH HOLD FROM (X,YOLD) FINDING YNEW, EST, AND
C  GAMMA.  CALCULATE HNEW AND TEST TO SEE IF YNEW IS ACCEPTABLE.
C
    1 CALL RKF(X,YOLD,YNEW,EST,HOLD)
      CALL GAMCAL(A,B,YOLD,EST,TOL,GAMMA)
      HNEW=.8*GAMMA*HOLD
      IF(GAMMA.GE.1.)  GO TO 2
C
C  YNEW IS NOT ACCEPTABLE.  RESTART FROM (X,YOLD) IF HNEW IS NOT
C  LESS THAN HMIN.
C
      IF(HNEW.LT.HOLD/10.)  HNEW=HOLD/10.
      IF(HNEW.LT.HMIN)  GO TO 3
      HOLD=HNEW
      GO TO 1
C
C  YNEW IS ACCEPTABLE.  TAKE NEXT STEP FROM (X+HOLD,YNEW) IF HNEW
C  IS NOT LESS THAN HMIN.
C
    2 IF(HNEW.LT.HMIN)  GO TO 3
      IF(HNEW.GT.5.*HOLD)  HNEW=5.*HOLD
      IF(HNEW.GT.HMAX)  HNEW=HMAX
      IF(X+HOLD.GE.B)  GO TO 4
      X=X+HOLD
      HOLD=HNEW
      YOLD=YNEW
      GO TO 1
C  ALGORITHM FAILED.  LAST POINT REACHED WAS (X,YOLD).
C
    3 IFLAG=0
      B=X
      YB=YOLD
      RETURN
C
C  ALGORITHM SUCCEEDED.  INITIALIZE HSTART FOR FURTHER INTEGRATIONS
C  AND EXECUTE STEP THAT REACHES EXACTLY TO X=B.
C
    4 IFLAG=1
      HSTART=HNEW
      HOLD=B-X
      CALL RKF(X,YOLD,YNEW,EST,HOLD)
      YB=YNEW
      RETURN
      END


      SUBROUTINE GAMCAL(A,B,YOLD,EST,TOL,GAMMA)
      ABSEST=ABS(EST)
      IF(ABSEST.EQ.0.)  GO TO 1
      IF(YOLD.EQ.0.)  GO TO 2
      GAMMA=(TOL*ABS(YOLD)/(ABSEST*(B-A)))**.25
      RETURN
    1 GAMMA=6.25
      RETURN
    2 GAMMA=(TOL*TOL/(ABSEST*(B-A)))**.25
      RETURN
      END
```

Figure 7.9 Subroutines DESORV, GAMCAL, and RFK.

```
SUBROUTINE RKF(X,YOLD,YNEW,EST,H)
REAL K1,K2,K3,K4,K5,K6
B21=1./4.
B31=3./32.
B32=9./32.
B41=1932./2197.
B42=-7200./2197.
B43=7296./2197.
B51=439./216.
B52=-8.
B53=3680./513.
B54=-845./4104.
B61=-8./27.
B62=2.
B63=-3544./2565.
B64=1859./4104.
B65=-11./40.
A1=25./216.
A3=1408./2565.
A4=2197./4104.
A5=-1./5.
R1=1./360.
R3=-128./4275.
R4=-2197./75240.
R5=1./50.
R6=2./55.
YT=YOLD
XT=X
CALL YPRIME(XT,YT,K1)
YT=YOLD+H*B21*K1
XT=X+H/4.
CALL YPRIME(XT,YT,K2)
YT=YOLD+H*(B31*K1+B32*K2)
XT=X+3.*H/8.
CALL YPRIME(XT,YT,K3)
YT=YOLD+H*(B41*K1+B42*K2+B43*K3)
XT=X+12.*H/13.
CALL YPRIME(XT,YT,K4)
YT=YOLD+H*(B51*K1+B52*K2+B53*K3+B54*K4)
XT=X+H
CALL YPRIME(XT,YT,K5)
YT=YOLD+H*(B61*K1+B62*K2+B63*K3+B64*K4+B65*K5)
XT=X+H/2.
CALL YPRIME(XT,YT,K6)
YNEW=YOLD+H*(A1*K1+A3*K3+A4*K4+A5*K5)
EST=R1*K1+R3*K3+R4*K4+R5*K5+R6*K6
RETURN
END
```

Figure 7.9 (continued)

Before explaining the program, we should consider some of the ways that a differential-equation solver might be used in practice. First of all, if we are interested only in the value of the solution at $x = b$, a single call to DESOLV is sufficient. On the other hand, we might want to produce a plot of the solution across the interval $[a, b]$ using increments of Δx. That is, for points $t_0, t_1, \ldots, t_i, \ldots$ where $t_i = a + i\Delta x$, we want to tabulate $y(t_i)$. In this case, we would call DESOLV repeatedly to solve $y' = f(x, y)$ across $[t_i, t_{i+1}]$, $i = 0, 1, \ldots$. There are other ways in which a differential-equation solver might be used. For instance, we might be interested in estimating the first point x at which $y(x)$ is

```
      DELX=.25
      NINCR=12
      YA=1.
      TOL=.25E-06
      HSTART=.1
      HMIN=1.E-04
      HMAX=.25
      DO 1 I=1,NINCR
        A=(I-1)*DELX
        B=I*DELX
        CALL DESOLV(A,B,YA,YB,TOL,HSTART,HMIN,HMAX,IFLAG)
        PRINT 100,B,YB,HSTART
        IF(IFLAG.EQ.0)  STOP
        YA=YB
    1 CONTINUE
  100 FORMAT(' ',3E20.7)
      STOP
      END


      SUBROUTINE YPRIME(X,Y,YPRIM)
      YPRIM=2.*X*Y
      RETURN
      END
```

Figure 7.10 Main program and subroutine YPRIME for the problem $y' = 2xy$, $y(0) = 1$.

equal to some prescribed value. As another example, applications areas such as control theory sometimes consider differential equations that contain parameters; the problem is then to estimate these parameters so that the solution has some desired property, perhaps matching an observed trajectory at several points, minimizing some function of the solution, etc. The program in Fig. 7.9 can be used for many of these applications and is simple enough so that it can be modified to handle others.

To make the program more readable, we will describe some of the structure below rather than in the comment sections of the listing. The basic mathematical problem is to solve $y' = f(x, y)$, $y(a) = y_0$ on the interval $[a, b]$; and the goal is to estimate the solution at $x = b$ with a relative error of ε or less.

The required inputs for DESOLV are A and B, the endpoints of the interval $[a, b]$; YA, the initial condition $y(a)$; TOL, the accuracy criterion; HSTART, a guess at an appropriate step size to use at $(x_0, y_0) = (a, y(a))$; HMIN, the minimum allowable step size; and HMAX, the maximum allowable step size. Note that the choice of HSTART is not critical but some value must be chosen to get the method started. Subroutine DESOLV will adjust HSTART until an acceptable step is found. If the program succeeds in solving $y' = f(x, y)$ across $[a, b]$, the last step used in getting to b is stored in HSTART and can be used to restart the algorithm if it is desired to continue the solution beyond $x = b$.

Subroutine DESOLV returns YB, which is the estimate to $y(b)$; and we anticipate that

$$\frac{|\text{YB} - y(b)|}{|y(b)|} \leq \text{TOL}.$$

DESOLV returns a value of 0 or 1 for IFLAG also where IFLAG = 0 indicates that DESOLV sensed that a step size smaller than HMIN would have been required in order to advance beyond some point x_F and still maintain accuracy. Thus IFLAG = 0 signals a failure to solve the problem, and in this case B is set to x_F and YB to the last good estimate; namely, the estimate to $y(x_F)$. If DESOLV is able to reach $x = b$ without using steps smaller than HMIN, then IFLAG is set to 1.

The computational details of DESOLV are fairly simple as the listing shows. Suppose the algorithm is at a point (x, y) with an acceptable estimate to $y(x)$ and with an estimate HOLD of the next step size to use. A step is made (using HOLD) by calling RKF. Subroutine RKF returns YNEW (an estimate to the solution at X+HOLD) and EST. The quantity EST is passed to GAMCAL, which returns a value of γ; and then a new step HNEW is computed. A decision is made as to whether YNEW is acceptable or not by testing $|\text{est}|$ against $\varepsilon|y|/(b - a)$; for convenience, we test $\gamma \geq 1$ to measure a success. The remainder of DESOLV merely follows the steps (1) – (6) given previously, constraining HNEW between .1*HOLD and 5.*HOLD and also testing to see that HNEW is between HMIN and HMAX. Finally, we want to advance exactly to $x = b$ so that if a step is successful, we test to see if X+HOLD ≥ B. If the inequality is satisfied, we make one last step from X using HOLD = B−X and return with IFLAG = 1 and HSTART = HNEW. (HNEW is the step size we would have used if we were to continue; so HSTART = HNEW is a very good estimate to the step we would want to use if we wish to continue the computation beyond $x = b$; in order, for example, to tabulate the solution should we wish to plot it.)

Subroutine RKF implements (7.29) in a straightforward fashion. The implementation can be made more efficient (some constants can be precomputed in DESOLV and passed to RKF) but we are more interested in clarity than efficiency. Similarly, the evaluation of $f(x, y)$ could be done by a function statement or by a function subprogram, but instead we evaluate $f(x, y)$ in a subroutine because it is easier to see how to modify the program for systems.

The calculation of γ in GAMCAL is not quite so obvious. Recalling that γ is defined by

$$\gamma = \left[\frac{\varepsilon|y_i|}{|\text{est}|(b - a)}\right]^{\frac{1}{4}},$$

we see that est = 0 can cause problems (and est = 0 will occur occasionally when the accuracy criterion, ε, is small). If est = 0, the step was successful and we arbitrarily set γ equal to 6.25 so that $.8\gamma = 5$. The other problem occurs when $y_i = 0$; and in that event we arbitrarily set $\gamma = [\varepsilon^2/|\text{est}|(b - a)]^{\frac{1}{4}}$. A logical alternative, which is frequently employed, is to use a mixed relative error/absolute error test, defining γ by

$$\gamma = \left[\frac{\varepsilon_R|y_i| + \varepsilon_A}{|\text{est}|(b - a)}\right]^{\frac{1}{4}}$$

With this, the test of $\gamma \geq 1$ is purely an absolute-error test when $\varepsilon_R = 0$ and purely a relative-error test when $\varepsilon_A = 0$. If $\varepsilon_R \neq 0$ and $\varepsilon_A \neq 0$, then the test is a mixed one—if $|y_i|$ is very small, ε_A is the dominant term and we are essentially controlling the absolute error; if $|y_i|$ is large, we are essentially controlling the relative error. For most applications, a mixed test is quite satisfactory; but it should be noted that if we try to follow a trajectory like $y(x) = e^{-5x}$ with a mixed test, the numerical solution will be small but will likely bear little other relation to $y(x)$.

In Fig. 7.10 we list a simple calling program (and the requisite subroutine YPRIME) that uses DESOLV to tabulate the solution of $y' = 2xy$, $y(0) = 1$ in steps of .25 from $x = 0$ to $x = 3$ (see also, Example 7.7 below). All the examples that follow were run in double precision on an IBM 370/158.

EXAMPLE 7.7. Figure 7.10 illustrates a calling program that uses DESOLV to tabulate the solution of $y' = 2xy$, $y(0) = 1$ in increments of .25 from $x = 0$ to $x = 3$. The accuracy criterion is TOL = .25E−06; HSTART = .1, HMIN = 1.E−04, HMAX = .25. Since we know the solution to $y' = 2xy$, we can calculate the relative errors and compare them with what we expect (the results are listed in Table 7.3). In Table 7.3 we also tabulate the HSTART values that were generated at each call to DESOLV. If $y' = 2xy$ is written in the form $y' = \lambda y$, it is obvious that $\lambda = 2x$ and so λ gets larger as x gets larger. As we show in a later section, when λ grows in magnitude, smaller steps are required to track the solution as is reflected in the table in which $h = .164 \ldots$ was the last predicted full step in [0, .25] and $h = .0143 \ldots$ was the last predicted full step in [2.75, 3.].

In this example, we made an accuracy request of TOL = .25E−06 and we anticipate that the program will make an error of about this size at each call to DESOLV. Since the errors may add, we can expect an error of about 1.E−06 at $x = 1$ (after four calls) and an error of about 3.E−06 at $x = 3$ (after twelve calls). This expectation is also reflected in the table. (There is a distinction between the way absolute errors grow and the way relative errors grow, and we will discuss this distinction in Section 7.6.) For reference, we note that the solution is $y(x) = e^{x^2}$ and that $y(3) = 8103.084$; DESOLV estimates the solution at $x = 3$ to be 8103.094. A tighter tolerance would produce better answers.

EXAMPLE 7.8. In this example we briefly illustrate the sorts of behavior that can be expected from a variable-step method. By way of example we use the program in Fig. 7.9 for three simple but representative problems.

First, the problem $y' = 2y$, $y(0) = 1$ was run using the program in Fig. 7.10. The relative errors ranged from 0.7694E−07 at $x = .25$ to 0.9147E−06 at $x = 3$. The solution is $y(x) = e^{2x}$; so $y(3) = 403.4288$; the program returned an estimate of 403.4292. A more interesting observation is that HSTART remained constant at all steps, with HSTART = 0.5682705E−01, to the places printed. The explanation for this behavior (see Section 7.6) is that when a fourth-order Runge-Kutta method is applied to $y' = \lambda y$, the local error at (x_i, y_i) is given by $y_i[e^{\bar{h}} - p(\bar{h})]$ where $\bar{h} = h\lambda$ and where

$$p(\bar{h}) = 1 + \bar{h} + \frac{\bar{h}^2}{2!} + \frac{\bar{h}^3}{3!} + \frac{\bar{h}^4}{4!}.$$

TABLE 7.3

x_i	y_i	Relative error	HSTART
.25	0.1064495E+01	0.3938 E−07	0.1643E+00
.50	0.1284026E+01	0.1289 E−06	0.1330E+00
.75	0.1755055E+01	0.2974 E−06	0.8715E−01
1.00	0.2718283E+01	0.4748 E−06	0.5636E−01
1.25	0.4770736E+01	0.5820 E−06	0.4367E−01
1.50	0.9487742E+01	0.6821 E−06	0.3392E−01
1.75	0.2138096E+02	0.7770 E−06	0.2832E−01
2.00	0.5459820E+02	0.8674 E−06	0.2376E−01
2.25	0.1579851E+03	0.9585 E−06	0.2043E−01
2.50	0.5180134E+03	0.1049 E−05	0.1804E−01
2.75	0.1924653E+04	0.1137 E−05	0.1598E−01
3.00	0.8103094E+04	0.1224 E−05	0.1431E−01

Thus, controlling the relative local error to within a tolerance of ε means that h is selected so that $|e^{\bar{h}} - p(\bar{h})| \le \varepsilon$ and the proper choice of h is independent of x in [0, 3]. [Note that if we were controlling absolute errors instead of relative errors, h would have to shrink as we advance since we would be insisting that $|y_i|\,|e^{\bar{h}} - p(\bar{h})| \le \varepsilon$ be satisfied at each step.] We should also note that if we write the problem solved in Example 7.7 ($y' = 2xy$) in the form $y' = \lambda y$, then $\lambda = 2x$. Thus for x near 1 we have $\lambda \simeq 2$ and so the algorithm applied to $y' = 2xy$ is nearly solving $y' = 2y$ when x is near 1. In particular, we would expect the algorithm to be using about the same step size near $x = 1$ whether we are solving $y' = 2y$ or $y' = 2xy$. This expectation is correct as a check of Table 7.3 will show; at $x = 1$ the last predicted full step is 0.5636E−01, which is close to being the constant step 0.5683E−01 that is used in solving $y' = 2y$.

Next the equation $y' = -10y$, $y(0) = 1$ was run, and the relative errors ranged from 0.1143E−06 at $x = .25$ to 0.1372E−05 at $x = 3$. The solution is $y(x) = e^{-10x}$; so $y(3) =$ 0.9357623E−13; the program returned an estimate of 0.9357610E−13 at $x = 3$. By tightening the tolerance, we could have achieved more places of agreement; it is also worthwhile to note that if we had used an absolute-error test or a mixed-error test with ε_A about 10^{-7} or so, we would have had a much larger relative error. As with the problem $y' = 2y$, this problem runs with a constant step, namely $h = 0.7466472\text{E}{-02}$ (a smaller step must be used since the factor λ is larger and we cannot maintain $|e^{\bar{h}} - p(\bar{h})| \le \varepsilon$ without having h smaller).

As a final example, the problem $y' = y^2$, $y(0) = 1$ was run; and since the solution is $y(x) = 1/(1 - x)$, we expect a failure as we near $x = 1$. The results returned were

x_i	y_i	Relative error	HSTART
.25	0.1333333E+01	0.1248E−06	0.5514E−01
.50	0.2000001E+01	0.3425E−06	0.3256E−01
.75	0.4000003E+01	0.8705E−06	0.1396E−01

The program signaled a failure at $x_F = 0.9950695$ and returned an estimate of 202.8277 to $y(x_F) = 202.8192$. In fact, the relative error at $x = x_F$ is 0.4726E−04, which exceeds the input tolerance by several orders of magnitude. The size of the actual relative error shows that we have already passed the point at which the machine estimates can be trusted.

EXAMPLE 7.9. In this example, we illustrate the problems that can arise when $f(x, y)$ is not smooth. In particular, we attempt to solve $y' = |x - 1|y$, $y(0) = 1$ on $[0, 2]$ with one call to DESOLV (we want the solution at $x = 2$ and are not interested in tabulating the intermediate steps). Note that $y'(x)$ is continuous on $[0, 2]$; but $y''(x)$ is not, having a point of discontinuity at $x = 1$. We cannot go into the details of why an equation whose solution is not smooth will cause difficulties for a method except to say that the derivation of a fourth-order method (and the error-monitoring machinery that goes with it) depended on the solution being smooth.

For the problem $y' = |x - 1|y$, we made one call to DESOLV with A = 0, B = 2, TOL = .25E−06, HSTART = .1, HMIN = 1.E−04, and HMAX = .25. The results are not good as we might expect. By inserting in DESOLV a print statement that lists each step, we find that the algorithm marches quickly along to $x = .90408$ and is predicting that a step of $h = .11936$ will be successful as the next step. It is not successful; so the step is adjusted and some minor advancement takes place until the algorithm stalls at $x_F = .9999576$ when DESOLV senses that a step less than HMIN must be taken. Subroutine DESOLV returns with an estimate of 1.648721 to $y(x_F)$. The solution to the problem is

$$y(x) = \begin{cases} e^{x - x^2/2}, & 0 \le x \le 1 \\ e^{x^2/2 - x + 1}, & 1 \le x \le 2 \end{cases}$$

and so the estimate to $y(x_F)$ is correct to the places listed. The algorithm can advance across the discontinuity of $y''(x)$; but in order to do so, we have to decrease HMIN or increase TOL. Taking the latter course, we reset TOL to TOL = .25E−03 and ran the problem again. This time the algorithm runs at a step of $h = .25$ (this is HMAX) up to $x = .85$, but the step from $x = .85$ with $h = .25$ fails. The step is adjusted, progress is made up to $x = .9780985$ where more failures occur. Steps are adjusted again and finally the algorithm is able to advance to 1.05312 by using $h = .0750214$. After passing $x = 1$, the algorithm quickly moves h back to HMAX and reaches $x = 2$ with an estimate of 2.718011. [The actual solution is $y(2) = 2.718282$; so the program was successful in meeting the accuracy criterion of .25E−03.] This example illustrates that $y' = f(x, y)$ can be solved even when the solution is not smooth, but the steps will have to be reduced substantially near discontinuities of higher derivatives.

PROBLEMS, SECTION 7.2.5

1. The equation $y' = 4x^3$, $y(0) = 1$ is easily solved. Solve the equation and calculate the local truncation error directly from (7.15) for Euler's method, Heun's method, and the modified Euler's method. Note that τ has the form $\tau = O(h^k)$ in each case and determine the appropriate value of k for each method.
2. Use (7.15) to calculate the local truncation error for Heun's method applied to $y' = 2x$, $y(0) = 0$. Argue from this calculation that Heun's method will (theoretically) track the solution exactly.

3. Verify that the local truncation error of a kth order Taylor's series method is given by (7.18).
4. Using (7.13a) and (7.18), show that the local truncation error of a kth order Runge-Kutta method satisfies $\tau = O(h^k)$.
5. To illustrate how integral curves may spread out, consider the problem $y' = y$, $y(0) = \sqrt{2}$. Now, $\sqrt{2}$ is not a machine number; so we must solve a slightly different problem $u' = u$, $u(0) = u_0$ where $u_0 \simeq \sqrt{2}$. If $u_0 - \sqrt{2} = 10^{-5}$, how large must x be in order that $|u(x) - y(x)| \geq 1$? While the absolute errors grow, the relative errors stay constant—what is $|u(x) - y(x)|/|y(x)|$?
6. Since a variable-step Runge-Kutta method attempts to control the local error, it is instructive to consider the form of this error and Heun's method is representative. If Heun's method is applied to $y' = \lambda y$, then $y_{i+1} = p(\bar{h})y_i$ where $\bar{h} = h\lambda$, $p(\bar{h}) = 1 + \bar{h} + \bar{h}^2/2$. If $u(x)$ solves $u' = \lambda u$, $u(x_i) = y_i$, show that

$$u(x_{i+1}) - y_{i+1} = y_i[e^{\bar{h}} - p(\bar{h})] \quad \text{and} \quad \frac{u(x_{i+1}) - y_{i+1}}{u(x_{i+1})} = \frac{e^{\bar{h}} - p(\bar{h})}{e^{\bar{h}}}.$$

7. Following Example 7.6, suppose that y_{i+1} is given by Euler's method and $\tilde{y}_{i+1}$ by the modified Euler's method. Suppose that $u(x)$ satisfies $u' = x - u$, $u(x_i) = y_i$; and write $u(x_{i+1}) = u(x_i) + u'(x_i)h + u''(x_i)h^2/2! + u'''(\gamma)h^3/3!$ Verify that the leading terms in $\tilde{y}_{i+1} - y_{i+1}$ and $u(x_{i+1}) - y_{i+1}$ agree.
8. Repeat Problem 7 for the equation $y' = y(2 - y)$.
9. Use DESOLV for the problem $y' = y^2$, $y(0) = -1$. Print the estimates and the relative errors in increments of .5 for $0 \leq x \leq 10$. Use a minimum step 10^{-4} and run with accuracy requests of 10^{-r}, $r = 4, 6, 8, 10$.
10. Repeat Problem 9 for the equations in Problems 9 and 10 of Section 7.1 and Problems 1 and 2 of Section 7.2.3.
11. If your computer center supports an adaptive differential-equation solver (such as DVERK in the IMSL or a variable-step Adams method), use that package to solve the equations in Problems 9 and 10.
12. Modify DESOLV to use a mixed relative-error/absolute-error test. Test your modification on $y' = 10y$, $y(0) = 1$ and $y' = -10y$, $y(0) = 1$; use $\varepsilon_R = 10^{-6}$, $\varepsilon_A = 10^{-6}$. Solve these over [0, 10]; print the estimates and both the relative and the absolute errors in increments of .5.

*7.3 CONVERGENCE AND ERROR ANALYSIS FOR ONE-STEP METHODS

In order for a one-step method $y_{i+1} = y_i + h\phi(x_i, y_i; h)$ to be useful, the method should have the property that the results get better as h gets smaller. In fact, if we apply the method to $y' = f(x, y)$, we would hope that $y_n \to y(x_n)$ as $h \to 0$. To put this idea into mathematical terms, we consider the initial-value problem $y' = f(x, y)$, $y(a) = y_0$ and fix a point x^* where $x^* > a$. For any n, we can define h from $h = (x^* - a)/n$; and if we apply the one-step method using this (fixed) step size h, then y_n is the approximation to $y(x^*)$. Thus we would like to say that the method is convergent if $y_n \to y(x^*)$ as $n \to \infty$ where h and n are constrained so that $x^* = a + nh$. Note that convergence is a mathematical idea; in reality, in any given machine with a fixed word length, we cannot actually use a value of h that is smaller than the underflow value. While convergence is a mathematical ideal, it is still useful, for if we can prove a method is convergent in the theoretical sense, then we know that it is a reasonable model for $y' = f(x, y)$ and we would expect to get better answers by decreasing h (although we might have to use extended-precision arithmetic actually to decrease h).

Before we formally define convergence, we see that there is another complication that must be dealt with. That is, the initial-value problem $y' = f(x, y)$, $y(a) = y_0$ might not have a solution at $x = x^*$; or if it does, solutions might not be unique. Because of this, it is customary to classify a method as being convergent if it converges (in a theoretical sense) for all initial-value problems that are "nice." For numerical methods, it is usual to define a nice initial-value problem as being one that satisfies the hypotheses of Theorem 7.1. In particular, a one-step method $y_{i+1} = y_i + h\phi(x_i, y_i; h)$ is said to be *convergent* if given any x^* in $[a, b]$, then

$$\lim_{\substack{n\to\infty \\ x^* = a+nh}} y_n = y(x^*)$$

whenever $y(x)$ is the solution of $y' = f(x, y)$, $y(a) = y_0$ and where $f(x, y)$ satisfies the hypotheses of Theorem 7.1 on $[a, b]$. To paraphrase this definition, a method is called convergent if it always works (in a theoretical sense) for a certain class of very well-behaved initial-value problems.

It is not hard to prove that Runge-Kutta methods are convergent and it is easy to analyze their rates of convergence; but a less specific theorem is given below as Theorem 7.2. To develop the hypotheses of Theorem 7.2, suppose that $y_{i+1} = y_i + h\phi(x_i, y_i; h)$ is a one-step method applied to $y' = f(x, y)$, $y(a) = y_0$ and suppose also that $\phi(x, y; h)$ satisfies the Lipschitz condition

$$|\phi(x, u; h) - \phi(x, v; h)| \leq K\,|u - v| \tag{7.32}$$

for all h, $0 < h \leq b - a$, for all x in $[a, b]$, and for all u and v, $-\infty < u, v < \infty$. We next fix a step size $h = (b - a)/n$; and if $y(x)$ is the solution of the initial-value problem, we let e_j denote the global error at x_j, $e_j = y(x_j) - y_j$. To derive a relation for the global errors, we use

$$y_{i+1} = y_i + h\phi(x_i, y_i; h) \tag{7.33a}$$

and

$$y(x_{i+1}) = y(x_i) + h\phi(x_i, y(x_i); h) + h\tau_i \tag{7.33b}$$

where $h\tau_i$ is the local truncation error at x_i. Subtracting (7.33a) from (7.33b), we obtain

$$e_{i+1} = e_i + h[\phi(x_i, y(x_i); h) - \phi(x_i, y_i; h)] + h\tau_i.$$

Employing the Lipschitz condition (7.32) on the equation above, we have

$$|e_{i+1}| \le (1 + hK)|e_i| + h|\tau_i|.$$

If we let $\tau = \max|\tau_i|$, $0 \le i \le n - 1$, then

$$|e_{i+1}| \le (1 + hK)|e_i| + h\tau. \tag{7.34}$$

The inequality (7.34) can be used to prove Theorem 7.2 below.

Theorem 7.2

If $\phi(x, y; h)$ satisfies (7.32), then for $0 < j \le n$

$$|e_j| \le \frac{\tau}{K}[e^{K(x_j - x_0)} - 1].$$

Proof. By the inequality (7.34), we have

$$|e_{i+1}| \le (1 + hK)\ |e_i| + h\tau;$$

and throwing the ith error back on the $(i - 1)$-st error, we see $|e_{i+1}| \le |e_{i-1}|$ $(1 + hK)^2 + h\tau[1 + (1 + hK)]$. Continuing, we finally obtain

$$|e_{i+1}| \le |e_0|\ (1 + hK)^{i+1} + h\tau[1 + (1 + hK) + \cdots + (1 + hK)^i].$$

Since $e_0 = y(x_0) - y_0$, it follows that $e_0 = 0$. Using the familiar formula for a finite geometric summation, we then have

$$|e_{i+1}| \le h\tau \frac{(1 + hK)^{i+1} - 1}{(1 + hK) - 1} = \frac{\tau}{K}[(1 + hK)^{i+1} - 1].$$

We next estimate $(1 + hK)^{i+1}$ by observing that $(1 + \alpha)^{i+1} \le e^{(i+1)\alpha}$ for $\alpha > 0$ (see Problem 8). Therefore, $(1 + hK)^{i+1} \le e^{(i+1)hK} = e^{K(x_{i+1} - x_0)}$ and this proves the theorem. ■

From Theorem 7.2, it follows that a one-step method is convergent if $\tau \to 0$ as $h \to 0$ and this condition is usually called the *consistency* condition. Note from (7.33b) that

$$\frac{y(x_i + h) - y(x_i)}{h} = \phi(x_i, y(x_i); h) + \tau_i;$$

so if $\tau \to 0$ as $h \to 0$, then $y'(x_i) = \phi(x_i, y(x_i); 0)$ if it is assumed that $\phi(x, y; h)$ is continuous. Therefore since $y'(x_i) = f(x_i, y(x_i))$, consistency means that $\phi(x_i, y(x_i); 0) = f(x_i, y(x_i))$. Note that the Taylor's series methods and all the Runge-Kutta methods we have discussed satisfy this condition. Next, in order to apply Theorem 7.2, it is necessary that the increment function $\phi(x, y; h)$ satisfy the Lipschitz condition (7.32). For the Runge-Kutta methods we have discussed, it is easy to show that (7.32) holds whenever $f(x, y)$ satisfies the hypotheses of Theorem 7.1; the Lipschitz condition on $\phi(x, y; h)$ can be derived from a Lipschitz condition on $f(x, y)$. (See Problem 9).

Finally we observe that Theorem 7.2 provides mainly qualitative rather than quantitative information. The inequalities leading up to the conclusion of the theorem are fairly crude unless h is very small. In addition, values for τ and K have to be known if the theorem is to be applied in a computational setting. However, the theorem does tell us what rate of convergence we should expect from a pth order method. Since $\tau = O(h^p)$, we expect the global error at $x = x^*$ to satisfy

$$y(x^*) - y_k = O(h^p), \qquad x^* = a + kh.$$

So for instance if we cut h in half, we expect the error to be reduced by a factor of 2^{-p}.

The proof of Theorem 7.2 implicitly assumes that the iteration $y_{i+1} = y_i + h\phi(x_i, y_i; h)$ is run using exact arithmetic and this assumption is unrealistic from a computational point of view. In the remainder of this section, we consider the limitations that roundoff errors will impose on the accuracy of the numerical solution. If we run the one-step method

$$y_{i+1} = y_i + h\phi(x_i, y_i; h)$$

on the machine, we will usually make errors in evaluating $\phi(x_i, y_i; h)$. Even if x_i, y_i, and h are exact machine numbers, the machine evaluation of $\phi(x_i, y_i; h)$ will not usually be exact. In addition, y_i is normally large with respect to $h\phi(x_i, y_i; h)$ and this comparative size causes a loss of significance in going from y_i to y_{i+1}. For example, consider Euler's method applied to $y' = x \sin(y)$; $y_{i+1} = y_i + hx_i \sin(y_i)$. Even if x_i and y_i are machine numbers, the machine representation of $x_i \sin(y_i)$ is probably not exact; and also, forming $y_i + h\, x_i \sin(y_i)$ will probably produce another rounding error.

From the discussion above, we see that in a machine implementation of $y_{i+1} = y_i + h\phi(x_i, y_i; h)$, the computer will actually calculate numbers of the form

$$u_{i+1} = u_i + h\phi(x_i, u_i; h) + \varepsilon_i \tag{7.35}$$

where we use ε_i to denote the "local rounding error" that is committed when we pass from the machine number u_i to the next machine number u_{i+1}. To analyze the effect of roundoff error, let $\delta_i = u_i - y_i$. Thus, δ_i represents the difference between what is calculated in the machine and what the method would produce if we could use exact arithmetic. Using (7.35), we see that

$$u_{i+1} - y_{i+1} = u_i - y_i + h[\phi(x_i, u_i; h] - \phi(x_i, y_i; h)] + \varepsilon_i; \tag{7.36a}$$

and subtracting (7.33b) from (7.33a), we also have

$$y_{i+1} - y(x_{i+1}) = y_i - y(x_i) + h[\phi(x_i, y_i; h) - \phi(x_i, y(x_i); h)] - h\tau_i. \tag{7.36b}$$

Adding (7.36a) to (7.36b) produces

$$u_{i+1} - y(x_{i+1}) = u_i - y(x_i) + h[\phi(x_i, u_i; h) - \phi(x_i, y(x_i); h)] + \varepsilon_i - h\tau_i;$$

and if we use the Lipschitz condition (7.32) on the equation above, we have for $0 \le i \le n - 1$

$$|E_{i+1}| \le (1 + hK)|E_i| + \varepsilon + h\tau \tag{7.37}$$

where $E_j = u_j - y(x_j)$; $\tau = \max|\tau_j|$, $0 \le j \le n - 1$; and $\varepsilon = \max|\varepsilon_j|$, $0 \le j \le n - 1$. Clearly, we are more concerned (computationally) with $E_j = u_j - y(x_j)$ than we are with $e_j = y_j - y(x_j)$ since the values u_j are what is actually calculated while the values y_j are what would be calculated if we could run the method using exact arithmetic.

Now, following the same steps as in the proof of Theorem 7.2, we can employ (7.37) to prove that

$$|E_{i+1}| \le (1 + hK)^{i+1}|E_0| + (h\tau + \varepsilon)[1 + (1 + hK) + \cdots + (1 + hK)^{i+1}].$$

If we assume that the initial condition $y(x_0) = y_0$ is such that y_0 is exactly representable in the machine, then $E_0 = 0$; and we can deduce that for all j

$$|E_j| \le \frac{e^{K(x_j - x_0)} - 1}{K}\left(\tau + \frac{\varepsilon}{h}\right). \tag{7.38}$$

The inequality (7.38) shows the dimensions of the problems caused by roundoff error. In particular although we expect that $\tau \to 0$ as $h \to 0$, we cannot expect that $\varepsilon \to 0$ as $h \to 0$. Hence although τ decreases as h decreases, the term ε/h *increases* with decreasing h. This situation agrees with our intuition, for although we expect the method to yield smaller truncation errors as h decreases, we also take more steps as h decreases and hence expect a larger influence from rounding. For a given machine, a given program, and a given differential equation, there is clearly an optimal value of h for which the quantity $(\tau + (\varepsilon/h))$ is a minimum. Unfortunately, in all but the simplest cases, we cannot expect to do more than make an educated guess at what this optimal step size might be.

EXAMPLE 7.10. As a demonstration of the effects of a decreasing step size, consider the equation $y' = 2xy$, $y(0) = 1$, which has $y(x) = e^{x^2}$ as the solution. In this example, we use Euler's method with step size $h = 1/n$ and find u_n as an estimate to $e = y(1) = 2.71828$ (to six places). Some representative results are tabulated below and come from a program run in single precision on an IBM 370.

n	u_n
100	2.72706
200	2.72238
300	2.72061
500	2.71894
600	2.71845

n	u_n
700	2.71797
1100	2.71666
2200	2.71386
3300	2.71140
4400	2.70894

As the table shows (for this particular differential equation and this particular machine), the results improve as h decreases to about $h = 1/600$. As h continues to decrease, the answers become progressively worse until for $h = 1/4400$ the estimate at $x = 1$ is actually worse than that computed with $h = 1/100$.

PROBLEMS, SECTION 7.3

1. If Euler's method is applied to $y' = 2xy$, $y(0) = 1$ on the interval $[0, 1]$, then a bound for the error at $x = 1$ is given by Theorem 7.2. Calculate the value of this bound for $h = 1/100, 1/200, 1/300$; and next calculate the actual errors from the table in Example 7.10 and compare. [Note: Roundoff errors are not significant for these values of h; and τ can be found from (7.17).]
2. Run Euler's method on $[0, 4]$ for the problem $y' = x - y$, $y(0) = 3$ with $h = k \times 10^{-2}$, $k = 1, 2, 3, 4$. Repeat the analysis of Problem 1 for this example.
3. Suppose that we use a one-step method to solve $y' = f(x, y)$, $y(x_0) = y_0$, employing a fixed step size h. If $y_n(h)$ is the estimate to $y(x^*)$ where $x^* = x_0 + nh$ and if the method is a pth-order method, then we expect $e_n(h) \simeq Ch^p$ where $e_n(h) = y(x^*) - y_n(h)$. Use this expected equation to establish the approximation
$$\frac{y_{4n}(h/4) - y_{2n}(h/2)}{y_{2n}(h/2) - y_n(h)} \simeq 2^{-p}.$$
4. Run Euler's method on $y' = 2xy$, $y(0) = 1$; and estimate $y(x^*)$, $x^* = 1$. Use step size $h = 2^{-k}$, $1 \le k \le 12$. For $n = 2^k$, $1 \le k \le 10$, print the ratios $(y_{4n} - y_{2n})/(y_{2n} - y_n)$ defined in Problem 3.
5. Using Heun's method instead of Euler's method, repeat Problem 4.
6. Using the third-order method from Problem 5, Section 7.2.3, repeat Problem 4.
7. Recall that consistency of a one-step method means that $\phi(x, y(x); 0) = f(x, y(x))$. Show that the m-stage Runge-Kutta method in (7.13) is consistent if $A_1 + A_2 + \cdots + A_m = 1$. Also, verify that Heun's method and the modified Euler's method are consistent.
8. Using the Maclaurin's series expansion for e^x, show that $(1 + \alpha)^j \le e^{j\alpha}$ where $\alpha \ge 0$.
9. Suppose that $f(x, y)$ satisfies the Lipschitz condition $|f(x, u) - f(x, v)| \le L|u - v|$ for $-\infty < u, v < \infty$, and $a \le x \le b$. Show that $\phi(x, y; h)$ given by (7.13b) satisfies (7.32) for $m = 2, 3, 4$. [Hint: Use (7.13c) to show that $K_j(x, y; h)$ satisfies a Lipschitz condition.]

7.4 *n*th-ORDER DIFFERENTIAL EQUATIONS AND SYSTEMS OF DIFFERENTIAL EQUATIONS

So far, we have considered only first-order differential equations, $y' = f(x, y)$, $y(x_0) = y_0$. In fact, most differential equations encountered in scientific and engineering problems are equations that are second order or higher. To be specific, the general nth-order initial-value problem takes the form

$$y^{(n)} = f(x, y, y', \ldots, y^{(n-1)}), \qquad y(x_0) = y_0, y'(x_0) = y_0', \ldots, y^{(n-1)}(x_0) = y_0^{(n-1)}.$$

The nth-order equation above can be converted into an equivalent system of n first-order equations by setting $u_1(x) = y(x)$, $u_2(x) = y'(x), \ldots, u_n(x) = y^{(n-1)}(x)$. When we do this, we obtain the system

$$\begin{aligned} u_1'(x) &= u_2(x) \\ u_2'(x) &= u_3(x) \\ &\vdots \\ u_n'(x) &= f(x, u_1(x), u_2(x), \ldots, u_n(x)). \end{aligned}$$

We next set

$$\mathbf{u}(x) = \begin{bmatrix} u_1(x) \\ u_2(x) \\ \vdots \\ u_n(x) \end{bmatrix}, \qquad \mathbf{F}(x, \mathbf{u}(x)) = \begin{bmatrix} u_2(x) \\ u_3(x) \\ \vdots \\ f(x, u_1(x), u_2(x), \ldots, u_n(x)) \end{bmatrix}$$

to obtain an equivalent first-order system of equations in vector form:

$$\mathbf{u}'(x) = \mathbf{F}(x, \mathbf{u}(x)), \ \mathbf{u}(x_0) = \mathbf{u}_0$$

where the vector $\mathbf{u}_0$ of initial-conditions is given by $\mathbf{u}_0 = [y_0, y_0', \ldots, y_0^{(n-1)}]^T$. In this reformulation, $\mathbf{u}'(x) = \mathbf{F}(x, \mathbf{u}(x))$, we are using differentiation and integration componentwise; so for example

$$\int_{x_0}^{x} \mathbf{F}(t, \mathbf{u}(t))\, dt = \begin{bmatrix} \displaystyle\int_{x_0}^{x} u_2(t)\, dt \\ \vdots \\ \displaystyle\int_{x_0}^{x} u_{n-1}(t)\, dt \\ \displaystyle\int_{x_0}^{x} f(t, u_1(t), \ldots, u_n(t))\, dt \end{bmatrix}.$$

To solve the problem $\mathbf{u}'(x) = \mathbf{F}(x, \mathbf{u}(x))$, we could apply any of the preceding numerical schemes with the only difference being that we calculate vector quantities instead of scalar quantities. For example, Euler's method becomes

$$\mathbf{u}_{i+1} = \mathbf{u}_i + h\mathbf{F}(x_i, \mathbf{u}_i), \qquad i = 0, 1, 2, \ldots.$$

In general, a one-step method $y_{i+1} = y_i + h\phi(x_i, y_i; h)$ can be converted to vector form:

$$\mathbf{u}_{i+1} = \mathbf{u}_i + h\boldsymbol{\phi}(x_i, \mathbf{u}_i; h).$$

A slight complication is that methods that are pth order for scalar differential equations are not necessarily pth order in their vector form [see Lambert (1973), p. 225]. However, Taylor's series methods and the specific Runge-Kutta methods we have discussed retain their original order when formulated for systems of equations.

Coding the vector form of a Runge-Kutta method is easy when a modular structure is used. That is, to solve $\mathbf{u}' = \mathbf{F}(x, \mathbf{u})$, a subroutine should be written that accepts x and the vector $\mathbf{u}$ and returns the vector $\mathbf{u}'$ where $\mathbf{u}' = \mathbf{F}(x, \mathbf{u})$. (The alternative, coding the vector form of a Runge-Kutta method componentwise, will lead to a very complicated program.) As a specific illustration, suppose we want to use the vector form of the method

$$y_{i+1} = y_i + \frac{h}{2}[f(x_i, y_i) + f(x_{i+1}, y_i + hf(x_i, y_i))] \tag{7.39}$$

to solve $y''' = x^2y'' + y^2y'$. In vector form, the differential equation is

$$u_1' = u_2$$

$$u_2' = u_3$$

$$u'_3 = x^2u_3 + u_1^2u_2.$$

The subroutine UPRIME, written for the system $\mathbf{u}' = \mathbf{F}(x, \mathbf{u})$ above, will evaluate $\mathbf{F}(x, \mathbf{u})$ for any x and $\mathbf{u}$:

```
SUBROUTINE UPRIME(X,U,UPRIM)
DIMENSION U(3),UPRIM(3)
UPRIM(1)=U(2)
UPRIM(2)=U(3)
UPRIM(3)=U(3)*X**2+U(2)*U(1)**2
RETURN
END
```

Without our reproducing an entire program that implements the method (7.39), the basic lines of code that are required to execute a step from x to $x + h$ using (7.39) are

```
      REAL K1(3),K2(3),UOLD(3),UNEW(3),UTEMP(3)
                          .
                          .
                          .
      CALL UPRIME(X,UOLD,K1)
      X=X+H
        DO 1 I=1,3
        UTEMP(I)=UOLD(I)+H*K1(I)
1       CONTINUE
      CALL UPRIME(X,UTEMP,K2)
        DO 2 I=1,3
        UNEW(I)=UOLD(I)+H*(K1(I)+K2(I))/2.
2       CONTINUE
                          .
                          .
                          .
```

Variable-step methods, such as the Runge-Kutta-Fehlberg method, can be put in vector form very easily; and we ask the reader to modify the program listed in Fig. 7.9 so that it can be used to solve a first-order system $\mathbf{u}' = \mathbf{F}(x, \mathbf{u})$. Note that a change must be made in subroutine GAMCAL since EST is now a vector quantity (see 7.29 d). In particular, if $\mathbf{u}$ is a k-dimensional vector, then a step of length h from $(x_i, \mathbf{u}_i)$ can be regarded as successful only if each component of EST satisfies (for a relative error test)

$$|\text{est}(j)| \le \frac{\varepsilon|\mathbf{u}(j)|}{b-a}, \qquad j = 1, 2, \cdots, k.$$

Thus γ must be defined by

$$\gamma = \min_{1 \le j \le k} \{r_j\}, \qquad r_j = \left[\frac{\varepsilon|\mathbf{u}(j)|}{(b-a)|\text{est}(j)|}\right]^{\frac{1}{4}}$$

This is a fairly significant modification since it means that the step size is limited by the components of $\mathbf{u}$ that are most difficult to track. In a later section, we consider "stiff" equations, which present a substantial problem for a differential-equation solver. As we will see, these equations have the unfortunate property that the component that is most difficult to track is also the one that is least significant to the solution. In other words, that part of the solution that least interests us is exactly the one that imposes a severe restriction on the size of the steps that we can take, and causes the method to run inordinately slowly (see Problems 5 through 13 for an example).

EXAMPLE 7.11. In this example, the Runge-Kutta-Fehlberg program in Fig. 7.9 was modified to solve first-order systems of differential equations. A run was made for van der Pol's equation:

$$y'' = (1 - y^2)y' - y, \qquad y(0) = .2, \; y'(0) = .2;$$

and the solution was tabulated in increments of $\Delta x = .1$ for $0 \le x \le 25$. The parameters for DESOLV were TOL $= (1 \times 10^{-10})/25$, HMIN $= 1 \times 10^{-4}$, HMAX $= 1 \times 10^{-1}$, and (initially) HSTART $= 1 \times 10^{-2}$. Some representative values returned by DESOLV are listed below.

x	y	y'
1.0	0.3510211E+00	0.5153074E−01
2.0	0.1993819E+00	−0.4014032E+00
5.0	−0.8316866E+00	0.9060111E+00
15.0	0.9067951E+00	−0.1242702E+01
25.0	−0.9004823E+00	0.1248455E+01

For reference, here are some values of HSTART: at $x = 5$, HSTART = 0.6429999E−02; and at $x = 25$, HSTART = 0.4680028E−02.

PROBLEMS, SECTION 7.4

1. Convert the following equations to the form $\mathbf{u}' = \mathbf{F}(x, \mathbf{u})$, $\mathbf{u}(0) = \mathbf{u}_0$.
   a) $y'' = -\sin(y)$, $y(0) = \pi/2$, $y'(0) = 0$
   b) $y'' - y = x$, $y(0) = 1$, $y'(0) = 1$
   c) $y'' + xy' + x^2y = 0$, $y(0) = 1$, $y'(0) = 0$
2. Find the exact solution of $y''' + 5y'' + 6y' = 2x$, $y(0) = 3$, $y'(0) = -3$, $y''(0) = 13$. Convert the equation to the form $\mathbf{u}' = \mathbf{F}(x, \mathbf{u})$, $\mathbf{u}(0) = \mathbf{u}_0$.
3. Code Euler's method so that it can be employed on a system $\mathbf{u}' = \mathbf{F}(x, \mathbf{u})$ where $\mathbf{u}$ is n-dimensional, $n \le 10$. Test your program on the equation in Problem 2; use a step size $h = 2^{-k}$, $k = 2, 3, \ldots, 8$. Print the estimates and the errors in increments of .25 for $0 \le x \le 4$ and verify that the errors are (approximately) proportional to h.
4. Code the third-order Runge-Kutta method in Problem 5, Section 7.2.3, for systems $\mathbf{u}' = \mathbf{F}(x, \mathbf{u})$. Test your program as in Problem 3 and verify that the errors are nearly proportional to h^3.
5. Find the exact solution to $\mathbf{u}' = A\mathbf{u}$, $\mathbf{u}(0) = \mathbf{u}_0$ where
$$A = \begin{bmatrix} -30 & 28 \\ 0 & -2 \end{bmatrix} \qquad \mathbf{u}_0 = \begin{bmatrix} 2 \\ 1 \end{bmatrix}.$$
   Verify that $\mathbf{u}(x) \to \boldsymbol{\theta}$ as x grows.
6. Repeat Problems 3 and 4 for the equation in Problem 5. Note the size of the errors for $h = \frac{1}{4}$ and $h = \frac{1}{8}$; see Problems 7 through 13.
7. Consider the equation $\mathbf{u}' = A\mathbf{u}$, $\mathbf{u}(0) = \mathbf{u}_0$ where A is an $(n \times n)$ constant matrix. Show that Euler's method applied to $\mathbf{u}' = A\mathbf{u}$ will generate the vector sequence $\mathbf{u}_{i+1} = p(hA)\mathbf{u}_i$, $i = 0, 1, \ldots$ where $p(hA) = I + hA$.
8. Use (7.10) and (7.12) to show that any two-stage second-order Runge-Kutta method applied to $\mathbf{u}' = A\mathbf{u}$ will generate the vector sequence $\mathbf{u}_{i+1} = p(hA)\mathbf{u}_i$, $i = 0, 1, \ldots$ where
$$p(hA) = I + hA + \frac{h^2}{2!}A^2.$$
9. Show that the third-order Runge-Kutta method in Problem 4 generates the vector sequence $\mathbf{u}_{i+1} = p(hA)\mathbf{u}_i$ when applied to $\mathbf{u}' = A\mathbf{u}$, where
$$p(hA) = I + hA + \frac{h^2}{2!}A^2 + \frac{h^3}{3!}A^3.$$
10. In Problems 7 through 9, if the matrix $p(hA)$ has even one eigenvalue that is larger than one in magnitude, then we can probably expect some of the components of $\mathbf{u}_k$ to be growing as k grows. Justify this statement by expressing $\mathbf{u}_0$ as a linear combination of the eigenvectors of $p(hA)$; assume that $p(hA)$ has n linearly independent eigenvectors.
11. Suppose that B is an $(n \times n)$ matrix and $p(x) = a_0 + a_1x + \cdots + a_rx^r$. The matrix polynomial $p(B)$ is defined by $p(B) = a_0I + a_1B + \cdots + a_rB^r$ where I is the $(n \times n)$ identity. Prove that if $B\mathbf{x} = \lambda\mathbf{x}$, $\mathbf{x} \ne \boldsymbol{\theta}$, then $p(B)\mathbf{x} = p(\lambda)\mathbf{x}$ [that is, if λ is an eigenvalue of B, then $p(\lambda)$ is an eigenvalue of $p(B)$].

12. What are the eigenvalues of the matrix A in Problem 5? What are the eigenvalues of hA? For each matrix polynomial $p(hA)$ in Problems 7 through 9 and for A as in Problem 5, answer this question: For what range on h will all the eigenvalues of $p(hA)$ be less than one in magnitude? [Hint: Use Problem 11 and Problems 10 and 11, Section 7.2.2.]
13. The equation in Problem 5 is an example of a mildly stiff problem; the solution $\mathbf{u}(x)$ contains terms $e^{\lambda_1 x}$ and $e^{\lambda_2 x}$ where λ_1 and λ_2 are the eigenvalues of A. Since $\lambda_1 < \lambda_2 < 0$, $\mathbf{u}(x) \to \boldsymbol{\theta}$ as $x \to \infty$; but the numerical solution will tend to $\boldsymbol{\theta}$ only if the eigenvalues of $p(hA)$ are less than one in magnitude. Thus, λ_2 is more significant with respect to the actual solution, and λ_1 determines the range of h for which the numerical solution tends to $\boldsymbol{\theta}$. As the ratio $|\lambda_1|/|\lambda_2|$ grows, the equation becomes "stiffer." Change the (1, 1) entry of A in Problem 5 to -120 and rework Problem 6. Before running the problem, determine which values of h should give reasonable answers.
14. Modify the program in Fig. 7.9 so that the program is suitable for systems $\mathbf{u}' = \mathbf{F}(x, \mathbf{u})$. Test the program on the equations in Problems 2 and 5 and on the equation in Example 7.11. For the first two equations, print the estimates, the actual relative errors, and HSTART, in increments of .25 for $0 \le x \le 5$; use accuracy requests of 10^{-6} and 10^{-10}.

7.5 LINEAR MULTISTEP METHODS

Linear multistep methods constitute a powerful class of numerical procedures for solving $y' = f(x, y)$. The multistep methods employ a philosophy that is somewhat different from that of the one-step methods. In particular, a multistep method uses "back values" $y_n, y_{n+1}, \ldots, y_{n+k-1}$ to determine y_{n+k} where $y_{n+j} \simeq y(x_{n+j})$, $0 \le j \le k$. In other words, a multistep method uses the past history (or trends) of the numerical solution at equally spaced points $x = x_n, x_{n+1}, \ldots, x_{n+k-1}$ in order to estimate the solution at $x = x_{n+k}$. By contrast, the one-step methods we have considered estimate the solution at $x = x_{n+k}$ solely in terms of the numerical solution at $x = x_{n+k-1}$ (as in Taylor's series methods) or by sampling the flow field between $x = x_{n+k-1}$ and $x = x_{n+k}$ (as in Runge-Kutta methods). The approximation philosophy of multistep methods (emphasizing past history) is clearly exhibited by the general form of a multistep method for $y' = f(x, y)$:

$$y_{n+k} = -\sum_{j=0}^{k-1} \alpha_j y_{n+j} + h \sum_{j=0}^{k} \beta_j f(x_{n+j}, y_{n+j}). \tag{7.40}$$

More precisely, the method in (7.40) is called a *linear k-step method*—linear because y_{n+k} is a linear combination of y_{n+j} and $f(x_{n+j}, y_{n+j})$; and k-step because k back values $y_n, y_{n+1}, \ldots, y_{n+k-1}$ are required to calculate y_{n+k}.

Linear multistep methods arise quite naturally in various schemes to approximate the solution of $y' = f(x, y)$. For example, one approximation to $y'(x)$ is the centered-difference formula

$$y'(x) \simeq \frac{y(x+h) - y(x-h)}{2h}.$$

Given equally spaced points x_n, x_{n+1}, x_{n+2}, this approximation can be written as $y'_{n+1} \simeq (y_{n+2} - y_n)/2h$. If we use this derivative approximation in $y'(x) = f(x, y(x))$, we obtain a finite-difference model for $y' = f(x, y)$:

$$\frac{y_{n+2} - y_n}{2h} = f(x_{n+1}, y_{n+1}).$$

Thus the model above leads to the iteration

$$y_{n+2} = y_n + 2hf(x_{n+1}, y_{n+1}), \tag{7.41}$$

and we note that this method has the form of (7.40); y_{n+2} is estimated in terms of y_n and y_{n+1} and (7.41) is a linear two-step method.

Similarly, *quadrature methods* also lead to iterations of the form of (7.40). For example, if we observe that

$$y(x_{n+2}) - y(x_n) = \int_{x_n}^{x_{n+2}} y'(x)dx,$$

then we can use Simpson's rule to approximate the integral

$$y(x_{n+2}) - y(x_n) \simeq \frac{h}{3}[y'(x_n) + 4y'(x_{n+1}) + y'(x_{n+2})].$$

Since $y'(x) = f(x, y(x))$, the approximation above is the same as

$$y(x_{n+2}) - y(x_n) \simeq \frac{h}{3}[f(x_n, y(x_n)) + 4f(x_{n+1}, y(x_{n+1})) + f(x_{n+2}, y(x_{n+2}))];$$

and if $y_{n+j} \simeq y(x_{n+j})$ for $j = 0, 1, 2$, then we are led to the iteration

$$y_{n+2} - y_n = \frac{h}{3}[f(x_n, y_n) + 4f(x_{n+1}, y_{n+1}) + f(x_{n+2}, y_{n+2})]. \tag{7.42}$$

Clearly, (7.42) is also a linear two-step method.

The most attractive feature of a multistep method is almost apparent from (7.40); we can save the past evaluations $f(x_{n+j}, y_{n+j})$ and thus need only make several evaluations of $f(x, y)$ to obtain y_{n+k}. To elaborate on this and several other aspects of the linear k-step method, we rewrite (7.40) in the form

$$\sum_{j=0}^{k} \alpha_j y_{n+j} = h \sum_{j=0}^{k} \beta_j f_{n+j} \tag{7.43}$$

where $f_{n+j} = f(x_{n+j}, y_{n+j})$. The notation for a linear k-step method in (7.43) is standard in the literature, and it is also conventional to insist that $\alpha_k = 1$ and that $|\alpha_0| + |\beta_0| > 0$. If $\beta_k = 0$, then the method (7.43) is called *explicit*; if $\beta_k \neq 0$, then the method is *implicit*. [Note: If $\beta_k \neq 0$, then y_{n+k} appears on both sides of the equation (7.43); on the right-hand side, y_{n+k} is in the term $h\beta_k f(x_{n+k}, y_{n+k})$.]

Now, if the method is explicit and if $f_n, f_{n+1}, \ldots, f_{n+k-2}$ have been saved, then a single calculation of f_{n+k-1} will produce y_{n+k}. Thus, only one evaluation per step is required to execute an explicit multistep method. In practice, however, explicit and implicit methods are usually used in pairs since (as we demonstrate later) the difference $y_{n+k} - \tilde{y}_{n+k}$ can be used to estimate the local (or per-step) errors. Here, y_{n+k} comes from an implicit method and $\tilde{y}_{n+k}$ comes from an explicit method. Two such methods used in conjunction form what is called a "predictor-corrector" pair; we will discuss some specific examples of predictor-corrector methods in the next section.

Several other computational aspects of (7.43) deserve some mention. First of all, in order to start the iteration (7.43) when we apply it to $y' = f(x, y)$, $y(x_0) = y_0$, we need to supply back values $y_0, y_1, y_2, \ldots, y_{k-1}$. These initial values can be obtained from a one-step method (these are "self-starting" methods) or by using a variable step/variable-order family of multistep methods (see Section 7.5.1). Next, we mentioned above that explicit and implicit methods are usually used together as predictor-corrector pairs in order to estimate and control the errors. This use poses two problems.

1. How do we calculate y_{n+k} when we use an implicit method?
2. If the step size h is changed in order to reduce the per-step error, how do we obtain the new equally spaced back values that are necessary to execute (7.43)?

For example, if we are using (7.42), then we need a way to calculate y_{n+2} since the method is implicit. Next, suppose we are using (7.42) and decide somehow that the step size h is too large to calculate y_{n+2} safely. If we reduce h to (say) $h/2$, then we will need back values at $x = x_{n+1}$ and $x = x_{n+1} + h/2$; but we do not have a value for $f(x, y)$ at $x = x_{n+1} + h/2$; and in particular, we do not have an estimate to $y(x)$ at $x = x_{n+1} + h/2$ to use to evaluate $f(x, y)$. We will treat these two problems in later sections.

Another computational problem is associated with multistep methods. For example, both the methods (7.41) and (7.42) have undesirable characteristics and cannot safely be used to solve $y' = f(x, y)$. This situation is somewhat paradoxical since (7.41) is derived from an excellent approximation to $y'(x_{n+1})$ and (7.42) is derived from an equally excellent approximation to $\int_{x_n}^{x_{n+2}} y'(x)dx$. The resolution of this apparent paradox must await a deeper theoretical analysis of linear multistep methods (see Section 7.5.4).

7.5.1. Adams Methods and Their Implementation as Predictor-Corrector Methods

Adams methods are the most popular multistep methods and are used in many of the best codes that implement variable-step/variable-order schemes for solving initial-value problems. Adams methods are quadrature methods that are based on interpolatory approximations to the integral of the form

$$\int_{x_{n+k-1}}^{x_{n+k}} g(x)dx \simeq h[A_0 g(x_n) + A_1 g(x_{n+1}) + \cdots + A_k g(x_{n+k})].$$

Thus an explicit Adams method (usually called an Adams-Bashforth method) has the form

$$y_{n+k} - y_{n+k-1} = h[A_0 f_n + A_1 f_{n+1} + \cdots + A_{k-1} f_{n+k-1}];$$

an implicit Adams method (called an Adams-Moulton method) has the form

$$y_{n+k} - y_{n+k-1} = h[A_0 f_n + A_1 f_{n+1} + \cdots + A_k f_{n+k}].$$

While the Adams methods are just a special case of the general multistep methods, the analyses of most multistep methods proceed exactly as they do for the Adams methods. Therefore, for simplicity, we describe the computational and theoretical aspects of the Adams methods first and do not go into so much detail in later sections that treat general multistep methods.

The Adams methods are normally employed in predictor-corrector pairs, using both an explicit and an implicit method to calculate y_{n+k}. Several explicit and implicit Adams methods are listed below, along with their associated truncation errors, $h\tau$, which we will discuss later. The first four Adams-Bashforth methods are

$$y_{n+1} - y_n = hf_n, \qquad h\tau = \frac{h^2}{2} y'' \tag{7.44a}$$

$$y_{n+2} - y_{n+1} = \frac{h}{2}[3f_{n+1} - f_n], \qquad h\tau = \frac{5h^3}{12} y''' \tag{7.44b}$$

$$y_{n+3} - y_{n+2} = \frac{h}{12}[23f_{n+2} - 16f_{n+1} + 5f_n], \qquad h\tau = \frac{3h^4}{8} y^{(4)} \tag{7.44c}$$

$$y_{n+4} - y_{n+3} = \frac{h}{24}[55f_{n+3} - 59f_{n+2} + 37f_{n+1} - 9f_n], \qquad h\tau = \frac{251h^5}{720} y^{(5)}. \tag{7,44d}$$

The first four Adams-Moulton methods are

$$y_{n+1} - y_n = hf_{n+1}, \qquad h\tau = -\frac{h^2}{2} y'' \tag{7.45a}$$

$$y_{n+2} - y_{n+1} = \frac{h}{2}[f_{n+2} + f_{n+1}], \qquad h\tau = -\frac{h^3}{12} y''' \tag{7.45b}$$

$$y_{n+3} - y_{n+2} = \frac{h}{12}[5f_{n+3} + 8f_{n+2} - f_{n+1}], \qquad h\tau = -\frac{h^4}{24} y^{(4)} \tag{7.45c}$$

$$y_{n+4} - y_{n+3} = \frac{h}{24}[9f_{n+4} + 19f_{n+3} - 5f_{n+2} + f_{n+1}], \qquad h\tau = -\frac{19h^5}{720} y^{(5)}. \tag{7.45d}$$

In listing the implicit methods (7.45 b, c, and d), we have not strictly followed the notational convention of (7.43) for multistep methods; for example, (7.45b) should have been recorded as $y_{n+1} - y_n = h[f_{n+1} + f_n]/2$. This deviation in notation is in accord with standard practice, however, since we will frequently want to pair corresponding formulas in (7.44) and (7.45). Note also that (7.44a) is already familiar as Euler's method. The method in (7.45a) is closely related to Euler's method and is called the backward Euler method; (7.45b) is called the trapezoidal method.

One reason that Adams methods are normally used in pairs is that the pair can be used to monitor and control the local errors. An example should serve to clarify this point; and we use (7.44c), the three-step Adams-Bashforth method, and (7.45c), the two-step Adams-Moulton method, as an illustration. However, even before considering an error monitor, we must decide how we can practically execute a step of (7.45c) since this method is implicit. That is, to use (7.45c), we must solve a nonlinear equation; and to exhibit this problem precisely, we rewrite (7.45c) as

$$y_{n+3} = \frac{5h}{12} f(x_{n+3}, y_{n+3}) + y_{n+2} + \frac{h}{12}[8f_{n+2} - f_{n+1}]. \tag{7.46}$$

Given back values y_{n+1} and y_{n+2}, (7.46) is a nonlinear equation where the unknown is y_{n+3}. In the form given above, it is natural to consider fixed-point iteration to solve (7.46) for y_{n+3}:

$$y_{n+3}^{(i+1)} = \frac{5h}{12} f(x_{n+3}, y_{n+3}^{(i)}) + y_{n+2} + \frac{h}{12}[8f_{n+2} - f_{n+1}], \qquad i = 0, 1, \ldots .$$

From the theory of fixed-point methods given in Section 4.3.1, we know that the iteration above will converge if the derivative of the iterating function (evaluated at the fixed point) is less than one in magnitude. In the context of (7.46), this condition on the derivative is equivalent to

$$\frac{5h}{12} \left|f_y(x_{n+3}, y_{n+3})\right| < 1$$

or $|5\bar{h}/12| < 1$ where $\bar{h} = h\lambda$ and $\lambda = f_y(x_{n+3}, y_{n+3})$. Normally, the step size h will be selected to be small enough (by the local error monitor, which is controlling the accuracy of the numerical solution) that the condition $|5\bar{h}/12| < 1$ is automatically satisfied (one exception occurs for "stiff" equations). Thus, iteration will usually be a successful method for determining y_{n+3} in (7.46).

For iteration to be successful, we also need a good initial guess $y_{n+3}^{(0)}$ to the fixed point y_{n+3}; as we explain later, one appropriate initial guess is the estimate provided by (7.44c), which we label $\tilde{y}_{n+3}$:

$$\tilde{y}_{n+3} = y_{n+2} + \frac{h}{12}\left[23f_{n+2} - 16f_{n+1} + 5f_n\right].$$

In addition, a potential advantage of a multistep method (as contrasted with a Runge-Kutta method) is that only a few evaluations of $f(x, y)$ are necessary to execute a step. Thus, we do not want to iterate very many times with (7.46) or we will lose this advantage. Frequently, only one iteration is made (although the rate of convergence can be monitored and more iterations performed if necessary). Assuming one iteration in (7.46), we are led to a method, based on (7.44c) and (7.45c), that estimates y_{n+3} in terms of the back values y_n, y_{n+1}, and y_{n+2}:

$$\tilde{y}_{n+3} = y_{n+2} + \frac{h}{12}\left[23f_{n+2} - 16f_{n+1} + 5f_n\right] \tag{7.47a}$$

$$y_{n+3} = y_{n+2} + \frac{h}{12}\left[5f(x_{n+3}, \tilde{y}_{n+3}) + 8f_{n+2} - f_{n+1}\right]. \tag{7.47b}$$

For obvious reasons, (7.47a) is called the predictor; and (7.47b), the corrector. The pair in (7.47) constitutes a predictor-corrector method that estimates y_{n+3} in terms of y_n, y_{n+1}, and y_{n+2} and uses two evaluations per step.

Generally, a predictor-corrector method based on the Adams formulas obtains a predicted value $\tilde{y}_{n+k}$ from one of (7.44) and then uses one iteration of a corresponding corrector formula in (7.45) to estimate y_{n+k}. There are, however, variations of this procedure; for example, although we have listed the formulas in (7.44) and (7.45) so that it appears natural to pair a k-step predictor with a $(k - 1)$-step corrector, it is also possible and very desirable to pair a k-step predictor witha k-step corrector—we comment on this pairing later. For multistep methods in general, the form of a predictor-corrector method follows the discussion above. That is, an explicit method is used (as a predictor) to produce $\tilde{y}_{n+k}$, and then an implicit method (a corrector) is iterated once or several times to produce the final estimate y_{n+k}.

Having seen the general form of a predictor-corrector method as illustrated in (7.47), we are ready to discuss an error-control mechanism for Adams methods. As with one-step methods the construction of an error monitor is based on knowing the theoretical form of the truncation error. As with a one-step method, the truncation error is defined (loosely) to be the discrepancy between the actual value of the solution $y(x_{n+k})$ and the value y_{n+k} that would be produced by the method if the back values $y_n, y_{n+1}, \ldots, y_{n+k-1}$ were exactly correct. (We will give the formal definitions for order, truncation error, etc. in the next section.) Given the theoretical form of the truncation error, it is easy to see how we can monitor and control the local errors.

We begin our discussion of truncation errors by considering an example, namely the predictor-corrector pair (7.47). First, consider the three-step predictor (7.47a):

$$\tilde{y}_{n+3} - y_{n+2} = \frac{h}{12}[23f_{n+2} - 16f_{n+1} + 5f_n].$$

The three-step method above is based on the interpolatory quadrature formula

$$\int_{x_{n+2}}^{x_{n+3}} g(x)dx \simeq \frac{h}{12}[23g(x_{n+2}) - 16g(x_{n+1}) + 5g(x_n)], \tag{7.48a}$$

and it is easy to show (see Problem 13) that the error term in (7.48a) is given by

$$\int_{x_{n+2}}^{x_{n+3}} g(x)dx = \frac{h}{12}[23g(x_{n+2}) - 16g(x_{n+1}) + 5g(x_n)] + E \tag{7.48b}$$

where $E = 3h^4 g'''(\eta)/8$ and where η is some point in $[x_{n+2}, x_{n+3}]$. If $u(x)$ is any solution to $y' = f(x, y)$ so that $u'(x) = f(x, u(x))$, then we can set $g(x) = u'(x)$ in (7.48b) and obtain

$$\begin{aligned} u(x_{n+3}) - u(x_{n+2}) = \frac{h}{12}[&23f(x_{n+2}, u(x_{n+2})) - 16f(x_{n+1}, u(x_{n+1})) \\ &+ 5f(x_n, u(x_n))] + \frac{3h^4}{8}u^{(4)}(\eta). \end{aligned} \tag{7.49}$$

Note that the error term in (7.49) is the same as the term $h\tau$ in (7.44c) as is true for all the methods in (7.44) and (7.45); the term $h\tau$, which we will identify later as the truncation error, is the same as the quadrature error that results from integrating $y'(x)$ on $[x_{n+k-1}, x_{n+k}]$. Moreover, we note that if the back values

y_n, y_{n+1}, y_{n+2} were exactly equal to $u(x_n)$, $u(x_{n+1})$, $u(x_{n+2})$, then (7.49) is the same as saying

$$u(x_{n+3}) - \tilde{y}_{n+3} = \frac{3h^4}{8} u^{(4)}(\eta). \tag{7.50}$$

In the one-step methods, error control was based on monitoring the local error $u(x_{n+k}) - y_{n+k}$ where $u(x)$ satisfies $y' = f(x, y)$ with $u(x_{n+k-1}) = y_{n+k-1}$. The same idea of controlling local errors in order to control the global errors is valid also for predictor-corrector methods. For the three-step predictor we are considering now, an expression for the local error $u(x_{n+3}) - y_{n+3}$ would be given by (7.50) if *all* the back values agreed with $u(x)$. While the definition of the local error assumes that $y_{n+2} = u(x_{n+2})$, we cannot expect that $y_n = u(x_n)$ and $y_{n+1} = u(x_{n+1})$ as well; so we cannot use (7.50) directly. However, if we assume that $u(x_n) - y_n = O(h^4)$ and $u(x_{n+1}) - y_{n+1} = O(h^4)$, then we can show that the leading term of $u(x_{n+3}) - \tilde{y}_{n+3}$ is given by (7.50). Moreover, we will show that this assumption on the back values is reasonable.

In particular, suppose $u(x)$ satisfies $y' = f(x, y)$ with $u(x_{n+2}) = y_{n+2}$; and suppose also that $u(x_n) - y_n = O(h^4)$ and $u(x_{n+1}) - y_{n+1} = O(h^4)$. The three-step predictor is

$$\tilde{y}_{n+3} - y_{n+2} = \frac{h}{12}[23f(x_{n+2}, y_{n+2}) - 16f(x_{n+1}, y_{n+1}) + 5f(x_n, y_n)]; \tag{7.51}$$

and if we subtract (7.51) from (7.49) and use the fact that $y_{n+2} = u(x_{n+2})$, we obtain

$$\begin{aligned} u(x_{n+3}) - \tilde{y}_{n+3} = \frac{h}{12}[&-16[f(x_{n+1}, u(x_{n+1})) - f(x_{n+1}, y_{n+1})] \\ &+ 5[f(x_n, u(x_n)) - f(x_n, y_n)]] + \frac{3h^4}{8} u^{(4)}(\eta). \end{aligned}$$

Using the mean-value theorem on the right-hand side of the equation above, we have

$$\begin{aligned} u(x_{n+3}) - \tilde{y}_{n+3} = {}& \frac{h}{12}[-16\lambda_1(u(x_{n+1}) - y_{n+1}) + 5\lambda_0(u(x_n) - y_n)] \\ &+ \frac{3h^4}{8} u^{(4)}(\eta) \end{aligned}$$

where $\lambda_1 = f_y(x_{n+1}, \theta_1)$ and $\lambda_0 = f_y(x_n, \theta_0)$. Under the assumption that $u(x_{n+1}) - y_{n+1} = O(h^4)$ and $u(x_n) - y_n = O(h^4)$, we thus have an expression for the local error of the predictor:

$$u(x_{n+3}) - \tilde{y}_{n+3} = \frac{3h^4}{8} u^{(4)}(\eta) + O(h^5). \tag{7.52}$$

Observe that the leading term for the local error of the predictor (7.44c) is the same as the term $h\tau$ in (7.44c); this is true of the local errors for the other three predictors in (7.44) as well, as an analogous argument will show.

Since we are considering the predictor-corrector pair (7.47), we are really interested in controlling the local error of the corrector, namely $u(x_{n+3}) - y_{n+3}$. That is, we are going to correct $\tilde{y}_{n+3}$ to y_{n+3} and then use y_{n+3} as our next estimate; so we are not directly interested in controlling $u(x_{n+3}) - \tilde{y}_{n+3}$. To analyze $u(x_{n+3}) - y_{n+3}$, we note first, as in (7.48) and (7.49), that it is easy to see that when we insert $u(x)$ into (7.45c) we have

$$\begin{aligned} u(x_{n+3}) - u(x_{n+2}) = \frac{h}{12}[5f(x_{n+3}, u(x_{n+3})) + 8f(x_{n+2}, u(x_{n+2})) \\ - f(x_{n+1}, u(x_{n+1}))] - \frac{h^4}{24}u^{(4)}(\gamma) \end{aligned} \tag{7.53a}$$

where γ is in $[x_{n+2}, x_{n+3}]$. [Note: In (7.53a), $u(x)$ has not changed since the derivation leading to (7.52). The function $u(x)$ still satisfies $y' = f(x, y)$, $u(x_{n+2}) = y_{n+2}$; and we still assume that $u(x_{n+1}) - y_{n+1} = 0(h^4)$.] For reference, the two-step corrector (7.47b) is

$$y_{n+3} - y_{n+2} = \frac{h}{12}[5f(x_{n+3}, \tilde{y}_{n+3}) + 8f(x_{n+2}, y_{n+2}) - f(x_{n+1}, y_{n+1})]. \tag{7.53b}$$

Subtracting (7.53b) from (7.53a) and using the mean-value theorem and that $y_{n+2} = u(x_{n+2})$, we obtain

$$\begin{aligned} u(x_{n+3}) - y_{n+3} = \frac{h}{12}[5\mu_1(u(x_{n+3}) - \tilde{y}_{n+3}) - \mu_0(u(x_{n+1}) - y_{n+1})] \\ - \frac{h^4}{24}u^{(4)}(\gamma) \end{aligned} \tag{7.54}$$

where $\mu_1 = f_y(x_{n+3}, \xi_1)$ and $\mu_0 = f_y(x_{n+1}, \xi_0)$. By assumption, $u(x_{n+1}) - y_{n+1} = O(h^4)$ and by (7.52) $u(x_{n+3}) - \tilde{y}_{n+3} = O(h^4)$. Therefore, the local error for one step of the predictor-corrector method (7.47) is given by

$$u(x_{n+3}) - y_{n+3} = -\frac{h^4}{24}u^{(4)}(\gamma) + O(h^5). \tag{7.55}$$

We can use (7.52) and (7.55) to estimate the local error $u(x_{n+3}) - y_{n+3}$. Specifically if we subtract (7.52) from (7.55), we obtain

$$\tilde{y}_{n+3} - y_{n+3} = -\frac{3h^4}{8}u^{(4)}(\eta) - \frac{h^4}{24}u^{(4)}(\gamma) + O(h^5).$$

If we now neglect terms of order h^5 and if we assume that $u(x)$ is smooth enough relative to the size of h that $u^{(4)}(\eta) \simeq u^{(4)}(\gamma)$, then we can make the approximation that

$$\tilde{y}_{n+3} - y_{n+3} \simeq -\frac{10h^4}{24} u^{(4)}.$$

From (7.55), this approximation leads to an estimate for the local error of the predictor-corrector method:

$$u(x_{n+3}) - y_{n+3} \simeq \frac{1}{10}[\tilde{y}_{n+3} - y_{n+3}]. \tag{7.56}$$

As was shown in Section 7.2.4, whenever we can estimate the local errors, we have the possibility of controlling the global errors (in either a relative or an absolute sense) by varying the step size h. Thus, (7.47) together with the local error monitor (7.56) could form the basis for an adaptive (variable-step) differential-equation solver. At one level, the basic principles of error control are the same whether we are using a one-step method or a multistep method. If the local error is too large, we want to use a smaller step; if the local error is small, we want to use a larger step; and we do not need to know whether the local error estimate came from a one-step or a multistep method in order to know in which direction the step size should be changed. In addition, if we know that the local error has the form $Ch^P + 0(h^{P+1})$ as in (7.55), then we know also how to calculate the size of the step we should be using [the analysis is exactly the same as in (7.30) and (7.31)]. At the practical computational level, however, we know that changing the step size in a multistep method means that we may have to obtain some additional back values. We will discuss this problem a bit later. For now, we wish to make several observations about the derivations leading up to (7.56).

First of all, although we used the predictor-corrector method (7.47) as an illustrative example, it should be clear that exactly analogous arguments could have been made for any pair consisting of a k-step Adams-Bashforth predictor and a $(k - 1)$-step Adams-Moulton corrector. (In practice, values of k up to about $k = 12$ are frequently used.) Precisely, the local error of such a predictor-corrector method is $u(x_{n+k}) - y_{n+k} = h\tau + 0(h^{k+2})$ where $h\tau$ is the truncation error for the corrector. Next if such a method is used, and if the truncation error for the predictor is $h\tau = \tilde{C}_{k+1}h^{k+1}y^{(k+1)}$ and the truncation error for the corrector is $h\tau = C_{k+1}h^{k+1}y^{(k+1)}$, then the local error can be estimated by

$$u(x_{n+k}) - y_{n+k} \simeq \frac{C_{k+1}}{C_{k+1} - \tilde{C}_{k+1}}(\tilde{y}_{n+k} - y_{n+k}). \tag{7.57}$$

The estimate in (7.57) is known as *Milne's device*. It can also be used in the context of a general predictor-corrector method and is not limited to Adams methods. In addition, there are other ways to estimate local errors; see Lambert (1973) or Shampine and Gordon (1975).

Another observation of some worth can be deduced from (7.54). That is, the local error of a predictor-corrector method is limited by the truncation error of the corrector. For example, even if the predictor in (7.54) were very accurate, say $u(x_{n+3}) - \tilde{y}_{n+3} = O(h^6)$, the local error $u(x_{n+3}) - y_{n+3}$ is still of the order $O(h^4)$, the order of the truncation error of the corrector. On the other hand, if an "inferior" predictor is used, then the corrector can be iterated to damp out the effects of a "bad" estimate. Next, we note that the derivation of Milne's device as an error monitor for (7.47) required that we assume $u(x_n) - y_n = O(h^4)$ and $u(x_{n+1}) - y_{n+1} = O(h^4)$. In (7.55), we also obtained $u(x_{n+3}) - y_{n+3} = O(h^4)$ on the basis of these assumptions. In order to continue the analyses inductively, we need to show that $v(x_{n+1}) - y_{n+1} = O(h^4)$ and $v(x_{n+2}) - y_{n+2} = O(h^4)$ where $v(x)$ satisfies $y' = f(x, y)$ with $v(x_{n+3}) = y_{n+3}$. This demonstration can be done if $f(x, y)$ satisfies the hypotheses of Theorem 7.1.

If we monitor local errors, we should also be prepared to change step sizes. There are several ways to change the step size in a multistep method in order to construct a variable-step predictor-corrector method [see Krogh (1973) for a list of ten]. If we limit step increases to doubling, then we can increase the step easily if we save enough back values. However, to decrease the step, we need back values for $f(x, y)$ and/or y that have not been calculated. One way to get the required back values is by calling a one-step method, such as a Runge-Kutta method, and taking care that both the one-step and multistep methods have comparable orders of accuracy. Another way to get back values is by interpolation, but Shampine and Gordon (1975) note that numerical instabilities may occur when interpolation is used in k-step methods if k is larger than about 6 or so; the source of these instabilities is not well understood at present. An alternate strategy for obtaining back values is given by Shampine and Gordon (1975), but it is a bit too complicated to describe here. (This valuable reference is devoted to variable-step, variable-order predictor-corrector methods based on the Adams's formulas). A variable-step, variable-order method is self-starting; typically, the solution starts at (x_0, y_0) and uses an Euler-method predictor and a trapezoidal-method corrector. As more back values become available, higher-order predictors and correctors are used. As the solution proceeds, decisions on raising or lowering the order and increasing or decreasing the step size are made on the basis of local error monitors; but again it is not possible here to describe adequately the various strategies that are used in practice; see Shampine and Gordon (1975) and Gear (1971).

PROBLEMS, SECTION 7.5.1

1. Code (7.44a) and (7.45a) as a predictor-corrector pair in the form

$$\tilde{y}_{n+1} = y_n + hf(x_n, y_n)$$

$$y_{n+1} = y_n + hf(x_{n+1}, \tilde{y}_{n+1}).$$

Test your program on $y' = 3x^2y^2$, $y(0) = -1$. Run this problem on $[0, 4]$; use step sizes $h = .2, .1, .02, .01$; and print the predicted, corrected, and true values in steps of .4.

2. Repeat Problem 1 for the equation $y' = \lambda y$, $y(0) = 1$ where $\lambda = 20$ and $\lambda = -20$.
3. Code (7.44d) and (7.45d) as a predictor-corrector pair and test your program as in Problems 1 and 2. Use exact starting values y_0, y_1, y_2, and y_3.
4. Modify the program in Problem 1 in order to run a first-order system $\mathbf{u}' = \mathbf{F}(x, \mathbf{u})$. Test your program as in Problem 1; use the equation $\mathbf{u}' = A\mathbf{u}$ in Problem 5, Section 7.4.
5. Suppose that either of the implicit one-step methods (7.45a) or (7.45b) is applied to the problem $y' = \lambda y$, $y(0) = y_0$. Suppose that $\lambda < 0$ and that the method is used with a fixed step size h. Show that the estimates y_k tend to zero as k grows (note that iteration is not required to execute a step of an implicit method when the differential equation is linear). Apply these two methods to $y' = -20y$, $y(0) = 1$ and contrast the results with those in Problem 2.
6. Apply the methods in Problem 5 to the equation $y' = 20y$, $y(0) = 1$ and contrast the results with those of Problem 2.
7. For $h = .2$ and $h = .02$, give the closed-form expressions for y_k in terms of y_0 when the methods in Problems 1 and 5 are applied to $y' = 20y$ and to $y' = -20y$. Use these expressions to explain the results of Problems 2, 5, and 6.
8. Code vector versions of (7.45a) and (7.45b) for the system $\mathbf{u}' = A\mathbf{u}$ as in Problem 4. Run these programs and contrast the results with those of Problem 4.
9. Suppose that A is an $(n \times n)$ matrix with a set of n linearly independent eigenvectors and also suppose that all the eigenvalues of A are real and negative. Suppose that (7.45a) or (7.45b) is applied to $\mathbf{u}' = A\mathbf{u}$, $\mathbf{u}(0) = \mathbf{u}_0$, and a fixed step size h is used. Show that the estimates $\mathbf{u}_i$ tend to $\boldsymbol{\theta}$. [Hint: If $p(x)$ and $q(x)$ are polynomials and if $A\mathbf{x} = \lambda\mathbf{x}$, then $p(A)\mathbf{x} = p(\lambda)\mathbf{x}$, $q(A)\mathbf{x} = q(\lambda)\mathbf{x}$, and $q(A)^{-1}p(A)\mathbf{x} = r(\lambda)\mathbf{x}$ where $r(x) = p(x)/q(x)$.]
10. The predictor-corrector pair in Problem 1 is derived from the quadratures

a) $\displaystyle\int_{x_n}^{x_{n+1}} g(x)dx = hg(x_n) + E_a$ and b) $\displaystyle\int_{x_n}^{x_{n+1}} g(x)dx = hg(x_{n+1}) + E_b$.

Show that the quadrature errors are

$$E_a = \frac{h^2}{2}g''(\eta) \quad \text{and} \quad E_b = -\frac{h^2}{2}g''(\gamma).$$

[Hint: For (a), write $g(x) = g(x_n) + g'(\theta_x)(x - x_n)$ where $x_n < \theta_x < x_{n+1}$ and then use Theorem 1.4. For (b), start with $g(x) = g(x_{n+1}) + g'(\theta_x)(x - x_{n+1})$.]

11. Suppose that $u(x)$ solves $u' = f(x, u)$, $u(x_n) = y_n$. Use Problem 10 to show that the method in Problem 1 satisfies

a) $u(x_{n+1}) - \tilde{y}_{n+1} = \dfrac{h^2}{2}u''(\eta)$

b) $u(x_{n+1}) = u(x_n) + hf(x_{n+1}, u(x_{n+1})) - \dfrac{h^2}{2}u''(\gamma)$.

Recall that $y_{n+1} = y_n + hf(x_{n+1}, \tilde{y}_{n+1})$ and use (b) and then (a) to show that

c) $u(x_{n+1}) - y_{n+1} = -\frac{h^2}{2} u''(\gamma) + 0(h^3)$.

Using (c) and (a) and supposing that $u''(\eta) = u''(\gamma) + 0(h)$, argue that the local error satisfies

d) $u(x_{n+1}) - y_{n+1} \simeq \frac{\tilde{y}_{n+1} - y_{n+1}}{2}$.

12. Using the local error estimator derived in (d) of Problem 11, code a variable-step method using a error-per-unit-step criterion. Test your program on the equation in Problem 1; use an accuracy request of 10^{-3}. (Note: This programming assignment is intended to illustrate some of the aspects of error control and does not reflect the design of a variable-step, variable-order method.)
13. Show that (7.48b) is an interpolatory quadrature formula by showing that the error is zero for $g(x) = 1$, $g(x) = x$, and $g(x) = x^2$. Next use (6.8a) and Theorem 1.4 to show that $E = 3h^4 g'''(\eta)/8$. Carry out a similar analysis for the corrector (7.53a).

7.5.2. Linear Difference Equations

For an adequate analysis of linear multistep methods, it is necessary to develop some of the elementary theory for linear difference equations. To see what is required, consider the linear k-step method

$$\sum_{j=0}^{k} \alpha_j y_{n+j} = h \sum_{j=0}^{k} \beta_j f_{n+j}.$$

Recall that we can start analyzing a method applied to $y' = f(x, y)$ by applying the method to $y' = \lambda y$. For the test equation $y' = \lambda y$, we have $f_{n+j} = \lambda y_{n+j}$; so the linear k-step method reduces to

$$\sum_{j=0}^{k} \alpha_j y_{n+j} = h\lambda \sum_{j=0}^{k} \beta_j y_{n+j}. \tag{7.58a}$$

If one sets $\bar{h} = h\lambda$, (7.58a) is the same as

$$\sum_{j=0}^{k} (\alpha_j - \bar{h}\beta_j) y_{n+j} = 0. \tag{7.58b}$$

Given starting values $y_0, y_1, \ldots, y_{k-1}$, the approximations $y_k, y_{k+1}, \ldots$ are generated from (7.58b) when the method is applied to $y' = \lambda y$. It turns out that for the sequence $\{y_i\}$ there is a closed-form expression that satisfies (7.58b) and that can be used to predict the quality of the approximations y_i to $y(x_i)$.

An equation of the form

$$a_k z_{n+k} + a_{k-1} z_{n+k-1} + \cdots + a_0 z_n = b_n, \qquad n = 0, 1, \ldots \tag{7.59}$$

is called a kth order *linear difference equation*. In (7.59), the constants $a_0, a_1, \ldots, a_k$ are given with $a_0 \neq 0$ and $a_k \neq 0$. The sequence $\{b_n\}$ is also given; and if

$b_n = 0$ for all n, the difference equation is called *homogeneous*. A solution to the difference equation is any sequence $\{w_i\}$ that satisfies the relation (7.59) for all i. Note that (7.58b) has the form of a kth order homogeneous difference equation. Note also that the existence of solutions to (7.59) is not an issue since we can select any starting values $z_0, z_1, \ldots, z_{k-1}$ and generate a solution from the iteration

$$z_i = \frac{1}{a_k}[b_i - (a_{k-1}z_{i-1} + \cdots + a_0 z_{i-k})], \qquad i = k, k+1, \ldots .$$

Conditions for uniqueness of solutions to (7.59) are also obvious; if $\{u_i\}$ and $\{v_i\}$ are sequences that satisfy (7.59) and if $u_i = v_i$ for $0 \le i \le k-1$, then $u_i = v_i$ for all $i = k, k+1, \ldots$ as well (see Problem 9). As a concrete example, consider the second-order homogeneous difference equation

$$z_{n+2} + z_{n+1} - 6z_n = 0. \tag{7.60}$$

It is easy to verify that the sequence $\{w_i\}$ given by $w_i = 2^i$, $i = 0, 1, \ldots$ satisfies the simple equation above since inserting $\{w_i\}$ into (7.60) gives

$$2^{n+2} + 2^{n+1} - 6 \cdot 2^n = 2^n[4 + 2 - 6] = 0, \qquad n = 0, 1, 2, \ldots .$$

Similarly, we can verify that $\{v_i\}$ given by $v_i = (-3)^i$ also satisfies (7.60) as does any sequence $\{z_i\}$ of the form

$$z_i = \gamma_1 2^i + \gamma_2(-3)^i, \qquad i = 0, 1, \ldots \tag{7.61}$$

where γ_1 and γ_2 are constants. In fact we see below that every sequence that satisfies (7.60) has precisely the form of (7.61) so that (7.61) is the *general solution* of (7.60).

The solution procedure for a kth order, linear, homogeneous difference equation is straightforward. In particular, consider the homogeneous equation

$$a_k z_{n+k} + a_{k-1}z_{n+k-1} + \cdots + a_0 z_n = 0. \tag{7.62}$$

If we define the *characteristic polynomial* for (7.62) as $p(x) = a_k x^k + a_{k-1}x^{k-1} + \cdots + a_1 x + a_0$ and if r is any root of $p(x) = 0$, then it is easy to verify (Problem 4) that the sequence $\{w_i\}$ defined as $w_i = r^i$, $i = 0, 1, \ldots$ will satisfy (7.62). Therefore if we write $p(x)$ in factored form as

$$p(x) = a_k(x - r_1)(x - r_2)\ldots(x - r_k),$$

then each of the sequences $\{(r_1)^i\}, \{(r_2)^i\}, \ldots, \{(r_k)^i\}$ satisfies (7.62). Next it is easy to show (Problem 3) that if $\{w_i\}$ and $\{v_i\}$ are solutions of (7.62), then so is the sequence $\{u_i\}$ where $u_i = \gamma_1 w_i + \gamma_2 v_i$, $i = 0, 1, \ldots$. Thus, given the roots of $p(x) = 0$, the sequence $\{z_i\}$ defined below will satisfy (7.62):

$$z_i = \gamma_1(r_1)^i + \gamma_2(r_2)^i + \cdots + \gamma_k(r_k)^i. \tag{7.63}$$

If the roots of $p(x) = 0$ are distinct, then it is easy to show that every solution of (7.62) can be expressed in the form of (7.63), and so we call (7.63) the general solution of (7.62). To prove that every solution has the form of (7.63), suppose that $\{w_i\}$ is some sequence that satisfies (7.62). Then, given w_0, $w_1, \ldots, w_{k-1}$, the terms of the sequence $\{w_i\}$ must satisfy

$$w_i = \frac{-1}{a_k}[a_{k-1}w_{i-1} + \cdots + a_0 w_{i-k}], \qquad i = k, k+1, \ldots.$$

That is, the specific solution $\{w_i\}$ is completely determined by the starting values $w_0, w_1, \ldots, w_{k-1}$. Since $r_1, r_2, \ldots, r_k$ are distinct, there are unique constants $\mu_1, \mu_2, \ldots, \mu_k$ such that for $0 \le j \le k-1$

$$\mu_1(r_1)^j + \mu_2(r_2)^j + \cdots + \mu_k(r_k)^j = w_j. \tag{7.64a}$$

[To see this, note that the k equations in (7.64a) can be written in matrix terms as $R\boldsymbol{\gamma} = \mathbf{w}$ where the coefficient matrix R is a Vandermonde matrix. Hence, R is nonsingular.] Given $\mu_1, \mu_2, \ldots, \mu_k$ that satisfy (7.64a), we note as in (7.63) that the infinite sequence $\{v_i\}$ given by

$$v_i = \mu_1(r_1)^i + \mu_2(r_2)^i + \cdots + \mu_k(r_k)^i, \qquad i = 0, 1, \ldots \tag{7.64b}$$

also satisfies (7.62). Since $\{v_i\}$ and $\{w_i\}$ both satisfy (7.62) and since $v_j = w_j$ for $0 \le j \le k-1$, it follows from Problem 8 that $v_i = w_i$ for all i. Since $\{v_i\}$ has the form of (7.63), it follows that any solution of (7.62) can be expressed in the closed form given in (7.63). As a specific example, the equation (7.60) has the associated polynomial $p(x) = x^2 + x - 6$ or $p(x) = (x-2)(x+3)$. From the discussion above, the general solution of (7.60) is given by (7.61).

If the roots of $p(x) = 0$ are not distinct, then the argument above breaks down. However if $p(x) = (x-r)^m q(x)$ where $q(r) \neq 0$, then it is not difficult to show that the m sequences

$$\{r^i\}, \{ir^i\}, \{i^2 r^i\}, \ldots, \{i^{m-1}r^i\}$$

all satisfy (7.62). Thus for each root of $p(x) = 0$ we can obtain m distinct "fundamental" solutions of (7.62) when the root has multiplicity m. Hence we can find k distinct "fundamental" solutions to (7.62). These k distinct solutions are "linearly independent" and can be used to express the general solution (the proofs are left to the problems). As an example, suppose that a sixth-order difference equation has the characteristic polynomial $p(x) = (x-r_1)^3(x-r_2)(x-r_3)^2$. The general solution of such a difference equation has the form

$$z_i = \gamma_1(r_1)^i + \gamma_2 i(r_1)^i + \gamma_3 i^2(r_1)^i + \gamma_4(r_2)^i + \gamma_5(r_3)^i + \gamma_6 i(r_3)^i$$

for $i = 0, 1, \ldots$.

In summary, if $\{z_i\}$ is a sequence satisfying (7.62), then $\{z_i\}$ has the form

$$z_i = \gamma_1 v_{1i} + \gamma_2 v_{2i} + \cdots + \gamma_k v_{ki}, \qquad i = 0, 1, \ldots$$

where the fundamental sequences $\{v_{ji}\}_{i=0}^{\infty}$, $1 \le j \le k$, all have this form: $v_{ji} = i^q(r)^i$, $i = 0, 1, \ldots$ where r is a root of $p(x) = 0$ of multiplicity m and where $0 \le q \le m-1$.

7.5.3. Order and Convergence for Multistep Methods

We are now in a position to begin an elementary analysis of multistep methods and we start with the definition of the truncation error for a k-step method $\sum_{j=0}^{k} \alpha_j y_{n+j} = h \sum_{j=0}^{k} \beta_j f_{n+j}$. If $y(x)$ is a solution of $y' = f(x, y)$, then the local truncation error, $h\tau$, is defined by

$$\sum_{j=0}^{k} \alpha_j y(x + jh) - h \sum_{j=0}^{k} \beta_j y'(x + jh) = h\tau. \tag{7.65}$$

For $x = x_n$, the local truncation error is given by $h\tau = \sum_{j=0}^{k} \alpha_j y(x_{n+j}) - h \sum_{j=0}^{k} \beta_j y'(x_{n+j})$; and since $y'(x_{n+j}) = f(x_{n+j}, y(x_{n+j}))$, $h\tau$ measures how much the method misses $y(x_{n+k})$ if the method is given exact back values $y(x_n), \ldots, y(x_{n+k-1})$.

If $y(x)$ is sufficiently differentiable, we can express $h\tau$ in the form

$$h\tau = C_0 y(x) + C_1 h y'(x) + C_2 h^2 y''(x) + \cdots + C_q h^q y^{(q)}(x) + \cdots. \tag{7.66}$$

To show this point, we first write

$$y(x + jh) = y(x) + y'(x)jh + \frac{y''(x)}{2!}(jh)^2 + \cdots$$

$$y'(x + jh) = y'(x) + y''(x)jh + \frac{y'''(x)}{2!}(jh)^2 + \cdots.$$

Inserting these expressions into (7.65) and collecting like power of h, we obtain (7.66) where the constants C_i are

$$C_0 = \alpha_0 + \alpha_1 + \cdots + \alpha_k$$

$$C_1 = \alpha_1 + 2\alpha_2 + 3\alpha_3 + \cdots + k\alpha_k - (\beta_0 + \beta_1 + \cdots + \beta_k);$$

and in general for $q \geq 1$

$$C_q = \frac{1}{q!}(\alpha_1 + 2^q\alpha_2 + 3^q\alpha_3 + \cdots + k^q\alpha_q) - \frac{1}{(q-1)!}(\beta_1 + 2^{q-1}\beta_2 + \cdots + k^{q-1}\beta_k). \tag{7.67}$$

If $C_0 = C_1 = \cdots = C_p = 0$ and $C_{p+1} \neq 0$, then the linear k-step method is said to be *pth order* and the constant C_{p+1} is called the *error constant*. Thus for a pth-order method, we expect that $h\tau = C_{p+1}h^{p+1}y^{(p+1)}(x) + O(h^{p+2})$ when $y(x)$ is smooth enough. Note that the classification of multistep methods according to order is completely specified by the coefficients α_j and β_j. For example, it is easy to see that the two-step Adams-Bashforth method (7.44b) is second order. In detail, $\alpha_0 = 0$, $\alpha_1 = -1$, $\alpha_2 = 1$, $\beta_0 = -1/2$, $\beta_1 = 3/2$, $\beta_2 = 0$; so

$$C_0 = \alpha_0 + \alpha_1 + \alpha_2 = 0$$

$$C_1 = \alpha_1 + 2\alpha_2 - (\beta_0 + \beta_1 + \beta_2) = 0$$

$$C_2 = \frac{1}{2}(\alpha_1 + 4\alpha_2) - (\beta_1 + 2\beta_2) = 0$$

$$C_3 = \frac{1}{6}(\alpha_1 + 8\alpha_2) - \frac{1}{2}(\beta_1 + 4\beta_2) = \frac{5}{12}.$$

Thus the method is second order and $h\tau = 5h^3y'''/12 + 0(h^4)$, which also agrees with (7.44b). Recall also from Section 7.5.1 that knowing the form of $h\tau$ is crucial to the error monitor and control features of multistep methods.

We next discuss the theoretical convergence properties of a linear k-step method. As before, we fix x^* and define h and n so that $x^* = x_0 + (n + k)h$. We want to determine the properties that a method must possess in order to be convergent; that is, so that $y_{n+k} \to y(x^*)$ as $h \to 0$ when the k-step method is applied to a well-behaved problem $y' = f(x, y)$. [The assumption is also that the starting values $y_0(h), y_1(h), \ldots, y_{k-1}(h)$ all tend to $y(x_0) = y_0$ as $h \to 0$.] As with the one-step methods, we first show that a convergent method necessarily has order one or larger. First, consider the rather trivial problem $y' = 0$, $y(0) = 1$ where the exact solution is the constant function $y(x) = 1$. This equation satisfies the hypotheses of Theorem 7.1 and the k-step method applied to this problem reduces to

$$\sum_{j=0}^{k} \alpha_j y_{n+j} = 0. \tag{7.68}$$

For any convergent k-step method, we clearly have (since k is fixed) that $y_{n+k} \to y(x^*)$, $y_{n+k-1} \to y(x^*)$, $\ldots$, $y_n \to y(x^*)$. Therefore, we can write $y_{n+j} = y(x^*) + \varphi_j(h)$, $0 \le j \le k$ where $\varphi_j(h) \to 0$ as $h \to 0$. Using this equation in (7.68) gives

$$\sum_{j=0}^{k} \alpha_j y(x^*) + \sum_{j=0}^{k} \alpha_j \varphi_j(h) = 0;$$

and since the second sum tends to zero with h, we are forced to conclude that $0 = \sum_{j=0}^{k} \alpha_j y(x^*) = y(x^*) \sum_{j=0}^{k} \alpha_j$ or that $C_0 = 0$ in (7.66).

To see that C_1 in (7.66) is also zero, consider the problem $y' = 1$, $y(0) = 0$, which has solution $y(x) = x$. The method applied to this problem reduces to

$$\sum_{j=0}^{k} \alpha_j y_{n+j} = h \sum_{j=0}^{k} \beta_j,$$

which is a (nonhomogeneous) difference equation. It is easy to show (see Problem 11) that one solution of this equation is the sequence $\{y_\nu\}$ defined by $y_\nu = \nu hM$, $\nu = 0, 1, \ldots$ where

$$M = \frac{\sum_{j=0}^{k} \beta_j}{\sum_{j=0}^{k} j\alpha_j}. \tag{7.69}$$

(We will see shortly that for a convergent method, $\sum_{j=0}^{k} j\alpha_j \neq 0$; so M is well defined.) Now, if the starting values $y_i(h)$ are defined to be $y_i(h) = ihM$, then the method will produce the sequence $y_\nu = \nu hM$ and in particular $y_{n+k} = (n + k)hM$. Since the starting values satisfy the restriction $y_i(h) \to y(0) = 0$, and since the method was assumed convergent, we must conclude that $y_{n+k} \to y(x^*)$. Now, $y_{n+k} = (n + k)hM = x^*M$; and since $y(x^*) = x^*$, we have $M = 1$ and thus $C_1 = 0$ [see (7.67)].

Roughly put, the condition $C_0 = 0$ guarantees that y_{n+k} converges to *some* value $y(x^*)$ as $h \to 0$; $C_1 = 0$ guarantees that $y(x^*)$ is actually the value of the solution to $y' = f(x, y)$ at $x = x^*$. [See Lambert (1973) for more details.] A method that is at least order one is called *consistent*; so a necessary condition for convergence is that the method be consistent. For a one-step method, consistency is also sufficient for convergence when the increment function satisfies some mild continuity conditions; see Henrici (1962). However for a multistep method to be convergent, an additional condition (sometimes called the "root condition") is required. To see why this condition is necessary, we consider another simple problem, $y' = 0$, $y(0) = 0$. The solution to this equation is $y(x) = 0$; the method applied to $y' = 0$ will reduce to

$$\sum_{j=0}^{k} \alpha_j y_{n+j} = 0. \tag{7.70}$$

Now, (7.70) describes a homogeneous difference equation, and it is easy to see that the sequence $\{y_m\}$ satisfies (7.70) where

$$y_m = h(r_i)^m, \qquad m = 0, 1, \ldots .$$

In the definition of $\{y_m\}$, r_i can be any one of the zeros of $\rho(r)$, the characteristic polynomial for (7.70), where

$$\rho(r) = r^k + \alpha_{k-1}r^{k-1} + \cdots + \alpha_1 r + \alpha_0.$$

Not only does the sequence $\{y_m\}$ above satisfy (7.70), but the starting values y_0, $y_1, \ldots, y_{k-1}$ tend to $y(0) = 0$ as $h \to 0$. If we assume the method is convergent, then we know that $y_{n+k} \to y(x^*) = 0$ as $h \to 0$. Since y_{n+k} is given by

$$y_{n+k} = h(r_i)^{n+k}$$

and since $n \to \infty$ as $h \to 0$, we see that we can have $y_{n+k} \to y(x^*) = 0$ only if $|r_i| \leq 1$. Thus, convergence implies that every zero r_i of $\rho(r)$ must satisfy $|r_i| \leq 1$. Moreover, if r_i is a multiple zero of $\rho(r)$, then we also obtain a solution $\{y_j\}$ of (7.70) by $y_j = hj^q(r_i)^j$, $j = 0, 1, \ldots$ where $q \leq m - 1$ when r_i is a zero of multiplicity m. For $j = n + k$, we have

$$y_{n+k} = h(n + k)^q(r_i)^{n+k}. \tag{7.71}$$

However, $h(n + k) = x^*$; so $y_{n+k} \to 0$ as $n \to \infty$ only if $|r_i| < 1$. To summarize, if a multistep method is convergent and if $r_1, r_2, \ldots, r_k$ are the zeros of $\rho(r)$, then we must have the *root condition*:

1. $|r_i| \leq 1$ if r_i is simple zero of $\rho(r)$;
2. $|r_i| < 1$ if r_i is a multiple zero of $\rho(r)$.

It can be shown that the root condition together with consistency is sufficient as well as necessary for convergence.

Theorem 7.3

A linear multistep method is convergent if and only if the method is consistent and satisfies the root condition.

A proof of this result can be found in Henrici (1962). Note that the roots of $\rho(r) = 0$ might be complex; so the root condition means that simple roots must lie in the closed unit disk in the complex plane and multiple roots must be in the interior of the disk. Note also that $C_0 = 0$ implies that $\rho(1) = 0$; and it is conventional to let r_1 denote this root: $r_1 = 1$. If the root condition holds, then $\rho'(r_1) \neq 0$ since $|r_1| = 1$. Furthermore, $\rho'(1)$ is precisely the term in the denominator of M in (7.69); so if the method is convergent, then M is well defined. As an example, the methods (7.41) and (7.42) both have $\rho(r) = r^2 - 1$ and hence satisfy the root condition. According to Problem 12, the methods are also consistent; and thus according to Theorem 7.3, they are convergent. As we have mentioned, neither of these methods is suitable for practical computation despite their being convergent. We will see the reason for unsuitability in the next section. As a final example, we note that all the Adams methods satisfy the root condition (the only roots are $r_1 = 1$ and $r_2 = 0$; the root r_2 is a multiple root). Moreover, the k-step Adams-Bashforth methods have order k and the k-step Adams-Moulton methods have order $k + 1$; by Theorem 7.3, all of the Adams methods are convergent.

PROBLEMS, SECTION 7.5.3

1. Use (7.63) to find the general solution of these difference equations.

   a) $z_{n+2} + z_{n+1} - 6z_n = 0$ b) $z_{n+2} - 4z_n = 0$

   c) $z_{n+2} - 3z_{n+1} + 2z_n = 0$ d) $z_{n+3} - 6z_{n+2} + 11z_{n+1} - 6z_n = 0$

2. For each difference equation in Problem 1, give the closed form for the solution $\{z_i\}$ which has these starting values.

   a) $z_0 = 5,\ z_1 = -10$ b) $z_0 = 4,\ z_1 = -4$

   c) $z_0 = 5,\ z_1 = 7$ d) $z_0 = 0,\ z_1 = 2,\ z_2 = 8$

   [Hint: Given the general solutions from Problem 1, set up the systems (7.64a).] Verify that your closed form for (a) is correct by comparing the terms with z_2, z_3, z_4 where $z_i = -z_{i-1} + 6z_{i-2}$, $i = 2, 3, 4$; $z_0 = 5$; $z_1 = -10$.

3. Suppose that the sequences $\{u_i\}$ and $\{v_i\}$ both satisfy $a_3z_{n+3} + a_2z_{n+2} + a_1z_{n+1} + a_0z_n = 0$. Verify that the sequence $\{w_i\}$ defined by $w_i = au_i + bv_i$, $i = 0, 1, \ldots$ also satisfies the difference equation.

4. Let $p(x) = a_3x^3 + a_2x^2 + a_1x + a_0$ be the characteristic polynomial for the difference equation in Problem 3. Show that if r is a root of $p(x) = 0$, then the sequence $\{u_i\}$ defined by $u_i = r^i$, $i = 0, 1, \ldots$ is a solution to the difference equation. [Hint: Substitute the assumed solution into the difference equation.]

5. Using the same notation as that in Problems 3 and 4, suppose that $p(r) = 0$ and $p'(r) = 0$. Let $\{v_i\}$ be the sequence defined by $v_i = ir^i$, $i = 0, 1, \ldots$. Show that $\{v_i\}$ is a solution to the difference equation. If $p(r) = p'(r) = p''(r) = 0$, show that $\{w_i\}$ defined by $w_i = i^2r^i$, $i = 0, 1, \ldots$ is a solution. (Note: The results in Problems 3 through 5 can obviously be extended to kth-order difference equations.)

6. Consider $z_{n+3} - 6z_{n+2} + 12z_{n+1} - 8z_n = 0$ and note that the characteristic polynomial $p(x)$ has the form $p(x) = (x - 2)^3$. Thus the sequence $\{w_i\}$ given by

$$w_i = a2^i + bi2^i + ci^22^i,\ i = 0, 1, \ldots$$

   is a solution to the difference equation. Show that a, b, and c in the expression above can be chosen so that $\{w_i\}$ has given starting values w_0, w_1, w_2. [Hint: Set up a system similar to (7.64a) and show that the system is nonsingular.]

7. Solve the difference equations with the starting values given.

   a) $z_{n+3} - 5z_{n+2} + 7z_{n+1} - 3z_n = 0$; $z_0 = 2,\ z_1 = 5,\ z_2 = 12$

   b) $z_{n+3} - 2z_{n+2} - 4z_{n+1} + 8z_n = 0$; $z_0 = 3,\ z_1 = -4,\ z_2 = 20$

8. Show that if $\{w_i\}$ is a sequence satisfying the difference equation in Problem 3 and if $w_0 = w_1 = w_2 = 0$, then $w_i = 0$ for all i. [Hint: Use induction on i.]

9. Suppose that $\{u_i\}$ and $\{v_i\}$ are sequences that satisfy the nonhomogeneous difference equation

$$a_3z_{n+3} + a_2z_{n+2} + a_1z_{n+1} + a_0z_n = b_n,$$

   and suppose that $u_0 = v_0,\ u_1 = v_1,\ u_2 = v_2$. Show that $u_i = v_i$ for all i. [Hint: The sequence $\{w_i\}$ defined by $w_i = u_i - v_i$ satisfies a homogeneous difference equation.]

10. Calculate the coefficients $C_0, C_1, C_2, \ldots$ in (7.67) for the Adams methods (7.44 a through c) and (7.45 a through c) and determine the order of each method. If a method is pth order, calculate C_{p+1} and compare with the terms $h\tau$ listed in (7.44) and (7.45).

11. Show that the sequence $\{y_\nu\}$ defined by $y_\nu = \nu hM$ where M is given in (7.69) satisfies the difference equation $\sum_{j=0}^{k} \alpha_j y_{n+j} = h \sum_{j=0}^{k} \beta_j$.

12. Show that the methods (7.41) and (7.42) are consistent, and determine their order. Also determine the leading coefficient of $h\tau$; see (7.66).
13. Consider the two-step method

$$u_{n+2} + \alpha_1 y_{n+1} + a y_n = h(\beta_2 f_{n+2} + \beta_1 f_{n+1} + \beta_0 f_n)$$

where a is a parameter. Determine α_1, β_2, β_1, β_0 so that the method has order three. Find a value for a so that the method has order four. For what range of a is the root condition satisfied?

7.5.4. Relative Stability for Multistep Methods

Convergence, the topic of the preceding section, is an important theoretical concept. We would hardly consider using a method that was not convergent (in a theoretical sense) because we could not expect to improve our approximations to the solution of $y' = f(x, y)$ by decreasing the step size h. However, for practical computational purposes we require more from a method than just that it be convergent. To see why, consider the test equation $y' = \lambda y$, $y(0) = y_0$. (The initial condition is chosen for convenience to be at $x = 0$; this choice does not have any significant impact on the discussion.) If we apply the method $\sum_{j=0}^{k} \alpha_j y_{n+j} = h \sum_{j=0}^{k} \beta_j f_{n+j}$ to $y' = \lambda y$, we find that the approximations y_m to $y(x_m)$ satisfy the homogeneous difference equation

$$\sum_{j=0}^{k} (\alpha_j - \overline{h}\beta_j) y_{n+j} = 0, \; n = 0, 1, \ldots \tag{7.72}$$

where, as before, $\overline{h} = h\lambda$. Of course in writing (7.72), we are assuming that the method is using a constant step size h; and we will return to this point a bit later.

Given starting values $y_0, y_1, \ldots, y_{k-1}$, the approximations y_m generated by (7.72) can be expressed in this form (see Section 7.5.2):

$$y_m = \gamma_1(\overline{r}_1)^m + \gamma_2(\overline{r}_2)^m + \cdots + \gamma_k(\overline{r}_k)^m, \qquad m = 0, 1, \ldots \tag{7.73}$$

where $\overline{r}_1, \overline{r}_2, \ldots, \overline{r}_k$ are the zeros of the characteristic polynomial $\rho(r) - \overline{h}\sigma(r)$,

$$\rho(r) - \overline{h}\sigma(r) = (\alpha_k - \overline{h}\beta_k)r^k + \cdots + (\alpha_1 - \overline{h}\beta_1)r + (\alpha_0 - \overline{h}\beta_0).$$

The notation $\rho(r) - \overline{h}\sigma(r)$ for the characteristic polynomial of (7.72) is standard in the literature and comes from the designations

$$\rho(r) = \alpha_k r^k + \cdots + \alpha_1 r + \alpha_0$$

$$\sigma(r) = \beta_k r^k + \cdots + \beta_1 r + \beta_0.$$

The roots of $\rho(r) - \overline{h}\sigma(r) = 0$ are designated as $\overline{r}_j$ since they are partly functions of $\overline{h}$. Note also that as $\overline{h} \to 0$, $\overline{r}_j \to r_j$, $1 \le j \le k$, where the r_j are the roots of $\rho(r) = 0$. If the method is convergent, then for small $\overline{h}$ the roots $\overline{r}_j$ are, because of the root condition, in the unit disk if $|r_j| < 1$ or near the boundary if $|r_j| = 1$. Finally, in writing the expression (7.73) for y_m, we are tacitly assuming that the roots $\bar{r}_j$ of $\rho(r) - \bar{h}\rho(r) = 0$ are distinct. For the moment we will assume distinct roots to simplify the discussion.

Now, $\bar{r}_1 \to r_1 = 1$ as $\bar{h} \to 0$ and we will see below that the component $\gamma_1(\bar{r}_1)^m$ in (7.73) is the one that approximates $y(x_m)$; the other components are "parasitic" components and must be controlled to be small relative to $\gamma_1(\bar{r}_1)^m$ in order for the method to be effective. To develop this concept, we use the definition of the local truncation error (7.65) to get an expression for $\bar{r}_1$. According to (7.65) and (7.66), we have the following equation for a pth order method:

$$\sum_{j=0}^{k} \alpha_j y(x + jh) - h \sum_{j=0}^{k} \beta_j y'(x + jh) = C_{p+1}h^{p+1}y^{(p+1)}(x)$$
$$+ C_{p+2}h^{p+2}y^{(p+2)}(x) + \cdots.$$

If we use the function $y(x) = e^{\lambda x}$ in the equation above and observe that

$$y(x + jh) = y(x)e^{j\bar{h}}$$
$$hy'(x + jh) = \bar{h}y(x)e^{j\bar{h}}$$
$$h^q y^{(q)}(x) = \bar{h}^q y(x), \qquad q = 2, 3, \ldots,$$

then we are led to

$$y(x) \sum_{j=0}^{k} \alpha_j e^{j\bar{h}} - \bar{h}y(x) \sum_{j=0}^{k} \beta_j e^{j\bar{h}} = C_{p+1}\bar{h}^{p+1}y(x) + 0(\bar{h}^{p+2}).$$

Canceling $y(x)$, combining the sums, and using $e^{j\bar{h}} = (e^{\bar{h}})^j$ lead to

$$\sum_{j=0}^{k} (\alpha_j - \bar{h}\beta_j)(e^{\bar{h}})^j = C_{p+1}\bar{h}^{p+1} + 0(\bar{h}^{p+2}). \tag{7.74}$$

In words, (7.74) says that $\rho(r) - \bar{h}\sigma(r)$ evaluated at $r = e^{\bar{h}}$ produces the result $C_{p+1}\bar{h}^{p+1} + O(\bar{h}^{p+2})$; so $e^{\bar{h}}$ is nearly a root when $\bar{h}$ is small. If we write $\rho(r) - \bar{h}\sigma(r)$ in factored form as $\rho(r) - \bar{h}\sigma(r) = (\alpha_k - \bar{h}\beta_k)(r - \bar{r}_1)(r - \bar{r}_2) \ldots (r - \bar{r}_k)$, then we have for $r = e^{\bar{h}}$

$$(\alpha_k - \bar{h}\beta_k)(e^{\bar{h}} - \bar{r}_1)(e^{\bar{h}} - \bar{r}_2) \ldots (e^{\bar{h}} - \bar{r}_k) = 0(\bar{h}^{p+1}). \tag{7.75}$$

The order relation above is valid for small $\bar{h}$. Note that as $\bar{h} \to 0$, $e^{\bar{h}} - \bar{r}_j \to 1 - r_j$, which is nonzero for $j = 2, 3, \ldots, k$ by the root condition (since $r_j \neq 1$, $2 \leq j \leq k$). Since $\alpha_k = 1$ and since $e^{\bar{h}} - \bar{r}_j \neq 0$ for $2 \leq j \leq k$, we can rewrite the order relation (7.75) as $e^{\bar{h}} - \bar{r}_1 = 0(\bar{h}^{p+1})$, or

$$\bar{r}_1 = e^{\bar{h}} + 0(\bar{h}^{p+1}). \tag{7.76}$$

Knowing the form of $\bar{r}_1$, we now return to the approximations y_m to $y(x_m)$ given in (7.73) where $\{y_m\}$ is the result of applying the pth-order k-step method to the problem $y' = \lambda y$, $y(0) = y_0$; see (7.72). Now, the solution to the initial-value problem at $x = x_m$ is $y(x_m) = y_0 e^{\lambda x_m}$; and since $x_m = mh$, we have

$$y(x_m) = y_0 e^{m\bar{h}}. \tag{7.77a}$$

On the other hand, the approximation y_m in (7.73) satisfies [by (7.76)]

$$y_m = \gamma_1 e^{m\bar{h}} + \gamma_2(\bar{r}_2)^m + \cdots + \gamma_k(\bar{r}_k)^m + 0(\bar{h}^{p+1}). \tag{7.77b}$$

Comparing (7.77a) and (7.77b), we see clearly that y_m is a good approximation to $y(x_m)$ if

1. $\gamma_1 \simeq y_0$, $\gamma_i \simeq 0$, $2 \leq i \leq k$;
2. $(\bar{r}_i)$ is small with respect to $e^{\bar{h}}$, $2 \leq i \leq k$.

Regarding these two points, it is easy to show (Problem 6) that if the starting values $y_0, y_1, \ldots, y_{k-1}$ are good, then $\gamma_1 \simeq y_0$ and $\gamma_i \simeq 0$ for $2 \leq i \leq k$. The second point leads to the definition of relative stability.

A linear k-step method is said to be *relatively stable* for a particular $\bar{h}$ if the roots of $\rho(r) - \bar{h}\sigma(r) = 0$ satisfy

$$|\bar{r}_1| > |\bar{r}_i|, \; i = 2, 3, \ldots, k.$$

The *interval of relative stability* for the method is the largest interval (α, β), $\alpha \leq 0 \leq \beta$, such that the method is relatively stable for all $\bar{h}$ in (α, β). As is obvious from (7.77), if $\bar{h}$ is outside the interval of relative stability, then we cannot expect that the relative errors $|y_m - y(x_m)|/|y(x_m)|$ will be small even if the starting values are very good. The size of the interval (α, β) puts a restriction on the step size that we can safely use; for λ large in magnitude we may have to choose h quite small so that $\bar{h} = h\lambda$ is in (α, β). However, as is pointed out in the next several sections, we may be interested in controlling absolute errors rather than relative errors; and then a similar concept, absolute stability, will come into play. The concept of relative stability can be seen qualitatively in (7.73) if we rewrite (7.73) in the form

$$y_m = \gamma_1(\bar{r}_1)^m \left[1 + \frac{\gamma_2}{\gamma_1}\left(\frac{\bar{r}_2}{\bar{r}_1}\right)^m + \cdots + \frac{\gamma_k}{\gamma_1}\left(\frac{\bar{r}_k}{\bar{r}_1}\right)^m \right].$$

It is clear that the smaller the ratios $|\bar{r}_i/\bar{r}_1|$ are, the better y_m is as an approximation to $y(x_m)$. Finally, we see that the possibility of multiple roots of $\rho(r) - \bar{h}\sigma(r) = 0$ will not affect the argument since if y_m has a component of the form $\gamma_i m^q(\bar{r}_i)^m$, the ratio $m^q(\bar{r}_i/\bar{r}_1)^m$ will still tend to zero as $m \to \infty$ when $|\bar{r}_i|/|\bar{r}_1| < 1$.

Some examples should serve to clarify the ideas developed above. In particular, let us consider the method (7.41) given by

$$y_{n+2} = y_n + 2hf_{n+1}.$$

For this two-step method, the "stability equation" $\rho(r) - \bar{h}\sigma(r) = 0$ is given by

$$r^2 - 2\bar{h}r - 1 = 0.$$

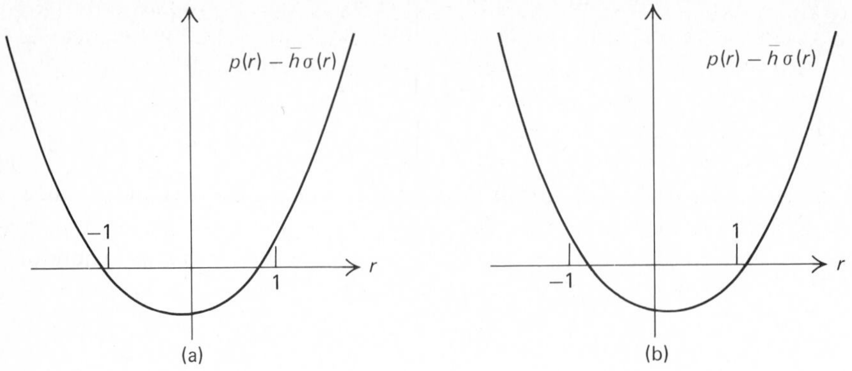

Figure 7.11 Graphs of $r^2 - 2\bar{h}r - 1$ for (a) $\bar{h} < 0$ and (b) $\bar{h} > 0$.

If we plot this polynomial, the representative graphs for $\bar{h} < 0$ and for $\bar{h} > 0$ are like those in Fig. 7.11. In Fig. 7.11, $\bar{r}_1$ is the root near $r = 1$ and $\bar{r}_2$ is the root near $r = -1$. The graph in the case $\bar{h} < 0$ indicates that $|\bar{r}_2| > |\bar{r}_1|$, and we see from the other that $|\bar{r}_1| > |\bar{r}_2|$ when $\bar{h} > 0$. Thus, the interval of relative stability for the method (7.41) is $(0, \infty)$; the method is relatively unstable for all $\bar{h} < 0$. The computational significance of the instability of (7.41) is easy to see. Recall that solving $y' = f(x, y)$ in a neighborhood of (x_i, y_i) is similar to solving $y' = \lambda y$ where $\lambda = f_y(x_i, y_i)$. Typically, $f_y(x, y)$ varies as we follow the solution $y(x)$, and it is quite possible that we will encounter regions in which $f_y(x, y)$ is negative and hence λ is negative. Of course, if $\lambda < 0$ then $\bar{h} < 0$; and thus we would expect the relative errors produced by (7.41) to grow in a neighborhood of (x, y) whenever $f_y(x, y) < 0$. An analysis similar to the one summarized in Fig. 7.11 will show that the interval of relative stability for the method (7.42) is also $(0, \infty)$. The conclusion is then that although they appear very reasonable in derivation, these methods are not suitable for general computation.

As a contrast, we consider the two-step Adams-Bashforth method:

$$y_{n+2} - y_{n+1} = \frac{h}{2}[3f_{n+1} - f_n]$$

whose stability polynomial is

$$\rho(r) - \bar{h}\sigma(r) = r^2 - \left(1 + \frac{3\bar{h}}{2}\right)r + \frac{\bar{h}}{2}.$$

A plot of this polynomial (as in Fig. 7.11) shows for $\bar{h} > 0$ that $0 < \bar{r}_2 < 1 < \bar{r}_1$ and hence that the method is relatively stable for any $\bar{h} > 0$. For $\bar{h} < 0$, the situation is slightly more complicated, but a plot reveals (see Problem 3) that $0 < \bar{r}_1 < 1$ for all $\bar{h} < 0$. On the other hand, $\bar{r}_2$ is near zero for $\bar{h}$ small; but $\bar{r}_2$ moves to the left of -1 as $2 + 2\bar{h}$ becomes negative [as $\bar{h} \to -\infty$, $\bar{r}_2 \to -\infty$; but $\bar{r}_1$ is always in (0, 1)]. To determine the value $\bar{h}$ where $|\bar{r}_2|$ first equals $|\bar{r}_1|$, we suppose that $\bar{r}_1 = -\bar{r}_2$ or $\bar{r}_1 + \bar{r}_2 = 0$. Since the coefficient of the linear term in

$\rho(r) - \bar{h}\sigma(r)$ is $-(\bar{r}_1 + \bar{r}_2)$, it follows that $|\bar{r}_2| < |\bar{r}_1|$ as $\bar{h}$ ranges to the left from zero up until the point that $1 + 3\bar{h}/2 = 0$. Thus, $|\bar{r}_2| < |\bar{r}_1|$ as long as $-2/3 < \bar{h}$ and so the interval of relative stability for the Adams-Bashforth method (7.44b) is given by $(-2/3, \infty)$. We conclude, as is illustrated in the example below, that this method is computationally practical.

From the preceding discussions, we can make an observation about the Adams methods. For a k-step Adams method, we have $r_1 = 1$ and $r_i = 0, 2 \le i \le k$. Thus since $\bar{r}_j \to r_j$ as $\bar{h} \to 0$, it follows that every Adams method has an interval of relative stability of the form (α, β) where $\alpha < 0 < \beta$. Therefore any of the Adams methods are computationally practical; moreover, we expect the ratios $|\bar{r}_i/\bar{r}_1|$ to be reasonably small for moderate $\bar{h}$ in (α, β).

EXAMPLE 7.12. To illustrate the preceeding discussion, we ran the problem $y' = -y$, $y(0) = 1$ with a step size $h = 0.02$, for the methods of both (7.41) and (7.44b). As expected, the approximations found by (7.41) became progressively worse. In the table below, the column headed A denotes the results found at x by (7.41) and those headed B denote the results found at x by (7.44b).

x	A	B	e^{-x}
4.00	0.18358E−01	0.18328E−01	0.18316E−01
8.00	0.23806E−02	0.33590E−03	0.33546E−03
16.00	0.60926E+01	0.11283E−06	0.11254E−06

We note that at $x = 16$ (after 800 steps), (7.41) is producing nonsense while (7.44b) is more nearly correct. Since $y(x) = e^{-x}$ is fairly small, relative errors are more meaningful than absolute errors. When these are listed, as below, the difference in the two methods becomes dramatically distinct where at $x = 16$, the relative error in column A is ten orders of magnitude larger than the relative error in column B. The columns headed EA and EB denote the relative errors:

x	EA	EB
4.00	0.229E−02	0.655E−03
8.00	0.610E+01	0.131E−02
16.00	0.541E+08	0.258E−02

As a final note on relative stability, recall that the analysis in this section is based on (7.72) where $\bar{h}$ is assumed constant. That $\bar{h} = h\lambda$ is constant means essentially that h and λ are assumed constant. As a practical matter, the step size h used is small enough so that $f_y(x, y)$ is not changing rapidly; so the assumption that λ is nearly constant should be valid for several steps. Most of the current codes that implement variable-step predictor-corrector methods do not vary h at each step as a variable-step Runge-Kutta method does. Instead, these codes attempt to hold h fixed for a block of steps (about ten steps) partly to reduce overhead costs of the code and partly to avoid stability problems that arise when the step size is changed frequently. From these comments, it is clear that the analysis in this section is reasonable over some block of steps. When h is changed, the analysis merely begins again with a new $\bar{h}$.

PROBLEMS, SECTION 7.5.4

1. To illustrate the effects of relative stability, consider the problem $y' = 3x^2y^2$, $y(1) = -1$. Run 30 steps of the methods below on this problem.

   a) $y_{n+2} = y_n + 2hf(x_{n+1}, y_{n+1})$

   b) $y_{n+2} = y_{n+1} + \dfrac{h}{2}[3f(x_{n+1}, y_{n+1}) - f(x_n, y_n)]$

   Use starting values $y_0 = -1$ and $y_1 = -.75$ and use $h = .1$. Print the estimates and the exact solution at each step for both methods.

2. To explain the output from Problem 1, note that $\lambda \simeq 6x^2y$ and so method (a) is relatively unstable when $y < 0$. Check the estimates to $y(2)$, $y(3)$, $y(4)$ for method (b); and verify that $\bar{h}$ is in the interval of relative stability.

3. As in Figure 7.11, graph the stability polynomial for method (b) in Problem 1. Verify that the method must have an interval of stability of the form (α, ∞) where $\alpha < 0$.

4. Show that the interval of relative stability for the method (7.42) is $(0, \infty)$.

5. Find the interval of relative stability for the method (7.45c).

6. Suppose [see (7.73) and (7.77b)] that a three-step method applied to $y' = \lambda y$ has starting values y_0, y_1, y_2 where

$$y_0 = \gamma_1 + \gamma_2 + \gamma_3$$

$$y_1 = \gamma_1 e^{\bar{h}} + \gamma_2 \bar{r}_2 + \gamma_3 \bar{r}_3 + O(\bar{h}^q)$$

$$y_2 = \gamma_1 e^{2\bar{h}} + \gamma_2 \bar{r}_2^{\,2} + \gamma_3 \bar{r}_3^{\,2} + O(\bar{h}^q).$$

   Suppose further that $y_i = y(x_i) + O(\bar{h}^q)$ for $0 \le i \le 2$. Show that $\gamma_1 = y_0 + O(\bar{h}^q)$ and $\gamma_i = O(\bar{h}^q)$ for $i = 2, 3$. [Hint: In the (3×3) system above, replace y_m by $y_0 e^{m\bar{h}} + O(\bar{h}^q)$, and rewrite the equations as $y_0 e^{m\bar{h}} = \gamma_1 e^{m\bar{h}} + \gamma_2 \bar{r}_2^{\,m} + \gamma_3 \bar{r}_2^{\,m} + O(\bar{h}^q), 0 \le m \le 2$. Use Cramer's rule to express $\gamma_1, \gamma_2, \gamma_3$.]

7.6 ABSOLUTE STABILITY FOR ONE-STEP METHODS AND MULTISTEP METHODS; STIFF EQUATIONS

As we mentioned in Section 7.5.4, controlling the absolute error in a numerical method is sometimes more desirable than controlling the relative error. For example, suppose that we are solving an equation that behaves like $y' = \lambda y$,

$y(0) = y_0$ where $\lambda < 0$. In this case, the solution $y(x) = y_0 e^{\lambda x}$ tends to zero very rapidly and we are normally not interested in finding an approximation that is correct to (say) 8 digits to a quantity that has a magnitude of 10^{-50}. Controlling absolute errors is often important when solving an m-dimensional system of equations $\mathbf{y}' = \mathbf{F}(x, \mathbf{y})$ where, as we shall see, the method behaves to some degree as though we were solving m uncoupled scalar equations:

$$\begin{aligned} y_1' &= \lambda_1 y_1, & y_1(0) &= c_1 \\ y_2' &= \lambda_2 y_2, & y_2(0) &= c_2 \\ &\vdots & &\vdots \\ y_m' &= \lambda_m y_m, & y_m(0) &= c_m. \end{aligned}$$

The solution of the uncoupled system above is

$$\mathbf{y}(x) = \begin{bmatrix} c_1 e^{\lambda_1 x} \\ c_2 e^{\lambda_2 x} \\ \vdots \\ c_m e^{\lambda_m x} \end{bmatrix},$$

and let us suppose that $\lambda_m \le \lambda_{m-1} \le \cdots \le \lambda_2 \le \lambda_1 < 0$ where λ_m is much larger in magnitude than λ_1. In this event, we do not care whether the method tracks the component $y_m(x) = c_m e^{\lambda_m x}$ accurately (in a relative sense); what we really care about is that the method follows $y_1(x) = c_1 e^{\lambda_1 x}$ accurately and that the approximations to $y_m(x)$ remain small relative to those for $y_1(x)$.

7.6.1. Absolute Stability for One-step Methods

The concept of absolute stability is related to some extent to the propagation of absolute errors and is easy to discuss in the context of one-step methods. Roughly, a method is "absolutely stable" if past errors are not magnified. To be more explicit, let us return to the test equation $y' = \lambda y$. For the one-step methods we have considered (Taylor's series methods and qth order, q-stage Runge-Kutta methods) we know that an explicit qth order, one-step method applied to $y' = \lambda y$ takes the form

$$y_{n+1} = p(\bar{h}) y_n$$

where $\bar{h} = h\lambda$ and $p(\bar{h}) = 1 + \bar{h} + \bar{h}^2/2 + \cdots + \bar{h}^q/q!$. Furthermore for the equation $y' = \lambda y$, we also know that $y(x) = y(x_n)e^{\lambda(x - x_n)}$; so for $x = x_{n+1}$ we have $y(x_{n+1}) = y(x_n)e^{\bar{h}}$ or

$$y(x_{n+1}) = p(\bar{h})\, y(x_n) + [e^{\bar{h}} - p(\bar{h})] y(x_n).$$

Subtracting these two equations we obtain an expression for the global error

$$e_{n+1} = p(\bar{h}) e_n + [e^{\bar{h}} - p(\bar{h})] y(x_n) \tag{7.78}$$

where $e_i = y(x_i) - y_i$, $i = 0, 1, \ldots$. Now if $|p(\bar{h})| < 1$, then the past errors are not magnified as we move from x_n to x_{n+1}, and this fact is the basis for the definition of absolute stability. A qth order Runge-Kutta method is *absolutely stable* for a particular value $\bar{h}$ if $|p(\bar{h})| < 1$.

Note that we always have $p(\bar{h}) > 1$ for $\bar{h} > 0$; so an explicit Runge-Kutta method cannot be absolutely stable when $\lambda > 0$. However, there is obviously a largest interval of the form $(\alpha, 0)$ for which $\bar{h}$ in $(\alpha, 0)$ implies that $|p(\bar{h})| < 1$; and this interval $(\alpha, 0)$ is called the *interval of absolute stability*. To see that $|p(\bar{h})| < 1$ for $\bar{h} < 0$ and small, one need note only that $p'(\bar{h}) = 1$ and $p(\bar{h}) = 1$ when $\bar{h} = 0$; therefore, $|p(\bar{h})|$ decreases from 1 as $\bar{h}$ moves to the left away from 0. As $\bar{h} \to -\infty$, $|p(\bar{h})| \to \infty$; so there must be an interval of the form $(\alpha, 0)$ in which the method is absolutely stable. Note also that $(\alpha, 0)$ depends only on $p(\bar{h})$, not on the particular choice of parameters. For example, given any second-order, two-stage Runge-Kutta method, $p(\bar{h})$ is always the polynomial

$$p(\bar{h}) = 1 + \bar{h} + \frac{\bar{h}^2}{2};$$

and thus the interval of absolute stability is always $(-2, 0)$.

To show the role that absolute stability plays, suppose we are using a second-order Runge-Kutta method to solve $u' = 100u + 100$, $u(0) = 2$. We can rewrite this equation as $y' = -100(u - 1)$ and make the substitution $y = u - 1$ to produce an equivalent equation:

$$y' = -100y, \qquad y(0) = 1 \tag{7.79}$$

[As a point of interest, it is easy to show that a second-order Runge-Kutta method applied to $y' = -100y$ will produce estimates y_k that run exactly parallel to the estimates u_k produced when the method is applied to $u' = -100u + 100$. That is, if $y_{n+1} = y_n + h\phi(x_n, y_n; h)$, $y_0 = 1$ and $u_{n+1} = u_n + h\phi(x_n, u_n; h)$, $u_0 = 2$, then $y_i = u_i - 1$ for $i = 0, 1, \ldots$ (see Problem 3). In general, the behavior of a qth order method for $u' = \lambda u + c$ is paralleled by the method applied to $y' = \lambda y$ where $u + c/\lambda = y$.] Continuing, let us apply a second-order Runge-Kutta method to (7.79). Since $\lambda = -100$, $\bar{h} = -100h$ and therefore the method will be absolutely stable only if $-2 < \bar{h} < 0$ or $h < 1/50$. To see the dramatic effect that $\bar{h} < -2$ can have, suppose $h = 1/25$; so $\bar{h} = -4$. As noted above, the method applied to $y' = \lambda y$ produces the iteration

$$y_{n+1} = p(\bar{h})y_n;$$

and for $\bar{h} = -4$, $p(\bar{h}) = 5$. With $y_0 = 1$, we are led to $y_k = 5^k$ and so after 25 steps, at $x_{25} = 1$, we have produced the approximation $y(1) \simeq y_{25} = 5^{25} \simeq 2198 \times 10^{17}$. The solution of $y' = -100\,y$ is $y(x) = e^{-100x}$; so $y(1) \simeq 3.72 \times 10^{-44}$ and we are 61 orders of magnitude off after 25 steps. Note also that we do not have to take very many steps before absolute instability becomes significant. For example, $y_2 = 5^2 = 25$ and $y_2 \simeq y(x_2) = y(.08) \simeq 3.36 \times 10^{-4}$; so after even two steps we have a large error.

By contrast if $h = 1/100$ so that $\bar{h} = -1$ and $p(\bar{h}) = .5$, we would produce

$y_k = 1/2^k$ and an approximation to $y(1)$ of $y_{100} = 1/2^{100} \simeq 7.89 \times 10^{-31}$. Note that this approximation is not accurate in a relative sense, but the absolute error $|y(x_{100}) - y_{100}|$ is small and the iterates y_k are at least moving in the right direction. As we noted above, the numerical solutions of $y' = -100y$ and $u' = -100u + 100$ move in parallel; so the same phenomenon is present when we solve $u' = -100u + 100$. [The actual solution to this problem is $u(x) = e^{-100x} + 1$; so the actual solution drops quickly to 1 and then stays nearly constant. The numerical solution will behave the same way when h is selected so that $|p(\bar{h})| < 1$.]

We wish to comment briefly on absolute stability in the context of a variable-step Runge-Kutta method. Suppose that we are solving a problem that looks (locally) like the test equation $y' = \lambda y$. As we noted above, absolute-stability is not an issue when $\bar{h} > 0$; so let us assume that $\lambda < 0$. Further, let us suppose that the step-selection decisions are made on the basis of an absolute-error test or a mixed test. In this case, the variable-step algorithm is controlling the absolute size of the local error; as observed in Section 7.2.5, such control amounts to controlling the quantity

$$|(e^{\bar{h}} - p(\bar{h}))y_n|$$

as we step from x_n to x_{n+1}. Now if we are running with $\bar{h}$ in the interval of absolute stability, then the sequence $\{y_n\}$ is tending rapidly to zero (since $y_{k+1} = p(\bar{h})y_k$, $k = 0, 1, \ldots$). Since $\{y_n\} \to 0$, the local error can be controlled to be small even with $\bar{h}$ large as long as $|y_n|$ is small [for large $\bar{h}$, $e^{\bar{h}} - p(\bar{h})$ may be large; but the local error, still small]. In fact as n increases, $|y_n|$ gets smaller as long as $|p(\bar{h})| < 1$; and we can use increasingly larger steps and still control the absolute local error. However, as h grows, so does $|\bar{h}|$; and as $\bar{h}$ leaves the interval of absolute stability, then $|p(\bar{h})|$ grows larger than 1 and causes $|y_n|$ to grow. As $|y_n|$ grows, so does the local error; and we are forced to cut the step down. Thus we would expect that a variable-step Runge-Kutta method (with an absolute-error test) would run with a step size h that keeps $\bar{h}$ near α where $(\alpha, 0)$ is the interval of absolute stability for the method. These intervals are tabulated below for orders one through four.

Order of the Runge-Kutta method	**Interval of absolute stability**
1	(−2, 0)
2	(−2, 0)
3	(−2.51, 0)
4	(−2.78, 0)

For example, the Runge-Kutta-Fehlberg method developed earlier would run on $y' = -100y$ with a step size h that is near .0278. This state of affairs is not desirable since the solution to $y' = -100y$ is almost constant for $x > 0$, and we should be able to track a slowly varying solution even with a large step h.

From the discussion above, an ideal method for solving $y' = \lambda y$ with $\lambda < 0$ appears to be one with an interval of absolute stability of the form $(-\infty, 0)$. If the Runge-Kutta methods could be absolutely stable for all $\bar{h} < 0$, then we could increase the step size with each iteration (and still control the absolute size of the local error) since we would have $\{y_n\} \to 0$ even though $e^{\bar{h}} - p(\bar{h})$ might be growing. A method that does have such a property is the backward Euler method:

$$y_{n+1} = y_n + hf_{n+1}.$$

This is an *implicit* one-step method; but for $y' = \lambda y$ the method reduces to

$$y_{n+1} = y_n + h\lambda y_{n+1}$$

or

$$y_{n+1} = \frac{1}{1 - \bar{h}} y_n.$$

For $\lambda < 0$, we see that $|y_{n+1}| < |y_n|$ no matter what the size of h. If we define $\varphi(\bar{h})$ by

$$\varphi(\bar{h}) = \frac{1}{1 - \bar{h}},$$

then as in the derivation of (7.78), we have

$$y_{n+1} = \varphi(\bar{h})y_n$$

$$y(x_{n+1}) = \varphi(\bar{h})y(x_n) + [e^{\bar{h}} - \varphi(\bar{h})]y(x_n);$$

and we are led to

$$e_{n+1} = e_n\varphi(\bar{h}) + [e^{\bar{h}} - \varphi(\bar{h})]y(x_n). \tag{7.80}$$

Now, $\varphi(\bar{h}) = 1 + \bar{h} + \bar{h}^2 + \cdots$; so

$$e^{\bar{h}} - \varphi(\bar{h}) = -\frac{\bar{h}^2}{2} + 0(\bar{h}^3);$$

and this means that we can control the local error

$$[e^{\bar{h}} - \varphi(\bar{h})]y_n$$

by choosing h small when y_n is large. However if y_n is small, then the local error will remain small even for large h. In addition, $|\varphi(\bar{h})| < 1$ for all $h < 0$; so there is no limit imposed on h by absolute stability considerations; we can use increasingly large h and still control the absolute local errors.

There is, however, a problem posed because the method is implicit. That is, we have to be able to solve $y_{n+1} = y_n + hf(x_{n+1}, y_{n+1})$ for y_{n+1}. For a linear equation $y' = p(x)y + q(x)$, this solution is easy; but in general $f(x, y)$ is nonlinear in y. Moreover, for $|\bar{h}|$ large, fixed-point iteration will not be successful since the convergence criterion (see Section 7.5.1) is $|hf_y(x_{n+1}, y_{n+1})| < 1$ or $|\bar{h}| < 1$. Thus for $|\bar{h}|$ large, a superlinearly convergent method such as Newton's method must be employed to solve $y_{n+1} = y_n + hf(x_{n+1}, y_{n+1})$ in order to execute each step of the backward Euler method. If a large step size h can be used, then the cost of solving the nonlinear equation may well be justified.

7.6.2. Absolute Stability for Multistep Methods

The concept of absolute stability for multistep methods parallels the same concept for one-step methods. In particular, if we apply the k-step method $\sum_{j=0}^{k} \alpha_j y_{n+j} = h \sum_{j=0}^{k} \beta_j f_{n+j}$ to $y' = \lambda y$, we obtain

$$\sum_{j=0}^{k} (\alpha_j - \bar{h}\beta_j) y_{n+j} = 0$$

where $\bar{h} = h\lambda$. On the other hand, using $y'(x) = \lambda y(x)$ in (7.65), we have for $x = x_n$:

$$\sum_{j=0}^{k} (\alpha_j - \bar{h}\beta_j) y(x_{n+j}) = h\tau_n.$$

If the method is pth order, then $h\tau_n = C_{p+1}\bar{h}^{p+1}y(x_n) + C_{p+2}\bar{h}^{p+2}y(x_n) + \cdots$. We subtract these two equations to get a linear difference equation for the global errors:

$$\sum_{j=0}^{k} (\alpha_j - \bar{h}\beta_j) e_{n+j} = h\tau_n$$

where $e_{n+j} = y(x_{n+j}) - y_{n+j}$. If we assume $h\tau_n$ is reasonably constant with varying n, then it is easy to show (Problem 5) that the global errors very nearly satisfy

$$e_m = \mu_1(\bar{r}_1)^m + \mu_2(\bar{r}_2)^m + \cdots + \mu_k(\bar{r}_k)^m - \frac{h\tau}{\bar{h} \sum_{j=0}^{k} \beta_j}$$

where the $\bar{r}_i$ are the roots of $p(r) - \bar{h}\sigma(r) = 0$. The coefficients μ_i are determined by the initial errors $e_0, e_1, \ldots, e_{k-1}$; and we note that if $|\bar{r}_i| < 1$ for $1 \le i \le k$, then the effect of the initial errors damps out and e_m is determined essentially by the local truncation error. If we do not vary h until a (small) block of steps has been executed, then we can expect this analysis to be valid over that block.

When we change h or when $h\tau$ changes substantially, we can restart the analysis with a new set of "initial" errors and different coefficients μ_i.

From the discussion above, we can expect that the effects of past errors will not be magnified if $|\bar{r}_i| < 1, 1 \le i \le k$. Thus, we are led to the definition: a linear k-step method is *absolutely stable* for a particular $\bar{h}$ if $|\bar{r}_i| < 1, 1 \le i \le k$. Recall that $\bar{r}_1 = e^{\bar{h}} + 0(\bar{h}^{p+1})$; so for $\bar{h} > 0$ and small, we must have $\bar{r}_1 > 1$; thus a multistep method cannot be absolutely stable for positive $\bar{h}$. [Similarly as we saw earlier, one-step methods are not absolutely stable for positive $\bar{h}$. The reason that one-step and multistep methods are absolutely unstable for $\bar{h} > 0$ is fairly clear. If we make even one small error in solving $y' = \lambda y$ with $\lambda > 0$, then that error will grow with increasing x since the integral curves for $y' = \lambda y$ are spreading out.] The *interval of absolute stability* for a linear k-step method is the largest interval $(\alpha, 0)$ such that the method is absolutely stable for all $\bar{h}$ in $(\alpha, 0)$. Some methods have no interval of absolute stability; for example (see Fig. 7.11), it is clear that the method (7.41) given by $y_{n+1} = y_n + 2hf_{n+1}$ has $\bar{r}_2 < -1$ when $\bar{h} < 0$. The same is true for the method (7.42); so not only are these methods relatively unstable for $\bar{h} < 0$, but also they are absolutely unstable for $\bar{h} < 0$.

On the other hand, the Adams methods clearly have a nondegenerate interval of absolute stability. To see this point, we note again that $\bar{r}_1 = e^{\bar{h}} + 0(\bar{h}^{p+1})$ for any pth order method. Thus for $\bar{h} < 0$ and small, the condition $|\bar{r}_1| < 1$ must hold for any multistep method, even for (7.41) and (7.42). However for a k-step Adams method, the other roots $\bar{r}_j, 2 \le j \le k$, all tend to zero as $\bar{h} \to 0$; so there must be a value $\alpha < 0$ such that $|\bar{r}_i| < 1$ for $\bar{h}$ in $(\alpha, 0)$. These intervals are listed below for the Adams methods (7.44) and (7.45) where the method is absolutely stable in $(\alpha, 0)$.

Method	α	Method	α
(7.44*a*)	-2	(7.45*a*)	$-\infty$
(7.44*b*)	-1	(7.45*b*)	$-\infty$
(7.44*c*)	$-6/11$	(7.45*c*)	-6
(7.44*d*)	$-3/10$	(7.45*d*)	-3

So far, we have discussed relative and absolute stability in the context of only a single multistep method. When two multistep methods are used as a predictor-corrector pair, then the stability analyses must be modified somewhat. It is tempting to assume that the stability characteristics of a predictor-corrector pair depends on the stability characteristics of the corrector alone since the predictor serves only to supply an initial guess for the corrector. This is not the case [see Lambert (1973)]. For example, the predictor-corrector pair (7.44d) and (7.45d) (using one iteration of the corrector) has an interval of absolute stability given by $(-1.25, 0)$ although from the table above, the corrector is absolutely stable on $(-3, 0)$. If the corrector is iterated to convergence, then the stability characteristics of the predictor-corrector pair are the same as those of the corrector alone. An elementary analysis of the stability of predictor-corrector pairs is not difficult, but the calculations are tedious and add little to understanding the basics of stability.

Finally we note that the ideas of stability are important mostly to the design and selection of numerical methods and to interpreting the behavior and output of the method. That is, a variable-step algorithm selects the appropriate step sizes h adaptively and uses a local error monitor of some sort—the user of the algorithm has little control of the step-selection strategy. If we are asking the algorithm to control absolute errors, then the absolute stability characteristics of the method being used will determine (in part) the step sizes that are used. If we ask the algorithm to control relative errors, then the relative stability characteristics of the method play a role in determining the steps.

7.6.3. Stability Considerations for Systems of Equations and the Problem of Stiffness

As noted in Section 7.4, in practice most problems that arise involving differential equations consist of systems of differential equations. If we can convert such a system to a first-order system $\mathbf{u}' = \mathbf{F}(x, \mathbf{u})$, then we can employ vector versions of Runge-Kutta methods or predictor-corrector methods to solve $\mathbf{u}' = \mathbf{F}(x, \mathbf{u})$. For these vector versions, stability considerations are quite important, and we briefly outline the dimensions of the problem in this section. The equation $\mathbf{u}' = \mathbf{F}(x, \mathbf{u})$ can be linearized in the same way in which a local approximation to $y' = f(x, y)$ was found of the form $y' = \lambda y + c$. In particular, suppose that $\mathbf{u}' = \mathbf{F}(x, \mathbf{u})$ is an m-dimensional problem and suppose that we have an approximation $(x_n, \mathbf{u}_n)$ to $(x_n, \mathbf{u}(x_n))$. Then a linearized approximation to $\mathbf{u}' = \mathbf{F}(x, \mathbf{u})$ takes the form $\mathbf{u}' = J\mathbf{u} + \mathbf{c}$ where J designates the Jacobian matrix $\partial\mathbf{F}/\partial\mathbf{u}$, evaluated at $(x_n, \mathbf{u}_n)$; J is an $(m \times m)$ constant matrix. If J is nonsingular, we can rewrite $\mathbf{u}' = J\mathbf{u} + \mathbf{c}$ in the form $\mathbf{u}' = J(\mathbf{u} + J^{-1}\mathbf{c})$; and if we make the change of variables $\mathbf{w} = \mathbf{u} + J^{-1}\mathbf{c}$, we obtain the equation $\mathbf{w}' = J\mathbf{w}$. Thus for an analysis of a method applied to $\mathbf{u}' = \mathbf{F}(x, \mathbf{u})$ we will consider the test equation $\mathbf{u}' = A\mathbf{u}$ where A is the Jacobian [the ijth entry of A is the derivative of the ith component function of $\mathbf{F}(x, \mathbf{u})$, differentiated with respect to the dependent variable u_j and evaluated at $(x_n, \mathbf{u}_n)$]. We will see that the stability characteristics of a method applied to $\mathbf{u}' = A\mathbf{u}$ are bound up with the eigenvalues of A.

We begin with an analysis of Runge-Kutta methods applied to $\mathbf{u}' = A\mathbf{u}$, $\mathbf{u}(0) = \mathbf{u}_0$. To simplify the discussion, assume that A is diagonalizable; so we assume that there is a matrix S such that $S^{-1}AS = D$ where D is diagonal (assuming that A can be diagonalized is not crucial but relieves us of the burden of using Jordan canonical form). We now make a change of variables in $\mathbf{u}' =$

$A\mathbf{u}$, and set $\mathbf{u}(x) = S\mathbf{y}(x)$ or $\mathbf{y}(x) = S^{-1}\mathbf{u}(x)$. With this change, $\mathbf{u}' = A\mathbf{u}$, $\mathbf{u}(0) = \mathbf{u}_0$ transforms to $S\mathbf{y}' = AS\mathbf{y}$, $\mathbf{y}(0) = \mathbf{y}_0 = S^{-1}\mathbf{u}_0$. Multiplying both sides of $S\mathbf{y}' = AS\mathbf{y}$ by S^{-1}, we obtain a new system $\mathbf{y}' = S^{-1}AS\mathbf{y}$ or $\mathbf{y}' = D\mathbf{y}$. The problem $\mathbf{y}' = D\mathbf{y}$, $\mathbf{y}(0) = S^{-1}\mathbf{u}_0$ is an uncoupled first-order system, and the solution of $\mathbf{y}' = D\mathbf{y}$ is related to the original system by

$$\mathbf{u}(x) = S\mathbf{y}(x).$$

We now observe that if a Runge-Kutta method is applied to $\mathbf{u}' = A\mathbf{u}$ and to $\mathbf{y}' = D\mathbf{y}$, $\mathbf{y}_0 = S^{-1}\mathbf{u}_0$, then the results produced will also be related by $\mathbf{u}_k = S\mathbf{y}_k$, $k = 0, 1, \ldots$. To see this relationship, suppose we apply a Runge-Kutta method to $\mathbf{u}' = A\mathbf{u}$. It is easy to verify that we will produce the sequence $\{\mathbf{u}_i\}$ where

$$\mathbf{u}_{n+1} = p(hA)\mathbf{u}_n, \qquad n = 0, 1, \ldots . \tag{7.81}$$

In (7.81) if the method is qth order, then $p(r) = 1 + r + r^2/2! + \cdots + r^q/q!$, and $p(hA)$ designates the matrix polynomial

$$p(hA) = I + hA + \frac{h^2}{2!}A^2 + \cdots + \frac{h^q}{q!}A^q.$$

If we apply the same method to $\mathbf{y}' = D\mathbf{y}$, $\mathbf{y}(0) = \mathbf{y}_0$, then we have

$$\mathbf{y}_{n+1} = p(hD)\mathbf{y}_n, \qquad n = 0, 1, \ldots . \tag{7.82}$$

Now since $\mathbf{y}_0 = S^{-1}\mathbf{u}_0$ and since $D = S^{-1}AS$, it is easy to show that $\mathbf{u}_i = S\mathbf{y}_i$, $i = 0, 1, \ldots$. Since $\mathbf{u}_i = S\mathbf{y}_i$ and $\mathbf{u}(x_i) = S\mathbf{y}(x_i)$ for all i, the global errors are clearly related by

$$\mathbf{u}(x_i) - \mathbf{u}_i = S(\mathbf{y}(x_i) - \mathbf{y}_i), \qquad i = 0, 1, \ldots . \tag{7.83}$$

Thus if we can determine how the errors behave when the method is applied to the diagonal system $\mathbf{y}' = D\mathbf{y}$, we can then deduce how the errors behave for $\mathbf{u}' = A\mathbf{u}$.

To treat the errors for $\mathbf{y}' = D\mathbf{y}$, let D be the diagonal matrix

$$D = \begin{bmatrix} \lambda_1 & 0 & \cdots & 0 \\ 0 & \lambda_2 & \cdots & 0 \\ \vdots & & & \vdots \\ 0 & 0 & \cdots & \lambda_m \end{bmatrix}.$$

(Note that the diagonal entries λ_i are the eigenvalues of A.) The solution of $\mathbf{y}' = D\mathbf{y}$, $\mathbf{y}(0) = \mathbf{y}_0$ is

$$\mathbf{y}(x) = \begin{bmatrix} c_1 e^{\lambda_1 x} \\ c_2 e^{\lambda_2 x} \\ \vdots \\ c_m e^{\lambda_m x} \end{bmatrix} \quad \text{where} \quad \mathbf{y}_0 = \begin{bmatrix} c_1 \\ c_2 \\ \vdots \\ c_m \end{bmatrix}. \tag{7.84a}$$

It is easy to verify from the form of the solution given above that

$$\begin{bmatrix} y_1(x_{n+1}) \\ y_2(x_{n+1}) \\ \vdots \\ y_m(x_{n+1}) \end{bmatrix} = \begin{bmatrix} y_1(x_n)e^{\lambda_1 h} \\ y_2(x_n)e^{\lambda_2 h} \\ \vdots \\ y_m(x_n)e^{\lambda_m h} \end{bmatrix} \quad \text{where} \quad \mathbf{y}(x) = \begin{bmatrix} y_1(x) \\ y_2(x) \\ \vdots \\ y_m(x) \end{bmatrix}. \tag{7.84b}$$

Defining a new diagonal matrix Q from

$$Q = \begin{bmatrix} e^{\bar{h}_1} & 0 & \cdots & 0 \\ 0 & e^{\bar{h}_2} & \cdots & 0 \\ \vdots & & & \vdots \\ 0 & 0 & \cdots & e^{\bar{h}_m} \end{bmatrix}$$

where $\bar{h}_i = h\lambda_i$, we can rewrite (7.84b) as

$$\mathbf{y}(x_{n+1}) = Q\mathbf{y}(x_n), \qquad n = 0, 1, \ldots. \tag{7.85}$$

(The matrix Q is the matrix exponential e^{hD}, but we do not need this detail here.) If we define the error vector $\mathbf{e}_j$ by $\mathbf{e}_j = \mathbf{y}(x_j) - \mathbf{y}_j$, then we can use (7.82) and (7.85) to obtain

$$\mathbf{e}_{n+1} = p(hD)\mathbf{e}_n + [Q - p(hD)]\mathbf{y}(x_n), \qquad n = 0, 1, \ldots. \tag{7.86}$$

Note that $p(hD)$ is a diagonal matrix and that $p(hD)$ has the form

$$p(hD) = \begin{bmatrix} p(\bar{h}_1) & 0 & \cdots & 0 \\ 0 & p(\bar{h}_2) & \cdots & 0 \\ \vdots & & & \vdots \\ 0 & 0 & \cdots & p(\bar{h}_m) \end{bmatrix}$$

The error relation (7.86) is a vector version of the already familiar (7.78) but has some quite interesting consequences. To explain, we first recall that Q and $p(hD)$ are diagonal matrices; and if we let $e_j^{(k)}$ denote the kth component of the error vector $\mathbf{e}_j$, then (7.86) leads (for $1 \le k \le m$) to

$$e_{n+1}^{(k)} = p(\bar{h}_k)e_n^{(k)} + [e^{\bar{h}_k} - p(\bar{h}_k)]y_k(x_n). \tag{7.87}$$

Thus the kth component of the error vector $\mathbf{e}_{n+1}$ is related to the kth component of $\mathbf{e}_n$ by (7.87), and thus the behavior of the errors $\mathbf{e}_i$ is jointly determined by all the values $\bar{h}_1, \bar{h}_2, \ldots, \bar{h}_m$. For instance, if we wish to control the absolute errors of a Runge-Kutta method applied to $\mathbf{y}' = D\mathbf{y}$, then we will have to use a step size h that takes into account all the values $h\lambda_1 = \bar{h}_1, h\lambda_2 = \bar{h}_2, \ldots, h\lambda_m = \bar{h}_m$ (if our step h is such that $|p(\bar{h}_j)| > 1$ for some j, $1 \le j \le m$, then the jth component of the error vectors is growing). Thus, the choice of a step size h is limited by the worst case of $\bar{h}_1, \bar{h}_2, \ldots, \bar{h}_m$; and (as we shall see) the worst case of $\bar{h}_j$ sometimes corresponds to the least significant component of the solution.

So that we do not lose track of the error analysis, we summarize what we have so far. If we apply a Runge-Kutta method to $\mathbf{u}' = \mathbf{F}(x, \mathbf{u})$, the method behaves (locally) as though we were solving $\mathbf{u}' = A\mathbf{u}$. The behavior of the method on $\mathbf{u}' = A\mathbf{u}$ is connected to the method applied to $\mathbf{y}' = D\mathbf{y}$ by $\mathbf{u}_i = S\mathbf{y}_i$ and $\mathbf{u}(x_i) - \mathbf{u}_i = S(\mathbf{y}(x_i) - \mathbf{y}_i)$. If we let $\hat{\mathbf{e}}_i = \mathbf{u}(x_i) - \mathbf{u}_i$, then the relation

$$\hat{\mathbf{e}}_i = S\mathbf{e}_i, \qquad i = 0, 1, \ldots, \tag{7.88}$$

tells us that if one component of $\mathbf{e}_i$ is growing absolutely, then some of the components (and perhaps all the components) of $\hat{\mathbf{e}}_i$ are growing absolutely. Since $\mathbf{e}_i$ can be controlled in an absolute sense only by restricting $\bar{h}_1, \bar{h}_2, \ldots, \bar{h}_m$, the same restrictions are necessary to control the errors $\hat{\mathbf{e}}_i$. Thus to control the errors (absolutely) when we solve $\mathbf{u}' = \mathbf{F}(x, \mathbf{u})$, we will need to choose h so that $|p(\bar{h}_j)| < 1$ for $1 \le j \le m$ where $\bar{h}_j = h\lambda_j$ and the λ_j are the eigenvalues of A.

Since the eigenvalues of A might be complex, the condition $|p(\bar{h}_j)| < 1$ has to be interpreted for complex numbers. This interpretation is straightforward, and we say a Runge-Kutta method has a *region of absolute stability* R if R is the largest region in the complex plane such that $|p(\bar{h})| < 1$ for all $\bar{h}$ in R. These regions are depicted in Fig. 7.12 for the q-stage explicit Runge-Kutta methods of order q for $1 \le q \le 4$.

From Fig. 7.12, if the λ_j are the (possibly complex) eigenvalues of A, then

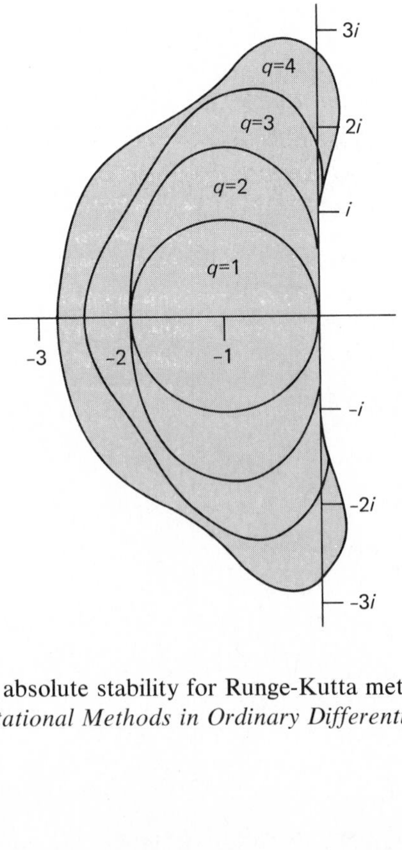

Figure 7.12 Regions of absolute stability for Runge-Kutta methods. (Adapted from J. D. Lambert, *Computational Methods in Ordinary Differential Equations*. New York: Wiley, 1973.)

$|p(\bar{h}_j)| < 1$ for $1 \le j \le m$ only if h is chosen so that the multiples $h\lambda_j$ are all in R, $1 \le j \le m$. From the picture, we also see that choosing h so that $\bar{h}_j$ is in R will not normally be possible unless λ_j has negative real parts (this condition corresponds to requiring $\lambda < 0$ for scalar equations $y' = \lambda y$). To see what $Re(\lambda_j) < 0$ means in terms of the solution of $\mathbf{u}' = A\mathbf{u}$, look again at $\mathbf{y}' = D\mathbf{y}$. The solution of $\mathbf{y}' = D\mathbf{y}$, $\mathbf{y}(0) = \mathbf{y}_0$ is given in (7.84a); and we recall the definition of $e^{\lambda x}$ for complex $\lambda = \alpha + i\beta$:

$$e^{\lambda x} = e^{(\alpha + i\beta)x} = e^{\alpha x}(\cos \beta x + i \sin \beta x). \tag{7.89}$$

In particular if $Re(\lambda) < 0$, then $e^{\lambda x} \to 0$ as $x \to \infty$. Therefore from (7.84a) and from $\mathbf{u}(x) = S\mathbf{y}(x)$, we conclude that all the components of $\mathbf{u}(x)$ tend to zero as x grows whenever $Re(\lambda_j) < 0$ for all j, $1 \le j \le m$. However, by (7.89), it is clear that $|e^{\lambda x}| \to 0$ very fast whenever $Re(\lambda) < 0$ and $|Re(\lambda)|$ is large. Thus when all the eigenvalues of A have negative real parts, the most significant contribution to a component of $\mathbf{u}(x)$ comes from the eigenvalue whose real part is smallest in magnitude. For example if we order the eigenvalues of A so that $Re(\lambda_m) \le \cdots \le Re(\lambda_2) \le Re(\lambda_1) < 0$, then [since $\mathbf{u}(x) = S\mathbf{y}(x)$] the ith component function of $\mathbf{u}(x)$ is

$$u_i(x) = s_{i1}c_1 e^{\lambda_1 x} + s_{i2}c_2 e^{\lambda_2 x} + \cdots + s_{im}c_m e^{\lambda_m x}. \tag{7.90}$$

As x grows, $e^{\lambda_m x}$ damps out very quickly and becomes insignificant compared with $e^{\lambda_1 x}$. However if we are trying to control the absolute errors, then λ_m is much more significant than λ_1 since we have to choose h so that $\bar{h}_m = h\lambda_m$ is in the region of absolute stability. In other words, the component of $\mathbf{u}_i(x)$, $e^{\lambda_m x}$, that interests us least is precisely the one that puts the most stringent constraints on the choice of the step size.

The definition of a stiff system is based on the discussion above. Loosely put, a system $\mathbf{u}' = \mathbf{F}(x, \mathbf{u})$ is called *stiff* in a neighborhood of $(x, \mathbf{u})$ if the eigenvalues of the Jacobian of F all have negative real parts where $Re(\lambda_m) \le Re(\lambda_{m-1}) \le \cdots \le Re(\lambda_2) \le Re(\lambda_1) < 0$ and where $Re(\lambda_m) << Re(\lambda_1)$. To be completely precise, we should specify also that the component of the solution determined by λ_m is insignificant [see Shampine and Gear (1979)]. If the component determined by λ_m is significant, then we will need to use small h to track this rapidly varying component and the system is not really stiff (yet). When this component becomes insignificant, then it is possible to control the absolute local error even with "large" h [see (7.86)]. However, with large h, absolute instability becomes a factor with explicit Runge-Kutta methods [again, see (7.86)]. With stiff systems, of course, we are usually concerned with controlling absolute errors rather than relative errors.

An exactly analogous analysis can be made for multistep methods, but we omit the details and say only that if we apply a multistep method to $\mathbf{y}' = D\mathbf{y}$, then the effect is to apply the method componentwise to the system using a common step size h for each component. This statement means that for relative stability in a k-step method we need $|r_1(\bar{h}_j)| > |r_i(\bar{h}_j)|$ for $2 \le i \le k$ and for all $\bar{h}_j$,

$1 \le j \le m$. For absolute stability, we will need $|r_i(h_j)| < 1$ for $1 \le i \le k$ and $1 \le j \le m$.

There are several ways that we can attempt to detect stiffness when we are solving $\mathbf{u}' = \mathbf{F}(x, \mathbf{u})$; see for example Shampine and Gordon (1975). If stiffness begins to limit the efficiency of the method being used, we can switch to a procedure designed for stiff equations. Without our going into all the details [see Gear (1971)], a method is called A-stable if it is absolutely stable for all $\bar{h}$ such that $Re(\bar{h}) < 0$. For example, the backward Euler method and the trapezoidal method are A-stable and are well suited for stiff systems except that these methods are low-order methods. Dahlquist has shown that no explicit multistep method can be A-stable and that no implicit multistep method can be A-stable unless it has order two or less. The R-stage implicit Runge-Kutta methods of order $2R$ are A-stable, but the vector version applied to an m-dimensional system requires the solution of a system of mR nonlinear equations in order to execute each step. Gear has defined a method to be stiffly stable if the method is (roughly) accurate for $|\bar{h}|$ small and is absolutely stable for $Re(\bar{h}) < a < 0$. See Gear (1971) or Lambert (1973) for a listing of some stiffly stable methods; these are implicit multistep methods of the form

$$\sum_{j=0}^{k} \alpha_j y_{n+j} = h\beta_k f_{n+k}.$$

Such stiffly stable methods represent a reasonable way of attacking a stiff problem.

PROBLEMS, SECTION 7.6.3

1. To illustrate absolute stability, apply Heun's method to $y' = -30y$, $y(0) = 1$. Use $h = .1$ and $h = .05$; print the estimates in increments of .1 for $0 \le x \le 2$.
2. Modify subroutine DESOLV in Fig. 7.9 to use an absolute-error test. Use the modified program on the equation in Problem 1 and print the estimates in increments of .2 for $0 \le x \le 5$. Also print the last full step used in each interval and the absolute and relative errors.
3. Rewrite the equation $u' = \lambda u + c$, $u(0) = u_0$ as $u' = \lambda(u + c/\lambda)$, $u(0) = u_0$. Make the change of variables $y(x) = u(x) + c/\lambda$, and consider the equivalent equation $y' = \lambda y$, $y(0) = u_0 + c/\lambda$. Show that any second-order Runge-Kutta method applied to these two problems will produce estimates u_n and y_n where $y_n = u_n + c/\lambda$, $n = 0, 1, \ldots$. [Hint: Use induction and recall that $y_{k+1} = p(h\lambda)y_k$, $k = 0, 1, \ldots$.]
4. Repeat Problems 2 and 3 for the equation $u' = -30u + 60$, $u(0) = 3$; and compare the results.
5. The global errors for a multistep method applied to $y' = \lambda y$ satisfy (approximately)

$$\sum_{j=0}^{k} (\alpha_j - \bar{h}\beta_j)e_{n+j} = h\tau.$$

This is a nonhomogeneous difference equation of the form

$$a_k z_{n+k} + \cdots + a_1 z_{n+1} + a_0 z_n = b. \tag{NHDE}$$

Show that the constant sequence $\{w_i\}$ defined by $w_i = b/(a_0 + a_1 + \cdots + a_k)$, $i = 0, 1, \ldots$, satisfies (NHDE), provided that $a_0 + a_1 + \cdots + a_k \neq 0$. Next show that every sequence $\{z_i\}$ satisfying (NHDE) has the form $z_i = u_i + w_i$, $i = 0, 1, \ldots$, where $\{u_i\}$ is a solution to the homogeneous equation $a_k z_{n+k} + \cdots + a_1 z_{n+1} + a_0 z_n = 0$.

6. Show that the method (7.42) is absolutely unstable for all $\bar{h} < 0$.
7. Verify that (7.44b) is absolutely stable for $-1 < \bar{h} < 0$; and (7.45b), for $-\infty < \bar{h} < 0$. [Hint: For (7.44b), what values of $\bar{h}$ will make $r = -1$ a root of $\rho(r) - \bar{h}\sigma(r) = 0$?]
8. Verify that (7.44c) is absolutely stable for $-6/11 < \bar{h} < 0$; and (7.45c), for $-6 < \bar{h} < 0$.

7.7 THE METHOD OF COLLOCATION

The method of collocation is a procedure unlike any of the single-step or multistep methods described in the previous sections. The output from a collocation method is a function (usually a polynomial) that approximates the solution of a differential equation throughout some range, $a \le x \le b$. This method is in contrast to single-step and multistep methods, which produce a discrete set of values $y_1, y_2, \ldots, y_n$. The basic ideas involved in collocation are fairly simple. As a starting point, let us consider the general initial-value problem

$$y'(x) = f(x, y(x)), \qquad y(0) = y_0, \qquad 0 \le x \le b. \tag{7.91}$$

We have chosen the initial condition to be specified at $x_0 = 0$ in order to simplify the development a bit although this choice of x_0 is not a requirement.

Suppose next that there is a polynomial $p(x) = a_0 + a_1 x + \cdots + a_n x^n$ that is a good approximation to the solution, $y(x)$, of (7.91). In order to determine $p(x)$, we consider the expression

$$e(x) = p'(x) - f(x, p(x)). \tag{7.92}$$

Our object is to find the coefficients $a_0, a_1, \ldots, a_n$ of $p(x)$ so as to make the error function, $e(x)$, as small as possible. These coefficients can be chosen several ways. For example we might ask that $a_0, a_1, \ldots, a_n$ be chosen so as to minimize either of the expressions (7.93a) or (7.93b):

$$\int_0^b [e(x)]^2 \, dx \tag{7.93a}$$

$$\sum_{j=0}^{m} [e(t_j)]^2, \qquad 0 \le t_0 < t_1 < \cdots < t_m \le b. \tag{7.93b}$$

Thus (7.93a) and (7.93b) represent least-squares problems, which (we note without further discussion) are related to methods that are termed *Galerkin procedures*. If $f(x, y)$ should be nonlinear in y [and if $f(x, y)$ is not too complicated], the parameters $a_0, a_1, \ldots, a_n$ in (7.93) can be found by solving the "normal equations" (see Problem 5).

Rather than using (7.93a) or (7.93b) as a criterion to determine the coefficients of $p(x)$, the method of collocation proceeds by asking that $p(x)$ be chosen so that $e(t_i) = 0$ at certain selected "collocation points," $t_0, t_1, \ldots, t_n$. (We note in passing that in the literature *collocation* is used interchangeably with *interpolation*.) Thus collocation leads to the system of $n + 1$ equations

$$e(t_i) = 0, \qquad i = 0, 1, \ldots, n.$$

In practice we normally ask that $p(x)$ satisfy the initial condition so that the approximate solution, $p(x)$, has the correct starting value. If $p(0) = y_0$, then clearly $a_0 = y_0$; so we are left with finding just n coefficients $a_1, a_2, \ldots, a_n$. We specify these by choosing n collocation (or interpolation) points $t_1, t_2, \ldots, t_n$ and insisting that

$$\begin{aligned} e(t_1) &= 0 \\ e(t_2) &= 0 \\ &\vdots \\ e(t_n) &= 0. \end{aligned} \tag{7.94}$$

Writing out a representative equation, $e(t_i) = 0$, for the system (7.94), we find the equation has the form

$$a_1 + 2a_2t_i + \cdots + na_nt_i^{n-1} - f(t_i, y_0 + a_1t_i + \cdots + a_nt_i^n) = 0. \tag{7.95}$$

If $f(x, y)$ is a nonlinear function of y, we see from (7.95) that (7.94) is a nonlinear system of n equations in n unknowns. Newton's method for several variables as given in Chapter 4 would then provide a tool for solving this system.

EXAMPLE 7.13. Consider the initial-value problem $y' = 2xy$, $y(0) = 1$, $0 \le x \le 1$. In this example, we will determine a cubic polynomial, $p(x)$, such that $p(0) = 1$ and

$$p'(t_i) - 2t_ip(t_i) = 0, \qquad i = 1, 2, 3.$$

As collocation points, we select the zeros of the Chebyshev polynomial $T_3(x)$, translated to the interval [0, 1]. Thus $t_1 = (2 - \sqrt{3})/4$, $t_2 = 0.5$, $t_3 = (2 + \sqrt{3})/4$. Since $p(x) = 1 + a_1x + a_2x^2 + a_3x^3$ and $p'(x) = a_1 + 2a_2x + 3a_3x^2$, we are led to the system:

$$\begin{aligned} (1 - 2t_1^2)a_1 + 2t_1(1 - t_1^2)a_2 + t_1^2(3 - 2t_1^2)a_3 &= 2t_1 \\ (1 - 2t_2^2)a_1 + 2t_2(1 - t_2^2)a_2 + t_2^2(3 - 2t_2^2)a_3 &= 2t_2 \\ (1 - 2t_3^2)a_1 + 2t_3(1 - t_3^2)a_2 + t_3^2(3 - 2t_3^2)a_3 &= 2t_3, \end{aligned}$$

which for this example is a (3×3) *linear* system of equations for the unknowns a_1, a_2, and a_3. [This system is linear since $f(x, y) = 2xy$ is a linear function in y.] Solving this linear system, we find

$$p(x) = 1 + 0.15529x - 0.33882x^2 + 1.88235x^3$$

as an approximation to the true solution $y(x) = e^{x^2}$. The following table lists representative values of $p(x)$ and $y(x)$.

x	$p(x)$	$y(x)$
0.25	1.04706	1.06449
0.50	1.22823	1.28403
0.75	1.72000	1.75505
1.00	2.69882	2.71828

Several observations should be made about collocation. First, there are many variations that can be useful. For example, we need not have used algebraic polynomials as an approximating form but could have chosen $p(x) = \Sigma_{i=0}^{n} a_i\phi_i(x)$ as our basic approximating form where the functions $\phi_i(x)$ might have been cubic splines, trigonometric functions, etc. Further, by differentiating the equation $y' = f(x, y)$ with respect to x, we could have insisted that $e(t_i) = 0$ and $e'(t_i) = 0$ at the collocation points (this would be a form of Hermite interpolation). In selecting the collocation points t_i in the example above, we chose a natural set of points with regard to the approximating form and assumed that the zeros of the Chebyshev polynomial $T_n(x)$ are well suited for algebraic polynomial interpolation. If we were using trigonometric polynomials as an approximating form, we might consider using equally spaced collocation points. Finally, collocation methods can be used in boundary-value problems (see Section 7.8) as well as in initial-value problems. In fact, this application of collocation is the most frequent.

In the example above, we see that the estimates to $y(x)$ are off by about as much as 4%. By choosing a polynomial of higher degree, we could have achieved a better approximation but at the expense of having to solve a larger system. However, even the collocation solution found in Example 7.13 is useful in that this solution provides an approximation to $y(x)$ *throughout* the interval. That is, collocation can be used to estimate roughly the behavior of the solution in the interval in question. This feature of collocation is useful in problems of engineering design in which a solution is sought that is optimal with respect to various engineering parameters. Finally, we note that the collocation solution provides an approximation $p(x)$ that can be evaluated at any point x. The utility of having a "closed form" approximation to the solution, $y(x)$, is rarely mentioned, and we wish to discuss this point briefly.

Suppose $p(x)$ is an approximation to $y(x)$ where $y(x)$ is the solution of $y' = f(x, y)$, $y(x_0) = y_0$. It seems intuitively plausible to set $y(x) = p(x) + c(x)$ and use the equation

$$p'(x) + c'(x) = f(x, p(x) + c(x)), \qquad p(x_0) + c(x_0) = y_0, \tag{7.96}$$

to solve for the "correction function," $c(x)$. If $p(x)$ is a known function, then (7.96) reduces to the initial-value problem

$$c'(x) = f(x, p(x) + c(x)) - p'(x), \qquad c(x_0) = y_0 - p(x_0). \tag{7.97}$$

It is, therefore, plausible to use any of the single-step or multistep methods developed previously to solve (7.97) for the correction, $c(x)$. Having $c(x)$, we could take $p(x) + c(x)$ as our approximation to $y(x)$. The hope in doing so is that the approximation, $p(x)$, would carry the bulk of the estimation to $y(x)$ and that $c(x)$ would be used to "fine-tune" the approximation. In this way, we might hope to use a larger step size in solving (7.97) than we could use for the same accuracy if we tried to solve $y' = f(x, y)$ directly. This technique is widely used in the area of celestial mechanics (e.g., Encke's method) and is demonstrated in Example 7.14.

EXAMPLE 7.14. In this example we return again to the equation $y' = 2xy$, $y(0) = 1$ used in Example 7.13. We use the collocation solution $p(x)$ found in Example 7.13 as an approximation, and then use Euler's method to solve

$$c'(x) = f(x, p(x) + c(x)) - p'(x), \qquad c(0) = 0. \tag{7.98}$$

In the table below, we tabulate the results of this experiment, in which we used $h = 0.0625$. In the column headed E, we list the results from solving $y' = 2xy$ with Euler's method. In the column headed CE, we list the estimates given by $c_i + p(x_i)$ where c_i is the result of solving (7.98) with Euler's method.

x_i	$y(x_i)$	CE	E
0.25	1.06449	1.05239	1.04755
0.50	1.28403	1.27368	1.23936
0.75	1.75505	1.75039	1.65010
1.00	2.71828	2.69672	2.46401

The program was rerun with $h = 0.02$ and yielded at $x = 0.5$, $CE = 1.28095$, $E = 1.26926$ and at $x = 1.$, $CE = 2.71109$, $E = 2.63082$.

Techniques such as the one illustrated above are useful when the evaluation of $f(x, y)$ is expensive. For example, problems in celestial mechanics typically involve solving the equations of motion for a body in vacuum, in which the only force present is gravity. An accurate evaluation of gravity at a point in space usually requires the evaluation of a *spherical-harmonic* series expansion, an evaluation which is time consuming. However, an approximate closed-form solution for the trajectory exists if we assume inverse-square gravity. This approximate trajectory, the Kepler solution, could play the same role in the equations of motion that $p(x)$ plays in (7.48). Thus we would be solving for the small perturbation that is necessary to correct the Kepler solution to the true trajectory.

7.8 BOUNDARY-VALUE PROBLEMS FOR ORDINARY DIFFERENTIAL EQUATIONS

The type of problem we will consider in this section is called a *boundary-value* problem, in which an interval $[a, b]$ is given and we seek a function $y(x)$ such that

$$y'' = f(x, y, y'); \qquad y(a) = \alpha, \qquad y(b) = \beta. \tag{7.99}$$

Thus in (7.99), instead of prescribing $y(a)$ and $y'(a)$ as in an initial-value problem, we prescribe values for $y(x)$ at $x = a$ and at $x = b$. Such problems are also called *two-point boundary-value* problems and the prescribed conditions $y(a) = \alpha$ and $y(b) = \beta$ are called the *boundary conditions*. Equations of this sort arise quite frequently in scientific and engineering applications and often pose a difficult challenge to the numerical analyst. The equation (7.99) may have no solution, a unique solution, or infinitely many solutions (see Problem 2). Problems less specific than (7.99) often occur. For example the boundary conditions in (7.99) need not be of the form $y(a) = \alpha$ and $y(b) = \beta$ but might take the form $k_1y(a) + k_2y'(a) = \alpha$ and $l_1y(b) + l_2y'(b) = \beta$ (or even more complicated forms). In addition, the function $f(x, y, y')$ may contain a parameter, λ, giving rise to an *eigenvalue problem,* in which we must also determine those values λ for which a solution exists and then find the solution. However, in order to keep our discussion to a reasonable length, we will restrict ourselves to the problem given in (7.99).

7.8.1. Finite-Difference Methods

A class of methods called *finite-difference methods* are frequently employed to solve (7.99) numerically. In a finite-difference procedure, we choose equally spaced points $a = x_0 < x_1 < x_2 < \cdots < x_{n-1} < x_n = b$ where $x_i = x_0 + ih$. We next replace the derivatives in (7.99) by suitably chosen difference quotients defined in terms of the points x_i. A typical choice is to use the *centered difference* approximations (see Section 6.4)

$$\begin{aligned} y'(x_i) &\approx \frac{y(x_{i+1}) - y(x_{i-1})}{2h} \\ y''(x_i) &\approx \frac{y(x_{i+1}) - 2y(x_i) + y(x_{i-1})}{h^2}. \end{aligned} \tag{7.100}$$

When these approximations are used in (7.99), we obtain a system of equations of the form

$$\frac{y_{i+1} - 2y_i + y_{i-1}}{h^2} = f\left(x_i, y_i, \frac{y_{i+1} - y_{i-1}}{2h}\right), \quad \begin{cases} i = 1, 2, \ldots, n-1 \\ y_0 = \alpha,\ y_n = \beta \end{cases} \tag{7.101}$$

In (7.101) we have designated $y_1, y_2, \ldots, y_{n-1}$ as the unknowns since any solution $y(x)$ of (7.99) will probably not satisfy (7.101) exactly. We think of the system (7.101) as an approximation to the equation in (7.99) and hope that for a sufficiently small spacing, h, the solutions y_i of (7.101) are good approximations to $y(x_i)$.

If the function $f(x, y, y')$ is nonlinear in y and y', we are faced with solving a nonlinear system of $(n - 1)$ equations in $(n - 1)$ unknowns. This circumstance would require that we use a scheme like Newton's method for several variables to determine $y_1, y_2, \ldots, y_{n-1}$. Note that Newton's method applied to (7.101) takes the form $\mathbf{Y}_{k+1} = \mathbf{Y}_k - J(\mathbf{Y}_k)^{-1}\mathbf{G}(\mathbf{Y}_k)$ where $\mathbf{Y}_k = [y_1^{(k)}, y_2^{(k)}, \ldots, y_{n-1}^{(k)}]^{\mathrm{T}}$ and where the ith component of the vector function $\mathbf{G}(\mathbf{Y})$ is given by

$$g_i(\mathbf{Y}) = y_{i+1} - 2y_i + y_{i-1} - h^2 f\left(x_i, y_i, \frac{y_{i+1} - y_{i-1}}{2h}\right).$$

Since $\dfrac{\partial g_i}{\partial y_j}$ is zero except for $j = i - 1, i, i + 1$, the $((n - 1) \times (n - 1))$ Jacobian matrix $J(\mathbf{Y}_k)$ is tridiagonal; and it is relatively easy to apply Newton's method to (7.101) even for large n (say for $n = 100$ or $n = 1000$). The starting vector $\mathbf{Y}_0$ could have components obtained by a linear interpolation of the boundary conditions:

$$y_i^{(0)} = \frac{x_i - a}{b - a}\beta + \frac{b - x_i}{b - a}\alpha, \qquad 1 \le i \le n - 1.$$

In practice $f(x, y, y')$ is frequently linear in y and y' so that (7.99) has the form

$$y'' - p(x)y' - q(x)y = r(x); \qquad y(a) = \alpha, \qquad y(b) = \beta. \tag{7.102}$$

The special case given in (7.102) is an important special case and contains, for example, many "Sturm-Liouville" problems. If we now make the finite difference approximations given in (7.100), we are led to

$$\frac{y_{i+1} - 2y_i + y_{i-1}}{h^2} - p(x_i)\frac{y_{i+1} - y_{i-1}}{2h} - q(x_i)y_i = r(x_i), \qquad 1 \le i \le n - 1, \tag{7.103a}$$

which is a *linear* system of equations. Multiplying (7.103a) by $-h^2/2$ and collecting like terms, we find for $1 \le i \le n - 1$ that (7.103a) becomes

$$\left(-\frac{1}{2} - p(x_i)\frac{h}{4}\right)y_{i-1} + \left(1 + q(x_i)\frac{h^2}{2}\right)y_i + \left(-\frac{1}{2} + p(x_i)\frac{h}{4}\right)y_{i+1} = -\frac{h^2}{2}r(x_i). \tag{7.103b}$$

Thus (7.103b) is a tridiagonal linear system of equations and can be solved quickly by the factorization method of Chapter 2. There is the question, of course, as to whether the system (7.103b) has a solution or not. It is easy to show (see Problem 3) that if $q(x)$ is continuous and positive for $a \le x \le b$ and if h is chosen so that $(h/2)|p(x)| \le 1$, then the coefficient matrix for (7.103b) is diagonally dominant and hence nonsingular. (See Theorem 2.3.)

7.8.2. Shooting Methods

A "shooting method," an aptly named procedure for solving (7.99), is applicable whether $f(x, y, y')$ is linear or nonlinear. The basic idea is to *guess* an initial value $y'(a) = \alpha_1$ to go along with the left-hand boundary condition $y(a) = \alpha$. Then the *initial-value* problem

$$y'' = f(x, y, y'); \qquad y(a) = \alpha, \qquad y'(a) = \alpha_1 \tag{7.104}$$

is solved using any of the single-step or multistep methods of this chapter. Naturally we do not expect that this solution will satisfy $y(b) = \beta$, but we can hope to adjust the condition on $y'(a)$ so that we converge to β for $x = b$. To see how this procedure is usually carried out, let $y_1(x)$ be the result of solving

$$y'' = f(x, y, y'); \qquad y(a) = \alpha, \qquad y'(a) = \alpha_1;$$

and let $y_2(x)$ be the result of solving

$$y'' = f(x, y, y'); \qquad y(a) = \alpha, \qquad y'(a) = \alpha_2.$$

For simplicity suppose that $y_1(b) < \beta < y_2(b)$. Then there is a number λ such that $\lambda y_1(b) + (1 - \lambda)y_2(b) = \beta$. We next try $\alpha_3 = \lambda\alpha_1 + (1 - \lambda)\alpha_2$ and find $y_3(x)$ to solve

$$y'' = f(x, y, y'); \qquad y(a) = \alpha, \qquad y'(a) = \alpha_3.$$

In the example above, $y_1(x)$ undershot the desired ending condition β and $y_2(x)$ overshot β. If $y_3(x)$ is still not near enough to β at $x = b$, we choose a new value α_4 for the initial conditions $y(a) = \alpha$, $y'(a) = \alpha_4$ and try again.

The shooting method described above is fairly crude and can be improved. To accomplish this improvement, it is convenient to introduce a parameter s and to rewrite (7.104):

$$y''(x, s) = f(x, y(x, s), y'(x, s)), \; y(a, s) = \alpha, \; y'(a, s) = s. \tag{7.105}$$

In (7.105), $y(x, s)$ denotes the solution of the differential equation $y'' = f(x, y, y')$ with initial conditions $y(a) = \alpha$, $y'(a) = s$. To solve (7.99), we want to choose a value s^* so that the solution $y(x, s^*)$ of (7.105) satisfies $y(b, s^*) = \beta$. Equivalently we want to solve the nonlinear equation

$$g(s) = 0$$

where $g(s) = y(b, s) - \beta$. Since Newton's method is an effective procedure for solving nonlinear equations, we are led to consider

$$s_{i+1} = s_i - \frac{g(s_i)}{\dot{g}(s_i)}, \qquad i = 0, 1, \ldots, \tag{7.106}$$

to solve $g(s) = 0$. [In (7.106), $\dot{g}(s)$ denotes dg/ds or differentiation of $g(s)$ with respect to s; $\dot{g}(s)$ indicates the rate of change of $y(b, s)$ with respect to changes in the initial slope s.] For a given value of s_i, $g(s_i)$ can be calculated by solving $y'' = f(x, y, y'')$, $y(a) = \alpha$, $y'(a) = s_i$; but we also need $\dot{g}(s_i)$ for (7.106); and a method for finding $\dot{g}(s_i)$ is given below.

We first define $w(x, s)$:

$$w(x, s) = \frac{\partial y(x, s)}{\partial s}.$$

Now if $y(x, s)$ is smooth enough (as a function of two variables), we have that

$$w'(x, s) = \frac{\partial w(x, s)}{\partial x} = \frac{\partial}{\partial x}\left[\frac{\partial y(x, s)}{\partial s}\right] = \frac{\partial}{\partial s}\left[\frac{\partial y(x, s)}{\partial x}\right] = \frac{\partial y'(x, s)}{\partial s}$$

where the prime denotes differentiation with respect to x. Similarly from the equation above we obtain

$$w''(x, s) = \frac{\partial w'(x, s)}{\partial x} = \frac{\partial}{\partial x}\left[\frac{\partial y'(x, s)}{\partial s}\right] = \frac{\partial}{\partial s}\left[\frac{\partial y'(x, s)}{\partial x}\right] = \frac{\partial y''(x, s)}{\partial s}.$$

Thus if we differentiate both sides of the identity (7.105) with respect to s, we find

$$\frac{\partial y''(x, s)}{\partial s} = f_y(x, y(x, s), y'(x, s))\,\frac{\partial y(x, s)}{\partial s} + f_z(x, y(x, s), y'(x, s))\,\frac{\partial y'(x, s)}{\partial s}.$$

Using $w' = \partial y'/\partial s$ and $w'' = \partial y''/\partial s$ in the equation above, we are led to an initial-value problem for $w(x, s)$:

$$\begin{aligned} w''(x, s) &= f_y(x, y(x, s), y'(x, s))w(x, s) \\ &\quad + f_z(x, y(x, s), y'(x, s))w'(x, s) \\ w(a, s) &= 0, \qquad w'(a, s) = 1. \end{aligned} \tag{7.107}$$

In (7.107) we regard s as a (fixed) parameter; the initial conditions come from (7.105) and the identification

$$w(x, s) = \frac{\partial y(x, s)}{\partial s}, \qquad w'(x, s) = \frac{\partial y'(x, s)}{\partial s}.$$

Equation (7.107) is called the *variational equation* [see Isaacson and Keller (1966)]. Note that $w(b, s)$ is precisely the quantity $\dot{g}(s)$ that we need for (7.106).

To summarize, we can execute one step of (7.106) as part of a shooting method by solving the coupled system of second-order equations

$$\begin{aligned} y'' &= f(x, y, y'); \qquad y(a) = \alpha, \qquad y'(a) = s_i \\ w'' &= f_y(x, y, y')w + f_z(x, y, y')w'; \qquad w(a) = 0, \qquad w'(a) = 1 \end{aligned}$$

and then forming

$$s_{i+1} = s_i - \frac{y(b) - \beta}{w(b)}.$$

EXAMPLE 7.15. Consider the two-point boundary-value problem

$$y'' = y^2(1 - .2 \sin 2x), \qquad y(0) = 1, \qquad y(3) = 0.$$

The associated variational equation is

$$w'' = 2y(1 - .2 \sin 2x)w, \qquad w(0) = 0, \qquad w'(0) = 1.$$

Writing these equations as a first-order system of four equations with $u_1 = y$, $u_2 = y'$, $u_3 = w$, $u_4 = w'$, we have (for the shooting method)

$$\begin{aligned} u_1' &= u_2, & u_1(0) &= 1 \\ u_2' &= u_1^2(1 - .2 \sin 2x), & u_2(0) &= s \\ u_3' &= u_4, & u_3(0) &= 0 \\ u_4' &= 2u_1(1 - .2 \sin 2x)u_3, & u_4(0) &= 1. \end{aligned}$$

Newton's method (7.106) was started with $s_0 = -1$, and a vector version of the Runge-Kutta-Fehlberg program in Figs. 7.9 through 7.10 was used to calculate $y(b)$ and $w(b)$ for each s_i. The program was executed with TOL = 1.3−08, HMIN = 1.E−10; and the Newton iteration was terminated when $|u_1(3)| \le 10^{-6}$. The results are tabulated below.

s_i	$u_1(3)$
−1.0000000	−.7873598E+00
− .5776113	.3489194E+01
− .6974514	.1061144E+01
− .7756662	.2182006E+00
− .8016254	.1645750E−01
− .8039210	.1156425E−03
− .8039373	.5820944E−08

7.8.3. Collocation Methods

Collocation methods are also suitable for problems of the form (7.99), and we describe them briefly. Suppose $l(x)$ is a linear polynomial such that $l(a) = \alpha$, $l(b) = \beta$; and suppose for each i that $q_i(x)$ is a polynomial of degree i such that $q_i(a) = q_i(b) = 0$. If we define $p(x)$ by

$$p(x) = l(x) + \sum_{i=2}^{n} a_i q_i(x) \tag{7.108}$$

where the coefficients a_i are as yet not specified, then we have $p(a) = \alpha$, $p(b) = \beta$. Choosing collocation points $t_2, t_3, \ldots, t_n$, we can determine $p(x)$ by insisting that

$$p''(t_i) = f(t_i, p(t_i), p'(t_i)), \qquad i = 2, 3, \ldots, n.$$

As in Section 7.7, we note that we need not restrict ourselves to polynomials in (7.108) but may choose any suitable approximating form.

PROBLEMS, SECTION 7.8.3

1. Write a program to use the method of collocation to estimate a solution to $y'' + y = 0$; $y(0) = 1$, $y'(0) = 0$ for $0 \le x \le 2$. Use $p(x) = 1 + a_2x^2 + \cdots + a_nx^n$ as an approximating form and insert this directly into the equation $y'' + y = 0$. [Note that $p(0) = y(0)$ and $p'(0) = y'(0)$ when $p(x)$ has the form given above.] Use the zeros of the shifted Chebyshev polynomials, $\tilde{T}_k(x)$, as collocation points for $k = 1, 2, \ldots, 10$.
2. Consider the boundary value problem $y'' + y = 0$; $y(0) = 0$ and $y(b) = \beta$. Recall that the most general solution of the equation $y'' + y = 0$ is $y(x) = A\cos(x) + B\sin(x)$. Find choices for b and β for which the boundary value problem has
   a) no solution,
   b) exactly one solution,
   c) infinitely many solutions.
3. Show that the coefficient matrix for (7.103b) is diagonally dominant when $q(x)$ is continuous and positive on $[a, b]$ and when h is chosen so that $(h/2)|p(x)| \le 1$ for $a \le x \le b$.
4. Program a shooting method and the finite-difference method for the problem $y'' + y = 0$; $y(0) = 0$, $y(1) = 0.841471$ [Note: $\sin(1) = 0.841471$].
5. Let $F(a_0, a_1, \ldots, a_n)$ equal either $\int_0^b [e(x)]^2\,dx$ or $\Sigma_{j=0}^{m} [e(t_j)]^2$ as in (7.93a) or (7.93b), respectively. As in the section on Newton's method in several variables in Chapter 4, the derivative of $F(\mathbf{a})$, $\mathbf{a} \equiv (a_0, a_1, \ldots, a_n)$, equals the $(1 \times (n + 1))$ Jacobian (or gradient),

$$\mathbf{F}'(\mathbf{a}) = \left(\frac{\partial F(\mathbf{a})}{\partial a_0}, \frac{\partial F(\mathbf{a})}{\partial a_1}, \ldots, \frac{\partial F(\mathbf{a})}{\partial a_n}\right).$$

   Solving the $(n + 1)$ equations, $\mathbf{F}'(\mathbf{a}) = \theta$, in the $(n + 1)$ unknowns, $a_0, a_1, \ldots, a_n$, is a condition for minimizing $F(\mathbf{a})$ and thus yielding a polynomial to minimize (7.93a) or (7.93b). The set of equations $\mathbf{F}'(\mathbf{a}) = \boldsymbol{\theta}$ is usually nonlinear and thus calls for a technique such as Newton's method. Perform the computations above on the specific problem given in Example 7.13.

8

Numerical Solution of Partial Differential Equations

8.1 INTRODUCTION

In this chapter we shall consider numerical approximations to the solution $u(x, y)$ of some commonly occurring special cases of the second-order partial differential equation

$$Au_{xx} + Bu_{xy} + Cu_{yy} + Du_x + Eu_y + Fu = G. \tag{8.1}$$

In (8.1) the coefficients $A(x, y)$, $B(x, y)$, . . . are continuous functions over some region W of the xy-plane; $u(x, y)$ and/or its derivatives are to satisfy prescribed conditions on the boundary of W. The theory of partial differential equations gives results on the existence and the uniqueness of solutions, but these results vary with features of the problem such as the type of equation, the boundary conditions, and the geometry of the domain.

Equation (8.1) is said to be (1) "elliptic" if $B^2 - 4AC < 0$ in W, (2) "hyperbolic" if $B^2 - 4AC > 0$ in W, and (3) "parabolic" if $B^2 - 4AC = 0$ in W. If $B^2 - 4AC$ changes sign in W, Eq. (8.1) is said to be of "mixed type." Typical examples follow.

1. **Elliptic:**

$$u_{xx} + u_{yy} = 0 \qquad \text{(Laplace's equation)}$$

$$u_{xx} + u_{yy} = f(x, y) \qquad \text{(Poisson's equation)}$$

   Both of these equations can be used, for example, to describe the steady-state temperature distribution in a plate.

2. **Parabolic:**

$$u_y = a^2u_{xx} \qquad \text{(the heat or diffusion equation)}$$

   This equation can be used to describe the transient temperature distribution in a rod or the diffusion of gases.

3. **Hyperbolic:**

$$u_{yy} = a^2u_{xx} \qquad \text{(the wave equation)}$$

$$u_{xx} = KLu_{yy} + (KR + LS)u_y + RSu, \quad K > 0, \quad L > 0 \qquad \text{(the telegraph equation)}$$

The wave equation can be used to describe the position of a vibrating string or longitudinal vibration in a bar. The telegraph equation describes either the electric potential or the current in a transmission line.

It can be shown [Garabedian (1964)] that (8.1) can be reduced by variable substitutions to generalized forms of Laplace's equation, the heat equation, or the wave equation, depending on whether (8.1) is elliptic, parabolic, or hyperbolic. The techniques that we consider shall be restricted to these three particular equations.

A boundary condition that prescribes the value of the solution $u(x, y)$ on the boundary of W is said to be of the *first type* or a *Dirichlet condition*. A condition that prescribes the normal derivative du/dn on the boundary is said to be of the *second type* or a *Neumann condition*. Among other kinds of boundary restrictions are those of the *third type,* which prescribe $\gamma u + du/dn$ on the boundary of W.

In this chapter we will consider techniques known as finite-difference methods and we will briefly introduce another technique known as the finite-element method. Finite-difference methods vary with the type of partial differential equation and the boundary conditions, but the central feature of each method is the approximation of derivatives by finite differences; these derivative approximations usually involve forward, backward, or central differences. In particular, for a given sufficiently differentiable function $f(x)$,

$$f(x \pm h) = f(x) \pm f'(x)h + f''(x)h^2/2 + f'''(x)h^3/3 + \cdots. \tag{8.2}$$

From (8.2) we can easily derive such expressions as

$$\begin{aligned} f'(x) &= [f(x + h) - f(x)]/h + 0(h) \\ f'(x) &= [f(x) - f(x - h)]/h + 0(h) \\ f'(x) &= [f(x + h) - f(x - h)]/2h + 0(h^2) \\ f''(x) &= [f(x + h) - 2f(x) + f(x - h)]/h^2 + 0(h^2). \end{aligned} \tag{8.3}$$

For example if $u(x, y)$ is sufficiently differentiable, we can use

$$u_{xx}(x, y) = [u(x + h, y) - 2u(x, y) + u(x - h, y)]/h^2 + 0(h^2). \tag{8.4}$$

PROBLEMS, SECTION 8.1

1. Determine the type of graph of the quadratic equation

$$Ax^2 + Bxy + Cy^2 + Dx + Ey + F = 0$$

when (a) $B^2 - 4AC > 0$, (b) $B^2 - 4AC = 0$, and (c) $B^2 - 4AC < 0$.

2. Verify that (8.1) can be written in terms of new variables $r(x, y)$ and $s(x, y)$ as $0 = Au_{xx} + Bu_{xy} + Cu_{yy} + Du_x + Eu_y + Fu - G = au_{rr} + bu_{rs} + cu_{ss} + \cdots$ where $a = Ar_x^2 + Br_xr_y + Cr_y^2$, $b = 2Ar_xs_x + Br_xs_y + Br_ys_x + 2Cr_ys_y$, and $c = As_x^2 + Bs_xs_y + Cs_y^2$.

3. a) Suppose that A, B, and C (with $A \neq 0$) are constants in (8.1) and that $B^2 - 4AC > 0$ (hyperbolic). Let $\alpha = (B + \sqrt{B^2 - 4AC})/2A$ and $\beta = (B - \sqrt{B^2 - 4AC})/2A$. If $r = y - \alpha x$ and $s = y - \beta x$, show that $a = c = 0$, $b = (4AC - B^2)/A$, and hence (8.1) can be written as $0 = u_{rs} + \cdots$.

   b) Show that where f and g are arbitrary differentiable functions, equation $u_{rs} = 0$ has the solution $u = f(r) + g(s) = f(y - \alpha x) + g(y - \beta x)$ (this is the d'Alembert solution).

   c) Under the rotation $\xi = (r - s)/\sqrt{2}$, $\eta = (r + s)/\sqrt{2}$, show that $u_{rs} = 0$ can be written as $u_{\xi\xi} - u_{\eta\eta} = 0$.

4. Suppose that A, B, and C (with $A \neq 0$) are constants in (8.1) and that $B^2 - 4AC = 0$ (parabolic). If $r = x$ and $s = y - (B/2A)x$, show that $a = A$, $b = c = 0$, and hence (8.1) can be written as $Au_{rr} + \cdots = 0$.

5. Suppose that A, B, and C (with $A \neq 0$) are constants in (8.1) and that $B^2 - 4AC < 0$ (elliptic). Let $\alpha = -B/2A$ and $\beta = \sqrt{4AC - B^2}/2A$. If $r = \alpha x + y$ and $s = \beta x$, show that $a = c = A\beta^2$, $b = 0$, and hence (8.1) can be written as $A\beta^2(u_{rr} + u_{ss}) + \cdots = 0$.

6. The reductions in Problems 3, 4, and 5 are called reductions to canonical form. Determine the type of the following equations and reduce them to canonical form. In the hyperbolic case give the d'Alembert solution.

   a) $u_{xx} + 2u_{xy} + u_{yy} + u_x - u_y = 0$

   b) $u_{xx} + 2u_{xy} + 5u_{yy} + 3u_x + u = 0$

   c) $3u_{xx} + 10u_{xy} + 3u_{yy} = 0$.

7. *(Separation of Variables)*

   a) Consider the wave equation $u_{tt} = u_{xx}$ where $u(0, t) = u(c, t) = 0$ and $u_t(x, 0) = 0$ for $0 \le x \le c$. Suppose that $u(x, t)$ can be written as $u(x, t) = X(x)T(t)$. Argue that $(X''(x)/X(x)) = (T''(t)/T(t)) = -\alpha^2$ (where α is a constant). Hence

$$X''(x) + \alpha^2 X(x) = 0, \qquad X(0) = X(c) = 0, \qquad \text{and}$$
$$T''(t) + \alpha^2 T(t) = 0, \qquad T'(0) = 0.$$

   Argue that for any positive integer n, $X(x) = \sin(n\pi x/c)$ and $T(t) = \cos(n\pi t/c)$ are solutions, and hence $u_N(x, t) = \sum_{n=1}^{N} b_n \sin(n\pi x/c)\cos(n\pi t/c)$ is a solution of the original problem for any positive integer N.

   b) As in part (a), show that $u_N(x, t) = \sum_{n=0}^{N} a_n \exp[-n^2\pi^2 t] \cos n\pi x$ is a solution of $u_t = u_{xx}$, $u_x(0, t) = u_x(1, t) = 0$.

   c) As in part (a), show that $u_N(x, y) = c_0 y + \sum_{n=1}^{N} c_n \cos(nx) \sinh(ny)$ is a solution of $u_{xx} + u_{yy} = 0$, $u_x(0, y) = u_x(\pi, y) = u(x, 0) = 0$.

8. Assuming sufficient differentiability of f, verify the following from (8.2).

   a) $f'(x) = [f(x + h) - f(x)]/h - f''(\xi)h/2$

   b) $f'(x) = [f(x) - f(x - h)]/h + f''(\xi)h/2$

   c) $f'(x) = [f(x + h) - f(x - h)]/2h - f'''(\xi)h^2/6$

   d) $f''(x) = [f(x + h) - 2f(x) + f(x - h)]/h^2 - f''''(\xi)h^2/12$

8.2 PARABOLIC EQUATIONS

For our study of finite-difference techniques for parabolic equations we consider the particular boundary-value problem (the heat equation) given by

$$\frac{\partial}{\partial t} u(x, t) = \alpha^2 \frac{\partial^2}{\partial x^2} u(x, t), \qquad 0 < x < L, \qquad t > 0 \tag{8.5}$$

where

$$u(x, 0) = f(x) \qquad \text{and} \qquad u(0, t) = u(L, t) = 0.$$

We let $h = L/M$ for some positive integer M, let k be the increment in time t, and define $x_i = ih$ for $0 \le i \le M$ and $t_j = jk$ for $j = 0, 1, 2, \ldots$. We use finite differences to approximate the solution u at the point (x_i, t_j) of the xt-plane in the region $W = \{(x, t) : 0 < x < L, t > 0\}$. We denote $u(x_i, t_j)$ as u_{ij}. Using a central difference in x and a forward difference in t, and assuming sufficient differentiability of u, we have

$$\begin{aligned} u_{xx}(x_i, t_j) &= \frac{u(x_i + h, t_j) - 2u(x_i, t_j) + u(x_i - h, t_j)}{h^2} \\ &\quad - \frac{\partial^4}{\partial x^4} u(x_i + \xi_i h, t_j)h^2/12 \end{aligned} \tag{8.6a}$$

$$u_t(x_i, t_j) = \frac{u(x_i, t_j + k) - u(x_i, t_j)}{k} - \frac{\partial^2}{\partial t^2} u(x_i, t_j + \zeta_j k)k/2 \tag{8.6b}$$

where $-1 < \xi_i < 1$ and $0 < \zeta_j < 1$. Thus using (8.5), we are led to

$$\begin{aligned} [u_{i,j+1} - u_{ij}]/k &= \alpha^2[u_{i+1,j} - 2u_{ij} + u_{i-1,j}]/h^2 \\ &\quad + \frac{\partial^2}{\partial t^2} u(x_i, t_j + \zeta_j k)k/2 - \alpha^2 \frac{\partial^4}{\partial x^4} u(x_i + \xi_i h, t_j)h^2/12. \end{aligned}$$

With higher-order terms neglected, an approximation w_{ij} for u_{ij} with a discretization error of order $(h^2 + k)$ can be found from

$$[w_{i,j+1} - w_{ij}]/k = \alpha^2[w_{i+1,j} - 2w_{ij} + w_{i-1,j}]/h^2. \tag{8.7}$$

Letting $r = \alpha^2 k/h^2$, we solve (8.7) for $w_{i,j+1}$ and obtain

$$w_{i,j+1} = rw_{i+1,j} + (1 - 2r)w_{ij} + rw_{i-1,j}. \tag{8.8}$$

In Fig. 8.1, we see that our designation of points (x_i, t_j) places a rectangular grid on W, and Eq. (8.8) specifies each approximation on the line $t = t_{j+1}$ in terms of the three approximations directly below it on line $t = t_j$.

From the boundary conditions we have that

$$w_{i,0} = u(x_i, 0) = f(x_i), \qquad w_{0,j} = u(0, t_j) = 0, \qquad \text{and} \qquad w_{M,j} = u(L, t_j) = 0.$$

Thus we see that we can use (8.8) with $j = 0$ to obtain w_{i1} for $1 \le i \le M - 1$. With these values, (8.8) yields w_{i2} for $1 \le i \le M - 1$. We continue, using (8.8) to calculate values $w_{i,j+1}$ from previously calculated values of w_{ij}. Since each

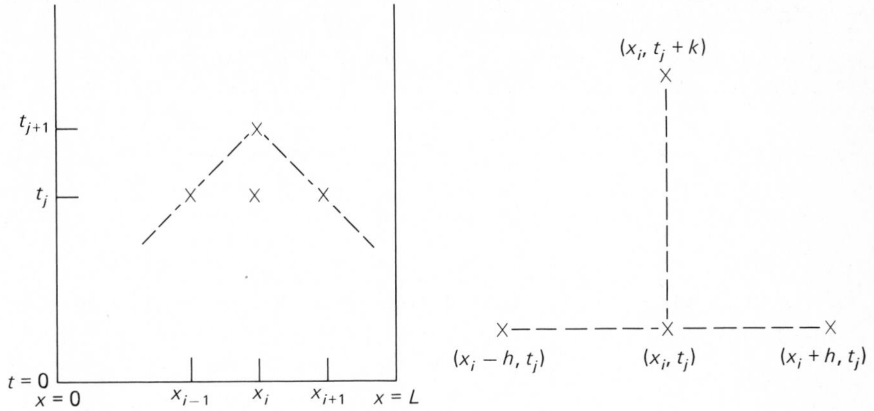

Figure 8.1

$w_{i,\,j+1}$ is individually calculated directly from (8.8) by use of known values from the previous t-step, this method is called an "explicit" method. In matrix form, this explicit method is given by $\mathbf{w}^{(j+1)} = A\mathbf{w}^{(j)}$ where the $((M-1) \times (M-1))$ tridiagonal matrix A and $((M-1) \times 1)$ vector $\mathbf{w}^{(j)}$ are given by

$$A = \begin{bmatrix} (1-2r) & r & 0 & \cdots & 0 \\ r & (1-2r) & r & & \vdots \\ 0 & r & (1-2r) & r & \\ \vdots & & & & r \\ 0 & \cdots & & r & (1-2r) \end{bmatrix}$$

$$\mathbf{w}^{(j)} = \begin{bmatrix} w_{1j} \\ w_{2j} \\ w_{3j} \\ \vdots \\ w_{M-1,\,j} \end{bmatrix}. \tag{8.9}$$

EXAMPLE 8.1. To illustrate the explicit method in (8.8), consider the parabolic equation in (8.5) with $\alpha^2 = 1$, $L = 1$, and $f(x) = \sin \pi x(1 + 6 \cos \pi x)$. The exact solution is $u(x, t) = \sin(\pi x)e^{-\pi^2 t} + 3 \sin(2\pi x)e^{-4\pi^2 t}$; and Table 8.1 lists the approximations at $t = .25$ for $x = .05, .10, \ldots, .90, .95$. All the approximations were made with $h = .05 = 1/20$; but to illustrate how different time steps affect the approximate solution, we used $k = 1/400$ $(r = 1)$, $k = 1/1200$ $(r = 1/3)$, and $k = 1/2400$ $(r = 1/6)$. Note that the approximations are clearly unacceptable for $r = 1$.

We will see that we cannot expect (8.8) to yield good approximations unless we have $r \leq 1/2$ as is illustrated in Table 8.1. To explain the results of Example 8.1, we need to consider the concepts of convergence and stability.

TABLE 8.1

		Approximate solution at (*x*, .25)		
x	*u*(*x*, .25)	*r*= 1	*r*= 1/3	*r*= 1/6
.05	0.1331437E−01	0.2002719E+36	0.1324321E−01	0.1331442E−01
.10	0.2629738E−01	−0.4020827E+36	0.2615708E−01	0.2629747E−01
.15	0.3862619E−01	0.6063715E+36	0.3842073E−01	0.3862632E−01
.20	0.4999469E−01	−0.8129329E+36	0.4972987E−01	0.4999485E−01
.25	0.6012134E−01	0.1019978E+37	0.5980453E−01	0.6012154E−01
.30	0.6875624E−01	−0.1223851E+37	0.6839614E−01	0.6875646E−01
.35	0.7568732E−01	0.1418941E+37	0.7529365E−01	0.7568755E−01
.40	0.8074553E−01	−0.1597812E+37	0.8032871E−01	0.8074577E−01
.45	0.8380883E−01	0.1751573E+37	0.8337965E−01	0.8380907E−01
.50	0.8480497E−01	−0.1870466E+37	0.8437428E−01	0.8480521E−01
.55	0.8371293E−01	0.1944641E+37	0.8329133E−01	0.8371316E−01
.60	0.8056312E−01	−0.1965956E+37	0.8016071E−01	0.8056332E−01
.65	0.7543625E−01	0.1924425E+37	0.7506242E−01	0.7543644E−01
.70	0.6846109E−01	−0.1818107E+37	0.6812432E−01	0.6846125E−01
.75	0.5981100E−01	0.1644842E+37	0.5951872E−01	0.5981114E−01
.80	0.4969954E−01	−0.1407239E+37	0.4945805E−01	0.4969965E−01
.85	0.3837512E−01	0.1111938E+37	0.3818951E−01	0.3837520E−01
.90	0.2611497E−01	−0.7694093E+36	0.2598909E−01	0.2611503E−01
.95	0.1321847E−01	0.3933908E+36	0.1315489E−01	0.1321850E−01

To examine convergence, we define the local truncation errors $e_{ij} = u_{ij} - w_{ij}$. Using $u_t - \alpha^2 u_{xx} = 0$ in (8.6a) and (8.6b), and combining this with (8.8) yield

$$e_{i,j+1} = re_{i+1,j} + (1 - 2r)e_{ij} + re_{i-1,j} + k\left[\frac{\partial^2}{\partial t^2} u(x_i, t_j + \zeta_j k)k/2 - \alpha^2 \frac{\partial^4}{\partial x^4} u(x_i + \xi_i h, t_j)h^2/12\right]. \tag{8.10}$$

Assuming sufficient differentiability of u, we see for the closed subregion of W bounded by $t \le T$ that the term in brackets in (8.10) is bounded by $k[M_1 k + M_2 h^2]$. For each j, let $E_j = \max|e_{ij}|$, and suppose that $r \le 1/2$. Then arguing recursively on (8.10), we have

$$\begin{aligned} E_{j+1} &\le E_j + k[M_1 k + M_2 h^2] \le E_{j-1} + 2k[M_1 k + M_2 h^2] \le E_{j-2} \\ &\quad + 3k[M_1 k + M_2 h^2] \le \cdots \le E_0 + (j+1)k[M_1 k + M_2 h^2] \\ &= t_{j+1}[M_1 k + M_2 h^2]. \end{aligned}$$

Letting $(h, k) \to (0, 0)$, we have $E_j \to 0$ or $e_{ij} \to 0$, and hence the explicit method is convergent for $r \le 1/2$.

It is clear that errors present in the jth t-step will affect the accuracy at the $(j + 1)$st step. Loosely speaking, the method is said to be *stable* if errors from whatever source at any t-step are not magnified as the iteration progresses and hence do not accumulate and destroy the accuracy indicated by the individual local truncation errors. For example, let us assume that an error $\mathbf{e}^{(0)}$ is made in obtaining $\mathbf{w}^{(0)} = [f(x_1), f(x_2), \ldots, f(x_{M-1})]^T$, and for simplicity assume that no other errors are made. We track this error through the iterations to see its effect on later stages. The initial vector is actually $\mathbf{w}^{(0)} + \mathbf{e}^{(0)}$, and so we have

$$\mathbf{w}^{(1)} = A(\mathbf{w}^{(0)} + \mathbf{e}^{(0)}),\ \mathbf{w}^{(2)} = A\mathbf{w}^{(1)} = A^2\mathbf{w}^{(0)} + A^2\mathbf{e}^{(0)}, \ldots, \mathbf{w}^{(n)} = A^n\mathbf{w}^{(0)} + A^n\mathbf{e}^{(0)}.$$

Hence at the nth step the contribution from $\mathbf{e}^{(0)}$ is $A^n\mathbf{e}^{(0)}$. In order for this error not to be magnified in the successive steps, we want $\|A^n\mathbf{e}^{(0)}\| \le \|\mathbf{e}^{(0)}\|$. Hence from a slight extension of Theorem 3.7, we need $\rho(A) \le 1$. [As in Section 3.4, $\rho(A)$ is the spectral radius or largest eigenvalue of A in magnitude.] From Exercise 4, we have that the eigenvalues of A in (8.9) are

$$\lambda_i = 1 - 4r\sin^2(i\pi/2M), \qquad 1 \le i \le M - 1.$$

Hence the explicit method is stable as described above only if $r \le 1/2$.

From the analyses of convergence and stability we must have $r \le 1/2$. However, this limitation on r places severe constaints on the explicit method. For example, the finite-difference approximation (8.6a) and (8.6b) for (8.5) is of order $(h^2 + k)$, and hence accuracy requires that h and k be fairly small. With $r = \alpha^2k/h^2 \le 1/2$ or $k \le h^2/2\alpha^2$, we see that the increments in t may have to be extremely small and thus often necessitate an enormous amount of computation to make any appreciable advance in the t-direction. This problem is countered by implicit methods, the topic of the next section.

8.2.1. Implicit Methods

We first note that for the explicit method (8.8) the approximation $w_{i,\,j+1}$ at (x_i, t_{j+1}) depends only on the information at the points below the slanted lines in Fig. 8.1. However, in a realistic problem, such as the temperature distribution in a homogeneous bar modeled by (8.5), the temperature $u(x_i, t_{j+1})$ depends also on the temperature at points above the slanted lines in Fig. 8.1, points such as (x_{i-1}, t_{j+1}) and (x_{i+1}, t_{j+1}). To include such information in the model, we consider the backward-difference approximation

$$u_t(x_i, t_j) = \frac{u(x_i, t_j) - u(x_i, t_{j-1})}{k} + \frac{\partial^2}{\partial t^2} u(x_i, t_j - \eta_j k)k/2 \tag{8.11}$$

where $0 < \eta_j < 1$. Using (8.11) and (8.6a) in (8.5), we have

$$(u_{ij} - u_{i,\,j-1})/k - \alpha^2(u_{i+1,\,j} - 2u_{ij} + u_{i-1,\,j})/h^2 = -\frac{\partial^2}{\partial t^2} u(x_i, t_j - \eta_j k)k/2 - \frac{\partial^4}{\partial x^4} u(x_i + \xi_i h, t_j)h^2/12.$$

Hence choosing v_{ij} to satisfy

$$(v_{ij} - v_{i,\,j-1})/k = \alpha^2(v_{i+1,\,j} - 2v_{ij} + v_{i-1,\,j})/h^2 \tag{8.12}$$

also yields a local truncation error of order $(h^2 + k)$ as in the explicit method. Letting $r = \alpha^2 k/h^2$ and solving (8.12) for $v_{i,\,j-1}$ yield

$$v_{i,\,j-1} = -rv_{i-1,\,j} + (1 + 2r)v_{ij} - rv_{i+1,\,j}. \tag{8.13}$$

Unlike the explicit method (8.8), the method given by (8.13) relates three approximations at the jth t-step to an approximation at the $(j - 1)$st step (see Fig. 8.2); and we cannot use (8.13) to solve for a single unknown in terms of previously calculated values as in the explicit method. Methods such as (8.13) are called *implicit*.

As before, the boundary conditions yield

$$v_{i,\,0} = u(x_i, 0) = f(x_i) \qquad \text{and} \qquad v_{0,\,j} = u(0, t_j) = 0,\ v_{M,\,j} = u(L, t_j) = 0.$$

Let the $((M - 1) \times (M - 1))$ matrix A and $((M - 1) \times 1)$ vector $\mathbf{v}^{(j)}$ be given by

$$A = \begin{bmatrix} (1 + 2r) & -r & 0 & 0 & \cdots & 0 \\ -r & (1 + 2r) & -r & & & \vdots \\ 0 & -r & & & & \\ \vdots & & & & & -r \\ 0 & \cdots & & & -r & (1 + 2r) \end{bmatrix} \qquad \mathbf{v}^{(j)} = \begin{bmatrix} v_{1j} \\ v_{2j} \\ \vdots \\ v_{M-1,\,j} \end{bmatrix}. \tag{8.14}$$

Then the method of (8.13) can be written as $A\mathbf{v}^{(j)} = \mathbf{v}^{(j-1)}$, $j = 1, 2, 3, \ldots$. Hence with the known $\mathbf{v}^{(0)}$ we solve for $\mathbf{v}^{(1)}$ and then use $\mathbf{v}^{(1)}$ to solve for $\mathbf{v}^{(2)}$, etc. Note that A is diagonally dominant for any $r > 0$; so A is nonsingular and $A\mathbf{v}^{(j)} = \mathbf{v}^{(j-1)}$ has a unique solution. Since a linear system of equations must be solved at each t-step, this method requires more computations per step than the explicit method requires. However as we shall see shortly, the ability to take

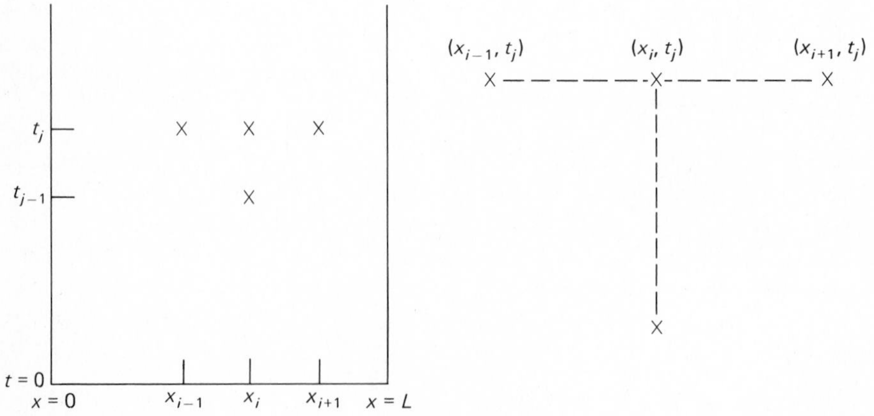

Figure 8.2

larger t-steps usually more than compensates for the additional per-step computations. Since A is tridiagonal, symmetric, and positive-definite, the Cholesky variant of the LU-decomposition in (2.33a) and (2.33b) would provide an efficient means of solving the linear systems.

Convergence and stability analyses for this implicit method can be made in much the same fashion as for the explicit method. By way of analysis one can show that the implicit method is convergent independent of the value of r. Similarly for stability we need $\rho(A^{-1}) \leq 1$. From Problem 10, the eigenvalues of A are given by

$$\lambda_i = 1 + 4r \sin^2(i\pi/2M), \qquad 1 \leq i \leq M - 1.$$

Thus $|\lambda_i| \geq 1$ and $\rho(A^{-1}) \leq 1$ for all $r > 0$. Hence this implicit method is convergent and stable for all positive h and k and is said to be *unconditionally stable*. We are able to advance in the t-direction with larger t-steps, and these larger steps are a definite computational advantage over the explicit method. [The size of h and k, however, are still limited in that the local truncation error is of order $(h^2 + k)$.]

EXAMPLE 8.2. To illustrate the implicit method in (8.13), we return to the parabolic equation of Example 8.1. Table 8.2 lists the approximations to $u(x, .25)$ for $x = .05, .10, \ldots, .90, .95$. All the approximations were made with $h = .05 = 1/20$ but with time steps $k = 1/200$ $(r = 2)$, $k = 1/400$ $(r = 1)$, and $k = 1/1200$ $(r = 1/3)$. Unlike the results of Example 8.1, the approximations for $r > 1/2$ are reasonable. The approximations for $r = 1/3$ are comparable for both the explicit and the implicit methods as is to be expected since stability is not an issue for $r = 1/3$ and both methods have truncation errors of order $(h^2 + k)$.

We conclude this section by considering an improvement of the implicit method above; this improvement has the same stability characteristics but has a local truncation error of order $(h^2 + k^2)$. For this case we consider the central difference approximation

$$u_t(x_i, t_j) = \frac{u(x_i, t_j + k) - u(x_i, t_j - k)}{2k} - \frac{\partial^3}{\partial t^3} u(x_i, t_j + \theta_j k)k^2/6 \qquad (8.15)$$

where $-1 < \theta_j < 1$. From (8.15) we see that by using $(v_{i,\,j+1} - v_{i,\,j-1})/2k$ as the left-hand side of (8.12), we can obtain a method of order $(h^2 + k^2)$. This method, however, has serious stability problems. These problems can be countered by averaging the forward-difference approximation at $t = t_j$ of (8.7) with the backward-difference approximation at $t = t_{j+1}$ of (8.12); that is, substitute v for w in (8.7) and average with (8.12) with $j + 1$ substituted for j. This procedure gives

$$[v_{i,\,j+1} - v_{ij}]/k = \alpha^2[(v_{i+1,\,j} - 2v_{ij} + v_{i-1,\,j}) + (v_{i+1,\,j+1} - 2v_{i,\,j+1} + v_{i-1,\,j+1})]/2h^2.$$

TABLE 8.2

		Approximate solution at $(x, .25)$		
x	$u(x, .25)$	$r = 2$	$r = 1$	$r = 1/3$
.05	0.1331437E−01	0.1426164E−01	0.1381907E−01	0.1352931E−01
.10	0.2629738E−01	0.2816321E−01	0.2729190E−01	0.2672103E−01
.15	0.3862619E−01	0.4135437E−01	0.4008135E−01	0.3924632E−01
.20	0.4999469E−01	0.5350395E−01	0.5186822E−01	0.5079358E−01
.25	0.6012134E−01	0.6430868E−01	0.6235954E−01	0.6107642E−01
.30	0.6875624E−01	0.7350110E−01	0.7129603E−01	0.6984095E−01
.35	0.7568732E−01	0.8085628E−01	0.7845859E−01	0.7687205E−01
.40	0.8074553E−01	0.8619720E−01	0.8367360E−01	0.8199865E−01
.45	0.8380883E−01	0.8939873E−01	0.8681693E−01	0.8509771E−01
.50	0.8480497E−01	0.9039006E−01	0.8781657E−01	0.8609691E−01
.55	0.8371293E−01	0.8915569E−01	0.8665386E−01	0.8497613E−01
.60	0.8056312E−01	0.8573491E−01	0.8336343E−01	0.8176741E−01
.65	0.7543625E−01	0.8021999E−01	0.7803167E−01	0.7655377E−01
.70	0.6846109E−01	0.7275310E−01	0.7079416E−01	0.6946678E−01
.75	0.5981100E−01	0.6352218E−01	0.6183184E−01	0.6068300E−01
.80	0.4969954E−01	0.5275594E−01	0.5136635E−01	0.5041941E−01
.85	0.3837512E−01	0.4071809E−01	0.3965443E−01	0.3892804E−01
.90	0.2611497E−01	0.2770092E−01	0.2698172E−01	0.2648979E−01
.95	0.1321847E−01	0.1401860E−01	0.1365600E−01	0.1340774E−01

[The left-hand side is actually a central difference at $(t_j + k/2)$ and thus accounts for $O(k^2)$ in t.] With $s = \alpha^2 k/2h^2$, this equation becomes

$$\begin{aligned} -sv_{i-1,\,j+1} + (1 + 2s)v_{i,\,j+1} - sv_{i+1,\,j+1} &= \\ sv_{i-1,\,j} + (1 - 2s)v_{ij} &+ sv_{i+1,\,j}. \end{aligned} \tag{8.16}$$

The method determined by (8.16) is usually called the Crank-Nicolson method; sometimes this method is called the modified Crank-Nicolson method and that given by (8.12) is called the Crank-Nicolson method. With $\mathbf{v}^{(0)}$ again determined by the boundary conditions with $t = 0$, the Crank-Nicolson method can be expressed in matrix form as $A\mathbf{v}^{(j+1)} = B\mathbf{v}^{(j)}$, $j = 0, 1, 2, \ldots$, where A and B are the $((M - 1) \times (M - 1))$ tridiagonal matrices given by

$$A = \begin{bmatrix} (1+2s) & -s & 0 & \cdots & 0 \\ -s & (1+2s) & -s & & \vdots \\ 0 & & & & 0 \\ \vdots & & & & -s \\ 0 & \cdots & 0 & -s & (1+2s) \end{bmatrix},$$

$$B = \begin{bmatrix} (1-2s) & s & 0 & \cdots & 0 \\ s & (1-2s) & s & & \vdots \\ 0 & & & & 0 \\ \vdots & & & & s \\ 0 & \cdots & 0 & s & (1-2s) \end{bmatrix}.$$

Once again the matrix A is diagonally dominant (and hence nonsingular); so each system $A\mathbf{v}^{(j+1)} = B\mathbf{v}^{(j)}$ has a unique solution. Since A is tridiagonal, symmetric, and positive-definite (Problem 10), an LU-decomposition may be efficiently used to solve each system. To have stability, we need $\rho(A^{-1}B) \le 1$. The eigenvalues of $A^{-1}B$ are (see Problem 12)

$$\lambda_i = \frac{2 - 8s \sin^2(i\pi/2M)}{2 + 8s \sin^2(i\pi/2M)}, \qquad 1 \le i \le M - 1.$$

Thus $|\lambda_i| \le 1$ and so the Crank-Nicolson method is stable for all $s > 0$.

A simple program that implements the Crank-Nicolson method for the problem (8.5) is listed in Fig. 8.3. The program uses a subroutine CNSTEP that executes one step of the Crank-Nicolson algorithm $A\mathbf{v}^{(j+1)} = B\mathbf{v}^{(j)}$ and a subroutine LU that finds an LU-decomposition for A. Briefly, a main program must set up the initial vector $\mathbf{v}^{(0)}$ for the equally spaced grid $0 = x_0 < x_1 < \cdots < x_M = L$ and calculate the parameter $s = \alpha^2 k/2h^2$ where $h = L/M$. Given s and M, an LU-decomposition for A is carried out in subroutine LU. Given any current state $\mathbf{v}^{(j)}$, subroutine CNSTEP finds $\mathbf{v}^{(j+1)}$ where $A\mathbf{v}^{(j+1)} = B\mathbf{v}^{(j)}$. In Fig. 8.3, $\mathbf{v}^{(j)}$ is stored in the array VOLD; and $\mathbf{v}^{(j+1)}$, in the array VNEW. An example of a calling program for CNSTEP and LU is listed in Fig. 8.3 in which the time step j is $j = 1/N$ and the approximate solution is printed at $t = .25$ (that is, at $t = t_i$ where $i = N/4$) and for $x = x_j$, $1 \le j \le M - 1$.

EXAMPLE 8.3. The Crank-Nicolson method as implemented in Fig. 8.3 was used on the parabolic equation in Example 8.1. Table 8.3 lists the approximations to $u(x, .25)$ for $x = .05, .10, \ldots, .90, .95$. Three runs were made with $h = 1/20$ and with time steps $k = 1/100$, $k = 1/48$, and $k = 1/16$. As a further illustration that accuracy is a function of both h and k, another run was made with $h = 1/40$ and $k = 1/100$; these approximations were the most accurate.

PROBLEMS, SECTIONS 8.2 and 8.2.1

1. Use Formula (8.2) and Problem 8, Section 8.1, to derive (8.6a), (8.6b), (8.11), and (8.15).
2. Consider the problem $u_t = u_{xx}$, $0 < x < .4$, $t > 0$, where $u(x, 0) = x(.4 - x)$, $u(0, t) = u(.4, t) = 0$. With $h = .1$ and $k = .005$, use the direct method (8.8) [or $\mathbf{w}^{(j+1)} = A\mathbf{w}^{(j)}$ from (8.9)] to calculate approximations for $u(x_i, .005)$ and $u(x_i, .01)$ for $x_i = \dfrac{i}{10}$; $i = 1, 2, 3$. [Answer: $u(x, t) = (1.28)\pi^{-3} \sum_{k=1}^{\infty} (2k - 1)^{-2} \sin((2k - 1)\pi x/.4)\exp(-(2k - 1)^2\pi^2 t/.16)$.]
3. Let $H = (h_{ij})$ be an $(n \times n)$ tridiagonal matrix with $h_{i,i+1} = h_{i+1,i} = 1/2$, $1 \le i \le n - 1$, and $h_{ii} = 0$, $1 \le i \le n$. Verify that the Sturm sequence (3.36) yields $p_0(x) = 1$, $p_1(x) = x$, and $p_k(x) = xp_{k-1}(x) - (1/4)p_{k-2}(x)$ for $2 \le k \le n$. From Example 6.9 and (6.38b), verify that $p_n(x) = V_n(x) = 2^{-n}U_n(x)$ where $U_n(x)$ is the Chebyshev polynomial of the second kind. Verify that $\lambda_j = \cos(j\pi/(n + 1))$, $1 \le j \le n$, are the zeros of $U_n(x)$ and hence are the eigenvalues of H.

```
      DIMENSION VOLD(40),VNEW(40),AM(40),D(40)
      M=20
      N=100
      H=1./M
      S=FLOAT(M*M)/(2.*FLOAT(N))
      MM1=M-1
        DO 1 I=1,MM1
        X=I*H
        VOLD(I)=F(X)
    1   CONTINUE
      CALL LU(S,AM,D,M)
      NSTOP=N/4
        DO 3 I=1,NSTOP
        CALL CNSTEP(S,AM,D,VOLD,VNEW,M)
          DO 2 J=1,MM1
          VOLD(J)=VNEW(J)
    2     CONTINUE
    3   CONTINUE
        DO 4 I=1,MM1
        PRINT 100,VOLD(I)
    4   CONTINUE
  100 FORMAT(' ',E20.7)
      STOP
      END


      FUNCTION F(X)
      PI=3.141592654
      F=SIN(PI*X)*(1.+6.*COS(PI*X))
      RETURN
      END


      SUBROUTINE LU(S,AM,D,M)
      DIMENSION AM(40),D(40)
C
C     GIVEN THE TRIDIAGONAL SYSTEM A*VNEW=B*VOLD TO SOLVE, THIS
C     SUBPROGRAM SETS UP A, REDUCES A TO TRIANGULAR FORM AND RECORDS
C     THE MULTIPLIERS AM(I) AND THE DIAGONAL D(I) OF THE EQUIVALENT
C     TRIANGULAR MATRIX.
C
      MM1=M-1
      DIAGA=1.+2.*S
        DO 1 I=1,MM1
        D(I)=DIAGA
    1   CONTINUE
        DO 2 I=2,MM1
        AM(I)=S/D(I-1)
        D(I)=D(I)-S*AM(I)
    2   CONTINUE
      RETURN
      END
```

Figure 8.3 Program for Example 8.3.

4. a) Let α and β be scalars and A be an $(n \times n)$ matrix. Show that if λ is an eigenvalue of A, then $\beta\lambda + \alpha$ is an eigenvalue of the matrix $\beta A + \alpha I$.

b) Let $A = (a_{ij})$ be an $(n \times n)$ tridiagonal matrix with $a_{i,\,i+1} = a_{i+1,\,i} = \gamma$ and $a_{ii} = \alpha$ for all i. With H as in Problem 3, verify that $A = 2\gamma H + \alpha I$, and hence that the eigenvalues of A are $2\gamma \cos(j\pi/(n+1)) + \alpha$, $1 \le j \le n$.

```
      SUBROUTINE CNSTEP(S,AM,D,VOLD,VNEW,M)
      DIMENSION AM(40),D(40),VOLD(40),VNEW(40),TEMP(40)
C
C  THIS SUBPROGRAM USES LU-DECOMPOSITION TO SOLVE A*VNEW=B*VOLD
C  IN ORDER TO EXECUTE ONE STEP OF THE CRANK-NICOLSON ALGORITHM.
C
      MM2=M-2
      MM1=M-1
      MP1=M+1
C
C  SET UP THE MATRIX B AND FORM THE PRODUCT B*VOLD.
C
      DIAGB=1.-2.*S
        DO 1 I=1,MM1
        TEMP(I)=VOLD(I)
    1   CONTINUE
      VOLD(1)=DIAGB*TEMP(1)+S*TEMP(2)
        DO 2 I=2,MM2
        VOLD(I)=S*TEMP(I-1)+DIAGB*TEMP(I)+S*TEMP(I+1)
    2   CONTINUE
      VOLD(MM1)=S*TEMP(MM2)+DIAGB*TEMP(MM1)
C
C  USE MULTIPLIERS AM(I) FROM LU-DECOMPOSITION TO MODIFY THE
C  RIGHT-HAND SIDE OF A*VNEW=B*VOLD.
C
        DO 3 I=2,MM1
        VOLD(I)=VOLD(I)+AM(I)*VOLD(I-1)
    3   CONTINUE
C
C  BACKSOLVE THE EQUIVALENT TRIANGULAR SYSTEM.
C
      VNEW(MM1)=VOLD(MM1)/D(MM1)
        DO 4 I=3,M
        K=MP1-I
        VNEW(K)=(VOLD(K)+S*VNEW(K+1))/D(K)
    4   CONTINUE
      RETURN
        END
```

Figure 8.3 (continued)

c) For the matrix A in (8.9), $n = M - 1$, $\gamma = r$, and $\alpha = 1 - 2r$. Verify that the eigenvalues of A are $\lambda_i = 1 - 4r \sin^2(i\pi/2M)$, $1 \leq i \leq M - 1$.

5. a) Repeat the calculations in Exercise 2 using $h = .1$ and $k = .01$ ($r = 1$) to obtain approximations for $u(x_i, 1)$, $i = 1, 2, 3$. Compare your results with the answers in Problem 2.

b) As in the text let an initial error $\mathbf{e}^{(0)}$ be given by $\mathbf{e}^{(0)} = \varepsilon[1, 1, 1]^T$. Calculate $A\mathbf{e}^{(0)}$ and $A^2\mathbf{e}^{(0)}$. Using $[\rho(A)]^n \leq \|A^n\|$, calculate $\rho(A)$ and give an estimate for $\|A^n\mathbf{e}^{(0)}\|$.

c) Derive formula (8.10) from (8.6a), (8.6b), and (8.8).

6. What modifications are necessary in (8.9) to treat the problem $u_t = \alpha^2 u_{xx}$, $u(x, 0) = f(x)$, $u(0, t) = g_0(t)$, $u(L, t) = g_1(t)$? [Hint: Consider the first and last equations.]

7. Modify (8.7) to be an approximation for the equation $u_t(x, t) = \alpha^2 u_{xx}(x, t) + F(x, t)u(x, t) + G(x, t)$. With boundary conditions as in (8.5), write the method in a matrix form similar to (8.9). If $F(x, t) = c$ (constant), how is the range of stability for r affected? (See Problem 4c.)

8. Write a program for the direct method and test the program against the results in

TABLE 8.3

	$h = 1/20$			$h = 1/40$
x	$k = 1/100$	$k = 1/48$	$k = 1/16$	$k = 1/100$
.05	0.1335303E−01	0.1324682E−01	0.1237273E−01	0.1329963E−01
.10	0.2637391E−01	0.2616538E−01	0.2443386E−01	0.2626862E−01
.15	0.3873901E−01	0.3843570E−01	0.3588014E−01	0.3858482E−01
.20	0.5014145E−01	0.4975416E−01	0.4642475E−01	0.4994269E−01
.25	0.6029892E−01	0.5984110E−01	0.5580486E−01	0.6006112E−01
.30	0.6896080E−01	0.6844781E−01	0.6378841E−01	0.6869046E−01
.35	0.7591431E−01	0.7536271E−01	0.7017990E−01	0.7561873E−01
.40	0.8098978E−01	0.8041648E−01	0.7482515E−01	0.8067678E−01
.45	0.8406465E−01	0.8348617E−01	0.7761475E−01	0.8374232E−01
.50	0.8506621E−01	0.8449808E−01	0.7848628E−01	0.8474269E−01
.55	0.8397316E−01	0.8342936E−01	0.7742521E−01	0.8365642E−01
.60	0.8081577E−01	0.8030841E−01	0.7446462E−01	0.8051340E−01
.65	0.7567479E−01	0.7521396E−01	0.6968367E−01	0.7539385E−01
.70	0.6867923E−01	0.6827295E−01	0.6320506E−01	0.6842610E−01
.75	0.6000287E−01	0.5965723E−01	0.5519149E−01	0.5978315E−01
.80	0.4985988E−01	0.4957929E−01	0.4584140E−01	0.4967833E−01
.85	0.3849949E−01	0.3828695E−01	0.3538391E−01	0.3835994E−01
.90	0.2619990E−01	0.2605730E−01	0.2407333E−01	0.2610524E−01
.95	0.1326154E−01	0.1319000E−01	0.1218319E−01	0.1321373E−01

Example 8.1. Also run the program to approximate $u(x_i, 1)$ for various values of h and k in Problem 2.

9. Use the implicit method of (8.13) or (8.14) for Problem 2 with $h = .1$ and $k = .01$.

10. a) For A in (8.14), use Problems 3 and 4 to verify that the eigenvalues are $\lambda_i = 1 + 4r\sin^2(i\pi/2M)$, $1 \le i \le M - 1$.

b) Since A is symmetric, it has a set of orthogonal eigenvectors $\{\mathbf{u}_i\}_{i=1}^{M-1}$ where $\mathbf{u}_i^T\mathbf{u}_j = 0$ for $i \ne j$. If $\mathbf{x}$ is an arbitrary vector in R^{M-1}, $\mathbf{x}$ can be written as $\mathbf{x} = c_1\mathbf{u}_1 + \cdots + c_{M-1}\mathbf{u}_{M-1}$. Use this fact and that each $\lambda_i > 0$ to show that $\mathbf{x}^TA\mathbf{x} > 0$ for $\mathbf{x} \ne \theta$, and hence A is positive definite.

11. Average the forward- and backward-difference operators to derive (8.16).

12. Let $C = (c_{ij})$ be an $((M - 1) \times (M - 1))$ tridiagonal matrix with $c_{i,\,i+1} = c_{i+1,\,i} = -1$ and $c_{ii} = 2$ for all i. From Problem 4 verify that the eigenvalues of C are $\mu_i = -2\cos(i\pi/2M) + 2$, $1 \le i \le M - 1$. If A and B are the Crank-Nicolson matrices ($A\mathbf{v}^{(j+1)} = B\mathbf{v}^{(j)}$), verify that $A = sC + I$ and $B = -sC + I$. In $A^{-1}B\mathbf{u}_i = \lambda_i\mathbf{u}_i$ or $(B - \lambda_iA)\mathbf{u}_i = \theta$, substitute for A and B in terms of C and verify that λ_i and μ_i are related by $\mu_i = (1 - \lambda_i)/s(1 + \lambda_i)$ or $\lambda_i = (1 - s\mu_i)/(1 + s\mu_i)$. Using the value above for μ_i, verify that $\lambda_i = (1 - 4s\sin^2(i\pi/2M))/(1 + 4s\sin^2(i\pi/2M))$, $1 \le i \le M - 1$.

13. Use the Crank-Nicolson method in Problem 2 with $h = .1$, $k = .01$.

14. Derive a Crank-Nicolson formula for these equations.

a) $u_t(x, t) = u_{xx}(x, t) + F(x, t)u(x, t) + G(x, t)$

b) $u_t(x, t) = H(x, t)u_{xx}(x, t) + F(x, t)u(x, t) + G(x, t)$

c) $u_t(x, t) = H(x, t)u_{xx}(x, t) + P(x, t)u_x(x, t) + F(x, t)u(x, t) + G(x, t)$

15. a) Use the program in Fig. 8.3 in Problem 2 and also for the problem $u_t = u_{xx}$, $u(0, t) = u(1, t) = 0$, $u(x, 0) = \sin \pi x$ with $h = .01$, $k = .01$ up to $t = 1$. [Answer: $u(x, t) = \exp(-\pi^2 t)\sin \pi x$.]

b) Modify this program to treat the equation in Problem 14(a) with boundary conditions $u(x, 0) = f(x)$, $u(0, t) = g_0(t)$, $u(0, L) = g_1(t)$.

c) Test this program on $u_t = u_{xx} - u - 2e^{-t}$, $u(x, 0) = x^2$, $u(0, t) = 0$, $u(1, t) = e^{-t}$. [Answer: $u(x, t) = x^2e^{-t}$.]

8.2.2. Derivative Boundary Conditions

Typically in heat-conduction problems one encounters boundary conditions that reflect Newton's law of cooling. A simple form of this law states that the rate of change of temperature across the boundary is directly proportional to the difference in temperatures of the object and the surrounding medium. Hence instead of the boundary conditions $u(0, t) = u(L, t) = 0$ in (8.5), the conditions have the form

$$u_x(0, t) = au(0, t) + b \quad \text{and} \quad u_x(L, t) = cu(L, t) + d. \tag{8.17}$$

A fairly simple procedure may be used to incorporate conditions of the form (8.17) into any of the three methods of the previous two sections. We first introduce two fictitious points in the x-direction, $x_{-1} = x_0 - h = -h$ and $x_{M+1} = x_M + h = L + h$, and assume that the equations determining the method [either (8.8) or (8.13) or (8.16)] hold also for $i = 0$ and $i = M$. Specifically for the Crank-Nicolson method we have

$$\begin{gathered} -sv_{-1, j+1} + (1 + 2s)v_{0, j+1} - sv_{1, j+1} = sv_{-1, j} + (1 - 2s)v_{0j} + sv_{1j} \\ -sv_{M-1, j+1} + (1 + 2s)v_{M, j+1} - sv_{M+1, j+1} = \\ sv_{M-1, j} + (1 - 2s)v_{Mj} + sv_{M+1, j}. \end{gathered} \tag{8.18}$$

We now use central-difference approximations for the derivatives in (8.17) to supply values for the fictitious approximations v_{ij} and $v_{i, j+1}$ in (8.18) when $i = -1$ and $i = M + 1$. On the assumption that $u(x, t)$ can be extended with sufficient differentiability to $\tilde{W} = \{(x, t): x_{-1} \le x \le x_{M+1}, t \ge 0\}$, then

$$\begin{aligned} u_x(0, t_j) &= \frac{u(x_1, t_j) - u(x_{-1}, t_j)}{2h} - \frac{\partial^3}{\partial x^3} u(\xi_0 h, t_j)h^2/6 \\ &= au(0, t_j) + b \\ u_x(L, t_j) &= \frac{u(x_{M+1}, t_j) - u(x_{M-1}, t_j)}{2h} - \frac{\partial^3}{\partial x^3} u(\xi_M h, t_j)h^2/6 \\ &= cu(L, t_j) + d. \end{aligned} \tag{8.19}$$

Thus $O(h^2)$ approximations at $t = t_j$ are supplied by

$$\frac{v_{1j} - v_{-1,j}}{2h} = av_{0j} + b \quad \text{and} \quad \frac{v_{M+1,j} - v_{M-1,j}}{2h} = cv_{Mj} + d$$

or

$$\begin{aligned} v_{-1,j} &= -2hav_{0j} + v_{1j} - 2hb \quad \text{and} \\ v_{M+1,j} &= v_{M-1,j} + 2hcv_{Mj} + 2hd. \end{aligned} \tag{8.20a}$$

Similarly

$$\begin{aligned} v_{-1,j+1} &= -2hav_{0,j+1} + v_{1,j+1} - 2hb \quad \text{and} \\ v_{M+1,j+1} &= v_{M-1,j+1} + 2hcv_{M,j+1} + 2hd. \end{aligned} \tag{8.20b}$$

Substituting (8.20a) and (8.20b) into (8.18) yields

$$\begin{aligned} &(1 + 2s + 2has)v_{0,j+1} - 2sv_{1,j+1} \\ &= (1 - 2s - 2has)v_{0j} + 2sv_{1j} - 4hbs, \end{aligned} \tag{8.21a}$$

$$\begin{aligned} &-2sv_{M-1,j+1} + (1 + 2s - 2hcs)v_{M,j+1} \\ &= 2sv_{M-1,j} + (1 - 2s + 2hcs)v_{Mj} + 4hds. \end{aligned} \tag{8.21b}$$

For this problem the values of $u(0, t_j)$ and $u(L, t_j)$ for $j \geq 1$ are unknown and are approximated by v_{0j} and v_{Mj}, respectively. For each t-step we solve for $v_{0,j+1}, v_{1,j+1}, \ldots, v_{M,j+1}$ in terms of the previously computed values of v_{0j}, $v_{1j}, \ldots, v_{Mj}$ from the $((M + 1) \times (M + 1))$ tridiagonal system whose first equation is (8.21a); the next $(M - 1)$ equations are given by (8.16) for $1 \leq i \leq M - 1$; and the last equation is (8.21b). The values $v_{0,0}, v_{1,0}, \ldots, v_{M,0}$ are once again supplied by the boundary condition $u(x, 0) = f(x)$ in (8.5).

PROBLEMS, SECTION 8.2.2

1. Modify the direct method (8.8) to treat the boundary conditions (8.17).
2. Use the result of Problem 1 and perform the required calculations in Problem 2, Section 8.2.1, with boundary conditions $u(0, t) = u(.4, t) = 0$ replaced by $u_x(0, t) = u(0, t) - 1$ and $u_x(.4, t) = 0$.
3. Repeat the required calculations in Problem 2 and use Crank-Nicholson modified by (8.21a) and (8.21b) with $h = .1$, $k = .01$.
4. Modify the program in Fig. 8.3 to treat derivative boundary conditions (8.17). Test the program on $u_t = u_{xx}$, $u(x, 0) = \sin \pi x + \cos \pi x$, $u_x(0, t) = \pi u(0, t)$, $u_x(1, t) = \pi u(1, t)$, up to $t = 1$ with $h = .1$, $t = .01$. [Answer: $u(x, t) = \exp(-\pi^2 t)(\sin \pi x + \cos \pi x)$.]
5. The value $r = 1/2$ may not be stable for the explicit method on a problem with derivative boundary conditions. Verify this statement with some sample computations on the problem $u_t = u_{xx}$, $u(x, 0) = 1$, $u_x(0, t) = u(0, t)$, $u_x(1, t) = 0$.

8.2.3. The Two-dimensional Heat Equation

In this section we discuss a particular form of a boundary-value problem that can be used to find the temperature distribution $u(x, y, t)$ of a homogeneous plate occupying a region R of the xy-plane. In this problem we are given values of u or its normal derivative on the boundary for all $t \geq 0$ and also an initial temperature distribution, $u(x, y, 0)$. For simplicity we assume that the region R is rectangular; and the specific problem we consider is

$$\frac{\partial}{\partial t} u(x, y, t) = \alpha^2\left(\frac{\partial^2}{\partial x^2} u(x, y, t) + \frac{\partial^2}{\partial y^2} u(x, y, t)\right)$$
$$0 < x < L,\ 0 < y < K,\ t \geq 0 \tag{8.22}$$

where $u(x, y, 0) = f(x, y)$ is the initial temperature distribution and $u(x, y, t)$ is specified on the boundary of R by $u(x, y, t) = g(x, y)$.

As with the one-dimensional problem (8.5), we let $\Delta x = L/M$ for some positive integer M and $x_i = i\Delta x$ for $0 \leq i \leq M$. We do the same in the y-direction: $\Delta y = K/N$ and $y_j = j\Delta y$ for $0 \leq j \leq N$. For simplicity of presentation we assume that M and N are chosen so that $\Delta x = \Delta y$, but this assumption is not necessary. We also choose an increment in t, Δt, and let $t_n = n\Delta t$ for $n = 0, 1, 2, \ldots$. We denote $u(x_i, y_j, t_n)$ as u_{ij}^n.

Using central differences as in (8.6a) for both u_{xx} and u_{yy} and the forward difference as in (8.6b) for u_t, we can derive the order $(\Delta x^2 + \Delta t)$ explicit method given by

$$\begin{aligned}(w_{ij}^{n+1} - w_{ij}^n)/\Delta t = \alpha^2[&(w_{i-1,j}^n - 2w_{ij}^n + w_{i+1,j}^n) \\ &+ (w_{i,j-1}^n - 2w_{ij}^n + w_{i,j+1}^n)]/(\Delta x)^2\end{aligned} \tag{8.23a}$$

or

$$w_{ij}^{n+1} = r(w_{i-1,j}^n + w_{i+1,j}^n + w_{i,j-1}^n + w_{i,j+1}^n) + (1 - 4r)w_{ij}^n \tag{8.23b}$$

where $r = \alpha^2\Delta t/(\Delta x)^2$. The points where approximations are related by (8.23) are shown in Fig. 8.4. At $n = 0$, the values w_{ij}^0 are given by the boundary condition $w_{ij}^0 = u(x_i, y_j, t_j, 0) = f(x_i, y_j)$. Inserting these values in (8.23) directly yields all values w_{ij}^1 at $t = t_1$. We then use the values w_{ij}^1 in (8.23) to obtain all w_{ij}^2, continue in this fashion, and advance in t. The role played by the coefficient $(1 - 2r)$ in (8.8) is analogous to the role of the coefficient $(1 - 4r)$ in (8.23) in a convergence and stability analysis [see (8.10)]; thus for the method of (8.23) we must have $r \leq 1/4$. [If $\Delta x \neq \Delta y$, we need $\alpha^2\Delta t/((\Delta x)^2 + (\Delta y)^2) \leq 1/8$.]

As in Section 8.2.1, we can obtain an order $((\Delta x)^2 + \Delta t)$ implicit method by using a backward-difference approximation for $u_t(x_i, y_j, t_n)$ as in (8.11). This procedure yields

$$\begin{aligned}(v_{i,j}^{n+1} - v_{ij}^n)/\Delta t = \alpha^2[&(v_{i-1,j}^{n+1} - 2v_{ij}^{n+1} + v_{i+1,j}^{n+1}) \\ &+ (v_{i,j-1}^{n+1} - 2v_{ij}^{n+1} + v_{i,j+1}^{n+1})](\Delta x)^2\end{aligned} \tag{8.24a}$$

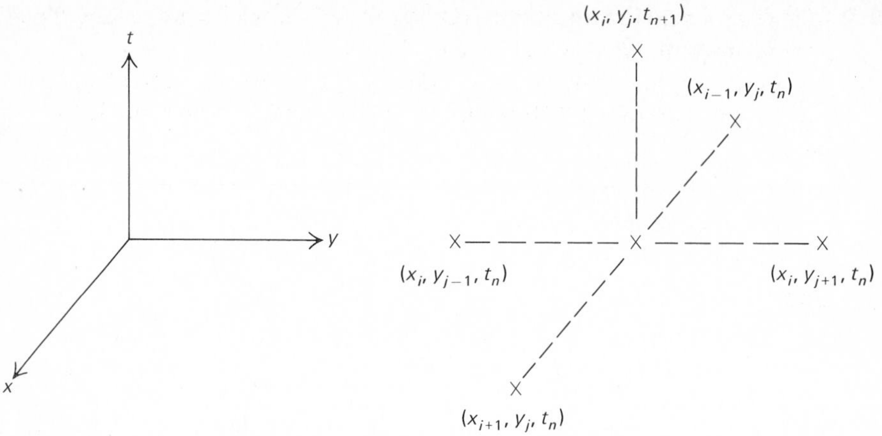

Figure 8.4

or

$$v_{ij}^n = -rv_{i-1,j}^{n+1} - rv_{i,j-1}^{n+1} + (1 + 4r)v_{ij}^{n+1} - rv_{i,j+1}^{n+1} - rv_{i+1,j}^{n+1}. \quad (8.24b)$$

To improve the order to $((\Delta x)^2 + (\Delta t)^2)$ as in Crank-Nicolson, we again average the forward-difference approximation (8.23a) with the backward-difference approximation (8.24a). This averaging yields

$$\begin{aligned} v_{ij}^{n+1} - v_{ij}^n = (r/2)[&(v_{i-1,j}^{n+1} - 2v_{ij}^{n+1} + v_{i+1,j}^{n+1}) + (v_{i-1,j}^n - 2v_{ij}^n + v_{i+1,j}^n) \\ &+ (v_{i,j-1}^{n+1} - 2v_{ij}^{n+1} + v_{i,j+1}^{n+1}) + (v_{i,j-1}^n - 2v_{ij}^n + v_{i,j+1}^n)]. \end{aligned} \quad (8.25)$$

The Crank-Nicholson method in (8.25) is unconditionally stable but requires the solution of an $[(M - 1)(N - 1) \times (M - 1)(N - 1)]$ linear system of equations to advance to $t = t_{n+1}$ from $t = t_n$. Each equation of (8.25) involves five unknowns $(v_{i-1,j}^{n+1}, v_{i,j-1}^{n+1}, v_{ij}^{n+1}, v_{i+1,j}^{n+1}, v_{i,j+1}^{n+1})$, and hence the system is no longer tridiagonal and can lead to unnecessary computation. This drawback is countered by a modification known as the *Peaceman-Rachford* or *alternating-direction implicit method*, or simply as ADI.

The ADI method of Peaceman and Rachford amounts to taking a time step $\Delta t/2$ using a backward-difference approximation that is implicit in only the x-direction. Specifically the backward-difference approximation

$$\frac{v_{ij}^{n+.5} - v_{ij}^n}{\Delta t/2} = \frac{\alpha^2}{(\Delta x)^2}[(v_{i-1,j}^{n+.5} - 2v_{ij}^{n+.5} + v_{i+1,j}^{n+.5}) + (v_{i,j-1}^n - 2v_{ij}^n + v_{i,j+1}^n)]$$

resembles (8.24a) except for using past values in the y-direction along the grid line $x = x_i$; the approximation is implicit in only the x-direction. The next step of ADI counters the bias introduced above by using

$$\frac{v_{ij}^{n+1} - v_{ij}^{n+.5}}{\Delta t/2} = \frac{\alpha^2}{(\Delta x)^2}[(v_{i-1,j}^{n+.5} - 2v_{ij}^{n+.5} + v_{i+1,j}^{n+.5}) + (v_{i,j-1}^{n+1} - 2v_{ij}^{n+1} + v_{i,j+1}^{n+1})].$$

The approximation above uses past values in the x-direction along the grid line $y = y_j$ and is implicit in only the y-direction.

With an * (asterisk) denoting the intermediate values calculated at $t = t_n + (\Delta t/2)$, the steps of the ADI method can be given in a form suitable for computation:

$$-v^*_{i-1,j} + \left(2 + \frac{2}{r}\right) v^*_{ij} - v^*_{i+1,j} = v^n_{i,j-1} + \left(\frac{2}{r} - 2\right) v^n_{ij} + v^n_{i,j+1} \quad (8.26)$$

and

$$-v^{n+1}_{i,j-1} + \left(2 + \frac{2}{r}\right) v^{n+1}_{ij} - v^{n+1}_{i,j+1} = v^*_{i-1,j} + \left(\frac{2}{r} - 2\right) v^*_{ij} + v^*_{i+1,j}. \quad (8.27)$$

To execute a complete step of ADI, we first use (8.26) to calculate all the intermediate values v^*_{ij}; this procedure is usually referred to as a ''horizontal sweep.'' In particular for each fixed index, j, $1 \le j \le N - 1$, we must solve the tridiagonal system (8.26) for $v^*_{1j}, v^*_{2j}, \ldots, v^*_{M-1,j}$. Thus in order to complete a horizontal sweep, we will solve $N - 1$ tridiagonal systems, each of size $M - 1$. Having the $(N - 1)(M - 1)$ intermediate values v^*_{ij}, we execute a vertical sweep by solving $M - 1$ tridiagonal systems in (8.27), one system for each index i, $1 \le i \le M - 1$. A horizontal sweep followed by a vertical sweep constitutes one step of the ADI method and produces an array of approximations v^{n+1}_{ij} to the solution $u(x_i, y_j, t_{n+1})$. For the two-dimensional heat equation on the rectangle (8.22), the ADI method has an accuracy comparable to the Crank-Nicolson method (8.25); see Young and Gregory (1973) or Ames (1977). The method is also unconditionally stable although the intermediate values v^*_{ij} have a bias and are not to be greatly trusted. We will see the philosophy of ADI employed also on elliptic problems in the next section.

EXAMPLE 8.4. As an illustration of the ADI method, we consider the equation

$$u_t = u_{xx} + u_{yy}, \qquad -1 \le x \le 1, -1 \le y \le 1$$

with $u(x, y, 0) = 0$ for $-1 < x < 1$, $-1 < y < 1$ and $u(x, y, t) = 1$ for (x, y) on the boundary. This equation could represent a long bar with a square cross section, where the temperature distribution is initially at 0° but then the surface is suddenly maintained at a temperature of 1°.

The ADI method was coded for this problem and a program is listed in Fig. 8.5. The program used $\Delta t = .05$, $\Delta x = .1$ and executed 20 time steps; given the current approximations v^n_{ij} to the solution at $t = t_n$ [stored in the array VOLD(I,J)], the main program calls Subroutine ADISTP to obtain v^{n+1}_{ij} [stored in the array VNEW(I,J)]. Subroutine ADISTP merely sets up the tridiagonal systems (8.26), (8.27) and then calls Subroutine TDSOLV to solve the systems.

The temperature distribution is symmetric in each quadrant of the square; and to illustrate the evolution of the temperature distribution with time, we printed the solution in the upper-right quadrant at (x_i, y_j) where $x_i = i/5$, $y_j = j/5$, $0 \le j \le 5$, $0 \le i \le 5$.

```
      DIMENSION VOLD(101,101),VNEW(101,101),EST(10)
      M=20
      N=M
      MP1=M+1
      NP1=N+1
      H=2./M
      DELT=.05
      R=DELT/(H*H)
        DO 2 I=2,M
          DO 1 J=2,N
          VOLD(I,J)=0.
    1     CONTINUE
    2   CONTINUE
        DO 3 K=1,MP1
        VOLD(1,K)=1.
        VOLD(MP1,K)=1.
        VOLD(K,1)=1.
        VOLD(K,NP1)=1.
    3   CONTINUE
      DO 8 ITIME=1,20
      CALL ADISTP(VOLD,VNEW,R,M,N)
        DO 5 JG=1,6
        JP=23-2*JG
          DO 4 IG=1,6
          IP=2*IG+9
          EST(IG)=VNEW(IP,JP)
    4     CONTINUE
        PRINT 100,(EST(K),K=1,6)
    5   CONTINUE
        DO 7 I=1,MP1
          DO 6 J=1,NP1
          VOLD(I,J)=VNEW(I,J)
    6     CONTINUE
    7   CONTINUE
    8 CONTINUE
  100 FORMAT(1H0,6F10.5)
      STOP
      END
```

Figure 8.5 (a) Main program for Example 8.4.

$t = .05$					
1.00000	1.00000	1.00000	1.00000	1.00000	1.00000
0.57943	0.58238	0.59646	0.64644	0.82028	1.00000
0.17262	0.17844	0.20613	0.30445	0.64644	1.00000
0.05566	0.06230	0.09391	0.20613	0.59646	1.00000
0.02271	0.02959	0.06230	0.17844	0.58238	1.00000
0.01579	0.02271	0.05566	0.17262	0.57943	1.00000

$t = .1$					
1.00000	1.00000	1.00000	1.00000	1.00000	1.00000
0.70958	0.71777	0.74858	0.81733	0.90697	1.00000
0.42973	0.44581	0.50631	0.64131	0.81733	1.00000
0.21508	0.23721	0.32048	0.50631	0.74858	1.00000
0.11890	0.14374	0.23721	0.44581	0.71777	1.00000
0.09333	0.11890	0.21508	0.42973	0.70958	1.00000

```
      SUBROUTINE ADISTP(VOLD,VNEW,R,M,N)
      DIMENSION VOLD(101,101),VINT(101,101),VNEW(101,101)
      DIMENSION RHS(101),TDSOL(101)
      MP1=M+1
      NP1=N+1
      RHO=2./R
C
C   SET BOUNDARY CONDITIONS ALONG TOP, BOTTOM, AND SIDES
C
        DO 1 I=1,MP1
        VINT(I,1)=1.
        VNEW(I,1)=1.
        VINT(I,NP1)=1.
        VNEW(I,NP1)=1.
    1   CONTINUE
        DO 2 J=1,NP1
        VINT(1,J)=1.
        VNEW(1,J)=1.
        VINT(MP1,J)=1.
        VNEW(MP1,J)=1.
    2   CONTINUE
C
C   EXECUTE HORIZONTAL SWEEP
C
        DO 5 J=2,N
          DO 3 I=2,M
          RHS(I)=VOLD(I,J-1)+VOLD(I,J+1)+(RHO-2.)*VOLD(I,J)
    3     CONTINUE
        RHS(2)=RHS(2)+VOLD(1,J)
        RHS(M)=RHS(M)+VOLD(MP1,J)
        CALL TDSOLV(RHS,TDSOL,RHO,M)
          DO 4 I=2,M
          VINT(I,J)=TDSOL(I)
    4     CONTINUE
    5   CONTINUE
C
C   EXECUTE VERTICAL SWEEP
C
        DO 8 I=2,M
          DO 6 J=2,N
          RHS(J)=VINT(I-1,J)+VINT(I+1,J)+(RHO-2.)*VINT(I,J)
    6     CONTINUE
        RHS(2)=RHS(2)+VINT(I,1)
        RHS(N)=RHS(N)+VINT(I,NP1)
        CALL TDSOLV(RHS,TDSOL,RHO,N)
          DO 7 J=2,N
          VNEW(I,J)=TDSOL(J)
    7     CONTINUE
    8   CONTINUE
      RETURN
      END
```

Figure 8.5 (b) Subroutine ADISTP.

		$t = .3$			
1.00000	1.00000	1.00000	1.00000	1.00000	1.00000
0.90174	0.90662	0.92053	0.94153	0.97374	1.00000
0.78116	0.79203	0.82301	0.86978	0.94153	1.00000
0.70258	0.71735	0.75946	0.82301	0.92053	1.00000
0.65051	0.66788	0.71735	0.79203	0.90662	1.00000
0.63224	0.65051	0.70258	0.78116	0.90174	1.00000

```
      SUBROUTINE TDSOLV(RHS,TDSOL,RHO,IDIM)
      DIMENSION RHS(101),TDSOL(101),D(101)
      DIAG=2.+RHO
      D(2)=DIAG
        DO 1 I=3,IDIM
        D(I)=DIAG-1./D(I-1)
        RHS(I)=RHS(I)+RHS(I-1)/D(I-1)
1       CONTINUE
      TDSOL(IDIM)=RHS(IDIM)/D(IDIM)
        DO 2 I=3,IDIM
        IEQN=IDIM+2-I
        TDSOL(IEQN)=(RHS(IEQN)+TDSOL(IEQN+1))/D(IEQN)
2       CONTINUE
      RETURN
      END
```

Figure 8.5 (c) Subroutine TDSOLV.

$t = 1.0$					
1.00000	1.00000	1.00000	1.00000	1.00000	1.00000
0.99650	0.99667	0.99717	0.99795	0.99895	1.00000
0.99317	0.99350	0.99447	0.99599	0.99795	1.00000
0.99057	0.99103	0.99237	0.99447	0.99717	1.00000
0.98892	0.98946	0.99103	0.99350	0.99667	1.00000
0.98835	0.98892	0.99057	0.99317	0.99650	1.00000

The exact solution has been tabulated at $(x, y) = (0, 0)$ for various values of time [see Carnahan, Luther, and Wilkes (1969)]. Listed below are values of the exact solution at (0, 0) and results from the ADI method with $\Delta t = .05$, $\Delta x = .1$ and with $\Delta t = .0125$, $\Delta x = .05$ ($r = 5$ in both cases).

Comparisons at $(x, y) = (0, 0)$

Time t	Exact value	$\Delta t = .0125$ $\Delta x = .05$	$\Delta t = .05$ $\Delta x = .1$
.1	.09883	.09907	.09333
.3	.63179	.63195	.63224
.5	.86252	.86259	.86282
.7	.94877	.94883	.94884

The methods of this section can be extended to three or more dimensions, but the size of the resultant systems can quickly become extremely large for realistic problems. In three dimensions a corresponding central difference for u_{zz} is included in the derivations and we proceed as indicated in this section. For the ADI method, three t-steps are required in an analogous fashion to that above. With $\Delta x = \Delta y = \Delta z$, this approach is unstable, however, for $r > 3/2$. Other modifications of ADI are available; for instance see Carnahan, Luther, and Wilkes (1969), Ames (1977), and Mitchell (1969).

PROBLEMS, SECTION 8.2.3

1. Using difference approximations as indicated in the text, derive (8.23b), (8.24b), and (8.25).
2. Consider $u_t = u_{xx} + u_{yy}$, $0 < x < 4, 0 < y < 3$, with $u(x, y, 0) = 0$, $u(0, y, t) = 10$, $u(4, y, t) = u(x, 0, t) = u(x, 3, t) = 0$.
   a) With $\Delta x = \Delta y = 1$ and $\Delta t = .25$ ($r = .25$), use the direct method (8.23b) to generate approximations w_{ij}^1 and w_{ij}^2, $1 \le i \le 3$, $1 \le j \le 2$.
   b) In matrix form display the Crank-Nicolson equations (8.25) applied to the problem above for $\Delta x = \Delta y = 1$ and $\Delta t = .5$.
   c) Apply the ADI method (8.26) and (8.27) for one t-step ($\Delta t/2 = .25$) to obtain approximations for u_{ij}^1, $1 \le i \le 3$, $1 \le j \le 2$.
3. Apply the program in Fig. 8.5 to Problem 2 with $\Delta x = \Delta y = .1$ and $\Delta t = .05$ to obtain approximations for $u(x_i, y_j, t_n)$, $1 \le i \le 39$, $1 \le j \le 29$, $1 \le n \le 10$.
4. Apply the program in Fig. 8.5 to $u_t = u_{xx} + u_{yy}$, $0 < x < 2, 0 < y < 1$, $u(x, y, 0) = \sin \pi y \sin 2\pi x$, $u(0, y, t) = u(2, y, t) = 0$, $u(x, 0, t) = u(x, 1, t) = 0$ for $\Delta x = \Delta y = .05$, $\Delta t = .05$, up to $t = 1$. (Answer: $u(x, y, t) = \exp[-5\pi^2 t]\sin 2\pi x \sin \pi y$.)
5. Perform sample calculations in Exercise 2a with $\Delta x = \Delta y = 1$, $\Delta t = .5$ to indicate instability for $r > 1/4$.
6. a) For $u_t = u_{xx} + u_{yy} + u_{zz}$, determine equations analogous to (8.25).
   b) Formulate the ADI algorithm for the equations in part (a).
7. Suppose that the boundary conditions for (8.22) are given by $u(x, y, 0) = f(x, y)$, $u_y(x, 0, t) = a_1 u(x, 0, t) + b_1$, $u_y(x, K, t) = a_2 u(x, K, t) + b_2$, $u_x(0, y, t) = a_3 u(0, y, t) + b_3$, and $u_x(L, y, t) = a_4 u(L, y, t) + b_4$. Modify both the direct method and the Crank-Nicolson method to treat this problem. [Hint: On the boundary u is unknown. Introduce fictitious points outside the boundary and use central differences to approximate derivatives.]
8. a) Consider the problem $u_t = u_{xx} + u_{yy}$, $u(x, y, 0) = F(x)G(y)$, $u(0, y, t) = u(L, y, t) = u(x, 0, t) = u(x, K, t) = 0$, $0 < x < L$, $0 < y < K$. Show that $u(x, y, t) = v(x, t)\, w(y, t)$ where

$$v_t = v_{xx}, \qquad v(x, 0) = F(x), \qquad v(0, t) = v(L, t) = 0$$

$$w_t = w_{yy}, \qquad w(y, 0) = G(y), \qquad w(0, t) = w(K, t) = 0.$$

   b) Suppose that one is interested in u only at a given xy-point, say $u(x_4, y_6, t)$. Devise an algorithm using the Crank-Nicolson method of Section 8.2.1 to obtain approximations for $u(x_4, y_6, t)$.

8.3 ELLIPTIC EQUATIONS

For our study of finite-difference techniques for elliptic equations, we restrict ourselves to the particular boundary-value problem

$$\frac{\partial^2 u}{\partial x^2}(x, y) + \frac{\partial^2 u}{\partial y^2}(x, y) + f(x, y)u(x, y) = g(x, y), \qquad 0 < x < L,\ 0 < y < K \tag{8.28}$$

where $u(x, y)$ is specified on the boundary by $u(x, y) = h(x, y)$ and $f(x, y)$ and $g(x, y)$ are given functions with $f(x, y) \leq 0$. When $f(x, y) = g(x, y) = 0$, the equation in (8.28) is called *Laplace's equation*; when $f(x, y) = 0$, Eq. (8.28) is called *Poisson's equation*; and if $f(x, y)$ is nonzero, (8.28) is called the *Helmholtz equation* or the *reduced wave equation*. Furthermore since u is prescribed on the boundary, (8.28) is a *Dirichlet* boundary-value problem.

As in previous sections, we let $h = L/M$ and $k = K/N$ for two positive integers M and N, and let $x_i = ih, 0 \leq i \leq M$, and $y_j = jk, 0 \leq j \leq N$. We denote $u(x_i, y_j)$ as u_{ij}, $f(x_i, y_j)$ as f_{ij}, and $g(x_i, y_j)$ as g_{ij}. We use central-difference approximations for u_{xx} and u_{yy} as in (8.6a); and with sufficient differentiability of u we obtain

$$\begin{aligned}(u_{i+1,j} - 2u_{ij} + u_{i-1,j})/h^2 + (u_{i,j+1} - 2u_{ij} + u_{i,j-1})/k^2 + f_{ij}u_{ij} - g_{ij}\\ = -\frac{\partial^4}{\partial x^4} u(x_i + \xi_i h, y_j)h^2/12 - \frac{\partial^4}{\partial y^4} u(x_i, y_j + \zeta_j k)k^2/12 \qquad (8.29)\end{aligned}$$

where $-1 < \xi_i, \zeta_j < 1$. For simplicity we assume that $h = k$, and so an approximation v_{ij} for u_{ij} with discretization error of order h^2 can be found from

$$-v_{i-1,j} - v_{i+1,j} + (4 - h^2 f_{ij})v_{ij} - v_{i,j-1} - v_{i,j+1} = -h^2 g_{ij}. \qquad (8.30)$$

The expression given by (8.30) is called a *five-point formula*. The coefficients and the relative positions of the points of approximation are given schematically in Fig. 8.6.

Since this approximation is made at each interior grid point, (8.30) gives rise to an $((M - 1)(N - 1) \times (M - 1)(N - 1))$ system of equations. In typical problems, the size of this system may be so large that solving the system

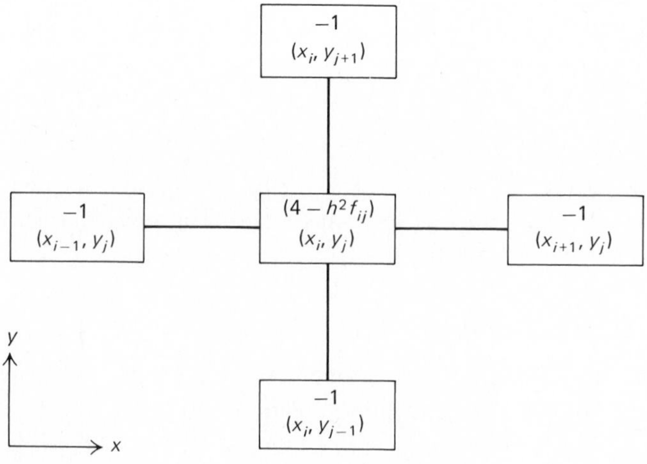

Figure 8.6

presents significant difficulties in terms of the number of operations and storage requirements. Note that if v_{ij} in (8.30) does not correspond to a point adjacent to the boundary, then the equation contains five variables and so the coefficient matrix is not (for example) tridiagonal. Hence in the remainder of this section, we consider properties of this particular system and numerical techniques for its solution.

The coefficient matrix for the $((M-1)(N-1) \times (M-1)(N-1))$ system in (8.30) has a natural banded structure when we order the equations from left to right, bottom to top; that is, if we express the system (8.30) as $A\mathbf{v} = \mathbf{g}$ where

$$\mathbf{v} = \begin{bmatrix} \mathbf{v}_1 \\ \mathbf{v}_2 \\ \vdots \\ \mathbf{v}_{N-1} \end{bmatrix}, \qquad \mathbf{g} = -h^2 \begin{bmatrix} \mathbf{g}_1 \\ \mathbf{g}_2 \\ \vdots \\ \mathbf{g}_{N-1} \end{bmatrix}, \qquad \mathbf{v}_j = \begin{bmatrix} v_{1j} \\ v_{2j} \\ \vdots \\ v_{M-1,j} \end{bmatrix}, \qquad \mathbf{g}_j = \begin{bmatrix} g_{1j} \\ g_{2j} \\ \vdots \\ g_{M-1,j} \end{bmatrix}.$$

An example of the coefficient matrix A is given in Fig. 8.7 where $M = N = 5$ and $f_{ij} = 0$, $1 \le i, j \le 5$. As is illustrated in Fig. 8.7, the diagonal entries of A contain the terms $4 - h^2 f_{ij}$; and no row nor column contains more than five nonzero terms. Even for modest values of M and N, the matrix A is quite large; for example, if $M = N = 100$, then A is a (9801×9801) matrix. Of course, the matrix A is sparse and the methods we will use to solve $A\mathbf{v} = \mathbf{g}$ take advantage of sparseness.

Before considering methods for solving the system $A\mathbf{v} = \mathbf{g}$ in (8.30), we first address the question of whether A is nonsingular or not. Now from (8.30) it follows that A is diagonally dominant (and hence nonsingular) if $f_{ij} < 0$, $1 \le$

$$A = \left[\begin{array}{cccc|cccc|cccc|cccc}
4 & -1 & 0 & 0 & -1 & 0 & 0 & 0 & 0 & 0 & 0 & 0 & 0 & 0 & 0 & 0 \\
-1 & 4 & -1 & 0 & 0 & -1 & 0 & 0 & 0 & 0 & 0 & 0 & 0 & 0 & 0 & 0 \\
0 & -1 & 4 & -1 & 0 & 0 & -1 & 0 & 0 & 0 & 0 & 0 & 0 & 0 & 0 & 0 \\
0 & 0 & -1 & 4 & 0 & 0 & 0 & -1 & 0 & 0 & 0 & 0 & 0 & 0 & 0 & 0 \\
\hline
-1 & 0 & 0 & 0 & 4 & -1 & 0 & 0 & -1 & 0 & 0 & 0 & 0 & 0 & 0 & 0 \\
0 & -1 & 0 & 0 & -1 & 4 & -1 & 0 & 0 & -1 & 0 & 0 & 0 & 0 & 0 & 0 \\
0 & 0 & -1 & 0 & 0 & -1 & 4 & -1 & 0 & 0 & -1 & 0 & 0 & 0 & 0 & 0 \\
0 & 0 & 0 & -1 & 0 & 0 & -1 & 4 & 0 & 0 & 0 & -1 & 0 & 0 & 0 & 0 \\
\hline
0 & 0 & 0 & 0 & -1 & 0 & 0 & 0 & 4 & -1 & 0 & 0 & -1 & 0 & 0 & 0 \\
0 & 0 & 0 & 0 & 0 & -1 & 0 & 0 & -1 & 4 & -1 & 0 & 0 & -1 & 0 & 0 \\
0 & 0 & 0 & 0 & 0 & 0 & -1 & 0 & 0 & -1 & 4 & -1 & 0 & 0 & -1 & 0 \\
0 & 0 & 0 & 0 & 0 & 0 & 0 & -1 & 0 & 0 & -1 & 4 & 0 & 0 & 0 & -1 \\
\hline
0 & 0 & 0 & 0 & 0 & 0 & 0 & 0 & -1 & 0 & 0 & 0 & 4 & -1 & 0 & 0 \\
0 & 0 & 0 & 0 & 0 & 0 & 0 & 0 & 0 & -1 & 0 & 0 & -1 & 4 & -1 & 0 \\
0 & 0 & 0 & 0 & 0 & 0 & 0 & 0 & 0 & 0 & -1 & 0 & 0 & -1 & 4 & -1 \\
0 & 0 & 0 & 0 & 0 & 0 & 0 & 0 & 0 & 0 & 0 & -1 & 0 & 0 & -1 & 4
\end{array}\right]$$

Figure 8.7 Coefficient matrix for (8.30) with $M = N = 5$, $f_{ij} = 0$.

$i \le M - 1, 1 \le j \le N - 1$. In the case $f_{ij} = 0$ (such as Laplace's equation or Poisson's equation), we can also show that A is nonsingular. To see this, suppose that $\mathbf{z}$ is any $(M - 1)(N - 1)$-dimensional vector satisfying $A\mathbf{z} = \theta$. Let p and q denote the index of the largest (in absolute value) component of $\mathbf{z}$ so that $\|\mathbf{z}\|_\infty = |z_{pq}|$. Since $A\mathbf{z} = \theta$, it follows from (8.30) that

$$-z_{p-1,\,q} - z_{p+1,\,q} + 4z_{pq} - z_{p,\,q-1} - z_{p,\,q+1} = 0$$

where we understand above that $z_{ij} = 0$ if $i = 0$, M or $j = 0$, N. Now since $|z_{pq}| \ge |z_{ij}|$ for all i and j, it follows also that $z_{p-1,\,q} = z_{p+1,\,q} = z_{pq} = z_{p,\,q-1} = z_{p,\,q+1} = \|\mathbf{z}\|_\infty$. Since $z_{p-1,\,q} = \|\mathbf{z}\|_\infty$, we can repeat the argument above and obtain $|z_{p-2,\,q}| = \|\mathbf{z}\|_\infty$. Continuing, moving to the left along the grid line $y = y_q$, we finally obtain $|z_{1,\,q}| = \|\mathbf{z}\|_\infty$. Since z_{1q} is adjacent to a boundary point, (8.30) becomes

$$-z_{2q} + 4z_{1q} - z_{1,\,q-1} - z_{1,\,q+1} = 0.$$

Since $|z_{1q}| = \|\mathbf{z}\|_\infty$, the equation above can hold only if $z_{1q} = 0$, which means $\mathbf{z} = \theta$. The only solution of $A\mathbf{x} = \theta$ is $\mathbf{z} = \theta$; so the system $A\mathbf{v} = \mathbf{g}$ in (8.30) always has a unique solution. Finally it can be shown [see Varga (1962)] that A is positive definite for either $f_{ij} < 0$ or $f_{ij} = 0$.

8.3.1. Iterative Methods for Solving Large Systems

For large systems $A\mathbf{v} = \mathbf{g}$ of the form (8.30), Gauss elimination may not be appropriate. Even though A is banded and Gauss elimination will not affect the zeros outside the bands, Gauss elimination will operate with the entries within the bands and can cause a large amount of fill in. Iterative methods, however, are quite effective for solving $A\mathbf{v} = \mathbf{g}$ and are also easy to understand and code.

Since A is symmetric and positive-definite, we have by Theorem 2.4 that the Gauss-Seidel iterative method (2.62) will converge. With the given ordering of equations, the Gauss-Seidel iteration applied to the system of (8.30) is given by

$$(4 - h^2 f_{ij})v_{ij}^{(n+1)} = v_{i-1,\,j}^{(n+1)} + v_{i,\,j-1}^{(n+1)} + v_{i+1,\,j}^{(n)} + v_{i,\,j+1}^{(n)} - h^2 g_{ij} \qquad (8.31)$$

where initial values $v_{ij}^{(0)}$ are initially given for all variables and the superscripts denote the iteration number [again see (2.62)]. Note that with the given ordering, updated values for $v_{i-1,\,j}$ and $v_{i,\,j-1}$ have already been computed and are used by Gauss-Seidel as soon as they are available. Note also that if v_{ij} corresponds to a point adjacent to the boundary, then either one or two terms on the right-hand side are given by the prescribed boundary conditions, and (8.31) must be modified accordingly.

Unfortunately, convergence of the algorithm given by (8.31) is often rather slow. To remedy this problem, successive overrelaxation (SOR) is usually employed. To develop the SOR algorithm from (8.31), we divide by $(4 - h^2 f_{ij})$, add and subtract the term $v_{ij}^{(n)}$, and obtain

$$v_{ij}^{(n+1)} = v_{ij}^{(n)} + [v_{i-1,j}^{(n+1)} + v_{i,j-1}^{(n+1)} + v_{i+1,j}^{(n)} + v_{i,j+1}^{(n)} - (4 - h^2 f_{ij})v_{ij}^{(n)} - h^2 g_{ij}]/(4 - h^2 f_{ij}). \tag{8.32}$$

Since the Gauss-Seidel method is convergent, $v_{ij}^{(n+1)} \to v_{ij}$ and $v_{ij}^{n} \to v_{ij}$ as $n \to \infty$, and so $|v_{ij}^{(n+1)} - v_{ij}^{(n)}| \to 0$ as $n \to \infty$. Hence we may think of the term in braces as a correction factor in going from $v_{ij}^{(n)}$ to $v_{ij}^{(n+1)}$, where the correction factor approaches zero as $n \to \infty$. The idea in the SOR method is to multiply this correction factor by a constant ω in order to accelerate the convergence. Hence the SOR iterations are given by

$$v_{ij}^{(n+1)} = v_{ij}^{(n)} + \omega\{[v_{i-1,j}^{(n+1)} + v_{i,j-1}^{(n+1)} + v_{i+1,j}^{(n)} + v_{i,j+1}^{(n)} - (4 - h^2 f_{ij})v_{ij}^{(n)}]/(4 - h^2 f_{ij})\} - \omega h^2 g_{ij}/(4 - h^2 f_{ij}). \tag{8.33}$$

[Once again modifications to (8.33) must be made if v_{ij} corresponds to a point adjacent to the boundary.]

The choice of a value for ω is a difficult matter. [Note that for $\omega = 1$, (8.33) is the Gauss-Seidel method.] If we write the original system of (8.30) as $A\mathbf{v} = \mathbf{g}$ with the horizontal ordering, then an equivalent system is $B\mathbf{v} = \mathbf{c}$ where $B = D^{-1}A$, $\mathbf{c} = D^{-1}\mathbf{g}$, and $D = (4 - h^2 f_{ij})I$. If one writes $B = I + L + U$ where L and U are the lower- and upper-triangular parts, respectively, of B, then (8.33) can be written in matrix form (see Problem 15, Section 2.4.2):

$$\mathbf{v}^{(n+1)} = \mathbf{v}^{(n)} - \omega L\mathbf{v}^{(n+1)} - \omega U\mathbf{v}^{(n)} - \omega\mathbf{v}^{(n)} + \omega\mathbf{c}$$

or

$$\mathbf{v}^{(n+1)} = (I + \omega L)^{-1}[(1 - \omega)I - \omega U]\mathbf{v}^{(n)} + \omega(I + \omega L)^{-1}\mathbf{c}. \tag{8.34}$$

Let $S_\omega = (I + \omega L)^{-1}[(1 - \omega)I - \omega U]$ and $\mathbf{e}^{(n)}$ denote the nth error vector; then (as in Section 2.4) $\mathbf{e}^{(n)} = S_\omega^n \mathbf{e}^{(0)}$. By Theorem 3.7 S_ω is convergent if and only if $\rho(S_\omega) < 1$. Furthermore the rate of convergence is governed largely by the size of $\rho(S_\omega)$, and so we wish to choose ω to minimize $\rho(S_\omega)$. Through a difficult and lengthy analysis [Varga (1962)] it can be shown that if ω^* is the value that minimizes $\rho(S_\omega)$, then $1 < \omega^* < 2$, and

$$\omega^* = 2/(1 + \sqrt{1 - \rho(M_J)^2})$$

where M_J is the matrix of the Jacobi method (2.59) applied to the system (8.30). For the case $f(x, y) = 0$ in (8.28), $\rho(M_J) = [\cos(\pi/M) + \cos(\pi/N)]/2$, [Young and Gregory (1973)]. Hence if $M = N$, $\omega^* = 2/(1 + \sin(\pi/M))$. For large M and for $f_{ij} < 0$ we would expect ω^* to be closer to two than to one. (See Young and Gregory for techniques to approximate ω^* in more general problems.) Both computational experience and further analysis indicate that it is better to overestimate ω^* than to underestimate ω^*. Often one finds a dramatic improvement in the rate of convergence as ω approaches ω^*.

EXAMPLE 8.5. As a simple illustration of SOR applied to Laplace's equation, consider

$$u_{xx} + u_{yy} = 0, \qquad 0 \le x \le \pi,\ 0 \le y \le \pi$$

where $u(x, \pi) = \sin(x)$, $u(x, 0) = u(0, y) = u(\pi, y) = 0$. The exact solution of this problem is given by

$$u(x, y) = \frac{\sinh(y)\sin(x)}{\sinh(\pi)}$$

as may be easily verified.

A simple program that uses SOR for this problem is listed in Fig. 8.8. The main program calculates the relaxation factor ω^* and a starting vector $\mathbf{v}_0$, and then calls Subroutine SOR. The SOR iteration is continued until $\|\mathbf{v}_{n+1} - \mathbf{v}_n\|_\infty$ is less than an input convergence criterion (TOL = .001 was used in the program in Fig. 8.8). The starting vector $\mathbf{v}_0$ [stored in the array VOLD(I, J)] is chosen by linearly interpolating the boundary values down the vertical grid lines of the $(\pi \times \pi)$ square. A square mesh of size $\pi/20$ was used, and the results from Subroutine SOR [stored in the array VNEW(I, J)] were printed at selected interior grid points; the results were printed at (x_i, y_j) where $x_i = i\pi/10$, $1 \le i \le 9$, $y_j = j\pi/10$, $1 \le j \le 9$.

The results are displayed below, in a pattern similar to the grid; the upper-right entry corresponds to the grid point $(.9\pi, .9\pi)$. At each location, the upper entry is the estimated solution and the lower entry is the actual solution. For these results, the convergence criterion was satisfied after 39 iterations.

0.22548	0.42891	0.59033	0.69397	0.72968	0.69396	0.59032	0.42889	0.22548
0.22534	0.42862	0.58994	0.69352	0.72921	0.69352	0.58994	0.42862	0.22534
0.16434	0.31256	0.43017	0.50568	0.53169	0.50565	0.43012	0.31250	0.16429
0.16408	0.31210	0.42957	0.50499	0.53098	0.50499	0.42957	0.31210	0.16408
0.11948	0.22721	0.31267	0.36751	0.38639	0.36746	0.31256	0.22708	0.11938
0.11915	0.22664	0.31195	0.36672	0.38559	0.36672	0.31195	0.22664	0.11915
0.08645	0.16435	0.22614	0.26577	0.27939	0.26568	0.22598	0.16417	0.08630
0.08608	0.16374	0.22537	0.26493	0.27857	0.26493	0.22537	0.16374	0.08608
0.06199	0.11777	0.16199	0.19034	0.20007	0.19022	0.16178	0.11752	0.06178
0.06158	0.11713	0.16121	0.18952	0.19927	0.18952	0.16121	0.11713	0.06158
0.04365	0.08284	0.11386	0.13373	0.14053	0.13359	0.11360	0.08251	0.04335
0.04320	0.08217	0.11310	0.13296	0.13980	0.13296	0.11310	0.08217	0.04320
0.02959	0.05606	0.07696	0.09033	0.09487	0.09016	0.07666	0.05567	0.02926
0.02912	0.05539	0.07624	0.08963	0.09424	0.08963	0.07624	0.05539	0.02912
0.01835	0.03471	0.04759	0.05580	0.05857	0.05563	0.04728	0.03433	0.01804
0.01794	0.03412	0.04697	0.05522	0.05806	0.05522	0.04697	0.03412	0.01794
0.00880	0.01661	0.02275	0.02665	0.02796	0.02655	0.02255	0.01637	0.00860
0.00855	0.01625	0.02237	0.02630	0.02765	0.02630	0.02237	0.01625	0.00855

```
      DIMENSION VOLD(101,101),VNEW(101,101),EST(9)
      M=20
      MP1=M+1
      PI=3.141592654
      H=PI/M
        DO 1 I=1,MP1
        VOLD(I,MP1)=SIN((I-1)*H)
 1      CONTINUE
        DO 3 J=1,M
        Q=(J-1)*H
          DO 2 I=1,MP1
          VOLD(I,J)=Q*VOLD(I,MP1)
 2        CONTINUE
 3      CONTINUE
      OMEGA=2./(1.+SIN(PI/M))
      TOL=.001
      CALL SOR(VOLD,VNEW,OMEGA,TOL,M)
        DO 4 J=1,9
        JP=21-2*J
          DO 5 I=1,9
          IP=2*I+1
          EST(I)=VNEW(IP,JP)
 5        CONTINUE
        PRINT 100,(EST(K),K=1,9)
 4      CONTINUE
100   FORMAT(1H0,9F10.5)
      STOP
      END
```

Figure 8.8 The program for Example 8.5.
(a) Main program

The same problem was run with a grid spacing of $\pi/100$; and as would be expected, the estimates were more accurate. However with the same convergence criterion of TOL = .001, the method required 195 iterations. The results are listed below.

0.22533	0.42858	0.58987	0.69343	0.72911	0.69342	0.58985	0.42856	0.22531
0.16411	0.31212	0.42954	0.50492	0.53087	0.50488	0.42947	0.31203	0.16405
0.11922	0.22675	0.31203	0.36674	0.38556	0.36666	0.31189	0.22660	0.11913
0.08621	0.16394	0.22556	0.26508	0.27866	0.26497	0.22537	0.16373	0.08607
0.06178	0.11743	0.16152	0.18977	0.19945	0.18963	0.16127	0.11715	0.06158
0.04347	0.08260	0.11353	0.13332	0.14007	0.13314	0.11321	0.08222	0.04321
0.02944	0.05592	0.07678	0.09009	0.09458	0.08985	0.07638	0.05546	0.02914
0.01823	0.03464	0.04752	0.05570	0.05842	0.05546	0.04712	0.03420	0.01797
0.00868	0.01659	0.02277	0.02666	0.02794	0.02651	0.02250	0.01632	0.00857

Modifications of Gauss-Seidel and SOR are naturally suggested by the form of (8.30). For simplicity we illustrate this point by considering (8.30) for Laplace's equation (so $f_{ij} = g_{ij} = 0$). In this case, (8.30) reduces to

$$-v_{i-1,j} - v_{i+1,j} + 4v_{ij} - v_{i,j-1} - v_{i,j+1} = 0. \tag{8.35}$$

If we move $v_{i,j-1}$ and $v_{i,j+1}$ to the right-hand side of (8.35), we have

$$-v_{i-1,j} + 4v_{ij} - v_{i+1,j} = v_{i,j-1} + v_{i,j+1}. \tag{8.36}$$

```
      SUBROUTINE SOR(VOLD,VNEW,OMEGA,TOL,M)
      DIMENSION VOLD(101,101),VNEW(101,101)
      S=OMEGA/4.
      MP1=M+1
C
C   SET BOUNDARY CONDITIONS
C
        DO 1 K=1,MP1
        VNEW(1,K)=VOLD(1,K)
        VNEW(MP1,K)=VOLD(MP1,K)
        VNEW(K,1)=VOLD(K,1)
        VNEW(K,MP1)=VOLD(K,MP1)
    1   CONTINUE
C
C   EXECUTE SOR ITERATION AND CALCULATE TEST WHERE
C   TEST IS THE MAXIMUM COMPONENT OF VNEW-VOLD
C
    2 TEST=0.
        DO 4 J=2,M
          DO 3 I=2,M
          VNEW(I,J)=VOLD(I,J)+S*(VNEW(I-1,J)+VNEW(I,J-1)+
     *    VOLD(I+1,J)+VOLD(I,J+1)-4.*VOLD(I,J))
          DIFF=ABS(VNEW(I,J)-VOLD(I,J))
          IF(TEST.LT.DIFF)  TEST=DIFF
    3     CONTINUE
    4   CONTINUE

C   CONVERGENCE CHECK; CONTINUE ITERATING IF TEST .GT. TOL
C
      IF(TEST.LT.TOL)  RETURN
        DO 6 J=2,M
          DO 5 I=2,M
          VOLD(I,J)=VNEW(I,J)
    5     CONTINUE
    6   CONTINUE
      GO TO 2
      END
```

Figure 8.8 (b) Subroutine SOR

The form of (8.36) suggests that we apply the philosophy of the Gauss-Seidel method and consider the iteration ($\mathbf{v}^0$ is a starting vector)

$$-v_{i-1,j}^{n+1} + 4v_{ij}^{n+1} - v_{i+1,j}^{n+1} = v_{i,j-1}^{n+1} + v_{i,j+1}^{n}, \qquad n = 0, 1, \ldots. \tag{8.37}$$

The iteration (8.37) proceeds through the grid one horizontal line at a time and is called the line Gauss-Seidel method. For $j = 1$, (8.37) asks that we solve a tridiagonal system for $v_{11}^{n+1}, v_{21}^{n+1}, \ldots, v_{M-1,1}^{n+1}$. Given v_{i1}^{n+1} for $1 \leq i \leq M-1$, we solve another tridiagonal system for v_{i2}^{n+1}, $1 \leq i \leq M-1$, etc.

The same idea applied to SOR leads to the successive line overrelaxation method (SLOR) given by

$$\begin{aligned} -v_{i-1,j}^{n+1} + 4v_{ij}^{n+1} - v_{i+1,j}^{n+1} &= \omega(v_{i,j+1}^{n} + v_{i,j-1}^{n+1}) \\ &\quad + (1-\omega)(4v_{ij}^{n} - v_{i+1,j}^{n} - v_{i-1,j}^{n}). \end{aligned} \tag{8.38}$$

Again (8.38) requires us to solve a tridiagonal system for $j = 1, 2, \ldots, N-1$. Also note that for $\omega = 1$, SLOR reduces to line Gauss-Seidel. The choice of the relaxation factor ω is again a difficult matter, but the ω in SLOR is related to the spectral radius of the line Jacobi iteration matrix in the same fashion as explained previously for the choice of ω in SOR. For the problems we are considering, SLOR converges "asymptotically" faster than SOR and can be programmed to have the same number of arithmetic computations per mesh point as SOR has. [For analysis and detailed discussion of the comments above, see both Varga (1962) and Young and Gregory (1973).]

The final technique we present for solving the system (8.30) is one that we have essentially already discussed in Section 8.2.3, the *Peaceman-Rachford* or *alternating-direction implicit* (ADI) method. In Section 8.2.3, ADI was presented for the two-dimensional transient-state heat equation, $u_t = \alpha^2(u_{xx} + u_{yy})$. As the temperature reaches steady-state, u_t approaches zero and essentially the equation we are considering is $u_{xx} + u_{yy} = 0$. Hence the techniques of ADI are appropriate also for (8.28).

The ADI philosophy applied to solve the $(M - 1)(N - 1)$ equations in (8.30) suggests that we repetitively execute horizontal and vertical sweeps. For convenience of presentation, we assume $f_{ij} = 0$. Thus one step of ADI applied to (8.30) would amount to first using a horizontal sweep and solving the tridiagonal systems

$$-v^*_{i-1,j} + 2v^*_{ij} - v^*_{i+1,j} = v^n_{i,j-1} - 2v^n_{ij} + v^n_{i,j+1} - h^2 g_{ij} \tag{8.39}$$

for $j = 1, 2, \ldots, N - 1$. The ADI step is completed by a vertical sweep and solving the tridiagonal systems

$$-v^{n+1}_{i,j-1} + 2v^{n+1}_{ij} - v^{n+1}_{i,j+1} = v^*_{i-1,j} - 2v^*_{ij} + v^*_{i+1,j} - h^2 g_{ij} \tag{8.40}$$

for $i = 1, 2, \ldots, M - 1$. For the equations (8.30), we are regarding ADI as an iterative method for solving (8.30); in Section 8.2.3, one step of ADI was used to advance in time from t_n to $t_n + \Delta t$; in (8.39) and (8.40), the iteration is being used to solve $A\mathbf{v} = \mathbf{g}$ in (8.30).

Experience has shown that an iteration parameter should be used in (8.39) and (8.40). To include this parameter, we rewrite (8.39) and (8.40):

$$-v^*_{i-1,j} + (2 + \rho)v^*_{ij} - v^*_{i+1,j} = v^n_{i,j-1} - (2 - \rho)v^n_{ij} + v^n_{i,j+1} - h^2 g_{ij} \tag{8.41}$$

$$-v^{n+1}_{i,j-1} + (2 + \rho)v^{n+1}_{ij} - v^n_{i,j+1} = v^*_{i-1,j} - (2 - \rho)v^*_{ij} + v^*_{i+1,j} - h^2 g_{ij}. \tag{8.42}$$

It can be shown [see Varga (1962), Young and Gregory (1973), or Ames (1977)] that the "asymptotic" rate of convergence for ADI using the optimum parameter ρ^* is the same as that for SOR using the optimum relaxation parameter ω^*. Moreover, ADI requires more effort per step. However [see Ames (1977)] if the parameter ρ in (8.41) and (8.42) varies from step to step, then very rapid convergence is possible to achieve. Although the theoretical justification is not given here, how the iteration parameters can be chosen for the problem (8.28) is fairly easy to describe. In particular, let $s = \max\{M, N\}$ and define

$$a = 4\sin^2\frac{\pi}{2s} \qquad b = 4\sin^2\frac{(s-1)\pi}{2s}$$

$$c = a/b \qquad \epsilon = 3 - 2\sqrt{2}$$

Let m be the smallest integer such that

$$\epsilon^{m-1} \leq c;$$

and define $\rho_1, \rho_2, \ldots, \rho_m$ by

$$\rho_k = bc^{\lambda_k}, \ \lambda_k = (k - 1)/(m - 1), \qquad 1 \leq k \leq m.$$

The iteration parameters ρ_k given above are called the Wachspress parameters and are used in an ADI variation called the cyclic ADI method.

Given a starting vector $\mathbf{v}^0$, the cyclic ADI method uses (8.41) and (8.42) with $\rho = \rho_1$ to calculate $\mathbf{v}^1$; then ρ_2, to calculate $\mathbf{v}^2$; etc. Having reached $\mathbf{v}^m$, the cycle begins again; ρ_1 is used to generate $\mathbf{v}^{m+1}$; ρ_2, to generate $\mathbf{v}^{m+2}$, etc.

EXAMPLE 8.6. As an illustration of the cyclic ADI method, the simple program listed in Fig. 8.9 was used on the problem in Example 8.5. Subroutine CYCADI (WOLD,WNEW,HSQG,TOL,M,N) was called by a main program that is essentially the same as that in Fig. 8.8 for SOR. The only difference was that the values $h^2 g_{ij}$ required in (8.41), (8.42) were passed to CYCADI in the array HSQG and that the relaxation parameter OMEGA needed by Subroutine SOR was not needed for Subroutine CYCADI. Also, CYCADI used the same tridiagonal system solver, TDSOLV, as SOR used. For the results listed below, a convergence criterion of TOL = .0001 was used, and M and N were set to 100. In Example 8.5, 195 iterations were necessary to meet the convergence requirement of .001 for a 100×100 grid; the program in Fig. 8.9 needed only 7 iterations to satisfy a stronger convergence test. Thus for this problem, the cyclic ADI method performed well. As a point of interest, the value of m was 6 and hence one complete cycle and then one additional step were required before $\|\mathbf{v}^{i+1} - \mathbf{v}^i\|_\infty \leq .0001$ The results are displayed below in the same format as those of Example 8.5.

0.22539	0.42874	0.59012	0.69372	0.72941	0.69374	0.59012	0.42875	0.22540
0.16416	0.31226	0.42983	0.50530	0.53127	0.50528	0.42982	0.31228	0.16417
0.11923	0.22681	0.31221	0.36703	0.38589	0.36701	0.31220	0.22682	0.11924
0.08616	0.16390	0.22561	0.26523	0.27885	0.26520	0.22561	0.16390	0.08616
0.06164	0.11727	0.16142	0.18977	0.19952	0.18975	0.16142	0.11727	0.06164
0.04326	0.08229	0.11327	0.13317	0.14000	0.13315	0.11327	0.08230	0.04326
0.02917	0.05549	0.07638	0.08979	0.09439	0.08977	0.07638	0.05549	0.02916
0.01797	0.03419	0.04706	0.05533	0.05816	0.05532	0.04706	0.03419	0.01797
0.00856	0.01629	0.02242	0.02636	0.02771	0.02635	0.02242	0.01629	0.00856

PROBLEMS, SECTIONS 8.3 and 8.3.1

1. Use (8.6a) to establish (8.29).
2. Display equations (8.30) in matrix form $A\mathbf{v} = \mathbf{g}$ for $M = 3$, $N = 4$ with $h = 1$, $f(x, y) = -x^2y^2$, and $g(x, y) = y$.

```
      SUBROUTINE CYCADI(VOLD,VNEW,HSQG,TOL,M,N)
      DIMENSION VOLD(101,101),VNEW(101,101),VINT(101,101),HSQG(101,101)
      DIMENSION RHS(101),TDSOL(101)
      PI=3.141592654
      MP1=M+1
      NP1=N+1
C
C   SET BOUNDARY CONDITIONS
C
      DO 1 I=1,MP1
      VINT(I,1)=VOLD(I,1)
      VINT(I,NP1)=VOLD(I,NP1)
    1 CONTINUE
      DO 2 J=1,NP1
      VINT(1,J)=VOLD(1,J)
      VINT(MP1,J)=VOLD(MP1,J)
    2 CONTINUE
C
C   CALCULATE ITERATION PARAMETERS, MITR IS THE CYCLE LENGTH
C
      MAX=MAX0(M,N)
      E=3.-2.*SQRT(2.)
      A=4.*SIN(PI/(2.*MAX))**2
      B=4.*SIN((MAX-1)*PI/(2.*MAX))**2
      C=A/B
      MITR=1
      Q=E
    3 MITR=MITR+1
      IF(Q.LE.C)  GO TO 4
      Q=Q*E
      GO TO 3
    4 CONTINUE
      EX=1./FLOAT(MITR-1)
      FACTOR=C**EX
      RHO=B
      DO 13 ITR=1,MITR
C
C   EXECUTE HORIZONTAL SWEEP
C
      DO 7 J=2,N
        DO 5 I=2,M
        RHS(I)=VOLD(I,J-1)+VOLD(I,J+1)-(2.-RHO)*VOLD(I,J)-HSQG(I,J)
    5   CONTINUE
      RHS(2)=RHS(2)+VOLD(1,J)
      RHS(M)=RHS(M)+VOLD(MP1,J)
      CALL TDSOLV(RHS,TDSOL,RHO,M)
        DO 6 I=2,M
        VINT(I,J)=TDSOL(I)
    6   CONTINUE
    7 CONTINUE
C
C   EXECUTE VERTICAL SWEEP
C
      DO 10 I=2,M
        DO 8 J=2,N
        RHS(J)=VINT(I-1,J)+VINT(I+1,J)-(2.-RHO)*VINT(I,J)-HSQG(I,J)
    8   CONTINUE
      RHS(2)=RHS(2)+VINT(I,1)
      RHS(N)=RHS(N)+VINT(I,NP1)
      CALL TDSOLV(RHS,TDSOL,RHO,N)
        DO 9 J=2,N
        VNEW(I,J)=TDSOL(J)
    9   CONTINUE
   10 CONTINUE
```

Figure 8.9 Subroutine CYCADI used in Example 8.6.

```
C
C   CHECK FOR CONVERGENCE
C
      TEST=0.
        DO 12 I=2,M
          DO 11 J=2,N
          DIFF=ABS(VNEW(I,J)-VOLD(I,J))
          IF(TEST.LT.DIFF)  TEST=DIFF
          VOLD(I,J)=VNEW(I,J)
   11     CONTINUE
   12   CONTINUE
      IF(TEST.LT.TOL)  RETURN
      RHO=RHO*FACTOR
   13 CONTINUE
      GO TO 4
      END
```

Figure 8.9 (continued)

3. a) Verify that the Gauss-Seidel method applied to the system (8.30) yields (8.31).

b) Give the modification of (8.31) when v_{ij} corresponds to a point adjacent to the boundary.

c) Verify that (8.34) is the matrix form of SOR from (8.33).

4. Let A be the matrix in Fig. 8.7, $\mathbf{g} = [2,1,1,2,1,0,0,1,1,0,0,1,2,1,1,2]^T$, and $\mathbf{v}^{(0)} = [v_1, v_2, \ldots, v_{16}]^T$ where $v_i = (-1)^i$, $1 \le i \le 16$.

a) Letting $\mathbf{v}^{(0)}$ be the initial vector, calculate $\mathbf{v}^{(1)}$ and $\mathbf{v}^{(2)}$, the first two iterates of the Gauss-Seidel method (8.31) for solving $A\mathbf{v} = \mathbf{g}$. (Answer: $\mathbf{v} = [1,1,1, \ldots, 1]^T$.)

b) Repeat the calculations in part (a) for the SOR method (8.33) with $\omega = 5/4$.

c) Repeat the calculations in part (a) for the SLOR method (8.38) with $\omega = 5/4$.

d) Compute the optimal relaxation factor in part (b).

5. Derive the line Gauss-Seidel equations and the SLOR equations for (8.30) with $f_{ij} < 0$ and $g_{ij} \neq 0$.

6. a) For $A\mathbf{v} = \mathbf{g}$ in Problem 4, use $\mathbf{v}^{(0)}$ as given and perform one step of the ADI method with $\rho = 1$.

b) Calculate the cycle length m for the cyclic ADI on this problem and calculate $\rho_1, \rho_2, \ldots, \rho_m$ also.

7. Use the routines (modified where necessary) in this section to generate approximations for the following problems.

a) $u_{xx} + u_{yy} = 0$, $0 < x < 1$, $0 < y < 1$, $u(x, 0) = u(0, y) = 0$, $u(x, 1) = x$, $u(1, y) = y$. [Answer: $u(x, y) = xy$.]

b) $u_{xx} + u_{yy} - (x^2 + y^2)u = 0$, $0 < x < 1$, $0 < y < 1$, $u(x, 0) = u(0, y) = 1$, $u(1, y) = e^{-y}$, $u(x, 1) = e^{-x}$. [Answer: $u(x, y) = \exp[-xy]$.]

c) $u_{xx} + u_{yy} - u = x^2 - 2$, $0 < x < 1$, $0 < y < 1$, $u(x, 0) = -x^2$, $u(0, y) = y$, $u(1, y) = ey - 1$, $u(x, 1) = e^x - x^2$. [Answer: $u(x, y) = ye^x - x^2$.]

d) $(x^2 + 1)^2(u_{xx} + u_{yy}) - (x^2 - x + 1)u = 2(x^2 + 1)^{5/2}$, $0 < x < 1$, $0 < y < 1$, $u(x, 0) = 0$, $u(x, 1) = (x^2 + 1)^{\frac{1}{2}}$, $u(0, y) = y^2$, $u(1, y) = \sqrt{2}\, y^2$. [Answer: $u(x, y) = (x^2 + 1)^{\frac{1}{2}}y^2$.]

8. a) For $u_{xx} + u_{yy} + u_{zz} + f(x, y, z)u = g(x, y, z)$, $0 < x < L$, $0 < y < K$, $0 < z < H$, derive a system of difference approximations analogous to (8.30).

   b) Display the resultant system of equations for $u_{xx} + u_{yy} + u_{zz} = 0$, $0 < x, y, z < 1$, $u(0, y, z) = 1$, $u(x, 0, z) = 2$, $u(x, y, 0) = 3$, and $u = 0$ on the remaining faces. Partition $0 < x, y, z < 1$ into cubes with volume h^3, h = 1/3.

   c) For $f = 0$, derive the three-part ADI equations analogous to (8.41) and (8.42) for solving the system in part (a).

8.3.2. Mixed Boundary Data and Irregular Regions

Often the boundary conditions of (8.28) are given in terms of the normal derivative, du/dn, such as

$$\frac{du}{dn} = au + b \qquad \text{on the boundary.} \tag{8.43}$$

For a rectangular region parallel to the x- and y-axes, du/dn is either $\partial u/\partial x$ or $\partial u/\partial y$; and either of these cases can be treated in the same fashion as in the parabolic case (Section 8.2.2).

Suppose that the boundary conditions are given by

$$\begin{aligned} u_x(0, y) &= a_1 u(0, y) + b_1, \qquad & u_x(L, y) &= a_2 u(L, y) + b_2, \\ u_y(x, 0) &= a_3 u(x, 0) + b_3, \qquad & u_y(x, K) &= a_4 u(x, K) + b_4. \end{aligned} \tag{8.44}$$

In this case the values of u on the boundary are unknown, but we assume that u can be extended to the region $-h < x < L + h$, $-h < y < K + h$. We again introduce fictitious grid points (x_{-1}, y_j), (x_{M+1}, y_j), (x_i, y_{-1}), and (x_i, y_{N+1}) where $x_{-1} = -h$, $x_{M+1} = L + h$, $y_{-1} = -h$, and $y_{N+1} = K + h$. Since (8.30) is an $O(h^2)$ approximation, we use central differences to approximate u_x and u_y on the boundary. For example with sufficient differentiability of u we have for $-1 < \xi < 1$

$$u_x(0, y_j) = [u_{1j} - u_{-1,j}]/2h - \frac{\partial^3}{\partial x^3} u(\xi h, y_j)h^2/6 = a_1 u_{0,j} + b_1. \tag{8.45}$$

Hence, solving for $u_{-1,j}$ from (8.45) and substituting into (8.29) with $i = 0$ and $k = h$ yield

$$\begin{aligned} &[2u_{1j} - 2(1 + a_1 h)u_{0,j} - 2hb_1]/h^2 + [u_{0,j+1} - 2u_{0,j} + u_{0,j-1}]/h^2 + f_{0,j}u_{0,j} \\ &- g_{0j} = \frac{\partial^3}{\partial x^3} u(\xi h, y_j)h/3 - \frac{\partial^4}{\partial x^4} u(x_0 + \xi_0 h, y_j)h^2/12 - \frac{\partial^4}{\partial y^4} u(x_0, y_j + \zeta_j h)h^2/12. \end{aligned} \tag{8.46}$$

Neglecting the right-hand side of (8.46), we obtain approximations $v_{0,\,j}$ for $u_{0,\,j}$ with discretization error of order h from

$$-2v_{1j} + (4 + 2a_1h - h^2f_{0j})v_{0j} - v_{0,\,j-1} - v_{0,\,j+1} = -2hb_1 - h^2g_{0j}. \quad (8.47)$$

Three similar sets of equations can be found for the remaining boundary conditions of (8.44). These equations along with (8.30) yield an $((M + 1)(N + 1) \times (M + 1)(N + 1))$ linear system for the approximations of u at the grid points, including those on the boundary.

We conclude this section with a brief discussion of the difficulties involved when the domain for (8.28) is not rectangular. For the problem (8.28), rectangular regions are easiest to deal with since the grid spacing can be chosen so that grid points fall on the boundary. If the region has an irregular boundary, such as that in Fig. 8.10, then there will be points "adjacent" to the boundary whose horizontal and/or vertical distance to the boundary is less than the mesh size h (see Fig. 8.10). In Fig. 8.10 the equation at point 1 is determined by (8.30) in terms of points 2, 3, 5, and 12. At points 2 and 3, however, (8.30) does not apply. Hence we consider a modification to obtain approximations at points 2 and 3. For this purpose we consider a general picture (Fig. 8.10b) representing any five points where each P_1 is $\gamma_i h \equiv h_i$ in distance from $P_0 = (x_0, y_0)$, $0 < \gamma_i < 1$, $i = 1, 2, 3, 4$. Assuming sufficient differentiability of u and setting $u_i = u(P_i)$, $0 \le i \le 4$, give

$$u_1 = u(x_0 + h_1, y_0) = [u + u_xh_1 + u_{xx}h_1^2/2 + u_{xxx}h_1^3/6 + 0(h_1^4)]\big|_{(x_0,\, y_0)}$$
$$u_3 = u(x_0 - h_3, y_0) = [u - u_xh_3 + u_{xx}h_3^2/2 - u_{xxx}h_3^3/6 + 0(h_3^4)]\big|_{(x_0,\, y_0)}.$$

Multiplying the first equation by h_3 and the second by h_1, and adding the resulting equation yield

$$u_{xx}(x_0, y_0) - \frac{2(h_3u_1 + h_1u_3 - (h_1 + h_3)u_0}{h_1h_3(h_1 + h_3)} = \frac{(h_3 - h_1)}{3}\, u_{xxx}(x_0, y_0) + 0(h_1^2 + h_3^2). \quad (8.48a)$$

Similarly

$$u_{yy}(x_0, y_0) - \frac{2(h_4u_2 + h_2u_4 - (h_2 + h_4)u_0)}{h_2h_4(h_2 + h_4)} = \frac{(h_4 - h_2)}{3}\, u_{yyy}(x_0, y_0) + 0(h_2^2 + h_4^2). \quad (8.48b)$$

Neglecting the terms on the right-hand sides of (8.48a) and (8.48b) and substituting for u_{xx} and u_{yy} in (8.28) yield an $O(h)$ difference approximation analogous to (8.30). [Note that if $h_1 = h_3 = h$ and $h_2 = h_4 = k$, we have (8.29); and if $h = k$, (8.30) follows.] The fact that (8.48a) and (8.48b) yield only $O(h)$ approx-

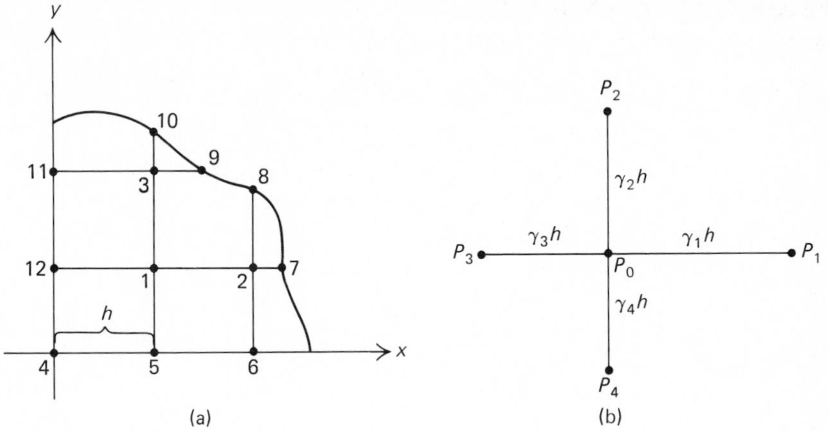

Figure 8.10

imations should be considered when selecting the mesh size. If the mesh is very fine, an alternative to the procedure above is to perturb the boundary so as to contain the ''exterior'' mesh points and at such a point assign a value of u that is the value of u at the nearest point on the boundary. (For example, point 2 in Fig. 8.10a would be assumed as a boundary point and u at point 2 would be given the value of u at point 7.)

EXAMPLE 8.7. Consider the problem $u_{xx} + u_{yy} = 0$ for the region in Fig. 8.10 where $u = 0$ along the x- and y-axes and $u = 20$ on the curved part of the boundary. Suppose that $h = 1$, $|P_2P_7| = 1/3$, $|P_2P_8| = 1/2$, and $|P_3P_9| = |P_3P_{10}| = 1/4$.

At points 2 and 3, respectively, (8.48a) and (8.48b) give the equations

$$(P_2)\quad 0 = u_{xx} + u_{yy} \simeq \frac{2\left(u_7 + \frac{1}{3}u_1 - \frac{4}{3}u_2\right)}{\frac{1}{3}\cdot\frac{4}{3}} + \frac{2\left(u_8 + \frac{1}{2}u_6 - \frac{3}{2}u_2\right)}{\frac{1}{2}\cdot\frac{3}{2}} \tag{8.49}$$

$$(P_3)\quad 0 = u_{xx} + u_{yy} \simeq \frac{2\left(u_9 + \frac{1}{4}u_{11} - \frac{5}{4}u_3\right)}{\frac{1}{4}\cdot\frac{5}{4}} + \frac{2\left(u_{10} + \frac{1}{4}u_1 - \frac{5}{4}u_3.\right)}{\frac{1}{4}\cdot\frac{5}{4}} \tag{8.50}$$

Letting v_i be the finite-difference approximations for u_i, $i = 1, 2, 3$, and using (8.30) at point 1, we have the following system:

$$\begin{aligned} 4v_1 - v_2 - v_3 &= 0 && \text{(Point 1)} \\ 9v_1 - 60v_2 &= 860 && \text{(Point 2)} \\ v_1 - 10v_3 &= -160 && \text{(Point 3).} \end{aligned}$$

PROBLEMS, SECTION 8.3.2

1. Consider $u_{xx} + u_{yy} = 0$ for $0 < x < 4$, $0 < y < 3$.
   a) For $h = 1$, display the five-point difference equations when $u(x, 0) = 0$, $u(0, y) = 0$, $u_x(4, y) = (.25)u(4, y)$, and $u_y(x, 3) = (1/3)u(x, 3)$.
   b) Solve the system in part (a). [Answer: $u(x, y) = xy$.]
2. Consider $u_{xx} + u_{yy} = xe^y$ for $0 < x < 4$, $0 < y < 3$.
   a) For $h = 1$, display the five-point difference equations when $u(0, y) = 0$, $u_x(4, y) = (.25)u(4, y)$, $u(x, 0) = x$, $u_y(x, 3) = u(x, 3)$.
   b) Solve the system in part (a). [Answer: $u(x, y) = xe^y$.]
3. In each of Problems 1 and 2, take $h = .1$ and solve the resulting system by SLOR and/or cyclic ADI. Make use of the routines in the previous section; use appropriate values of ω and ρ.
4. Verify that (8.48a) and (8.48b) yield (8.30) if $h_i = h$ for $1 \le i \le 4$.
5. With data as given in Example 8.7 for Fig. 8.10, give the system of difference approximations for these equations.
   a) $u_{xx} + u_{yy} = -2$
   b) $u_{xx} + u_{yy} - xyu = 0$

8.4 HYPERBOLIC EQUATIONS

For our study of finite-difference techniques for hyperbolic equations, we restrict ourselves to the particular boundary-value problem (the wave equation) given by

$$\frac{\partial^2 u}{\partial t^2}(x, t) = \alpha^2 \frac{\partial^2}{\partial x^2} u(x, t), \qquad 0 < x < L,\ 0 < t < T \tag{8.51}$$

where $u(0, t) = u(L, t) = 0$, $u(x, 0) = f(x)$, $(\partial/\partial t)u(x, 0) = g(x)$.

To solve this equation, a procedure known as the "method of characteristics" is usually recommended; see Ames (1977). However, a discussion of the method of characteristics would be quite involved and so we will restrict our discussion to finite-difference techniques; these are known to be suitable when (8.51) has solutions that are well behaved.

We let $h = L/M$ and $k = T/N$ for some positive integers M and N, and let $x_i = ih, 0 \le i \le M$, and $t_j = jk, 0 \le j \le N$. We denote $u(x_i, t_j)$ as u_{ij}. We again use the central-difference formula (8.6a) for $u_{xx}(x_i, t_j)$ and the corresponding central-difference formula for $u_{tt}(x_i, t_j)$. Assuming sufficient differentiability of u, we then have

$$\begin{aligned}(u_{i,j+1} - 2u_{ij} + u_{i,j-1})/k^2 &- \alpha^2(u_{i+1,j} - 2u_{ij} + u_{i-1,j})/h^2 \\ &= \frac{\partial^4}{\partial t^4} u(x_i, t_j + \zeta_j k)k^2/12 - \alpha^2 \frac{\partial^4}{\partial x^4} u(x_i + \xi_i h, t_j)h^2/12\end{aligned} \tag{8.52}$$

where $-1 < \zeta_j, \xi_i < 1$. Hence an approximation v_{ij} for u_{ij} with discretization error of order $(h^2 + k^2)$ can be found from

$$[v_{i,j+1} - 2v_{ij} + v_{i,j-1}]/k^2 = \alpha^2[v_{i+1,j} - 2v_{ij} + v_{i-1,j}]/h^2. \tag{8.53}$$

Letting $r = \alpha^2k^2/h^2$ in (8.53) and solving for $v_{i,j+1}$ yield

$$v_{i,j+1} = rv_{i-1,j} + 2(1 - r)v_{ij} + rv_{i+1,j} - v_{i,j-1}. \tag{8.54}$$

The boundary conditions yield $v_{0,j} = u(0, t_j) = 0$ and $v_{Mj} = u(L, t_j) = 0$. Hence (8.54), for $1 \le i \le M - 1$, is given in matrix form as

$$\mathbf{v}^{(j+1)} = A\mathbf{v}^{(j)} - \mathbf{v}^{(j-1)}, \qquad j = 1, 2, 3, \ldots, \tag{8.55}$$

where

$$A = \begin{bmatrix} 2(1-r) & r & 0 & \cdots & 0 \\ r & 2(1-r) & r & & \vdots \\ 0 & & & & 0 \\ \vdots & & & & r \\ 0 & & \cdots & r & 2(1-r) \end{bmatrix} \qquad \mathbf{v}^{(j)} = \begin{bmatrix} v_{1j} \\ v_{2j} \\ \vdots \\ v_{M-1,j} \end{bmatrix}. \tag{8.56}$$

The vector $\mathbf{v}^{(0)}$ is known from the boundary conditions since $v_{i,0} = u(x_i, 0) = f(x_i)$. In order to compute $\mathbf{v}^{(j+1)}$, however, we need both $\mathbf{v}^{(j)}$ and $\mathbf{v}^{(j-1)}$. Thus in order to use (8.55) we must derive an expression for $\mathbf{v}^{(1)}$. This derivation can be done in a manner analogous to that of Section 8.2.2 by assuming that the solution u can be extended with sufficient differentiability for $0 < x < L$ and $t > -k$. We introduce fictitious grid points (x_i, t_{-1}) with $t_{-1} = -k$ and use a central-difference approximation for the remaining boundary condition,

$$u_t(x_i, 0) = [u_{i1} - u_{i,-1}]/2k - \frac{\partial^3}{\partial t^3} u(x_i, \eta k)k^2/6 = g(x_i)$$

where $-1 < \eta < 1$. We solve this expression for $u_{i,-1}$ and substitute these values into (8.52) with $j = 0$. With $u_{i,0} = f(x_i)$, (8.52) becomes

$$\begin{aligned}(2u_{i1} - 2f(x_i) - 2kg(x_i))/k^2 - \alpha^2(f(x_{i+1}) - 2f(x_i) + f(x_{i-1}))/h^2 \\ = \frac{\partial^3}{\partial t^3} u(x_i, \eta k)k/3 + \frac{\partial^4}{\partial t^4} u(x_i, \zeta_0 k)k^2/12 + \frac{\partial^4}{\partial x^4} u(x_i + \xi_i h, 0)h^2/12.\end{aligned} \tag{8.57}$$

Hence an approximation v_{i1} for u_{i1} with discretization error of order $(k^3 + k^2h^2)$ is provided by

$$v_{i1} = (r/2)f(x_{i+1}) + (1 - r)f(x_i) + (r/2)f(x_{i+1}) + kg(x_i). \tag{8.58}$$

Using $\mathbf{v}^{(0)}$ as above and $\mathbf{v}^{(1)}$ from (8.58), we can now use (8.55) to generate $\mathbf{v}^{(2)}$, $\mathbf{v}^{(3)}, \ldots, \mathbf{v}^{(N)}$.

This method has a rather remarkable property when $r = 1$, u has a convergent Taylor expansion throughout the region, and when $\mathbf{v}^{(0)}$ and $\mathbf{v}^{(1)}$ contain exact values for u_{i0} and u_{i1}. In this situation, the remaining iterations $\mathbf{v}^{(j)}$ yield the actual values of u as well; that is, $v_{ij} = u_{ij}$ for all i and j. To prove this fact, we first note that since u has a convergent Taylor series, the right-hand side of (8.52) can also be written as

$$2\left\{\left(k^2 \frac{\partial^4}{\partial t^4} u_{ij} - \alpha^2 h^2 \frac{\partial^4}{\partial x^4} u_{ij}\right)/4! + \left(k^4 \frac{\partial^6}{\partial t^6} u_{ij} - \alpha^2 h^4 \frac{\partial^6}{\partial x^6} u_{ij}\right)/6! + \left(k^6 \frac{\partial^8}{\partial t^8} u_{ij} - \alpha^2 h^6 \frac{\partial^8}{\partial x^8} u_{ij}\right)/8! + \cdots\right\}. \tag{8.59}$$

Since $u_{tt} = \alpha^2 u_{xx}$, then

$$\frac{\partial^4 u}{\partial t^4} = \frac{\partial^2}{\partial t^2}[u_{tt}] = \frac{\partial^2}{\partial t^2}[\alpha^2 u_{xx}] = \alpha^2 \frac{\partial^2}{\partial x^2}[u_{tt}] = \alpha^4 \frac{\partial^4 u}{\partial x^4}. \tag{8.60}$$

With $r = 1$, $\alpha^2 k^2 = h^2$; and so by (8.60) the first term in (8.59) equals zero. A similar argument shows that each term in (8.59) is zero. Thus under the conditions above, (8.52) and (8.53) are the same, and so the only errors in $\mathbf{v}^{(j)}$ generated by (8.55) are due to rounding or machine representation.

Since $f(x)$ is given, then $\mathbf{v}^{(0)}$ will be mathematically correct. From (8.57) we see also that if $g(x)$ and u are such that $u_{ttt}(x, \Upsilon) = 0$ for $-k < \Upsilon < k$, then $\mathbf{v}^{(1)}$ is also mathematically correct. Unfortunately in realistic problems all of the stipulations above do not occur often. For example if $f(x)$ and/or $g(x)$ have discontinuities, the expansion (8.59) and subsequent analysis are not valid. Furthermore if $f(x)$ and/or $g(x)$ have a finite number of finite discontinuities [but for given values of h and k, $f(x_i)$ and $g(x_i)$ are still defined], then there are an infinite number of functions $\tilde{f}(x)$ and $\tilde{g}(x)$ where $\tilde{f}(x_i) = f(x_i)$, $\tilde{g}(x_i) = g(x_i)$, and $\tilde{f}(x)$ and $\tilde{g}(x)$ are continuous and differentiable. The method of (8.55) then will converge to the solution of the *different* boundary-value problem whose boundary conditions are prescribed by $\tilde{f}(x)$ and $\tilde{g}(x)$. In summary, this method tends to "smooth out" discontinuous boundary conditions; and even though it is stable for $r = 1$, it is not usually used for problems of this type. The method of characteristics is better suited to treat problems with discontinuous boundary data [see Gerald (1978)].

PROBLEMS, SECTION 8.4

1. Use (8.2) to derive (8.52) and (8.57).
2. a) Use Problems 3 and 4, Section 8.2.1, to find the eigenvalues of A in (8.56).
   b) Suppose that $\{\mathbf{u}^{(j)}\}$ are the exact solutions of (8.55); that is, $\mathbf{u}^{(j+1)} = A\mathbf{u}^{(j)} - \mathbf{u}^{(j-1)}$. Suppose simplistically that errors $\mathbf{e}$ and $\boldsymbol{\varepsilon}$ are made in obtaining $\mathbf{v}^{(0)}$ and $\mathbf{v}^{(1)}$ and

that no further errors are made; that is, $\mathbf{v}^{(0)} = \mathbf{u}^{(0)} + \mathbf{e}$ and $\mathbf{v}^{(1)} = \mathbf{u}^{(1)} + \boldsymbol{\varepsilon}$. Show that $\mathbf{v}^{(5)} = \mathbf{u}^{(5)} + A^4\boldsymbol{\varepsilon} - 3A^2\boldsymbol{\varepsilon} + 2A\mathbf{e} + \boldsymbol{\varepsilon}$.

3. Use Problem 3, Section 8.1, to give the d'Alembert solution for $u_{xx} = \alpha^2 u_{tt}$. How do the two functions in this solution relate to $f(x)$ and $g(x)$ in (8.51)?
4. Show that all of the terms in (8.59) are zero when $r = 1$.
5. In (8.51) let $\alpha = 1$, $L = 4$, $T = 2$, $f(x) = 0$, and $g(x) = (x - 4)$. With $h = 1$, use (8.54) to find $u(1, 2)$, $u(2, 2)$, and $u(3, 2)$. [Answer: $u(x, y) = (x - 4)t$.]
6. Modify (8.54) to treat $u_{xx} = \alpha^2 u_{tt} + F(x, t)u + G(x, t)$.

8.5 FINITE-ELEMENT TECHNIQUES

Finite-element methods, especially as applied to elliptic boundary-value problems have become increasingly popular in recent years. Rather than approximate derivatives by finite differences (as in the previous sections), finite-element methods attempt to approximate the solution from a family of spline functions and often use a family of first-degree splines. The general principles of the finite-element approach can be outlined in a fairly simple fashion, but details for specific problems can require a very sophisticated analysis. At the risk of oversimplifying some of this analysis, we present in this section a brief conceptual discussion of the finite-element approach in order to provide an introduction to the computational aspects and to serve as background for reading comprehensive finite-element texts such as Oden and Reddy (1976), Strang and Fix (1973), and Prenter (1975).

8.5.1. The Rayleigh-Ritz Method

The Rayleigh-Ritz procedure is a particular form of the finite-element method, and a particular application of the Rayleigh-Ritz procedure was presented in Section 5.3.3 for ordinary differential equations. The ideas underlying the Rayleigh-Ritz procedure are the same in a more general setting as outlined here. Suppose that L is a linear differential operator and f is a given function, and we wish to find the function u such that $L[u] = f$ where u satisfies certain given constraints. If X is a vector space of functions that contains u, and X_N is a subspace of X, then we wish to find in X_N a function u_N that approximates u. Suppose that $\langle \cdot, \cdot \rangle$ is an inner product on X. (As in Section 5.3, the inner product has these properties for all v, w, and z in X and all real a: $\langle v, v \rangle \geq 0$, $\langle v, v \rangle = 0$ if and only if $v = \theta$, $\langle v, w \rangle = \langle w, v \rangle$, $\langle av, w \rangle = a\langle v, w \rangle$, and $\langle v + w, z \rangle = \langle v, z \rangle + \langle w, z \rangle$.) The associated norm given by $\|v\| = (\langle v, v \rangle)^{\frac{1}{2}}$ can be used to measure the magnitude of any function v in X. Suppose also that every function w in X_N can be written as $w = c_1\phi_1 + c_2\phi_2 + \cdots + c_N\phi_N$ where $\{\phi_i\}_{i=1}^N$ is linearly independent. (That is, X_N has dimension N and a basis $\{\phi_i\}_{i=1}^N$.) Analogous to Corollary 1 of Theorem 5.10, function w_N in X_N for which $\|w_N - u\| \leq$

$\|w - u\|$ for all w in X_N (w_N is "closer" to u than any other w in X_N) satisfies $\langle w_N - u, \phi_i \rangle = 0, 1 \le i \le N$. Writing $w_N = \sum_{j=1}^{N} \gamma_j \phi_j$, we thus obtain the system of equations

$$\sum_{j=1}^{N} \langle \phi_i, \phi_j \rangle \gamma_j = \langle u, \phi_i \rangle, \qquad 1 \le i \le N. \tag{8.61}$$

If we knew the scalars $\langle u, \phi_i \rangle$, we could solve (8.61) for $\gamma_1, \gamma_2, \ldots, \gamma_N$ and we would then have the best approximation w_N. However since we are trying to solve $L[u] = f$, we normally do not have $\langle u, \phi_i \rangle$.

Since u is not known, the Rayleigh-Ritz procedure considers the following alternative. If the differential operator L is symmetric and positive definite—that is, if $\langle L[v], w \rangle = \langle L[w], v \rangle$ for all v and w in X and $\langle L[v], v \rangle > 0$ for $v \neq \theta$—then we can define on X a new inner product (satisfying the inner-product properties above): $\langle v, w \rangle_L = \langle L[v], w \rangle$. This new inner product has the corresponding norm $\|v\|_L = (\langle v, v \rangle_L)^{\frac{1}{2}} = (\langle L[v], v \rangle)^{\frac{1}{2}}$. Since $L[u] = f$, the function $u_N = \sum_{i=1}^{N} c_i \phi_i$ in X_N where $\|u_N - u\|_L \le \|w - u\|_L$ for all w in X_N is found by solving

$$\sum_{j=1}^{N} \langle \phi_i, \phi_j \rangle_L c_j = \langle u, \phi_i \rangle_L = \langle f, \phi_i \rangle, \qquad 1 \le i \le N. \tag{8.62}$$

The function u_N obtained from (8.62) is the Rayleigh-Ritz approximation to u. Although the basic concepts underlying (8.62) are fairly simple, some extensive theoretical and computational details are involved. These include the identification of a proper function space X and corresponding inner product so that L is symmetric and positive-definite. A further problem is the identification of the subspace X_N, in terms of both theoretical and computational features.

We now consider the Rayleigh-Ritz procedure for this elliptic boundary-value problem on a region D:

$$L[u] = -\frac{\partial}{\partial x}[pu_x] - \frac{\partial}{\partial y}[qu_y] + ru = f \tag{8.63}$$

$$u = 0 \text{ on the boundary } C \text{ of } D.$$

In (8.63), p and q are positive differentiable functions on D, and r is a positive function on D. We use as an inner product

$$\langle v, w \rangle = \iint_D v(x, y) w(x, y) dx\, dy. \tag{8.64}$$

To show that $L[u]$ is symmetric and positive definite, we will use Green's theorem, which states that if $S(x, y)$ and $T(x, y)$ are continuously differentiable functions on $D \cup C$, then

$$\int_D\int (S_x - T_y)dx\,dy = \int_C (T\,dx + S\,dy) \tag{8.65}$$

where the line integral on C is taken in the positive sense. Suppose that v and w are two functions in the domain of L such that $v = 0$ and $w = 0$ on C. By (8.63) and (8.64),

$$\langle L[v], w\rangle = -\int_D\int [(pv_x)_x + (qv_y)_y - rv]w\,dx\,dy. \tag{8.66}$$

We next observe that

$$\begin{aligned}&\int_D\int [(pv_x)_x w + (qv_y)_y w]\,dx\,dy\\ &= \int_D\int [(pv_x w)_x + (qv_y w)_y - pv_x w_x - qv_y w_y]\,dx\,dy \qquad (8.67)\\ &= \int_C -qv_y w\,dx + pv_x w\,dy - \int_D\int [pv_x w_x + qv_y w_y]\,dx\,dy.\end{aligned}$$

Since $w = 0$ on C, the line integral above is zero. Substituting (8.67) into (8.66) yields

$$\langle L[v], w\rangle = \int_D\int [pv_x w_x + qv_y w_y + rvw]\,dx\,dy. \tag{8.68}$$

From (8.68) it is clear that $\langle L[v], w\rangle = \langle L[w], v\rangle$ (symmetry), and $\langle L[v], v\rangle > 0$ for v differentiable and nonzero (positive-definite). Also it is immediate that $\langle L[av], w\rangle = a\langle L[v], w\rangle$, and $\langle L[v + w], z\rangle = \langle L[v], z\rangle + \langle L[w], z\rangle$. Hence in a function space where the operations above are valid, $\langle v, w\rangle_L = \langle L[v], w\rangle$ is an inner product with corresponding norm $\|v\|_L = (\langle v, v\rangle_L)^{\frac{1}{2}}$. (This norm is usually called the *energy norm*.)

The analysis above leaves many questions to be answered. Is there such a space X? Is $L[v]$ defined for all v in X; and if not, can (8.68) be extended to all of X? Is X large enough to contain functions (such as piecewise linear functions) that prove to be computationally convenient for the approximating subspace X_N? We shall not attempt to answer such questions here; we only state that the problem is usually embedded in a Sobolev function space where v, v_x, and v_y are square integrable on D in the Lebesque sense and $v = 0$ on C.

Now that we have an energy inner product $\langle v, w\rangle_L$ given by (8.68), we proceed as in the general pattern above for the Rayleigh-Ritz technique and select an approximating subspace X_N. In Section 5.3.3, X_N was a space of cubic splines. In two or more dimensions, however, it is often not feasible (because of the amount and difficulty of computation) to use approximating functions other than piecewise linear functions. Furthermore from (8.62) we see that we shall be computing terms of the form $\langle \phi_i, \phi_j \rangle_L$, and so we see that we can reduce the computation if we choose each $\phi_i(x, y)$ so that it is nonzero on as small a subregion of D as possible. To this end we choose the following procedure although other approaches are possible.

Suppose that the region D is subdivided into triangles as in Fig. 8.11a. Let $\{P_i\}_{i=1}^M$ denote the points of the triangular grid (that is, the vertices of the triangles) and suppose that $\{P_i\}_{i=1}^N$ are the interior grid points (not on the boundary). Let the triangles be denoted by $\{T_j\}_{j=1}^K$ where we assume that the triangles were constructed so that each T_i has at least one interior vertex. For each interior point P_i we wish to construct a continuous, piecewise, linear function $\phi_i(x, y)$ such that $\phi_i(P_i) = 1$ and such that ϕ_i is zero on any triangle for which P_i is not a vertex as in Fig. 8.11b. On each triangle for which P_i is a vertex, ϕ_i is given by the equation of the plane that is one at P_i and zero at the other two vertices. For example if $P_i = (x_0, y_0)$ and the other two vertices of the particular triangle under consideration are (x_1, y_1) and (x_2, y_2), then ϕ_i is given on that triangle by

$$\phi_i(x, y) = ax + by + c$$

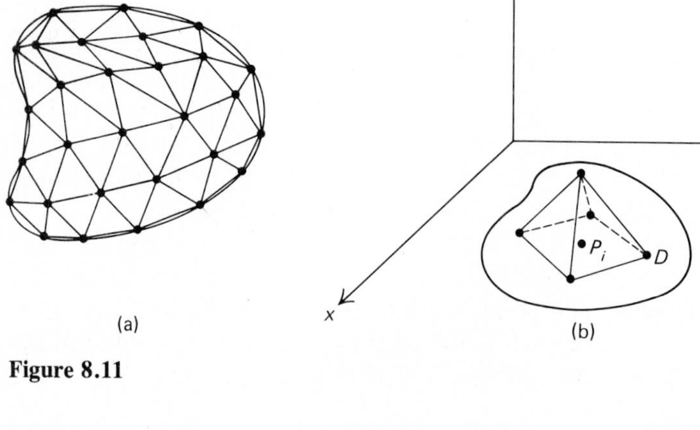

Figure 8.11

where a, b, and c are chosen to satisfy the three equations

$$\begin{aligned} ax_0 + by_0 + c &= 1 \\ ax_1 + by_1 + c &= 0, \qquad \text{or } A\mathbf{x} = \mathbf{e}_1. \\ ax_2 + by_2 + c &= 0 \end{aligned} \tag{8.69}$$

Thus to evaluate ϕ_i to a point P in one of the triangles with vertex P_i, we determine which triangle P is in and use the equation of the plane corresponding to that triangle.

A common procedure for coding and recording the functions ϕ_i is to label the vertices of each T_k as V_1^k, V_2^k, and V_3^k. If all three vertices are interior points, each will have a corresponding basis function ϕ_i. Hence as in (8.69), we solve the three systems $A\mathbf{x} = \mathbf{e}_1$, $A\mathbf{x} = \mathbf{e}_2$, and $A\mathbf{x} = \mathbf{e}_3$ where the rows of the coefficient matrix come from the coordinates of V_1^k, V_2^k, and V_3^k. The resulting functions are given as

$$p_m^k(x, y) = a_m^k x + b_m^k y + c_m^k, \qquad m = 1, 2, 3, \tag{8.70}$$

and are stored as arrays $A = (a_m^k)$, $B = (b_m^k)$, and $C = (c_m^k)$. If either one or two of the vertices of T_k is a boundary point, then there is no basis function ϕ_i associated with such a vertex and no $p_m^k(x, y)$ is constructed that equals one at that point.

We now complete the procedure by determining in X_N the function u_N that is nearest the solution u with respect to the energy norm. The subspace X_N is the set of all functions of the form

$$v = \gamma_1\phi_1 + \gamma_2\phi_2 + \cdots + \gamma_N\phi_N.$$

(Note that each v in X_N satisfies the boundary condition $v = 0$ on C.) We set $u_N = \sum_{j=1}^{N} c_j\phi_j$; and as in (8.62) solve the system

$$\sum_{j=1}^{N} \langle\phi_i, \phi_j\rangle_L c_j = \langle f, \phi_i\rangle, \qquad 1 \le i \le N, \qquad \text{or } S\mathbf{c} = \mathbf{b} \tag{8.71}$$

where S is the $(N \times N)$ matrix $S = (\langle\phi_i, \phi_j\rangle_L)$ and $\mathbf{b}$ is the $(N \times 1)$ vector $\mathbf{b} = (\langle f, \phi_i\rangle)$.

Computation of each $\langle\phi_i, \phi_j\rangle_L$ proceeds as follows. Since [see (8.68)]

$$\langle L[\phi_i], \phi_j\rangle = \int\!\!\int_D \left[p\,\frac{\partial\phi_i}{\partial x}\frac{\partial\phi_j}{\partial x} + q\,\frac{\partial\phi_i}{\partial y}\frac{\partial\phi_j}{\partial y} + r\phi_i\phi_j\right] dx\,dy,$$

the integration on D is done individually on the triangles $\{T_k\}_{k=1}^{K}$ and summed. In most cases ϕ_i and ϕ_j have no triangle in common where they are both nonzero, and so the integrals over such triangles are zero. If both are nonzero on a triangle T_k, then P_i and P_j are both vertices of T_k, say V_1^k and V_2^k, respectively. In this case $(P_i \neq P_j)$ we have by (8.70)

$$\int_{T_k}\int \left[p\frac{\partial\phi_i}{\partial x}\frac{\partial\phi_j}{\partial x} + q\frac{\partial\phi_i}{\partial y}\frac{\partial\phi_j}{\partial y} + r\phi_i\phi_j\right] dx\, dy$$

$$= a_1^k a_2^k \int_{T_k}\int p\, dx\, dy + b_1^k b_2^k \int_{T_k}\int q\, dx\, dy \tag{8.72a}$$

$$+ \int_{T_k}\int r[(a_1^k x + b_1^k y + c_1^k)(a_2^k x + b_2^k y + c_2^k)]dx\, dy.$$

Furthermore if $P_i \neq P_j$, then there are in the grid only two triangles T_k and $\hat{T}_k$ that have both P_i and P_j as vertices. Hence another calculation as in (8.72a) is done on the other triangle $\hat{T}_k$, and the sum of the results yields $\langle L[\phi_i], \phi_j\rangle = \langle L[\phi_j], \phi_i\rangle$. If $P_i = P_j$, then the calculations above are done on each triangle T_k having P_i as a vertex; and we use the $p_m^k\ (x, y)$ in (8.70) where m is chosen so that $P_i = V_m^k$. When $P_i = P_j$, it is convenient in this step also to calculate the right-hand side of (8.71):

$$\langle f, \phi_i\rangle = \int_D\int f\phi_i\, dx\, dy = \Sigma \int_{T_k}\int f\phi_i\, dx\, dy. \tag{8.72b}$$

Unless the integrals in (8.71) and (8.72) are simple enough to be generally calculated and programmed, they will be approximated numerically. If the mesh is sufficiently fine, a very simple formula, such as an average of the integrand times the area of T_k or the integrand evaluated at some point times the area of T_k, is appropriate.

An alternate approach to compute the entries of S and $\mathbf{b}$ is to perform all computations for each triangle, one triangle at a time, and then sort out which computations contribute to the terms $\langle L[\phi_i], \phi_j\rangle$ and $\langle f, \phi_i\rangle$. For example, if the vertices of T_k are interior grid points, the six different computations corresponding to $P_1^kP_1^k$, $P_1^kP_2^k$, $P_1^kP_3^k$, $P_2^kP_2^k$, $P_2^kP_3^k$, and $P_3^kP_3^k$ must be done (the computations for $P_2^kP_1^k$, $P_3^kP_1^k$, $P_3^kP_2^k$ are obtained by symmetry). In the case in which $u \neq 0$ on the boundary or mixed boundary conditions are given (the Galerkin method of Section 8.5.2), the computations above must be altered somewhat, and this approach is probably more appropriate.

We see in the work above that the labeling of the grid points has a significant influence on the form of the coefficient matrix S. Labeling strategies are still the subject of extensive research, but a common approach is to follow the technique explained earlier for finite-difference approximations and to label "horizontally" left to right, bottom to top. To illustrate, we consider the given triangulation of a portion of a rectangular region with uniform mesh as shown in Fig. 8.12, in which P_i and its "adjacent" mesh points are interior points of the

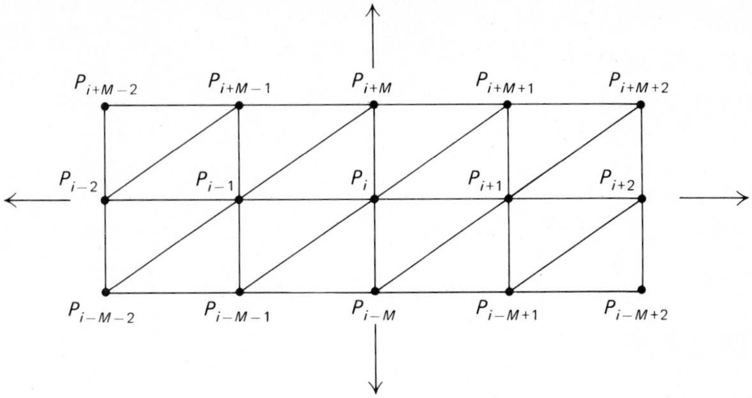

Figure 8.12

grid. There is an equation of (8.71) for every interior mesh point. We consider the ith equation of (8.71) and hence the entries s_{ij} of the ith row of S. Unless there is algebraic cancellation to zero from the integrations in (8.72) contributing to $s_{ij} = \langle L[\phi_i], \phi_j \rangle$, we see from Fig. 8.12 that the entries $s_{i,\,i-M-1}$, $s_{i,\,i-M}$, $s_{i,\,i-1}$, s_{ii}, $s_{i,\,i+1}$, $s_{i,\,i+M}$, and $s_{i,\,i+M+1}$ are nonzero, and that the remaining entries in the ith row are zero. Hence in this simplistic case, S would be a banded matrix with band width $2M + 2$; and the solution of (8.71) would probably be sought with block Gauss-Seidel or SLOR. (The same type of arguments can be made with respect to a "vertical" ordering, which would be appropriate for reducing band width when the region is "longer" in the y-direction.)

PROBLEMS, SECTION 8.5.1

1. With $\langle L[u], w\rangle$ as in (8.68), show that $\langle L[v], w\rangle = \langle L[w], v\rangle$ and that $\langle L[v], v\rangle > 0$ for v differentiable and nonzero.
2. Consider the given (4×4) symmetric matrix A with eigenvalues 1, 2, 5, 10.
   a) Show that the eigenvalues' being positive implies that $\mathbf{v}^T A \mathbf{v} > 0$ for all $\mathbf{v} \neq \theta$ in R^4.

$$A = \begin{bmatrix} 5 & 4 & 1 & 1 \\ 4 & 5 & 1 & 1 \\ 1 & 1 & 4 & 2 \\ 1 & 1 & 2 & 4 \end{bmatrix}$$

[Hint: Let $\mathbf{v} = c_1\mathbf{u}_1 + c_2\mathbf{u}_2 + c_3\mathbf{u}_3 + c_4\mathbf{u}_4$ where $\{\mathbf{u}_i\}_{i=1}^4$ are orthogonal eigenvectors (which need not be calculated for the proof).]

b) For any $\mathbf{v}$ and $\mathbf{w}$ in R^4 define $\langle \mathbf{v}, \mathbf{w} \rangle_L = \mathbf{v}^T A \mathbf{w}$. Show that $\langle \mathbf{v}, \mathbf{w} \rangle_L$ satisfies the four inner-product properties.

c) Consider the problem $L[\mathbf{u}] = \mathbf{b}$ where $\mathbf{b} = [8, 8, -4, 2]^T$ and $L[\mathbf{x}] = A\mathbf{x}$ for all $\mathbf{x}$ in R^4. Use formula (8.71) to find $\mathbf{u}_3 = c_1\mathbf{e}_1 + c_2\mathbf{e}_2 + c_3\mathbf{e}_3$, the Rayleigh-Ritz approximation to $\mathbf{u}$ from X_3 that is spanned by $\{\mathbf{e}_1, \mathbf{e}_2, \mathbf{e}_3\}$.

3. Let D be the square with vertices $P_3 = (0, 0)$, $P_4 = (3, 0)$, $P_5 = (3, 3)$, and $P_6 = (0, 3)$. Using $P_1 = (1, 1)$ and $P_2 = (2, 2)$ as interior grid points, divide D into six triangles.

a) Use (8.69) and (8.70) to give expressions for basis functions $\phi_1(x, y)$ and $\phi_2(x, y)$.

b) In (8.63) let $p = q = 1$, $r = 0$, and $f = -2$. Use (8.68) to calculate $\langle \phi_1, \phi_1 \rangle_L$, $\langle \phi_1, \phi_2 \rangle_L$, and $\langle \phi_2, \phi_2 \rangle_L$. Also, use (8.64) to calculate $\langle f, \phi_1 \rangle$ and $\langle f, \phi_2 \rangle$.

c) Let $u_2 = c_1\phi_1 + c_2\phi_2$ and solve for c_1 and c_2 from (8.62) or (8.71).

4. a) Let (x_i, y_i), $i = 0, 1, 2$, be the vertices of a triangle where the vertices are numbered clockwise. Show that the area of the triangle is given by $\left(\frac{1}{2}\right)$ det (A) where

$$A = \begin{bmatrix} x_0 & y_0 & 1 \\ x_1 & y_1 & 1 \\ x_2 & y_2 & 1 \end{bmatrix}.$$

b) Find the coordinates of the point P_c where the medians intersect.

c) Approximate the integrals in part (b) of Exercise 3; use the integrand evaluated at P_c times the area for each triangle.

5. Use Cramer's rule to give formulas for a_m^k, b_m^k, c_m^k, $m = 1, 2, 3$ in (8.70) for a given triangle T_k with vertices V_m^k, $m = 1, 2, 3$.

6. Suppose that a uniform square mesh of height and width $h = (1/5)$ is placed on the unit square. Connect the mesh points and let the triangulation of the unit square be accomplished by drawing the two diagonals of each mesh square. If the interior mesh points are labeled "horizontally," determine the band width of the matrix S in (8.71).

7. Write a program to find the Rayleigh-Ritz approximation for the solution of (8.63); use (8.71). You may wish to incorporate the ideas of Problems 4 and 5. In order to perform individual computations, you must establish an association between the grid points P_i (interior or exterior), the triangles T_k, and the vertices V_m^k, $m = 1, 2, 3$. Test your program on the following.

a) In (8.63) let $p = y^2 + 1$, $q = x^2 + 1$, $r = 2$, $f = -6$, and $C = \{(x, y)\colon x^2 + y^2 = 1\}$. [Answer: $u = x^2 + y^2 - 1$.]

b) In (8.63) let $p = q = 1$, $r = \pi^2$, and $f = 3\pi^2 \sin \pi x \sin \pi y$. Let $D = \{(x, y)\colon 0 < x, y < 1\}$. [Answer: $u = \sin \pi x \sin \pi y$.]

8. Given the figure below, determine a_k, b_k, c_k, and d_k so that if $q_k(x, y) = a_k + b_k x + c_k y + d_k xy$, then on rectangle k, $1 \le k \le 4$, $q_k(x_i, y_j) = 1$ and at the other three corners $q_k = 0$. Define the function $\phi_{ij}(x, y)$ so that $\phi_{ij}(x, y) = q_k(x, y)$ for (x, y) in rectangle k, $1 \le k \le 4$, and $\phi_{ij} = 0$ elsewhere. Briefly outline a Rayleigh-Ritz procedure for (8.63); use a rectangular mesh and basis functions ϕ_{ij} for the approximation. As in Problem 3 use the same region D with interior mesh points $P_1 = (1,$

1.5) and $P_2 = (2, 1.5)$ to form a six-part rectangular mesh. Using the basis functions above, repeat parts (b) and (c) of Problem 3.

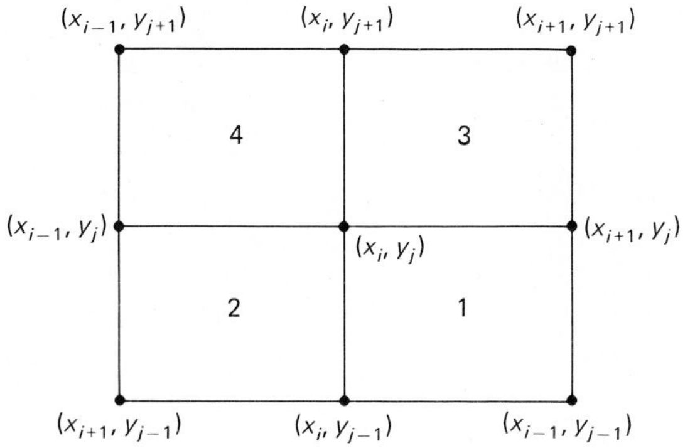

8.5.2. The Galerkin Method

The Rayleigh-Ritz method is a special case of the Galerkin method. Once again we wish to approximate the solution of an operator equation $L[u] = f$ from a subspace X_N of X, the inner-product space on which L is defined. Since $L[u] = f$, then $\langle (L[u] - f), v \rangle = 0$ for all v in X. In the general version of the Galerkin method, to approximate u by u_N in X_N, we find u_N such that $\langle (L[u_N] - f), v \rangle = 0$ for all v in a given subspace Y_N of X; that is, we approximate the relationship $\langle (L[u] - f), v \rangle = 0$ by restricting u_N to X_N and v to Y_N. Let $\{\phi_i\}_{i=1}^N$ and $\{\psi_i\}_{i=1}^N$ be bases for X_N and Y_N, respectively; that is, each x in X_N and each y in Y_N can be written thus:

$$x = \sum_{i=1}^{N} \beta_i \phi_i \qquad \text{and} \qquad y = \sum_{i=1}^{N} \gamma_i \psi_i.$$

(The functions ϕ_i are usually called "trial functions," and the functions ψ_i are usually called "test functions.") We then seek $u_N = \sum_{j=1}^{N} c_j \phi_j$ in X_N such that $\langle (L[u_N] - f), v \rangle = 0$ for all v in Y_N, or $\langle (L[u_N] - f), \psi_i \rangle = 0$ for $1 \le i \le N$ or

$$\sum_{j=1}^{N} \langle L[\phi_j], \psi_i \rangle c_j = \langle f, \psi_i \rangle, \qquad 1 \le i \le N. \tag{8.73}$$

We wish to choose X_N and Y_N such that (8.73) has a solution and thus u_N exists. Furthermore as X_N and Y_N are enlarged and become dense in X (as in the case of spaces of spline functions, with a refinement of the mesh and the resultant increase in N), the equations of (8.73) should more closely approximate the relationship $\langle (L[u] - f), v\rangle = 0$ for all v in X. We would then hope to be able to show that the sequence of approximations $\{u_N\}$ converges to u.

Usually we take $\psi_i = \phi_i$, $1 \le i \le n$. In this case, if $\langle v, w\rangle_L = \langle L[v], w\rangle$ is an inner product on X, then (8.73) becomes (8.62) and we have the Rayleigh-Ritz procedure. In the Rayleigh-Ritz situation, we know that (8.73) has a solution since u_N is the unique, best least-squares approximation in X_N to u in X with respect to the energy norm. (By analogy to Theorem 5.10, if $\{\tilde{\phi}_i\}_{i=1}^N$ is an orthonormal basis for X_N, $\langle \tilde{\phi}_i, \tilde{\phi}_j\rangle_L = 0$ if $i \ne j$, and $\langle \tilde{\phi}_i, \tilde{\phi}_i\rangle_L = 1$, then $u_N = \sum_{i=1}^{N} \langle u, \tilde{\phi}_i\rangle_L \tilde{\phi}_i$.) If the operator L is not symmetric and positive-definite (and hence $\langle L[u], w\rangle$ does not yield a new inner product), it is possible that the Galerkin technique may still be applicable. In this case we would like conditions on L, ϕ_i, and ψ_i to ensure that (8.73) has a solution. We will not pursue this point further except to show by a simple example that (8.73) need not always have a solution even in the case in which $\psi_i = \phi_i$.

EXAMPLE 8.8. Instead of function spaces, let $X = R^4$ and suppose that $A = (a_{ij})$ is a (4×4) nonsingular matrix and $\mathbf{b} = [b_1, b_2, b_3, b_4]^T$ is a given vector in R^4. Let $X_2 = Y_2$ be the subspace of R^4 consisting of all linear combinations of the unit vectors $\mathbf{e}_1$ and $\mathbf{e}_2$; that is, let $\phi_1 = \psi_1 = \mathbf{e}_1$ and $\phi_2 = \psi_2 = \mathbf{e}_2$. Suppose that we wish to find the Galerkin approximation $\mathbf{u}_2 = c_1\mathbf{e}_1 + c_2\mathbf{e}_2$ in X_2 for the solution $\mathbf{x}$ in R^4 of $A\mathbf{x} = \mathbf{b}$. With A written in column form, $A = [\mathbf{A}_1, \mathbf{A}_2, \mathbf{A}_3, \mathbf{A}_4]$, (8.73) yields

$$\begin{aligned} \langle A\mathbf{e}_1, \mathbf{e}_1\rangle c_1 + \langle A\mathbf{e}_2, \mathbf{e}_1\rangle c_2 &= \langle \mathbf{b}, \mathbf{e}_1\rangle \\ \langle A\mathbf{e}_1, \mathbf{e}_2\rangle c_1 + \langle A\mathbf{e}_2, \mathbf{e}_2\rangle c_2 &= \langle \mathbf{b}, \mathbf{e}_2\rangle. \end{aligned} \tag{8.74}$$

For all (4×1) vectors $\mathbf{v}$ and $\mathbf{w}$, $\langle \mathbf{v}, \mathbf{w}\rangle = \mathbf{v}^T\mathbf{w}$; and since $A\mathbf{e}_1 = \mathbf{A}_1$ and $A\mathbf{e}_2 = \mathbf{A}_2$, (8.74) becomes

$$\begin{aligned} \mathbf{e}_1^T(c_1\mathbf{A}_1 + c_2\mathbf{A}_2) &= \mathbf{e}_1^T\mathbf{b} & & a_{11}c_1 + a_{12}c_2 = b_1 \\ & & \text{or} \\ \mathbf{e}_2^T(c_1\mathbf{A}_1 + c_2\mathbf{A}_2) &= \mathbf{e}_2^T\mathbf{b} & & a_{21}c_1 + a_{22}c_2 = b_2. \end{aligned} \tag{8.75}$$

Even though $A\mathbf{x} = \mathbf{b}$ has a unique solution, if $\Delta = (a_{11}a_{22} - a_{12}a_{21}) = 0$, then (8.75) may have no solution; and hence there would be no Galerkin approximation, $\mathbf{u}_2 = c_1\mathbf{e}_1 + c_2\mathbf{e}_2$. However if A is a symmetric, positive-definite matrix, then $\Delta > 0$ (Exercise1), and (8.75) has a unique solution. This is now a Rayleigh-Ritz problem since $\langle \mathbf{v}, \mathbf{w}\rangle_A = \langle A\mathbf{v}, \mathbf{w}\rangle = \mathbf{w}^T A\mathbf{v}$ is a genuine inner product on R^4.

We return now to elliptic boundary-value problems and consider

$$\begin{aligned} L[u] &= -(pu_x)_x - (pu_y)_y + ru = f \\ u + \beta \frac{du}{dn} &+ \gamma = 0 \text{ on the boundary } C \text{ of } D. \end{aligned} \tag{8.76}$$

In (8.76), du/dn is the normal derivative; p, r, f, β, and γ are functions with p, r, and β positive and p differentiable. We again use the inner product (8.64). If v and w are two differentiable functions satisfying the boundary conditions of (8.76), and if we examine $\langle L[v], w\rangle$ as in (8.66) and (8.67) with $p = q$, then the results are the same except the line integral in (8.67) is no longer zero. With $dv/dn = (-v - \gamma)/\beta$, this line integral becomes [Seeley [1970], p 220]

$$\begin{aligned}\int_C -pv_y w\,dx + pv_x w\,dy &= \int_C (pw)\frac{dv}{dn}\,ds \\ &= -\int_C [(pw)(v+\gamma)/\beta]\,ds\end{aligned} \tag{8.77}$$

where s denotes the parameterization of C with respect to arc length. Hence

$$\langle L[v], w\rangle = \int_D\int [pv_x w_x + pv_y w_y + rvw]dx\,dy + \int_C [pw(v+\gamma)/\beta]ds. \tag{8.78}$$

Note that if $\gamma = 0$, then by (8.78) $\langle L[v], w\rangle = \langle L[w], v\rangle$ and $\langle L[v], v\rangle > 0$ for $v \neq \theta$; that is, L is symmetric and positive definite. Hence the Rayleigh-Ritz procedure is applicable and we use formulas (8.71) as in Section 8.5.1; because of the line integral in (8.78), a contribution is added to each inner product. If $\gamma \neq 0$ in (8.78), then in general $\langle L[v], w\rangle \neq \langle L[w], v\rangle$; and thus this operation does not yield an inner product. The more general Galerkin procedure may still be applicable in this case, however, and we proceed as before using (8.78) in (8.73) even though $\langle L[v], w\rangle$ is not an inner product. The computations are similar to those in Section 8.5.1; but before we develop them, we present a different approach to the Galerkin procedure; the results of that approach may seem simpler but actually turn out to yield the same approximation.

An equivalent form of the Galerkin method for (8.76) can be obtained through what is known as a variational principle. In general, for a given differential equation the basic idea of the variational technique is to find an integral form $F(w)$ such that $F(w)$ is minimized when $w = u$ where u is the solution of the differential equation. We can replace the problem of solving the differential equation by the problem of minimizing $F(w)$ for w in a given class of functions. Often $F(w)$ is difficult to identify (if it even exists), but for the problem given by (8.76) we can use (8.78) to deduce the form of $F(w)$. For motivation, we first consider the following simple example in R^n.

EXAMPLE 8.9. Let A be a symmetric, positive-definite, $(n \times n)$ matrix; then there is a unique solution $\mathbf{x}$ for the system $A\mathbf{x} = \mathbf{b}$. Furthermore the solution $\mathbf{x}$ satisfies

$$\langle (A\mathbf{x} - \mathbf{b}), \mathbf{x}\rangle = 0 \qquad \text{or} \qquad \mathbf{x}^{\mathrm{T}}A\mathbf{x} - \mathbf{x}^{\mathrm{T}}\mathbf{b} = 0.$$

To proceed as described above, we wish to find a functional $F(\mathbf{w})$ such that $F(\mathbf{w})$ is minimized when $\mathbf{w} = \mathbf{x}$. Consider the case in which $n = 1$. Then $A\mathbf{x} - \mathbf{b} = \theta$ becomes $q(x) = ax - b = 0$ where $a > 0$. Any antiderivative of $q(x)$ has a critical point at the root of $q(x) = 0$. Hence we let $F(x) = \left(\frac{1}{2}\right)ax^2 - bx$. If $x_0 = b/a$, then $q(x_0) = F'(x_0) = ax_0 - b = 0$, and $F''(x_0) = a > 0$. Thus the minimum of F occurs at the root of $q(x) = 0$.

Generalizing to R^n, we then choose $F(\mathbf{w})$ as

$$F(\mathbf{w}) = \left(\tfrac{1}{2}\right)\mathbf{w}^T A\mathbf{w} - \mathbf{w}^T\mathbf{b}, \tag{8.79}$$

and define $\boldsymbol{\varepsilon}$ in R^n such that $\mathbf{w} = \mathbf{x} + \boldsymbol{\varepsilon}$. Then

$$\begin{aligned} F(\mathbf{w}) &= \left(\tfrac{1}{2}\right)(\mathbf{x} + \boldsymbol{\varepsilon})^T A(\mathbf{x} + \boldsymbol{\varepsilon}) - (\mathbf{x} + \boldsymbol{\varepsilon})^T\mathbf{b} \\ &= \left(\tfrac{1}{2}\right)\mathbf{x}^T A\mathbf{x} + \left(\tfrac{1}{2}\right)(\mathbf{x}^T A\boldsymbol{\varepsilon} + \boldsymbol{\varepsilon}^T A\mathbf{x} + \boldsymbol{\varepsilon}^T A\boldsymbol{\varepsilon}) - \mathbf{x}^T\mathbf{b} - \boldsymbol{\varepsilon}^T\mathbf{b}. \end{aligned} \tag{8.80}$$

Since A is symmetric, $\mathbf{x}^T A\boldsymbol{\varepsilon} = \boldsymbol{\varepsilon}^T A\mathbf{x}$; and since $F(\mathbf{x}) = \left(\frac{1}{2}\right)\mathbf{x}^T A\mathbf{x} - \mathbf{x}^T\mathbf{b}$, then (8.80) yields

$$\begin{aligned} F(\mathbf{w}) &= F(\mathbf{x}) + \boldsymbol{\varepsilon}^T(A\mathbf{x} - \mathbf{b}) + \left(\tfrac{1}{2}\right)\boldsymbol{\varepsilon}^T A\boldsymbol{\varepsilon} \\ &= F(\mathbf{x}) + \left(\tfrac{1}{2}\right)\boldsymbol{\varepsilon}^T A\boldsymbol{\varepsilon}. \end{aligned} \tag{8.81}$$

Since A is positive definite, $\boldsymbol{\varepsilon}^T A\boldsymbol{\varepsilon} > 0$ for $\boldsymbol{\varepsilon} \neq \theta$; and so $F(\mathbf{w}) > F(\mathbf{x})$ for all $\boldsymbol{\varepsilon} \neq \theta$. Hence the functional F is minimized uniquely by $\mathbf{x}$, the solution of $A\mathbf{x} = \mathbf{b}$.

The analysis in the example above will extend directly to the operator equation $L[u] = f$ whenever L is a symmetric positive-definite operator on the inner-product space X, the domain of L. For example, the Rayleigh-Ritz procedure for (8.63) can be analyzed in terms of the variational principle by using the inner product (8.68) and proceeding as above. This analysis will be a special case of the following variational approach for the Galerkin method applied to (8.76).

We recall that the Galerkin method uses the fact that if $L[u] = f$, then $\langle (L[u] - f), v\rangle = 0$ for all v in X, and hence $\langle (L[u] - f), u\rangle = 0$. This fact is analogous to $(\mathbf{x}^T A\mathbf{x} - \mathbf{x}^T\mathbf{b}) = 0$ in Example 8.9. In Example 8.9, however, A was symmetric and positive definite. We see from (8.78) that L fails to be symmetric and positive definite only when γ is nonzero. This result indicates the possibility that the steps of Example 8.9 may be valid applied to (8.76) with $\langle L[u], u\rangle$ as given by (8.78).

In analogy to (8.79), we begin by choosing $F(w)$ so that the "quadratic" terms of $\langle (L[w] - f), w\rangle$ are given factors of $\frac{1}{2}$. Since

$$\begin{aligned} \langle (L[w] - f), w\rangle = &\int_D\!\!\int [pw_x^2 + pw_y^2 + rw^2 - fw]dx\,dy \\ &+ \int_C [pw(w + \gamma)/\beta]ds, \end{aligned} \tag{8.82}$$

then for all sufficiently differentiable functions w where $w + \beta\dfrac{dw}{dn} + \gamma = 0$, we choose

$$\begin{aligned} F(w) = &\left(\tfrac{1}{2}\right)\int_D\!\!\int [pw_x^2 + pw_y^2 + rw^2 - 2fw]dx\,dy \\ &+ \left(\tfrac{1}{2}\right)\int_C [pw^2 + 2p\gamma w]/\beta\, ds. \end{aligned} \tag{8.83}$$

Again as in Example 8.9 we choose the function ε such that $w = u + \varepsilon$. Then

$$F(w) = \left(\tfrac{1}{2}\right)\int_D\int [p(u_x^2 + 2u_x\varepsilon_x + \varepsilon_x^2) + p(u_y^2 + 2u_y\varepsilon_y + \varepsilon_y^2)$$
$$+ r(u^2 + 2u\varepsilon + \varepsilon^2) - 2f(u + \varepsilon)]dx\,dy + \left(\tfrac{1}{2}\right)$$
$$\int_C [p(u^2 + 2u\varepsilon + \varepsilon^2)/\beta]ds + \int_C [p\gamma(u + \varepsilon)/\beta]ds$$
$$= F(u) + \int_D\int [pu_x\varepsilon_x + pu_y\varepsilon_y]dx\,dy + \int_D\int [ru\varepsilon - f\varepsilon]dx\,dy$$
$$+ \left(\tfrac{1}{2}\right)\int_D\int [p\varepsilon_x^2 + p\varepsilon_y^2 + r\varepsilon^2]dx\,dy$$
$$+ \int_C [(pu\varepsilon + p\gamma\varepsilon)/\beta]ds + \left(\tfrac{1}{2}\right)\int_C [p\varepsilon^2/\beta]ds.$$

Using Green's theorem as in (8.67), we have

$$\int_D\int [pu_x\varepsilon_x + pu_y\varepsilon_y]dx\,dy = \int_C -pu_y\varepsilon\,dx + pu_x\varepsilon\,dy$$
$$- \int_D\int [(pu_x)_x\varepsilon + (pu_y)_y\varepsilon]dx\,dy$$
$$= \int_C \left[p\varepsilon\frac{du}{dn}\right]ds - \int_D\int [(pu_x)_x\varepsilon + (pu_y)_y\varepsilon]dx\,dy$$
$$= -\int_C [p\varepsilon(u + \gamma)/\beta]ds$$
$$- \int_D\int [(pu_x)_x\varepsilon + (pu_y)_y\varepsilon]dx\,dy.$$

To verify the second equality above, see Seeley (1970), p. 220. The third equality follows from $u + \beta\dfrac{du}{dn} + \gamma = 0$ on C. Substitution of this expression into the expression for $F(w)$ yields

$$F(w) = F(u) + \int_D\int [-(pu_x)_x\varepsilon - (pu_y)_y\varepsilon + ru\varepsilon - f\varepsilon]dx\,dy$$
$$- \int_C [p\varepsilon(u + \gamma)/\beta]ds + \left(\tfrac{1}{2}\right)\int_D\int [p\varepsilon_x^2 + p\varepsilon_y^2 + r\varepsilon^2]dx\,dy \qquad (8.84)$$
$$+ \int_C [(pu\varepsilon + p\gamma\varepsilon)/\beta]ds + \left(\tfrac{1}{2}\right)\int_C [p\varepsilon^2/\beta]ds.$$

The first integral in (8.84) is $\int_D\int [(L[u] - f)\varepsilon]dx\,dy = \langle (L[u] - f), \varepsilon\rangle = 0$ (since $L(u) = f$), and so

$$F(w) = F(u) + \left(\tfrac{1}{2}\right)\int_D\int [p\varepsilon_x^2 + p\varepsilon_y^2 + r\varepsilon^2]dx\,dy + \left(\tfrac{1}{2}\right)\int_C [p\varepsilon^2/\beta]ds. \quad (8.85)$$

The integrals in (8.85) are positive, and so $F(w) > F(u)$ when $\varepsilon \neq \theta$. Thus $F(w)$ is uniquely minimized by the solution u of (8.76). Furthermore, we can repeat the analysis above in X_N to show that if in X_N, w_N is the function that minimizes F over all functions in X_N, then $w_N = u_N$ where u_N is the Galerkin approximation from (8.73) with $Y_N = X_N$; that is, $\langle (L[u_N] - f), v\rangle = 0$ for all v in X_N. To see this result, let w be any function in X_N and define the function ε by $w = u_N + \varepsilon$. We substitute $w = u_N + \varepsilon$ into (8.83) and note that the arguments following (8.83) remain valid; therefore (8.84) is valid for u_N replacing u. Since $\varepsilon \in X_N$, u_N satisfies $\langle (L[u_N] - f), \varepsilon\rangle = 0$, the first integral in (8.84) is again zero and we obtain (8.85) with u_N replacing u. Thus we have $F(w) \geq F(u_N)$ for all w in X_N. So for (8.76) the Galerkin procedure using (8.78) in (8.73) with $Y_N = X_N$ yields the same solution u_N as in minimizing $F(w)$ in (8.85), if u_N exists. In other problems, however, the two procedures may not be the same. For instance, in Example 8.8 with $F(w) = \left(\tfrac{1}{2}\right)\mathbf{w}^T A\mathbf{w} - \mathbf{w}^T\mathbf{b}$ and $\mathbf{w} = c_1\mathbf{e}_1 + c_2\mathbf{e}_2$, the equations $\partial F/\partial c_1 = 0$ and $\partial F/\partial c_2 = 0$ yield

$$a_{11}c_1 + \left(\tfrac{1}{2}\right)(a_{12} + a_{21})c_2 = b_1$$
$$\left(\tfrac{1}{2}\right)(a_{12} + a_{21})c_1 + a_{22}c_2 = b_2,$$

which is not the same as (8.75) unless $a_{12} = a_{21}$ and even then may have no solution.

In the case of Dirichlet boundary conditions in (8.76), that is, $\beta = 0$ in (8.76) and u satisfies

$$u = -\gamma(x, y) \equiv g(x, y) \qquad \text{for } (x, y) \text{ on } C,$$

then a slight modification of the analysis above yields the functional $\tilde{F}(w)$ where

$$\tilde{F}(w) = \left(\tfrac{1}{2}\right)\int_D\int [pw_x^2 + pw_y^2 + rw^2 - 2fw]dx\,dy. \quad (8.86)$$

This analysis yields further that $\tilde{F}(u + \varepsilon) > \tilde{F}(u)$ when $\varepsilon \neq \theta$ (Problem 2).

If $\beta = \gamma = 0$ in (8.76), then (8.76) is the same as (8.63) with $q = p$. In the previous section we used the Rayleigh-Ritz principle to find u_N in X_N such that $\|u - u_N\|_L \leq \|u - w\|_L$ for all w in X_N. We now show that in X_N, u_N is also the function that minimizes $\tilde{F}(w)$ in (8.86); and hence the Rayleigh-Ritz procedure is equivalent to the variational procedure for the problem (8.76) with $\beta = \gamma = 0$.

From (8.68) with $q = p$ we have for all w in X_N,

$$
\begin{aligned}
\langle (L[w] - f), w\rangle &= \int_D\int [pw_x^2 + pw_y^2 + rw^2 - fw]dx\,dy \\
&= \int_D\int [pw_x^2 + pw_y^2 + rw^2 - 2fw]dx\,dy \qquad (8.87)\\
&\quad + \int_D\int fw\,dx\,dy.
\end{aligned}
$$

Then by (8.86) and (8.87),

$$\langle (L[w] - f), w\rangle = 2\tilde{F}(w) + \langle f, w\rangle. \qquad (8.88)$$

Thus for the solution u of (8.76) with $\beta = \gamma = 0$ and for any w in X_N, we have

$$
\begin{aligned}
\|w - u\|_L^2 &= \langle L[w - u], w - u\rangle = \langle (L[w] - f), w - u\rangle \\
&= \langle (L[w] - f), w\rangle - \langle L[w], u\rangle + \langle f, u\rangle \\
&= \langle (L[w] - f), w\rangle - \langle L[u], w\rangle + \langle f, u\rangle \qquad (8.89)\\
&= \langle (L[w] - f), w\rangle - \langle f, w\rangle + \langle f, u\rangle \\
&= 2\tilde{F}(w) + \langle f, w\rangle - \langle f, w\rangle + \langle f, u\rangle.
\end{aligned}
$$

Thus

$$\tilde{F}(w) = \left(\tfrac{1}{2}\right)\|w - u\|_L^2 - \left(\tfrac{1}{2}\right)\langle f, u\rangle \qquad (8.90)$$

for all w in X_N. Since $\|u_N - u\|_L \leq \|w - u\|_L$ for all w in X_n, (8.90) shows that $\tilde{F}(w) \geq \tilde{F}(u_N)$, and hence the Rayleigh-Ritz solution is the same as would be found by minimizing $\tilde{F}(w)$ on X_N. We note that (8.89) and (8.90) are valid for the Rayleigh-Ritz technique for approximating the solution u of the general linear operator equation $L[u] = f$ from a subspace X_N of the inner-product space X on which L is defined and where L is symmetric and positive definite. Further, the variational function $\tilde{F}(w)$ is given by (8.90). For example, in the problems the reader is asked to verify that (8.90) yields $\tilde{F}(w)$ in (8.86) for $\beta > 0$, $\gamma = 0$, and yields $\tilde{F}(w) = \left(\frac{1}{2}\right)\mathbf{w}^{\mathrm{T}}A\mathbf{w} - \mathbf{w}^{\mathrm{T}}\mathbf{b}$ for A symmetric and positive definite in Example 8.9. The reader is asked also to use (8.90) to identify $\tilde{F}(w)$ for the two-point boundary-value problem of Section 5.3.3.

The computations to minimize either $F(w)$ in (8.83) or $\tilde{F}(w)$ in (8.86) are as follows. Suppose that X_N is determined by the basis functions $\{\phi_i\}_{i=1}^N$ and that an arbitrary function w in X_N is given by

$$w = c_1\phi_1 + c_2\phi_2 + \cdots + c_N\phi_N. \qquad (8.91)$$

Then with $F(w)$ given by (8.83)

$$F(w) = \left(\tfrac{1}{2}\right)\int_D\int \left[p\left(\sum_{j=1}^N c_j(\phi_j)_x\right)^2 + p\left(\sum_{j=1}^N c_j(\phi_j)_y\right)^2 + r\left(\sum_{j=1}^N c_j\phi_j\right)^2 - 2f\sum_{j=1}^N c_j\phi_j\right]dx\,dy + \left(\tfrac{1}{2}\right)\int_C \left[p\left(\sum_{j=1}^N c_j\phi_j\right)^2 + 2p\gamma\sum_{j=1}^N c_j\phi_j\right]\Big/\beta\,ds. \tag{8.92}$$

Since each w in X_N can be obtained by a particular choice of its coefficients c_j in (8.91), F is also a function of $c_1, c_2, \ldots, c_N$. Hence a necessary condition for F to be minimized over X_N is that $\partial F/\partial c_i = 0$ for $1 \le i \le N$. However, if $\beta = 0$ and $\gamma \ne 0$ in (8.76) on some part of C, some of the coefficients (say, $c_{n+1}, c_{n+2}, \ldots, c_N$) are reserved to try to make each w satisfy the boundary conditions (see below in the case of triangulation of the region). We use the remaining coefficients as variables and set $\partial F/\partial c_i = 0$ for $1 \le i \le n$.

Differentiating (8.92) yields

$$\frac{\partial F}{\partial c_i} = \int_D\int \left[p\left(\sum_{j=1}^N c_j(\phi_j)_x\right)(\phi_i)_x + p\left(\sum_{j=1}^N c_j(\phi_j)_y\right)(\phi_i)_y + r\left(\sum_{j=1}^N c_j\phi_j\right)\phi_i - f\phi_i\right]dx\,dy + \int_C \left[p\left(\sum_{j=1}^N c_j\phi_j\right)\phi_i + p\gamma\phi_i\right]\Big/\beta\,ds, \qquad 1 \le i \le n. \tag{8.93}$$

If one sets each $\partial F/\partial c_j = 0$, $1 \le j \le n$, the linear system from (8.93) can be written as $A\mathbf{c} = \mathbf{b}$ where $A = (a_{ij})$, $\mathbf{c} = [c_1, c_2, \ldots, c_n]^{\mathrm{T}}$, $\mathbf{b} = [b_1, b_2, \ldots, b_n]^{\mathrm{T}}$ are given by

$$a_{ij} = \int_D\int [p(\phi_i)_x(\phi_j)_x + p(\phi_i)_y(\phi_j)_y + r\phi_i\phi_j]dx\,dy + \int_C [p\phi_i\phi_j/\beta]ds, \tag{8.94a}$$

$$b_i = \int_D\int [f\phi_i]dx\,dy - \int_C [p\gamma\phi_i/\beta]ds - \sum_{k=n+1}^N a_{ik}c_k. \tag{8.94b}$$

We now suppose that the region D has a boundary consisting of straight-line segments. (For curved boundaries either we can transfer the boundary data to each corresponding exterior line segment, or we must use other techniques such as *isoparametric transformations*, which we cannot discuss here.) We then divide D into triangles $\{T_j\}_{j=1}^K$ with vertices labeled $\{P_i\}_{i=1}^N$ as in Section 8.5.1. We again construct basis functions $\{\phi_i(x, y)\}_{i=1}^N$ such that $\phi_i(P_i) = 1$

and $\phi_i(x, y) = 0$ whenever (x, y) is not in a triangle that has P_i as a vertex. The construction proceeds exactly as in Section 8.5.1 via formulas (8.69) and (8.70). This time, however, unless β and γ are both zero on C, a ϕ_i is constructed for each boundary mesh point, as well as for the interior mesh points. Let $P_{n+1}, P_{n+2}, \ldots, P_N$ denote the boundary mesh points where $\beta = 0$ so that $u(P_i) = -\gamma(P_i)$, $n + 1 \le i \le N$. We choose $c_i = -\gamma(P_i)$ for $n + 1 \le i \le N$; and hence for $w = \sum_{j=1}^{N} c_j\phi_j$, $w(P_i) = -\gamma(P_i)$ for $n + 1 \le i \le N$. The remaining ϕ_i, $1 \le i \le n$, are used to satisfy $\partial F/\partial c_i = 0$, $1 \le i \le n$, as in (8.93). The coefficients and constants of this system given by (8.94a) and (8.94b) are computed as described in Section 8.5.1 with additional contributions from the line integrals for the remaining boundary points where $\beta > 0$. Note that a_{ik} for $n + 1 \le k \le N$ [with $\beta = 0$, and hence the line integral is taken as zero from $\tilde{F}$ in (8.86)] must also be computed for use in (8.49b). Again the ordering of points significantly affects the structure of A, and the previously described "horizontal" or "vertical" orderings are appropriate. As with finite-difference methods, when N is large, an iterative method for solving $A\mathbf{c} = \mathbf{b}$ is probably preferable. For an algorithm incorporating these details the reader is referred to Burden, Faires, and Reynolds (1981).

We conclude this section by considering another special case of the Galerkin procedure: the *method of least squares*. Instead of $\psi_i = \phi_i$, this method uses $\psi_i = L[\phi_i]$ in (8.73), which then becomes

$$\sum_{j=1}^{N} \langle L[\phi_j], L[\phi_i]\rangle c_j = \langle L[\phi_i], f\rangle, \qquad 1 \le i \le N. \tag{8.95}$$

In this case, each y in Y_N can be expressed as

$$y = d_1L[\phi_1] + d_2L[\phi_2] + \cdots + d_NL[\phi_N].$$

Hence if $u = \sum_{i=1}^{N} d_i\phi_i$ in X_N, then $y = L[u]$. Furthermore if in Y_N, y_N is the best least-squares approximation to the function f, that is,

$$\|y_N - f\| = \min_{y \in Y_N} \|y - f\|$$

where the norm is the inner-product norm on X, then similarly to Corollary 1 of Theorem 5.10, $\langle (y_N - f), y\rangle = 0$ for all y in Y_N, and hence

$$\langle (y_N - f), L[\phi_i]\rangle = 0, \qquad 1 \le i \le N. \tag{8.96}$$

Writing $y_N = \sum_{j=1}^{N} c_jL[\phi_j]$ and substituting in (8.96) yield (8.95). After solving (8.95) for $\{c_j\}_{j=1}^{N}$, we then take $u_N = \sum_{j=1}^{N} c_j\phi_j$ as the Galerkin approximation for u. We note that y_N always exists and is unique, and hence there is a unique solution for (8.95).

PROBLEMS, SECTION 8.5.2

1. a) In Example 8.8, suppose that A is symmetric and positive definite; that is, $A = A^T$ and $\mathbf{x}^T A\mathbf{x} > 0$ for $\mathbf{x} \neq \theta$. Show that $\mathbf{x} = \mathbf{e}_1$ yields $a_{11} > 0$. Show that $\mathbf{x} = (a_{12}/\sqrt{a_{11}})\mathbf{e}_1 - \sqrt{a_{11}}\,\mathbf{e}_2$ yields $\Delta = (a_{11}a_{22} - a_{12}^2) > 0$ and that hence (8.75) has a solution.

   b) Using $\psi_1 = A\mathbf{e}_1$ and $\psi_2 = A\mathbf{e}_2$, repeat the procedure in Example 8.8. Does the resulting (2×2) system have a solution? How does this result compare with (8.95)?

2. With $\tilde{F}(w)$ given by (8.86) and $\beta = 0$ in (8.76), verify that $\tilde{F}(u + \varepsilon) > \tilde{F}(u)$ for $\varepsilon \neq \theta$.

3. a) If $\beta > 0$ and $\gamma = 0$ in (8.76), verify that (8.90) yields $\tilde{F}(w)$ in (8.86).

   b) If A is symmetric and positive definite in Example 8.9, verify that (8.90) yields $\tilde{F}(\mathbf{w}) = \left(\frac{1}{2}\right)\mathbf{w}^T A\mathbf{w} - \mathbf{w}^T\mathbf{b}$.

   c) Use (8.90) to derive the variational functional $\tilde{F}(w)$ for the two-point boundary-value problem in Section 5.3.3.

4. Set up the system $A\mathbf{c} = \mathbf{b}$ of (8.94a) and (8.94b) for Problem 3, Section 8.5.1.

5. Program the Galerkin method as in (8.94a) and (8.94b) for Problem 7, Section 8.3.1.

6. Modify the program in Problem 5 to treat $-(e^{-x}u_x)_x - (e^{-x}u_y)_y + e^{-x}u = e^{-x-y}$, $0 < x, y < 1$, $u(x, 0) = x$, $u(0, y) = 0$, $u_x(1, y) + u(1, y) - 2e^{-y} = 0$, $u_y(x, 1) + u(x, 1) = 0$. [Answer: $u(x, y) = xe^{-y}$.]

Appendix: Optimization

A.1 INTRODUCTION

Optimization is a broad topic that treats some very modern and practical problems. These range from basic problems, such as a store manager's selection of products to maximize profits or a farmer's choice of crops, fertilizers, and pesticides to maximize returns, to modern control theory problems, such as determining a trajectory for a satellite rendezvous that uses a minimum amount of fuel. Optimization is thus a modern and ongoing topic of practical research. As one might suspect, however, such a broad and powerful tool is quite deep theoretically and full of pitfalls for one who tries to use it without a sound knowledge of its basic principles. We must therefore limit ourselves in this text to a brief and somewhat superficial survey of certain optimization techniques, with a clear warning that caution must be exercised in most cases. Some of the material in this chapter depends on the use of vectors and vector norms (Chapter 2); the reader should be familiar also with Newton's method in several variables (Chapter 4). The material on linear programming in Section A.5, however, is primarily self-contained and can be read without these prerequisites.

A.2 EXTENSIONS OF RESULTS FROM CALCULUS

We shall first consider the problem of finding the relative maxima or minima (relative extrema) of a real-valued function of n variables, $f(x_1, x_2, \ldots, x_n) \equiv f(\mathbf{x})$: $R^n \to R$ where $\mathbf{x} \equiv (x_1, x_2, \ldots, x_n)^T$. A vector $\mathbf{x}^*$ is a *relative maximum* for f on a domain $D \subseteq R^n$ if there exists an open sphere S centered at $\mathbf{x}^*$ such that $f(\mathbf{x}^*) > f(\mathbf{x})$ for every $\mathbf{x}$ in $D \cap S$. [Note that $f(\mathbf{x}^*) > f(\mathbf{x})$ is well defined since $f(\mathbf{x}^*)$ and $f(\mathbf{x})$ are real numbers.] A similar definition holds for a relative minimum. The problem of finding extrema is difficult in terms of both theory and computation. Letting $z = f(x_1, x_2, \ldots, x_n)$, we must find the relative minima and maxima for z as the vector $\mathbf{x} = (x_1, x_2, \ldots, x_n)^T$ ranges through some region D in R^n. For $n = 2$, $z = f(x_1, x_2)$ can be graphed as a surface above the x_1x_2-plane (see Fig. A.1). We often use this model for our intuition and think of $z = f(x_1, x_2, \ldots, x_n)$ as a surface in R^{n+1} although we realize that in the graph the replacement of the x_1x_2-plane by R^n is somewhat artificial (again see Fig. A.1).

Most often we like to visualize the surface $z = f(x_1, x_2, \ldots, x_n)$ in terms of a topographical map in which the mountain peaks represent the maxima and the valley bottoms represent the minima. We must be careful to keep in mind that the "surfaces" for $n > 2$ can be much more complicated than the familiar surfaces in $n = 2$. We must also realize that the techniques we develop probably will locate only relative extrema, and we hope that the absolute extrema are among the relative extrema that we are able to locate. This situation is often likened to a person walking in dense fog in mountainous terrain. The bottom of a valley can be located by a process of continuous descent, but the path to the minima may be far from optimal, and furthermore the person has no idea of whether the surrounding valleys (if any exist) are deeper.

In order to gain more insight and a better intuitive feeling, let us consider the basic problem in just one variable, that is, $n = 1$, and see which results from calculus extend to higher dimensions. Readers should recall that if they wish to minimize or maximize a real-valued function, $f(x)$, for $a \le x \le b$, then the typical procedure is to find the derivative, $f'(x)$, and then values x^* such that $f'(x^*) = 0$. Even in this seemingly elementary case we can encounter many theoretical and practical difficulties. First of all, we must know that $f'(x)$ exists for $a < x < b$ and be able to calculate it (which we may or may not be able to do in closed form). Next we must find the zeros of $f'(x)$; as we have seen in Chapter 4, this task can be formidable. Given such a zero x^* (a *critical point*), we must still determine whether it represents a minimum, a maximum, or an inflection point. All of this time we must be aware that an extremum can exist at the endpoint of the interval, and also we must check the points where the derivative does not exist. Thus the simple task of *setting the derivative equal to*

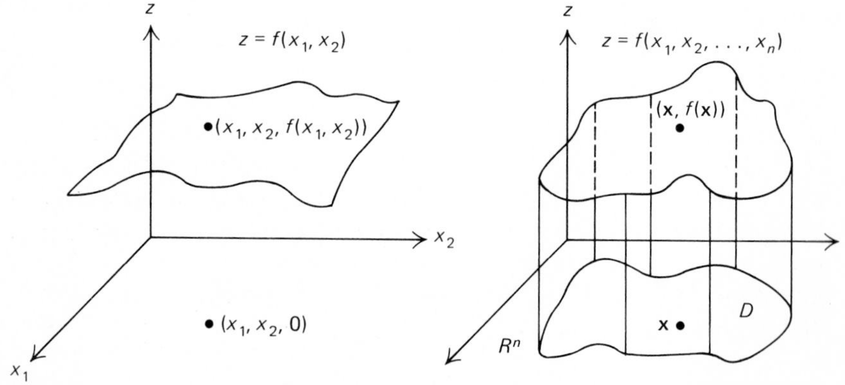

Figure A.1 Visualizing $z = f(x_1, x_2, \ldots, x_n)$.

zero may not be so "simple" after all. As an elementary example, consider the function $f(x) = |x|$ for $-1 \le x \le 1$. This function obviously has extrema at $x = -1$, 0, and 1; yet none of these represents a zero of the derivative.

Some calculus texts present methods for finding extrema of real-valued functions of two variables [see Thomas (1972)], but few texts present the problem for more than two variables. In the two-variable case, suppose $z = f(x, y)$. If the point (x^*, y^*) is an extremal point in the *interior* of some region in the xy-plane, and if tangent planes to the surface $z = f(x, y)$ exist at all points in a neighborhood of (x^*, y^*), then the tangent plane to the surface at $(x^*, y^*, f(x^*, y^*))$ will be parallel to the xy-plane. Hence the gradient vector,

$$\nabla f(x^*, y^*) \equiv \left(\frac{\partial f(x^*, y^*)}{\partial x}, \frac{\partial f(x^*, y^*)}{\partial y}\right),$$

equals (0, 0). [We recall from calculus (see Thomas [1972], pp. 683–6 for example) that the normal line of the tangent plane at $(x^*, y^*, f(x^*, y^*))$ has direction numbers

$$\left(\frac{\partial f(x^*, y^*)}{\partial x}, \frac{\partial f(x^*, y^*)}{\partial y}, -1\right).$$

Hence if $\nabla f(x^*, y^*) = (0, 0)$, the normal is parallel to the z-axis. Thus the tangent plane at this point is horizontal since (x^*, y^*) is extremal.] Once again we should be aware that $\nabla f(u, v) = \theta$ does not imply that the point (u, v) is extremal.

To see the analogy of this two-variable result to the familiar one-variable case, we recall from Chapter 4 that the derivative of the function $\mathbf{f}\colon R^n \to R^p$, that is,

$$\mathbf{f}(\mathbf{x}) \equiv \mathbf{f}(x_1, x_2, \ldots, x_n) = \begin{bmatrix} f_1(x_1, x_2, \ldots, x_n) \\ f_2(x_1, x_2, \ldots, x_n) \\ \vdots \\ f_p(x_1, x_2, \ldots, x_n) \end{bmatrix},$$

is the $(p \times n)$ Jacobian matrix

$$J(\mathbf{x}) = \left(\frac{\partial f_i(x_1, x_2, \ldots, x_n)}{\partial x_j}\right).$$

In this appendix we are considering the problem of finding the extremal points of a *real-valued* function, $f\colon R^n \to R$; so $p = 1$. In this case the Jacobian matrix is actually a $(1 \times n)$ vector that we commonly call the *gradient of f*; that is, for $p = 1$

$$\nabla f(\mathbf{x}) \equiv J(x_1, x_2, \ldots, x_n) = \left(\frac{\partial f(x_1, x_2, \ldots, x_n)}{\partial x_1}, \frac{\partial f(x_1, x_2, \ldots, x_n)}{\partial x_2}, \ldots, \frac{\partial f(x_1, x_2, \ldots, x_n)}{\partial x_n}\right). \quad \text{(A.1)}$$

Customarily in this text a vector $\mathbf{x} \in R^n$ has been regarded as a column vector; that is, $\mathbf{x}$ is $(n \times 1)$. Since the gradient $\nabla f(\mathbf{x})$ in (A.1) is a $(1 \times n)$ matrix, we must be a bit careful to avoid confusion with our usual notation. For this reason we shall hereafter in this chapter use the notation that $\mathbf{g}(\mathbf{x}) \equiv \mathbf{g}(x_1, x_2, \ldots, x_n) \equiv (\nabla f(\mathbf{x}))^{\mathrm{T}}$; that is, $\mathbf{g}(\mathbf{x})$ is an $(n \times 1)$ representation of the gradient. We shall use $\nabla f(\mathbf{x})$ and $\mathbf{g}(\mathbf{x})$ interchangeably when there is no danger of confusion. Hence, from above, we see that finding (x^*, y^*) such that $\nabla f(x^*, y^*) = (0, 0)$ is the direct vector analog of "setting the derivative equal to zero." Furthermore we can extend the procedure to higher dimensions as we shall show momentarily.

First, however, we note that we can think of the gradient $\mathbf{g}(\mathbf{x})$ in (A.1) as a function from $R^n \to R^n$; that is, for every n-tuple $\mathbf{x} = (x_1, x_2, \ldots, x_n)$, $\mathbf{g}(\mathbf{x})$ is also an n-tuple given by $\nabla f(\mathbf{x})^{\mathrm{T}}$ in (A.1). Thus it is possible for $\mathbf{g}(\mathbf{x})$ to have a derivative. If so, since $\mathbf{g}$: $R^n \to R^n$, the derivative of $\mathbf{g}(\mathbf{x})$ will be the $(n \times n)$ Jacobian matrix

$$H(\mathbf{x}) \equiv g'(\mathbf{x}) \equiv g'(x_i, x_2, \ldots, x_n) \equiv \left(\frac{\partial g_i(x_1, x_2, \ldots, x_n)}{\partial x_j}\right) \qquad \text{(A.2a)}$$

for $1 \leq i \leq n$ and $1 \leq j \leq n$. Since it is the derivative of $\mathbf{g}(\mathbf{x})$, $H(\mathbf{x})$ must also be the second derivative of $f(\mathbf{x})$ and is called the *Hessian* of $f(\mathbf{x})$. As we shall see, the Hessian will provide us with an analog of the second derivative test. Writing $H(\mathbf{x})$ in terms of $f(\mathbf{x})$, we have, since

$$g_i(\mathbf{x}) = \frac{\partial f(\mathbf{x})}{\partial x_i},$$

that

$$H(\mathbf{x}) \equiv f''(\mathbf{x}) = \left(\frac{\partial^2 f(x_1, x_2, \ldots, x_n)}{\partial x_i\, \partial x_j}\right). \qquad \text{(A.2b)}$$

In the single-variable case, a differentiable function can have an extremum at the endpoint of an interval without having its derivative equal to zero there. Similarly we must be careful also of boundary points of regions in higher-dimensional problems. Thus in this section we shall consider derivative techniques only for "nonboundary" points. Loosely speaking, we say that a vector $\mathbf{c}$ in a region D of R^n is an *interior point of D* if there exists a number ε and a sphere of radius $\varepsilon > 0$ and centered at $\mathbf{c}$ such that the sphere is contained in D. For example, if D is the unit disk in the xy-plane, $D = \{(x, y): x^2 + y^2 \leq 1\}$, then the point (x, y) is interior to D only if $x^2 + y^2 < 1$. Intuitively if (x, y) is anywhere inside the unit circle, $x^2 + y^2 = 1$, we can draw a sphere centered at (x, y) with a radius small enough that all points within this sphere lie in D.

With these preliminaries, we are now able to state (without proof) multivariable analogs of the familiar calculus derivative procedures for finding extrema of functions of a single variable.

Theorem A.1

Let $f(\mathbf{x})$ map D into the reals where $D \subseteq R^n$. If $\mathbf{c}$ is an interior point of D at which f is differentiable and at which f has a relative extremum, then $\nabla f(\mathbf{c})^T \equiv \mathbf{g}(\mathbf{c}) = \theta$.

Now if $\mathbf{c}$ is a relative minimum of $f(\mathbf{x})$, then $\mathbf{c}$ is a relative maximum of the function $(-f(\mathbf{x}))$. Thus a minimization problem can always be viewed as a maximization problem and vice versa. We do, however, need to be concerned with inflection or saddle points, which seem to be more prevalent and harder to distinguish in higher-dimensional problems. (As in the familiar single-variable case, an *inflection* or *saddle* point is a vector where the derivative equals zero but is not a relative extremum.) As a simple example, let $n = 3$, and let $f(x_1, x_2, x_3) = x_1x_2x_3$. Then $\nabla f(\mathbf{x}) = (x_2x_3, x_1x_3, x_1x_2)$ and $\mathbf{x} = \theta$ is a critical point of $f(\mathbf{x})$. However since $f(\mathbf{x})$ assumes both positive and negative values in every neighborhood of the origin, then $\mathbf{x} = \theta$ must be a saddle point. Another relatively simple example of a saddle point is the origin for the familiar hyperbolic paraboloid $z = x^2/a^2 - y^2/b^2$.

To determine whether a critical point is an extremal point or a saddle point, we have the following multidimensional analog of the second derivative test.

Theorem A.2

Let f: $D \to R$, $D \subseteq R^n$, have continuous second partial derivatives in the neighborhood of a critical point $\mathbf{c}$ in R^n. If the second derivative (the Hessian), $H(\mathbf{c})$ of (A.2b) satisfies

1. $\mathbf{x}^T H(\mathbf{c})\mathbf{x} > 0$ for all $\mathbf{x} \neq \theta$ in R^n, then $\mathbf{c}$ is a relative minimum;
2. $\mathbf{x}^T H(\mathbf{c})\mathbf{x} < 0$ for all $\mathbf{x} \neq \theta$ in R^n, then $\mathbf{c}$ is a relative maximum.

Finally if $\mathbf{x}^T H(\mathbf{c})\mathbf{x}$ has both positive and negative values as $\mathbf{x}$ ranges through R^n, then $\mathbf{c}$ is a saddle point.

If $(\partial^2 f(\mathbf{c}))/(\partial x_i\, \partial x_j) = (\partial^2 f(\mathbf{c}))/(\partial x_j\, \partial x_i)$ for all i and j (as is implied by the hypotheses above), then $H(\mathbf{c})$ in (A.2b) is symmetric. If $\mathbf{x}^T H(\mathbf{c})\mathbf{x} > 0$ (<0) for all $\mathbf{x} \neq \theta$ in R^n, then we recall from Chapters 2 and 3 that $H(\mathbf{c})$ is positive (negative) definite, which is equivalent (see Problem 7) to the eigenvalues of $H(\mathbf{c})$ being all positive (negative). In these cases, $H(\mathbf{c})$ is nonsingular.

A.3 DESCENT METHODS

Since maximization problems are easily transposed (as above) into minimization problems, we shall concentrate without loss of generality on the latter. We see by Theorem A.1 that we should try to find the zeros of $f'(\mathbf{x}) \equiv \nabla f(\mathbf{x}) \equiv \mathbf{g}(\mathbf{x})^T$ and then use Theorem A.2 to test them. Since $\nabla f(\mathbf{x})^T \equiv \mathbf{g}(\mathbf{x})$: $R^n \to R^n$, we immediately recall that Newton's method for several variables (Section 4.5) is specifically designed to solve a problem like $\nabla f(\mathbf{x})^T = \theta$. Since $g'(\mathbf{x}) = H(\mathbf{x})$ as in A.2b, then Newton's method [formula (4.51)] for this problem becomes

$$\mathbf{x}_0, \text{ an initial guess for the minimum point,}$$
$$\mathbf{x}_{i+1} = \mathbf{x}_i - (H(\mathbf{x}_i))^{-1}\mathbf{g}(\mathbf{x}_i), \qquad i \geq 0, \tag{A.3a}$$

or equivalently solve

$$H(\mathbf{x}_i)\mathbf{z}_i = -\mathbf{g}(\mathbf{x}_i), \qquad \text{where} \qquad \mathbf{z}_{i+1} \equiv \mathbf{x}_{i+1} - \mathbf{x}_i, \qquad i \geq 0. \tag{A.3b}$$

We immediately see difficulties with this approach. The computation of $H(\mathbf{x}_i)$ can often be quite cumbersome or impossible to perform for practical purposes, or $H(\mathbf{x}_i)$ may be singular or may not even exist. We note also that the sequence $\{\mathbf{x}_i\}$ in (A.3a) may not converge to a zero of $\mathbf{g}(\mathbf{x})$ unless the initial guess $\mathbf{x}_0$ is sufficiently close to that zero. Thus we see that a good strategy for finding a zero of $\mathbf{g}(\mathbf{x})$ might be to develop a rather crude method for obtaining an initial approximation to that zero and then use Newton's method (or a modification of it to ease the computational difficulties) to refine that initial approximation. [This suggestion is exactly analogous to our suggestions for root finding in Chapter 4, in which we advocated first using crude techniques to isolate zeros and then using Newton's method to converge quickly (quadratically) to them.] Thus we shall now turn to the task of developing this package, consisting of a technique called *steepest descent* to give first approximations to local minima and a modification of Newton's method to improve computational efficiency.

A.3.1. Steepest Descent

To describe intuitively the method of steepest descent, we return to our analogy of a person trying to descend to the bottom of a valley in heavy fog. At any particular point, the natural course would be to descend in the direction in which the slope of the land is steepest and to continue in that direction as long as descent continues. When a point is reached where the path would start to ascend if continued, the person stops, chooses a new direction of steepest descent, and repeats the process. If the valley were shaped like the bottom of a sphere, direction would never need to be changed and the bottom would be reached along the original path. However if the valley were shaped like a long narrow kidney bean and the descent were started at the end opposite the true bottom, it could take a considerable time along a very irregular path to reach the bottom. This course would clearly not be an optimal way of reaching the bottom, but the person would at least be assured of eventually getting arbitrarily close to the goal (under moderate assumptions about the walls of the valley).

We wish to develop our minimization technique to mathematically model the process above. In doing so, we must be able to answer two questions at each step of the process. (1) At any given point $\mathbf{x}$ in R^n, which direction represents the direction of steepest descent? (2) How far should one go in that direction before changing directions?

The first question is difficult to analyze for $n > 2$ because any actual graph of the surface is somewhat artificial. For $n = 2$, however, we can answer this question with basic calculus. Since we wish to proceed in the direction in which the slope is steepest, we must be able to examine the slope in any direction. This concept is precisely that of *directional derivative,* which we shall now briefly review. Given a surface $z = f(x, y)$, a fixed point $\mathbf{P}_0 \equiv (x_0, y_0, f(x_0, y_0))$ on the surface, and a vertical plane Π containing $\mathbf{P}_0$ where Π makes an angle ϕ with the x-axis, then the plane and surface have a curve of intersection that also contains $\mathbf{P}_0$. If the surface is sufficiently smooth, this curve will have at $\mathbf{P}_0$ a tangent, which is called the *directional derivative of* $z = f(x, y)$ *in the direction* ϕ *at the point* $\mathbf{P}_0$ (see Fig. A.2).

Since Π is a vertical plane, it has the same equation as its projection line ℓ in the xy-plane. If s represents the distance along the line ℓ from the point $(x_0, y_0, 0)$, then ℓ can be represented by the parametric equations:

$$x = x_0 + s\cos(\phi), \qquad y = y_0 + s\cos\left(\frac{\pi}{2} - \phi\right) = y_0 + s\sin(\phi). \quad \text{(A.4)}$$

The slope of the tangent line (the directional derivative) is therefore given by

$$\frac{df}{ds} = \lim_{s \to 0} \frac{f(x_0 + s\cos(\phi), y_0 + s\sin(\phi)) - f(x_0, y_0)}{s}. \quad \text{(A.5a)}$$

By the chain rule and the equations (A.4) for ℓ, (A.5a) becomes

$$\frac{df}{ds} = \frac{\partial f}{\partial x}\frac{dx}{ds} + \frac{\partial f}{\partial y}\frac{dy}{ds} \equiv \frac{\partial f(x_0, y_0)}{\partial x}\cos(\phi) + \frac{\partial f(x_0, y_0)}{\partial y}\sin(\phi). \quad \text{(A.5b)}$$

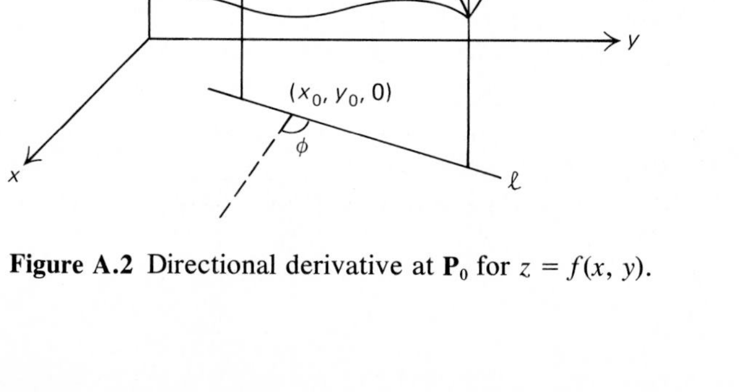

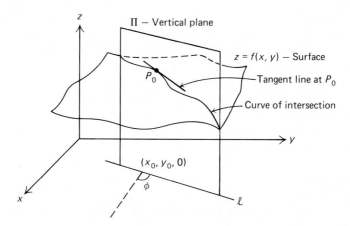

Figure A.2 Directional derivative at $\mathbf{P}_0$ for $z = f(x, y)$.

Letting $\mathbf{u} = (\cos(\phi), \sin(\phi))$, we can express (A.5b) symbolically as the inner product

$$\frac{df}{ds} = \langle \nabla f(x_0, y_0), \mathbf{u} \rangle. \tag{A.5c}$$

(To avoid unnecessary theoretical technicalities, we shall adopt the following notation. If $\mathbf{x}$ and $\mathbf{y}$ are m-tuples with components $\{x_j\}_{j=1}^m$ and $\{y_j\}_{j=1}^m$, respectively, we shall regard the inner product $\langle \mathbf{x}, \mathbf{y} \rangle$ as

$$\langle \mathbf{x}, \mathbf{y} \rangle = x_1 y_1 + x_2 y_2 + \cdots + x_m y_m$$

regardless of whether x and/or y are row vectors or column vectors.)

Now we wish to find the angle ϕ_0 that yields the direction for which $|df/ds|$ is maximized, i.e., where the slope is steepest. We can do this by differentiating (A.5b) with respect to ϕ, setting it equal to zero, and solving for ϕ_0. The steps are as follows:

$$\frac{d}{d\phi}\left(\frac{df}{ds}\right) = \frac{\partial f(x_0, y_0)}{\partial x}(-\sin(\phi_0)) + \frac{\partial f(x_0, y_0)}{\partial y}(\cos(\phi_0)) = 0.$$

Thus with

$$\lambda_1 = \frac{\partial f(x_0, y_0)}{\partial x} \quad \text{and} \quad \lambda_2 = \frac{\partial f(x_0, y_0)}{\partial y},$$

we see $\tan(\phi_0) = \lambda_2/\lambda_1$. For $r = \sqrt{\lambda_1^2 + \lambda_2^2}$, it follows that $\cos(\phi_0) = \lambda_1/r$ and $\sin(\phi_0) = \lambda_2/r$; so in (A.5c) the direction vector $\mathbf{u}$ that is in the direction of maximum increase is $\mathbf{u} = (1/r)\, \nabla f(x_0, y_0)$. Thus $\mathbf{u}$ is in the direction of the gradient when the directional derivative is maximized. Since the gradient represents the direction of the outward normal, and we wish the direction of steepest *descent*, we thus see that this direction is given by the negative gradient, $-\nabla f(x_0, y_0)$.

We pause here to give an intuitive argument that this result is true in higher dimensions as well. Suppose that we are at the vector $\tilde{\mathbf{x}}$ and wish to move to a vector of the form $\mathbf{x} = \tilde{\mathbf{x}} + \alpha \mathbf{e}$ where $\alpha > 0$ is small and $\mathbf{e}$ is a unit vector determining the direction in which we wish to move so that the object function f decreases most rapidly. If α is sufficiently small and ∇f is continuous, then by the mean-value theorem in several variables (see Section 4.5) we have the approximate equality

$$f(\mathbf{x}) \approx f(\tilde{\mathbf{x}}) + \langle \mathbf{x} - \tilde{\mathbf{x}}, \nabla f(\tilde{\mathbf{x}}) \rangle.$$

[To have equality, we should use $\nabla f(\mathbf{z})$ rather than $\nabla f(\tilde{\mathbf{x}})$ where $\mathbf{z}$ is on the line between $\mathbf{x}$ and $\tilde{\mathbf{x}}$.] If one defines the unit vector $\mathbf{u}$ to be $\mathbf{u} \equiv \nabla f(\tilde{\mathbf{x}})/\|\nabla f(\tilde{\mathbf{x}})\|_2$ and notes that $(\mathbf{x} - \tilde{\mathbf{x}}) = \alpha \mathbf{e}$, the approximation above becomes

$$f(\mathbf{x}) - f(\tilde{\mathbf{x}}) \approx \alpha \|\nabla f(\tilde{\mathbf{x}})\|_2 \langle \mathbf{e}, \mathbf{u} \rangle.$$

Thus the direction of maximum change to make the object function smaller forces $\langle \mathbf{e}, \mathbf{u} \rangle$ to be as large as possible negatively. From the Cauchy-Schwarz inequality (Chapter 2), $|\langle \mathbf{e}, \mathbf{u} \rangle| \le \|\mathbf{e}\|_2 \, \|\mathbf{u}\|_2$ with equality holding if and only if $\mathbf{u}$ is a multiple of $\mathbf{e}$. Thus to make $\langle \mathbf{e}, \mathbf{u} \rangle$ as large as possible negatively, we must choose $\mathbf{e} = -\mathbf{u}$, or $\mathbf{e}$ is in the direction of the negative gradient, $-\nabla f(\mathbf{x})$.

Thus we have the answer to our first question. If we are at a point $\mathbf{x}$ in R^n and wish to move in the direction of steepest descent of the function $f(\mathbf{x})$, we proceed in the direction of the negative gradient,

$$-\nabla f(\mathbf{x})^{\mathrm{T}} \equiv -\nabla f(x_1, x_2, \ldots, x_n)^{\mathrm{T}} \equiv -\mathbf{g}(x_1, x_2, \ldots, x_n).$$

Thus our algorithm for steepest descent will generate a sequence of vectors $\{\mathbf{x}_k\}$, with $\mathbf{x}_0$ an initial guess for the minimum point, by

$$\mathbf{x}_{k+1} = \mathbf{x}_k - \alpha_k \mathbf{g}_k, \qquad k \ge 0, \tag{A.6}$$

where $\mathbf{g}_k \equiv \mathbf{g}(\mathbf{x}_k)$ and where α_k is a positive constant telling how far to go in the direction $(-\mathbf{g}_k)$ in order to proceed from $\mathbf{x}_k$ to $\mathbf{x}_{k+1}$.

Thus, choosing a value for α_k for each k is equivalent to answering our second question. To see how this might be done, suppose that $\mathbf{x}_k$ is not the minimal point. Then for sufficiently small α, $f(\mathbf{x}_k - \alpha \mathbf{g}_k) < f(\mathbf{x}_k)$ since $-\alpha \mathbf{g}_k$ is in the direction of steepest descent from $\mathbf{x}_k$. Thus we wish to choose $\alpha = \alpha_k > 0$ to minimize the expression $f(\mathbf{x}_k - \alpha \mathbf{g}_k) \equiv h(\alpha)$. Since for a given k, $\mathbf{x}_k$ and $\mathbf{g}_k$ are fixed, $h(\alpha)$ is actually a real-valued function of the real variable α. Thus $h(\alpha)$ can be minimized by the familiar methods of single-variable calculus; i.e., set

$$h'(\alpha) \equiv \frac{d}{d\alpha} f(\mathbf{x}_k - \alpha \mathbf{g}_k) = 0 \tag{A.7}$$

and solve for α, usually taking the smallest positive value of α that satisfies (A.7). The computation of the derivative in (A.7) and the solution for its zeros may be so cumbersome as to be impractical. In this case we can resort to cruder methods such as calculating $h(\alpha)$ for various values of $\alpha > 0$. For small α, $h(\alpha)$ should initially decrease. (If it doesn't, α has not been chosen small enough initially.) The values $h(\alpha)$ will eventually start to increase as α gets larger; so after we locate two values of α (the first to the left of α_k and the second to the right of α_k), we can close in on α_k by a bisection technique, which proceeds exactly as in Chapter 4 except that now we are searching for a minimum of $h(\alpha)$ instead of for one of its zeros. As an alternative to the bisection approach, we can instead search until we find three values of α where $h(\alpha)$ is smaller at the middle α then at the other two. We can then fit a quadratic interpolating polynomial through these three values of $h(\alpha)$, minimize the quadratic, and accept the value of α where the minimum occurs as α_k in (A.6). This process then completes our description of how to use (A.6), the steepest descent algorithm for finding the relative minimum of a function f: $R^n \to R$.

Several variations of the method of (A.6), each called a *gradient method*, will not be pursued here. We wish to reiterate that the steepest descent method is often used to obtain a first approximation for a faster, but more local technique such as the Newton-type methods we will present in the next section. There are some problems, however, in which it is advantageous to use steepest descent as the entire procedure.

EXAMPLE A.1. Let us now reconsider the problem of an overdetermined linear system of equations (Chapter 2) in light of the material of this appendix. Let $m > n$, and let A be an $(m \times n)$ matrix and $\mathbf{b}$ an $(m \times 1)$ vector. Our problem is to find an $(n \times 1)$ vector $\mathbf{x} \equiv (x_1, x_2, \ldots, x_n)^T$ such that $f(\mathbf{x}) \equiv \|A\mathbf{x} - \mathbf{b}\|_2^2 = \mathbf{x}^T A^T A\mathbf{x} - 2\mathbf{x}^T A^T\mathbf{b} + \mathbf{b}^T\mathbf{b}$ is minimized. We leave to the reader to verify that $\nabla f(\mathbf{x})^T = 2A^TA\mathbf{x} - 2A^T\mathbf{b}$, and so $\nabla f(\mathbf{x})^T = \theta$ when $A^TA\mathbf{x} = A^T\mathbf{b}$. Thus we arrive at the "normal" equations, just as in Chapter 2. We can easily see also that $f''(\mathbf{x}) \equiv H(\mathbf{x}) = 2A^TA$; and so by Theorem A.2, the zero of $\nabla f(\mathbf{x})$ is unique if A^TA is positive definite. [From Chapter 3, A^TA is positive definite if and only if A has n linearly independent columns. We note in the special problem of obtaining the polynomial of degree $(n - 1)$ that provides the best least-squares fit to m data points, A is the $(m \times n)$ Vandermonde-type matrix that has rank n, and so the polynomial is unique.)

If we choose to solve this problem by the method of steepest descent using (A.6), instead of directly solving the normal equations (such as by Gauss elimination), then for each k, we must minimize

$$f(\mathbf{x}_k - \alpha\mathbf{g}_k) = \mathbf{x}_k^T A^T A\mathbf{x}_k - 2\alpha\mathbf{g}_k^T A^T A\mathbf{x}_k + \alpha^2\mathbf{g}_k^T A^T A\mathbf{g}_k - 2\mathbf{x}_k^T\mathbf{b} + 2\alpha\mathbf{g}_k^T\mathbf{b} + \mathbf{b}^T\mathbf{b} \equiv h(\alpha)$$

as a function of α. Now,

$$h'(\alpha) = -2\mathbf{g}_k^T A^T A\mathbf{x}_k + 2\alpha\mathbf{g}_k^T A^T A\mathbf{g}_k + 2\mathbf{g}_k^T\mathbf{b}.$$

Recalling that $\mathbf{g}_k = 2A^TA\mathbf{x}_k - 2A^T\mathbf{b}$ and setting $h'(\alpha_k) = 0$ yield $\alpha_k = \mathbf{g}_k^T\mathbf{g}_k/2\mathbf{g}_k^T A^T A\mathbf{g}_k$ for use in the steepest descent algorithm (A.6).

Suppose that the rows of the normal equations, $A^TA\mathbf{x} = A^T\mathbf{b}$, are divided by appropriate constants so that the diagonal entries of A^TA are all equal to 1; that is, $A^TA\mathbf{x} = A^T\mathbf{b}$ has the equivalent form $Q\mathbf{x} = \mathbf{c}$ with $q_{ii} = 1$, $1 \le i \le n$. Suppose also that we modify the steepest descent procedure on this problem and choose $\alpha_k = 1/2$ for each k. Under this modification the gradient becomes $\mathbf{g}_k = 2Q\mathbf{x}_k - 2\mathbf{c}$, and the iteration of (A.6) becomes

$$\mathbf{x}_{k+1} = \mathbf{x}_k - \frac{1}{2}\mathbf{g}_k = \mathbf{x}_k - Q\mathbf{x}_k + \mathbf{c} = (I - Q)\mathbf{x}_k + \mathbf{c},$$

which is precisely the Jacobi iteration technique on $Q\mathbf{x} = \mathbf{c}$ (see Chapter 2).

A.3.2. Quasi-Newton Methods

We earlier pointed out some difficulties in using Newton's method (A.3a) or (A.3b) as the entire minimization process. One of these difficulties was our lack of anything beyond a local convergence theorem for the method. This problem is due mainly to the possibility that $f(\mathbf{x}_{k+1}) > f(\mathbf{x}_k)$ [where $\mathbf{x}_{k+1}$ and $\mathbf{x}_k$ are generated by (A.3a)] if $\mathbf{x}_k$ is remote from the minimum point. Thus we are led to consider a modified Newton algorithm:

$$\begin{aligned} &\mathbf{x}_0, \text{ an initial guess for the minimum point,} \\ &\mathbf{x}_{k+1} = \mathbf{x}_k - \alpha_k (H(\mathbf{x}_k))^{-1}\mathbf{g}(\mathbf{x}_k), \qquad k \geq 0, \end{aligned} \tag{A.8}$$

where $H(\mathbf{x}_k)$ and $\mathbf{g}(\mathbf{x}_k)$ are again the Hessian matrix and the transpose of gradient vector, respectively. The scalar α_k is again chosen in the same manner as in the method of steepest descent, (A.6); that is, α_k is chosen to minimize $k(\alpha) \equiv f(\mathbf{x}_k - \alpha(H(\mathbf{x}_k))^{-1}\mathbf{g}(\mathbf{x}_k))$. Once again, since $\mathbf{x}_k$ is known, $k(\alpha)$ is a real-valued function of the real variable α. Hence α_k can be obtained from the solution of

$$k'(\alpha) \equiv \frac{d}{d\alpha} f(\mathbf{x}_k - \alpha H_k^{-1}\mathbf{g}_k) = 0; \tag{A.9}$$

and the remarks that follow (A.7) and concern the computation of α_k pertain to this equation as well.

We still have the usually formidable computational problems of calculating H_k^{-1} at each step. We now turn to quasi-Newton methods, which use an initial estimate and information from previous calculations to generate an estimate Q_k to the inverse Hessian H_k^{-1} at each step. The iteration for any such technique then becomes

$$\mathbf{x}_{k+1} = \mathbf{x}_k - \alpha_k Q_k \mathbf{g}_k \tag{A.10}$$

where α_k is again chosen to minimize f in the direction $-Q_k\mathbf{g}_k$ [see (A.7) and (A.9)].

The theory necessary to develop these methods rigorously is beyond the scope of this text; so we will use a more intuitive approach. In order to see how the approximations Q_k to the inverse Hessian H_k^{-1} might be formed, let us suppose that $f(\mathbf{x})$: $R^n \to R$ can be written as an n-dimensional quadratic form

$$f(\mathbf{x}) = \gamma + \mathbf{b}^{\mathrm{T}}\mathbf{x} + \frac{1}{2}\mathbf{x}^{\mathrm{T}}A\mathbf{x} \tag{A.11}$$

where γ is a scalar constant, $\mathbf{b}$ is a constant vector in R^n, and A is a symmetric positive definite $(n \times n)$ constant matrix. If $f(\mathbf{x})$ has this form, then the gradient $\mathbf{g}(\mathbf{x})$ is given by $\mathbf{g}(\mathbf{x}) = \mathbf{b} + A\mathbf{x}$ and the Hessian is the constant matrix $H(\mathbf{x}) = A$. [By Theorems A.1 and A.2, the unique critical point $\mathbf{x}^* = -A^{-1}\mathbf{b}$ is a minimum since $f''(\mathbf{x}^*) = A$ is positive definite. Thus for the simple function $f(\mathbf{x})$ given in (A.11), we already have the minimum. However, we are using (A.11) only as a model to lead us to minimization techniques for more complicated functions.]

For the function in (A.11), by the mean-value theorem we have that

$$\mathbf{g}(\mathbf{x}_{k+1}) = \mathbf{g}(\mathbf{x}_k) + H(\mathbf{x}_k)(\mathbf{x}_{k+1} - \mathbf{x}_k) \tag{A.12}$$

where (A.12) is an identity since $H(\mathbf{x}_k)$ is the constant matrix A. Thus we see that the inverse Hessian has the property that

$$H_k^{-1}(\mathbf{g}_{k+1} - \mathbf{g}_k) = \mathbf{x}_{k+1} - \mathbf{x}_k. \tag{A.13}$$

If we have an approximation Q_k to the inverse Hessian [as in the iteration defined by (A.10)], it seems plausible that we might choose an $(n \times n)$ matrix D_k to *correct* Q_k in the sense that the equation

$$(Q_k + D_k)(\mathbf{g}_{k+1} - \mathbf{g}_k) = \mathbf{x}_{k+1} - \mathbf{x}_k \tag{A.14}$$

is satisfied when $\mathbf{x}_{k+1}$ is given by (A.10). Setting $Q_{k+1} = Q_k + D_k$, we then proceed with the next step of (A.10) and go in the direction $-Q_{k+1}\mathbf{g}_{k+1}$.

This procedure condenses into the following algorithm for minimizing $f(\mathbf{x})$.

1. Choose an initial estimate $\mathbf{x}_0$ and an initial approximation Q_0 to the inverse Hessian. The usual choice for Q_0 is the identity matrix I; thus we proceed initially in the direction of the negative gradient.
2. For any k, choose $\alpha_k > 0$ to minimize $f(\mathbf{x}_k - \alpha Q_k \mathbf{g}_k)$ as a function of α.
3. Set $\mathbf{x}_{k+1} = \mathbf{x}_k - \alpha_k Q_k \mathbf{g}_k$.
4. Compute $\mathbf{g}_{k+1} \equiv \nabla f(\mathbf{x}_{k+1})$ and determine a matrix D_k so that
$$(Q_k + D_k)(\mathbf{g}_{k+1} - \mathbf{g}_k) = \mathbf{x}_{k+1} - \mathbf{x}_k.$$
5. Let $Q_{k+1} = Q_k + D_k$, and return to step (2).

As a further motivation for this procedure, we note that when we are sufficiently near a minimum of $f(x)$, then $f(x)$ can be well approximated by a quadratic form. Furthermore it can be shown that if the procedure outlined above is followed in the case that $f(\mathbf{x})$ *is* a quadratic form, then the minimum is *found* in n steps or less. Thus when we are near a minimum [so that $f(\mathbf{x})$ is very nearly a quadratic form], we expect that convergence will be rapid.

Several possible choices for the correction matrix D_k cannot be derived in this text. However, for a variety of substantial reasons, Shanno (1970) advocates the use of (A.15) below where $\mathbf{r}_k = \mathbf{x}_{k+1} - \mathbf{x}_k$ and $\mathbf{y}_k = \mathbf{g}_{k+1} - \mathbf{g}_k$. [In the following formulas the reader should be aware that if $\mathbf{w}$ is any vector in R^n (an $(n \times 1)$ column vector), then $(\mathbf{w}\mathbf{w}^T)$ is an $(n \times n)$ matrix, but $\mathbf{w}^T\mathbf{w}$ is a scalar.]

$$D_k = t\frac{\mathbf{r}_k\mathbf{r}_k^T}{\mathbf{r}_k^T\mathbf{y}_k} + \frac{((1-t)\mathbf{r}_k - Q_k\mathbf{y}_k)((1-t)\mathbf{r}_k - Q_k\mathbf{y}_k)^T}{((1-t)\mathbf{r}_k - Q_k\mathbf{y}_k)^T\mathbf{y}_k}. \tag{A.15}$$

In (A.15), t is an arbitrary scalar parameter. If $t = 0$, (A.15) yields

$$D_k = \frac{(\mathbf{r}_k - Q_k\mathbf{y}_k)(\mathbf{r}_k - Q_k\mathbf{y}_k)^T}{(\mathbf{r}_k - Q_k\mathbf{y}_k)^T\mathbf{y}_k}, \tag{A.16a}$$

which is known as the Barnes-Rosen method. If $t = 1$, (A.15) yields

$$D_k = \frac{\mathbf{r}_k\mathbf{r}_k^{\mathrm{T}}}{\mathbf{r}_k^{\mathrm{T}}\mathbf{y}_k} - \frac{Q_k\mathbf{y}_k\mathbf{y}_k^{\mathrm{T}}Q_k}{\mathbf{y}_k^{\mathrm{T}}Q_k\mathbf{y}_k}, \qquad \text{(A.16b)}$$

which is known as the Fletcher-Powell method. [We leave to the reader to verify the important fact that D_k in (A.15) satisfies (A.14).] Shanno (1970) advocates other choices for t, choices which we cannot explore here. However, Fletcher-Powell is usually preferred over Barnes-Rosen and has the theoretically important property that if Q_0 is positive definite, then the successive Q_k's remain positive definite.

PROBLEMS, SECTION A.3

1. Let $\mathbf{x} = (x_1, x_2, x_3)$ and $f(\mathbf{x}) = x_1x_2^2 + x_2x_3^2 + x_3^3$. Compute the gradient and Hessian of f, and find all values of $\mathbf{x}$ for which $\nabla f(\mathbf{x}) = (0, 0, 0)$. Is the Hessian singular or nonsingular at these values?
2. Restate Theorems A.1 and A.2 componentwise for $n = 1$ and $n = 2$. Compare your results with the usual theorems from calculus in these cases.
3. Let $f_1(x_1, x_2) = x_1^4 + x_2^4$ and $f_2(x_1, x_2) = x_1^4 - x_2^4$. Determine if the origin (0, 0) is a relative extremum or saddle point for each of these two functions.
4. a) If $f(\mathbf{x}) = \|A\mathbf{x} - \mathbf{b}\|_2^2$ as in Example A.1, verify that $\nabla f(\mathbf{x})^{\mathrm{T}} = 2A^{\mathrm{T}}A\mathbf{x} - 2A^{\mathrm{T}}\mathbf{b}$ and $H(\mathbf{x}) \equiv f''(\mathbf{x}) = 2A^{\mathrm{T}}A$.

   b) Write a computer program utilizing the steepest-descent method for minimizing this $f(\mathbf{x})$.

   c) Use this program to find the best cubic least-squares polynomial fitting the data points: $(-2, 1)$, $(-1, 2)$, $(0, 1)$, $(1, 1)$, and $(2, 2)$.
5. Let $f_1(x_1, x_2) = x_1^2 - x_1 + x_2 - 1$ and $f_2(x_1, x_2) = x_1^3 - x_1^2 + 2x_1x_2 - x_1 - x_2 - 2$. Define $F(x_1, x_2) \equiv f_1(x_1, x_2)^2 + f_2(x_1, x_2)^2$. Use an initial guess of $(\xi_0, \zeta_0) \equiv (0, 0)$ and apply four iterations of the steepest-descent method toward minimizing $F(x_1, x_2)$. Note that any absolute minimum of $F(x_1, x_2)$ is a simultaneous solution of $f_1(x_1, x_2) = 0$, $f_2(x_1, x_2) = 0$. Loosely compare the amount of necessary computation in using Newton's method in two variables for the simultaneous solution of $f_1 = f_2 = 0$ versus the Newton method of this chapter in minimizing F. (This analysis should indicate why steepest descent is advocated above.)
6. a) Given a vector $\mathbf{x} \in R^n$, a function $f\colon R^n \to R$, its gradient $\mathbf{g}(\mathbf{x}) \equiv \nabla f(\mathbf{x})^{\mathrm{T}}$, and an $(n \times n)$ positive definite matrix Q, let $k(\alpha) \equiv f(\mathbf{x} - \alpha Q\mathbf{g}(\mathbf{x}))$, $\alpha > 0$. Write a computer program using a bisection technique to locate the smallest α minimizing $k(\alpha)$.

   b) Implementing the program in part (a), write a steepest-descent program for $n = 2$ and test the program on the Rosenbrock function $f(x_1, x_2) = 100(x_2 - x_1^2) + (1 - x_1)^2$ with the initial estimate $(1, -1.2)$.

c) Using a Fletcher-Powell program, repeat the problem in part (b).

7. Show that a real, symmetric matrix is positive definite if and only if all of its eigenvalues are positive.

A.4 LAGRANGE MULTIPLIERS

The minimization problems we have considered thus far are said to be *unconstrained*. This term means essentially that we have an object function $f: R^n \to R$, and we wish to minimize it without imposing any other conditions on it. There are many practical problems that arise, however, in which we want to minimize a function $f: R^n \to R$ subject to certain restraints or limits on the vectors in the domain of f that we wish to consider. Perhaps we can best illustrate this situation by example.

EXAMPLE A.2. Suppose that we wish to minimize the object function $f(\alpha, \beta, \gamma) = \alpha^2 + (\beta - 2)^2 + \gamma^2$ subject only to the constraints that we wish to minimize $f(\alpha, \beta, \gamma)$ for vectors (α, β, γ) that satisfy $g_1(\alpha, \beta, \gamma) = \alpha^2 + \beta^2 - 1 = 0$ and $g_2(\alpha, \beta, \gamma) = \alpha^2 + \gamma^2 - 1 = 0$. Geometrically, this problem is equivalent to finding the minimum distance from the curve of intersection of $g_1(\alpha, \beta, \gamma) = g_2(\alpha, \beta, \gamma) = 0$ to the point $\mathbf{P} \equiv (0, 2, 0)$ (see Fig. A.3). From the constraint equations, we have $\beta^2 = 1 - \alpha^2$ and $\gamma^2 = 1 - \alpha^2$; so the object function under these two constraints becomes

$$\begin{aligned} f(\alpha, \beta, \gamma) &= \alpha^2 + (1 - \alpha^2) - 4\sqrt{1 - \alpha^2} + 4 + (1 - \alpha^2) \\ &= -\alpha^2 - 4\sqrt{1 - \alpha^2} + 6 \equiv h(\alpha). \end{aligned}$$

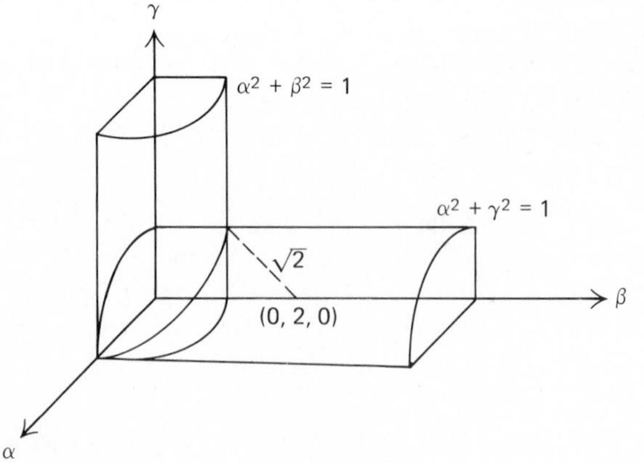

Figure A.3 Geometry of Example A.2.

Now, $h'(\alpha) = -2\alpha + 4\alpha(1 - \alpha^2)^{-1/2}$; so $h'(0) = 0$ and $\alpha = 0$ is a critical point. Furthermore, $h''(\alpha) = -2 + 4(1 - \alpha^2)^{-1/2} + 4\alpha^2(1 - \alpha^2)^{-3/2}$, and so $h''(0) = 2 > 0$, and thus $h(\alpha)$ is minimized for $\alpha = 0$ [and $h(0) = 2$]. [Therefore since $f(\alpha, \beta, \gamma)$ equals the square of the distance, the minimum distance is $\sqrt{h(0)} = \sqrt{2}$ as we can clearly see from Fig. A.3.]

We chose the example above because of its simplicity in that we were able to use the equations $g_1(\alpha, \beta, \gamma) = g_2(\alpha, \beta, \gamma) = 0$ to eliminate the variables β and γ explicitly from the object function $f(\alpha, \beta, \gamma)$ and thereby were able to minimize the object function in terms of the single variable α. More often than not, however, the constraint equations will be too complex to eliminate variables explicitly as we were able to do in the example above. For instance, if we had had $g_1(\alpha, \beta, \gamma) = e^{\alpha\beta} - \alpha\beta^2\gamma + \cos(\gamma) = 0$ and $g_2(\alpha, \beta, \gamma) = \cos(\alpha\beta) - \sin(\beta\gamma) = 0$, we would probably not have been able to eliminate β and γ explicitly from the object function. However, the two simultaneous equations in three variables, $g_1(\alpha, \beta, \gamma) = 0$ and $g_2(\alpha, \beta, \gamma) = 0$, can still be envisioned as two restrictions (the equations) in three free parameters (the three variables). Using two of the free parameters to fulfill the two restrictions, we are still left with one free variable with which we can theoretically minimize the object function. Generalizing this notion to higher dimensions, suppose that we have an object function of n variables, $f(x_1, x_2, \ldots, x_n)\colon R^n \to R$, to minimize with respect to m constraint equations; $g_i(x_1, x_2, \ldots, x_n) = 0$, $1 \le i \le m$; with $m < n$ (fewer restrictions than free parameters). We can visualize the m constraint equations as implicitly eliminating m of the variables of the object function, and our problem is then equivalent to minimizing an *unconstrained* functional of $(n - m)$ variables for which purpose the theory and techniques of the previous sections are applicable. For instance, in the example above we had $m = 2$ constraint equations for an object function of $n = 3$ variables, and the problem reduced to minimizing an object function of $(n - m) = (3 - 2) = 1$ variable; and we used the familiar minimization techniques of calculus. [In order to make the statements above completely rigorous, we must call upon the implicit function theorem (see Bartle [1964], p. 260). Such rigor would go far beyond the level of this text, however; and so the point will be pursued no further than the intuitive approach above.]

Since we cannot usually expect to be able to use the m constraint equations explicitly to eliminate m variables in the object function (as we did in the example above), we must search for another procedure. An elegant, yet still very practical alternate method was developed by Lagrange. In Lagrange's method we introduce $(m + 1)$ additional variables, $\{\lambda_0, \lambda_1, \lambda_2, \ldots, \lambda_m\}$, called *Lagrange multipliers*, and consider the new function

$$F(x_1, x_2, \ldots, x_n) \equiv \lambda_0 f(x_1, x_2, \ldots, x_n) + \sum_{i=1}^{m} \lambda_i g_i(x_1, x_2, \ldots, x_n). \qquad \text{(A.17)}$$

Loosely speaking, if $\mathbf{x}^* = (x_1^*, x_2^*, \ldots, x_n^*)$ is a minimum point of the object function subject to the constraint equations, then there exist values $\lambda_0^*, \lambda_1^*, \ldots, \lambda_m^*$ for the Lagrange multipliers such that

$$\lambda_0^* \nabla f(x_1^*, x_2^*, \ldots, x_n^*) + \sum_{i=1}^{m} \lambda_i^* \nabla g_i(x_1^*, x_2^*, \ldots, x_n^*) = \theta. \qquad \text{(A.18)}$$

Furthermore if the m vectors $\{\nabla g_i(\mathbf{x}^*)\}_{i=1}^m$ are linearly independent, then we may take $\lambda_0^* = 1$, and the remaining multipliers are unique and not all zero. In the following material we shall assume that this condition is true, but the reader must be aware of the possibility that it is not; in that case the variable λ_0 must be incorporated in the equations. Thus $\mathbf{x}^*$ can be found by solving the following $(n + m)$ equations in the $(n + m)$ unknowns, $\{x_1, x_2, \ldots, x_n, \lambda_1, \lambda_2, \ldots, \lambda_m\}$:

$$\nabla f(x_1, x_2, \ldots, x_n) + \sum_{i=1}^{m} \lambda_i \nabla g_i(x_1, x_2, \ldots, x_n) = \theta.$$

and (A.19)

$$g_i(x_1, x_2, \ldots, x_n) = 0, \qquad 1 \le i \le m.$$

Once again we emphasize that the rigorous derivation of the equations of (A.19) or even the conditions under which they will yield a solution to our problem are beyond the scope of this text; thus the reader must be very cautious in their utilization. We note also that (A.19) usually represents a system of $(n + m)$ *nonlinear* simultaneous equations in the $(n + m)$ unknowns, and so a nonlinear method for their solution must be employed. We could possibly use Newton's method in $(n + m)$ variables to solve this system, but once again this process involves computing the Hessian at each step. An alternate way to attempt to solve this system is to define the functional

$$F(\mathbf{y}) \equiv F_1(\mathbf{y})^2 + F_2(\mathbf{y})^2 + \cdots + F_{n+m}(\mathbf{y})^2$$

where each $F_j(\mathbf{y})$ is the jth component of the equations (A.19) where $\mathbf{y} \equiv (x_1, x_2, \ldots, x_n, \lambda_1, \lambda_2, \ldots, \lambda_m)$. Since $F(\mathbf{y})$ is minimized at the solution vector $\mathbf{y}^*$ of (A.19), that is, $F(\mathbf{y}^*) = 0$, we can apply the method of steepest descent to $F(\mathbf{y})$ for determining $\mathbf{y}^*$. We note further that the Lagrange multipliers $\{\lambda_1, \lambda_2, \ldots, \lambda_m\}$ play the role of auxiliary variables, but their values are necessary in determining the minimizing vector $\mathbf{x}^*$.

As motivation for the derivation of the equations in (A.19), consider the differentials of the object function and the constraint equations (recall the definition of "differential" from calculus):

$$df = \frac{\partial f}{\partial x_1} dx_1 + \frac{\partial f}{\partial x_2} dx_2 + \cdots + \frac{\partial f}{\partial x_n} dx_n = 0,$$

and

$$dg_i = \frac{\partial g_i}{\partial x_1}\,dx_1 + \frac{\partial g_i}{\partial x_2}\,dx_2 + \cdots + \frac{\partial g_i}{\partial x_n}\,dx_n = 0, \qquad 1 \le i \le m.$$

Multiplying each dg_i by λ_i and adding yield

$$\left\{\frac{\partial f}{\partial x_1} + \sum_{i=1}^{m} \lambda_i \frac{\partial g_i}{\partial x_1}\right\} dx_1 + \left\{\frac{\partial f}{\partial x_2} + \sum_{i=1}^{m} \lambda_i \frac{\partial g_i}{\partial x_2}\right\} dx_2 + \cdots + \left\{\frac{\partial f}{\partial x_n} + \sum_{i=1}^{m} \lambda_i \frac{\partial g_i}{\partial x_n}\right\} dx_n = 0. \tag{A.20}$$

If each of the quantities in brackets in (A.20) vanishes regardless of any variation dx_j, $1 \le j \le n$, then we can think of this equation as representing a horizontal tangent and thus a candidate for a minimum. We leave to the reader to verify that setting each bracket equal to zero *plus* satisfying $g_i(x_1, x_2, \ldots, x_n) = 0$, $1 \le i \le m$, yields precisely the equations of (A.19).

We conclude this section with two examples of the use of the Lagrange multiplier technique. In the first example we shall rework the problem of Example A.2 by this technique and in the second example we shall illustrate the technique in terms of a data-fitting problem.

EXAMPLE A.3. With $f(\alpha, \beta, \gamma)$, $g_1(\alpha, \beta, \gamma)$, and $g_2(\alpha, \beta, \gamma)$ given as in Example A.2, we have from (A.19)

$$\nabla f + \lambda_1 \nabla g_1 + \lambda_2 \nabla g_2 = \begin{bmatrix} 2\alpha(1 + \lambda_1 + \lambda_2) \\ 2\beta(1 + \lambda_1) - 4 \\ 2\gamma(1 + \lambda_2) \end{bmatrix}^{\mathrm{T}} = \begin{bmatrix} 0 \\ 0 \\ 0 \end{bmatrix}^{\mathrm{T}}$$

with $\gamma = \sqrt{1 - \alpha^2}$ and $\beta = \sqrt{1 - \alpha^2}$. Algebraically solving these five equations, we obtain one solution: $\alpha = 0$, $\beta = 1$, $\gamma = 1$, and $\lambda_1 = 1$, $\lambda_2 = -1$. Thus $f(0, 1, 1) = 2$ as before. There are other solutions to these five equations. We leave it to the reader to find them and to interpret their significance via Fig. A.3 (Exercise 3).

EXAMPLE A.4. (A piecewise discrete least-squares data-fit with continuity constraints.) We have mentioned before that polynomial data-fitting leads to an undesirable oscillatory behavior in the approximation. Cubic spline interpolation is one way of overcoming this difficulty, but for a large number of data points this process can become very cumbersome. Another approach is to partition the data points into several different sets and obtain a separate low-degree polynomial least-squares fit on each individual set. For example, let the data points $\{(x_i, y_i)\}_{i=1}^{5}$ be the first set S_1; let S_2 consist of $\{(x_i, y_i)\}_{i=6}^{10}$, and S_3 consist of $\{(x_i, y_i)\}_{i=11}^{20}$. Let $p_1(x)$ be the best quadratic least-squares polynomial on S_1, $p_2(x)$ be the best quadratic least-squares polynomial on S_2, and $p_3(x)$ be the best cubic least-squares polynomial on S_3. Then we could take $P(x)$ given by

$$P(x) = \begin{cases} p_1(x), & \text{for } x_1 \le x < x_6 \\ p_2(x), & \text{for } x_6 \le x < x_{11} \\ p_3(x), & \text{for } x_{11} \le x \le x_{20} \end{cases}$$

as our approximating curve. The difficulty with this approach is that usually $p_1(x_6) \neq p_2(x_6)$ and $p_2(x_{11}) \neq p_3(x_{11})$; and thus $P(x)$ is discontinuous at x_6 and x_{11}, usually an undesirable condition in practice. We can use the Lagrange multipler technique to remedy this deficiency by still requiring a least-squares polynomial fit on S_1, S_2, and S_3 as above, but also imposing the side constraints that $p_1(x_6) = p_2(x_6)$ and $p_2(x_{11}) = p_3(x_{11})$. [Note that by altering the problem in this way we will not obtain the same polynomials $p_1(x)$, $p_2(x)$, and $p_3(x)$ as above.] Our variables are the coefficients of the three polynomials plus the two Lagrange multipliers corresponding to the side constraints. Thus we let $p_1(x) = \alpha_1 + \alpha_2 x + \alpha_3 x^2$, $p_2(x) = \alpha_4 + \alpha_5 x + \alpha_6 x^2$, and $p_3(x) = \alpha_7 + \alpha_8 x + \alpha_9 x^2 + \alpha_{10} x^3$ where $\{\alpha_i\}_{i=1}^{10}$ are the variables to be determined subject to the constraints $p_1(x_6) = p_2(x_6)$ and $p_2(x_{11}) = p_3(x_{11})$. The constraint equations expressed in these variables are

$$g_1(\alpha_1, \ldots, \alpha_{10}) \equiv (\alpha_1 + \alpha_2 x_6 + \alpha_3 x_6^2) - (\alpha_4 + \alpha_5 x_6 + \alpha_6 x_6^2) = 0$$

and

$$g_2(\alpha_1, \ldots, \alpha_{10}) \equiv (\alpha_4 + \alpha_5 x_{11} + \alpha_6 x_{11}^2) - (\alpha_7 + \alpha_8 x_{11} + \alpha_9 x_{11}^2 + \alpha_{10} x_{11}^3) = 0.$$

Thus

$$\nabla g_1(\alpha_1, \ldots, \alpha_{10}) = (1, x_6, x_6^2, -1, -x_6, -x_6^2, 0, 0, 0, 0)$$

and

$$\nabla g_2(\alpha_1, \ldots, \alpha_{10}) = (0, 0, 0, 1, x_{11}, x_{11}^2, -1, -x_{11}, -x_{11}^2, -x_{11}^3).$$

As in Example A.1, we take $f(\alpha_1, \alpha_2, \ldots, \alpha_{10}) = \|A\mathbf{x} - \mathbf{b}\|_2^2$ where $\mathbf{x}^T \equiv (\alpha_1, \alpha_2, \ldots, \alpha_{10})^T$, $\mathbf{b}^T \equiv (y_1, y_2, \ldots, y_{20})^T$, and A is the (20×10) Vandermonde-type matrix shown in Fig. A.4. Again, as in Example A.1, $\nabla f(\alpha_1, \alpha_2, \ldots, \alpha_{10}) = 2A^T A\mathbf{x} - 2A^T\mathbf{b}$. Thus to minimize $f(\alpha_1, \ldots, \alpha_{10})$ under the constraints $g_1(\alpha_1, \ldots, \alpha_{10}) = 0$ and $g_2(\alpha_1, \ldots, \alpha_{10}) = 0$, we must satisfy the equations of formula (A.19):

$$\begin{aligned} \nabla f(\alpha_1, \ldots, \alpha_{10}) + \lambda_1 \nabla g_1(\alpha_1, \ldots, \alpha_{10}) + \lambda_2 \nabla g_2(\alpha_1, \ldots, \alpha_{10}) &= 0 \\ g_1(\alpha_1, \ldots, \alpha_{10}) &= 0 \\ g_2(\alpha_1, \ldots, \alpha_{10}) &= 0. \end{aligned} \tag{A.21}$$

Since the bottom two equations (A.21) are linear in the variables $\{\alpha_1, \alpha_2, \ldots, \alpha_{10}\}$, (A.21) represents a linear system of twelve equations in twelve unknowns, $\{\alpha_1, \alpha_2, \ldots, \alpha_{10}, \lambda_1, \lambda_2\}$. In matrix form these equations become

$$\left[\begin{array}{c|cc} A^T A & B^T & \\ \hline B & 0 & 0 \\ & 0 & 0 \end{array}\right] \begin{bmatrix} \alpha_1 \\ \alpha_2 \\ \vdots \\ \alpha_{10} \\ \tilde{\lambda}_1 \\ \tilde{\lambda}_2 \end{bmatrix} = \begin{bmatrix} A^T\mathbf{b} \\ 0 \\ 0 \end{bmatrix} \tag{A.22}$$

$$A = \begin{bmatrix} 1 & x_1 & x_1^2 & 0 & 0 & 0 & 0 & 0 & 0 & 0 \\ 1 & x_2 & x_2^2 & 0 & 0 & 0 & 0 & 0 & 0 & 0 \\ \vdots & & & & & & & & & \vdots \\ 1 & x_5 & x_5^2 & 0 & 0 & 0 & 0 & 0 & 0 & 0 \\ 0 & 0 & 0 & 1 & x_6 & x_6^2 & 0 & 0 & 0 & 0 \\ 0 & 0 & 0 & 1 & x_7 & x_7^2 & 0 & 0 & 0 & 0 \\ \vdots & & & & & & & & & \vdots \\ 0 & 0 & 0 & 1 & x_{10} & x_{10}^2 & 0 & 0 & 0 & 0 \\ 0 & 0 & 0 & 0 & 0 & 0 & 1 & x_{11} & x_{11}^2 & x_{11}^3 \\ 0 & 0 & 0 & 0 & 0 & 0 & 1 & x_{12} & x_{12}^2 & x_{12}^3 \\ \vdots & & & & & & & & & \vdots \\ 0 & 0 & 0 & 0 & 0 & 0 & 1 & x_{20} & x_{20}^2 & x_{20}^3 \end{bmatrix}$$

Figure A.4

where $\tilde{\lambda}_1 = \lambda_1/2$, $\tilde{\lambda}_2 = \lambda_2/2$, and B is the (2×10) matrix

$$B = \begin{bmatrix} 1 & x_6 & x_6^2 & -1 & -x_6 & -x_6^2 & 0 & 0 & 0 & 0 \\ 0 & 0 & 0 & 1 & x_{11} & x_{11}^2 & -1 & -x_{11} & -x_{11}^2 & -x_{11}^3 \end{bmatrix}.$$

Using the expressions for ∇f, ∇g_1, and ∇g_2, we leave to the reader to verify componentwise that (A.22) is precisely (A.21).

We have presented this method of data-fitting in a very specialized and simplified form to illustrate more clearly the derivation of the linear system (A.22). It should be clear to the reader that any number of data points can be used, there can be a different number of partitions, they can be separated at different points, the degrees of the polynomials on each partition can be altered, etc. We can put even further conditions on the degree of continuity between the polynomials of each partition. [For example, we could require that $p_1'(x_6) = p_2'(x_6)$, etc.] For each additional constraint we have another Lagrange multiplier; however, these types of constraints are still linear with respect to the polynomial coefficients, and we still obtain a partitioned matrix form similar to (A.22) to solve for the polynomial coefficients and Lagrange multipliers. A data-fitting technique of this type is called a *smoothing* procedure since it reduces the oscillatory behavior of the approximation.

PROBLEMS, SECTION A.4

1. In formula (A.19) let $n = 5$ and $m = 3$. Write (A.19) in componentwise form yielding the eight equations in eight unknowns that must be solved.
2. Show that the constraints, $g_i(\mathbf{x}) = 0$, $1 \leq i \leq m$, plus setting the quantities in parentheses in (A.20), yield the equations of formula (A.19).
3. Find the other solutions to the equations in Example A.3, and interpret their geometric significance to Fig. A.3.

4. Verify that the matrix representation in (A.22) is equivalent on the equations of formula (A.21).
5. a) In Example A.4 specify numerical values for $\{(x_i, y_1)\}_{i=1}^{20}$; and find $p_1(x)$, $p_2(x)$, and $p_3(x)$ by solving (A.22).

   b) Modify the problem of Example A.4 by imposing the additional restriction that $p_2'(x_{11}) = p_3'(x_{11})$. Display the modifications this restriction makes on the matrix system (A.22).
6. Use Lagrange's method to solve the following.
   Given the plane $\alpha + 2\beta + \gamma = 1$, find the point on this plane that is nearest the origin. [Hint: Minimize $f(\alpha, \beta, \gamma) = \alpha^2 + \beta^2 + \gamma^2$ with respect to the constraint $\alpha + 2\beta + \gamma - 1 = 0$.]
7. Find the maximum of $f(x_1, x_2, \ldots, x_n) = (x_1x_2 \ldots x_n)^2$ subject to the constraint $x_1^2 + x_2^2 + \cdots + x_n^2 = 1$.

A.5 LINEAR PROGRAMMING

The minimization problem considered in Section A.4 had *equality constraints* of the form $g_i(x_1, x_2, \ldots, x_n) = 0$, $1 \le i \le m$. A more complicated problem, in general, is to replace the equality constraints above by inequalities of the form $g_i(x_1, x_2, \ldots, x_n) \le 0$, $1 \le i \le m$. This general problem is beyond the level of this text; so we shall be content to consider a specialized but important problem of the latter type. The problem is called the *linear programming* problem and seeks to minimize an objective function that is *linear* in the unknowns, subject to constraints, which consist of *linear* equalities and *linear* inequalities. In standard form the inequalities are changed to equalities and the problem can be expressed thus:

$$\text{minimize } f(x_1, \ldots, x_n) = c_1x_1 + c_2x_2 + \cdots + c_nx_n$$

$$\begin{aligned} \text{subject to } a_{11}x_1 + a_{12}x_2 + \cdots + a_{1n}x_n &= b_1 \\ a_{21}x_1 + a_{22}x_2 + \cdots + a_{2n}x_n &= b_2 \\ \vdots \qquad\qquad\qquad \vdots \quad & \\ a_{m1}x_1 + a_{m2}x_2 + \cdots + a_{mn}x_n &= b_m \end{aligned} \tag{A.23}$$

and

$$x_1 \ge 0, \quad x_2 \ge 0, \ldots, x_n \ge 0$$

where the a_{ij}'s, c_i's and b_i's are known values, and the x_i's are the unknowns to be determined. We shall also require that in (A.23) each $b_i \ge 0$ [since this condition can be achieved by multiplication by (-1) whenever necessary]. Furthermore, we note that minimizing $f(x_1, x_2, \ldots, x_n)$ is equivalent to maximizing

$$(-f(x_1, x_2, \ldots, x_n)).$$

If $A = (a_{ij})$ is an $(m \times n)$ matrix, $\mathbf{x} = (x_1, x_2, \ldots, x_n)^T$, $\mathbf{c} \equiv (c_1, c_2, \ldots, c_n)^T$, and $\mathbf{b} \equiv (b_1, b_2, \ldots, b_m)^T$, then the linear programming problem, stated above, can be expressed in more compact vector form:

$$\begin{aligned} &\text{minimize } f(\mathbf{x}) \equiv \mathbf{c}^T\mathbf{x} \\ &\text{subject to } A\mathbf{x} = \mathbf{b} \quad \text{and} \quad \mathbf{x} \geq \theta. \end{aligned} \tag{A.24}$$

Often a practical problem will not occur directly in the standard form of (A.23) or (A.24); a procedure to convert it to this form is introducing what are known as *slack variables*. For instance, consider the following simple example in which we wish to determine x_1 and x_2 such that $x_1 \geq 0$, $x_2 \geq 0$; $z = -3x_1 - 4x_2$ is minimized, and x_1 and x_2 must satisfy these three constraints:

$$-x_1 + 2x_2 \geq -1, \qquad 5x_1 + 2x_2 \leq 7, \qquad \text{and} \qquad 2x_1 + 3x_2 \leq 8.$$

We recognize that the first constraint is equivalent to $x_1 - 2x_2 \leq 1$. We then introduce three new "slack variables": $x_3 \geq 0$, $x_4 \geq 0$, $x_5 \geq 0$, such that the constraints become

$$x_1 - 2x_2 + x_3 = 1, \qquad 5x_1 + 2x_2 + x_4 = 7, \qquad \text{and} \qquad 2x_1 + 3x_2 + x_5 = 8.$$

This problem is now in the standard form given by (A.23) with respect to the nonnegative variables, x_1, x_2, x_3, x_4, and x_5. We observe that the original problem involves only two unknowns whereas the auxiliary problem (in standard form) involves five unknowns. It can be shown, however, that a solution of the latter problem will determine a corresponding solution of the original problem as illustrated later. The following example shows where the linear-programming problem above could occur in a hypothetical real-world setting.

EXAMPLE A.5. Let x_1 and x_2 represent the respective amounts of two types of pain-killing liquid that are made in one day; each type contains certain amounts of drug A and drug B. Each unit of the first type of pain-killing liquid requires $\frac{5}{7}$ unit of drug A, and each unit of the second type of pain-killing liquid requires $\frac{2}{7}$ unit of drug A. If only one unit of drug A is available in one day, then

$$\frac{5}{7}x_1 + \frac{2}{7}x_2 \leq 1 \qquad \text{or} \qquad 5x_1 + 2x_2 \leq 7.$$

Each unit of the first type requires $\frac{1}{4}$ unit of drug B, and each unit of the second type requires $\frac{3}{8}$ unit of drug B. If only one unit of drug B is available in one day, then

$$\frac{1}{4}x_1 + \frac{3}{8}x_2 \leq 1 \qquad \text{or} \qquad 2x_1 + 3x_2 \leq 8.$$

Moreover FDA regulations require that the amount x_1 cannot exceed $1 + 2x_2$; that is,

$$x_1 \leq 1 + 2x_2 \qquad \text{or} \qquad x_1 - 2x_2 \leq 1.$$

If three dollars and four dollars are the respective profit margins for one unit of the two pain-killing liquids, then $3x_1 + 4x_2$ represents the total profit per day. We want to determine the values of x_1 and x_2 that will maximize the profit.

In order to obtain some geometric insight into the linear programming problem, we consider a geometric approach to the particular problem above. The first constraint demands that $x_1 - 2x_2 \le 1$. Therefore our solution, (x_1^*, x_2^*), must be a point lying below the line $x_1 - 2x_2 = 1$ in the x_1x_2-plane. Similarly the other two constraints insist that (x_1^*, x_2^*) lie below the lines $5x_1 + 2x_2 = 7$ and $2x_1 + 3x_2 = 8$. Since we must also have $x_1 \ge 0$ and $x_2 \ge 0$, the solution (x_1^*, x_2^*) must lie in the area denoted *feasible solutions* in Fig. A.5.

We observe that minimizing $z = -3x_1 - 4x_2$ is equivalent to maximizing $z = 3x_1 + 4x_2$. After a few lines of the form $z = 3x_1 + 4x_2$ are graphed for various values of z, it becomes readily apparent that the largest value that z can attain and still have the line intersect the area of feasible solutions comes from the line that passes through the point $C = (5/11, 26/11)$. The value of z for this line is $z^* = 3(5/11) + 4(26/11) = 119/11$.

Obviously a similar graphical procedure would be very difficult to execute in a problem involving several variables. For example, in a problem involving three variables, the area of feasible solutions becomes a three-dimensional region bounded by planes; and in the general case of n variables the area becomes an n-dimensional solid called a *polytope* bounded by *hyperplanes*. A hyperplane in R^n is the set of all vectors $\mathbf{x} = (x_1, x_2, \ldots, x_n)$ whose coordinates satisfy an equation of this form: $a_1x_1 + a_2x_2 + \cdots + a_nx_n = b$. A hyperplane of this form thus divides R^n into two parts or *half-spaces*. One half-space is the set of all vectors $\mathbf{x} = (x_1, x_2, \ldots, x_n)$ such that $a_1x_1 + a_2x_2 + \cdots + a_nx_n \ge b$, and the other half-space is the set of all vectors $\mathbf{x} = (x_1, x_2, \ldots, x_n)$ such that $a_1x_1 + a_2x_2 + \cdots + a_nx_n \le b$. Clearly the hyperplane is the intersection of these two half-spaces. In Fig. A.5, $n = 2$, and each line (say, $x_1 - 2x_2 = 1$) is a hyperplane that divides the x_1x_2-plane into two parts or half-spaces. If $n = 3$, a hyperplane is the set of all vectors $\mathbf{x} = (x_1, x_2, x_3)$ such that $a_1x_1 + a_2x_2 + a_3x_3 = b$. The reader should recall from calculus that this equation is the general equation of a

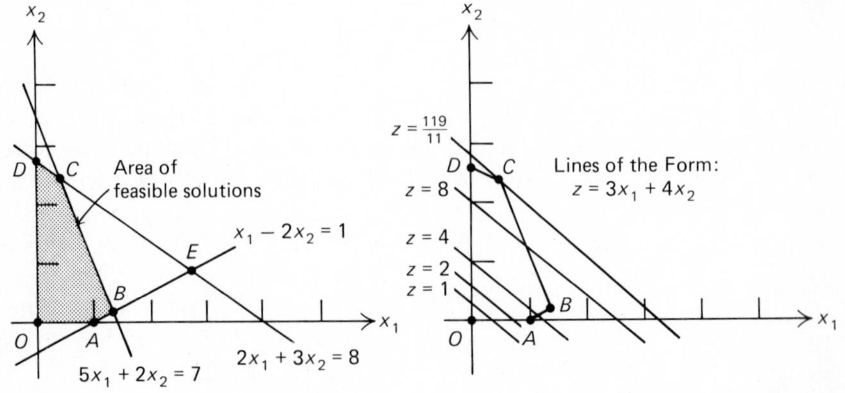

Figure A.5 Area of feasible solutions and the graphical solution.

plane and any plane divides R^3 into two parts. If $n > 3$, we lose this type of graphical perspective, but we see from above that a hyperplane in R^n still divides R^n into two parts. Thus we are led to suspect that some of the geometrical properties of the two-variable problem above carry over into the n-dimensional problem. Any vector (if any exists) that lies in the set of feasible solutions is called a *feasible point* and its coordinates satisfy all constraints of the problem. Furthermore the n-dimensional polytope (set of feasible points) has vertices. In Fig. A.5, $n = 2$ and the points O, A, B, C, and D are vertices. Even though E is the intersection of $x_1 - 2x_2 = 1$ and $2x_1 + 3x_2 = 8$, the point E is not a vertex of the polytope because the coordinates of E do not satisfy the constraint $5x_1 + 2x_2 \leq 7$ and hence E is not a feasible point. A vertex of a polytope defined by the constraints can be regarded geometrically as a *feasible* point where the boundaries of at least n of the half-spaces, determined by the constraints, intersect. These vertices are called *extreme feasible points*. It can be shown that a solution point is always an extreme feasible point (such as the point C in Fig. A.5). If, as can happen, two extreme feasible points are solutions, the entire edge connecting them consists of solution points. The same is true in higher dimensions.

The method we shall discuss for solving the linear programming problem is known as the *simplex method*, which we shall introduce by illustrating its use on the problem above. The first step is to introduce nonnegative slack variables to put the problem in standard form. For example, the problem above now becomes this problem.

Determine $x_i \geq 0$, $1 \leq i \leq 5$, such that $z \equiv f(x_1, x_2, \ldots, x_5) = 3x_1 + 4x_2 + 0x_3 + 0x_4 + 0x_5$ is maximized subject to

$$x_1 - 2x_2 + x_3 \qquad\qquad = 1 \tag{A.25a}$$

$$5x_1 + 2x_2 \qquad + x_4 \qquad = 7 \tag{A.25b}$$

$$2x_1 + 3x_2 \qquad\qquad + x_5 = 8, \tag{A.25c}$$

$$x_i \geq 0, \qquad 1 \leq i \leq 5.$$

We recall that a solution must occur at an extreme feasible point of the polytope. Since we do not know which extreme feasible point yields the solution, however, we proceed in a systematic fashion, which will amount to choosing a vertex and moving along an edge from that vertex to another vertex in a manner such that z will be increased or unchanged. This process will be repeated again and again until the solution is found. Several complications can arise in this procedure: *degeneracy*, contradictory constraint equations, multiple solutions, etc.; but it is beyond the scope of this text to investigate them [see Luenberger (1973)]. These complications are the primary reason why we consider only the simple problem above to illustrate the technique.

Once the problem is in standard form, our first step is to set $x_1 = x_2 = 0$, solve (A.25) for x_3, x_4, x_5, and obtain $x_3 = 1$, $x_4 = 7$, and $x_5 = 8$. For this choice

$z = 0$ and is not maximized since z can be increased by increasing either x_1 or x_2. Since x_2 has the larger coefficient in $z = 3x_1 + 4x_2 + 0x_3 + 0x_4 + 0x_5$, we allow x_2 to increase and keep x_1 at zero. We next select one of the three remaining variables, x_3, x_4, or x_5, to be set equal to zero in place of x_2. To determine which one, we consider the equations of (A.25) with $x_1 = 0$:

$$-2x_2 + x_3 = 1 \qquad \text{(A.26a)}$$

$$2x_2 + x_4 = 7 \qquad \text{(A.26b)}$$

$$3x_2 + x_5 = 8. \qquad \text{(A.26c)}$$

Since x_3, x_4, x_5 cannot be negative, (A.26b) yields $x_2 \le \frac{7}{2}$ and (A.26c) yields $x_2 \le \frac{8}{3}$. Since the coefficient of x_2 in (A.26a) is negative, x_3 will be positive for any $x_2 \ge 0$, and so (A.26a) places no restriction on x_2. The two restrictions we do have on x_2 must hold simultaneously; and so we let x_2 equal the smaller of the two, $x_2 = \frac{8}{3}$, which requires $x_5 = 0$ because of (A.26c). Now solving equations (A.26) yields $x_1 = 0$, $x_2 = \frac{8}{3}$, $x_3 = \frac{19}{3}$, $x_4 = \frac{5}{3}$, and $x_5 = 0$. With these values we have $z = 3x_1 + 4x_2 = \frac{32}{3}$, which is substantially better than our first value $z = 0$. (We note here that the work above is precisely equivalent to moving from the origin to the vertex D in Fig. A.5). Before we can repeat this step and move to another vertex, we must reformulate equations (A.25) so that x_2, x_3, and x_4 play the roles of the slack variables. This requirement means that we wish to convert the equations of (A.25) to the form $a_k x_1 + \beta_k x_5 + x_k = \gamma_k$ for $k = 2, 3$, and 4, and that we wish to express z as a function of x_1 and x_5 rather than of x_1 and x_2. We can achieve this expression by use of (A.25c) since it can be used to solve for x_2 in terms of x_1 and x_5. (A.25c) becomes $\frac{2}{3}x_1 + x_2 + \frac{1}{3}x_5 = \frac{8}{3}$. Now we use it to eliminate x_2 from the first two equations and get

$$\frac{7x_1}{3} + x_3 + \frac{2x_5}{3} = \frac{19}{3} \qquad \text{(A.27a)}$$

$$\frac{11x_1}{3} + x_4 - \frac{2x_5}{3} = \frac{5}{3} \qquad \text{(A.27b)}$$

$$\frac{2x_1}{3} + x_2 + \frac{x_5}{3} = \frac{8}{3}. \qquad \text{(A.27c)}$$

Using (A.27c) to eliminate x_2 from $z = 3x_1 + 4x_2$, we obtain $z = \frac{32}{3} + \frac{1}{3}x_1 - \frac{4}{3}x_5$.

We have now completed one step of the simplex method and are ready to repeat the same process using equations (A.27) and $z = \frac{32}{3} + \frac{1}{3}x_1 - \frac{4}{3}x_5$. Since x_1 has the larger coefficient in z $\left(\frac{1}{3} > -\frac{4}{3}\right)$, and the coefficient is positive, the value of z can be increased by increasing the value of x_1. Therefore we allow x_1 to increase while we keep x_5 at zero. We now repeat the process above to select

x_2, x_3, or x_4 to take the place of x_1 and be set equal to zero. From (A.27) with $x_5 = 0$ we obtain

$$\frac{7x_1}{3} + x_3 \qquad = \frac{19}{3} \tag{A.28a}$$

$$\frac{11x_1}{3} \qquad + x_4 = \frac{5}{3} \tag{A.28b}$$

$$\frac{2x_1}{3} \qquad + x_2 = \frac{8}{3}. \tag{A.28c}$$

Since x_2, x_3, x_4 are nonnegative, (A.28a) yields $\frac{7}{3}x_1 \leq \frac{19}{3}$, (A.28b) yields $\frac{11}{3}x_1 \leq \frac{5}{3}$, and (A.28c) yields $\frac{2}{3}x_1 \leq \frac{8}{3}$. Since these relations must hold simultaneously, we have from (A.28b) that $x_1 \leq \frac{5}{11}$ since this is the smallest of the three bounds on x_1. Setting $x_1 = \frac{5}{11}$ yields $x_4 = 0$ from (A.28b). Now solving equations (A.28) yields $x_1 = \frac{5}{11}$, $x_2 = \frac{26}{11}$, $x_3 = \frac{58}{11}$, $x_4 = 0$, and $x_5 = 0$. We calculate $z = \frac{32}{3} + \frac{1}{3}x_1 - \frac{4}{3}x_5$ with these values and obtain $z = \frac{119}{11}$, which is larger than our previous value of $z = \frac{32}{3}$. (Note that this step is precisely equivalent to moving from vertex D to vertex C in Fig. A.5.) Now we reformulate equations (A.27) so that x_1, x_2, and x_3 play the role of the slack variables. Using (A.27b) to eliminate x_1 in the remaining two equations yields

$$x_1 \qquad + \frac{3x_4}{11} - \frac{2x_5}{11} = \frac{5}{11} \tag{A.29a}$$

$$x_2 \qquad - \frac{2x_4}{11} + \frac{5x_5}{11} = \frac{26}{11} \tag{A.29b}$$

$$x_3 - \frac{7x_4}{11} + \frac{12x_5}{11} = \frac{58}{11}. \tag{A.29c}$$

Using (A.29a) to eliminate x_1 from $z = \frac{32}{3} + \frac{1}{3}x_1 - \frac{4}{3}x_5$ yields $z = \frac{119}{11} - \frac{1}{11}x_4 - \frac{14}{11}x_5$. Since increasing either x_4 or x_5 in this expression for z will decrease the value of z, this expression results in the maximum value for z. (Moving to any other vertex decreases z and thus we are at the extreme feasible point where z is maximized.) Thus our final answer is $z = \frac{119}{11}$, which occurs at $x_1 = \frac{5}{11}$ and $x_2 = \frac{26}{11}$, and agrees with the answer achieved by the graphical method in Fig. A.5.

We notice that each step of the simplex method involves solutions of linear systems of equations and selection of largest *pivotal elements,* i.e., the choice of which variable to be increased to maximize the increase in z in going from one vertex to another. These procedures can be organized and put in an algorithmic form through what is known as a *simplex tableau.* This process primarily involves an extension of the familiar Gauss-Jordan elimination method with pivoting to solve a set of simultaneous linear equations. Since this section is intended as only a brief introduction to the linear programming problem, and because of the many complications that must be considered to treat the problem thoroughly, we shall not pursue this concept further. Numerous texts are devoted primarily to the theory of linear programming and the efficient utilization of the simplex method. An excellent reference is the above-mentioned text by Luenberger. We call attention also to the text by Kuo (1972). In particular we reference Kuo's computer program for the simplex method for up to 34 constraints and 85 variables on pages 360–363 and his flow chart on pages 364–366.

PROBLEMS, SECTION A.5

1. Consider the problem of maximizing $z = 3x_1 + 2x_2$ subject to the constraints: $2x_1 - x_2 \leq 4$, $2x_1 + x_2 \leq 8$, $x_1 + x_2 \leq 6$, and $-3x_1 + x_2 \leq 2$, with $x_1 \geq 0$ and $x_2 \geq 0$.
   a) Graph the region of feasible solutions in the x_1x_2-plane and identify the extreme feasible points.
   b) Solve this problem graphically.
   c) Solve this problem by putting it in standard form and using the simplex method. Identify your path from vertex to vertex as you proceed.
2. Consider the problem of maximizing $z = 2x_1 + 3x_2 + x_3$ subject to these constraints: $-2x_1 + x_3 \leq 0$, $2x_1 + 3x_2 \leq 9$, $x_1 - x_2 \leq 2$, and $x_1 + x_3 \leq 3$ with $x_1 \geq 0$, $x_2 \geq 0$, $x_3 \geq 0$. Repeat the steps of parts (a), (b), and (c) of Problem 1 for this three-dimensional problem. (The solution is found after three steps of the simplex method. Note, however, that the results of the first and second steps yield the same values of the unknowns and do not increase the value of z. This phenomenon is known as *degeneracy*. It is theoretically possible that this situation could cycle indefinitely without reaching a maximum. This situation, however, is not common; and modifications can be made to the simplex method to treat the situation.)
3. Mr. H. H. Schmidtke makes doll houses and toy barns as a hobby and sells them for profit. He has six days of vacation to work on his hobby and wishes to work at most eight hours each day. He furthermore wishes to spend two days cutting pieces and four days for assembly and finishing work. Each doll house requires thirty minutes to cut pieces and one hour and twenty minutes to assemble and finish. Each toy barn requires twenty minutes to cut pieces and thirty minutes to assemble and finish. His retail outlet demands that he make thirteen more toy barns than doll houses. His net profit is $20 for each doll house and $12 for each toy barn. How many of each should he make in order to maximize his profit over the vacation, provided he is willing to leave a doll house and a toy barn only partially done?

Answers to Selected Exercises

CHAPTER 1

Section 1.3

1. $1023 = (3FF)_{16} = (1111111111)_2$, $1025 = (401)_{16} = (10000000001)_2$, $278.5 = (116.8)_{16} = (100010110.1)_2$, $14.09375 = (E.18)_{16} = (1110.00011)_2$, $0.1240234375 = (.1FC)_{16} = (.0001111111)_2$
2. a) 4131
   b) 256
   c) 420.75
   d) 107.109375
   e) 4095.068359375
3. a) $(.3243F7)_{16} \times 16$, $(.2B7E15)_{16} \times 16$, $(.555555)_{16}$
   b) $(.3243F6)_{16} \times 16$, $(.2B7E15)_{16} \times 16$, $(.555555)_{16}$
5. 0.122×10^{-4}, 0.88×10^{-5}, 0.157×10^{-4}, 0.322×10^{-4}
8. 9001

CHAPTER 2

Section 2.1

3. n^3
4. $(n^2 + n)/2$
6. If T_k is the number of multiplications for a $(k \times k)$ determinant, then $T_{k+1} = (k + 1)(T_k + 1)$, $k \geq 2$. $T_2 = 2$; then $T_3 = 9$, $T_4 = 40$, $T_5 = 205$, $T_n > n!$ for $n > 2$.
7. In $L\mathbf{x} = \theta$ with $\ell_{11} \neq 0$, the first equation yields $x_1 = 0$. Continuation shows $\mathbf{x} = \theta$.
10. Assume that $i > j$. $(TS)_{ij} = \sum_{k=1}^{n} t_{ik}s_{kj}$. For $k < i$, $t_{ik} = 0$. For $k \geq i > j$, $s_{kj} = 0$. Then $(TS)_{ij} = 0$.
12. Consider $AB\mathbf{x} = \theta$. Let $\mathbf{y} = B\mathbf{x}$. A nonsingular in $A\mathbf{y} = \theta$ yields $\mathbf{y} = \theta$. B nonsingular in $B\mathbf{x} = \mathbf{y} = \theta$ yields $\mathbf{x} = \theta$.

Sections 2.2.1-2.2.4

4. $y = (2 - 10^n)/(1 - 10^n)$, $x = -10^n/(1 - 10^n)$. For $n = 3$, computation yields $y = .999$, $x = 1.00$. From Example 2.8 with $n = 4$, $y = 1.00$ and $x = 0.00$. For $n \geq 4$, computations are unreliable.

5. $x_1 = 10^n/(10^n + 1)$, $x_2 = (3 \cdot 10^n + 2)/(10^n + 1)$. Computational values: $n = 1$, $x_1 = .900$, $x_2 = 2.91$; $n = 2$, $x_1 = 1.00$, $x_2 = 2.99$; $n = 3$, $x_1 = 0.00$, $x_2 = 3.00$; unreliable for $n \geq 3$.

10. Let $A = \begin{bmatrix} 1 & 2 & 1 \\ 3 & 10 & 0 \\ 2 & 3 & 3 \end{bmatrix}$ Partial pivoting: interchange row_1 and row_2, Implicit scaling: interchange row_1 and row_3.

12. The kth row of I is $\mathbf{e}_k^T$ and $\mathbf{e}_k^T\mathbf{x} = x_k$. Hence $P\mathbf{x}$ interchanges ith and jth components of $\mathbf{x}$. Since $PA = [P\mathbf{A}_1, P\mathbf{A}_2, \ldots, P\mathbf{A}_n]$, PA interchanges ith and jth rows of A. Using $A = P$ yields $PP = I$ or $P^{-1} = P$.

14. i) Assume that A and B are nonsingular; consider $AB\mathbf{x} = \theta$. Since A is nonsingular, $A(B\mathbf{x}) = \theta$ implies $B\mathbf{x} = \theta$. Since B is nonsingular, then $\mathbf{x} = \theta$; AB, nonsingular.

ii) Assume that AB is nonsingular and consider $B\mathbf{x} = \theta$. Then $A(B\mathbf{x}) = \theta = (AB)\mathbf{x}$, and so $\mathbf{x} = \theta$ and B is nonsingular. Now consider $A\mathbf{x} = \theta$ and let $\mathbf{y} = B^{-1}\mathbf{x}$ or $\mathbf{x} = B\mathbf{y}$. Then $A\mathbf{x} = AB\mathbf{y} = \theta$; so $\mathbf{y} = \theta$; and so $\mathbf{x} = \theta$ and A is nonsingular.

16.

$$A = \begin{bmatrix} 1 & 0 & 0 & 0 \\ \frac{7}{5} & 1 & 0 & 0 \\ \frac{6}{5} & -2 & 1 & 0 \\ 1 & 0 & \frac{3}{2} & 1 \end{bmatrix} \begin{bmatrix} 5 & 7 & 6 & 5 \\ 0 & .2 & -.4 & 0 \\ 0 & 0 & 2 & 3 \\ 0 & 0 & 0 & .5 \end{bmatrix} = LU$$

$L\mathbf{y} = \mathbf{b}$, $\mathbf{y} = [-1, .4, -6, -1]^T$; $U\mathbf{x} = \mathbf{y}$, $\mathbf{x} = [-1, 2, 0, -2]^T$

17. $\text{Row}_k(L) = \mathbf{e}_k^T$, $k \leq j$; and $\text{row}_k(L) = [0, \ldots, 0, M_{kj}, 0, \ldots, 0, 1, 0, \ldots, 0]$, $k > j$. $\text{Col}_k(L^{-1}) = \mathbf{e}_k$, $k \neq j$; $\text{col}_j(L^{-1}) = [0, \ldots, 0, 1, -M_{j+1,j}, \ldots, -M_{nj}]^T$ Clearly, $\text{row}_i(L)\text{col}_k(L^{-1}) = \delta_{ik}$.

20.

$$A = \begin{bmatrix} 1 & 0 & 0 \\ -1 & 1 & 0 \\ 2 & 1 & 1 \end{bmatrix} \begin{bmatrix} 1 & 0 & 1 \\ 0 & 1 & 0 \\ 0 & 0 & -2 \end{bmatrix}$$

Sections 2.3.1-2.3.4

1. $\mathbf{x} = [1, 1, 1, 1]^T$, $\mathbf{y} = \mathbf{e}_3$

3. a) $\|\mathbf{x}\|_1^2 = \|\mathbf{x}\|_2^2 + |x_1|\,|x_2| + |x_1|\,|x_3| + \cdots \geq \|\mathbf{x}\|_2^2 = x_1^2 + x_2^2 + \cdots + x_n^2 \geq \|\mathbf{x}\|_\infty^2$; and $n\|\mathbf{x}\|_\infty \geq |x_1| + |x_2| + \cdots + |x_n| = \|\mathbf{x}\|_1$.

b) For any p and q, $\|\mathbf{x}\|_p \geq c_{pq}\|\mathbf{x}\|_q$, c_{pq} independent of $\mathbf{x}$ and positive. Then $\|\mathbf{x}^{(j)}\|_p < \varepsilon c_{pq}$ implies that $\|\mathbf{x}^{(j)}\|_q < \varepsilon$.

c) $\|\mathbf{x}^{(j)} - \mathbf{x}\|_p < \varepsilon/\|A\|_p$ implies that $\|A\mathbf{x}^{(j)} - A\mathbf{x}\|_p < \varepsilon$.

6. $\|I\|_m = \|AA^{-1}\|_m \leq \|A\|_m\|A^{-1}\|_m$ and $0 < \|I\|_m = \|I^2\|_m \leq \|I\|_m\|I\|_m$; so $\|I\|_m \geq 1$.

7. $.10127 \times 10^{-12} \le \|\mathbf{x}_t - \mathbf{x}_c\|_\infty / \|\mathbf{x}_t\|_\infty \le .2034 \times 10^{-2}$.

9. $A_n^{-1} = \begin{bmatrix} 4n^2+1 & -2n^2 \\ -2n^2 & n^2 \end{bmatrix}$. Let $B = \begin{bmatrix} 1 & 2 \\ 2 & 4 \end{bmatrix}$; then

$(\|A_n - B\|_\infty / \|A_n\|_\infty - 1/K(A_n)) = (5n^2+1)/(6n^2+1)^2 < 1/n^2$.

10. $\det(D_n) = (.1)^n$, $\|D_n\|_\infty \|D_n^{-1}\|_\infty = 1$, $\min[\|D_n - B\|_\infty, B \text{ singular}] = .1$

11. a) $[1, -2, -1]^T$ b) $[2, 0, 3]^T$ c) $[1, -1, 1]^T$

12. a) $\mathbf{x} = \begin{bmatrix} 3 \\ -\frac{8}{3} \\ -\frac{11}{12} \end{bmatrix}$ b) $\mathbf{y} = \begin{bmatrix} -\frac{7}{2} \\ -6 \\ -\frac{3}{4} \end{bmatrix}$ d) $A = \begin{bmatrix} -\frac{1}{2} & -\frac{1}{2} & 2 \\ \frac{1}{2} & -\frac{1}{4} & -2 \\ 1 & -1 & 4 \end{bmatrix}$

13, 14. $\|H_3^{-1}\|_1 = 408$

18. 1) Solve $U^T\mathbf{z} = \mathbf{b}$ 2) Solve $L^T\mathbf{y} = \mathbf{z}$

Section 2.4

5. a) $\|N^{-1}P\|_\infty = 1$, $\|N^{-1}P\|_1 = .9$, convergent
   b) $\|M_J\|_\infty = .8$, $\|M_J\|_1 = 17/30$

6. $N = \begin{bmatrix} 4 & 0 & 0 \\ 0 & 6 & 0 \\ 0 & 0 & -4 \end{bmatrix}$, $P = \begin{bmatrix} -1 & -1 & 1 \\ 2 & 1 & -1 \\ -1 & 1 & -1 \end{bmatrix}$, $b = \begin{bmatrix} 2 \\ -1 \\ 4 \end{bmatrix}$, $A = N - P$, $\|N^{-1}P\|_\infty =$ .75(convergent), $\mathbf{x} = [1/9, 1/9, -4/3]^T$

7. n^2, $k \simeq n/3$

8. If α and β are the (1, 1) and (1, 2) entries, respectively, of the altered matrix $\tilde{A}$, then $\tilde{A}\mathbf{x} = \theta$ where $\mathbf{x} = -\beta\mathbf{e}_1 + \alpha\mathbf{e}_2$.

11. For $\mathbf{y} \ne \theta$, let $\mathbf{z} = A\mathbf{y} \ne \theta$. Then $\mathbf{y}^T(A^TA)\mathbf{y} = (A\mathbf{y})^TA\mathbf{y} = \mathbf{z}^T\mathbf{z} > 0$.

13. $M_J^{2n} = \begin{bmatrix} \left(\frac{9}{14}\right)^n & 0 \\ 0 & \left(\frac{9}{14}\right)^n \end{bmatrix}$, $\|M_J^{2n}\|_1 = (9/14)^n$, $\|M_J^{2n+1}\|_1 \le \|M_J^{2n}\|_1 \|M_J\|_1 = (3/2)(9/14)^n$.

Hence $\lim_{k\to\infty} \|M_J^k\|_1 = 0$, $\|\mathbf{e}^{(k+1)}\|_1 \le \|M_J^k\|_1 \|\mathbf{e}^{(0)}\|_1 \to 0$ as $k \to \infty$.

Section 2.5

1. $\mathbf{x}^* = [-.5, 2]^T$

3. $\mathbf{x}^* = [.4 + 2a, 8/35 - 3a, a]^T$

4. $\mathbf{x}^TA^TA\mathbf{x} = 0$ for $\mathbf{x} = a[2, -3, 1]^T$

5. a) $y = (1/29)(12x + 15)$ b) $y = (13x + 11)/20$ c) $y = (-7x + 5)/10$

6. a) $y = (x^2 + 2)/3$ b) $y = (5x^2 - x - 1)/20$

7. $y = (5 - 4\cos x - 2\sin x)/8$

8. $\mathbf{x}^TA^TA\mathbf{x} = (A\mathbf{x})^TA\mathbf{x} \ge 0$

CHAPTER 3

Section 3.1

1. A: $\lambda_1 = -2$, $\mathbf{u}_1 = a[1, -1]^T$, $\lambda_2 = 3$, $\mathbf{u}_2 = b[1, 4]^T$

   B: $\lambda_1 = 1$, $\mathbf{u}_1 = a[1, 0, 1]^T$, $\lambda_2 = \lambda_3 = 2$, $\mathbf{u}_2 = b[1, 1, 0]^T$
2. $\mathbf{x} = a[(-3 + 7\sqrt{7}i), (-18 - 6\sqrt{7}i), 16]^T$
6. $A = \begin{bmatrix} 1 & 1 \\ 0 & 1 \end{bmatrix}$, $\lambda = 1$, $u = \begin{bmatrix} 1 \\ 0 \end{bmatrix}$; $A^T = \begin{bmatrix} 1 & 0 \\ 1 & 1 \end{bmatrix}$, $\lambda = 1$, $u = \begin{bmatrix} 0 \\ 1 \end{bmatrix}$
7. $\mathbf{x}_i^T(c_1\mathbf{x}_1 + c_2\mathbf{x}_2 + \cdots + c_k\mathbf{x}_k) = c_i\mathbf{x}_i^T\mathbf{x}_i = 0$, $c_i = 0$, $1 \le i \le k$.
8. A_1: $\{\mathbf{e}_1\}$; A_2: $\{\mathbf{e}_1, \mathbf{e}_3\}$; A_3: $\{\mathbf{e}_1, \mathbf{e}_2, \mathbf{e}_3\}$
9. If $c_1\mathbf{u}_1 + \cdots + c_r\mathbf{u}_r + c_{r+1}\mathbf{u}_{r+1} = \theta$, not all $c_i = 0$; then $c_{r+1} \ne 0$ by linear independence of $\{\mathbf{u}_i\}_{i=1}^r$. Multiply $-c_{r+1}\mathbf{u}_{r+1} = c_1\mathbf{u}_1 + \cdots + c_r\mathbf{u}_r$ by λ_{r+1} and by A, subtract and obtain a contradiction.
10. $A\overline{\mathbf{x}} = \overline{A\mathbf{x}} = \overline{\lambda\mathbf{x}} = \overline{\lambda}\overline{\mathbf{x}}$

Section 3.2

3. $(k - 1)n^3 + n^2$ versus kn^2
4. a) From (3.12), $\lim_{k\to\infty}\beta_k = \lambda_1(\alpha_1\gamma_1/\alpha_1\gamma_1) = \lambda_1$

   b) With $\lambda_1 = \lambda_2$, $|\lambda_2| > |\lambda_3|$ in (3.12), $\lim_{k\to\infty}\beta_k = \lambda_1(\alpha_1\gamma_1 + \alpha_2\gamma_2)/(\alpha_1\gamma_1 + \alpha_2\gamma_2) = \lambda_1$
5. $\mathbf{x}_k^T\mathbf{x}_k = (\alpha_1\lambda_1^k\mathbf{u}_1 + \cdots + \alpha_n\lambda_n^k\mathbf{u}_n)^T(\alpha_1\lambda_1^k\mathbf{u}_1 + \cdots + \alpha_n\lambda_n^k\mathbf{u}_n) = \alpha_1^2\lambda_1^{2k} + \cdots + \alpha_n\lambda_n^{2k}$
6. Multiply $A\mathbf{u}_j = \lambda_j\mathbf{u}_j$ by $\mathbf{u}_i^T$ and $\mathbf{u}_i^TA^T = \mathbf{u}_i^TA = \lambda_i\mathbf{u}_i^T$ by $\mathbf{u}_j$, subtract, and obtain $0 = (\lambda_j - \lambda_i)\mathbf{u}_i^T\mathbf{u}_j$.
8. Establish $\mathbf{y}_k = A^k\mathbf{w}_0/(\|\mathbf{y}_1\|\cdots\|\mathbf{y}_{k-1}\|) = \mathbf{w}_k/D_{k-1}$, $k > 1$. Then $\mathbf{x}_{k-1} = \mathbf{y}_{k-1}/\|\mathbf{y}_{k-1}\| = \mathbf{w}_{k-1}/D_{k-1}$. Hence $(\mathbf{v}^T\mathbf{y}_k/\mathbf{v}^T\mathbf{x}_{k-1}) = (\mathbf{v}^T\mathbf{w}_k/D_{k-1})/(\mathbf{v}^T\mathbf{w}_{k-1}/D_{k-1}) = \mathbf{v}^T\mathbf{w}_k/\mathbf{v}^T\mathbf{w}_{k-1}$.
10. $A\mathbf{x} = \lambda B\mathbf{x}$ is the same as $B^{-1}A\mathbf{x} = \lambda\mathbf{x}$. If $\mathbf{y}_{k+1} = B^{-1}A\mathbf{x}_k$, $\mathbf{y}_{k+1}$ can be found by solving $B\mathbf{y}_{k+1} = A\mathbf{x}_k$.

Sections 3.2.1-3.2.4

1. $\begin{bmatrix} 3.2 & -1.4 \\ .6 & -1.2 \end{bmatrix}$
2. For $k > 1$, $A_1\mathbf{u}_k = A\mathbf{u}_k - \lambda_1\mathbf{u}_1\mathbf{u}_1^T\mathbf{u}_k = A\mathbf{u}_k = \lambda_k\mathbf{u}_k$, and $A_1\mathbf{u}_1 = A\mathbf{u}_1 - \lambda_1\mathbf{u}_1(\mathbf{u}_1^T\mathbf{u}_1) = A\mathbf{u}_1 - \lambda_1\mathbf{u}_1 = \theta$.
11. $\lambda = 12$, $1 \pm 5i$, 2
12. Any real α is equidistant from λ_n and $\lambda_{n-1} = \overline{\lambda}_n$
13. $P = [\mathbf{e}_1, 9\ldots, \mathbf{e}_{r-1}, \mathbf{e}_s, \mathbf{e}_{r+1}, \ldots, \mathbf{e}_{s-1}, \mathbf{e}_r, \mathbf{e}_{s+1}, \ldots, \mathbf{e}_n]$; then clearly $P = P^T$. Since $A\mathbf{e}_j = \mathbf{A}_j$, then $AP = [\mathbf{A}_1, \ldots, \mathbf{A}_{r-1}, \mathbf{A}_s, \mathbf{A}_{r+1}, \ldots, \mathbf{A}_{s-1}, \mathbf{A}_r, \mathbf{A}_{s+1}, \ldots, \mathbf{A}_n]$. Since $\mathbf{P}_r = \mathbf{e}_s$ and $\mathbf{P}_s = \mathbf{e}_r$, then $PP = [\mathbf{e}_1, \mathbf{e}_2, \ldots, \mathbf{e}_n] = I$. $(PA)^T = A^TP$, interchanging rth and sth columns of A^T, and hence PA interchanges rth and sth rows of A.

14. $(\mathbf{x}_1)_1 = (\mathbf{u}_1)_2 \neq 0$. $P^{-1} = P$; so $PAP = PAP^{-1}$. $\mathbf{u}_1 = P^{-1}\mathbf{x}_1 = P\mathbf{x}_1$, $PAP\mathbf{x}_1 = PA\mathbf{u}_1 = P(\lambda_1\mathbf{u}_1) = \lambda_1\mathbf{x}_1$.

15. $\mathbf{x}^T\mathbf{x} = \alpha_1^2 + \alpha_2^2 + \cdots + \alpha_n^2$, $\mathbf{x}^TA\mathbf{x} = \alpha_1^2\lambda_1 + \alpha_2^2\lambda_2 + \cdots + \alpha_n^2\lambda_n$; then $\lambda_n(\alpha_1^2 + \cdots + \alpha_n^2) \le \mathbf{x}^TA\mathbf{x} \le \lambda_1(\alpha_1^2 + \cdots + \alpha_n^2)$.

16. $r_1 = r_2 = 6$, $r_3 = r_4 = 4$, $|\lambda - 5| \le 6$, $|\lambda - 4| \le 4$; then $-1 \le \lambda \le 11$.

17. For A: $r_1 = r_2 = r_3 = r_4 = 9$, $|\lambda - 6| \le 9$, $-3 \le \lambda \le 15$

For B: $r_1 = 8$, $r_2 = 7$, $r_3 = 6$, $r_4 = 11$, $|\lambda - 2| \le 8$, $|\lambda + 3| \le 7$, $|\lambda - 6| \le 6$, $|\lambda + 1| \le 11$; then $-12 \le \lambda \le 12$

Section 3.3

1. $\mathbf{y}_1 = [2, 3, 9]^T$, $\mathbf{y}_2 = [17, 41, 57]^T$, $\mathbf{y}_3 = [90, 251, 321]^T$, $15\mathbf{y}_0 + 7\mathbf{y}_1 - 7\mathbf{y}_2 + \mathbf{y}_3 = \theta$, $p(\lambda) = \lambda^3 - 7\lambda^2 + 7\lambda + 15$

2. $\mathbf{y}_1 = [1, 2, 0, 0]^T$, $\mathbf{y}_2 = [3, 2, -2, 0]^T$, $\mathbf{y}_3 = [3, 4, 2, -4]^T$, $\mathbf{y}_4 = [5, 4, 0, -4]^T$, $\mathbf{y}_1 - \mathbf{y}_2 - \mathbf{y}_3 + \mathbf{y}_4 = \theta$, $p(\lambda) = \lambda^4 - \lambda^3 - \lambda^2 + \lambda$

3. The coefficient of λ^n is always 1, but the other coefficients vary. The resulting polynomials are not scalar multiples.

5. Fails if $\beta = 0$ or if $\alpha = \beta$.

6. d)
$$PHP = \begin{bmatrix} 0 & 2 & 1 & 1 \\ 1 & 1 & 1 & 1 \\ -1 & 0 & -2 & -2 \\ 0 & 0 & 2 & 2 \end{bmatrix}$$

8. $N = n^3 + (n^2 + n)/2$ multiplications/divisions. For $n = 20$, $N = 8210$.

Section 3.3.1

2. For $k > 2$, the kth row of S_1A is $(-a_{k1}/a_{21})$ times the second row of A plus the third row of A, creating zeros in the $(k, 1)$ positions. Applying this rule with S_1^{-1} as in (3.29b) shows $S_1S_1^{-1} = I$. The first column of S_1^{-1} is $\mathbf{e}_1$ and hence $(S_1AS_1^{-1})_1 = (S_1A)_1$.

3. a)
$$H = \begin{bmatrix} -7 & -8 & -3 \\ 8 & 9 & 3 \\ 0 & 1 & 1 \end{bmatrix}$$
b)
$$H = \begin{bmatrix} -6 & 31 & -14 \\ -1 & 6 & -2 \\ 0 & 2 & 1 \end{bmatrix}$$
c)
$$H = \begin{bmatrix} 1 & 1 & 3 \\ 1 & 3 & 1 \\ 0 & 4 & 2 \end{bmatrix}$$
d)
$$H = \begin{bmatrix} 6 & 33 & 8 & 4 \\ 1 & 38 & 8 & 4 \\ 0 & -120 & -25 & -15 \\ 0 & 0 & 0 & 5 \end{bmatrix}$$
e)
$$H = \begin{bmatrix} 1 & -3 & 4 & -2 \\ 1 & 9 & -3 & 2 \\ 0 & 14 & -5 & 4 \\ 0 & 0 & 2 & 2 \end{bmatrix}$$
f)
$$H = \begin{bmatrix} 1 & -5 & 10 & 4 & 3 \\ -3 & 2 & 6 & 1 & 2 \\ 0 & -1 & 0 & 3 & 1 \\ 0 & 0 & -8 & 7 & 0 \\ 0 & 0 & 0 & -11 & -4 \end{bmatrix}$$

5. $(L_1A)_1 = [1, -a_{21}/a_{11}, \ldots, -a_{n1}/a_{11}]^T$, $L_1 + L_1^{-1} = 2I$

7. a) $\mathbf{w}_0 = \mathbf{e}_1$, $\mathbf{w}_1 = [2, -1, 0]^T$, $\mathbf{w}_2 = [2, -1, -2]^T$, $\mathbf{w}_3 = [4, -3, -4]^T$, $p(\lambda) = \lambda^3 - 2\lambda^2 - \lambda + 2$.

b) $\mathbf{w}_0 = \mathbf{e}_1$, $\mathbf{w}_1 = [3, -1, 0]^T$, $\mathbf{w}_2 = [17, -1, -6]^T$, $\mathbf{w}_3 = [35, -3, -12]^T$, $p(\lambda) = \lambda^3 - 2\lambda^2 - \lambda + 2$

c) $\mathbf{w}_0 = \mathbf{e}_1$, $\mathbf{w}_1 = [2, 2, 0, 0]^T$, $\mathbf{w}_2 = [10, 18, 2, 0]^T$, $\mathbf{w}_3 = [32, 64, 8, 8]^T$, $\mathbf{w}_4 = [64, 144, 16, 64]^T$, $p(\lambda) = \lambda^4 - 8\lambda^3 + 24\lambda^2 - 32\lambda + 16$

10. If the program does not require multiplications by the normal zeros and ones in each S_k and S_k^{-1} and operations to create zeros are ignored, total operations are $\sum_{k=1}^{n-2} (n - k - 1)(2n - k + 1) = (n - 2)(7n^2 - 4n - 3)/6$ to obtain H and $n(n + 1)/2$ to backsolve.

11. a) $\lambda = 2$, $\mathbf{x} = [1, -1, 0, 0]^T$; $\lambda = 1$, $\mathbf{x} = [-9, 5, 1, 1]^T$; $\lambda = 3$, $\mathbf{x} = [3, -9, 1, -1]^T$

b) $\lambda = -1$, $\mathbf{x} = [2, 0, -1, 0]^T$; $\lambda = 0$, $\mathbf{x} = [-1, 1, 1, 0]^T$; $\lambda = 2$, $\mathbf{x} = [1, 15, 1, 6]^T$

Section 3.3.2

1. Exercise 3d, Section 3.3.1 provides an example.

2. a) i) $\mathbf{u} = \begin{bmatrix} 0 \\ 4 \\ 4 \end{bmatrix}$, $Q = \begin{bmatrix} 1 & 0 & 0 \\ 0 & 0 & -1 \\ 0 & -1 & 0 \end{bmatrix}$ ii) $\mathbf{u} = \begin{bmatrix} 0 \\ 8 \\ 4 \end{bmatrix}$, $Q = \begin{bmatrix} 1 & 0 & 0 \\ 0 & -.6 & -.8 \\ 0 & -.8 & .6 \end{bmatrix}$

iii) $\mathbf{u} = \begin{bmatrix} 0 \\ 25 \\ -5 \end{bmatrix}$ iv) $\mathbf{u} = \begin{bmatrix} 0 \\ 5 \\ 1 \\ -2 \end{bmatrix}$, $Q = (1/15)\begin{bmatrix} 15 & 0 & 0 & 0 \\ 0 & -10 & -5 & 10 \\ 0 & -5 & 14 & 2 \\ 0 & 10 & 2 & 11 \end{bmatrix}$

v) $\mathbf{u} = [0, 10, 2, -6]^T$ vi) $\mathbf{u} = [0, 9, 2, 1, 2]^T$

3. a) $\begin{bmatrix} 2 & .4 & 2.2 \\ 5 & .68 & .24 \\ 0 & 2.24 & -.68 \end{bmatrix}$ b) $\begin{bmatrix} 1 & 3 & 3 & 1 \\ 2 & 2 & 1 & 1 \\ 0 & -1 & 0 & 2 \\ 0 & 0 & 2 & -1 \end{bmatrix}$ c) $\begin{bmatrix} 1 & 2 & 1 & 2 \\ 3 & 3 & 2.2 & .4 \\ 0 & -5 & -1.6 & -.2 \\ 0 & 0 & -.2 & .6 \end{bmatrix}$

7. $c = -2$: {+, +, +, +, +}, $c = 5$: {+, −, +, −, +}, $c = 0$: {+, +, −, −, −}, $c = 1$: {+, +, −, −, +}, $c = .5$: {+, +, −, −, +}, $c = .25$: {+, +, −, −, +} One eigenvalue in $(-2, 0)$, one in $(0, .25)$, two in $(1, 5)$

9. i) $Q_1 = \begin{bmatrix} -.6 & 0 & -.8 \\ 0 & 1 & 0 \\ -.8 & 0 & .6 \end{bmatrix}$, $Q_2 = \begin{bmatrix} 1 & 0 & 0 \\ 0 & -.6 & -.8 \\ 0 & -.8 & -.6 \end{bmatrix}$,

$$Q_2Q_1A = \begin{bmatrix} -5 & -7 & -3 \\ 0 & -5 & -.2 \\ 0 & 0 & 1.4 \end{bmatrix} = R$$

CHAPTER 4

Section 4.2

3. There is a sign change in [99, 100]. By the mean-value theorem, $f(\alpha) - f(s) = f'(\theta)$ $(\alpha - s)$ for some θ between α and s. Therefore, $f(\alpha) = f'(\theta)(\alpha - s)$; and since $f'(\theta)$ is on the order of $2 + 10^6$, we can expect $f(\alpha) \simeq 200$ if $\alpha - s \simeq 10^{-4}$.
4. The interval [1, 2] contains the smallest positive root.
6. $f(j\pi/n) = \sin(nj\pi/n) - \alpha \sin((n - 1)j\pi/n)$
$= -\alpha \sin(j\pi - j\pi/n) = \alpha \sin(j\pi/n)\cos(j\pi)$.
Since $\sin(j\pi/n)$ is positive, we see that $f(j\pi/n)$ is positive for j even and negative for j odd.
7. By Problem 6, $f(x)$ is negative at $x = \pi/n$. However, $f(x)$ must be positive for x in $(0, \varepsilon)$ for some small ε since $f(0) = 0$ and $f'(0) > 0$; that is, $f(x)$ is an increasing function as x moves through $x = 0$.

Section 4.3.1

2. For (a), $g'(x) = 3x^2/13$ and thus $g'(x) \geq 0$ for all x. Next, $g'(x) < 1$ if and only if $x^2 < 13/3$. So, for instance, $|g'(x)| \leq k < 1$ for $-\infty < x < 2.08$ and an interval [1.92, 2.08] will satisfy the conditions of Problem 2. For (d), the condition $|g'(x)| < 1$ is $-1 < 3x^2 - 12 < 1$ or $11 < 3x^2 < 13$. Again, the interval [1.92, 2.08] will be satisfactory.
5. To show that $g(x)$ is in $[s, b]$ whenever x is in $[s, b]$, note first that $g(x)$ is increasing; so $x \geq s$ implies $g(x) \geq g(s) = s$. Next, $g(x) - g(s) = g'(\theta)(x - s) \leq x - s \leq b - s$; and therefore since $g(s) = s$, $g(x) \leq b$. The same argument shows that $g(x_i) - g(s) \leq x_i - s$ or $x_{i+1} \leq x_i$ for all i.
7. If $|g'(s)| > 1$, then there is an interval $I = [s - \varepsilon, s + \varepsilon]$ where $|g'(x)| \geq \kappa > 1$ for all x in I. If x_0 is in I, then $g(x_0) - g(s) = g'(\theta)(x_0 - s)$ or $|x_1 - s| = |g'(\theta)|\,|x_0 - s| \geq \kappa\,|x_0 - s| > |x_0 - s|$.
9. For $-1 < x_0 < 1$, $\{x_i\} \to 0$. For $x_0 = 1$ or $x_0 = -1$, $\{x_i\} \to 1$. For $x_0 < -1$ or $1 < x_0$, the iteration does not converge.

Section 4.3.2

1. Since $g'(x) = (x - 2)/2$, it follows that $-1 < g'(x) < 1$ when $-2 < x - 2 < 2$ or $0 < x < 4$. Clearly, $|g'(x)| \leq 1 - .5\varepsilon$ when $\varepsilon < x < 4 - \varepsilon$. The rate of convergence is linear since $g'(s) \neq 0$.
4. $g'(x) = 6/x^2$; so $0 < g'(x) \leq 2/3$ when $3 \leq x < \infty$.
5. The fixed points are $s = 1$ and $s = 2$. An iteration can converge only to $s = 1$, and convergence is quadratic to $s = 1$.

7. For (a), $\lim_{n\to\infty} \frac{t_{n+1} - 2}{t_n - 2} = 1$; so Aitken's Δ^2-method cannot work. For (b), $\lim_{n\to\infty} \frac{t_{n+1} - 1}{t_n - 1} = \frac{7}{8}$; so the Δ^2-method will work.

9. Note that $\{S_n\} \to 1/(1 - r) = S^*$ and hence $S_n - S^* = -r^{n+1}/(1 - r)$. Thus it follows that $\lim_{n\to\infty} \frac{S_{n+1} - S^*}{S_n - S^*} = r$; and so the Δ^2-method can be applied.

Section 4.3.3

8. If $g(x) = x - f(x)/f'(x)$ and if $f(s) = 0$ with $f'(s) \neq 0$, then $g'(s) = 0$. Thus by Theorem 4.4 there is an $\varepsilon > 0$ such that $|x_0 - s| < \varepsilon$ implies that $\{x_i\} \to s$. From the hint, $e_{n+1} = e_n + (s - x_n) + f''(\theta_n)e_n^2/2!f'(x_n) = f''(\theta_n)e_n^2/2!f'(x_n)$ and hence $e_{n+1}/e_n^2 = f''(\theta_n)/2f'(x_m)$. Now since $\{x_n\} \to s$, $\{\theta_n\} \to s$ and therefore $\kappa = f''(s)/2f'(s)$.

9. From l'Hospital's rule, it follows that $\lim_{x\to s} g(x) = s$. Setting $g(s)$ equal to s, we have $[g(s + h) - g(s)]/h = 1 - f(s + h)/hf'(s + h)$. But, $f(s + h)/hf'(s + h) = f^{(p)}(\theta)(p - 1)!/f^{(p)}(\beta)p!$; and thus, letting $h \to 0$, we obtain $g'(s) = 1 - 1/p$.

11. Clearly if $x_i > r_n$, then $x_{i+1} < x_i$ since $f(x_i)/f'(x_i) > 0$. To see $x_{i+1} > r_n$, consider

$$x_{i+1} - r_n = x_i - r_n - \frac{f(x_i)}{f'(x_i)} = -\frac{[f(x_i) + f'(x_i)(r_n - x_n)]}{f'(x_i)}.$$

But, $f(r_n) = f(x_i) + f'(x_i)(r_n - x_i) + f''(\theta_i)(r_n - x_i)^2/2!$; so $x_{i+1} - r_n = f''(\theta_i)(r_n - x_i)^2/2f'(x_i) > 0$ since $r_n < \theta_i < x_{i+1}$ and $f''(x) > 0$ for $x > r_n$.

Section 4.3.5

3. $\gamma^{n+1} = \gamma^n + \gamma^{n-1}$ holds for all $n \geq 1$ if and only if $\gamma^2 = \gamma + 1$ or $\gamma^2 - \gamma - 1 = 0$; we assume $\gamma = 0$ is not interesting. Thus $\gamma_1 = (1 + \sqrt{5})/2$ and $\gamma_2 = (1 - \sqrt{5})/2$.

Section 4.4

1. Using $x - r$ for $Q(x)$ in (4.16) gives $p(x) = (x - r)S(x) + R(x)$ where $R(x)$ is a constant polynomial. Since (4.16) is an identity in x, $p(r) = R(r)$, or $R(r) = 0$; and since $R(x)$ is a constant polynomial, $R(x) \equiv 0$.

2. If r_1 and r_2 are zeros of $p(x)$, so are $\bar{r}_1$ and $\bar{r}_2$. From (4.14), $p(x) = 2(x - r_1)(x - \bar{r}_1)(x - r_2)(x - \bar{r}_2)$ or $p(x) = 2(x^2 - 4x + 13)(x^2 + 4x + 13)$.

4. By (4.16), $p(x) = (x - r)^2Q(x) + R(x)$ where $R(x)$ has degree 1 or less, $R(x) = ax + b$. Since $P'(r) = 0$, it follows that $R'(r) = 0$ and thus $a = 0$. Next since $P(r) = 0$, we see that $b = 0$ as well.

CHAPTER 5

Section 5.1

2. For any x, $|f(x) - p_k(x)| \le |f^{(k+1)}(\xi)|\,|x|^{k+1}/(k+1)!$ where ξ is somewhere between 0 and x. For $f(x) = \cos(x)$, $|f^{(k+1)}(\xi)| \le 1$; so we have $|f(x) - p_k(x)| \le |x|^{k+1}/(k+1)!$ For $x = 3$ and $k = 16$, $|3|^{17}/17! < 3.64 \times 10^{-7} < 10^{-6}$; so $k = 16$ will suffice.

4. $\left|\int_0^4 \frac{p_k(t)}{t}\,dt - S_i(4)\right| = \left|\int_0^4 \frac{p_k(t)}{t}\,dt - \int_0^4 \frac{\sin(t)}{t}\,dt\right| \le \int_0^4 \left|\frac{p_k(t) - \sin(t)}{t}\right|\,dt$

Now, $p_k(t) - \sin(t) = f^{(k+1)}(\eta_t)t^{k+1}/(k+1)!$ where $0 < \eta_t < t$ for any t in $[0, 4]$, and where $f(t) = \sin(t)$. Therefore the integrand is bounded by

$$\left|\frac{p_k(t) - \sin(t)}{t}\right| \le \frac{t^k}{(k+1)!} \le \frac{4^k}{(k+1)!}$$

for all t in $[0, 4]$. Consequently

$$\left|\int_0^4 \frac{p_k(t)}{t}\,dt - S_i(4)\right| \le 4^{k+1}/(k+1)!$$

and $k = 19$ is sufficient to estimate $S_i(4)$ to within 10^{-6}.

Section 5.2.1

2. For part (a), if $p(x) = a_0 + a_1x + a_2x^2 + a_3x^3$, then the system to be solved is

$$\begin{aligned} a_0 - a_1 + a_2 - a_3 &= -1 \\ a_0 &= 3 \\ a_0 + 2a_1 + 4a_2 + 8a_3 &= 11 \\ a_0 + 3a_1 + 9a_2 + 27a_3 &= 27. \end{aligned}$$

The solution is $a_0 = 3$, $a_1 = 2$, $a_2 = -1$, $a_3 = 1$; and thus $p(x) = 3 + 2x - x^2 + x^3$ will interpolate the data.

3. For part (a), the divided difference table is

$$\begin{array}{r|rrrr} -1 & -1 & & & \\ & & 4 & & \\ 0 & 3 & & 0 & \\ & & 4 & & 1 \\ 2 & 11 & & 4 & \\ & & 16 & & \\ 3 & 27 & & & \end{array}$$

Thus $p(x) = -1 + 4(x+1) + 0(x+1)x + 1(x+1)x(x-2)$ or $p(x) = -1 + 4(x+1) + (x+1)x(x-2)$. Clearly $p(-2) = -1 - 4 - 8 = -13$.

5. For part (a), $p(x) = 1 + 0(x + 1) + 1(x + 1)x = 1 + (x + 1)x$. For part (e), $p(x) = 1 + 2x + 10x(x - 1) + 19x(x - 1)(x - 2) + 9x(x - 1)(x - 2)(x - 3)$.

6. Let $p(x) = a_0 + a_1x + a_2x^2$ and note that $p'(x) = a_1 + 2a_2x$. The constraints $p(1) = 0$, $p'(1) = 7$, $p(2) = 10$ lead to this system:

$$\begin{aligned} a_0 + a_1 + a_2 &= 0 \\ a_1 + 2a_2 &= 7 \\ a_0 + 2a_1 + 4a_2 &= 10. \end{aligned}$$

The solution is $a_0 = -4$, $a_1 = 1$, $a_2 = 3$ and so $p(x) = -4 + x + 3x^2$.

Section 5.2.2

1. For $f(x) = x^2$, $\Delta f(x) = (x + h)^2 - x^2 = x^2 + 2xh + h^2 - x^2$ or $\Delta f(x) = 2xh + h^2$. Next, $\Delta^2 f(x) = (2(x + h)h + h^2) - (2xh + h^2) = 3h^2 - h^2 = 2h^2$. Since $\Delta^2 f(x)$ is a constant function, $\Delta^3 f(x)$ and $\Delta^4 f(x)$ are each the zero function.

2. The forward difference table is

x	$f(x)$			
0	1			
		8		
2	9		48	
		56		48
4	65		96	
		152		
6	217			

The data spacing is $h = 2$ with $x_0 = 0$; so with $x = x_0 + rh = 2r$,

$$p(x) = p(2r) = 1 + 8\binom{r}{1} + 48\binom{r}{2} + 48\binom{r}{3}.$$

Written out at length,

$$p(x) = p(2r) = 1 + 8r + 24r(r - 1) + 8r(r - 1)(r - 2).$$

In terms of x, this reduces to the divided difference form for $p(x)$:

$$p(x) = 1 + 4x + 6x(x - 2) + x(x - 2)(x - 4).$$

4. For $n = 3$, $p(x_0 + rh) = f(x_0) + \Delta f(x_0)r + \Delta^2 f(x_0)\dfrac{r(r - 1)}{2} + \Delta^3 f(x_0)\dfrac{r(r - 1)(r - 2)}{6}$.

Therefore $\dfrac{dp}{dx} = \dfrac{1}{h}\dfrac{dp}{dr}$ means that

$$\frac{dp}{dx} = \frac{1}{h}\left[\Delta f(x_0) + \Delta^2 f(x_0)\,\frac{2r - 1}{2} + \Delta^3 f(x_0)\,\frac{3r^2 - 6r + 2}{6}\right].$$

For $r = 3/2$, we are led to the formula

$$p'(x_0 + 1.5h) \doteq \frac{1}{h}\left[\Delta f(x_0) + \Delta^2 f(x_0) - \frac{\Delta^3 f(x_0)}{24}\right]$$

when $p(x)$ is a cubic polynomial interpolating $f(x)$ at $x_i = x_0 + ih$, $0 \le i \le 3$.

Section 5.2.4

1. For $n = 5$, (5.28) is $|e(x)| \le \kappa_5\, |W(x)|/6!$ where $W(x) = (x - .1)(x - .2)(x - .3)(x - .4)(x - .5)(x - .6)$. For part (b), $f(x) = \ln(x)$; so $f^{(6)}(x) = 120x^{-6}$. For $.1 \le x \le .6$, $|f^{(6)}(x)| \le 120 \times 10^6$; so $|e(x)| \le 10^6|W(x)|/6$. For $x = .15$, $|W(x)| \simeq 1.48 \times 10^{-5}$; so the best we can say from (5.28) is that $|e(.15)| \le 296$. For $f(x) = \cos(x)$, $\kappa_5 = 1$ and we have $|e(.15)| \le 1.48 \times 10^{-5}/6! = 2.06 \times 10^{-8}$.
2. For $f(x) = \ln(x)$, $\kappa_5 = 120 \times 10^6$; and (5.36) reduces to

$$|e(x)| \le \frac{120 \times 10^6}{32 \cdot 6!}(.25)^6 \simeq 1.27$$

while for $f(x) = \cos(x)$, $|e(x)| \le 1.06 \times 10^{-8}$.

Section 5.2.6

3. The system to be solved is

$$\begin{aligned} a_0 - 2a_1 + 4a_2 - 8a_3 + 16a_4 &= 2 \\ a_1 - 4a_2 + 12a_3 - 32a_4 &= -1 \\ a_0 &= 0 \\ a_0 + 2a_1 + 4a_2 + 8a_3 + 16a_4 &= 2 \\ a_1 + 4a_2 + 12a_3 + 32a_4 &= 1. \end{aligned}$$

The polynomial is $p(x) = (12x^2 - x^4)/16$.

10. If equality is to hold in (5.50), then it must be that $\int_a^b [g''(x) - S''(x)]^2 dx = 0$ or $g''(x) - S''(x) \equiv 0$ on $[a, b]$. Therefore $g(x) - S(x) \equiv rx + s$ on $[a, b]$, for some constants r and s. However, $g(a) - S(a) = 0$ and $g(b) - S(b) = 0$; so $r = s = 0$.

Section 5.2.7

3. The basis functions are, respectively,

$(0, 0) : (1 - x)(1 - y)$ $(1, 0) : x(1 - y)$
$(0, 1) : (1 - x)y$ $(1, 1) : xy$

4.
$$p(.5, .5) = .25f(0, 0) + .25f(0, 1) + .25f(1, 0) + .25f(1, 1)$$
$$= \frac{\frac{f(0, 0) + f(0, 1)}{2} + \frac{f(1, 0) + f(1, 1)}{2}}{2}$$

CHAPTER 6

Section 6.2.2

1. $A_0 = \int_0^{2h} \frac{x - h}{-h}\, dx = 0; \qquad A_1 = \int_0^{2h} \frac{x}{h}\, dx = 2h$

Thus, $\int_0^{2h} f(x)dx \simeq Q(f) = 2hf(h)$

2. The system is

$$A_0 + A_1 + A_2 = 3h$$
$$A_1h + A_2 2h = \frac{9h^2}{2}$$
$$A_1h^2 + A_2 4h^2 = 9h^3.$$

The solution is $A_0 = 3h/4$, $A_1 = 0$, $A_2 = 9h/4$.

Thus $\int_0^{3h} f(x)dx \simeq Q(f) = \frac{h}{4}[3f(0) + 9f(2h)]$.

6. $\ell_0(x) = (x - \alpha)(x - \alpha - h)/2h^2$; so $\ell_0'(x) = [(x - \alpha - h) + (x - \alpha)]/2h^2$. Therefore $A_0 = \ell_0'(\alpha) = -h/2h^2 = -1/2h$. Similarly, $A_1 = 0$ and $A_2 = 1/2h$. This result leads to the numerical differentiation formula

$$f'(\alpha) \simeq \frac{f(\alpha + h) - f(\alpha - h)}{2h}.$$

7. The system is

$$A_0 + A_1 + A_2 = 0$$
$$A_0(\alpha - 2h) + A_1(\alpha - h) + A_2\alpha = 1$$
$$A_0(\alpha - 2h)^2 + A_1(\alpha - h)^2 + A_2\alpha^2 = 2\alpha.$$

The solution is $A_0 = 1/2h$, $A_1 = -4/2h$, $A_2 = 3/2h$; and the differentiation formula is

$$f'(\alpha) \simeq \frac{1}{2h}[f(\alpha - 2h) - 4f(\alpha - h) + 3f(\alpha)]$$

Section 6.2.4

2. A convenient form of (6.21b) for this problem is $e_N^T = -f''(\eta)(b-a)^3/(12N^2)$. For (a), $f''(x) = 2/x^3$ and an upper bound for $|f''(x)|$ on $[1, 2]$ is 2. Thus we want N such that $2/(12N^2) \le 10^{-3}$ or $1000/6 \le N^2$ or $17 \le N$. For (b), $f''(x) = (6x^2 - 2)/(1 + x^2)^3$; a crude upper bound for $|f''(x)|$ is 8; the actual maximum is 2. Using 2 and noting $b - a = 2$, we require $16/(12N^2) \le 10^{-3}$ or $N = 37$. For (c), $N = 26$ is sufficient.

4. $I(f) - T_{2N}(f) = \dfrac{(h/2)^3}{12} \displaystyle\sum_{k=0}^{2N-1} f''(\beta_k)$ where $y_k \le \beta_k \le y_{k+1}$. Thus $I(f) - T_{2N}(f) = -\dfrac{h^3}{8 \cdot 12} \displaystyle\sum_{j=0}^{N-1} [f''(\beta_{2j}) + f''(\beta_{2j+1})]$ where β_{2j} and β_{2j+1} are both in $[x_j, x_{j+1}]$. If h is small enough so that

$$\frac{f''(\beta_{2j}) + f''(\beta_{2j+1})}{2} \simeq f''(\eta_j),$$

then we can rewrite $I(f) - T_{2N}(f)$ as

$$I(f) - T_{2N}(f) \simeq \frac{-h^3}{4 \cdot 12} \sum_{j=0}^{N-1} f''(\eta_j) = \frac{1}{4}[I(f) - T_N(f)].$$

This approximation is equivalent to

$$4[I(f) - T_{2N}(f)] \simeq [I(f) - T_N(f)] \text{ or } 3[I(f) - T_{2N}(f)] \simeq [T_{2N}(f) - T_N(f)].$$

6. The first test is $|T_{16}(f) - T_8(f)| \le \text{TOL}$. At the nodes $x_j = j\pi/8$ and $y_k = k\pi/16$, the integrand has the value 1; $T_{16}(f) = T_8(f)$ because the composite trapezoid rules believe they are integrating $f(x) \equiv 1$ on $[0, 1]$.

8. $S_N(f) = \displaystyle\sum_{j=0}^{N-1} \frac{h}{6}[f(x_j) + 4f(x_j + h/2) + f(x_{j+1})]$ where $x_{j+1} - x_j = h = (b-a)/N$ and $x_i = a + ih$, $0 \le i \le N$. By Lemma 6.1,

$$\frac{h}{6}[f(x_j) + 4f(x_j + h/2) + f(x_{j+1})] = hf(\eta_j)$$

for some η_j in $[x_j, x_{j+1}]$. Thus $S_N(f)$ can be written as $S_N(f) = \displaystyle\sum_{j=0}^{N-1} f(\eta_j)h$ and this is a Riemann sum for $f(x)$. Therefore $S_N(f) \to I(f)$ as $N \to \infty$.

Section 6.4.1

1. The table, to six places, is

.361172			
.355353	.353413		
.353997	.353545	.353554	
.353664	.353553	.353554	.353554

3. $f(\alpha + h) = f(\alpha) + f'(\alpha)h + f''(\alpha)h^2/2! + f'''(\alpha)h^3/3! + f''''(\alpha)h^4/4! + \cdots$ and $f(\alpha - h) = f(\alpha) - f'(\alpha)h + f''(\alpha)h^2/2! - f'''(\alpha)h^3/3! + f''''(\alpha)h^4/4! + \cdots$. Thus

$$\frac{f(\alpha + h) - f(\alpha - h)}{2h} = \frac{2hf'(\alpha) + 2h^3f'''(\alpha)/3! + 2h^5f^{(5)}(\alpha)/5! + \cdots}{2h},$$

which establishes (6.28).

6. For any j, $A_{0,j} = A(r^jh) = a_0 + a_k(r^jh)^k + 0(h^{k+1})$
Thus $A_{0,j} - A_{0,j-1} = a_k(r^jh)^k - a_k(r^{j-1}h)^k + 0(h^{k+1}) - [r^k - 1]\,a_k(r^{j-1}h)^k + 0(h^{k+1})$.
Consequently

$$\frac{A_{0,m} - A_{0,m-1}}{A_{0,m+1} - A_{0,m}} = \frac{[r^k - 1]a_k(r^{m+1}h)^k + 0(h^{k+1})}{[r^k - 1]a_k(r^mh)^k + 0(h^{k+1})}.$$

Neglecting the terms $0(h^{k+1})$ in the numerator and the denominator, then canceling $[r^k - 1]a_k(r^{m-1}h)^k$, we obtain

$$\frac{A_{0,m} - A_{0,m-1}}{A_{0,m+1} - A_{0,m}} \simeq \frac{1}{r^k}.$$

Section 6.5.3

6. Let $p(x) = p_{n+1}(x)p_{n+1}(x)$ and note that $Q_n(p) = 0$ since the nodes x_j are roots of $p_{n+1}(x) = 0$. However, $I(p) = \langle p_{n+1}, p_{n+1}\rangle \neq 0$.

11. a) $P_m(x) = \sum_{j=0}^{m} \langle f, p_j\rangle p_j(x) = \sum_{j=0}^{m} \left(\left(\int_a^b w(t)f(t)p_j(t)dt\right)p_j(x)\right)$. So $P_m(x) = \int_a^b w(t)f(t)$

$$\left(\sum_{j=0}^{m} p_j(t)p_j(x)\right)dt = \int_a^b w(t)f(t)G_m(x, t)dt.$$

c) For $r(x) \equiv 1$, $P_m(x) \equiv 1$. Thus from (a), $1 = \int_a^b w(t)G_m(x, t)dt$. Therefore $f(x) =$

$f(x) \cdot 1 = f(x) \int_a^b w(t)G_m(x, t)dt$; and since t is the variable of integration, $f(x)$ can be moved inside the integral.

e) By (d), $f(x_0) - P_m(x_0) = \int_a^b w(t)[f(x_0) - f(t)]G_m(x, t)dt$; and applying (b),

$$G_m(x, t) = \frac{\lambda_m[p_{m+1}(x)p_m(t) - p_{m+1}(t)p_m(x)]}{\lambda_{m+1}(x - t)}.$$

Therefore,

$$f(x_0) - P_m(x_0) = \gamma_m \int_a^b w(t)g(t)p_{m+1}(x_0)p_m(t)dt - \gamma_m \int_a^b w(t)g(t)p_m(x_0)p_{m+1}(t)dt$$

where $\gamma_m = \lambda_m/\lambda_{m+1}$. Equivalently,

$$f(x_0) - P_m(x_0) = \gamma_m[p_{m+1}(x_0)\langle p_m, g\rangle - p_m(x_0)\langle p_{m+1}, g\rangle];$$

so $P_m(x_0) \to f(x_0)$ as $m \to \infty$.

CHAPTER 7

Section 7.1

4. a) $L = 1$

   b) $f_y(x, y) = -2xye^{-y^2}$ and for any fixed x, the maximum of $f_y(x, y)$ is at $y = .5$. Thus $L = |f_y(1, .5)| \simeq .7788$.

6. $w(x) = (y_0 + c/\lambda)e^{\lambda x}$; so $y(x) = (y_0 + c/\lambda)e^{\lambda x} - c/\lambda$. For $\lambda < 0$, $y(x) \to -c/\lambda$ as $x \to \infty$.

11. $y(x) = y_0\exp[2x - x^2/2]$; so $y(9) = y_0\exp(-22.5)$ and we need $y_0 \leq 10^{-3}e^{22.5} \simeq 5.9 \times 10^6$.

Section 7.2.2

4. Euler's method will follow the trivial solution $y(x) \equiv 0$.

10. $y_{k+1} = y_k + hf(x_k, y_k) = y_k + h\lambda y_k = (1 + h\lambda)y_k$ for $k = 0, 1, 2, \ldots$; and therefore $y_i = (1 - h\lambda)^i y_0$. Clearly we can have $|y_i| \to 0$ only if $|1 + h\lambda| < 1$. Hence h must be chosen so that $-1 < 1 + h\lambda < 1$ or $-2 < h\lambda < 0$. For $\lambda = -25$, we need $h < 2/25 = .08$ in order for the numerical solution to tend to zero.

11. Since $y' = \lambda y$, $y'' = \lambda y' = \lambda^2 y$; $y''' = \lambda y'' = \lambda^3 y, \ldots y^{(k)} = \lambda^k y$. Thus $y_{j+1} = p(\bar{h})y_j$ for $j = 0, 1, \ldots$. For $k = 2$, $p(\bar{h}) = 1 + \bar{h} + \bar{h}^2/2$; and so let us consider the polynomial $p(t) = 1 + t + t^2/2$. For $t > 0$, $p(t) > 1$; but note that $p(0) = 1$ while $p'(0) > 0$. Thus there is a value $a < 0$ such that $|p(t)| < 1$ for t in $(a, 0)$. The minimum of $p(t)$ is at $t = -1$ and $p(-1) = 1/2$; so $p(t)$ is always positive and concave up. Therefore $|p(t)| < 1$ in $(a, 0)$ where $p(a) = 1$ or $a = -2$. For $k = 3$, α is about -2.51.

Section 7.2.3

7. Heun's method can be expressed as

$$y_{i+1} = y_i + h[K_1(x_i, y_i; h + K_2(x_i, y_i; h)]/2;$$

   and for $f(x, y) = \lambda y$,

$$K_1(x_i, y_i; h) = f(x_i, y_i) = \lambda y_i$$

$$K_2(x_i, y_i; h) = f(x_i + h, y_i + hf(x_i, y_i)) = \lambda[y_i + hf(x_i, y_i)]$$
$$= \lambda y_i + h\lambda^2 y_i.$$

   Thus $y_{i+1} = y_i + h[\lambda y_i + (\lambda y_i + h\lambda^2 y_i)]/2 = p(\bar{h})y_i$.

8. From $f(x, y) = \lambda y$, (7.10) reduces to $y_{i+1} = y_i + h[A_1\lambda y_i + A_2\lambda(y_i + \alpha h\lambda y_i)]$ or $y_{i+1} = y_i + h\lambda y_i[A_1 + A_2(1 + \alpha h\lambda)]$. By (7.12), $A_1 + A_2 = 1$, $A_2\alpha = 1/2$; so we obtain $y_{i+1} = y_i + h\lambda y_i[1 + h\lambda/2] = p(\bar{h})y_i$.

Section 7.2.5

1. The solution to $y' = 4x^3$ is $y(x) = x^4$; so in (7.15) we have $y(x + h) = (x + h)^4$. Thus (7.15) reduces to $x^4 + 4x^3h + 6x^2h^2 + 4xh^3 + h^4 = x^4 + h\phi(x, y(x); h) + h\tau$. For

Euler's method, $\phi(x, y(x); h) = f(x, y(x)) = 4x^3$; and we find $6x^2h^2 + 4xh^3 + h^4 = h\tau$. Therefore, $\tau = h[6x^2 + 4xh^2 + h^3]$ and $k = 1$ or Euler's method is first order. For the improved Euler's method, $\phi(x, y(x); h) = f(x + h/2, y(x) + hy'(x)/2) = 4(x + h/2)^3$. Thus $\phi(x, y(x); h) = 4x^3 + 6x^2h + 9xh^2 + h^3/2$, and (7.15) reduces to $4xh^3 + h^4 = 9xh^3 + h^4/4 + h\tau$. Therefore $\tau = h^2(-5x + .75h^3)$.

5. $u(x) - y(x) = (u_0 - y_0)e^x = 10^{-5}e^x$. Therefore $|u(x) - y(x)| \geq 1$ when $e^x \geq 10^5$ or when x is larger than about 11.5. The relative difference between $u(x)$ and $y(x)$ is

$$\frac{u(x) - y(x)}{y(x)} = \frac{(u_0 - y_0)e^x}{y_0 e^x} = \frac{u_0 - y_0}{y_0} = \frac{10^{-5}}{\sqrt{2}}.$$

6. If $u(x)$ satisfies $u' = \lambda u$, $u(x_i) = y_i$, then $u(x) = u(x_i)e^{\lambda(x-x_i)}$ and so $u(x_{i+}) = y_i e^{\bar{h}}$. Therefore $u(x_{i+1}) - y_{i+1} = y_i e^{\bar{h}} - y_i p(\bar{h}) = y_i[e^{\bar{h}} - p(\bar{h})]$.

Section 7.3

1. The constants required for Theorem 7.2 are K and τ. Clearly, $K = 2$ while τ is the maximum of $h|y''(x)|/2$. Since $y(x) = e^{x^2}$, $y''(x)/2 = (2x^2 + 1)e^{x^2}$ and hence $\tau = 3he$. Thus $|e_j| \leq 1.5\ he[e^{2x_j} - 1]$.

3. $e_{2n}(h/2) \simeq ch^p/2^p$ and thus $e_{2n}(h/2) - e_n(h) = y_{2n}(h/2) - y_n(h) \simeq ch^p[2^{-p} - 1]$. Similarly, $y_{4n}(h/4) - y_{2n}(h/2) \simeq ch^p 2^{-p}[2^{-p} - 1]$, which establishes the result.

7. From (7.13c), $K_j(x, y(x); 0) = f(x, y(x))$; and thus by (7.13b), $\phi(x, y(x); 0) = f(x, y(x))$ when $A_1 + A_2 + \cdots + A_m = 1$.

Section 7.4

1. c) $u_1' = y_2 \quad ; \quad u_1(0) = 1$

$u_2' = xu_2 + x^2u_1 \quad ; \quad u_2(0) = 0$

5. When a (2×2) matrix A has a set of linearly independent eigenvectors (say $\mathbf{v}_1$, $\mathbf{v}_2$ where $A\mathbf{v}_1 = \lambda_1\mathbf{v}_1$, $A\mathbf{v}_2 = \lambda_2\mathbf{v}_2$), then for $\lambda_1 \neq \lambda_2$ the solution of $\mathbf{u}' = A\mathbf{u}$, $\mathbf{u}(0) = \mathbf{u}_0$ is given by

$$\mathbf{u}(x) = a_1 e^{\lambda_1 x}\mathbf{v}_1 + a_2 e^{\lambda_2 x}\mathbf{v}_2$$

where $a_1\mathbf{v}_1 + a_2\mathbf{v}_2 = \mathbf{u}_0$. Since the eigenvalues of A are $\lambda_1 = -30$ and $\lambda_2 = -2$, it follows that any solution $\mathbf{u}(x)$ of $\mathbf{u}' = A\mathbf{u}$ tends to θ as x grows.

11. If $B\mathbf{x} = \lambda\mathbf{x}$ then $B(B\mathbf{x}) = B(\lambda\mathbf{x})$ or $B^2\mathbf{x} = \lambda B\mathbf{x} = \lambda^2\mathbf{x}$. Similarly, $B^j\mathbf{x} = \lambda^j\mathbf{x}$ for $j = 3, 4, \ldots, r$ and thus $p(B)\mathbf{x} = p(\lambda)\mathbf{x}$.

12. If A has eigenvalues λ_1 and λ_2, then hA has eigenvalues $h\lambda_1$ and $h\lambda_2$.

CHAPTER 8

Section 8.1

1. a) hyperbola b) parabola c) ellipse

2. $u_x = u_r r_x + u_s s_x$

$u_{xx} = (u_{rr}r_x + u_{rs}s_x)r_x + u_r r_{xx} + (u_{sr}r_x + u_{ss}s_x)s_x + u_s s_{xx}$, etc.

6. a) parabolic, $u_{rr} + u_r - 2u_s = 0$
b) elliptic, $4u_{rr} + 4u_{ss} - 3u_r + 6u_s + u = 0$
c) hyperbolic, $-(64/3)u_{rs} = 0$, $u = f(y - 3x) + g(y - x/3)$

Section 8.2

2. $u(x, t) = (1.28)\pi^{-3} \sum_{k=1}^{\infty} [(2k - 1)^{-2} \sin((2k - 1)\pi x/.4)\exp(-(2k - 1)^2\pi^2 t/.16)]$

5. b) $\rho(A) = |1 - 4\sin^2(3\pi/8)| \simeq 2.4142$. $\|A^n\mathbf{e}^{(0)}\| \le \|A^n\|\,\|\mathbf{e}^{(0)}\|$, $\|A^n\|_1 \ge (\rho(A))^n \simeq (2.4142)^n$, $\|\mathbf{e}^{(0)}\|_1 = 3\varepsilon$. $A^n\mathbf{e}^{(0)} = (-1)^n\varepsilon[\alpha_{n-1}, \beta_{n-1}, \alpha_{n-1}]^{\mathrm{T}}$, $\alpha_i = [(1 + \sqrt{2})^i - (1 - \sqrt{2})^i]/2\sqrt{2}$, $\beta_i = \alpha_i + \alpha_{i-1}$. $A\mathbf{e}^{(0)} = \varepsilon[1, 0, 1]^{\mathrm{T}}$, $A^2\mathbf{e}^{(0)} = \varepsilon[1, -1, 1]^{\mathrm{T}}$.

14. c) $(v_{i,j+1} - v_{ij})/k = (1/2h^2)(H_{ij}\delta_x^2 + hP_{ij}\mu\delta_x + h^2F_{ij})(v_{i,j+1} + v_{ij}) + (G_{i,j+1} + G_{ij})$ where $\delta_x v_{ij} = v_{i+\frac{1}{2},j} - v_{i-\frac{1}{2},j}$, $\mu v_{ij} = (v_{i+\frac{1}{2},j} + v_{i-\frac{1}{2},j})/2$.

15. a) $u(x, t) = \exp(-\pi^2 t)\sin \pi x$
c) $u(x, t) = x^2 e^{-t}$

Section 8.2.2

1. For $1 \le i \le M - 1$, equations same as (8.8). $w_{0,j+1} = 2rw_{1j} + (1 - 2r - 2har)w_{0j} - 2hbr$, $w_{M,j+1} = (1 - 2r + 2hcr)w_{Mj} + 2rw_{M-1,j} + 2hdr$.

4. $u(x, t) = \exp(-\pi^2 t)(\sin \pi x + \cos \pi x)$

Section 8.2.3

2. b) Let $\mathbf{v}^{(k)} = [v_{11}^k, v_{12}^k, v_{13}^k, v_{21}^k, v_{22}^k, v_{23}^k]^{\mathrm{T}}$ and $\mathbf{r} = [20, 0, 0, 20, 0, 0]^{\mathrm{T}}$; then $A\mathbf{v}^{(n+1)} = B\mathbf{v}^{(n)} + \mathbf{r}$ where

$$A = \begin{bmatrix} 8 & -1 & 0 & -1 & 0 & 0 \\ -1 & 8 & -1 & 0 & -1 & 0 \\ 0 & -1 & 8 & 0 & 0 & -1 \\ -1 & 0 & 0 & 8 & -1 & 0 \\ 0 & -1 & 0 & -1 & 8 & -1 \\ 0 & 0 & -1 & 0 & -1 & 8 \end{bmatrix} \qquad B = \begin{bmatrix} 0 & 1 & 0 & 1 & 0 & 0 \\ 1 & 0 & 1 & 0 & 1 & 0 \\ 0 & 1 & 0 & 0 & 0 & 1 \\ 1 & 0 & 0 & 0 & 1 & 0 \\ 0 & 1 & 0 & 1 & 0 & 1 \\ 0 & 0 & 1 & 0 & 1 & 0 \end{bmatrix}$$

4. $u(x, y, t) = \exp[-5\pi^2 t)\sin 2\pi x \sin \pi y$.

6. b) A modified ADI method with increased range of stability is given by

$$(v^* - v^n)/\Delta t = \tfrac{1}{2}\delta_x^2(v^* + v^n) + \delta_y^2 v^n + \delta_z^2 v^n$$

$$(v^{**} - v^n)/\Delta t = \tfrac{1}{2}\delta_x^2(v^* + v^n) + \tfrac{1}{2}\delta_y^2(v^{**} + v^n) + \delta_z^2 v^n$$

$$(v^{n+1} - v^n)/\Delta t = \tfrac{1}{2}[\delta_x^2(v^* + v^n) + \delta_y^2(v^{**} + v^n) + \delta_z^2(v^{n+1} + v^n)]$$

where δ_x^2, δ_y^2, δ_z^2 are central-difference operators with respect to x, y, and z, respectively, analogous to (8.4).

Section 8.3.1

2. $A\mathbf{v} = \mathbf{g}$, $\mathbf{v} = [v_{11}, v_{21}, v_{12}, v_{22}, v_{13}, v_{23}]^{\mathrm{T}}$ where

$$A = \begin{bmatrix} 5 & -1 & -1 & 0 & 0 & 0 \\ -1 & 8 & 0 & -1 & 0 & 0 \\ -1 & 0 & 8 & -1 & -1 & 0 \\ 0 & -1 & -1 & 20 & 0 & -1 \\ 0 & 0 & -1 & 0 & 13 & -1 \\ 0 & 0 & 0 & -1 & -1 & 40 \end{bmatrix}, \quad \mathbf{g} = \begin{bmatrix} -1 \\ -1 \\ -2 \\ -2 \\ -3 \\ -3 \end{bmatrix}$$

3. b) Let $h_{ij} = h(x_i, y_j)$, $(4 - h^2 f_{1j})v_{1j}^{(n+1)} = h_{0j} + v_{1,\,j-1}^{(n+1)} + v_{2j}^{n} + v_{1,\,j+1}^{n} - h^2 g_{ij}$ and $(4 - h^2 f_{M-1,\,j})v_{M-1,\,j}^{(n+1)} = v_{M-1,\,j-1}^{(n+1)} + h_{M,\,j} + v_{M-1,\,j+1}^{(n)} - h^2 g_{M-1,\,j} + v_{M-2,\,j}^{(n+1)}$, $0 \le j \le N - 2$. $(4 - h^2 f_{i1})v_{i1}^{(n+1)} = v_{i-1,\,1}^{(n+1)} + v_{i+1,\,1}^{(n)} + h_{i0} - h^2 g_{i1} + v_{i2}^{(n)}$ and $(4 - h^2 f_{i,\,N-1})v_{i,\,N-1} = v_{i-1,\,N-1}^{(n+1)} + v_{i,\,N-2}^{(n)} + v_{i+1,\,N-1}^{(n)} + h_{iN} - h^2 g_{i,\,N-1}$ for $2 \le i \le M - 2$. Modify at points $(1, 1)$, $(M - 1, 1)$, $(1, N - 1)$, and $(M - 1, N - 1)$.

4. a) $\mathbf{v} = [1, 1, 1, \ldots, 1]^{\mathrm{T}}$ d) $\omega^* = 1.259616184$

6. b) $a = 4\sin^2(\pi/10) = .3819660112$, $b = 4\sin^2(4\pi/10) = 3.618033989$, $c = a/b$, $(3 - 2\sqrt{2})^{m-1} \le c$ for $m = 3$, $\rho_1 = bc^0 = b$, $\rho_2 = bc^{.5} = 1.175570505$, $\rho_3 = bc = a$.

7. a) $u(x, y) = xy$ b) $u(x, y) = \exp(-xy)$ c) $u(x, y) = ye^x - x^2$
d) $u(x, y) = (x^2 + 1)^{\frac{1}{2}} y^2$

8. a) $-v_{i-1,\,j,\,k} - v_{i+1,\,j,\,k} - v_{i,\,j-1,\,k} - v_{i,\,j+1,\,k} - v_{i,\,j,\,k-1} - v_{i,\,j,\,k+1} + (6 - h^2 f_{ijk})u_{ijk} = -h^2 g_{ijk}$.
c) Use Exercise 6b, Section 8.2.3, with $u_t = 0$.

Section 8.3.2

1. a) $4v_{11} - v_{21} - v_{12} = 0$, $-v_{11} + 4v_{21} - v_{31} - v_{22} = 0$, $-v_{21} + 4v_{31} - v_{41} - v_{32} = 0$, $-2v_{31} + (7/2)v_{41} - v_{42} = 0$, $-v_{11} + 4v_{12} - v_{22} - v_{13} = 0$, $-v_{21} - v_{12} + 4v_{22} - v_{32} - v_{23} = 0$, $-v_{31} - v_{22} + 4v_{32} - v_{42} - v_{33} = 0$, $-v_{41} - 2v_{32} + (7/2)v_{42} - v_{43} = 0$, $-2v_{12} + (10/3)v_{13} - v_{33} = 0$, $-2v_{22} - v_{13} + (10/3)v_{23} - v_{33} = 0$, $-2v_{32} - v_{23} + (10/3)v_{33} - v_{43} = 0$, $-2v_{42} - 2v_{33} + (17/6)v_{43} = 0$. $(u(x, y) = xy)$.

2. b) $u(x, y) = xe^y$

5. a)
$$\begin{aligned} 4v_1 - v_2 - v_3 &= -2 \\ 4v_1 - 40v_3 &= -645 \\ 9v_1 - 60v_2 &= -872 \end{aligned}$$
b)
$$\begin{aligned} 3v_1 - v_2 - v_3 &= 0 \\ 9v_1 - 72v_2 &= -860 \\ 4v_1 - 45v_3 &= -640 \end{aligned}$$

Section 8.4

2. a) With H as in Exercise 3, Section 8.2.1, $A = 2rH + 2(1 - r)I$.
$\lambda_j(H) = \cos(j\pi/M)$; so $\lambda_j(A) = 2r\cos(j\pi/M) + 2(1 - r)$
$= 2r(\cos(j\pi/M) - 1) + 2 = 2 - 4r\sin^2(j\pi/2M)$

3. Let $r = x - t/\alpha$, $s = x + t/\alpha$. Then $u_{xx} - \alpha^2 u_{tt} = 4u_{rs} = 0$, and $u = f(r) + g(s) = f(x - t/\alpha) + g(x + t/\alpha)$.

5. $u(x, y) = (x - 4)t$.

Section 8.5.1

2. a) $\mathbf{v}^T A \mathbf{v} = \sum_{i=1}^{4} c_i^2 \lambda_i (\mathbf{u}_i^T \mathbf{u}_i) > 0$

   c) $\langle \phi_i, \phi_j \rangle_L = (A\mathbf{e}_i)^T \mathbf{e}_j = a_{ij}$, $\langle \phi_i, b \rangle = b_i$, $1 \le i, j \le 3$. $\mathbf{u}_3 = [-(10/17), -(286/17), (258/17)]^T$.

3. a) $\phi_1(x, y) = y$ on T_1, $(-2/3)x - (1/3)y + 2$ on T_2, $(-1/3)x - (2/3)y + 2$ on T_5, x on T_6. $\phi_2(x, y) = (1/3)x + (2/3)y - 1$ on T_2, $-x + 3$ on T_3, $-y + 3$ on T_4, $(2/3x + (1/3)y - 1$ on T_5.

5. Let $V_m^k = (x_m, y_m)$, $1 \le m \le 3$; and let A be the (3×3) matrix whose mth row is $[x_m, y_m, 1]$. Then $[a_m^k, b_m^k, c_m^k]^T$ is the solution of $A\mathbf{x} = \mathbf{e}_m$, $1 \le m \le 3$, and hence the mth column of A^{-1}, which can be supplied by Gauss elimination or Cramer's rule.

6. band-width = 18

7. a) $u(x, y) = x^2 + y^2 - 1$ b) $u(x, y) = \sin \pi x \sin \pi y$

8. For $q_1(x, y)$, $a_1 = (x_{i-1}y_{j-1})/h^2$, $b_1 = (h - y_j)/h^2$, $c_1 = (h - x_i)/h^2$, $d_1 = 1/h^2$.

References

Ames, W. F. (1977), *Numerical Methods for Partial Differential Equations*, New York: Academic Press.

Atkinson, K. E. (1978), *An Introduction to Numerical Analysis*, New York: John Wiley and Sons.

Bartle, R. G. (1964), *The Elements of Real Analysis*, New York: John Wiley and Sons.

Burden, R. L., J. D. Faires, and A. C. Reynolds (1981), *Numerical Analysis,* Boston: Prindle, Weber, and Schmidt.

Carnahan, B., H. A. Luther, and J. O. Wilkes (1964), *Applied Numerical Methods,* New York: John Wiley and Sons.

Cheney, E. W. (1966), *Introduction to Approximation Theory*, New York: McGraw-Hill.

Clenshaw, C. W., and A. R. Curtis (1960), "A method for numerical integration on an automatic computer," *Numerische Mathematik* **2**.

Cline, A. K., C. B. Moler, G. W. Stewart, and J. H. Wilkinson (1979), "An estimate for the condition number of a matrix," *SIAM J. Num. Anal.* **16**, pp. 368–375.

Coddington, E. A., and N. Levinson (1955), *Theory of Ordinary Differential Equations*, New York: McGraw-Hill.

Conte, S. D., and C. deBoor (1980), *Elementary Numerical Analysis*, New York: McGraw-Hill.

Dahlquist, G., and A. Bjork (1974), *Numerical Methods*, Englewood Cliffs, N.J.: Prentice-Hall.

Davis, P. J. and P. Rabinowitz (1975), *Methods of Numerical Integration*, New York: Academic Press.

Davis, P. J. and P. Rabinowitz (1978), *A First Course in Numerical Analysis*, New York: McGraw-Hill.

Dongarra, J. J., C. B. Moler, J. R. Bunch, and G. W. Steward (1979), *LINPACK User's Guide*, Philadelphia: SIAM.

Faddeev, D. K., and V. N. Faddeeva (1963), *Computational Methods of Linear Algebra*, San Francisco: Freeman.

Forsythe, G. E. (1967), "Today's computational methods of linear algebra," *SIAM Review*.

Forsythe, G. E., and C. Moler (1967), *Computer Solution of Linear Algebraic Systems*, Englewood Cliffs, N.J.: Prentice-Hall.

Fox, L. (1965), *Introduction to Numerical Linear Algebra*, New York: Oxford University Press.

Fröberg, C. E. (1969), *Introduction to Numerical Analysis*, Reading, Mass.: Addison-Wesley.

Garabedian, P. R. (1964), *Partial Differential Equations*, New York: John Wiley and Sons.

Gear, C. W. (1971), *Numerical Initial Value Problems in Ordinary Differential Equations*, Englewood Cliffs, N.J.: Prentice-Hall.

Gerald, C. F. (1978), *Applied Numerical Analysis*, Reading, Mass.: Addison-Wesley.

Givens, W. (1954), "Numerical computation of the characteristic values of a real symmetric matrix," *Rep. ORNL 1574*, Oak Ridge, Tenn.: Oak Ridge National Laboratory.

Hall, C. A. (1968), "On error bounds for spline interpolation," *J. Approx. Theory* **1**.

Henrici, P. (1962), *Discrete Variable Methods in Ordinary Differential Equations*, New York: John Wiley and Sons.

Henrici, P. (1974), *Applied and Computational Complex Analysis*, Vol. 1, New York: John Wiley and Sons.

Householder, A. S. (1970), *The Numerical Treatment of a Single Nonlinear Equation*, New York: McGraw-Hill.

Householder, A. S., and F. L. Bauer (1959), "On certain methods for expanding the characteristic polynomial," *Numerische Mathematick* **1**, pp. 29–37.

Isaacson, E., and H. B. Keller (1966), *Analysis of Numerical Methods*, New York: John Wiley and Sons.

Krogh, F. T. (1973), "Algorithms for changing the step size," *SIAM J. Num. Anal.* **10**, pp. 949–965.

Kuo, S. S. (1972), *Computer Applications of Numerical Methods*, Reading, Mass.: Addison-Wesley.

Lambert, J. D. (1973), *Computational Methods in Ordinary Differential Equations*, London: John Wiley and Sons.

Luenberger, D. G. (1973), *Introduction to Linear and Nonlinear Programming*, Reading, Mass.: Addison-Wesley.

Maxfield, J. E., and M. W. Maxfield (1971), *Abstract Algebra and Solution by Radicals*, Philadelphia: W. B. Saunders.

Natanson, I. P. (1965), *Constructive Function Theory*, Vol. III, New York: Frederick Ungar.

Oden, J. T., and J. N. Reddy (1976), *An Introduction to the Mathematical Theory of Finite Elements*, New York: John Wiley and Sons.

Ostrowski, A. M. (1966), *Solution of Equations and Systems of Equations*, New York: Academic Press.

Prenter, P. M. (1975), *Splines and Variational Methods*, New York: John Wiley and Sons.

Ralston, A. (1965), *A First Course in Numerical Analysis*, New York: McGraw-Hill.

Ralston, A., and P. Rabinowitz (1978), *A First Course in Numerical Analysis*, New York: McGraw-Hill.

Rivlin, T. J. (1969), *An Introduction to the Approximation of Functions*, Waltham, Mass.: Blaisdell.

Rowland, J. H., and Y. L. Varol (1972), "Exit criteria for Simpson's compound rule," *Mathematics of Computation* **26**, no. 119.

Schoenberg, I. J. (1946), "Contributions to the problem of approximation of equidistant data by analytic functions," *Quart. Appl. Math.* **4**.

Seeley, R. T. (1970), *Calculus of Several Variables, an Introduction*, Glenview, Ill.: Scott, Foresman and Company.

Shampine, L. F., and R. C. Allen (1973), *Numerical Computing: An Introduction*, Philadelphia: Saunders.

Shampine, L. F., and C. W. Gear (1979), "A user's view of solving stiff ordinary differential equations," *SIAM Review* **21**, pp. 1–17.

Shampine, L. F., and M. Gordon (1975), *Computer Solution of Ordinary Differential Equations*, San Francisco: W. H. Freeman and Company.

Shanno, D. F. (1970), "Conditioning of quasi-Newton methods for function minimization," *Mathematics of Computation* **24**, no. 111.

Strang, G., and G. Fix (1973), *An Analysis of the Finite Element Method*, Englewood Cliffs, N.J.: Prentice-Hall.

Stroud, A. H. (1971), *Approximate Calculation of Multiple Integrals*, Englewood Cliffs, N.J.: Prentice-Hall.

Stroud, A. H., and D. Secrest (1966), *Gaussian Quadrature Formulas*, Englewood Cliffs, N.J.: Prentice-Hall.

Thomas, G. B. (1972), *Calculus and Analytic Geometry*, 4th ed., Reading, Mass.: Addison-Wesley.

Todd, J. (1962), *Survey of Numerical Analysis*, New York: McGraw-Hill.

Varga, R. S. (1962), *Matrix Iterative Analysis*, Englewood Cliffs, N.J.: Prentice-Hall.

Wilkinson, J. (1963), *Rounding Errors in Algebraic Processes*, Englewood Cliffs, N.J.: Prentice-Hall.

Wilkinson, J. (1965), *The Algebraic Eigenvalue Problem*, Oxford, England: Oxford University Press.

Wilkinson, J. and C. Reinsch, ed. (1971), "Handbook for automatic computation," Vol. II, *Linear Algebra*, New York: Springer-Verlag.

Index